Mike Ellis

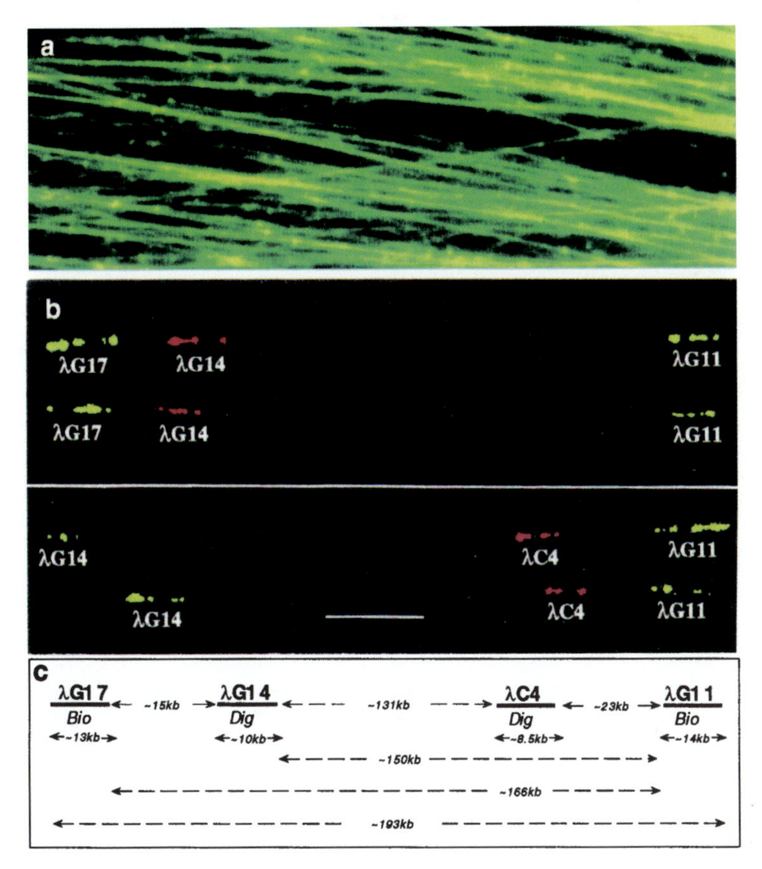

Plate 1. Fibre-fluorescence *in situ* hybridization (FISH): (a) extended DNA fibres on glass slides after 6-diamidino-2-phenylindole (DAPI) staining (pseudo-coloured green); (b) multicolour hybridization on extended DNA fibres with three lambda clones delineates the order of the three major histocompatibility complex (MHC) loci; (c) clone size and interclone distances estimated using the Watson-Crick approach. (Courtesy of A. Sjöberg *et al.*, 1997.)

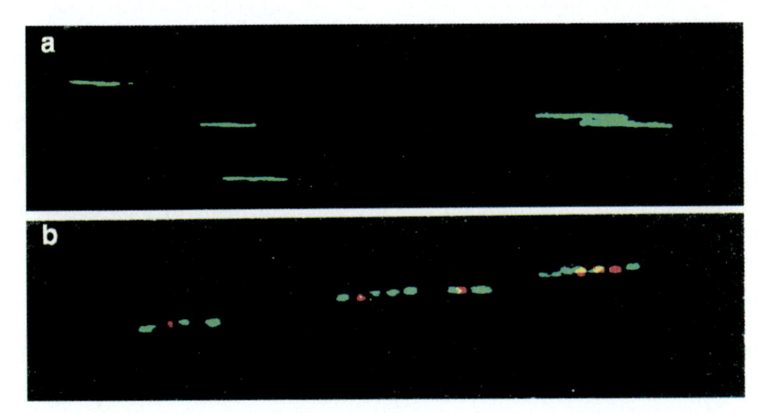

Plate 2. DNA-combing: (a) non-mechanically stretched whole lambda clones stained with DAPI; (b) multicolour FISH showing the order/arrangement of subclones derived from these clones. (Courtesy of Liu *et al.*, 1998.)

The assistance of PIC UK, Abingdon, Oxfordshire, towards the printing of the colour plates is gratefully acknowledged.

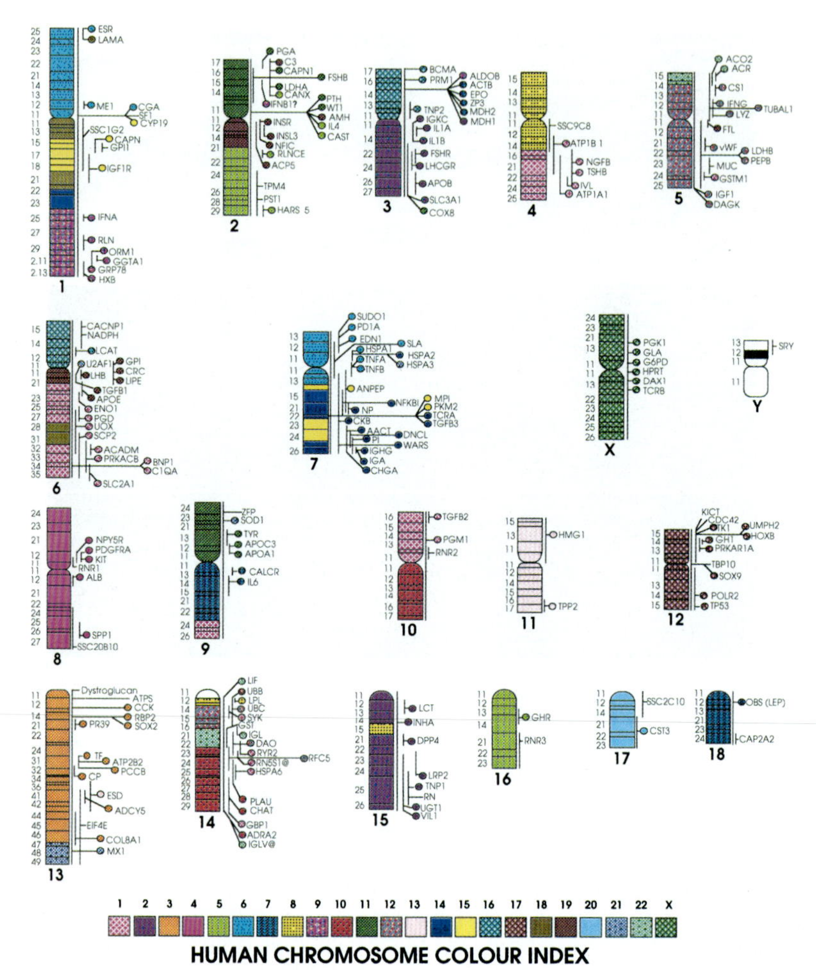

Plate 3. A comparative map of the pig and human genomes. An index of colour patterns represents individual human chromosomes. Homologies between human and porcine karyotypes are shown by these colour patterns on the pig chromosomes. Loci mapped to individual pig chromosomes are arranged on the right-hand side of each chromosome and colour indexed corresponding to the human chromosome they map to. *In situ* hybridization (ISH) mapped loci are represented with a horizontal bar defining the localization. Syntenic loci which map to the same human chromosome are connected to a vertical line spanning the entire length of the chromosome. Where available, segmental locations for loci mapped by somatic cell hybrids are also presented. Loci not showing correspondence between the comparative painting and gene mapping data are distinguishable by their colour index. The porcine gene mapping data were obtained from Pigbase (http://www.ri.bbsrc.ac.uk/pigmap/pigbase/pigbase.html) and some recent data from Chaudhary *et al.* (1998a), Chowdhary *et al.* (unpublished data). Corresponding human locations for the genes can be searched through the genome database (GDB) (http://gdb.gdbnet.ad.jp/gdb-bin/gdb/lite/bin/genes), where the details of the locus symbols can also be found.

THE GENETICS *of the Pig*

These efforts are dedicated to our loving families, supportive colleagues and to the pig industry that employs and feeds hundreds of millions of people worldwide. We also dedicate these efforts to our friend and colleague Dr Larry Young who died during the writing of this book.

MFR
AR

THE GENETICS
of the Pig

Edited by

M.F. ROTHSCHILD

Department of Animal Science
Iowa State University
USA

and

A. RUVINSKY

Department of Animal Sciences
University of New England
Australia

CAB INTERNATIONAL

CAB INTERNATIONAL
Wallingford
Oxon OX10 8DE
UK

Tel: +44 (0)1491 832111
Fax: +44 (0)1491 833508
E-mail: cabi@cabi.org

CAB INTERNATIONAL
198 Madison Avenue
New York, NY 10016-4314
USA

Tel: +1 212 726 6490
Fax: +1 212 686 7993
E-mail: cabi-nao@cabi.org

A catalogue record for this book is available from the British Library, London, UK

Library of Congress Cataloging-in-Publication Data
The genetics of the pig/edited by F. Rothschild and A. Ruvinsky.
p. cm.
Includes index.
ISBN 0-85199-229-3 (alk. paper)
1. Swine—Genetics. 2. Swine—Breeding. I. Rothschild, Max Frederick, 1952–
II. Ruvinsky, Anatoly. III. C.A.B. International.
SF396.9.G45 1998
636.4′0821—dc21 97-34917
CIP

ISBN 0 85199 229 3

Typeset in Garamond by AMA Graphics Ltd
Printed and bound in the UK at the University Press, Cambridge

Contents

Preface

Genetic theory and practice have evolved enormously over the past two decades. The separate fields of classical and quantitative genetics have now been joined by immunogenetics and molecular genetics creating new methods and insights into understanding biological processes. Today, genetics serves as one of the pillars of modern agricultural sciences, medicine and the biotechnology industry. Animal breeders and geneticists more widely than ever before are using knowledge from the different fields of genetics for manipulation and improvement of their livestock. Given these enormous changes in our understanding of genetics it seems an appropriate time to publish a series of books devoted to the genetics of the domestic livestock species which serve humans.

Domestication of the pig occurred some 5000 years ago and it has been a tremendously important food source in several civilizations. Today, the pig continues to be a valued source of food worldwide. Modern biological discoveries and technological improvements in management practices have revolutionized pork production. Approximately one billion pigs are raised worldwide and pork is the dominant meat source representing 40% of all the red meat eaten. In addition, the pig, because of physiological and genetic similarities with man, serves as an excellent animal model for medical research. Animal scientists, geneticists, veterinarians, livestock producers, medical researchers and students are interested in the pig. An improved understanding of the genetics of the pig is imperative for the pig to serve as a primary food source, as an animal model for human research or as a future organ donor.

The purpose of this book is to present in one location a complete, comprehensive and updated description of the modern genetics of the pig. It is our intention to combine essential knowledge from the various fields of genetics integrated with livestock management in order to provide a useful and informative guide on the genetic improvement of the pig. Recent genetic improvements over the past 10 years in the pig industry have been dramatic. Tremendous progress has also been made in understanding the genetic control of the pig. An excellent example of these advancements is the success in porcine genome mapping. In 1990 only about 50 genes and markers had been mapped or assigned to individual porcine chromosomes. At the time of this publication, the number of genes and markers mapped is approaching 1800 loci.

This book is addressed to a diverse audience including students, researchers, veterinarians and pig breeders. The initial chapters are devoted to taxonomy and domestication and overview several pig breeds. Chapters 3 and 4 extensively cover the genetics of coat colour, morphological characteristics and inherited disorders. Molecular, biochemical and immunological genetics are described in Chapters 5, 6 and 7 and present considerable new information combined with classical data. Genetic linkage and cytological genetic maps are presented in Chapters 8 and 9, and demonstrate the substantial recent advancements in pig gene mapping. In Chapters 10–13 the new areas of developmental and behavioural genetics are explored, as is the genetics of reproduction and new emerging technologies that combine genetics and reproductive biology.

Chapter 14 addresses genetic diversity and concerns for maintaining exotic and rare local breeds. The genetics of performance traits and carcass and meat quality traits are included in Chapters 15 and 16. Chapter 17 is devoted to overall genetic improvement.

Considerable genetics research with the pig is ongoing at the time of publication of this book. Therefore, it is inevitable that some recent research will fail to be cited. We hope that errors or omissions will be noted and brought to our attention. Also at the time during which this book has been written there have been several changes in the nomenclature related to genes and alleles. Many authors were not aware of these changes and while the editors tried to develop consistency between chapters in using nomenclature this was not always possible. The editors apologise for this problem. Finally, in no way is this book meant to replace the many fine textbooks devoted to the theory of animal breeding.

Further it should be noted that *The Genetics of the Pig* is the second publication in a series of monographs devoted to genetics of mammalian species published by CAB International. The previous book in this series is *The Genetics of Sheep*, and the next one *The Genetics of Cattle* will be published in the near future.

This book is a result of international efforts. The editors wish to express their appreciation to the authors for accepting the challenges of writing each chapter. In addition, many scientists helped to review, edit and critique all the materials presented in the book. Their contributions are gratefully appreciated. Our particular thanks are due to L. Andersson, T. Baas, R. Beiharz, N. Cameron, K. Campbell, P.Chardon, L. Christian, A. Conley, S. Dooling, R. Fernando, J. Gibson, C. Groves, S. Hermesch, B. Kinghorn, S. Lamont, N. Larsen, D. Lay, J. Mabry, L. Messer, L. Ollivier, C. Rexroad Jr, A. Stratil, S. Sun, M. Vaiman and L. Wang. The assistance of PIC UK, Abingdon, Oxfordshire, towards the printing of colour plates is gratefully acknowledged. The assistance of the publisher, CAB INTERNATIONAL, is also acknowledged and appreciated. It is our hope that this text will serve as a useful resource for all those people who study or work with pigs.

Max F. Rothschild
Anatoly Ruvinsky
December 1997

Systematics and Evolution of the Pig

1

A. Ruvinsky[1] and M.F. Rothschild[2]

[1]*Department of Animal Science, University of New England, Armidale 2351, NSW, Australia;* [2]*Department of Animal Science, 225 Kildee Hall, Iowa State University, Ames, IA 50011, USA*

Introduction

According to the traditional classification system the order Artiodactyla (even-toed ungulates), to which the pig belongs, has a complicated and diversified taxonomy. It is comprised of about ten families with approximately 88 genera and more than 200 species. In addition, recent investigations of mammalian phylogeny show a certain degree of similarity between Artiodactyla and Cetacea (whales), even though they are not deeply nested within the artiodactyl phylogenetic tree. Cetacea appear to be more closely related to the members of the suborder Ruminantia, than to the two other traditionally defined suborders, Tylopoda and Suiformes (Graur and Higgins, 1994). The limited molecular genetic studies of artiodactyls have shown good agreement with the basic structure of their phylogeny supported by morphological data (Novacek, 1992). Current estimates indicate that the common ancestor for artiodactyla (including Cetacea) may have existed around 65 million years ago (MYA) (Graur and Higgins, 1994). It seems likely that future molecular genetic research will provide additional evidence of the phylogenetic structure of this mammalian group.

The suborder Tylopoda (camelids) consists of two genera: camels of the Old World and American llamas. These species have a number of very distinctive differences with other representatives of Artiodactyla. They lack typical hoofs, have a stomach with three compartments which is very different from ruminants, and differences exist in the structure and functioning of the placenta, extra embryonic tissues and sexual organs. Many adaptations of these animals are unique and are absent in other Artiodactyla. It has even been suggested that camels and llamas should be considered as a separate order (Bannikov, 1980). Whether or not this is acceptable, a reasonable level of similarity between camelid species does exist and the molecular data from restriction site patterns in ribosomal DNA seem to support their common origin (Semorile *et al.*, 1994). A common ancestor for the pig and camel lived approximately more than 50 MYA (Jermann *et al.*, 1995).

The suborder Ruminantia is certainly the most advanced and numerous group in the order Artiodactyla. It includes six families, of which the largest are Cervidae (deer) and Bovidae (oxen). These two most recent suborders are represented worldwide and were enormously successful and competitive during the last 20–25 million years. The most important features that appeared during the evolution of the suborder Ruminantia were changes in the digestive physiology and morphology of the digestive organs. This made them able to ferment cellulose with increased efficiency. The estimated time of divergence between the Ruminantia and Suiformes was about 55–60 MYA (Graur and Higgins, 1994).

Taxonomy and Phylogeny of the Suborder Suiformes

The Suiformes are the third and probably the more primitive lineage among Artiodactyla. Their differences relate to the structure and function of the stomach and the morphology of their teeth. The three living families within the Suiformes include Hippopotamidae (hippos), Tayassuidae (peccaries) and Suidae (pigs).

Family Hippopotamidae

There are only two genera, each containing one species, which are known as hippos. The first one is *Hippopotamus amphibius*, which was very common in most of Africa at the beginning of the 19th century. Presently these giant animals, whose mature weight may reach 3000 kg, live in very limited areas of Africa and mostly in the national parks. Hippos can be distinguished by their lack of the snout disc and by the fact that the collateral digits on each foot reach the ground (Groves, 1981). The digestive tract is very long (about 60 metres) and the stomach has three compartments. These features increase its capacity to assimilate cellulose quite effectively. Hippos were a good source of meat (one animal can produce more than 500 kg of pure meat), which was one of the primary reasons for the dramatic reduction of this species. Their diploid chromosome number is 36 (Gernike, 1965).

The other genus is represented by *Hexaprotodon liberiensis*, the Pigmy Hippopotamus, which is only 250–260 kg, and has several differences in body composition and morphology of the teeth. The typical habitats of this species are restricted to swamp forests of West Equatorial Africa. The species is now quite rare in its natural habitat.

The two genera can be traced back to the latest part of the Miocene era, while the family itself possibly originated about 11 MYA (Coryndon, 1977 as cited by Groves, 1981). The latest reconstruction, based on the investigation of the molecular evolution of the ribonuclease superfamily, predicts a later date of about 35 MYA for the separation of Hippopatamidae and Suidae (Jermann *et al*., 1995)

Family Tayassuidae

The family Tayassuidae (peccaries) was separated from pigs not later than in the Oligocene era. Their fossils have been found in Eurasia and even in Africa. However, their modern species live only in the Americas. The peccaries, like pigs, have a snout disc, but the differences between the two families are very significant. The stomach in peccaries is subdivided into three compartments. In several features of the digestive system peccaries resemble ruminants, though whether the features developed independently or in parallel is not known.

Peccaries have three hoofs on their back legs. In addition their upper canine teeth point downwards and they have a total of 38 teeth. They have a scent-producing gland located on their backs. Peccaries are significantly smaller than pigs with an average body size of approximately 30 kg.

The Tayassuidae family, according to most recent classification (Grubb, 1993a) includes three genera and three species. The Collared Peccary (*Pecari tajacu*) spread across South and Central America and the southern part of North America. Their diploid chromosome number is 30. It is a very common species, reproducing well and used widely for hunting, because of its good meat and leather quality. The other species is the White-Lipped Peccary (*Tayassu pecari*) which is bigger than the Collared Peccary. Their diploid chromosome number is 26. Its range is from southern Mexico to the south in Central America. Hybridization between both species is probably possible, but no hybrids with pigs have been described (Bannikov, 1980). The third species, the Chacoan Peccary (*Catagonus wagneri*), was known only through fossil records until recently when it was discovered on the Paraguay–Bolivia–Argentina border (Wetzel, 1977). This species, as shown by Benirschke *et al.* (1985), differs significantly from the other two peccaries in chromosome number ($2n = 20$). This may indicate a considerable evolutionary distance between *C. wagneri* and other Tayassuidae.

Phylogenetic relationships

Tentative phylogenetic relationships between previously described genera and Suidae, which are discussed in the following sections, are represented in Fig. 1.1. A lack of data prevents, at the moment, construction of a completely justified phylogenetic tree. However, the latest information on the molecular evolution of 12S rRNA (Douzery and Catzeflis, 1995) and numerous classical data have demonstrated the monophyly of the suborder of Suiformes and provide an opportunity for rough phylogenetic reconstruction. This phylogenetic reconstruction shows three groups of species corresponding to three families within Suiformes. A recent comparative cytogenetic study confirms zoological data that Tayassuidae significantly lost homology with the representatives of Suidae and the separation might have occurred quite long ago (Bosma *et al.*, 1995). Additional data using SINE (short interspersed repetitive elements) sequence analysis have suggested that separation of Tayassuidae and Suidae occurred less than 43.2 MYA (Yasue and Wada, 1996).

Taxonomy and Phylogeny of the Suidae Family

Systematics

The Suidae family includes the most widely spread species of non-ruminant even-toed ungulates. All of them have an elongated muzzle with a snout disc and four-toe extremities with well developed side toes. The canine teeth are

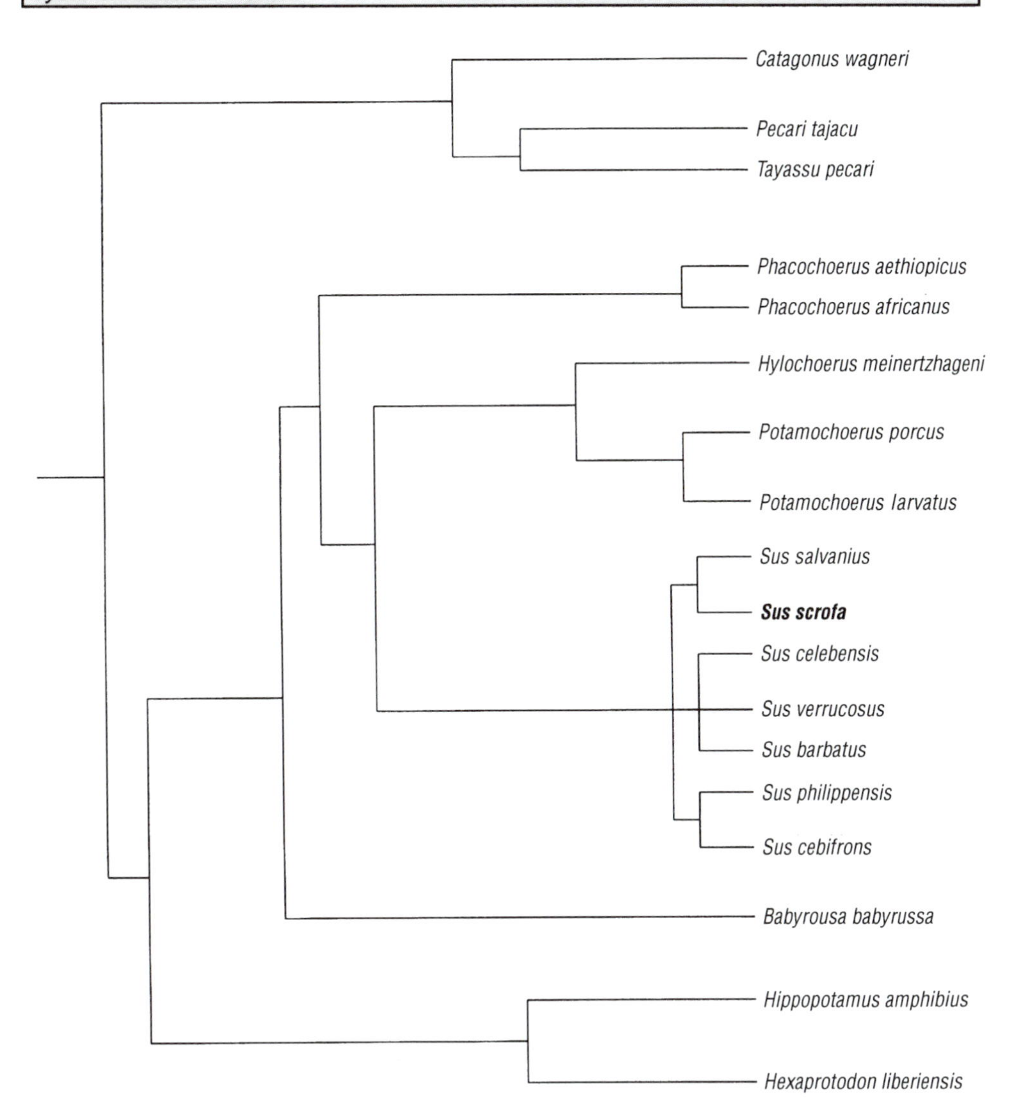

Fig. 1.1. Possible phylogenetic relationships in the suborder Suiformes. The length of branches is not necessarily proportional to real time and genetic distances between taxons. The position of internal nodes as they relate to *Sus* in particular are not finally resolved.

large and the upper ones are curved. The stomach is simple with an additional sack. The Suidae are omnivores. This family traces back to the Oligocene era in Europe. More than a dozen species, which belong to five genera and three subfamilies are living now and have been grouped into subfamilies and tribes (Thenius, 1970; Grubb, 1993a,b):

Subfamily Babyrousinae
  Genus *Babyrousa*
    Species *Babyrousa babyrussa* (the Babirusa)

Subfamily Phacochoerinae
Genus *Phacochoerus*
Species *Phacochoerus africanus* (the Common Warthog)
Species *Phacochoerus aethiopicus* (the Cape and Somali Warthog)
Subfamily Suinae
Tribe Potamochoerini
Genus *Potamochoerus*
Species *Potamochoerus porcus* (the Red River Hog)
Species *Potamochoerus larvatus* (the Bushpig)
Genus *Hylochoerus*
Species *Hylochoerus meinertzhageni* (the Giant Forest Hog)
Tribe Suini
Genus *Sus*
Species *Sus scrofa* (the Eurasian Wild Boar)
Species *Sus salvanius* (the Pygmy Hog)
Species *Sus verrucosus* (the Javan Warty Pig)
Species *Sus barbatus* (the Bearded Pig)
Species *Sus celebensis* (the Sulawesi Warty Pig)
Species *Sus philippensis* (the Philippine Warty Pig)
Species *Sus cebifrons* (the Visayan Warty Pig)

The general taxonomy of the family seems reasonably justified. However, the taxonomy of some genera, *Sus* in particular, may be the subject of future reconsiderations. There are indications concerning the existence of at least two more species: *Sus bucculentis* and *Sus heureni*, which may inhabit South Vietnam and Flores Islands (Indonesia), respectively. However, the data are not sufficient at the moment to draw any further conclusions.

Subfamily Babyrsousinae

The genus Babyrousa

The Babirusa ('babi-rusa' Indonesian, babi = pig, rusa = deer), *Babyrousa babyrussa,* differs significantly from other Suinae in that it has long legs, a relatively small head and a body free of most hair. The stomach is more complex. Grass comprises the main source of food and the typical digging or rooting behaviour of pigs is not known. The canine teeth in males are considerably more developed and are very large and curved, sometimes creating a spiral. The most unusual feature is that the mandibular canine and the maxillary canine teeth are both pointed upward, which is uncommon for mammals. The Babirusa spread across Sulawesi and nearby islands. The animals swim well and can cross rivers and even sea harbours. Females usually deliver two offspring. The adults are large in size. Local people in the past used to tame the Babirusa and transport them from place to place.

The diploid chromosome number in the Babirusa is 38. Eleven of the autosome pairs and the X chromosome look almost identical to chromosomes of the domestic pig (Bosma *et al.*, 1991a). Future molecular phylogenetic investigations may give a clearer answer concerning the origin of the Babirusa. According to Groves (1981) it is possible that the Babirusa could have phylogenetic connections with the Anthracotheriidae, the source group of the hippos. If this were the case, the proper systematic position for the Babirusa might be closer to the Hippotamidae family. However, a recent morphological study of the placenta and heart anatomy indicates that the Babirussa has more characters in common with pigs than with hippos (MacDonald, 1994).

Subfamily Phacochoerinae

The genus Phacochoerus

The Common Warthog (*Phacochoerus africanus*) is widespread in a majority of African countries. The name is derived from the large warts that are located on the muzzle and whose function is not known. The shape of the skull of the Warthog differs essentially from other pigs and the number of teeth is greatly reduced. The canines are large, sharp and represented in both sexes. Adults are approximately 145–190 cm in length, 65–85 cm in height and weigh 50–150 kg. Average litter size is three or four piglets. The offspring are susceptible to the cold immediately after birth and therefore do not leave a burrow, where the temperature is around 30°C constantly. Their main food is grass and the animal grazes on its knees which causes it to have callouses. Adults, in contrast to their offspring, enter the burrow backwards. The species is spread widely in Sub-Saharan Africa, and not in the rain forests of Western Africa. In southern Africa this species has been reintroduced again for hunting. The diploid chromosome number of the Warthog is 34 (Melander and Hansen-Melander, 1980).

The Cape and Somali Warthog, *Phacochoerus aethiopicus*, was recognized by zoologists as a separate species quite recently (Grubb, 1993b). However, palaeontologists have recognized these two as different species for a long time, mainly because of the lack of functional incisors in *P. aethiopicus*. The Common Warthog has two incisors in the upper jaw and usually six in the low jaw. Further comparative investigations of both species are desirable.

Subfamily Suinae

The genus Potamochoerus

Potamochoerus porcus (Red River Hog) is one of the smallest African pigs. This species is widely spread in the central and southern parts of the continent. It is variable in colour and size and may be made up of several subspecies or may be even two to three separate species. The animals that live in West Equatorial Africa, for instance, are usually very bright and are red in colour with a white bar on the back, white hairs on the muzzle and brushes of long hairs on the ears. Males from a majority of habitats are characterized by the canine apophyses (located between the ears and the nose), which in older males looks like two small horns directed backwards. Length of the body varies from 100 to 150 cm. The height varies from 55 to 80 cm and the weight may reach 80 kg (Bannikov and Flint, 1989). The number of offspring ranges from three to four. The skull of *P. porcus* is very much like that of *Sus*. Groves (1981) considered that the skull structure was an indication that these genera may be more closely related or both have changed little from their more distant common ancestor. The diploid chromosome set of the Red River Hog is 34 (Melander and Hansen-Melander, 1980). However, an absence of comparative cytogenetic data (Bosma *et al.*, 1991a) and molecular data makes it difficult to estimate the phylogenetic distance from other genera.

The range of the Bushpig (*Potomochoerus larvatus*) is mainly to the east and south of that of *P. porcus* and occurs not only in East and South Africa, but even in Madagascar. It is also bristly, but bristly pelage extends from the head over the whole body and gives the live animal a shaggy, crested appearance, different from the river hog. Both species are, for the most part, separated territorially but in some places their areas may overlap (Grubb, 1993b). An introgression between the species is assumed (Kingdon, 1979). The limited amount of information about the biology and evolution of both of these species needs to be rectified.

The genus Hylochoerus

The Giant Forest Hog, *Hylochoerus meinertzhageni,* is one of the largest wild pigs (length 155–190 cm, height up to 110 cm, weight up to 250 kg). However, it overlaps in size with the *Potamochoerus* species. The Giant Forest Hog of East Africa is particularly large (Grubb, 1993b). The head and muzzle are very broad and the snout is big, and well developed for extensive digging. This species and genus is a relatively recent discovery. It was found in Kenya earlier in the 20th century and is now known to be found throughout the tropical forest region of Africa. The external appearance of this animal is that it is covered with long black hair. The biology of this species is still under investigation. The diploid chromosome number of the Giant Forest Hog is 32 (Melander and Hansen-Melander, 1980).

Concluding remarks

Palaeontology data trace the Suidae to the Oligocene (Thenius, 1970). Molecular phylogenetic data are very limited and therefore a comprehensive molecular phylogeny is still absent. It is known, however, that evolutionary rates differ significantly in different Suidae lineages. According to Groves (1981) *Potamochoerus* and *Sus* had already developed into their modern forms in the Pliocene, while at the beginning of the Pleistocene the ancestors of the Warthog and Giant Hog were not different enough to be placed in different genera.

Bosma *et al.* (1991a) assume that further cytogenetic analysis is appropriate to point out differences and similarities between chromosomes of the African pigs. It appears that current data support the close phylogenetic links between the African species (i.e. *Phacochoerus aethiopicus*) and the domestic pig. Figure 1.1 contains a tentative description of the phylogeny of Suidae. The three branches of the family represent modern pigs of African and Eurasian origin.

Taxonomy of the Genus *Sus*

Introductory remarks

Groves (1981) has presented a very comprehensive and complete analysis of the taxonomy and phylogeny of the genus *Sus* based on morphological, palaeontological, biogeographical and other data. Recently Groves and Grubb (1993) made a revision of *Sus* systematics. This analysis is summarized in the following description. At present, seven species comprise the genus *Sus*. Investigations for the past 150 years of this genus have taken different approaches and the number of discriminated species has varied from very few to a total of 37. It is very likely that future classifications, based on classical and molecular phylogenetic data, may change our current knowledge concerning *Sus* taxonomy and evolution.

Wild pigs spread naturally through vast territories, covering most of Europe, Asia and some areas of northern Africa. They were introduced into North and South America, Australia and Oceania. Domestic pigs are very common in a majority of countries worldwide, except in those where religious restrictions exist. Several features, including teeth and skull morphology, external proportions, hair and colour patterns, biochemical and molecular polymorphisms, ecology and behaviour, reproductive isolation and natural areas, are used for discrimination of the many species in the genus.

Sus scrofa (the Eurasian Wild Boar)

This is one of the most widespread mammalian species. *Sus scrofa* is the main ancestor for domesticated pigs and this issue is discussed in Chapter 2. *S. scrofa* is extremely variable in the majority of traits studied. The number of subspecies is uncertain and depends upon the definition of the subspecies. However, it is possible to discriminate at least 16 more or less distinct subspecies (Groves, 1981; Groves and Grubb, 1993).

S. s. scrofa Western, Central and parts of Southern Europe
S. s. attila East Europe, northern slope of Caucasus, parts of Western Siberia, Central and Western Asia.
S. s. meridionalis South Spain, Corsica and Sardinia.
S. s. algira North-west Africa
S. s. libica Asia Minor, Middle East, southern part of Eastern Europe
S. *s. nigripes* Southern Siberia, Central Asia
S. s. sibiricus Eastern Siberia, Mongolia
S. s. ussuricus Russian Far East, Korea

S. s. moupinensis Eastern China, South East Asia
S. s. leucomystax Japan
S. s. riukiuanus Ryukyu Islands
S. s. taivanus Taiwan
S. s. davidi Western India
S. s. cristatus Eastern India, western part of Indochina
S. s. affinis Southern India, Sri Lanka
S. s. vittatus Malaysia, southern Indonesian Islands

Areas of these subspecies are close and the level of discriminating differences may be quite small, involving size, colour, proportions, skull characters and in several cases chromosome numbers. The variation in chromosome number is a result of two distinct Robertsonian translocations, which were found in the different geographic areas of the species (Tikhonov and Troshina, 1974). The usual number of chromosomes in *S. scrofa* is 38 chromosomes (Bosma, 1976). However, translocation I (T I) involving chromosomes 16 and 17 and translocation II (T II) involving chromosomes 15 and 17 were found in Kirgizian and European Boars (Tikhonov and Troshina, 1978), reduces the number to 37 chromosomes in heterozygotes and to 36 chromosomes in homozygotes. A full description of these translocations is presented in Chapter 8.

Adaptations of these animals to different food and climatic conditions have been amazing. The flexible behaviour of the wild boar is perhaps one of the important features providing this adaptability. *S. scrofa* are well adapted to Siberian winters, tropical conditions, mountains and semi-deserts. Pigs can tolerate temperature conditions from −50°C to +50°C due to well developed thermoregulation and nest building behaviour. Despite being under significant human and predator pressure, there are populations of wild boar that are quite numerous in many parts of the world.

Body size variations are significant among subspecies. The largest are *S. s. ussuricus* (males up to 300 kg) and *S. s. attila* (males up to 275 kg). Generally, mature weight is quite variable depending on age, sex, food availability, season and habitat. Body and head length is about 130–175 cm and height ranges up to 100 cm. The smallest forms of wild boar are from Asia. Detailed description of skulls, and osteometrical studies are published elsewhere (Groves, 1981; Endo *et al.*, 1994).

Current knowledge concerning the evolution of *S. scrofa* is still limited and fragmented data do not create a full-scale picture of the origin and phylogeny of the species. It is known that a fully evolved *S. scrofa* lived in the Biharian fauna in Europe and replaced the previously existing lineages of *S. minor* and *S. strozzi* (Hünermann, 1969). The facial shortening that occurred in *S. celebensis* and in *S. scrofa* has been used as a possible argument in favour of their common origin from perhaps southern or the south-eastern regions of Asia. The spread of the two above mentioned Robertsonian translocations does not contradict this possibility. However, it is obvious that an extensive molecular genetics study of the problem would be necessary before any clear conclusions can be drawn.

Sus salvanius (the Pygmy Hog)

This is the smallest representative of the genus. It is the second non-warty pig of the genus *Sus,* besides *S. scrofa.* Body and head length is 66–71 cm in males and 55–62 cm in females. The shoulder height is correspondingly 23–30 cm and 20–22 cm and weight is 9–10 kg and 6–7 kg in males and females, respectively (Mallinson, 1978). The basic colour is dark brown. Structure of the skull differs significantly from that of *S. scrofa.* The number of pairs of teats is three, instead of six pairs typical for other *Sus* species, and the number of piglets born is usually three to four. Ears are large and rounded. The tail is very short and the inner toes are short compared with other species in the genus (Groves, 1981). *S. salvanius* is distributed currently in quite a narrow part of northern Assam (India) in the long-grass belt. The number of animals in the area is very small. This species is therefore considered to be endangered. Diploid chromosome number is 38. Comparative analysis of G-bands shows that the pygmy hog chromosomes are very similar to those of domestic pig and those wild *S. scrofa* that possess $2n = 38$ chromosomes (Bosma *et al.*, 1983). Except for small size of the body there are relatively few characters which may serve as diagnostic for discrimination between this species and *S. scrofa.*

Sus verrucosus (the Javan Warty Pig)

This species lives now mainly in Java. Two subspecies have been described (Groves and Grubb, 1993). The most typical common feature is three warts on a specific location of the muzzle strongly developed in the adult males. Colour varies from overall black to a pale red. Size also varies from relatively large to small. Sexual dimorphism in size is greater than in other species. Diploid chromosome number is 38. In spite of some differences in G-banding and the structure of Y chromosome with *S. scrofa vittatus* and *S. celebensis,* similarity is significant (Bosma *et al.*, 1991b). Closeness of *S. verrucosus* and *S. scrofa vittatus* is supported by the observation and precise description of interspecies hybrids in nature (Blouch and Groves, 1990). Several morphological features make *S. verrucosus* close to the other South East Asian species, *S. barbatus.*

Sus barbatus (the Bearded Pig)

The common name of this pig is due to the elongated whiskers around the muzzle from the mouth to the ears. A few warts on the muzzle are very typical. Mature size varies significantly between several subspecies and is on average close to that of *S. scrofa.* The length ranges from 100 to 160 cm and the weight is approximately 100 kg. Some males are much bigger. The Bearded Pig inhabits the Malaysian peninsula, Sumatra, Java, Borneo, Palau, Bangka, Palawan and some other islands. *S. barbatus* sometimes migrates and these migrations involve thousand of animals. Fertile hybrids with *S. scrofa* obtained in captivity are known (Blouch and Groves, 1990).

Sus celebensis (the Sulawesi Warty Pig)

This wild pig from Sulawesi and several other islands including possibly Timor has been recognized as separate from the *S. verrucosus* species by Groves (1981). Recent cytogenetic analysis strongly supported this point. The diploid chromosome number is 38, but the structure of the Y chromosome is different from *S. verrucosus* (Bosma *et al.*, 1991b). Animals are usually black with a few white or yellowish hairs intermixed and they have crown tufts of hair. Other colour types have been described. The muzzle is short, like the Eurasian wild pig, and is of small size. Legs are also short, and it has small short ears with a relatively large head. It occurs on Sulawesi and other offshore islands. There are indications that *S. celebensis* was domesticated during the early Holocene and spread as far as Roti, a medium sized island south-west of New Guinea, where they are living now (Groves, 1981). In other places they probably have been replaced by domestic *S. scrofa vittatus.* The several forms of wild pigs in New Guinea could be a result of hybridization between *S. scrofa vittatus* and *S. celebensis* (Groves, 1981). It appears that strong reproductive isolation has not been developed by this species.

Sus philippensis (the Philippine Warty Pig)

According to the latest information there are sufficient arguments to discriminate this species from *S. celebensis* and *S. barbatus* (Groves, 1981; Groves and Grubb, 1993). It occurs on several islands of the eastern Philippines. The colour is black, sometimes with a pale snout-band and red-brown patches in the mane. This pig is smaller than *S. barbatus.* Further investigations of this species are desirable.

Sus cebifrons (the Visayan Warty Pig)

This small pig occurs allopatrically to *S. philippensis* on the west central islands of the Philippines (Groves and Grubb, 1993). The data about the biology of this species are very limited.

Interrelationships of the species in the genus Sus

A high level of morphological similarities between all species of the genus is an argument in favour of their relatively recent origin from a common ancestor. The same chromosome number and their high level of homology support this conclusion. However, these close relationships complicate phylogenetic reconstruction. A possible phylogeny of the genus *Sus* is presented at Fig. 1.1.

It follows from the previous description that, in several cases, different *Sus* species coexist in the same area yet have maintained significant differences in morphology, ecology and behaviour. This may be reasonably explained by a reproductive isolation that appears to exist between the species that may have contact. This is applicable to *S. scrofa* and *S. salvanius* in northern India and *S. scrofa* and the 'Indonesian' species: *S. barbatus, S. verrucosus* and *S. celebensis*. However, interspecies hybridization with *S. scrofa* can occur and this probably indicates a limited reproductive isolation (Groves, 1983). Fertile hybrids between a European Wild Boar and Bearded Pig sows (*S. barbatus*) have been reported (Lotsy, 1922). Hybridization between *S. scrofa* and *S. verricosus* has been recorded fairly recently in Java (Blouch and Groves, 1990). According to Groves (1996, personal communication) in some parts of the Philippines the indigenous wild pigs (especially *S. cebifrons*) are in danger of being hybridized out of existence by crossing with feral domestic pigs. Groves (1981) assumed that the ancestors of these species were separated at least in Middle Pleistocene, but this may have happened much longer ago.

Discriminant analysis based on 16 morphological traits was used to evaluate the level of differences and similarities among the *Sus* species (Groves, 1981). Results from this analysis are given in Fig. 1.2 and reveal that there are

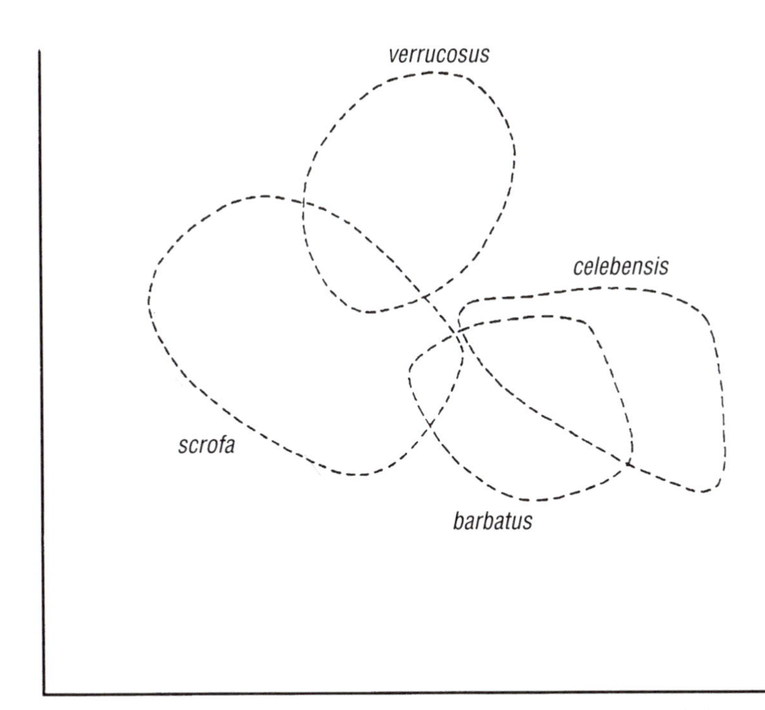

Fig. 1.2. Representation of morphological similarities based on discriminant analysis between *S. scrofa*, *S. verrucosus*, *S. celebensis* and *S. barbatus*. (From Groves, 1981, with permission.)

some overlaps of the four species based on several factors in the discriminant functions.

A hypothetical scenario of *Sus* evolution has been suggested (Groves, 1981). He proposed that the *verrucosus–barbatus* lineage, which has been present from the beginning of the Pliocene in Europe, entered Indonesia about 2 MYA. It appears that these animals cohabited with the older *celebensis* lineage. *S. scrofa* may possibly have evolved out of the *celebensis* lineages and entered Europe about 700,000 years ago, where it replaced *verrucosus*-like pigs. The high level of similarity in chromosome structure of *S. scrofa* and *S. celebensis* does not contradict this hypothesis. Several independent sets of data support a Far East origin of *S. scrofa* and steady spreading in a westerly direction. The previously mentioned Robertsonian translocations, which are typical for some Siberian, Central Asian and European populations, probably appeared and became fixed in the populations after or during their western movement. Numerous investigations have been devoted to comparing geographic distribution of alleles for blood group antigens, isozymes and other proteins, but have not directly used these data as arguments in resolution of the problem of the species origins (Gorelov, 1994). Further study of this problem may yield more information. It seems very likely that South East Asia and the Indonesian Islands may have played a special role in the *Sus* speciation and evolution.

Conclusions

The information presented in this chapter gives the general overview of the systematic position and phylogeny of the wild ancestors of the domestic pig. The family Suidae appeared during the evolution of the early Oligocene relatively soon after separation of the suborder Suiformes from other Artiodactyla. From a morphological point of view the Suiformes are more primitive and less specialized.

Suidae has spread widely across Africa, Europe and Asia. *Sus* itself appeared in the Lower Pliocene at least 3–5 MYA in Europe and Indonesia. A distribution of these ancestors' species through the Indonesian Islands was possibly essential for speciation within the genus.

One of these species, *S. scrofa*, was tremendously successful and spread through Asia and Europe replacing previous species. A number of more or less distinctive subspecies emerged and some of them were independently involved in the domestication process which began over 5000 years ago.

Acknowledgements

We wish to express our gratitude to Dr Colin P. Groves from the Australian National University, whose critical comments were very important and useful and also for permission to reproduce Fig. 1.2.

References

Bannikov, A.G. (1980) Order Tylopoda. In: Sokolov, V.E. (ed.) *Life of Animals. Mammals.* Prosveshenie, Moscow, pp. 422–426 (in Russian).

Bannikov, A.G. and Flint, V.E. (1989) Order Artiodactyla. In: Sokolov, V.E. (ed.) *Life of Animals. Mammals.* Prosveshenie, Moscow, pp. 426–522 (in Russian).

Bernirschke, K., Kumamoto, A.T. and Merrit, D.A. (1985) Chromosomes of the Chacoan peccary, *Catagonus wagneri* (Rusconi). *Journal of Heredity* 76, 95–98.

Blouch, R.A. and Groves, C.P. (1990) Naturally occurring suid hybrid in Java. *Zeitschrift für Säugetierkunde* 55, 270–275.

Bosma, A.A. (1976) Chromosomal polymorphism and G-banding patterns in the wild boar (*Sus scrofa* L.) from The Netherlands. *Genetica* 46, 391–399.

Bosma, A.A., Oliver, W.L.R. and MacDonald, A.A. (1983) The karyotype, including G- and C-banding patterns, of the pigmy hog *Sus (Porcula) silvanius* (Suidae: Mammalia). Genetica 61, 99–106.

Bosma, A.A., de Haan, N.A., Blouch, R.A. and MacDonald A. A. (1991a) Comparative cytogenetic studies in *Sus verrucosus*, *Sus celebensis* and *Sus scrofa vittatus* (Suidae: Mammalia). *Genetica* 83, 189–194.

Bosma, A.A., de Haan, N.A. and MacDonald, A.A. (1991b) The current status of cytogenetics of the Suidae: a review. *Frädrich-Jubiläumsband. Bongo, Berlin* 18, pp. 258–272.

Bosma, A.A., de Haan, N.A., Mellink, C.N.M., Yerle, M. and Zijstra, C. (1995) Study of chromosome homology in *Sus*, *Babyrousa* and *Tayassu* by chromosome painting. *Proceedings 9th North American Colloquium on Domestic Animals. Cytogenetics & Gene Mapping.* Texas A&M University, College Station, Texas, USA.

Coryndon, S.C. (1977) The taxonomy and nomenclature of the Hippopatamidae (Mammalia: Artiodactyla) and description of two fossil species. *Proceedings konilijke Nederlanse Akademie van Wetenschap. Amsterdam B* 80, 61–88.

Douzery, E. and Catzeflis, F.M. (1995) Molecular evolution of the mitochondrial 12S rRNA in Ungulata (Mammalia). *Journal of Molecular Evolution* 41, 622–636.

Endo, H., Kurohmaru, M. and Hayashi, Y. (1994) An osteometrical study of the cranium and mandible of Ryukyu wild pig in Iriomote Island. *Journal of Veterinary and Medical Science* 56, 855–860.

Gernike, W.H. (1965) The chromosomes and neutrophil nuclear appendages of *Hippopotamus amphibicus* Linnaeus 1758. *Onderstepport Journal of Veterinary Research* 32, 181.

Gorelov, I.G. (1994) *Biology of Siberian Boar.* Institute Cytology and Genetics, Novosibirsk, 82 pp (in Russian).

Graur, D. and Higgins, D.G. (1994) Molecular evidence for the inclusion of cetaceans within the order Artiodactyla. *Molecular Biology and Evolution* 11, 357–364.

Groves, C. (1981) Ancestor for the pigs: taxonomy and phylogeny of the genus *Sus. Technical Bulletin No 3. Department of Prehistory, Research School of Pacific Studies.* Australian National University, Canberra, 96 pp.

Groves, C.P. (1983) Pigs east of the Wallace line. *Journal de la Société des Océanistes* 77(39), 105–119.

Groves, C.P. and Grubb, P. (1993) The Eurasian Suids: *Sus* and *Babyrousa*. In: Oliver, W.L.R. (ed.) *Pigs, Peccaries and Hippos.* IUCN, The World Conservation Union, pp. 107–111.

Grubb, P. (1993a) Order Artiodactyla. In: Wilson, D.E. and Reeder D.M. (eds.) *Mammalian Species of the World: A Taxonomic and Geographic Reference.* Smithsonian Institution Press, Washington DC, and London, pp. 374–414.

Grubb, P. (1993b) The Afrotropical suids: *Phacochoerus, Hylochoerus* and *Potamochoerus.* In: Oliver, W.L.R. (ed.) *Pigs, Peccaries and Hippos.* IUCN, The World Conservation Union, pp. 66–75.

Hünermann, K.A. (1969) *Sus scrofa priscus* Goldfuss im Pleistozän von Süssenborn bei Weimar. *Paläontologische Abhandlungen* A 3, 611–616.

Jermann, T.M., Opitz, J.G., Stackhouse, J. and Benner S.A. (1995) Reconstructing the evolutionary history of the artiodactyl ribonuclease superfamily. *Nature* 374, 57–59.

Kingdon, J. (1979) *East African Mammals.* Vol. III (B). Academic Press, London.

Lotsy, J.P. (1922) Die Aufarbeitung des Kühn'schen Kreuzungs materials im Institut für Tierzucht der Universität Halle. *Genetica* 4, 32–61.

MacDonald, A.A. (1994) The placenta and cardiac foramen oval of the babirusa (*Babyrousa babyrussa*). *Anatomy and Embryology* 190, 489–494.

Malinson, J.J.C. (1978) Breeding of the pigmy hog *Sus salvanius* (Hogson) in northern Assam. *Journal of the Bombay National History Society* 74, 288–297.

Melander, Y. and Hansen-Melander, E. (1980) Chromosome studies in African wild pigs (Suidae, Mammalia). *Hereditas* 92(2), 209–216.

Novacek, M.J. (1992) Mammalian phylogeny: shaking the tree. *Nature* 356, 121–125.

Semorile, L.C., Crisci J.V. and Vidal-Rioja L. (1994). Restriction patterns in the ribosomal DNA of Camelidae. *Genetica* (Dordrecht) 92, 115–122.

Thenius, E. (1970) Zur Evolution und Verbreitungsgeschichte der Suidae (Artiodactyla: Mammalia). *Zeitschrift für Säugetirkunde* 35, 321–342.

Tikhonov, V.N. and Troshina, A.I. (1974) Identification of chromosomes and their aberrations in karyotypes of subspecies of *Sus scrofa* L. by differential staining. *Doklady Akademii Nauk SSSR* 214, 932–935.

Tikhonov, V.N. and Troshina, A.I. (1978) Production of hybrids of wild and domestic pigs with two chromosomal translocations. *Doklady Akademii Nauk SSSR* 239, 713–716.

Wetzel, R.M. (1977) The Chacoan peccary *Catagonus wagneri* (Rusconi). *Bulletin of the Carnegie Museum of Natural History* 3, 1–36.

Yasue, H. and Wada, Y. (1996) A swine SINE (PRE-1 sequence) distribution in swine-related animal species and its phylogenetic analysis in swine genome. *Animal Genetics* 27, 95–98.

Genetic Aspects of Domestication, Common Breeds and their Origin

2

G.F. Jones
Department of Agriculture, Western Kentucky University, 1 Big Red Way, Bowling Green, KY 42101-3576, USA

Domestication

Based upon archaeological findings, many volumes have been written outlining the chronology and history of swine domestication. Jonsson (1991) presented a thorough discussion of the scientific aspects of domestication as well as a chronological history of the development of modern-day domesticated swine. However, the purpose of this chapter is to discuss domestication, breed formation and breeds of swine with respect to worldwide swine genetic resources and how these concepts relate to past and future selection schemes around the world.

The reasons for domestication of different animal species are many. There is general agreement that the primary reason for domestication of swine was to provide a source of food protein. However, it could be argued that a more logical reason was the unique organoleptic characteristics of pork rather than its nutritive value. This is especially true in China where the native swine population appears to have been changed little by artificial selection programmes. In China pork is the meat of choice and pork fat is used as shortening in the majority of dishes; Chinese cuisine has actually evolved around pork. In addition to swine being domesticated as a primary source of protein, one must appreciate that the unique flavour of pork could have contributed significantly to the domestication of swine.

Meat scientists and pork industry enthusiasts often extol the nutritive value of pork as being a source of high quality protein, B vitamins and trace minerals. However, domestication may have resulted due to the fact that pork can be very fatty and thus provide an excellent source of energy. During the first half of the 20th century, pork was the primary meat source in many parts of the world because of its high fat content. In human populations that have high energy demands, pork fat is an excellent source of highly palatable energy. From a genetic standpoint, one may assume that the reasons for domestication are irrelevant to selection schemes for today's swine population. However, if the unique culinary qualities of pork were a major reason for domestication, then selection programmes which emphasize pork quality traits that result in improved palatability may have a major impact upon consumer demand for pork products.

Domestication seems to have caused several changes in physical type. In this context, type refers to that combination of characteristics that make an animal useful for a specific purpose. Bokonyi (1974) suggested that one of the earliest results of domestication was a decrease in skeletal size. However, this author also points out the existence of archaeological remains which dispute this theory. There is some suggestion that the smaller skeletal size of domesticated pigs lasted until breed formation was beginning when a few of the developing breeds began to approach the size of their wild ancestors. Based upon early breed development, it appears that some of the earliest swine breeds developed in Europe were much larger than Asian breeds. Size differences in various areas of the world may have resulted because of diversity in feed resources. A logical explanation for this size difference could be that the feed resources in Asia were much higher in fibre and lower in energy content. This difference in diet may also aid in explaining the apparent difference in size of the belly region between the European and Asian types.

Another observed difference between wild and domestic swine is the differential development of body parts. Hammond (1962) concluded that nutrients are first utilized for the development of the head, brain and spinal column, and with limited nutrient availability, these body parts will become more fully developed relative to other body parts. Likewise, loin muscle and fat deposition have a lower priority for development. This concept may explain a major difference in phenotypic appearance of wild and domesticated pig types. Furthermore, this pattern of growth has implications with reference to modern day selection programmes. In order to produce maximum fat and muscle variation and facilitate accurate evaluations for selection purposes, appropriate nutrient levels should be used to allow full expression of genetic potential. Furthermore, findings by Stahly *et al.* (1994) suggest that maintaining pigs in a 'high health' status may be equally important in allowing pigs to optimize genetic potential.

According to Darwin (1868), the Chinese claimed to have domestic pigs for 4900 years. The Chinese pigs were markedly different from those in Europe having wider and shorter heads, dished faces, short legs and considerable fatness. Because of the large human population in China, pigs were confined to small households or farms much earlier than in Europe. This type of confinement led pigs to evolve into more docile, lethargic animals that were more dependent upon man for their food supply. During this long process, pigs lost their natural roaming and rooting instincts. Specifically, the Chinese pigs have been described as 'lazy' beasts that convert more of the energy consumed into fat rather than using the energy for roaming and searching for food. Therefore, over time the belly region became large enough to decrease the ability of the pigs to move swiftly and easily.

From a theoretical standpoint, the domestication of animals functions to slow or stop natural evolution since mates are no longer selected based on the 'survival of the fittest' concept. Once domestication of a species has occurred, mate selection becomes the choice of man. However, this basic change does not imply that 'evolutionary-like' changes have not occurred in animal populations since the time of domestication. For example, some of the 'evolutionary-like' changes have most likely resulted from dietary conditions which influenced alterations in size and possibly the physiology of the digestive systems. An important question relative to this issue concerns the size of the abdomen or the 'pot-bellied' appearance of Chinese pigs when compared with most other swine of the world. Is the large abdomen of the Chinese pigs the result of artificial selection by man, or is the 'pot-belly' the result of 'evolutionary-like' changes in the Chinese pig populations resulting from a diet high in fibre content? The omnivorous characteristic of pigs has allowed them to be fed a variety of diets. Consequently, variations in available feed resources have influenced differences in the digestive systems of pigs around the world. Using imported Chinese germplasm, US and European scientists are investigating evolutionary differences between the Chinese breeds and US or European breeds.

Concerning the 'evolutionary-like' changes that have occurred during domestication, Dr Jay L. Lush (1945) gave the following explanation:

> Domestication merely intensified forces or processes which already existed in nature. Increased inbreeding alternating with wider outcrossing, more intense selection devoted toward a wider variety of goals, and mating like to like wherever one man or tribe was breeding the same species for two or more different goals, all had the net effect of tremendously speeding up the slow process of evolution as it occurs in nature, until remarkably large changes were made in animals under domestication during what was a very short period in terms of geologic time. That the changes thus brought about were at the maximum rate possible seems highly unlikely. If further changes are desired, it is probable that the possibilities in most directions are by no means exhausted and that intelligent use of these same processes can result in much faster progress than has been averaged during the long (in terms of human lifetimes) history of domestication.

Even though the use of ultrasonic evaluation and Best Linear Unbiased Prediction (BLUP) statistical procedures may have resulted in a much faster rate of genetic change in recent years than Dr Lush had observed in 1945, it is probable that further improvements are possible for most economically important traits. However, these traits continue to operate based on the concept of genetic homeostatis introduced by Lerner (1954). A classic example is the divergent line selection project conducted by Hetzer and Miller (1972) for a high and low backfat line of Duroc pigs (Fig. 2.1). Selection resulted in steady progress within both lines; however, the rate of genetic change was greater for the fat line. This indicates a 'natural' tendency for swine populations to regress towards a more favourable homeostatic condition, in this case a regression towards increased fat deposition which is often thought to be beneficial for many fitness traits.

Darwin (1868) stated 'domestication implies almost complete fertility under new and changed conditions of life and this is far from being invariably the case'. It is possible that the new environments and selection may have adversely affected reproductive cycles and reproductive rates. Bassett (1996) suggests that the problems associated with summer infertility in swine may be the result of

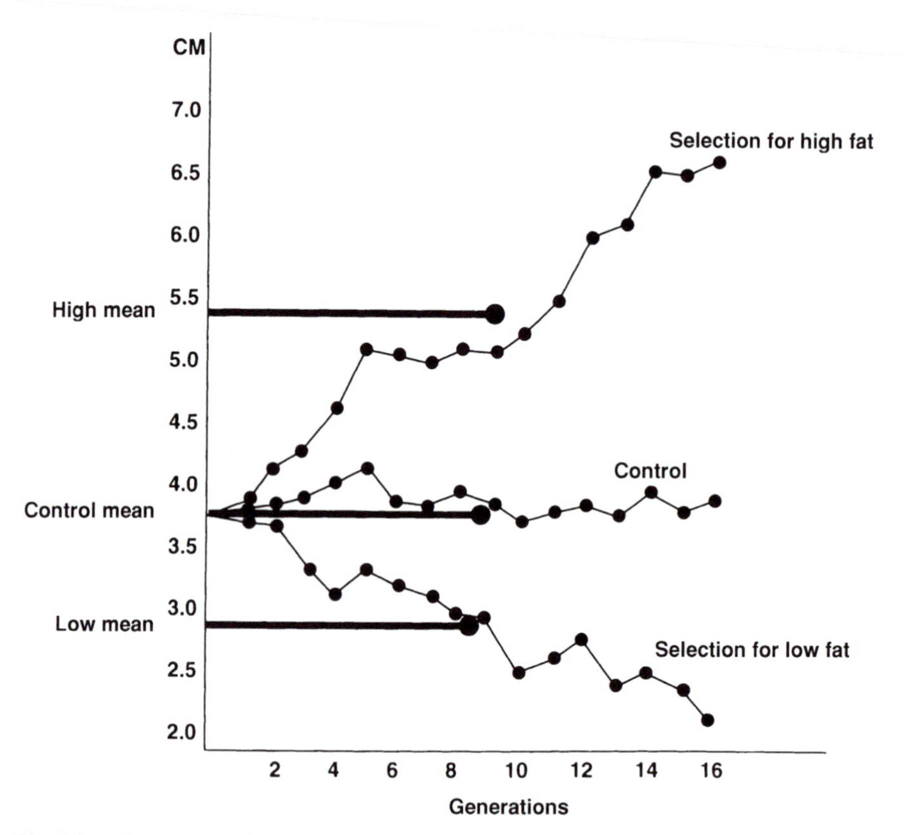

Fig. 2.1. Duroc generation means of backfat thickness (Hetzer and Miller, 1972)

residual reproductive seasonality inherited from their European Wild Boar ancestry. He further suggests that selective breeding against reproductive seasonality may provide the best long-term solution to what has been perceived by many swine producers and researchers alike as purely an environmental phenomenon caused by elevated atmospheric temperatures during the summer months.

Breed Development

Numerous authors have attempted to describe the history and characteristics of the various swine breeds of the world. Mason (1969) listed 330 non-extinct breeds of swine, 89 which he considered to be 'important and recognized'. McLaren (1990) listed 201 pig breeds in an appendix to the chapter on world breed resources; however, many of these breeds are extinct and had little or no lasting influence on the swine industry. For those interested in the total array of genetic resources of the world, Porter (1993) presented a thorough coverage of the various breeds. The historical development of breeds may be important to understand the attributes and problems of today's breeds. Even though much of the early breed history is highly speculative, there are common theories and ideas which are logical based upon type differences that developed among the breeds. The common classroom or textbook definition of a breed as 'a group of animals of common origin that possess certain true breeding distinguishable characteristics that make them different from other members of the species' will suffice to explain the basic idea of breeds in this chapter.

Most authors agree that the major centres of breed development were England, China and the US. There is little doubt that domestication of swine occurred earlier in China than in Europe, and consequently the evolution of the different Chinese breeds must have occurred earlier than that of the English breeds. Records of breed formation in China are quite vague because breed societies have never become a part of Chinese culture. In McLaren's account of the breeds, no Chinese breed was listed as having a breed society.

The Large White or Yorkshire

There has been confusion about the names Yorkshire and Large White. Much of the early breed development occurred in the county of Yorkshire, and many of the early importations to other countries were from Yorkshire. For this reason, the name Yorkshire has been used in many countries around the world. However, in later years, Large Whites from other counties were exported as well. Therefore, over the years the terms have become interchangeable. The Large White breed was first recognized as a distinct breed in England in 1868 and the first herd book was published in 1884 (Plager, 1975). Breed associations and herd books have been established in several other countries (Table 2.1), with both Yorkshires and Large Whites accepted into these registries. Movement of

Table 2.1. A list of Large White (Yorkshire) national breeds derived from the British Large White.

Breed Society	Country	Period of development	Breed
Canadian Yorkshire	Canada	Mid-19th C.	1889
American Yorkshire	USA	Late 19th C.	1893
Danish Yorkshire	Denmark	Late 19th C.	1895
German Yorkshire (Edelschwein)	Germany	Late 19th C.	1904
Dutch Yorkshire	The Netherlands	Late 19th C.	1913
French Large White	France	Early 20th C.	1927

Source: Sellier and Rothschild, 1991.

this breed from country to country has certainly exemplified the idea of Lush (1945) who suggested that within breeds there is often intense selection with inbreeding followed by wider outcrossing, in this case by genetic migrations from other countries. No doubt animal health restrictions have limited the rate of migration. Limiting migration may have been desirable because the resulting changes in gene frequencies have been quite large when migrations have occurred.

During the latter half of the 19th century and the early part of the 20th century, the Large Whites or Yorkshires were exported from England to many other countries around the world. Vaughan (1937) found that Sanders Spencer, a leading English breeder, had exported Large White pigs to 46 different countries in Europe, Asia, Africa, Australia and North and South America. The popularity of the breed has continued to increase around the world, and it is clearly one of the two major maternal breeds in the world. In addition, Yorkshires are continuing to be developed as a terminal sire breed in many countries. The Large White and Yorkshire breeds have been selected to meet the needs of the swine industries of each country, but over the years a great deal of genetic migration has occurred between countries.

Weir and Coleman (1877) concluded that the Old English Pig was 'more or less white in colour, and that our dark breeds owe their colour to the influence of foreign blood'. They used these facts to conclude that Large White pigs were uncrossed with foreign breeds. Weir and Coleman state that the Large White pigs were 'cultivated principally in the counties of Yorkshire, Lancashire, Lincolnshire and Leicestershire, and it is probable that they are descended pretty directly from the Old English Pig'. They quoted Mr Rowlandson who had written a prize essay on the breeding and management of pigs as saying:

> There are good grounds for supposing that 'the Old English Hog' with flop ears was originally the only domestic animal of its kind throughout the kingdom. The genuine old English breed was coarse-boned, long in limb, narrow in the back, and low-shouldered – a form to which they were most probably predisposed from the fact of having to travel far and labor hard for their food, and undergo considerable privations during winter. [He proceeded to say he had seen] surprising improvements resulting from better care, shelter, and food, where no fresh blood was introduced. In these cases the thick flop ears became fine and thin, the bones

a moderate size, bristles gave place to hair, and the white skin became fine and ruddy.

Weir and Coleman (1877) described the white pigs of the mid 1800s as being enormous specimens 'often weighing as much as a small bullock'. These animals were noted for their quality of bacon and for their hardy and prolific character with litter size often as large as 16 to 18 piglets. However, these hogs were apparently very 'late maturing' and consumed large volumes of feed in preparation for slaughter. Before the Yorkshires or Large Whites were 'cultivated', they were described as being 'of large size, gaunt, greedy, and unthrifty; coarse in the quality of meat, flat-sided, and huge boned'. Plager (1975) cited Cully who wrote in 1774 that 'a Large White hog measured nine feet, eight inches and was four feet, five and one-half inches tall and weighed 1400 pounds'.

Weir and Coleman (1877) gave credit to the famous English stockman, Robert Bakewell, for improving the Leicestershire strain of the Large Whites. Most early authors suggest that Bakewell made great contributions in the moulding of the Old English Hog into the Large White or Yorkshire breed. Bakewell likely used the same principles he had used so successfully with other species, 'selection, discarding the large, coarse animals, and selecting such as were more symmetrical and finer of bone'. Perhaps he was selecting for earlier maturing animals that would be much easier to 'fatten' for slaughter. Plager (1975), based upon findings from research in England, concluded that most, if not all, the outstanding strains of Yorkshires descended from Bakewell's herd which began in 1761. Porter (1993) cited several sources suggesting that Bakewell's contributions to the White Leicester pigs may have been insignificant. However, based on numerous accounts, it seems quite evident that during the latter half of the 18th century the Large White breed became more consistent in type and more widely recognized as one of the major swine breeds of the era. This trend towards consistency is consistent with Bakewell's principles of breeding.

Porter (1993) suggested 1770 as the date when major importations of Chinese pigs were made into England. From 1770 to approximately 1860, there was a popular 'fad' of introducing Chinese, Siamese or Neapolitan pigs into the English pig populations. It is generally agreed these introductions resulted in decreasing the age at slaughter from 18 months to 9 months or less. Although the Neapolitan pigs probably contained some Chinese breeding, these pigs were more moderate in size than the extremely small Chinese breeds. Porter (1993) concluded that 'the general effect of these exotic introductions was to bring earlier maturity and quicker fattening (with plenty of fat and good pork), finer bones, finer skin and hair, smaller shorter carcasses, shorter heads, smaller prick ears and often dished faces, and a general air of refinement'. The contribution of Chinese pigs to breed development in England appears quite conclusive; however, the extent to which these imported pigs contributed to the various breeds has been widely debated. Many sources claim the Large White or Yorkshire pigs were free from 'foreign blood' as evidenced by their white colour; however, there is little doubt that the Chinese breeds contributed significantly to the development of the Berkshire, Small White and the Middle White breeds. Porter (1993) related the story about a prize winning Middle Yorkshire sow that reportedly had 13 crosses of Chinese breeding. Anderson (1931) suggested that Large White hogs were improved by the addition of Berkshire blood as early as 1842 and by the incorporation of White Leicester hogs in later years. He also credited the Leicester for the erect ears, refinement of head and bone and improved fattening qualities. Anderson (1931) further concluded that Small Whites were used to improve the fattening qualities. Concerning the Small White breed, Weir and Coleman (1877) wrote 'it is impossible to keep them poor; the tendency to lay on fat is remarkable; hence, they are well suited for porking purposes, but lack the lean meat so desirable for bacon'. Weir and Coleman (1877) suggested that Middle Whites farrowed from nine to twelve piglets per litter and that Small Whites seldom farrowed more than seven to nine piglets. These data suggest that the Chinese breeds which had been incorporated were not the prolific Taihu breeds, but rather breeds from the Canton region of southern China.

The Small and Middle Yorkshire breeds gained some popularity during the late 1800s and even during the early 1900s, but these breeds have essentially become extinct today. The major contribution of these breeds to the swine industry is through the crosses that were made with the Large White breed. Based upon the characteristics of the Old English Hog and the general characteristics of the Chinese breeds, there is little doubt that 'present day' Yorkshires or Large Whites were derived by blending the three distinct white breeds. Plager (1975) concluded that the Large White, Middle White and Small White each contributed significantly in the formation of the Yorkshire breed in Canada and the United States. His opinion was that 'the Middle and Small Whites did more harm than good on this side of the water', presumably because he was associated with Yorkshires during the period of change to 'meat type' pigs in the US. Because of the greater Chinese influence, the Small and Middle White breeds were much smaller framed and earlier maturing. This prompted Plager to conclude that the Yorkshire lines with the influence of these smaller statured breeds became fatter and showed decreased growth rate at much lighter weights.

The Landrace

The lop-eared pigs which were developed in the northern regions of Spain, Portugal, France, Italy, and nearly all of Northern Europe are generally referred to as Celtic pigs. Porter (1993) described the Celtic as 'a large, coarse, late-maturing meat type with those long lop ears, long legs, flat sides, and a dirty or yellowish white colour with or without varying degrees of spotting or belting'. Of course, this description applies to native populations before the infusion of Asian breeds. In the 18th and 19th centuries, these 'land pigs' were widespread across the northern European countries with the common characteristic of large drooping ears hanging over the eyes (Porter, 1993).

The degree of infusion of outside genetics on the native Danish 'land pigs' is debatable. There are reports of importations from China and other parts of the world in the early 1800s and there was certainly use of British breeds during the late 1800s. Porter (1993) points out that some authors claim the Danish Landrace originated by crossing the Large White with local Celtic types. Even though the English Large White had become popular in Denmark during the late 1800s, the Danes deny that the Large White had any influence on the famous Danish Landrace.

In Denmark, a national pig breeding scheme was established in 1896 for the improvement of the Large White and the 'land pig'. The national scheme was implemented in order to develop superior breeding stock for crossbreeding programmes designed to produce pigs suited for the export market in Great Britain (Porter 1993). During this time the 'land pig' was first called the Danish Landrace. The type of pig that emerged from this Danish breeding effort was a long, lean pig that was ideal as a 'baconer' to produce Wiltshire sides for the market in Britain. The famous Wiltshire sides in Great Britain were produced by curing the entire carcass halves. The bacon cut from the mid region of the side included the loin and belly being sliced together. Most other continental pigs of this era were too short and fat for this special market. Improvements in the Danish Landrace occurred rapidly, and the breed was soon exported to become the foundation of other Landrace breeds (Table 2.2).

After World War II, the Danes elected to restrict the exportation of Landrace breeding stock. This action opened the door for other countries and Sweden became a major exporter of Landrace breeding stock. Therefore, in recent years, the Swedish Landrace have had much greater influence on worldwide Landrace populations than the Danish Landrace. In more recent years, Norway and Finland have also made major exportations of Landrace seedstock. Figure 2.2 outlines the basic manner in which Landrace populations have migrated around the world. This chart was prepared by Porter (1993) and modified to reflect other significant genetic migrations.

The Landrace breed is extremely variable in type with each country having distinct types which characterize different selection goals. The extremes in type range from the muscular, high cutability Belgian Landrace to the extremely large

Table 2.2. A list of Landrace national breeds derived from the Danish Landrace.

Breed	Country	Period of development	Breed Society	Remarks
Swedish Landrace	Sweden	Early 20th C.	1907	
Dutch Landrace	The Netherlands	Early 20th C.	1933	
American Landrace	USA	Mid-20th C.	1950	
French Landrace	France	Mid-20th C.	1951	
British Landrace	UK	Mid-20th C.	1953	
Canadian Landrace	Canada	Mid-20th C.	1956	
Belgian Landrace	Belgium	Mid-20th C.	1966	France (BS 1971)

Source: Sellier and Rothschild, 1991.

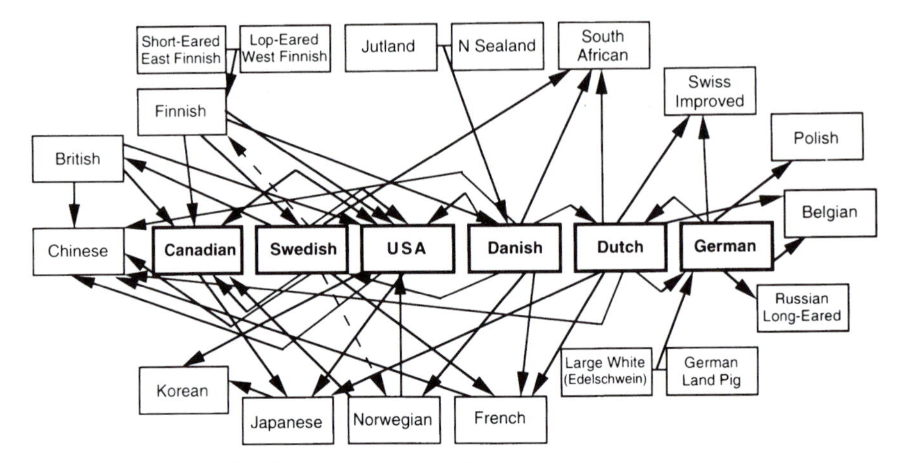

Fig. 2.2. Landraces: their influences on each other.

framed, late maturing Norwegian Landrace. The Landrace breed appears to have had an abundance of inherent genetic variability which continues to allow it to be changed to different types when subjected to diverse breeding schemes. The use of the different 'land pigs' from various European countries in breed formation contributed to the genetic diversity of the breed. In the US, where Landrace from several countries have been used in the development of the population, some breeders have created Landrace lines with high carcass lean percentage while others have concentrated on improving maternal characteristics.

The Duroc

During the last quarter of a century, the Duroc has been the fastest growing breed in the world. The Duroc breed was developed in the US, and for many years, all the growth and development of the breed occurred in the US. The Duroc has become one of the most important terminal sires in Canada, Denmark, Japan, China, Thailand, Taiwan and many other countries around the world. In most countries, the Duroc has been used primarily as a terminal breed; however, in more recent years, the Duroc has been exploited as a breed to produce more durable maternal-line females. Some of the major breeding stock companies have incorporated the Duroc into maternal lines to improve appetite during lactation and, consequently, to improve the ability of sows to maintain body condition, wean heavier litters and decrease the number of days from weaning to rebreeding. Including Duroc in the maternal line has improved structural soundness and increased sow longevity. Furthermore, the Duroc has been incorporated into maternal lines in some countries to enhance the sows' ability to adapt and thrive under outdoor rearing conditions.

The first history of the red hogs of America was recorded in 1872 at the first National Swine Breeders Convention held at Indianapolis, Indiana. The Jersey

Reds of New Jersey and the Durocs of New York were described as two families of red swine having similar type characteristics. This history is recorded in the first volume of the *Duroc–Jersey Record* (Curtis, 1885). In the preface of the first volume, a sound rationale for including a wide range of red hogs in the breed was presented.

> We have kept out of our Record all foreign crosses, for we have a pride in the Duroc-Jerseys not only praiseworthy but strictly American. In grouping all of the families of the same color and original blood into one name, standard and breed; we take national ground with broader and more comprehensive views than to try to confine our efforts to a single family narrowed by the selfishness of a few individuals or one locality.
>
> This is not all, we are by this union of blood giving our hogs the benefit of the strength and constitutional vigor which comes from the commingling of blood not akin; thus avoiding the debilitating effects of the close inbreeding which follows breeding all in one family or with limited stock. Animals thus bred speedily deteriorate.
>
> We are satisfied that it is better for all of us to be able to draw fresh blood from both these families than to restrict our breeding to one, especially since they are all alike in their general characteristics and origin.

A claim in the preface is that 'Duroc–Jersey swine will make greater weights in the same time and on the same food than any other breed'. For many years the advertising slogan of the Duroc breed was 'Durocs Grow Faster on Less Feed'. The preface also claims the Duroc–Jerseys to be outstanding for 'docility, prolificness, hardiness, and lean meat, and they seem particularly adapted to give vitality and great growth to their progeny with whatever breed they are crossed'.

The first record association was the Duroc or Jersey Red Swine Club organized by a group of nine Wisconsin breeders in 1882. There was another meeting in Chicago in 1883 and at this meeting the name was changed to the American Duroc–Jersey Swine Breeders' Association. At this time, 36 animals had been recorded by six members (Vaughan, 1937). When the first volume of the *Duroc–Jersey Record* was published in 1885, the Association had 246 members in 20 different states. The first volume contained pedigrees of 1300 animals and there were already 500 additional entries on file for recording. The breed name was changed to Duroc in 1934 (Porter, 1993).

The history states that the origin of the Jersey Reds was unknown, but that the breed had been in existence for more than 50 years in the New Jersey area. It is suggested that the Jersey Reds descended from importations of old fashioned Berkshires. The Jersey Reds resembled the old Berkshires in many respects, but the red hogs were much coarser than the improved swine of the Berkshire breed. This statement could probably be translated to mean that the early Berkshire importations to the US (in the early 1800s) were less influenced by the Chinese breeds than the Berkshires imported to the US after 1860. Porter (1993) stated that by 1840, the Berkshire in England already had prick (or erect) ears, and, of course, the Duroc–Jerseys had drooping or lop ears similar to Old Berkshires of England.

The hogs that were united into one group to form the Duroc–Jersey breed not only included the Jersey Reds from New Jersey and the Duroc from New York but also the Red Berkshires from Connecticut and the Red Rocks from Vermont. Anderson (1931) suggested the following ancestry of the Duroc–Jersey along with some dates of importation or breed recognition:

	Jersey Red (1832)	Red Guinea Pigs (1804)
		Spanish Red Pigs (1837)
Duroc–Jersey (1877)	Duroc (1823)	Portuguese Pigs (1852)
	Vermont Red Rocks	Berkshires
		Tamworths
	Red Berkshires	Red Berkshires

The only debatable part of this ancestry is the Tamworth. In the original history, F.D. Curtis (1885) suggested that Tamworth had never been imported into the US at the time the Duroc–Jerseys were established. If the Tamworth was involved, it probably would have been only as an ancestor of the Red Berkshires.

The Duroc family of red hogs was established in Saratoga County, New York. Mr Isaac Frink took a mare to the farm of Mr Harry Kelsey to be bred to a noted trotting stallion whose name was Duroc. Mr Kelsey simply called his pigs 'red pigs', but Mr Frink called them Duroc – a compliment to Mr Kelsey's famous horse. Mr Frink purchased a Duroc boar from Mr Kelsey and the boar's services were much sought after because his progeny were 'very growthy with a quiet disposition and produced pork of excellent quality'. This first 'Duroc' boar and red hogs imported from Connecticut to New York gave the foundation animals for the Duroc breed in New York.

The wide genetic base of the Duroc breed has provided genetic variability to give the breed a high degree of genetic elasticity. During different eras of the swine industry in the US, vast changes in type have occurred. The Duroc breed has quickly responded to differential selection pressures for a variety of traits. In the late 1800s and prior to World War I, the Duroc breed emphasized extremely large size and Duroc boars were often the winners of heaviest hog competitions. In the 1930s, a trend towards much taller, trimmer bacon-type hogs occurred and Durocs led this trend. At the outbreak of World War II, there was again increased demand for lard and the Duroc breed quickly responded by producing extremely large, fat hogs with high lard yields. In the 1950s, the demand for lard declined and the so-called 'meat-type' era began. The Duroc breed quickly converted to lean, muscular hogs with a high percentage of carcass lean content. Even though some breeders were slow to emphasize carcass cutability as the main criteria for selection, there has been a steady progression in changing the Duroc to the lean, high cutability pigs demanded in most countries today. The migration of Durocs to other countries, particularly to Canada and Denmark, has allowed this breed to be selected under the national genetic evaluation and selection schemes of those countries. Many present-day Durocs found in the US and other countries around the world are descendants of high cutability lines developed in Denmark and Canada. Durocs are now being exported from Sweden.

From the early years of development, the Duroc breed emphasized pork quality as one of the strong attributes of the breed. The two Duroc lines tested in the US National Genetic Evaluation Program (NGEP), completed in 1995 by The National Pork Producers' Council (NPPC), were clearly superior for muscle quality traits (Table 2.3). Two other very important questions remain concerning muscle quality of Durocs:

1. Can the quality of muscle tissue be maintained if total carcass leanness continues to be improved?
2. Do Durocs have some inherent quality advantage independent of the degree of total fat in the carcass?

In Denmark there has been a steady decline in the percentage of intramuscular fat (IMF) in the M. longissimus for the four major breeds since 1978 (Table 2.4). It should be noted that the selection criterion for selecting herd replacements during this period was based on a selection index heavily weighted on meat content of carcasses (muscling and leanness). In Britain, during a 20-year period, average backfat was reduced by 33% and separable carcass fat by 35%, but the percentage IMF in the loin muscle was reduced by 27% (Warkup, 1993). In both the British and Danish industries there was a high correlated response of reducing percentage IMF while total carcass fatness was being reduced.

Both Barton-Gade (1990) and Warkup (1993) reported consumer concerns about pork quality. Barton-Gade (1990) suggested 2–3% IMF as ideal for fresh pork products; many Chinese suggest 3–4% as minimum levels. Barton-Gade (1990) also reported a genetic correlation of –0.30 between per cent IMF and meat content for the Danish data. She also reported heritabilities of per cent IMF to range from 0.50 to 0.76 based upon four separate studies. Based upon data from 2111 pigs of eight breeds, Goodwin (1994) estimated the heritability of per cent IMF to be 0.61 and the genetic correlation between per cent IMF and loin muscle area (LMA) to be –0.37. In the US NGEP study (NPPC, 1995), which involved 3216 pigs from ten genetic lines, the estimated genetic correlation between lipid content of the M. longissimus and LMA was –0.25. Low genetic correlations and high heritabilities indicate it should be possible to select for increased per cent IMF and meat content simultaneously. In the late 1980s per cent IMF was incorporated into the Swiss swine breeding scheme, and it has now been incorporated into the Danish programme.

Since many swine populations produce carcasses with less than 2% IMF, it seems logical that per cent IMF should be considered in breeding schemes of other countries as well. Based upon the British and Danish data, if selection to increase per cent IMF is not incorporated into breeding programmes, the trend

Table 2.3. US national genetic evaluation programme terminal sire line trait summary (1995).**

	Berkshire	Danbred HD	Duroc	Hampshire	NGT Large White	NE SPF Duroc	Newsham Hybrid	Spotted	Yorkshire
Average daily feed intake (kg)	2.66cd	2.47a	2.61c	2.57bc	2.55ab	2.62c	2.51ab	2.69d	2.54b
Feed conversion (29.5–113.6 kg)	3.07c	2.88ab	2.91b	2.92b	2.94b	2.89ab	2.83a	3.14d	2.93b
Lean efficiency	9.42c	7.81a	8.26b	8.03a	8.51b	8.37b	8.02a	9.41c	8.29b
Days to 113.6 kg	175.0b	176.5b	169.7a	175.4b	173.3ab	167.9a	172.0ab	174.2b	174.6b
Average daily gain on test (g) (29.5–113.6 kg)	840c	831c	885a	849bc	849bc	894a	863ab	835c	835c
Lean gain per day on test (g)	286c	327a	318ab	322a	295c	331a	331a	286c	309b
Carcass dressing (killing out) %	73.6	74.4	73.5	73.8	73.6	74.2	73.8	73.8	73.6
Carcass lean %	47.0c	52.0a	49.0b	51.2a	47.7c	49.8b	51.3a	47.4c	49.9b
Carcass length (cm)	82.0d	82.8bc	82.0d	82.6c	84.3a	82.3cd	83.6ab	81.8d	82.8bc
Tenth rib backfat thickness (mm)	31.8d	24.4a	28.7c	25.7a	29.7cd	28.2bc	24.4a	31.5d	26.7ab
Last rib backfat thickness (mm)	32.3d	27.9ab	30.0c	29.0bc	29.04bc	29.5bc	27.4a	32.3d	29.0bc
Loin muscle area (cm^2)	37.0c	43.5a	39.6b	42.4a	36.2c	41.0ab	41.6a	37.6c	39.8b
Ultimate pH of loin muscle	5.91a	5.75cd	5.85ab	5.70d	5.84ab	5.88ab	5.82bc	5.83bc	5.84ab
Loin muscle drip loss %	2.43a	3.34cd	2.75ab	3.56d	2.92bc	2.81ab	2.99bc	2.88b	2.85b
Loin muscle minolta reflectance – in plant cooler	22.6a	23.0a	23.2a	25.3b	23.4a	23.1a	22.7a	23.3a	23.0a
Loin muscle minolta hunter L. colour – in plant cooler	47.3a	47.7a	47.9a	49.9b	48.2a	47.7a	47.4a	47.9a	47.7a
Loin muscle lipid content %	2.41bc	2.33c	3.03a	2.57b	2.15c	2.71ab	2.25c	2.35c	2.33c
Loin cooking loss %	22.5a	24.3c	23.1ab	26.0d	22.9ab	22.5a	24.2bc	23.4ab	23.5bc
Cooked loin instron tenderness	5.74ab	5.81ab	5.65a	5.86ab	6.09c	5.78ab	6.12c	5.92bc	6.13c
Value on Index 1 ($/Pig)*	–4.14b	2.12a	0.64a	1.34a	–0.97ab	1.50a	3.53a	–4.77b	0.74a
Value on Index 2 ($/Pig)*	–4.05bc	1.52b	10.51a	1.55b	–8.87c	8.25a	0.89b	–8.20c	–1.70b

*Index 1: Includes feed conversion, days to 250 lb, 10th rib backfat.
*Index 2: Includes Index 1 traits plus loin muscle area, pH, intramuscular fat, drip loss and instron tenderness score.
**Means followed by the same letter are not statistically different ($P < 0.05$).
HD = Hampshire × Duroc F_1; NGT = National Genetic Technology; NE = Nebraska.

Table 2.4. Percentage intramuscular fat (% IMF) of loins for Danish breeds.

Breed	1978	1988	1992
Landrace	1.89	1.58	1.00
Large White	1.93	1.59	1.00
Duroc	4.15	3.21	2.05
Hampshire	2.47	1.86	1.20

Source: Barton-Gade, 1990 and Goodwin, 1994.

towards lower percentage IMF content will likely continue to levels that are unacceptable to consumers. The Duroc may therefore play an important role in improving IMF.

The Hampshire

The Hampshire, black with a white belt encircling the body at the shoulders, is another US breed that has gained worldwide prominence since the 1960s. The Hampshire was the undisputed leader of the 'meat-type' era of the 1950s in the US. The Meat Certification programme developed in the early 1950s was a progeny testing programme for all US swine breeds. During the years of this programme, the Hampshire breed qualified more Certified Meat Litters and Certified Meat Sires than any other breed. Because of superior carcass cutability, Hampshire hogs were exported to numerous countries. Hampshires have been popular as a terminal sire breed, but, like the Duroc, Hampshires have also been used in maternal lines in selected environments. Hampshire crossbred females have been popular in the Upper Midwest of the US and the breed has been incorporated into some European female lines utilized in outdoor production systems. Although in recent years the popularity of the Hampshire breed has increased rapidly in many European countries and in Canada, the breed has had difficulty in adapting to tropical environments. In many countries Hampshire × Duroc crossbred terminal sires are popular, but in tropical regions Hampshires are not frequently utilized in commercial breeding programmes because of fertility problems.

The origin of the Hampshire breed was not clearly recorded. However, there is general agreement that the ancestors of the Hampshire originated in the Hampshire area of England. Two breeds in the Hampshire area of England, the Wessex Saddleback and the Essex, have similar colour patterns to Hampshires, but those breeds are lop-eared breeds and the early Hampshire hogs had relatively small erect ears. Some authorities have suggested that the Hampshire is a direct descendant of the Old English Hog which had been crossed with the Berkshire. Depending upon the date of the Berkshire introductions, this could be the source of the erect ears.

Anderson (1931) suggested that hogs of a belted colour pattern were reported in Massachusetts as early as 1820 and in New York in 1830. A ship owner named McKay had reportedly imported belted hogs from England to Massachusetts about 1820. Those hogs spread to various areas of the US and were known as 'McKay', 'McGee', 'Ring Middle' or 'Saddle Back' hogs. Fifteen head of belted hogs from Pennsylvania were taken to Boone County, Kentucky, in 1835 by Major Joel Garnett, and these were the foundation animals for the 'Thin Rind' breed. Their numbers increased rapidly and, supposedly, improvements in the breed came about by selection within the breed with very little migration of outside genetics. The American Thin Rind Association was organized in 1893, and foundation stock from 12 herds located in Boone County, Kentucky, were allowed registration in this association. The name 'Thin Rind' was given to the breed by packers in Cincinnati, Ohio, probably because the backfat and/or the skin was thinner than for other hogs. The name of the breed was officially changed to the American Hampshire Swine Record Association in 1904 to reflect the assumed origin of the breed. Unlike the apparent broad genetic base of the Duroc, the Hampshire breed was established from those original 15 animals shipped from Pennsylvania to Kentucky in 1835.

The Hampshire breed grew in size and became one of the most influential breeds in the US by the early 1950s, with the popularity of the breed expanding rapidly from 1955 until 1980. Even though the genetic base had appeared to be quite variable during the successful years of the swine certification programme, the pedigrees became very similar in the late 1970s and early 1980s. A large majority of the pedigrees could be traced to animals from the herd of A. Ruben Edwards at Middletown, Missouri. Almost all the influential herd boars were bred by Mr Edwards or were descendants of boars from his farm. In the July 1980 *Hampshire Herdsman*, the annual herd sire influence rankings revealed that five of the top 53 sires were Edward's herd sires and 17 others were bred by his farm. Only two of the top 53 most influential sires were outcross to the Edwards programme; the other 51 had pedigrees heavily influenced by the Edwards breeding programme. This resulted in most Hampshires having linebred pedigrees with a relatively high degree of inbreeding. During this era, major reproductive problems occurred in the breed, the worst being poor libido and fertility in the boars. Inferior reproductive performance of Hampshire boars led many commercial producers desiring to use Hampshire genetics to use Hampshire crossbred boars. Hence, the reproductive problems among Hampshires contributed to the increased use of crossbred boars in US commercial swine operations. Poor reproductive performance may have resulted from reduced fitness or inbreeding depression or possibly occurred because a few very influential sires in the breed simply possessed genetically inferior reproductive performance. The answer to this question is not clear; however, the reproductive performance of the Hampshire breed has been markedly improved since this era. There were major migrations of Hampshires to Canada in the 1960s and some of those bloodlines were later exported from Canada to Sweden and from Sweden back to Canada. Some of these Canadian–Swedish bloodlines have made major contributions to the improvement of reproductive efficiency and the carcass cutability of Hampshires. Additionally, there were a few US Hampshire breeders who maintained the fertile, aggressive, high cutability bloodlines which had been so successful in earlier years. Fortunately, some superior genetic variability for these important reproductive traits was preserved.

Although the Hampshire breed has become popular as a terminal sire breed, presently there is major concern regarding the muscle quality characteristics of Hampshire-sired carcasses. The results of the US NGEP Study (NPPC, 1995) showed the Hampshire breed to be inferior for ultimate muscle pH, loin drip loss, loin cooking loss, loin colour score, loin marbling score and loin firmness score (Table 2.3). These inferior characteristics most likely resulted because of the high frequency of the *RN* (Rendement Napole) gene in the Hampshire-sired pigs (Rendement: yield in French language; *Napole: Na* veau, *Po*mmeret, *Le*chaux, names of the authors of the method of evaluating technological yield). This 'Hampshire effect' was first described by Monin and Sellier (1985) as a condition resulting in very low ultimate pH and high cooking loss during processing. The Napole technological yield (RTN) is a measure of cooked or cured weight in comparison to fresh weight, and the gene responsible for a lower yield is often referred to as the 'Napole' gene. Milan *et al.* (1995) found this condition to be the result of an autosomal dominant gene located on chromosome 15. Enfalt *et al.* (1994) reported the frequency of the *RN* gene in purebred Swedish Hampshires to be 0.72. (Further discussion is in Chapter 16.)

The Berkshire

The Berkshire breed is included because of its contribution to the development of other breeds and renewed interest in US Berkshires resulting from the breed's superior muscle quality characteristics. The Berkshire breed has gradually declined in popularity around the world since the 1940s until the only sizeable population remained in the US. Interest in Berkshire hogs has been rekindled because some major pork packers in the US have begun to pay a premium for Berkshire sired pigs (Berkshire Gold Program). The carcasses and/or wholesale cuts from these pigs are being exported to Japan to meet the Japanese demand for the superior muscle quality. The Japanese perception of the superior meat quality of Berkshires is supported by results of the US NGEP study (NPPC, 1995). Carcasses from Berkshire-sired pigs were superior for loin meat and eating quality traits (Table 2.3). This same study also revealed reasons which may explain the decline in popularity of Berkshires around the world. The breed was inferior for average daily gain, feed conversion ratio, backfat depth, loin muscle area and carcass per cent lean (Table 2.3).

According to Vaughan (1937), the Berkshire breed was the first breed of hogs selected and improved in the world. The breed originated in Berkshire and adjoining counties in England, and the foundation for the breed was the Old English Hog. The Berkshire became a recognized breed in the early 1800s, and some authorities suggest that the early Berkshire hogs played a role in development of all British swine breeds. The Chinese and Siamese hogs were used to improve the Berkshire as early as 1780 and may account for the variation in size, colour and type observed during the 1800s. The infusion of Asian pigs into the Berkshire breed probably continued until about 1850 (Porter, 1993). The Chinese genes decreased the skeletal or mature body size of the Berkshires and improved the fattening qualities of the breed. The black colour, white points and the slightly dished face had become well established by about 1869. Berkshires were exported to many other countries before and after the present-day colour and ear characteristics were established. Berkshires were first imported into the US in 1823, and the American Berkshire Society was established in 1875. The English Berkshire society was established in 1884, but the popularity of the breed has declined and now it is considered a rare breed in the UK. The American Berkshire Association had 1938 litters and 13,400 individual pigs recorded in 1995, and numbers have increased for the first half of 1996.

The Berkshire, much like the Small White and Middle White breeds in Britain, must have continued incorporating Chinese or Siamese genes into the breed for an extended period of time until there was little resemblance of the old Berkshires of the early 19th century. The Berkshires of the late eighteenth and early 19th centuries were described as large, lop eared pigs with red and black markings (Porter, 1993). It is not clear whether the continued decrease in size and increased fatness of the Berkshire resulted from additional Chinese introductions or simply from selection for the 'fancy, show type' Berkshires. There is little doubt of the Chinese contribution to the genetic make-up of the Berkshire breed. The phenotypic change from the lop-eared appearance of the Old English Hog and the early Berkshire to the prick-ears of the late 19th century is additional evidence of the change in breed composition. But similar to the situation with the Small and Middle Whites, the Chinese influence must have come from breeds other than the Taihu group because Berkshires have been noted as lacking in prolificacy. The Chinese perceive many native breeds to possess superior muscle quality. If this perception is correct, this may explain the superior eating quality characteristics of present-day Berkshires. Unfortunately, the extra fatness of today's Berkshires may be traced to the same origin. Improving carcass cutability while maintaining the breed's superior meat quality is a huge challenge for Berkshire breeders.

The Piétrain

The Piétrain may be described as the most interesting and controversial swine breed of the 20th century. Around 1920, an obscure breed of uncertain ancestry originated around the small village of Piétrain near the town of Brabant, Belgium. This breed quickly became recognized as one of the leanest and heaviest muscled breeds in existence. Although the ancestry of the Piétrain is not clear, there are several theories about the composition of the breed. Some have suggested that the Berkshire, Normand and Large White breeds were used along with local Belgian pigs in the breed development. Because the spots are sometimes rusty red in appearance the Tamworth has been suggested as one of the possible ancestors. It is also possible that the greyish-black Perigord variety of the Limousin was included in the ancestry (Porter, 1993). It is interesting to note that the Piétrain breed became almost extinct during World War II because of the increased demand for fat during that period.

After World War II, the demand for leaner fresh pork products grew rapidly in countries worldwide, but particularly in Belgium. Because of the increased demand for lean, muscular pigs, the Piétrain breed society was formed in 1952. Interest in this unique greyish-white breed with black spots quickly spread around the world. The breed was being exported to France by 1955 and a French herdbook was established in 1958. The breed was exported to Germany in 1960 and the German selection scheme produced Piétrain pigs with even greater musculature.

Much of the popularity of the Piétrain breed was due to its extreme muscling, which is due in part to the high frequency of the halothane gene in the breed. Selection schemes that emphasized leanness and muscularity also increased the halothane gene frequency. This increased gene frequency resulted in the breed being more extreme for the 'double muscled' appearance and improved the carcass cutability traits that were responsible for the rapid increase in popularity of the breed. The breed has been noted for high carcass yields, low fat content, large loin eye muscle area, more carcass weight in the hams and a very high muscle to bone ratio in the carcass (Porter, 1993). However, the detrimental effects of the halothane or porcine stress syndrome (PSS) gene have resulted in tremendous economic loss to the swine industry. When in the homozygous state, there is a tendency for pigs suddenly to collapse and die when subjected to various environmental stressors. In addition, the PSS gene, either in the heterozygous or homozygous state, has been shown to have a detrimental effect upon pork quality.

From the standpoint of a purebred breed, the Piétrain has never attained the worldwide popularity of the Large White, Landrace, Duroc or Hampshire breeds. However, the breed has been exploited by many breeding stock companies to meet increased packer demands for higher carcass cutability. Undoubtedly, the impact of the halothane gene, whether from the Piétrain or other sources, has resulted in deterioration of pork quality in most countries of the world.

Results from a study by Barton-Gade (1990) comparing halothane-positive and halothane-negative pigs are presented in Table 2.5. Because of the goal of reducing the incidence of the pale, soft and exudative (PSE) condition in pork carcasses, the Danish industry was the first in the world to attempt to eliminate the stress gene from the swine breeding populations. As a result of this strategy which began in 1986, Barton-Gade (1990) reported that the incidence of PSE meat in pigs from the Danish breeding centres had declined and was no longer a problem for Duroc, Hampshire and Large White (Table 2.6). The Landrace continued to have a small problem only in the M. longissimus. In 1996, the US National Pork Producers Council ratified a resolution to take aggressive measures to eliminate the halothane gene from the US swine population. Introductions of the Piétrain and other halothane-positive pigs into China have been

Table 2.5. Within-litter comparison of halothane-positive and halothane-negative Landrace and Large White pigs.

	Description			
	Landrace		Large White	
	Hal +	Hal –	Hal +	Hal –
No. of animals:	22	22	12	12
KK-index[1]	2.6	5.4	1.7	5.6
% PSE[2]-biceps femoris	54.5	9.1	75.0	0
% PSE-semimembranosus	57.1	4.5	91.7	0
% PSE-longissimus dorsi	90.9	27.3	91.7	33.3
% PSE-pig	90.9	36.4	91.7	33.3

Source: Barton-Gade, 1990.
[1]Danish meat quality index (scale, 0–10, values below 6.5 are unacceptable).
[2]PSE – pale, soft and exudative.

Table 2.6. PSE-incidence (%) in Danish breeding pigs from 1972 to 1989.

Year	Landrace	Large White	Duroc	Hampshire
1972–1973	39.2	—	—	—
1976–1977	26.5	—	—	—
1980–1981	14.5	9.8	3.6	5.6
1984–1985	17.0	6.2	2.3	6.2
1988–1989	7.7	2.3	2.0	2.8

From Barton-Gade, 1990.
Pigs were considered PSE if one or more of muscles measured (biceps femoris, semimembranosus and longissimus dorsi) was PSE.

few, and these breeding animals were quickly eliminated due to consumer rejection of pale, watery pork products. The wet pork market in China and Hong Kong is intolerant of any pork products that exhibit PSE-like qualities.

Even though the Piétrain is an extremely muscular breed that is relatively free of fat at weights up to 80–100 kg, there has been a problem among the Piétrain-sired pigs when heavier market weights of around 120 kg or greater are desired. Because of the small stature or skeletal size of the Piétrain, there is a marked tendency for the pigs to become slow growing and deposit subcutaneous fat rapidly at weights beyond 100 kg. This early maturity pattern is particularly true of castrates. To produce fresh pork in Italy, the Piétrain and Belgian Landrace are used extensively as terminal sires for pigs slaughtered at 90–100 kg. However, these small statured breeds are not used to sire pigs which are slaughtered at heavier weights of 160–170 kg to produce Parma ham and salami products.

Other US, Canadian and European Breeds

Many other breeds have continued to be propagated in the US, in Canada and in some European countries. Several breeds have maintained active breed societies. However, most of the minor breeds have been unable to maintain active breed societies. Breeds without breed societies have become classified as 'rare' or 'extinct' in some countries. Efforts to preserve breed genetic resources will be discussed in Chapter 14.

Other breeds that continue to make some contribution to commercial swine production include the Chester White, Spotted and Poland China breeds in the US; the Lacombe in Canada; and the Welsh and Wessex Saddleback in England. There are probably other minor breeds that continue to have some influence on commercial swine production in isolated areas. Unique genetic resources among these breeds may be a needed asset in future genetic development.

The Chester White has been promoted as the 'rugged white breed' in the US and has been used primarily as a maternal breed. Schneider *et al.* (1982b) reported an additive genetic effect of +0.2 and a maternal effect of +0.96 for number of pigs born live for the Chester White breed. However, Schneider *et al.* (1982a) also reported an additive genetic effect of +3.4 for age at 100 kg. Although the Chester White breed has gained some popularity because of its apparent superior maternal ability, the lack of growth ability has also been quite detrimental. Chester Whites tend to be smaller framed and earlier maturing than Landrace and Yorkshires; therefore, field reports among commercial producers are that the major problem with using Chester White in breeding programmes is slower growth and deposition of excess fat, particularly at heavier slaughter weights. Schneider *et al.* (1982a) reported an additive genetic effect of 0.101 for carcass backfat. The Chester White has received only limited use outside the US.

In the early 1970s, the Spotted breed opened its herd book to Poland China breeders, and for a few years, the breed gained enough popularity to become the fourth most popular breed in the US. However, the popularity of this breed has steadily declined in recent years. Results of the US NGEP study (NPPC, 1995) showed the Spotted breed to be inferior for most traits. The Poland China breed has continued to decline in popularity and is now the smallest US breed.

The Welsh is a British breed that had decreased to only 33 registered boars in 1949, but by opening the herdbook to Swedish Landrace, the breed began to gain in popularity (Porter, 1993). By 1981, the Welsh had become the third most popular breed in Great Britain. The Welsh has been exported and promoted for both terminal and maternal use in the US. Some imported Welsh boars have been heterozygous for the halothane gene.

Another white breed that has continued to see limited use in Canada is the Lacombe. The Lacombe is a synthetic breed derived from Danish Landrace, Berkshire and Chester White (Porter, 1993). The breed continues to be used as a maternal breed in Canada and has been exported to a few other countries.

The Wessex Saddleback continues as a relatively small purebreeding population in England. However, as producers have moved from intensive to more outdoor production systems, this breed has begun to regain popularity because of its adaptability to outdoor rearing conditions (Porter, 1993). The breed has also been used in the US in sow lines adapted to outdoor conditions.

The Chinese Swine Industry and Chinese Breeds

The Chinese swine industry is by far the largest of any country in the world. Pork comprises about 80% of the total meat consumed in China and pork is included in almost every Chinese dish. Based upon discussions in 1995 with Madam Wang at the Chinese Ministry of Agriculture, 412 million head of pigs were slaughtered in 1994. Other sources reported the 1994 standing population to be 330 million head and the slaughter at 290 million. Because so many of the pigs in China are slaughtered by local farmers and small operators, it is extremely difficult to get an accurate assessment of Chinese pig numbers. However, there is little doubt that China has a swine population that is at least five times as large as that of the US and possibly seven times as large. The pigs in China grow much more slowly than pigs in most other countries and the average slaughter weight is considerably lower. These facts help explain why the standing population in China is probably larger than the number slaughtered annually in the US which has a standing population of about 55 million and results in an annual slaughter of around 90 million. Several European countries have standing populations only one-half the size of the annual number slaughtered.

The breed composition of the breeding herds in China is difficult to determine, but it is very clear that there is a rapid transition from the native Chinese breeds to US and/or European breeds. Modernized swine farms use 100% imported breeding stock, and the most popular breeding scheme utilizes Landrace × Yorkshire F_1 sows bred to a terminal Duroc boar. Hampshires have also been used in terminal sire lines, but the Chinese are well aware of pork quality problems created by the use of Hampshires. Piétrain and Piétrain hybrids have also sired carcasses that were unacceptable to Chinese consumers. Madam Wang suggested that 30% of the total swine production was produced by 100% native Chinese breeds, but most small farm commercial production involves two-way or three-way crosses with local breeds. The two-breed cross utilizes a native breed for females and US or European breeds for terminal sires. The three-breed cross utilizes an F_1 female produced by crossing the native sow with a Landrace or Yorkshire boar and a boar such as a Duroc for the terminal sire. For farms using these two-way or three-way crossing schemes, Madam Wang suggested that about one-third of them use the two-breed scheme and about two-thirds use the three-breed scheme. Based upon discussions with many experts in the Chinese swine industry and upon farm visits, it appears there is a rapid transition to using a high percentage US or European breeding stock.

Even though the Meishan breed is being analysed in Europe as a possible maternal line, Chinese swine industry officials seem to be promoting the use of 100% imported genetic material. There is an effort by the Chinese government to preserve native breeds on state-owned farms. One such preservation farm is in the Dongguan area where two local breeds are being preserved, but the production potential of these breeds is so poor that most of the male pigs are destroyed at birth. Preservation efforts will be discussed in Chapter 14. This author visited several large commercial farms in four provinces in 1995 and no local breeds were utilized on these farms. In addition, discussions with large companies planning to build major swine production facilities revealed that none of these has plans for using local Chinese breeds in their breeding programmes. Although the Chinese breeds, particularly those in the Taihu group, seem to be quite prolific, the kilograms of pork produced per sow per year for those breeds appears to be considerably lower than US/European breeds. In an unpublished paper by Shi Qishun (1988, personal communication) from Hunan Agricultural College, there is discussion of 13 new breeds that have been produced by crossing Large White and Landrace with local breeds. The Hubei White was reported to be the most popular of these with 1000 purebred sows representing five lines being used in a breeding system to produce 50,000 hogs annually. Most Chinese officials discount the potential of the new swine breeds.

In his paper titled 'Chinese pig breed resources and their improvement', Shi Qishun (Hunan, China; 1988, personal communication) concluded that more than 100 local swine breeds were in existence in China. He classified the pigs into six distinct types based upon location of origin. These locations are shown in Fig. 2.3 (North China type, Central China type, Southwest type, River–Sea type, South China type, and Plateau type). Qishun's description of the six types follows:

1. *North China type* This type is distributed in the large areas in the north of the Huai River and the Qin Mountains. These pigs have large and high bodies, narrow backs and loins, strong legs, long and straight heads and mouths, big and drooping ears, thick skins, black, rough and long hairs. In winter, they grow thick, downy brown hairs. There are about eight pairs of teats. This type of pig is cold resistant and able to eat all kinds of feed. There is much abdominal fat. The litter size is over 12 piglets and there is much lean meat, like the Northeast Min Pig and the Huang-Huai-Hai Black pig.

2. *South China type* This type is distributed in the tropical mountainous areas of Yunnan, Guangdong, Guangxi, Fujian and Taiwan provinces. These pigs have short, wide and round bodies, concave backs, big and drooping bellies, plump hams and rumps. The hairs are black or black and white spot. The head is small and the skin and hair are thin. The ears are small, straight or stretched out. This type has the characteristics of early maturity, easy fattening, thick back fat, much abdominal fat, fine and tender meat, and rather low reproductive ability. There are five to six pairs of teats in each pig, like the Small Ear Pig of South Yunnan and the Liang Guang Small Spotted Pig.

3. *Central China type* This type is mainly distributed in the large areas between the Changjiang and the Pearl River. The characteristics of this type are rather wide back and loin which are usually concave, big and drooping belly, drooping ears of medium size, thin hair of black and white spot, rapid growth, fine and tender meat, and early maturity. The litter size is usually 10–13, and there are six to eight pairs of teats, like the Jinhua pig and the Ning Xiang pig.

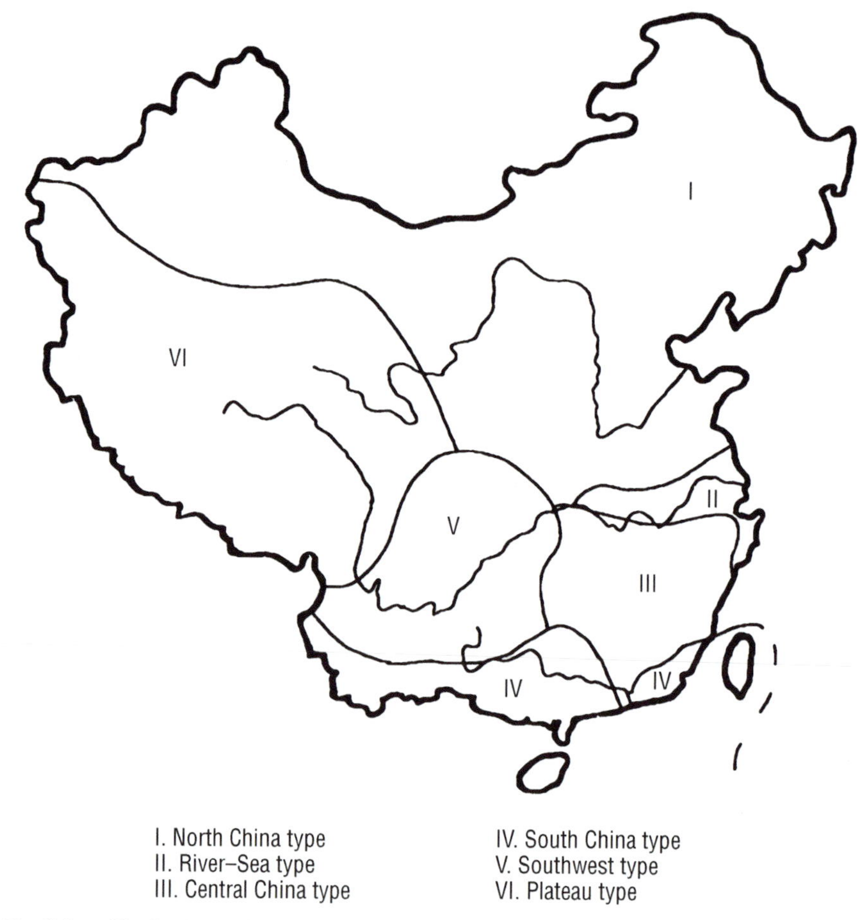

Fig. 2.3. Distribution of Chinese native pig breed types.

4. *River–Sea type* (Lower Changjiang Basin type) This type is distributed near the Han River, the middle and lower sources of the Changjiang and along the southeast coast. It is developed by the crossing of the North China type and the Central China type. It is famous for its high reproductive ability. The litter size is generally over 13, sometimes over 20. There are eight pairs of teats. The wrinkles on the forehead are deep and usually in a rhombus shape. Ears are big and drooping. The hairs are black or some of them have white spots, like the Taihu pig and Jinangquhai pig.

5. *South-west type* This type is distributed over large areas of Sichuan, Yunnan and Guizhou provinces. The characteristics of this type are bigger bodies and big heads, short and strong necks. There are wrinkles on the forehead. The hairs are all black or 'six white', some have black and white spots or brown colour. The litter size is generally eight to ten, like the Neijiang pig and Guanling pig.

6. *Plateau type* This type is distributed on the plateaux of Qing Hai and Tibet, over 3000 m above sea level. The characteristics are small bodies, narrow, long and straight heads, small and straight ears, narrow backs, small bellies, and thick, long, black hairs. This type is of late maturity and adapts to life in the plateau areas. The litter size is five to six. There are five pairs of teats, like the Zang pig.

Qishun described the Chinese native pig breeds as having high reproductive rates, fine meat quality, good adaptability for extensive feeding and management and the ability for high utilization of green-crude feed. But the undesirable traits are slow growth rate, low dressing percentage and low lean meat percentage. The following are Qishun's descriptions of the general evolutionary characteristics of the Chinese pigs:

1. The body size is large in the north and small in the south. For example, the average body weight of brood sow, Min pig is 148 kg, Huang-Huai-Hai Black pig is 115 kg, Jinhua pig is 97 kg, Liang Guang Small Spotted pig is 92 kg, and South Yunnan Small Ear pig is 69 kg.
2. The hair colour is black in the north and spotted in the south. For example, North China type pigs are usually black, but those in the south of the Changjiang River are black in the head and rump, or black and white spot, or black back bar.
3. The litter size of Taihu pig is one of the largest, but towards north, west and south, the litter size reduces gradually. Taking the litter sizes in the third and the later litters as examples, Taihu pig is 15–16; towards north, Jiangquhai pig is 13–14; Huai pig is 12–13; Min pig is 13–14; towards west, Wai pig is 13–14; NingXiang pig is 11–12; Neijiang pig is 10–11; Zang pig is 6–7; towards south, Jinhua pig is 13–14; Liang Guang Small Spotted pig is 10–11; South Yunnan Small Ear pig is 9–10.

There is little question regarding the superior reproductive performance of most Chinese breeds, and in particular, the prolificacy of the Taihu breeds. Cheng (1983) reported that Taihu boars can mate and fertilize females as early as 3 months of age and generally are 100% fertile by 4 months of age. In China, first service of young boars begins at 6–8 months of age. Xue (1991) presented reproductive performance of some traditional pig breeds in China (Table 2.7). The unusually young age at puberty and sexual maturity appear quite consistent among the Chinese breeds and may be related to the inherent fatness of the breeds. Litter size difference between sows and gilts appears to be greater than for US and European breeds (Table 2.8). The likely explanation is that the Chinese gilts farrow the first litter at a much younger age than US or European breeds. There also appears to be a major difference in level of production in the rural areas as compared with the pig breeding farms (Table 2.9). Whether this is related to genetics or the environment is difficult to determine, but the improved level of feeding and management on the larger state-owned farms is a likely explanation. The excellent longevity of the Chinese breeds (Table 2.10) is an extremely desirable attribute that has often been overlooked. The average litter size for some of these breeds is much larger than for US or European breeds. Xue (1991) reported finding a Dingyuan black sow that had farrowed 27 litters with no significant decrease in reproductive rate.

Table 2.7. Reproductive performance of some traditional Chinese pig breeds.

Breed	Female			Male		
	Puberty (days)	1st service		1st mounting behaviour (days)	1st mating	
		Age (days)	Body wt (kg)		Age (days)	Body wt (kg)
Min	90–120	240	—	—	270	—
Large Black-White	90	—	—	40	90	—
Daweizi	120	150	25–35	90	120	30
Wei	60–90	180	30–40	60–90	180	30–40
Jinhua	65	90–130	20–40	54	300	—
Jiangquhai	100	120–150	—	—	120–150	—
Taihu	60–70	120–150	20–30	—	—	—
Dongchuan	90–120	120–150	30–35	90	—	—

Source: Xue, 1991.

Table 2.8. Litter size of some traditional Chinese pig breeds.

Breed	Place of origin	Gilt		Sow	
		No. of litters	No. of piglets	No. of litters	No. of piglets
Min	Heilongjiang, Jilin & Liaoning	78	11.40	273	13.26
Huai	Huai River Valley	397	9.16	1139	12.98
Shenzhou	Hebei	194	10.10	159	12.80
Hainan	Hainan	65	9.50	96	12.20
Large Black-White	Guangdong	116	12.10	587	13.40
Jinhua	Zhejiang	125	10.50	270	13.78
Longyou	Zhejiang	409	8.31	1594	12.91
Qingping	Hubei	31	9.42	269	12.12
Taihu	Taihu Park Valley	1218	12.14	2862	15.83
Jiangquhai	Jiangsu	162	9.96	608	15.83
Dongchuan	Jiangsu	299	10.63	769	14.19
Wei	Anhui	80	8.31	366	12.12

Source: Xue, 1991.

Research efforts in Europe and the US have been aimed at explaining the unique reproductive characteristics of the Chinese breeds. Lower embryonic mortality has been theorized as one of the explanations for the increased litter size of the Chinese breeds. Bolet *et al.* (1986) compared Meishan, Large White and 'hyperprolific' Large White sows for litter size born, ovulation rate and embryonic mortality rate (Table 2.11). This study reported no difference between the Meishan and Large White, but the 'hyperprolific' Large White sows ovulated more eggs. Embryonic mortality was higher in the Large White than Meishan and much higher among the hyperprolific Large White sows.

Gueblez *et al.* (1986) compared one-half and one-quarter Chinese sows with Large White × Landrace F_1 sows in a field study in France (Table 2.12). The Meishan sows in this study were one-half Large White and three-quarters Large White. He reported an advantage of 26% and 8% in pigs/sow/year for the one-half Meishan and one-quarter Meishan sows, respectively, compared with the Large White × Landrace controls. The study also reported a decreased dressing and carcass lean percentage for the Meishan crossbred pigs. The researchers concluded that the increase in sow productivity was insufficient to compensate for the inferior carcass cutability of the Meishan crosses. In trial one, the

Table 2.9. Comparison of litter size (number of piglets born) between rural areas and breeding units in China.

	Gilt		Sow	
Breed	Rural area	Pig unit	Rural area	Pig unit
Ningxiang	8.3	8.0	9.7	11.5
Jinhua	—	—	11.9	14.3
Jiangquhai	7.6	10.2	11.9	14.2
Taihu	7.4	11.9	—	15.0
Rongchang	6.9	9.7	9.9	12.0
Chenghua	7.5	9.0	9.4	10.4
Yanan	8.1	8.4	9.9	12.4
Taoyuan Black	6.5	7.5	9.4	11.5

Source: Xue, 1991.

Table 2.10. Number of piglets at each farrowing of some traditional Chinese pig breeds.[1]

	Parity											
Breed	1	2	3	4	5	6	7	8	9	10	11	12
Min	30	21	23	21	14	6	10	6	7	5	4	3
	11.97	14.86	14.74	14.14	16.36	14.62	14.00	17.36	17.86	17.00	18.25	15.67
Large	116	111	111	99	84	56	56	47	45	33	29	18
Black-White	12.10	12.90	13.40	13.80	13.40	14.10	14.30	12.90	13.04	13.60	12.90	11.40
Daweizi	82	71	61	49	41	29	18	11	10	10	10	10
	8.56	10.55	10.11	10.41	11.00	12.70	12.10	12.20	11.10	10.90	10.60	11.30
Wei	23	17	14	18	17	11	6					
	9.35	10.88	11.50	11.98	11.94	12.45	13.00					
Jinhua	174	161	150	136	90	62	34	20	13	6	7	61
	7.25	8.57	11.14	11.21	12.01	12.35	13.50	13.55	12.70	13.74	13.74	12.71
Jangquhai	116	103	74	58	14	39	26	16	12	10		
	7.60	9.20	10.80	11.90	11.90	11.80	12.00	12.00	14.10	12.50		
Taihu	216	197	84	83	74	46	27					
	7.37	10.70	12.83	13.60	16.87	14.15	14.00					

Source: Xue, 1991.

[1]The first line represents number of statistical litters and the second line is number of piglets per litter at birth for each breed.

one-quarter Meishan crossbred pigs had an advantage in meat quality index, but there was no significant difference in trial two.

The oestrogen receptor gene has been identified in the Meishan by Rothschild *et al.* (1994). The presence of this favourable allele helps to explain the superior litter size of the Taihu breeds. The same allele has been found in some other breeds. Identification and selection to increase the gene frequency is warranted for maternal lines. Locating this allele stimulates interest in the development of hyperprolific strains in the future, but it may be possible that

Table 2.11. Comparison of Meishan (MS), Large White (LW) and 'Hyperprolific' Large White (HLW) sows for litter size born (LSB), ovulation rate (OR) and embryonic mortality rate (EMR).

	No. of sows			Mean ± standard error		
Genetic type	LSB	OR	EMR (%)	LSB	OR	EMR (%)
MS ± 9	16	17	16	15.7 ± 1.5	17.2 ± 1.2	16
LW ± 7	20	15	13	12.1 ± 1.1	17.6 ± 0.8	26
HLW ± 7	25	18	17	13.1 ± 1.3	22.9 ± 1.0	41

Source: Bolet *et al.*, 1986.

Table 2.12. Field evaluation of crossbred Large White–Meishan Sows.

	Trial 1		Trial 2	
Trait	1/2-MS[1]	Control	1/4-MS Control	Control
No. of sows	234	—	219	3183
No. of litters	662	—	477	7098
Litter size born	13.9	—	11.8	11.0
Litter size born alive	13.2	—	11.2	10.3
Litter size weaned	10.8	—	9.4	9.0
Weaning to rebreeding interval (days)	7.9	—	9.0	14.1
Pigs weaned sow^{-1} $year^{-1}$	26.4	—	22.6	21.0
No. of pigs[2]	76	77	101	100
No. of sires[3]	36	36	54	54
Average daily gain (g day^{-1})	836	858	825*	852
Feed-to-gain ratio	3.07	3.00	2.98	2.94
Dressing per cent	71.1**	71.8	72.3**	72.9
Carcass lean (%)	49.4**	53.6	51.7**	54.2
Meat quality index	10.7*	9.9	10.3	10.6

Source: Gueblez *et al.*, 1986

[1]1/2-MS = Large White × Meishan; Control = Large White × French Landrace; single control for prolificacy means.

[2]Pigs produced by part Chinese sows are 1/4 or 1/8 Meishan.

[3]Part Chinese and control pigs sired by the same European boars.

*Part Meishan and control means differ ($P < 0.05$).

**Part Meishan and control means differ ($P < 0.01$).

quantitative selection procedures are doing an effective job of identifying animals with this gene. (For additional details see Chapter 11.)

The Meishan is resistant to the K88ac strain of *Escherichia coli* (Michaels *et al.*, 1991). This resistance exists because of a lack of receptor or attachment sites for the bacteria on the lining of the small intestine. Some degree of resistance was also found among the Minzhu and Fengjing breeds. The possibility of transgenic transfer of this resistance could also be a reality in the future if the gene can be isolated.

The extent of use of Chinese germplasm in breeding programmes outside China will likely depend on the degree of success in increasing the carcass cutability of the new synthetics while maintaining the desirable attributes of the Chinese breeds. Widespread use of the Chinese breeds may also depend upon how well the carcasses sired by ultra high cutability boars are accepted by consumers in the future. The Meishan breed has already been incorporated into hyperprolific female lines by some European seedstock companies. In addition, an Irish firm (Hermitage A.I. and Pedigree Pigs) offers semen of Meishan boars and of the Hermitage Hysan, an F_1 resulting from Meishan × Landrace.

Migration – Past and Future

Since the 1700s when modern swine breed development was occurring in England, migration of genetic material between countries and within countries has provided the genetic variability necessary to allow major phenotypic and genetic changes in breed populations. Health restrictions of the 20th century have complicated the movement of genetic material. However, migrations have significantly contributed to changes in the economically important traits of world pig populations. Even though most European countries have breeds derived from native swine populations, the majority of the breeds utilized today have been developed by the use of genetic material migrated from other European countries, Canada or the US. Likewise, European genetic lines have had a major impact upon the swine populations of the US, Canada and most Asian countries. The Asian breeds were influential in the early development of the European breeds.

These migrations in the past have been conducted primarily between privately owned pedigreed seedstock producers who have been eager to share their genetic material for swine industry improvement. These migrations have generally followed the pattern described by Lush (1945) with a period of inbreeding or selection followed by migrations of animals with 'outcross' pedigrees. During recent years, a significant amount of genetic material has become controlled by 'breeding stock companies' who have the policy of selling only 'non-reproducible' genetic material. In other words, these large companies are able to access the genetic material from private sectors of the world, but they generally do not allow reverse migration of reproducible genetic materials to the private sector.

Because of this outward movement of superior genetics with no opportunity to share in future genetic developments from the companies, breeding schemes such as those used in Sweden, Finland and Denmark may need to be considered by other countries of the world. These countries use across herd breeding value estimations and/or progeny testing to identify the outstanding young sires. These sires are then made available to the entire seedstock industry through artificial insemination centres that sell semen from these boars. This procedure effectively increases breed population size and allows greater selection intensity for males. This method offers the private sector an advantage of increased population size over the largest of the breeding stock companies. These types of programmes provide an opportunity for private pedigreed breeders to continue to compete in the production of superior seedstock for use in commercial swine production. Identification of outstanding young sires and widespread use of fresh and frozen semen may make worldwide genetic migrations an even greater factor in breed improvements of the future.

Breed Purity

Since the early days of breed development, attempts to develop 'pure' or 'true' breeding groups of animals have been a major goal of livestock breeders. The idea of 'fixing' characters in breeds of livestock began with the great English stockman, Robert Bakewell, and continued with many of the apprentice breeders who used Bakewell's principle of 'mating the best to the best regardless of relationship'. A major component of breed society formation was to establish criteria for registration in each respective society or association. The criteria generally used are those characteristics of the breed that make them different from other members or breeds in the species.

Over the years the criteria have been revised for some breeds as the breeds have strived to develop procedures to ensure purity. Most Yorkshire and Landrace breed societies of the world have recognized that their breeds were developed from white animals with some degree of spotting occurring in the rump region (see Chapter 3). Most breed societies originally allowed animals with some black spotting in the rump area to be recorded. Since the 1960s, the degree of spotting has increased in some countries and there have been numerous reports of black spots or 'masking' around the eyes of some purebred animals. It is difficult to ascertain if these deviant colour patterns resulted from segregation of the original genes that were present in the breeds at the time of formation or if there have been illegal migrations of coloured or spotted breeds into these white breeds. The Hampshire breed originated as a black hog possessing erect ears and a white belt encircling the body at the shoulders. Even though the 'belt' was recognized as a characteristic of the breed, it was only included as a requirement for registry in 1932. Since the 1960s, the recessive red gene has been found in the Hampshire breed. Although the red gene could have resulted from a mutation, there is speculation that the gene was illegally or accidentally introduced from the Duroc breed. In the first volume of the *Duroc–Jersey Record*, there is mention of pigs with white feet or white legs with occasional black spots. However, the rules for registry require that the Duroc be solid red in colour with no black spots larger than a US quarter.

Because of speculation that migration of genes from other breeds may have occurred, the Yorkshire, Landrace and Hampshire breeds in the US require that boars in artificial insemination stations be 'test mated' for colour before AI registration certificates will be issued. The Yorkshire or Landrace boars must be mated to two coloured females and have two litters resulting in at least eight white pigs and no coloured, masked or spotted pigs. The Hampshire boars must be mated to at least two red sows and have two litters farrowed with at least eight black pigs and no red pigs in the litters. Since this policy has been implemented, several boars have failed to pass the 'colour test', and therefore, their AI sired progeny are not eligible for registration. The majority of the white boars that have sired coloured pigs have been imported from Europe. Debate continues as to whether the colour or spotting present in the white breeds is residual from the foundation animals. (See Chapter 3 for discusssions on coat colour.)

It is possible that ineligible colour patterns resulted from the illegal introductions of breeds such as Piétrain or Hampshire in an effort to accelerate the rate of genetic progress towards high cutability carcasses. If the latter is true, the crossbred animals with high cutability ancestors would more likely be selected because most selection schemes around the world are designed to select the high cutability animals regardless of other traits such as reproductive efficiency and maternal ability.

The major concern about the migration of outside genes relates to expected heterosis of future offspring, to the uniformity of animals and to the loss of natural superiority for traits in a breed that may not be included in trait evaluation and selection schemes. There is concern in the white breeds relative to incorporating breeds that may later be used as terminal sires. This could result in a decrease in the degree of hybrid vigour. With regard to the loss of uniformity, this phenomenon could possibly play a role in prolonging the 'close out' period for finishing units if the 'mixing' contributed to more variability in the genetic potential for growth rate. Probably the most serious concern for the white breeds is a rapid increase in the percentage composition of the 'outside' breed. In countries such as Canada where the Estimated Breeding Value (EBV) index is based upon backfat thickness and growth rate, animals with an infusion of 'terminal sire' genes will most likely have higher index values than the true purebreds. Therefore, if selection over time is based upon this EBV index, it is possible that a large proportion of the genes in a breed population would become 'outside' genes. In the absence of selection pressure for age at puberty, fertility, litter size, mothering ability and longevity, a breed that has been noted for being outstanding for maternal traits could lose the advantage of inherent maternal superiority in just a few generations.

Breed Societies – Past and Future

The primary function of all breed societies (registries or associations) is to record and maintain records of ancestry. Sellier and Rothschild (1991) listed the following as functions of swine breed societies:

- maintenance of breed purity by preventing migration of outside genes
- registration of pedigrees on all animals (litter recording)
- definition and maintenance of 'true type' characteristics
- sponsorship of livestock shows largely based on phenotypic excellence as judged by eye
- assistance of breeders in advertising and merchandising.

For many breed societies, publication and distribution of the breed magazine has been an important means of providing information to purebred breeders and promotion of the breed to prospective customers. Several US breed associations publish an annual herdsire reference edition which provides a source of pedigree information for breeders, educators and others interested in pedigrees of the various breeds. In the 1990s, the US breed associations have attempted to incorporate performance information as part of the criteria for judging during live shows. In fact, this has been used as a means of encouraging participation in the genetic evaluation programme by breeders and as an educational tool for both seedstock and commercial producers.

The relationship between breed societies and governments varies greatly around the world. In the US there is no formal government input into the function of breed associations; however, in many countries, there are often close ties with the respective ministries of agriculture. In the US there is often input from the state universities with active swine breeding and extension programmes.

The US meat certification program conducted from 1953 to 1973 was a highly successful progeny testing programme which could be given major credit for converting the US swine industry from the production of 'lard-type' pigs to 'meat-type' pigs. This programme was implemented solely by the breed associations with no government involvement. The success of this programme indicates the kind of genetic progress that can be made when members of a breed society work together to accomplish a breeding objective.

In the future, breed societies have an even greater opportunity for encouraging genetic change in the swine industry. By the use of BLUP procedures for conducting across-herd genetic evaluations, the opportunity to make genetic change appears greater than ever before. The Canadian estimated breeding value programme was the first of its nature to be implemented. Now both seedstock producers and commercial producers alike make widespread use of the data from this programme. In Canada, the sire summaries are published quarterly and the early summaries of the US have been published semi-annually. Dr John Mabry and his associates at the University of Georgia made a programme available to US breed associations which allows across-herd analyses to be calculated daily. Recent developments in computer software and hardware technology make it possible to calculate up-to-date across herd expected progeny differences (EPDs) on each contemporary group of animals as their phenotypic test data are accumulated. Because genetic evaluations are conducted daily, more timely results can be made available to seedstock breeders and commercial producers.

Consequently, if breed associations are successful in encouraging private breeders to use these new programmes, breed associations could play a much greater role in swine genetic improvement of the future. In order for genetic evaluations to have significant impact, purebred breeders must actively participate in the programmes and use the results (EPDs) to guide their breeding programmes. In addition, the breed associations must actively promote the use of EPDs to commercial producers. Breed associations may also be effective by implementing progeny testing programmes, which often prove useful in evaluating traits such as muscle quality.

References

Anderson, A.L. (1931) *Swine Enterprises.* J.B. Lippincott Company, Chicago, 458 pp.

Barton-Gade, P.A. (1990) Pork quality in genetic improvement programs – the Danish Experience. *Proceedings National Swine Improvement Federation.* Annual Meeting, Des Moines, Iowa, vol. 15, pp. 10–19.

Bassett, J.M. (1996) Reacting to the rhythm of summer infertility. In: Evans, A. (ed.) *Pigs.* Missett. Vol. 12:2 pp. 8–10.

Bokonyi, S. (1974) *History of Domestic Mammals in Central and Eastern Europe.* Akademiai Kiado, Budapest, 597 pp.

Bolet, G., Martinat-Botte, F., Locatelli, A., Gruand, J., Terqui, M. and Berthelot, F. (1986) Components of prolificacy in hyperprolific Large White sows compared with the Meishan and Large White breeds. *Genetics, Selection, Evolution* 18, 333.

Cheng, P.L. (1983) A highly prolific breed of China – the Taihu pig, I and II. Commonwealth Agricultural Bureaux. *Pig News and Information,* 4, 407–425.

Curtis, F.D. (1885) History of Duroc–Jersey swine of the United States. In: *American Duroc–Jersey Record,* Springfield, Illinois, pp. 16–20.

Darwin, C. (1868) *The Variation of Animals and Plants under Domestication,* Vols. I, II and III. John Murray, London.

Enfalt, A.C., Lundstrom, K., Lundkvist, L., Karlsson, A. and Hansson, I. (1994) Technological meat quality and the frequency of the RN-gene in purebred Swedish Hampshire and Yorkshire pigs. *Proceedings of 40th International Congress of Meat Science and Technology,* The Hague, The Netherlands.

Goodwin, R.N. (1994) Genetics of pork quality. *Proceedings of National Swine Improvement Federation.* Annual Meeting, Des Moines, Iowa, Vol. 19, pp. 15–23.

Gueblez, R. Legault,, C., Bruel, L. and Le Henaff, G. (1986) Reproduction and production results of crossbred Chinese pigs under field conditions in France. *Proceedings of 37th Annual Meeting of European Association Animal Production,* Budapest, Hungary, vol. 1, p. 121

Hammond, J. (1962) Some changes in the form of sheep and pigs under domestication. *Zeitschrift für Tierzucht und Züchtungsbiologie* 77,156–158.

Hetzer, H.O. and Miller, R.H. (1972) Rate of growth as influenced by selection for high and low fatness in swine. *Journal of Animal Science* 35, 730–742.

Jonsson, P. (1991) Evolution and domestication, an Introduction. In: Maijala, K (ed.) *Genetic Resources of Pig, Sheep and Goat.* Elsevier, Amsterdam, Oxford, New York and Tokyo, pp. 1–10.

Lerner, I.M. (1954) *Genetic Homeostasis.* Oliver and Boyd, Edinburgh and London, 134 pp.

Lush, J.L. (1945) *Animal Breeding Plans.* The Iowa State College Press, Ames, Iowa, 443 pp.

McLaren, D.G. (1990) World breed resources. In: Young, L.D. (ed.) *Genetics of Swine.* NC-103 Regional Research Report (Meat Animal Research Center, Nebraska) pp. 11–36.

Mason, I.L. (1969) *A World Dictionary of Livestock Breeds, Types and Varieties.* 2nd Edition. CAB International, Farnham Royal, UK, 348 pp.

Michaels, R.D., Whipp, S.C. and Rothschild, M.F. (1991) *Resistance to K88 E. coli in Chinese Pigs.* 1991 Iowa State University Swine Research Report ASL-R859. Iowa State University, Ames 50011.

Milan, D., LeRoy, P., Woloszyn, N., Caritez, J.C., Elsen, J.M., Sellier, P. and Gellin, J. (1995) The RN locus for meat quality maps to pig chromosome 15. *Genetics, Selection, Evolution* 27, 195–199.

Monin, G. and Sellier, P. (1985) Pork of low technological quality with a normal rate of muscle pH fall in the immediate postmortem period: the case of the Hampshire breed. *Meat Science* 13, 49–63.

NPPC (1995) *Genetic Evaluation: Terminal Line Program Results.* National Pork Producers Council Publication. Des Moines, Iowa.

Plager, W. L. (1975) *History of Yorkshires and American Yorkshire* Club. American Yorkshire Club, Lafayette, Indiana, 333 pp.

Porter, V. (1993) *Pigs: A Handbook to the Breeds of the World.* Cornell University, Ithaca, New York, 256 pp.

Rothschild, M.F., Jacobson, C., Voske, D.A. and Tuggle, C.K. (1994) *Discovery of a Major Gene for Litter Size in Pigs.* 1994 Iowa State University Swine Research Report ASL–R1178. Iowa State University, Ames 50011.

Schneider, J.F., Christian, L.L. and Kuhlers, D.L. (1982a) Crossbreeding in swine: genetic effects on pig growth and carcass merit. *Journal of Animal Science* 54, 747.

Schneider, J.F., Christian, L.L. and Kuhlers, D.L. (1982b) Crossbreeding in swine: Genetic effects on litter performance. *Journal of Animal Science* 54, 739.

Sellier, P. and Rothschild, M.F. (1991) Breed identification and development in pigs. In: Maijala, K, (ed.) *Genetic Resources of Pig, Sheep and Goat.* Elsevier, Amsterdam, pp. 125–141.

Stahly, T.S., Williams, N.H. and Swenson, S.G. (1994) *Interactive Effects of Immune System Activation and Lean Growth Genotype on Growth of Pigs.* 1994 Iowa State University Swine Research Report ASL-R1165. Iowa State University, Ames 50011.

Vaughan, H.W. (1937) *Breeds of Livestock in America.* R.G. Adams and Company, Columbus, Ohio, 780 pp.

Warkup, C. (1993) Improvement of pigmeat quality. *Proceedings National Swine Improvement Federation.* Annual Meeting, St Louis, Missouri, vol. 18, pp. 16–24.

Weir, H. and Coleman, J. (1877) *The Sheep and Pigs of Great Britain. A series of articles on the various breeds of sheep and pigs of the United Kingdom, their history, management, etc.* 'The Field' Office, London, 214 pp.

Xue, J. (1991) *Highly Prolific Swine Breeds of China. World Review of Animal Production,* vol. 2, pp. 55–57.

Genetics of Colour Variation 3

C. Legault
INRA, Station de Génétique Quantitative et Appliquée,
78352 Jouy-en-Josas cedex, France

Introduction

Most of the work on colour inheritance in pigs occurred during the first half of the century. Wright (1918) undertook a first overall review dealing with mammals. This was followed by other surveys that dealt especially with pig coat colour (Wentworth and Lush, 1923; Kosswig and Ossent, 1931, 1932, 1934; Smith *et al.*, 1938). The main body of experimental data, however, remains the series of eight papers published by Hetzer from 1945 to 1954 (Hetzer,

1945a,b,c,d, 1946, 1947, 1948, 1954). Since this time, little work has been done on the genetics of pig coat colour.

Searle (1968) discussed the possible homologies between pig colour genes and those of other mammalian species. Fourteen years later, Ollivier and Sellier (1982) published an exhaustive review of 'pig genetics' including a chapter on coat colour. The most recent review by Rempel and Marshall (1990) contained new hypotheses concerning *red-dilution* and *white-head* genes.

As was concluded in these two most recent publications (Ollivier and Sellier, 1982; Rempel and Marshall, 1990) and following the work of Hetzer, genetics of coat colour seems to be clearly established for several aspects including the 'dominant white' (*I* locus), the colouration extension (*E* locus) and the white belt (*Be* locus). Unfortunately, genetics of many colour variations remains obscure and inconclusive. In fact, with the exception of Hetzer's experiments, most observations concerning colour are secondary observations derived from experiments that were designed to study predominantly economic traits in pigs (reproductive ability, growth rate and carcass quality). In addition, the size of the samples and the number of observed generations (or crosses) were not large enough to lead to definite conclusions. Most of these studies were concerned only with a limited number of European and North-American breeds. The Asian breeds (in particular, the Chinese breeds) were not included in these earlier investigations. However, due to the past importance of breed standard in selection, pure breeds have the advantage of being homozygous for the most common or well identified alleles involved in coat colour. This is not true for traits such as white marks which can be confused with other coloration patterns.

There has recently been renewed interest in this area, however, as colour genes are being considered as possible markers in addition to those already in use, such as blood, biochemical and molecular markers. Colour variants may also be useful in identifying the components of some specific crossbreeding schemes as well as contributing to the image associated with high quality regional products.

This review updates the two previous surveys on genetics of pig coloration (Ollivier and Sellier, 1982; Rempel and Marshall, 1990) and includes some previously unpublished material in this area which deals with a larger number of European native breeds as well as three Chinese breeds that were imported into France in 1979 (Legault, 1997). This chapter will also take into account some of the recent advances in molecular genetics and mapping which have led to the chromosomal mapping of two coloration loci (Johansson *et al.*, 1992; Kijas *et al.*, 1996; Mariani *et al.*, 1996).

Inventory of the Main Colour Types and Breeds

Wild type

This type was described by Ollivier and Sellier (1982) as follows:

The wild colour, similar to agouti in rodents, is characterized by a yellow subterminal band on an otherwise dark dorsal hair and also often characterized by a variable colour intensity according to the body region. A peculiarity of the wild pig is that the piglets at birth exhibit longitudinal stripes which gradually disappear later in life. This may also occur in some domestic breeds (e.g. Mangalitza or, to a lesser extent, red breeds like Duroc).

Longitudinal stripes also segregate at a low frequency in primary populations (i.e. Corsican pig, Caribbean créole pig) as well as in crosses or composite lines including red, black and white pigs.

Uniform black

This type is illustrated by the English Large Black, the French Gascon, the German Cornwall and specific varieties of the Iberian breed. The uniform black coat is the most common type among the numerous native breeds from China (Epstein, 1969; Zheng, 1984; Zhang, 1986) and Vietnam. This type segregates with a high frequency in the various native populations of Iberian origin such as Corsican, African and Caribbean créole pigs.

Uniform red

Three breeds show this colour, namely the Duroc, Tamworth and Minnesota No.1. A red variety has also been reported in the Mangalitza and Iberian breeds. This type of coloration, which is common in Mediterranean native populations as well as in African and American native populations of Iberian origin, seems to be absent from the Asian native breeds.

'Domino' black spotting

Black spots generally occur on a white background, but red hair in variable amounts may be mixed with the white, up to a uniformly red background. Such a pattern is found in Spotted Poland China (now called Spotted), Gloucester Old Spot, Piétrain, Bayeux, Polish Zlotnicka and several native breeds from central Europe and Russia. It is also segregating into various native populations of Iberian origin. The term 'domino' was proposed by Lauvergne and Canope (1979) to signify a large number of black spots of medium and irregular size spread over the whole body, except generally on the legs, the forehead and the extremity of the tail (Fig. 3.1).

The term 'dalmatian spotting' could be used to signify a large number of small black spots on a red background usually observed in Piétrain × Duroc and Piétrain × Minnesota No.1 crosses.

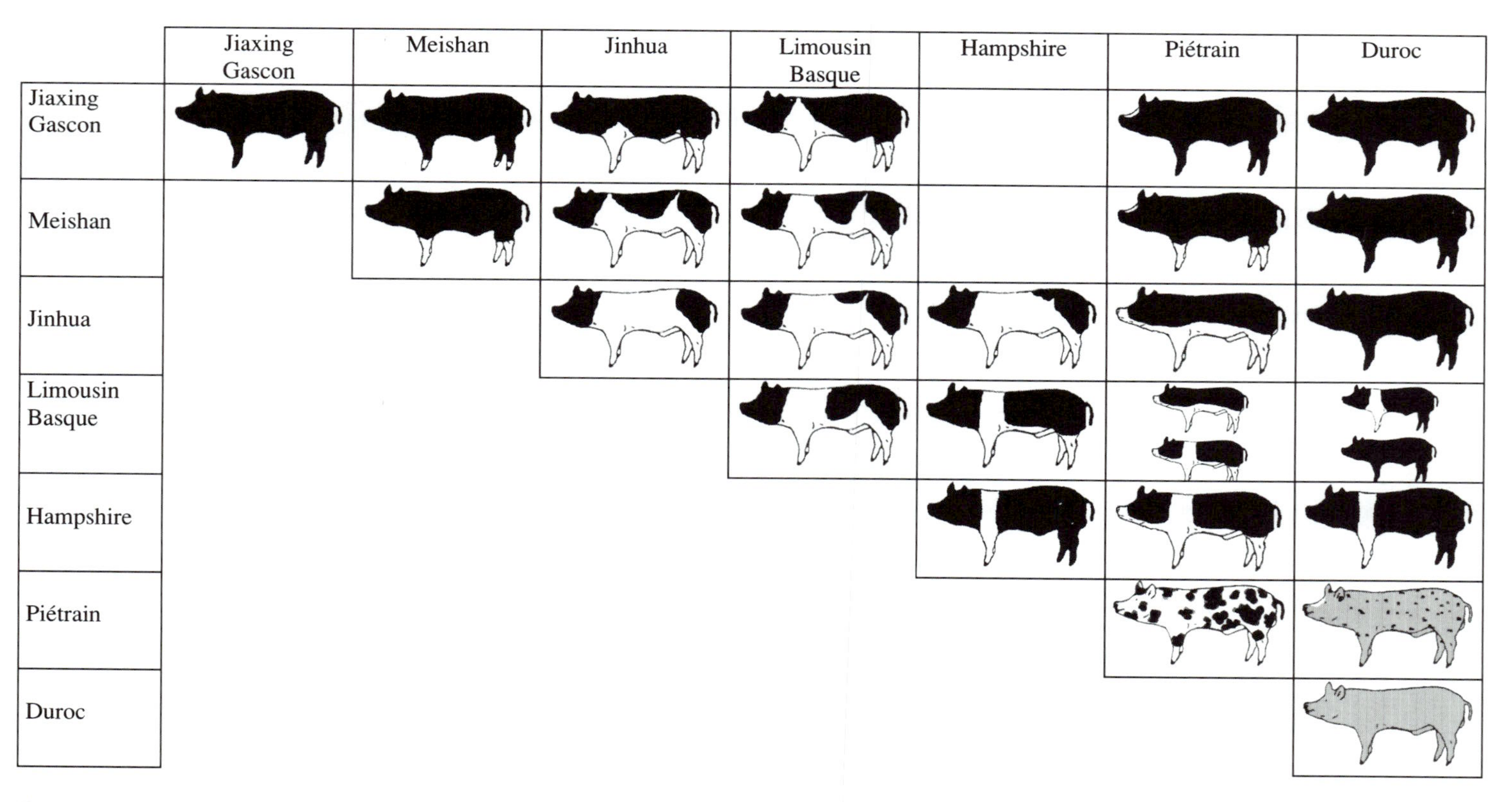

Fig. 3.1. Dominant colour pattern in some coloured pure breeds (on the diagonal) and their crosses.

Black and red piebald

As opposed to 'domino' represented by a large number of fairly small black spots located everywhere on the body, the piebald pattern consists of a small number of large black or red spots located primarily on the head and on the rump with possibly intermediate spots on the top of the back. Three variants may occur (Fig. 3.1).

Black pied with black head
This is probably the most common type represented by the French native breeds, Limousin and Basque, and by 13 Chinese native breeds including Meishan and Jinhua described by Zhang (1986).

Black pied with white marks on head
This type is illustrated by 10 Chinese native breeds described by Zhang (1986), including the Dahuabei (Pearl River breed) and by the Moncaï Vietnamese native breed. When the background of the coat is red, we obtain the Hereford type.

White belt
Pig breeds having a white belt on a black background are the Hampshire, Essex and Wessex British breeds. When the background is red, we have the Bavarian Landschwein. As will be discussed later, the white belt can be confused with pie patterns in case of a large extension of the white as in the Chinese Jinhua breed.

All these types and their combinations can occur in the various native populations of Mediterranean origin as well in the foundation stock of the composite strains including Duroc, Hampshire, Piétrain and Large White.

Black with white points

The coat is uniformly black with the exception of six white points (feet, tail and snout) in Berkshire and Poland-China, or with the exception of four white socks in the Meishan breed.

White

Two types of white coat exist, a shiny white with a usually white skin as in Large White (Yorkshire), Middle White, Chester White, Lacombe, Blanc de l'Ouest and Landrace of various European countries, and a 'dirty' white on a pigmented skin as in Mangalitza.

The Longchang pig is the only white breed considered by Zheng (1984) to be native to China. Recessive white pigs (dirty white) usually segregate in the Caribbean créole and Corsican coloured populations (Lauvergne and Canope, 1979).

Over 61 native breeds from China are described with colour pictures by Zhang (1986) in a book. Of these 37 are black, 13 are black pied with a black head, 10 are black pied with a white head and one is white. No cases of red or domino-spotting are mentioned.

Rarer colour types have occasionally been reported and may be worth mentioning: sepia hair; sepia coat which is a mixture of white, dark and banded hair; and roan or grey, a mixture of white and black hair. The blue colour, which consists of white hair on a black skin, has also sometimes been referred to as roan or grey.

Genetics of Coat Colour

The order of presentation for this section follows that suggested by Berge (1961), Searle (1968) and Ollivier and Sellier (1982).

Agouti locus

In their review, Ollivier and Sellier (1982) concluded that segregation results observed in crosses between wild and red or black breeds are complex and that various hypotheses have to be put forward in order to explain them. This is a consequence of epistatic relations, not yet well understood, between the *Agouti* locus (*A*) and the other colour loci.

It can reasonably be assumed, as suggested by Berge (1961), Searle (1968) and Lauvergne and Canope (1979), that most domestic breeds carry the recessive *non-agouti* (*a*) allele, although the wild allele (*A*) may by present in some red breeds. For example, the light-belly agouti pattern was shown to appear in an F_2 Berkshire × Duroc pig by Lush (1921). The Mangalitza breed is another exception as wild-type F_1 piglets may appear when the Mangalitza is crossed with various other domestic breeds (Kosswig and Ossent, 1931; Constantinescu, 1933; Teodoreanu, 1935). For this reason, Kosswig and Ossent (1931) assume that the *A* allele, which would be responsible for the juvenile striping pattern, is present at a high frequency in the Mangalitza population. This hypothesis is not accepted by Constantinescu (1933), Teodoreanu (1935) and Hetzer (1945a), who consider that the juvenile stripes, a constant characteristic of Mangalitza, and the wild pattern are determined by genes at two different loci. They suggest that hair structure rather than pigmentation is involved in the phenomenon. Comparing wild coat colour between Papua New Guinea village pigs and the French wild boar, Lauvergne *et al.* (1982) suggested that another agouti pattern may exist besides that of the light-belly agouti. This is a kind of badger-face pattern, with black belly and snout. The symbol they propose for the *light-belly agouti* is A^w (w for white) and they proved that this allele is dominant to a. The tentative symbol for the *badger face* variant could be A^b (b for badger face).

The recent contributions concerning crosses between wild pigs and domestic breeds (Thielscher, 1986; Johansson *et al.*, 1992) have not made it possible to choose clearly between the different hypotheses.

Brown (B), Albinism (C) and Dilution (D) loci

No members of the B and C series were reported until recently, but according to Lauvergne *et al.* (1982), a brown variant of eumelanin could exist in Papua New Guinea village pigs. Albinism is unknown in the pig, but Searle (1968) considers the dirty white colour of Mangalitza as being possibly due to an allele of the C series, homologous with the extreme dilution c^e in other mammals.

A likely member of the *Dilution* (D) locus is, according to Searle (1968), the recessive *sepia* factor observed by McPhee *et al.* (1931), who describe this colour as being due to a partial dilution of the black pigment and to a mixture of white, pigmented and banded hair. Berge (1961) attributes it to an allele a^s of the *Agouti* locus. The analysis of Piétrain × Minnesota No.1 crosses led Rempel and Marshall (1990) to hypothesize that a high frequency dilution gene is carried by the Piétrain breed but they did not offer any conclusion about the locus involved. This gene would be recessive to red colour.

Extension (E) locus

After Hetzer's investigations, a series of three alleles at the E locus was considered to be definitely established. These alleles are, in Hetzer's nomenclature, E for *uniform black*, E^P for *black-spotting* and e for *uniform red*. The allelism of E and e has been shown in crosses involving Hampshire, Cornwall, Duroc, Bavarian and Minnesota No.1 breeds (Kronacher, 1924; Bushnell, 1943; Rempel and Marshall, 1990). Allelism between E and E^P may also be inferred from Large Black × Berkshire crosses (Carr-Saunders, 1922), Berkshire × Cornwall crosses (Kosswig and Ossent, 1931), from crosses involving Meishan, Large White and Piétrain (Legault, 1997) and from comparisons between Landrace × Large Black and Landrace × Poland China (or Berkshire) crosses by Hetzer (1945b,c,d). Hetzer (1946) also confirmed the order of dominance $E/E^P/e$. It may thus be assumed that Berkshire and Poland-China are E^PE^P. The same genotype is found in the Piétrain breed (Lauvergne and Ollivier, 1966; Milojic, 1966; Rempel and Marshall, 1990; Legault, 1997). Thus the Berkshire (Poland China) black is merely an extended form of black spotting. This was first suggested by Wright (1918) and was later confirmed by Hetzer (1954) who showed that domino black spotting may be experimentally extended by selection. According to

Hanset (1959), the origin of the Piétrain coat is the Berkshire, which was originally a 'domino' spotted breed. Recent observations dealing with Piétrain × Minnesota No.1 and Piétrain × Duroc crosses (Rempel and Marshall, 1990; Legault, 1997) give F_1 animals exhibiting a 'dalmatian' pattern instead of a 'domino' pattern. Therefore, the complete dominance of E^P/e seems to be doubtful. The 'orange' background which is intermediate between the red of Duroc and the white of Piétrain would be associated with heterozygous F_1 animals (E^Pe) and could be related to the Piétrain *Dilution* gene.

The results of observations made during a crossbreeding experiment (Bidanel *et al.*, 1989; Legault, 1997) between Meishan (MS), Large White (LW) and Piétrain (PT) can be summarized as follows (Legault, 1997). As expected, (MS × LW) F_1 pigs were white with generally blue spots concentrated on the rump and the shoulder. The (MS × LW) F_2 pigs were white, black or domino in the proportion of 12 : 3 : 1 respectively, while MS × (MS × LW) backcross pigs were white or black in the ratio 1 : 1. In the second step of this experiment in which different genotypes of sows were crossed with Piétrain terminal boars, white, black and domino colour patterns were observed in the following proportions: 8 : 4 : 4 in the PT × (MS × LW) as well as in the PT × F_2 crosses; 4 : 9 : 3 in the PT × [MS × (MS × LW)] crosses and 12 : 1 : 3 in the PT × [LW × (MS × LW)] cross. These results confirm the epistasy of *I* locus over *E* locus, the genotypes $II\ E^PE^P$, $ii\ EE$ and $ii\ E^PE^P$ for LW, MS and PT respectively and also the dominant allelism of *E* over E^P.

Further results illustrated in Figs 3.1 and 3.2, implying various crosses between three Chinese breeds and various coloured Western breeds, seem to indicate that uniform black breeds (Gascon, Jiaxing) and black pied breeds (Meishan, Jinhua, Limousin) have the same *ii EE* genotype. Observations from China on the Meishan (black with four white socks) and black Fengjing breeds indicated the possibility of obtaining uniform black by selecting in favour of black extension. Selecting in favour of white extension applied in Meishan,

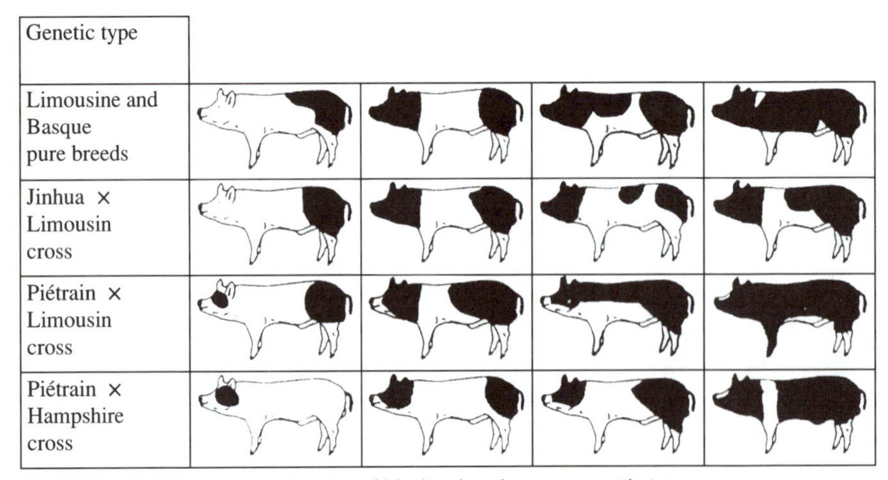

Genetic type
Limousine and Basque pure breeds
Jinhua × Limousin cross
Piétrain × Limousin cross
Piétrain × Hampshire cross

Fig. 3.2. Variation in the extension of black colour in some genetic types.

however, led to a black pied and finally to a Jinhua pattern (black on the two extremities). Such a selection is facilitated by the large variability of black extension illustrated in Fig. 3.2 and by the high heritability of this trait as demonstrated by Hetzer (1954) in Beltsville No.1 line.

Crosses between the Chinese breeds and Duroc gave a large proportion of uniform black pattern (nearly 90%) except when the white belt was present or segregating in the partner breed. Thus, white belt would be absent from Jinhua in spite of a large apparent white belt in this breed, while the *white belt* gene would segregate at a medium frequency in the Limousin breed. In such crosses, coat colour is frequently dark red at birth and becomes black or black with possible red shades by two months of age.

The existence of a fourth allele at the *E* locus is likely. Kosswig and Ossent (1931) distinguish between a hypostatic black (*E*) in Cornwall (or Large Black) and a dominant black (E^d), in Hannover–Braunschweig (or Hampshire). This series of four alleles, $E^d/E/E^P/e$, would be thus similar to that found in the guinea pig and the rabbit (Searle, 1968). On the basis of homology with cattle (pie-black × uniform red giving uniform black), the existence of an epistatic gene for uniform pattern (black or red), may also be suggested. This gene could be the same as E^d and the one that Kosswig and Ossent (1931) and Searle (1968) suggested would be present in Hampshire and Hannover pigs and which would segregate at a high frequency in Duroc (Legault, 1997). Whether it exists at the *E* locus or not, however, is far from being confirmed. The half-red coloured Bavarian pig, which is recessive to uniform colour (Berge, 1961), could belong to the pie-red category with a large extension of white.

The extension coat colour *E* has been mapped on the short arm of chromosome 6p near the most distal marker *S0035* (Kijas *et al*., 1996; Mariani *et al.*, 1996). Comparative mapping information reveals that pig chromosome 6p shares homology with mouse chromosome 8 and cattle chromosome 18. This supports homology for extension loci in the three species, in particular that it is the gene for melanocyte stimulating hormone receptor (*MSHR*).

Pink or Red-eye series (P and R) loci

As there have been no new results in this area, this section repeats the conclusions of Ollivier and Sellier (1982): 'According to Searle (1968), no member of the *P* series is known, but the analogous (and linked) *R* locus (red eye) of the rat seems to have an equivalent in the recessive autosomal gene (*r*) shown by Roberts and Krider (1949) to be responsible for the red-eye and dilution of the black pigment into a sepia colour, found in the Hampshire breed.'

White (I) locus

White is the most frequently occurring colour among the current domestic breeds of pigs, and, not surprisingly, crosses between white and coloured breeds have been the most thoroughly investigated. As early as 1906, Spillman established the dominance of the white colour in a Tamworth × Yorkshire cross. On the basis of those results, Wright (1918) assumed that white was due to two dominant genes but, from later results, Wentworth and Lush (1923) put forward the hypothesis of a single dominant gene. This was confirmed by Hetzer (1945b,c,d) who called the gene *I* (inhibition of colour). Later results also demonstrated the independence between the *I* and the *E* colour loci (Hetzer, 1946). This hypothesis was confirmed by the more recent studies of Lauvergne and Ollivier (1966), Rempel and Marshall (1990) and Legault (1997). White breeds, such as Yorkshire, Large White and Landrace, are generally homozygous for *I*, an allele which inhibits both black and yellow pigment production. Coloured breeds, such as Berkshire, Poland-China, Large Black, Duroc, Piétrain and coloured Chinese breeds are homozygous recessive *ii*. A third allele, I^d, was postulated by Hetzer (1948) in order to explain the occurrence of grey-roans in some crosses between Landrace and Hampshire. The I^d allele would be recessive and would have the same inhibitory effect on pigments as *I* when E^P is present, and it would give a grey-roan phenotype (mixture of black and white hair) when *E* is present. The 'sapphire hog' described by McLean (1914) probably presented the same roan character, originating from the white breeds used in its foundation. This I^d allele, also present in the Créole pig of Guadeloupe (Lauvergne and Canope, 1979), is possibly homologous to the allele responsible for roan found in cattle and horses (Searle, 1968). According to Lauvergne and Canope (1979), the $I^d i\ E^P E^P$ genotype corresponds to the recessive dirty white (locally called 'chabin'). A fourth allele i^m is assumed by Berge (1961) to be responsible for the recessive white of Mangalitza, the order of dominance at the *I* locus being, then, $I/I^d/i/i^m$. However this assumption is not accepted by Searle (1968) who considers the Mangalitza white to be due to an allele at the *C* locus (see above).

The black patches occasionally observed in F_1 crosses between white breeds (Large White, Landrace) and European wild pigs (Johansson *et al.*, 1992) or black Chinese breeds (Legault, 1997) may have two explanations. For Johansson *et al.* (1992), it could be due to a recessive allele at the *I* locus, called I^P and segregating at a low frequency in European white breeds. Because the black patches are generally absent in pure white breeds, except at a very low frequency in Scandinavia, Legault (1997) suggests the hypothesis of an incomplete penetrance of the dominant white allele in the heterozygous state.

As mentioned previously, animals obtained by crossing silky white breeds (Large White, Landrace) with black or black and white breeds (Large Black, Gascon, Saddleback, Hampshire, Chinese breeds) generally exhibit large blue patches (white hair on black skin) while white breeds × Piétrain crossbreds are generally perfectly white. The fact that blue spots (associated with the *E* allele) are systematically rejected from white breed standards offers an explanation for most Large White and Landrace being currently homozygous $E^P E^P$. Another possible explanation for the blue patches could be the occurrence of a mutation within an *ii* $E^P E^P$ genotype.

Johansson *et al.* (1992) have analysed the segregation of the dominant white *I* gene in a reference pedigree for gene mapping developed by crossing European wild pigs and Large White. The gene for dominant white colour was shown to be closely linked to the *ALB* (Albumin) and *PDGFRA* genes (platelet-derived growth factor receptor alpha) on chromosome 8. The *ALB–PDGFRA–I* linkage group was found to share homology with part of mouse chromosome 5, human chromosome 4 and horse linkage group II containing dominant genes for white or white spotting. The gene is now known to be the *KIT* gene and a result of a duplication event in pigs (Johansson Moller *et al.*, 1996).

White Belt (Be) locus

This pattern was first studied genetically by Spillman (1907) who assumed it to be due to the complementary action of two genes. Durham (1921) favours a major dominant gene, as do both Olbrycht (1941) and Donald (1951). The latter authors, however, disagree over the explanation for the colour polymorphism observed in the Wessex Saddleback and Essex breeds, where the belt is variable in width and black pigs are reported to remain at a noticeable frequency in spite of their regular elimination by breeders. Olbrycht (1944) believes that belt width is essentially a polygenic character, whereas for Donald (1951) the narrow belt selected for by breeders is an 'unfixable' heterozygous genotype ($Be^{w}be$), black piglets being mainly *bebe*. According to Berge (1961), the wide belt of the Hannover-Braunschweig would correspond to the homozygous $Be^{w}Be^{w}$, and the extension of the belt towards the front found in the half-coloured Bavarian Landschwein, which is recessive to uniform colour, would be due to a third allele be^{b} at the same locus, the order of dominance being $Be^{w}/be/be^{b}$. Searle (1968), however, considers the dominant white belt of Hampshire (analogous to the belt in cattle) and the recessive half-coloured pattern (similar to belted in mice) as two separate entities which have not been proven to be allelic. This opinion is somewhat reinforced by the suggestions of a close linkage between the *E* locus and the half-coloured pattern on one hand (Kronacher, 1924) and of a loose linkage between the white belt factor and the *E* locus on the other (Bushnell, 1943). Searle (1968) also considers the white face pattern of the 'Hereford hog' described by Smith *et al.* (1938) as probably being due to the same gene as the half-coloured pattern, whereas Berge (1961) assumes that an allele of the *E* series is responsible for the white face.

Recent observations illustrated in Figs 3.1 and 3.2, involving various crosses between Limousin, Meishan, Jinhua, Hampshire and Duroc breeds (Legault, 1997), have suggested that apparent white belts can be obtained by selecting in favour of extension of the white pattern within black pied breeds such as Meishan and Jinhua. Crosses with uniform red Duroc appear to be a good way to reveal the presence of the *Be* gene in black pied populations. Thus, the Be^{w} allele seems to be absent from Meishan and Jinhua breeds, but is segregating with a medium frequency in the Limousin breed. The half-black coloured animals that are occasionally observed in pure Limousin breed 'cul-noir de Saint Yriex' as well as in Jinhua × Limousin or Limousin × Hampshire crossbreds are probably similar to half-red coloured Bavarian pigs. The *white head* gene, which will be described below, could also favour the production of half-coloured animals.

White head or Hereford locus

The white face pattern described by Smith *et al.* (1938) is probably the same as the white head pattern studied by Rempel and Marschal (1990) in Piétrain × Minnesota No.1 cross, and observed by Legault (unpublished observations) in Chinese × Piétrain and Duroc × Piétrain crosses. As was discussed in the previous section, several inconclusive hypotheses have been put forward to explain this colour particularity (Smith *et al.*, 1938; Berge, 1961; Searle, 1968).

In a series of observations made on various crosses involving Piétrain, Minnesota No.1 and Hampshire breeds, Rempel and Marshall (1990) clearly established that white head spots were due to a single dominant allele. The gene frequency for the white head spot in this specific herd of Piétrain (Rosemont, Minnesota) was very high and varied from 0.92 to 0.98 depending on the method of estimation.

This hypothesis seems to be confirmed by 15 years of observations in France on various crosses between Chinese breeds (Meishan, Jiaxing and Jinhua) and several Western breeds including Piétrain, Hampshire, Duroc, Limousin and Large White (Legault, unpublished observations). The observation of the progeny of Piétrain boars used in AI centres leads to an estimation of the gene frequency of 0.87 (four heterozygous among 15 boars). White head spots appear on both black backgrounds (Piétrain × Chinese breeds) and red ones (Piétrain × Duroc cross). As observed by Rempel and Marshall (1990) and confirmed by Legault (1997), the size of the spot varies from a few white hairs to a large triangular mark (Fig. 3.3). The size of white spot in crosses with Piétrain also depends on the extension of the white pattern in the partner breed. Thus, crosses with uniform black or red breeds (i.e. Duroc, Jiaxing, Gascon) give white spots that are generally of reduced size (Fig. 3.3d and 3.3e). When Piétrain is crossed with black pied breeds such as Jinhua or Limousin, the white pattern is continuous from the front of the head to the throat and belly (Figs 3.1, 3.2 and 3.3c). As illustrated in Figs 3.1, 3.2 and 3.3b, crosses with white-belted breeds

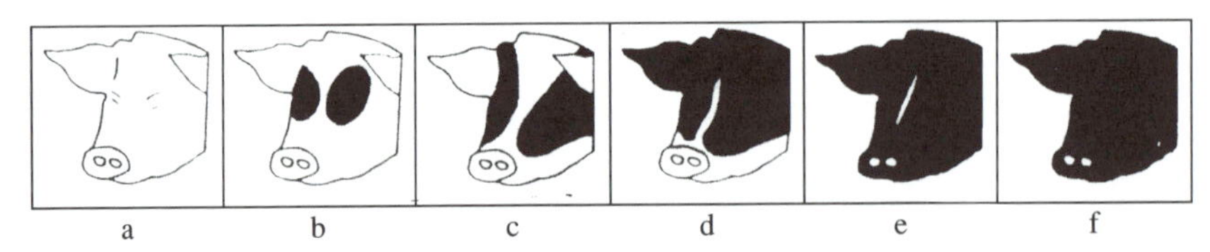

Fig. 3.3. Variation in the extension of white coloration on the face.

(Hampshire) can lead to black spots surrounding the eyes (called Panda pigs in China).

The observation of black patterned F_2 pigs from two independent Meishan × Large White crossbreeding experiments (Bidanel *et al.*, 1989; Bidanel *et al.*, 1996) revealed 9% and 7% of white head spots respectively. In the absence of a linkage with the *E* locus, these figures lead to a low frequency of the white head allele in the Large White breed (somewhere between 0.07 and 0.10). The white face dominant allele could be called Hereford (*He*) because of its first description by Smith *et al.* (1938). Hereford pigs correspond to strains that are homozygous *HeHe* with a red colour pattern. The native Chinese breeds that are black pied with a white head can also be considered as homozygous at the *He* locus. They include the so-called Panda varieties having a large extension of the white pattern.

As illustrated in Figs 3.1 and 3.2, the high correlation generally observed for the surface area of white marks (or white background) between the average parental breeds and their F_1 progenies is an indicator of the common polygenic determination of white extension in any pattern (i.e. piebald, domino, white belt, white head).

Hair Mutations

The hereditary basis for hairlessness (hypotrichosis) was established by Roberts and Carroll (1931). This condition, which should not be confused with a similar one that is due to a deficiency of iodine, is due to a single autosomal recessive gene which reduces the number of hair follicles.

Another type of hypotrichosis has been described by Meyer and Drommer (1968). In this case an autosomal dominant gene is involved. This character is lethal, as homozygous hypotrichotic piglets die within 10 days. The vitality of heterozygous individuals is also reduced.

The woolly hair condition, which occurs frequently in the native Brazilian Canastrao breed, the Min breed of northern China, and in the Mangalitza breed was studied by Rhoad (1934), who showed it to be determined by a single autosomal dominant gene, segregating independently from genes for coat colour and pattern. The data were insufficient to establish its independence from the recessive type of hairlessness, a factor which also was present in the Brazilian breed. Disturbances in the arrangement of the hair, known as whorls, swirls or 'roses', occur mainly along the spinal column. They have been explained by the complementary action of two dominant genes (Nordby, 1932).

General Discussion and Conclusions

It is relatively well recognized that the domestication of wild pigs started about 7000 years ago in two independent areas located in the Near East and in Asia, respectively (Epstein and Bichard, 1984; Zheng, 1984). Western Europe cannot be included as a possible domestication centre due to the specific karyotype ($2n = 36$) of the majority of native wild pigs in this area. With respect to coat colour, a general survey of native breeds shows a contrast between the mutations common to the Asian breeds (i.e. *E* for black or piebald and *He* for white head) and a larger number of mutations common in Near East and Mediterranean populations. In particular, black, domino spotting, red, and white belts have occurred in the Near East and in the Mediterranean area while the dominant white mutation (*I*) was present during the Middle Ages in Northern Europe. Visible characters, and more particularly coat colour and pattern, have generally been the first criterion used by pig breeders to create new breeds and make them of more homogeneous appearence. The generalization of breed standards then led to homozygous populations for alleles directly responsible for visible traits as well as for alleles associated with a greater homogeneity such as E^P in white breeds, the recessive dilution gene in Piétrain and possibly the gene controlling colour highlights in red breeds such as Duroc. The presence of the *He* gene in European breeds, particularly in Berkshire, can be explained by the introduction of Asian pigs into England in the early 19th century, originating from South China (an area where most native breeds belong the 'black pied with white head' type). Taking into account the hypothesis that this discrete gene is independant of economic traits, its low freqency in Large White (less than 10%) could give an indication of the proportion of genes of Asian origin in such a breed.

The present review supports the findings of earlier geneticists with respect to the dominant white (*I* locus), the extension of coloration (*E* locus), the white belt (*Be* locus) and the white head (*He* locus). However, the number of alleles segregating at each of these loci as well as their specific properties are not clearly established. Information concerning other loci such as wild-*Agouti* (*A*) locus, *Dilution* (*D*) locus and possibly the *Albino* (*C*) locus, remains quite confused. Recently, an increasing number of investigations involving new breeds such as Piétrain, native European breeds of Iberian origin and Chinese breeds have updated the conclusions of Ollivier and Sellier (1982) (Table 3.1), and suggested several hypotheses to explain the different observations. Table 3.2 proposes the most probable genotypes for representative breeds as well as for specific colour phenotypes.

New hypotheses essentially concern the white head, which is probably associated with white under jaw and white belly (Rempel and Marshall, 1990; Legault, unpublished observations), *Dilution* genes (Lauvergne and Canope, 1979; Rempel and Marshall, 1990) and the *Extension* locus in which uniform black and piebald could be determined by the same *E* allele (Legault, unpublished observations). Also discussed are a possible dominant gene determining uniform (solid) black or red colour segregating at a high frequency in Duroc and the role of polygenic systems for explaining the extension of white in different colour patterns (Hetzer, 1954; Legault, unpublished observations).

It remains necessary to undertake many experiments specifically designed to study coat colour in order to establish clearly the allelism of the presumed

Table 3.1. A list of coat colour genes in the pig.

Locus	Alleles	
A	A^w	*agouti white belly*
	A^b	*agouti badger face*[1]
	a	*non agouti*
	a^s	*sepia*[1]
C	*C*	normal
	c^e	extreme dilution[1] *(= dirty white)*
D	*D*	*normal*
	d^s	*sepia*[1]
	d^p	*recessive dilution* gene[1]
E	E^d	*dominant black*[1]
	E	*black or pie-black spotting*
	E^p	*domino spotting and black with six white points*
	e	*red*
	e^h	*white face*[1]
He	*He*	*white face*
	he	*normal*
I	*I*	*inhibition of colour*
	I^d	*roan*
	I^p	*black patches*
	i	*coloured*
	i^m	*dirty white*[1]
Be	Be^w	*belt*
	be	*self*
	be^b	*half-coloured* and possibly *white face*[1]
R	*R*	*normal*
	r	*red-eye*

Source: Adapted from Berge, 1961; Searle, 1968; Ollivier and Sellier, 1982.
[1]Allelism not proven.

genes such as I^P (black patches), E^d [(uniform (or solid) pattern dominant or epistatic over piebald pattern], *He* (dominant white face), as well as to be able to identify possible linkage situations between colour loci.

Divergent selection experiments for heritable black extension, as initiated by Hetzer (1954) in Berkshire, would confirm the high heritability of this trait for any pattern (piebald, domino, white belt or Hereford) and would definitely determine the influence of polygenic effects. Such a selection could be carried out on the width of white belt in Hampshire, on the overall area of black spots in pie-black animals such as Meishan or Limousin. The latter experiment could lead either to a uniform black pattern or to an apparent large white belt or half-coloured pattern. It would also be interesting to repeat Hetzer's selection

Table 3.2. Presumed genotype in some domestic breeds of specific coloration phenotypes.

Breeds or specific phenotypes	Coat colour phenotype	Genotype locus				
Breeds		***A***	***I***	***E***	***He***	***Be***
Large White, Landrace	White	*a a*	*I I*	$E^p E^p$	*he he*	*be be*
Gascon, Jiaxing (Cornwall)	Black	*a a*	*i i*	*E E*	*he he*	*be be*
Meishan, Jinhua	Pie-black, black head	*a a*	*i i*	*E E*	*he he*	*be be*
Limousin, Saddleback	Pie-black, black head	*a a*	*i i*	*E E*	*he he*	Be^w *be*
Dahua bei, Moncaï	Pie-black, white head	*a a*	*i i*	*E E*	*He He*	*be be*
Piétrain, Berkshire, Poland-China	Domino black spotted	*a a*	*i i*	$E^p E^p$	*He He*	*be be*
Duroc, Tamworth, Minnesota No.1	Red	*a a*	*i i*	*e e*	*he he*	*be be*
Hampshire, Hannover	Black, white belt	*a a*	*i i*	$E^d E^d$	*he he*	$Be^w Be^w$
Particular phenotypes						
Dalmatian domino spots		*a a*	*i i*	$E^p e$	—	—
Grey-roan (sapphire)		*a a*	$I^d i$	*E E*	—	—
Dirty white, recessive white, chabin		*a a*	$I^d i$	$E^p E^p$	—	—

experiment on the 'domino' spotting pattern of foundation stocks that either carried the white face gene or not.

Currently, development in this area is also characterized by the progress in molecular genetics which led to the mapping of the *I* locus (Johansson *et al*., 1992), its identification of the *KIT* locus as the *I* locus (Johansson Moller *et al*., 1996) and the discovery of the *E* locus (Kijas *et al*., 1996; Mariani *et al*., 1996). In the near future, this kind of investigation will be facilitated by the continuing advances in molecular biology, either through a rapid mapping of the *A*, *Be* and *He* loci, or by identifying genes through homologies between species.

References

Berge, S. (1961) Heredity of colour in pigs (in Norwegian). *Tidssskiftr. Norske Landbruk.* 68, 159–188.

Bidanel, J.P., Caritez, J.C. and Legault, C. (1989) Estimation of crossbreeding parameters between Large White and Meishan porcine breeds. I – Reproductive performance. *Genetics, Selection, Evolution* 24, 507–526.

Bidanel, J.P., Bonneau, M., Chardon, P., Elsen, J.M., Gellin, J., Le Roy, P., Milan, D. and Ollivier, L. (1996) Etablissement et utilisation de la carte génétique porcine. *INRA Productions Animales* 9, 299–310.

Bushnell, R.L. (1943) Linked color factors in Hampshire swine. Linkage of black and the basic white of the white belt pattern. *Journal of Heredity* 34, 303–306.

Carr-Saunders, A.M. (1922) Note on inheritance in swine. *Science* 55, 19.

Constantinescu, G.K. (1933) Vererbungsversuche an Schweinen unter besonderer Berücksichtigung des Mangalitza Schweines. *Zeitschrift für Züchtung. B.* 26, 395–427.

Donald, H.P. (1951) Genetic variation in colour pattern of Wessex Saddleback pigs. *Journal of Agricultural Science, Cambridge* 41, 214–221.

Durham, G.B. (1921) Inheritance of belting spotting in cattle and swine. *American Naturalist* 55, 476–477.

Epstein, H. (1969) *Domestic Animals of China.* Commonwealth Agricultural Bureaux, Farnham Royal, UK.

Epstein, H. and Bichard, M. (1984) Pig. In: Mason, I.L. (ed.) *Evolution of Domesticated Animals.* Longman, pp. 145–162.

Hanset, R. (1959) Un aperçu de la génétique des robes chez le porc. *Annales de Médecine Vétérinaire* 103, 53–66.

Hetzer, H.O. (1945a) Inheritance of coat color in swine. I. General survey of major color variations in swine. *Journal of Heredity* 36, 121–128.

Hetzer, H.O. (1945b) Inheritance of coat color in swine. II. Results of Landrace by Poland China crosses. *Journal of Heredity* 36, 187–192.

Hetzer, H.O. (1945c) Inheritance of coat color in swine. III. Results of Landrace by Berkshire crosses. *Journal of Heredity* 36, 255–256.

Hetzer, H.O. (1945d) Inheritance of coat color in swine. IV. Analysis of hybrids of Landrace and Large Black. *Journal of Heredity* 36, 309–312.

Hetzer, H.O. (1946) Inheritance of coat color in swine. V. Results of Landrace by Duroc–Jersey crosses. *Journal of Heredity* 37, 217–224.

Hetzer, H.O. (1947) Inheritance of coat color in swine. VI. Results of Yorkshire by Duroc–Jersey crosses. *Journal of Heredity* 38, 121–124.

Hetzer, H.O. (1948) Inheritance of coat color in swine. VII. Results of Landrace by Hampshire crosses. *Journal of Heredity* 39, 123–128.

Hetzer, H.O. (1954) Effectiveness of selection for extension of black-spotting in Beltsville no.1 swine. *Journal of Heredity* 45, 215–223.

Johansson-Moller, M., Chowdhary, R., Hellmen, E., Höyheim, B., Chowdhary, B. and Andersson, L. (1996) Pigs with the dominant white coat color phenotype carry a duplication of the KIT gene encoding the mast/stem cell growth factor receptor. *Mammalian Genome* 7, 822–830.

Johansson, M., Ellegren, H., Marklund, L., Gustavsson, U., Ringmar-Cederberg, E., Anderson, K., Edfors-Lilja, I. and Andersson, L. (1992) The gene for dominant white color in the pig is closely linked to ALB and PDGFRA on chromosome 8. *Genomics* 14, 965–969.

Kijas, J., Moller, M., Marklund, L., Wales, R. and Andersson, L. (1996) The extension coat color locus maps to pig chromosome 6. *25th International Conference on Animal Genetics, 21–25 July 1996, Tours, France,* 126 (abstract).

Kosswig, C. and Ossent, H.P. (1931) Die Vererbung der Haarfarben beim Schwein. *Zeitschrift für induktive Abstammungs – und Vererburgslehre. Abstamm. -u. Vererblehre B.* 22, 297–381.

Kosswig, C. and Ossent, H.P. (1932) Ein Beitrag zur Vererbung der Haarfarben beim Schwein. *Züchter* 4, 225–230.

Kosswig, C. and Ossent, H.P. (1934). Weitere Ergebnisse über die Vererbung der Haarfarben beim Schwein. *Züchter* 6, 306–308.

Kronacher, C. (1924) Vererbungsversuche und Beobachtungen an Schweinen. *Z. induktive Abstamm. -u. Vererblehre* 34, 1–120.

Lauvergne, J.J. and Canope, I. (1979) Etude de quelques variants colorés du porc Créole de la Guadeloupe. *Annales de Génétique et de Sélection Animale* 11, 381–390.

Lauvergne, J.J. and Ollivier, L. (1966) A propos de colorations observées lors de croisements entre porcs de Piétrain et porcs Large White. *Annales de Génétique* 9, 39–41.

Lauvergne, J.J., Malynicz, G.L. and Quartermain, A.R. (1982) Coat colour variants of village pigs in Papua New Guinea. *Annales de Génétique et de Sélection Animale* 14, 29–42.

Lush, J.L. (1921) Inheritance in swine. *Journal of Heredity* 12, 57–71.

McLean, J.A. (1914) The sapphire hog. *Journal of Heredity* 5, 301–304.

McPhee, H.C., Russel, E.Z. and Zeller, J. (1931) An inbreeding experiment with Poland-China swine. *Journal of Heredity* 22, 393–403.

Mariani, P., Johansson-Moller, M.H., Hoyheim, B., Marklund, L., Davies, W., Ellegren, H. and Andersson, L. (1996) The extension coat color locus and the loci for blood group O and tyrosine aminotransferase are on pig chromosome 6. *Journal of Heredity* 87, 272–276.

Meyer, H. and Drommer, W. (1968) Erbliche Hypothrichie beim Schwein. *Dt. Tierärztl. Wschr.* 75, 13–17.

Milojic, M. (1966) Color inheritance when crossing certain pig breeds (in Serbo-Croat). Zbornik *radova Poljoprivrednog Fakulteta, Universitet u Beogradu* 14, 1–13.

Nordby, J.E. (1932) Inheritance of whorls in the hair of swine. *Journal of Heredity* 23, 397–404.

Olbrycht, T.M. (1941) Statistical analysis of black colour in Wessex Saddleback breed. *Annals of Eugenics* 11, 80–88.

Olbrycht, T.M. (1944) Behaviour of some external characteristics in Essex pigs. *Journal of Agricultural Science, Cambridge* 34, 16–21.

Ollivier, L. and Sellier, P. (1982) Pig genetics: a review. *Annales de Génétique et de Sélection Animale* 14, 481–544.

Rempel, W.E. and Marshall, M.L. (1990) Inheritance of coat color in swine. In: *Genetics of Swine.* Agricultural Research Service, Roman L. Hruska U.S. Meat Animal Research Center, Clay Center, Nebraska.

Rhoad, A.O. (1934) Woolly hair in swine. *Journal of Heredity* 25, 371–375.

Roberts, E. and Carroll, W.E. (1931) The inheritance of 'hairlessness' in swine. Hypotrichosis II. *Journal of Heredity,* 22, 125–132.

Roberts, E. and Krider, J.L. (1949) Inheritance of red-eye in swine. *Journal of Heredity* 40, 306.

Searle, A.G. (1968) *Comparative Genetics of Coat Colour in Mammals.* Logos Press, London.

Smith, A.D.B., Robinson, O.J. and Bryant, D.M. (1938) The genetics of the pig. *Bibliographia Genetica* 12, 1–160.

Spillman, W.J. (1906) Inheritance of coat colour in swine. *Science* 24, 441–443.

Spillman, W.J. (1907) Inheritance of the belt in Hampshire swine. *Science* 25, 541–543.

Teodoreanu, N.I. (1935) Vererbungsbeobachtungen über die Farbe des roten und des schwarzen Mangaliczaschweines. *Analele Academici Române (Menoriile) Sectianii Stiitifice* 10, Men. 11, 20 pp.

Thielscher, H.H. (1986) Colour variations in crosses of wild boars. *Veterinary Medical Review, German Federal Republic* (*1*), 183–188.

Wentworth, E.N. and Lush, J.L. (1923) Inheritance in swine. *Journal of Agricutural Research* 23, 557–582.

Wright, S. (1918) Color inheritance in mammals. VIII. Swine. *Journal of Heredity* 9, 33–38.

Zhang, Z. (1986) Pig breeds in China. In: *Breeds of Domestic Animal and Poultry in China.* Shanghai Scientific & Technical Publishers, Shanghai, 236 pp.
Zheng, P.L. (1984) *Livestock Breeds of China.* FAO Animal Production and Health, paper 46, Roma, 217 pp.

Genetics of Morphological Traits and Inherited Disorders

F.W. Nicholas
Department of Animal Science, University of Sydney, Sydney, NSW 2006, Australia

Introduction

Morphological traits have not been of much importance in pigs. In contrast, inherited disorders have been of considerable importance, from an agricultural point of view (e.g. malignant hyperthermia) and in terms of animal models for inherited human diseases. This chapter, therefore, concentrates on inherited disorders. However, it does include mention of morphological traits that have been documented.

With the molecular revolution now in full swing and, in particular, with the development of gene markers covering all regions of all pig chromosomes (see Chapters 8 and 9), knowledge of morphological traits and inherited disorders in pigs will increase rapidly in the decades ahead. As described below, regularly-updated information is available on the Internet. By this means, it will be possible for readers throughout the world to obtain the latest information on any trait or disorder mentioned in this review.

The Range of Possibilities

The spectrum of morphological traits and inherited disorders ranges from those that are definitely due to the action of just one gene, to those that are due to the combined action of many genes and many non-genetic (environmental) factors. In between these two extremes are many traits and disorders which appear to run in families, but for which there is insufficient information to enable a conclusion to be drawn about whether one or more genes are involved. Unfortunately, the literature abounds with examples of traits and disorders that have been claimed to be due to just one gene, despite the data being so sparse that such a claim cannot be justified. Similar problems exist with claims of inheritance being recessive or dominant; in most cases, there is insufficient information to justify the claims that have been made. In the fullness of time, of course, additional data might support the initial claims. But we must be careful not to jump the gun.

This scarcity of reliable data on the inheritance of traits and disorders poses a challenge to those who are asked to compile lists of such traits – as required for this chapter. As will be seen below, the present author's solution has been to follow McKusick's (1997) well-established practice of providing a full list of all traits and disorders for which claims of inheritance have been made, using an asterisk to indicate those for which reasonable evidence of single-locus inheritance has been published.

During the 1980s and 1990s, there have been considerable advances made in identifying the molecular basis of inherited disorders in domestic animal species, and the pig has been at the forefront of such studies. Indeed, the discovery of the molecular basis of malignant hyperthermia by Fujii *et al.* (1991) provides a marvellous example of the power of molecular genetics to solve important, practical and economically important problems.

Previous Reviews

Several comprehensive reviews of inherited traits and disorders in pigs have been published over the years. The first major summary was by Smith *et al.* (1936). More recently, there have been comprehensive surveys of inherited disorders by Johansson (1964), Ollivier and Sellier (1982), Hanset (1991) and Robinson (1991), together with more wide-ranging reviews by Koch *et al.* (1957), Huston *et al.* (1978) (congenital disorders), Wiesner and Willer (1974) and Hamori (1983). It should be noted that some of these reviews are concerned with congenital traits and disorders, i.e. traits and disorders that are present at birth. Not all such traits and disorders are inherited.

In compiling the list presented in this chapter, the information provided by Ollivier and Sellier (1982) and Robinson (1991) has been used as a convenient starting point. It is interesting to compare the conclusions drawn in these two reviews. Ollivier and Sellier discussed 76 traits and disorders, and concluded that 29 are determined by a single locus. Nearly a decade later, Robinson listed 64 traits and disorders, of which 33 were regarded as being due to a single locus. Of the 64 in Robinson's list, only five were first reported after the publication of Ollivier and Sellier's review. Thus, Robinson excluded 17 of the disorders described by Ollivier and Sellier. This is consistent with Robinson's declaration that he had 'been deliberately more stringent in accepting that a defect could be genetically determined'.

Among the 33 traits and disorders included in Robinson's single-locus list, only three were first reported after publication of Ollivier and Sellier's review. Thus, despite being more stringent, Robinson actually declared one more pre-1982 trait/disorder to be due to a single locus than did Ollivier and Sellier. To what extent do the two single-locus lists coincide? Somewhat less than one might have thought. In fact, only 20 traits/disorders are common to the two lists: nine of Ollivier and Sellier's single-locus traits/disorders were excluded from Robinson's list; and ten of Robinson's single-locus traits/disorders were excluded by Ollivier and Sellier.

Of course, these statistics are simply a reflection of the difficulties involved in evaluating inadequate evidence concerning inheritance, as described above. Both lists were compiled with great care by very competent geneticists. The fact that their conclusions differ is a cautionary tale for anyone charged with the task of determining whether a particular trait/disorder is due to a single gene.

Current Sources of Information

While a list of reviews is useful, it is even more useful to have a single catalogue of morphological traits and inherited disorders that is regularly updated, and which is made available both electronically (on the Internet) and in hardcover (book) form on a regular basis. Human geneticists have long had access to such a resource, McKusick's *Mendelian Inheritance in Man* (MIM). Now in its 12th edition as a book (McKusick, 1997), and accessible on the Internet as Online Mendelian Inheritance in Man (OMIM) at http://www3.ncbi.nlm.nih.gov:80/omim/, this catalogue contains a wealth of information on thousands of morphological traits and inherited disorders in humans. It also contains a surprising quantity of information on pigs, because McKusick has always been interested in potential animal models of human disorders.

In 1978, the present author commenced compiling a catalogue of inherited traits and disorders in a wide range of animal species. Being modelled on, and complementary to, McKusick's catalogue, this catalogue is called Mendelian Inheritance in Animals (MIA). It is accessible on the Internet as Online Mendelian Inheritance in Animals (OMIA) at http://www.angis.su.oz.au/databases in the same format as OMIM, and at http://probe.nalsasda.gov:8300/animal/omia.html in a different format.

The OMIA includes entries for all inherited disorders in pigs, together with other traits in pigs for which single-locus inheritance has been claimed, however dubiously. Each entry consists of a list of references arranged chronologically, so as to present a convenient history of knowledge about each disorder or trait. For some entries, there is additional information on inheritance or molecular genetics. If the disorder or trait has a human homologue, the relevant MIM numbers are included, providing a direct hyperlink to the relevant entry in McKusick's online catalogue OMIM.

Table 4.1. Inherited traits and disorders in pigs grouped by type of disorder. Columns represent traits and disorders, brief summary of current knowledge and earliest and recent references. An asterisk (*) implies convincing evidence for single-locus inheritance. (Clinical descriptions are substantially drawn from Blood and Studdert, 1988.)

Traits and disorders	Current knowledge	Reference
Hair and skin		
Coat colour, general	Determined by several loci, which are discussed in detail in Chapter 3 of this book.	Spillman (1906); Johansson Moller *et al.* (1996)
*Dermatosis vegetans	A well-characterized syndrome comprising skin lesions on the body, swollen feet (club foot), and multinucleate giant cells (MGC) in the lungs associated with fatal pneumonia. Occurs only in the Landrace breed. Breeding data support single-locus autosomal recessive inheritance (Flatla *et al.*, 1961). Evensen (1993) showed that the giant cells in the lungs of affected pigs are derived from mesenchymal cells, with a monocyte/macrophage origin.	Hjarre (1953); Flatla *et al.*, (1961); Evensen (1993)
*Epitheliogenesis imperfecta	Congenital absence of the areas of skin. Also known as aplasia cutis. Affected animals usually die within 3 days, but some survive to adulthood, the chance of survival is indirectly proportional to the area of skin that is absent. A single-locus autosomal recessive disorder.	Nordby (1929b); Sailer (1955)
*Hypotrichosis, dominant	An autosomal dominant form of hairlessness (or hypotrichie). The listing here of two separate loci for hypotrichosis is based on precedent (Ollivier and Sellier, 1982; Robinson, 1991) and on supposition of the present author. Breeding trials to test non-allelism have not been reported.	Meyer and Drommer (1968)
*Hypotrichosis, recessive	An autosomal recessive form of hairlessness (or hypotrichie).	Roberts and Carroll (1931)
Melanoma, congenital	A congenital disorder involving tumours arising from melanocytes, dendritic cells of neuroectodermal origin, or melanoblasts. The disorder occurs at a high frequency in the Munich Miniature Swine (MMS) Troll. Muller *et al.* (1995) conducted a segregation analysis of data from the F_2 of a cross between MMS and another breed, but could find no good evidence for single-locus inheritance.	Nordby (1933); Muller *et al.* (1995)
Nipples, asymmetrical numbers	Mayer and Pirchner (1995) documented substantial asymmetry in the number of teats in German sows, and a positive correlation between the extent of asymmetry and the occurrence of inverted nipples (see following entry). However, the heritability of asymmetry was zero.	Mayer and Pirchner (1995)
Nipples, inverted	Also known as inverted teats or cratered teats. They occur most commonly in the umbilical and anterior regions, being much less frequent towards the rear. Early claims of single-locus inheritance have not withstood close scrutiny. Clayton *et al.* (1981) showed that this disorder has to be regarded as multifactorial, and estimated heritability to be approximately 20%.	Nordby (1934b); Clayton *et al.* (1981)

Pityriasis rosea	A transient skin disease that is often mistaken for ringworm. Lesions generally appear first on the ventral (underneath) surface, and have usually disappeared completely by puberty. Affected pigs do not show any ill effects. There is convincing evidence for familial occurrence, but no convincing data to support single-locus inheritance.	Wellmann (1953); Done (1964)
Teat number	A discontinuous but multifactorial trait, with a heritability of between 10 and 20%.	Enfield and Rempel (1961); McKay and Rahnefield (1990)
Whorls	The occurrence of hairs in a spiral arrangement.	Nordby (1932)
*Woolly hair	Curly coat, as seen in the Canastrao breed. An autosomal dominant trait, which may be allelic with recessive hypotrichosis.	Rhoad (1934)
Skeletal system: axial		
Cephalothoracophagus	A twin 'monster' united at the head, neck and thorax.	Halnan (1970)
Miniature	Several strains of 'miniature' pigs have been developed by selection for small body size, maturing at less than 70 kg. Their main use is as an experimental animal (see, for example, Friedman *et al.*, 1994). The genetic basis of their smallness has not been elucidated, but it is certain to be multifactorial. 'Native' populations of small pigs have existed for many decades or even centuries in various parts of the world. One of the smallest is the Mexican Cuino, with a mature weight of only 12 kg (Anon., 1991).	Anon. (1991); Friedman *et al.*, 1994)
Osteochondrosis	A disorder characterized by abnormal differentiation of growth cartilage. Also called dyschondroplasia. Definitely familial but no good evidence for single-locus inheritance.	Reiland *et al.* (1980); Goedegbuure *et al.* (1988); Uhlhorn *et al.* (1995)
*Pulawska factor	A combination of skeletal disorders, including malformation of cranium bones and ribs, fusion of vertebrae, enlarged liver, pancreas, kidney and intestine, and rudimentary lungs. Citing the evidence of Dabczewski (1949), Ollivier and Sellier (1981) were convinced that this was a single-locus autosomal recessive disorder.	Dabczewski (1949); Ollivier and Sellier (1982)
Tail, kinky	Kinked, flexed or crooked tail, as distinct from the usual porcine spiral. Results from fusion of the caudal vertebrae. Donald's (1949) substantial segregation analysis of the disorder in British pigs showed that it is definitely not due to a single locus, but is certainly familial.	Nordby (1934a); Donald (1949)
Taillessness	Brooksbank (1958) reported two case studies of congenital tail abnormalities in pigs, one of which resembled the trademark tailless trait of Manx cats. The family history of this case provided anecdotal evidence for hereditary influence, but no hard data.	Brooksbank (1958)

Table 4.1. *continued*

Traits and disorders	Current knowledge	Reference
Twinning, conjoined	Congenital disorder in which twins are partly joined. Also known as Siamese twins.	McKay (1955); McManus *et al.* (1994)
Vertebral anomalies	The most common form of vertebral anomaly involves fusion of vertebrae, giving rise to a shortened body, and the appearance of a reduction in the actual number of vertebrae. Familial, but no good evidence for single-locus inheritance.	Fredeen and Newman (1962a,b); Gluhovschi *et al.* (1967)
Skeleton system: head and appendages		
Agnathia	Congenital absence of the lower jaw.	Kruger (1964)
*Aplasia of tongue	Congenital absence of the median portion of the apex of the tongue. Nes (1958) presented data from several matings supporting single-locus autosomal recessive inheritance.	Nes (1958)
Arthritis	Inflammation of a joint. Used interchangeably with arthrosis, which refers to degenerative disease of joints. A familial disorder with a low heritability.	Sabec (1963); Lingaas and Ronningen (1991)
Brachydactyly	Abnormal shortness of the digits.	Meyer (1964)
Brachygnathia	Congenital abnormal shortness of the mandible (lower jaw), resulting in protrusion of the maxilla (upper jaw). Also called overshot jaw or parrot mouth.	Hamori (1941)
Brachygnathia superior	Congenital abnormal shortness of the maxilla (upper jaw), resulting in protrusion of the mandible (lower jaw). Also called monkey mouth.	Lutz (1996)
Cerebellar hypoplasia	Underdevelopment of the cerebellum. Familial in other species, very limited data available in pigs.	Gitter and Bowen (1962); Kronevi *et al.* (1975)
Cleft palate	Congenital fissure (split) that involves the hard or soft palate (roof of the mouth). Also called palatoschisis.	Nes (1958); Titze (1963)
Cranioschisis	Also known as cranium bifidum and fronto-parietal skeletal defect, this disorder comprises a cleft in the fronto-parietal bones of the skull. If meninges (the membranes covering the brain) protrude through the cleft, the disorder is sometimes called meningocele. If meninges and brain tissue protrude, it is sometimes called encephalocele or meningo-encephalocele. The most definitive studies of its inheritance, by Wijeratne *et al.* (1974) and Vogt *et al.* (1986), showed that the disorder is familial, but is not due to a single gene. Vogt *et al.* (1986) presented extensive incidence data, showing variation between breeds and between years, with an overall incidence of 0.12% in the University of Missouri-Columbia herd.	Nordby (1929a); Vogt *et al.* (1986)
Dicephalus	Developmental disorder resulting in two heads. No genetic data in isolated reports.	Thielscher (1994)
Harelip	A congenital disorder comprising a cleft in the upper lip, resulting from failure of fusion between the maxillary and medial nasal processes. Also called cleft lip or cheiloschisis. A familial disorder, but insufficient data to justify any conclusion as to mode of inheritance.	Nes (1958); Titze (1963)

Holoprosencephaly	Developmental failure of cleavage of the forebrain (prosencephalon); with a deficit in midline facial development, and with cyclopia (see entry in this table) in the severe form. No data available on inheritance.	Fisher *et al.* (1989)
Hydrocephalus	Enlargement of the cranium caused by accumulation of fluid. Also called 'water on the brain'. Despite several claims of single-locus inheritance, the data are not convincing. It is, however, definitely familial.	Blunn and Hughes (1938); Smith and Stevenson (1973)
*Oedema disease, resistance to	Not to be confused with oedema, oedema disease is a fatal infectious disease of weaners and growers, characterized by uncoordination, a hoarseness of voice, weakness, flaccid paralysis and blindness. Although oedema of the eyelids, face and ears is diagnostic, this trait is seldom visible during clinical examination. Also called gut oedema or bowel oedema. Caused by opportunistic colonization by certain strains of *Escherichia coli*. Bertschinger *et al.* (1993) provided substantial evidence from breeding experiments for single-locus autosomal recessive inheritance of resistance to experimental challenge with a causative strain of *E. coli*.	Bertschinger *et al.* (1993)
*Wattles	Appendages suspended from the head. Also known as tassles or bells, they are fleshy masses of cartilaginous material covered with normal skin and suspended from the mandibular area. Roberts and Morrill (1944) provided good evidence for single-locus autosomal recessive inheritance.	Kronacher (1924); Roberts and Morrill (1944)
Skeletal system: limbs		
Arthritis deformans	Erosion of articular cartilage and destruction of subchondral bone. Regarded as an auto-immune disease. Also known as arthrosis deformans, erosive arthritis/arthrosis or deforming arthritis/arthrosis.	Sabec *et al.* (1961); Glawischnig (1965)
Hip dysplasia	Laxity of the hip joint, resulting from a shallow acetabulum and/or a small, misshapen head of the femur. Also called congenital dislocation of the hip. In severe cases, affected animals are lame.	Gotoh and Ando (1993)
Hyperostosis	Also known as thick forelegs. Affected piglets have a congenital enlargement of the forelegs below the elbow joint, due to abnormal development of bone and connective tissue: the bone is thick, the periosteum is rough and there is extensive oedema. The resultant difficulty in standing and moving about results in death from starvation in the first few days of life. Claims of single-locus inheritance have been made, but the data are not convincing.	Morrill (1947); Cargill and Riley (1971)
Leg weakness	Congenital defect of the hindlimbs that prevents standing, because the hindlimbs are splayed. Also called splayleg or spraddle leg or myofibrillar hypoplasia. A classic example of a multifactorial disorder with intermediate heritability.	Smith (1966); Rothschild *et al.* (1988); Antalikova *et al.* (1996)
*Legless	The absence of all four legs. Also called streamlined. Johnson and Lush (1939) and Johnson (1940) provided convincing evidence of autosomal recessive inheritance of this lethal disorder.	Johnson and Lush (1939); Johnson (1940)

Table 4.1. *continued*

Traits and disorders	Current knowledge	Reference
Monobrachia	Having only one forelimb.	Ilancic and Fuks (1962)
Osteoarthritis	A non-inflammatory degenerative joint disease characterized by degeneration of the articular cartilage, hypertrophy of bone at the margins, and changes in the synovial membrane. Also called degenerative joint disease. One of the causes of leg weakness.	Hill (1990a,b)
Paresis	Slight or incomplete paralysis. There has been only one case report (McClymont, 1954), with no data on inheritance.	McClymont (1954)
Polydactyly	Presence of extra digits. Definitely familial but, with the exception of the form of polydactyly described in the following entry, there is no convincing evidence of single-locus inheritance.	Hughes (1935); Ptak (1963)
*Polydactyly with otocephalic monster	In village pigs of Papua New Guinea, Malynicz (1982) reported an autosomal dominant disorder in which heterozygotes are polydactylous, and homozygotes are 'monsters', having club foot (dactylomegaly, or abnormally large digits) and otocephaly (absence of the lower jaw, with the ears united below the face).	Malynicz (1982)
*Progressive myopathy	A progressive degeneration of the hind legs, resulting in collapse. Also known as creeper syndrome. Wells *et al.* (1980) provided evidence for autosomal recessive inheritance.	Wells and Bradley (1978); Wells *et al.* (1980)
*Syndactyly	Fusion of the digits. Convincing evidence for autosomal dominant inheritance was provided by Simpson and Simpson (1908), soon after the re-discovery of Mendelism. Breeding data reported by Detlefsen and Carmichael (1921) confirmed this conclusion. There have been no recent reports of breeding data for this disorder.	Auld (1889); Leipold and Dennis (1972)
*Three-legged	Citing evidence reviewed by Koch *et al.* (1957), Ollivier and Sellier (1981) were convinced that this disorder is due to an autosomal recessive gene.	Koch *et al.* (1957); Ollivier and Sellier (1982)
Muscular system		
*Meat quality	The *RN* (Rendement Napole) gene was first documented by Leroy *et al.* (1990). In longissimus dorsi muscles, carriers of this dominant gene show lower pH, higher surface and internal reflectance values, lower protein extractability, lower water-holding capacity, lower Napole yield (yield after curing and cooking), and greater cooking loss. On the positive side, carriers have a lower shear force value, a stronger taste and smell, and greater acidity. The primary cause of these differences is that the mutant allele results in higher stored glycogen content in muscle. The gene is located on the short arm (q21–22) of chromosome 16.	Leroy *et al.* (1990); Lundstrom *et al.* (1996)

Muscular hypertrophy	Abnormal increase in muscular tissue caused entirely by enlargement of existing cells (in contrast to muscular hyperplasia, in which the abnormal increase in muscular tissue is due to the formation and growth of new, normal muscle cells). The Belgian Pietrain breed of pig shows considerable muscular hypertrophy, not unlike the 'double muscling' seen in Belgian Blue cattle. From the results of a cross between Pietrain and another breed, Ollivier and Lauvergne (1967) concluded that a major gene was responsible for muscular hypertrophy in pigs. However, it was later concluded (Ollivier, 1980) that this gene is actually the gene for malignant hyperthermia.	Ollivier and Lauvergne (1967); Ollivier (1980)
Eye		
Anophthalmos	Congenital absence of one or both eyes, or the presence of rudimentary eyes. Also called anophthalmia.	Hale (1935); Bendixon (1944)
Blindness	Congenital blindness.	Maneely (1951)
Cyclopia	Congenital developmental disorder characterized by a single orbital fossa. Named after the race of one-eyed giants in Greek mythology. In humans, this disorder is often associated with a chromosomal abnormality. In the only reported case in pigs, however, the affected animal, which died within an hour of birth, was shown to have a normal karyotype (Arakaki and Vogt, 1976). No evidence on inheritance.	Arakaki and Vogt (1976)
*Heterochromia iridis	Difference in colour of the iris in the two eyes, or in different areas of the one iris. In pigs, part (partial heterochromia) or all (complete heterochromia) of the iris lacks pigment. The latter form is called glass-eye. Reconciling data from Durr (1937) and Gelati *et al.* (1973), Ollivier and Sellier (1982) concluded that bilateral complete heterochromia iridis is autosomal recessive, and that heterozygotes show unilateral and partial heterochromia.	Durr (1937); Gelati *et al.* (1973)
Macrophthalmia	Abnormal enlargement of the eyeball. There are insufficient genetic data to enable any conclusion on inheritance to be drawn.	Darcel *et al.* (1960); Krajsa (1961)
Microphthalmia	Abnormal smallness in all dimensions of one or both eyes.	Roberts (1948); Krajsa (1961)
Neurological and neuromuscular systems		
*Arthrogryposis	Persistent flexion of a joint. Also known as bent-stiff-legged, bentleg, and congenital articular rigidity (CAR). Can be caused by various non-genetic factors such as ingestion of Jimsonweed or tobacco by pregnant sows. However, there is convincing evidence of a single-locus autosomal recessive form of this disorder (Lomo, 1985).	Hallqvist (1933); Lomo (1985)

Table 4.1. *continued*

Traits and disorders	Current knowledge	Reference
Asymmetric hindquarter syndrome	Asymmetry of the hindquarters. The disorder is familial, but there is no evidence for single-locus inheritance.	Done *et al.* (1975)
*Ataxia, progressive	A progressive failure of muscle coordination, resulting in perverse movements. Also known as congenital motor defect or congenital ataxia. The central nervous system appears normal at birth, but older pigs show dysplasia of the cerebellar cortex. Rimaila-Parnanen (1982) provided convincing evidence of autosomal recessive inheritance.	Rimaila-Parnanen (1982)
Cerebellar anomaly, congenital	Uncoordination, ataxia, dizziness, unthriftiness, resulting in death. Associated with misshapen nuclei in some of the Purkinje cells of the cerebellum.	Seibold and Roberts (1957)
*Hindlimb paralysis	Evidence for autosomal recessive inheritance of hindlimb paralysis was presented by Berge (1941). Ollivier and Sellier (1982) cited a study by Ludvigsen *et al.* (1963) as providing evidence for autosomal recessive inheritance of paralysis of the hind limbs, associated with abnormal lumbar vertebrae. Because the Berge disorder did not involve vertebral abnormalities, it is possible that more than one locus can give rise to hindlimb paralysis.	Berge (1941); Ludvigsen *et al.* (1963)
Hypomyelinogenesis, congenital	Congenital deficiency of myelin, especially in the cerebellum and brainstem, includes failure of formation of myelin, plus incomplete and delayed myelination of axons. Clinical signs include inability to rise, and severe muscle tremor with periods of spasticity.	Emerson and Delez (1965); Patterson and Sweasey (1970)
*Malignant hyperthermia	A progressive increase in body temperature, muscle rigidity and metabolic acidosis, leading to rapid death. In pigs, malignant hyperthermia (MH) leads to rapid postmortem changes in muscle, resulting in pale soft exudative (PSE) meat. MH can be triggered by a minor stress, such as loading, transport, sexual intercourse, high ambient temperature, or exposure to the anaesthetic halothane. Susceptibility to halothane-induced MH is an autosomal recessive trait in pigs. Together, sudden death and PSE constitute porcine stress syndrome (PSS), which became a major economic problem in many countries in the 1970s. In part, this was due to strong selection for increased leanness, which is associated with susceptibility to PSS. In 1991, a major breakthrough occurred when a Canadian research team led by David MacLennan (Fujii *et al.*, 1991) showed that MH is due to a base substitution (C → T) in the 1843rd nucleotide of the *RYR* gene for the calcium release channel in the sarcoplasmic reticulum of skeletal muscle. The base substitution causes an amino-acid substitution (arginine → cysteine) in the 615th position of the calcium release channel, resulting in altered calcium flow. A PCR genotyping test based on the causative mutation has been used extensively in many countries, leading to the elimination of the mutation from many herds (see Chapter 16).	Briskey (1964); Mickelson and Louis (1996)

Melanoma, Sinclair swine cutaneous malignant	A disorder involving tumours arising from melanocytes, dendritic cells of neuroectodermal origin, or melanoblasts, that occurs spontaneously in young pigs of the Sinclair strain, which are a valuable model for human melanoma. Spontaneous regression is common. This familial disorder shows association with certain alleles at the major histocompatibility complex.	Tissot *et al.* (1987); Blangero *et al.* (1996)
*Motor neurone disease, lower	A distinctive locomotor disorder of weaners, characterized by progressive ataxia and paresis of variable severity. Histological examination reveals significant degenerate changes in lower motor neurones, lower (motor) spinal nerve roots, and myelinated axons of peripheral nerves and of ventral and lateral spinal columns (O'Toole *et al.*, 1994b). Lipid-like inclusions and mitochondrial swelling suggest an underlying defect in lipid metabolism and/or mitochondrial metabolism (O'Toole *et al.*, 1994b). Limited breeding data are strongly suggestive of autosomal dominant inheritance, and a breeding colony has been established at the Wyoming State Veterinary Laboratory (O'Toole *et al.*, 1994a).	Wells *et al.* (1987); O'Toole *et al.* (1994b)
Myositis ossificans	A generalized inflammation of involuntary muscle, associated with dysptrophic ossification of muscle. The only known publication (Seibold and Davis, 1967) reports the disorder in approximately one-quarter of the pigs of an affected sire, but wisely avoids any single-locus attribution.	Seibold and Davis (1967)
Recumbent piglet trait	According to O'Toole *et al.* (1994b), this disorder was reported by Korovetzkaya (1938) to be inherited as a recessive trait in Large White pigs in the Soviet Union. Although definite evidence is lacking, it is possible that this disorder is the same as motor neurone disease, lower.	Korovetzkaya (1938); O'Toole *et al.* (1994b)
Spongiform encephalopathy	Spongiform encephalopathies are a class of fatal neurological diseases. Clinical signs are characteristic of a progressive degeneration of the central nervous system, they include pruritis, abnormalities of gait and recumbency. Death is inevitable. On postmortem, brain histopathology shows a characteristic spongy appearance. The infectious agent is a modified form of a protein encoded by a gene in the host. The name given to this infectious particle is prion. The host gene is called the prion protein (*PrP*) gene, which is a normal part of the genome of mammals and chickens. Its polypeptide product, called PrP^{C}, is a naturally-occurring protein attached to the outer surface of neurones and some other cells. PrP^{C} appears to play a role in maintaining the Purkinje cells of the cerebellum, which are essential for balance and muscular function. The infectious agent, called PrP^{Sc}, is a modifed form of PrP^{C}, where the modifications involve glycosylation and the creation of intra-strand disulphide bonds. It is important to realize that these modifications involve no change in amino acid sequence. When PrP^{Sc} molecules enter a previously uninfected host, they convert the naturally occurring PrP^{C} molecules, produced by the host gene, into infectious PrP^{Sc} particles, which ultimately cause clinical signs in that animal, and which can spread to other animals, both horizontally (by infection) and vertically (by maternal transmission). There have been several reports of the transmission of bovine spongiform encephalopathy (BSE) to pigs.	Dawson *et al.* (1990); Pearce (1996)

Table 4.1. *continued*

Traits and disorders	Current knowledge	Reference
*Tremor type A III, congenital	There are several different forms of congenital tremor syndrome, also known as dancing pig disease, trembles or myoclonia congenita. The common clinical feature comprises rhythmic tremors of the head and limbs, being most severe when standing, and being absent when sleeping. Except when provoked by external stimulus, e.g. handling or sudden noise, tremors are rarely seen after the first few months of life, and usually disappear by the second or third week. Affected pigs have decreased life expectancy due to decreased ability to nurse and to avoid being crushed by the sow. Done and Harding (1967) distinguished between those forms that show lesions in the central nervous system (type A) and those that do not (type B) Within type A, there are two forms caused by infectious agents (I and II) and two inherited forms (III and IV) (Done, 1968). Congenital tremor type III A, also known as cerebrospinal hypomyelinogenesis, is an X-linked recessive disorder.	Harding *et al.* (1973); Blakemore *et al.* (1974)
*Tremor type A IV, congenital	Background details are given in the previous entry. Congenital tremor type A IV, also known as cerebrospinal dysmyelinogenesis, is an autosomal recessive disorder.	Patterson *et al.* (1973); Blakemore and Harding (1974)
*Tremor, high-frequency	This disorder was described by Richter *et al.* (1995), from whom the following description is taken. The disorder is characterized by muscular weakness and a very intense tremor of the legs when standing and walking but not when at rest in a lying position. The intensity of tremor and muscular weakness progressively increases with age, resulting in pronounced postural instability. Also known as the Campus syndrome, after the boar whose offspring first showed the disorder. Breeding data show autosomal dominant inheritance.	Richter *et al.* (1995)
Haematological and cardiac systems		
Anaemia	Abnormally low number of red blood cells (erythrocytes) or abnormally low quantity of haemoglobin.	Gabris (1974)
Artery, anomaly of	Congenital abnormality of an artery. The only case reported in pigs (Gregg, 1946) involved the right subclavian artery arising as a side branch of the pulmonary trunk on its dorso-lateral surface, observed in a preserved nearly-full-term fetus.	Gregg (1946)
Atherosclerosis	A common form of arteriosclerosis in which deposits of yellowing plaques (atheromas) containing cholesterol and other lipid material are formed within the arteries. The disorder is familial in pigs. A strain that spontaneously develops atherosclerosis has been developed by selection, and is used as an animal model for medical studies of the same disease in humans.	Skold and Getty (1961); Rapacz and Hasler-Rapacz (1990)
Blood group system, general	Each blood group system consists of a set of blood types, each of which corresponds to a particular antigen (usually a glycoprotein) on the surface of red blood cells. The different types within a system are the result of the action of different alleles at a locus that usually encodes an enzyme that catalyses the creation of the unique feature of the glycoprotein (see Chapter 4).	Andresen *et al.* (1960); Hojny *et al.* (1994)

*Blood group system J	This system has at least 3 alleles.	Andresen (1957); Hradecky *et al.* (1985)
*Blood group system I	This system has at least 2 alleles.	Andresen (1957); Andresen (1966)
*Blood group system H	This system has at least 7 alleles. This locus is linked to the gene for malignant hyperthermia, forming part of what is commonly called the Hal linkage group.	Andresen (1957); Zeveren *et al.* (1988)
*Blood group system G	This system has at least 3 alleles.	Andresen (1957); Erhard *et al.* (1988)
*Blood group system F	This system has at least 3 alleles.	Andresen (1957); Erhard *et al.* (1988)
*Blood group system E	This system has at least 15 alleles.	Andresen and Irwin (1959a); Hojny and Nielsen (1992)
*Blood group system N	This system has at least 3 alleles.	Hala and Hojny (1964)
*Blood group system O	This system has at least 2 alleles.	Hojny and Hala (1965); Mariani *et al.* (1996)
*Blood group system K	This system has at least 6 alleles.	Andresen and Irwin (1959b); Erhard *et al.* (1988)
*Blood group system L	This system has at least 6 alleles.	Andresen (1962); Marklund *et al.* (1993)
*Blood group system M	This system has at least 18 alleles.	Nielsen (1961); Nielsen (1991)
*Blood group system D	This system has at least 2 alleles.	Andresen (1962); Erhard *et al.* (1988)
*Blood group system A	This system has at least 2 alleles. Also known as the A–O blood group system.	Sprague (1958); Erhard *et al.* (1988)
*Blood group system C	This system has at least 2 alleles.	Andresen and Baker (1964); Rasmusen (1982)
*Blood group system B	This system has at least 2 alleles.	Baker and Andresen (1964); Erhard *et al.* (1988)
*Suppression of blood group system A	This locus controls the expression of blood group antigens of the A blood group system, with suppression being recessive. This locus is in the HAL linkage group.	Rasmusen (1964); Zeveren *et al.* (1988)
Cardiomyopathy, dilated	A disorder characterized by cardiac enlargement, especially of the left ventricle, poor myocardial contractility, and congestive heart failure. Also called congestive cardiomyopathy.	Hendrick *et al.* (1990)
Cardiomyopathy, hypertrophic	Increase in volume of the muscle tissue of the heart, due to an increase in the size of muscle cells, primarily in the left ventricle and ventricular septum. A familial, multifactorial disorder, with a moderate heritability (around 30%). A hypertrophic cardiomyopathy (HCM) strain of pigs has been developed (Huang *et al.*, 1996).	Liu *et al.* (1994); Huang *et al.* (1996)

Table 4.1. *continued*

Traits and disorders	Current knowledge	Reference
Fragile site	A site on a chromosome that does not stain, at which a break in the chromosome often occurs. In cultured cells, chromosomal breakage at fragile sites can be induced by the addition of caffeine or aphidicolon or bromodeoxyuridine to the culture medium. Some fragile sites in pigs correspond to breakpoints in known reciprocal translocations (Yang and Long, 1993). In humans, fragile sites are sometimes associated with tandem repeats of three nucleotides (triplet microsatellites) which, if the number of repeats increases, can cause inherited disorders. No such examples have yet been documented in domesticated animals.	Riggs and Chrisman (1991); Ronne (1995)
Hypercholesterolaemia, spontaneous	An excess of cholesterol in the blood. Also called cholesterolaemia or hypercholesteraemia. A strain of pigs that shows spontaneous development of hypercholesterolaemia has a mutation in the gene for apolipoprotein B (apoB), designated Lpb5. This mutant apoB allele is associated with low density lipoprotein (LDL) particles which are deficient in their ability to bind to the LDL receptor (Purtell *et al.*, 1993), and which are therefore cleared from the circulation more slowly. The resultant hypercholesterolaemia, however, is not a single-locus trait. Indeed, there appears to be at least one other gene of large effect involved in this disorder (Aiello *et al.*, 1994).	Lee *et al.* (1990); Hasler-Rapacz *et al.* (1995)
Hyposelenaemia–hyperselenaemia	Stowe and Miller (1985) showed that selenium concentration is a multifactorial trait with a non-zero (but undetermined) heritability.	Stowe and Miller (1985)
*Lymphosarcoma	A malignant neoplastic disorder of lymphoid tissue. Also called lymphoma. Sometimes also called leukaemia, although strictly speaking, this is a different disorder (namely a malignant disorder of the blood-forming tissue) Clinical signs of lymphosarcoma in pigs include stunted growth, pot-belly, enlargement of superficial lymph nodes, an increase in circulating lymphocytes, and death before 15 months of age. Postmortem reveals sarcoma involving all lymph nodes, but primarily those draining the gut and lung (Head *et al.*, 1974). McTaggart *et al.* (1979) provided convincing evidence that this disorder is autosomal recessive in pigs.	McTaggart *et al.* (1971); McTaggart *et al.* (1979)
Persistent truncus arteriosus	The truncus arteriosus is the main arterial trunk arising from the embryonic heart. It normally develops into the aortic and pulmonary arches. This disorder is the result of the failure of this development. A familial trait in other species.	Vitums (1972)

*Porphyria	Porphyria is a general term for disorders resulting from a deficiency of any one of the six enzymes involved in the biosynthesis of protoporphyrin from aminolaevulinic acid. Some of the intermediates (loosely called porphyrins) are extremely photoreactive. Because a deficiency of any one of these enzymes results in a buildup of intermediates, photosensitivity is a common clinical sign in most species, but not in pigs. In this species, clinical diagnosis is based on discoloration (pink to brown) of the teeth. On autopsy, bones, kidneys and lymph nodes show a brown discoloration. Another major clinical sign is haemolytic anaemia, due to a deficiency of haemoglobin, of which protoporphyrin is a vital component. There have been only isolated reports in pigs, with only one study describing the likely enzyme deficiency: Roels *et al.* (1995) attributed their cases of porphyria to a combined deficiency of the third (uroporphyrinogen-III-cosynthetase) and fourth (uroporphyrinogen decarboxylase) enzymes. Their cases would therefore be a 'mixed' porphyria, comprising aspects of congenital erythropoeitic porphyria and porphyria cutaneous tarda.	Clare and Stephens (1944); Roels *et al.* (1995)
Subaortic stenosis	Congenital obstruction to the outflow of blood from the left ventricle into the aorta, characterized by stenosis (narrowing) of the aorta just below the semilunar valves. Familial but not due to a single gene.	Emsbo (1956); Baker (1976)
*Thrombopathia	A blood coagulation disorder due to failure of ADP release from platelets following stimulation by aggregation factors such as thromboplastin. Also called storage pool deficiency. Characterized by mild to moderate bleeding. Shown to be autosomal recessive by Thiele *et al.* (1986).	Daniels *et al.* (1986); Thiele *et al.* (1986)
Tachycardia	Abnormally rapid heart rate.	Hendrick *et al.* (1990)
Ventricular septal defect	A congenital heart defect characterized by persistent patency (openness) of the ventricular septum, permitting flow of blood directly between ventricles, bypassing the pulmonary circulation and resulting in various degrees of cyanosis (blue discoloration of the skin) due to oxygen deficiency. Clinical signs include systolic murmur and a palpable thrill on both sides of the chest, dyspnoea and poor tolerance of exercise.	Swindle *et al.* (1990); Johnson *et al.* (1993)
*von Willebrand disease	von Willebrand factor (vWF) is a multimeric form of a plasma protein encoded by an autosomal gene (not yet mapped in pigs). vWF plays a vital role in platelet adhesion and clot formation. It also combines with factor VIIIC (the product of the X-linked haemophilia *A* locus), forming factor VIII. vWF accounts for 99% of the mass of factor VIII, its role is to protect factor VIIIC from degradation. von Willebrand disease (also called pseudohaemophilia or vascular haemophilia) is an autosomal bleeding disorder resulting from deficient or defective vWF. The molecular basis of this disorder in pigs has not yet been determined.	Hogan *et al.* (1941); Roussi *et al.* (1996)

Table 4.1. *continued*

Traits and disorders	Current knowledge	Reference
Endocrine and metabolic systems		
*Dwarfism	There have been several isolated reports of dwarfism in pigs, with the most convincing single-locus evidence being provided by Jensen *et al.* (1984). The two reports with the greatest clinical detail are those by Kaman *et al.* (1991) and Shirota *et al.* (1995).	Petrov (1974); Shirota *et al.* (1995)
*Gangliosidosis, GM2	A lysosomal storage disease in which there is a build-up (storage) of GM2 gangliosides (a type of glycolipid) in various tissues, due to the lack of the enzyme hexosaminidase, whose task is to break down the GM2 ganglioside into its constituents. Characterized by progressive neuromuscular dysfunction and impaired growth from an early age.	Read and Bridges (1968); Kosanke *et al.* (1978)
Goitre, familial	Enlargement of the thyroid gland, causing a swelling in the front of the neck. Familial forms of this disorder have been identified in several species. But only in cattle and goats has the molecular basis been determined.	Bokori (1956)
*Nucleoside transport defect	Defect in the transport of nucleosides (purine or pyrimidine base attached to a ribose or deoxyribose sugar) across erythrocyte membranes. This disorder has not been recorded in pigs *in vivo*. However, in a kidney cell line, Aran and Plagemann (1992) created several different mutants that resulted in failure to transport thymidine and uridine. The faulty gene has not yet been identified.	Aran and Plagemann (1992)
*Oedema	Abnormal accumulation of fluid in tissues and/or body cavities. Also called myxoedema, dropsy or hydrops. In reviewing the evidence summarized by Koch *et al.* (1957), Ollivier and Sellier (1982) concluded that oedema is an autosomal recessive disorder, possibly to do with a thyroid defect.	Young (1952); Neeteson (1964)
*Pseudovitamin D deficiency rickets	Vitamin D (cholecalciferol) is synthesized in the skin from 7-dehydrocholesterol by the action of UV radiation from sunlight. Cholecalciferol, however, has very little biological activity: it requires two hydroxylations in order to become (biologically) active. The first hydroxylation, catalysed by cholecalciferol 25hydroxylase, occurs in the liver. The second of these hydroxylations occurs in the kidney under the action of the enzyme 25alpha-hydroxy-cholecalciferol 1-hydroxylase. The resultant active form of vitamin D (called 1,25-dihydroxycholecalciferol or 1,25$(OH)_2$D) is a steroid hormone that plays a vital role in whole-body calcium homeostasis. Pseudo-vitamin D deficiency rickets is an inherited deficiency of the 1-hydroxylase enzyme. As expected, this deficiency results in clinical signs indistinguishable from those seen in individuals suffering from non-genetic lack of vitamin D (vitamin D deficiency rickets), most commonly resulting from a dietary deficiency of calcium or insufficient exposure to sunlight. The clinical signs of rickets (inherited and non-genetic) arise from defects in calcium homeostasis. The most noticeable effects include a failure of calcification of bones (leading to bowing of limbs) and delayed dentition. Pseudo-vitamin D deficiency rickets is an autosomal recessive disorder. The molecular basis of this disorder in pigs has not yet been determined.	Meyer and Plonait (1968); Barckhaus *et al.* (1994)

*Respiratory distress syndrome	Also known as barker syndrome, because of the resemblance between affected piglets and barker foals (Gibson *et al.*, 1976). Characterized by very small thyroid glands, hairlessness, retarded ossification, and delayed haemopoiesis, and death soon after birth, due to acute respiratory distress (Wrathall, 1976). Breeding data presented by Wrathall (1976) provide convincing evidence of autosomal recessive inheritance.	Gibson *et al.* (1976); Wrathall (1976); Wrathall *et al.* (1977)
Digestive tract		
Agenesia of anal sphincter	Congenital absence of anal sphincter. Also known as aplasia of the anal sphincter.	Hamori (1966)
Atresia ani	Congenital absence of the anus, causing a build-up of faeces and consequent distension of the abdomen. A large German segregation analysis (Stigler *et al.*, 1991, 1992) showed that this disorder is definitely familial, but is also definitely not due to a single gene.	Berge (1941); Stigler *et al.* (1992)
Atresia ilei	Congenital absence of the ileum (the distal portion of the small intestine, extending from the jejunum to the caecum).	Stevenson and Walker (1954)
Cloaca	Absence of the anus, causing faeces and urine to be voided through the vulva. Named after the structure in birds through which faeces and urine are normally voided together.	Knilans (1943)
Hernia	Protrusion of part of an organ or tissue through the structures normally containing it. Often shows familial occurrence, but convincing evidence for single-locus inheritance is lacking.	Warwick (1926); Stigler *et al.* (1992)
Hernia, diaphragmatic	A congenital opening in the thoracic diaphragm, permitting the displacement of abdominal organs into the thorax. Shows familial occurrence, but convincing evidence for single-locus inheritance is lacking.	Griffin (1965); Gregory (1971)
Hernia, inguinal	Protrusion of part of the intestine through the abdominal wall. Often shows familial occurrence, but convincing evidence for single-locus inheritance is lacking.	Lewis (1965); Wensing (1976)
Hernia, scrotal	An inguinal hernia that has passed into the scrotum. Often shows familial occurrence, but convincing evidence for single-locus inheritance is lacking.	Berge (1941); Lingaas and Ronningen (1991)
Hernia, umbilical	Protrusion of part of the intestine or other abdominal organs through the abdominal wall at the umbilicus, the protrusion being covered with skin and subcutaneous tissue. Often shows familial occurrence, but convincing evidence for single-locus inheritance is lacking.	Hamori (1962); Hayes (1974)
Megacolon	A disorder in which the large intestine (caecum, colon and sometimes rectum) undergoes a large dilation and fills with faecal mass. Also known as Hirschsprung disease. In other species, this disorder is due to incomplete migration of nerve cells to the intestine during embryonic development, resulting in lack of peristalsis and hence a build-up of faecal material. A single-locus disorder in some other species, but no good family data have been published for pigs.	Kernkamp and Kanning (1955); Osborne *et al.* (1968)

Table 4.1. *continued*

Traits and disorders	Current knowledge	Reference
*Neonatal diarrhoea, K88	Neonatal diarrhoea in piglets is often caused by strains of *Escherichia coli* bacteria having a cell-surface antigen called K88, which combines with a glycoprotein receptor on the wall of a piglet's intestines, enabling the bacteria to attach themselves to the intestine. The receptor is a type of transferrin. Once attached, the bacteria proliferate, releasing enterotoxins and thus producing diarrhoea, which can lead to high mortality. Certain piglets lack the intestinal receptor to K88, and are therefore resistant to K88 bacteria and hence to diarrhoea caused by K88 strains. Lack of the K88 receptor is a single-locus autosomal recessive trait. In fact, there are several different antigenic variants of K88 (ab, ac, ad), and it seems that there is a separate receptor for each. The presence or absence of at least two of these receptors (for K88ab and K88ac) is determined for each receptor by one of two closely-linked genes on chromosome 13. The immunological and population-genetics implications of segregation at these loci are very interesting: the result is selection against heterozygotes (Nicholas, 1996, pp. 135–136).	Gibbons *et al.* (1977); Grange and Mouricout (1996)
Neonatal diarrhoea, K99	It seems likely that resistance to the K99 strains of *Escherichia coli* will also turn out to be determined by a single gene. At present, however, the evidence for this is lacking.	Teneberg *et al.* (1990); Yuyama *et al.* (1993)
Neonatal diarrhoea, F4	It seems likely that resistance to the F_4 strains of *Escherichia coli* will also turn out to be determined by a single gene. At present, however, the evidence for this is lacking.	Runnels *et al.* (1992)
Ulcer, stomach	An excavation of the surface (mucosal epithelium) of the stomach. Assessible only by physical examination of stomach tissue after slaughter. Scored on a scale from 0 (normal) to 4 (severe), or as a zero-one trait. Multifactorial rather than single locus. In the most thorough study reported to date, Berruecos and Rabinan (1972) reported an incidence of 29% in Durocs and 12% in Yorkshires, with a heritability of 52% on the underlying scale. Genetic correlations with production traits were favourable for daily gain and food conversion efficiency, but were unfavourable for backfat thickness.	Kowalczyk *et al.* (1960); Berruecos and Robison (1972)
Urogenital system		
Cryptorchidism	Undescended testes. Can be either unilateral or bilateral. Both disorders are familial, but there is no convincing evidence for single-locus inheritance.	McPhee and Buckley (1934); Rothschild *et al.* (1988); Vandenbroeck and Maghuinrogister (1995)
Cystic bile ducts and renal tubules	Affected piglets appear normal at birth, but develop distended abdomens and show signs of distress between 24 and 48 hours. Death follows within 10 days. Autopsy revealed numerous cysts (up to 1 cm in diameter) in the liver and on the surface of the kidney. The only published report (Groth and Rosdail, 1955) provides limited evidence of familial occurrence, and no genetic data.	Groth and Rosdail (1955)

Epididymal aplasia	Total or partial failure of development of the epididymis. Konig *et al.* (1972) provided substantial inheritance data in cattle, but less decisive data for pigs.	Konig *et al.* (1972)
Infertility	Inability to conceive. Has a wide variety of causes, some of which are at least partly genetic.	Wilson *et al.* (1949); Adams (1970)
Intersex	A mixture of male and female characteristics. Affected animals are also called hermaphrodites. Often due to an abnormality of the sex chromosomes.	Kingsbury (1909); Hunter and Greve (1996)
Knobbed acrosome	The presence of a protrusion (knob) on the acrosome (cap) of a sperm. Associated with sterility. The disorder is certainly familial, but the limited evidence for single-locus inheritance is not convincing.	Rob (1963); Toyama and Itoh (1993)
*Membranoproliferative glomerulonephritis type II	A progressive inflammation of the capillary loops in the glomeruli of the kidney in which the glomeruli become enlarged as a result of proliferation of mesangial cells and irregular thickening of the capillary walls, due to massive glomerular depositis of complement component C3 and the terminal C5b-9 complement complex. This disorder is due to a deficiency of complement factor H, whose task is to restrict the formation of C3 convertase. The hypermetabolism of the excess C3 results in the deposits of complement in the kidney, and in hypocomplementaemia. The disorder is autosomal recessive in pigs. Affected piglets die at around 5 weeks.	Jansen (1993); Jansen *et al.* (1995)
Mortality	Wijeratne *et al.* (1970) presented evidence of genetic variation (a sire effect) in mortality.	Wijeratne *et al.* (1970)
Nephropathy	Disease of the kidneys.	Olsen *et al.* (1993)
Ovarian aplasia	Defective development, or complete absence, of the ovaries.	Bosma *et al.* (1975)
Persistent frenulum praeputii	A close attachment of the prepuce to the ventral surface of the body by a stretch of mucous membrane (a frenulum), resulting in inadequate protrusion of the penis, and subsequent inability to breed.	Aamdal and Nes (1958)
*Protamine-2 deficiency	In most mammals, protamine-2 constitutes the major component of basic protein in sperm nuclei. It binds to DNA during the elongation of spermatids. Pigs and cattle have the protamine-2 gene, but produce very little protamine-2. In the case of pigs, this is due to a deletion of 27 bases in the middle of the gene. It appears that all pigs are homozygous for this mutation. Presumably, the mutation occurred before the splitting of the porcine evolutionary lineage.	Maier *et al.* (1990)
*Renal cysts	The occurrence in the kidney of closed epithelium-lined sacs containing a liquid or semi-solid substance. Wijeratne and Wells (1980) reported breeding data that provided convincing evidence of autosomal dominant inheritance.	Wijeratne and Wells (1980)
*Renal hypoplasia, bilateral	Failure of development of the mesonephric mesenchyme, resulting in renal insufficieny and consequent death within 2 months of birth. Cordes and Dodd (1965) presented sufficient breeding data to justify a conclusion of single-locus autosomal recessive inheritance.	Cordes and Dodd (1965)

Table 4.1. *continued*

Traits and disorders	Current knowledge	Reference
Renal hypoplasia, unilateral	As above, but only one kidney affected. Insufficient data in pigs to enable any conclusion about mode of inheritance.	Brody and Bailey (1939)
SME sperm head defect	In 1973, Blom reported the presence of a rounded cyst-like body about 2 μm in diameter, in about 30% of sperms from a Danish Landrace boar called SME. The boar's semen had poor motility. The body is hypochromatic (eosin–nigrosin stain) and, by electron microscopy, has moderate electron density. It occurs in the front half of the sperm head, within the region of the acrosomal cap (Blom and Jensen, 1977). The same defect was later observed in a Norwegian Landrace boar and in a Yorkshire boar. Blom and Jensen (1977) reported the results of matings between offspring of this latter boar. Although the results were consistent with autosomal recessive inheritance, the numbers were too small to enable any firm conclusion to be reached.	Blom (1973); Blom and Jensen (1977)
Testicular feminization	This is an abnormality of sexual development in which affected individuals have an XY chromosomal constitution, undescended testes and female secondary sexual characteristics (including female external genitalia). Also, instead of normally developed Mullerian duct derivatives (Fallopian tubes, uterus, cervix, and upper portion of the vagina), they have underdeveloped Wolffian duct derivatives (epididymis, vas deferens, and seminal vesicle). In humans and mice, this disorder is known to be due to a deficiency of an androgen receptor encoded by a gene on the X chromosome. In pigs, the sole report of testicular feminization (Lojda, 1975) is insufficient to enable firm conclusions to be drawn.	Lojda (1975)
Uterus didelphys	Complete duplication of the cervix and uterus.	
*Uterus aplasia	Incomplete development of the uterus. King and Linares (1980) provided convincing evidence for autosomal recessive inheritance.	King and Linares (1980)
*Wilms' tumour	A rapidly developing malignant mixed tumour of the kidneys which has been recognized in pigs for many decades (Parish and Done, 1962). A gene for Wilms' tumour in human and mouse was cloned in the early 1990s. Since its peptide is involved in the development of the urogenital tract and the gonads, the gene is of considerable interest in relation to the genetic basis of sexual development. Using primers based on human and mouse sequence, Vaiman *et al.* (1995) amplified a 150-bp fragment of the porcine Wilms' tumour gene by PCR. The fragment contained a TG microsatellite which also exists in the same fragment in cattle, sheep and goats. This microsatellite appears to be monomorphic in pigs.	Parish and Done (1962); Vaiman *et al.* (1995)

Respiratory system		
Atrophic rhinitis	A disease of young pigs, resulting in permanent deformities of the facial and turbinate bones. Caused by *Pasteurella multocida* serotype D and *Bordetella bronchiseptica*. Resistance to atrophic rhinitis is multifactorial, with heritability of the order of 20%, and negligible phenotypic and genetic correlation with production traits (Lundeheim, 1979). A useful summary of the genetic aspects of resistance to atrophic rhinitis is provided by Straw and Rothschild (1992, pp. 710–711).	Koch *et al.* (1958); Gatlin *et al.* (1996)
Immotile cilia syndrome	Congenital defect in functioning of the cilia, which are minute hair-like processes that extend from cell surfaces, beating rhythmically to move fluid or mucus over the surface. Also called ciliary dyskinesia and Kartagener syndrome. The abnormal functioning results in bronchiectasis (chronic dilatation of the bronchi and bronchioles, with associated infection) and sinusitis (inflammation of one or more of the paranasal sinuses). Is also associated with dextrocardia (location of the heart in the right side of the thorax, rather than the usual left side). This disorder appears to be a good animal model for the disorder in humans. There is too little data on the disorder in pigs to enable any conclusion as to mode of inheritance. In humans, the disorder shows definite single-locus inheritance.	Roperto *et al.* (1993)

An Overview

In Table 4.1 is provided a list of a total of 161 traits and disorders that have been reported in pigs together with the earliest report plus the most recent reference, extracted from MIA. A complete set of up-to-date references for each entry can be obtained by accessing OMIA on the Internet. Those traits and disorders for which there is reasonable evidence of single-locus inheritance (57 in total, including 15 blood group systems) are indicated by an asterisk (*).

Many of the traits and disorders are of only minor interest, having been the subject of only one or a few poorly-documented reports. Only a few words are devoted to such entries. Some traits have been discussed in detail in other chapters of the present book, e.g. coat colour in Chapter 3, blood groups in Chapter 5, and malignant hyperthermia in Chapter 16. Readers are referred to these chapters for more detailed information.

Conclusions

The list of inherited morphological traits and disorders presented in this chapter provides an indication of the range of such traits and disorders that have been observed and studied in pigs. The molecular and gene-mapping revolutions now underway will lead to an explosion of knowledge in this area in the years ahead. To exploit fully the genetic variation that does occur, breeders and researchers need to be continually on the lookout for unusual animals, saving them where possible. If DNA can be sampled from several generations of a family in which a particular morphological trait or disorder occurs, and if careful records on the occurrence of the trait or disorder in that family have been kept, it will be an increasingly straightforward matter to identify the gene responsible.

References

Aamdal, J. and Nes, N. (1958) Persistent frenulum praeputii as a cause of mating impotence in boars. Preliminary communication. *Nordisk Veterinaermedicin* 10, 444–446.

Adams, W.M. (1970) Hormonal and anatomical causes of infertility in swine: effect of disease and stress of reproductive efficiency in swine. In: Lucus, L.E. and Wagner, W.C. (eds) *Proceedings of a Symposium Sponsored by the National Pork Producers Council.* University of Nebraska, College of Agriculture, pp. 16–21.

Aiello, R.J., Nevin, D.N., Ebert, D.L., Uelmen, P.J., Kaiser, M.E., Maccluer, J.W., Blangero, J., Dyer, T.D. and Attie, A.D. (1994) Apolipoprotein B and a second major gene locus contribute to phenotypic variation of spontaneous hypercholesterolemia in pigs. *Arteriosclerosis and Thrombosis* 14, 409–419.

Andresen, E. (1957) Investigations on blood groups of the pig. *Nordisk Veterinaermedicin* 9, 274–284.

Andresen, E. (1962) Blood groups in pigs. *Annals of the New York Academy of Sciences* 97, 207–225.

Andresen, E. (1966) Blood groups of the I system in pigs: association with variants of serum amylase. *Science* 153, 1660–1661.

Andresen, E. and Baker, L.N. (1964) The C blood group system in pigs and the detection and estimation of linkage between the C and J systems. *Genetics* 49, 379–386.

Andresen, E. and Irwin, M.R. (1959a) The E blood system of the pig. *Nordisk Veterinaermedicin* 11, 540–547.

Andresen, E. and Irwin, M.R. (1959b) The K blood group system of the pig. *Acta Agriculturae Scandinavica* 9, 253–260.

Andresen, E., Hojgaard, N., Jylling, B., Larsen, B., Moustgaard, J. and NeimannSorensen, A. (1960) [Blood and serum-group investigations on cattle, pig and dog in Denmark]. *Zuchtungskunde* 32, 306–323.

Anon. (1991) Micropigs. In: National Research Council (ed.) *Microlivestock: Little-Known Small Animals with a Promising Economic Future.* National Academy Press, Washington DC, pp. 63–72.

Antalikova, L., Horak, V. and Matolin, S. (1996) Ultrastructural demonstration of glucose-6-phosphatase activity and glycogen in skeletal muscles of newborn piglets with the splayleg syndrome. *Reproduction, Nutrition, Development* 36, 205–212.

Arakaki, D.T. and Vogt, D.W. (1976) A porcine cyclops with normal female karyotype. *American Journal of Veterinary Research* 37, 95–96.

Aran, J.M. and Plagemann, P.G.W. (1992) Nucleoside transport-deficient mutants of PK15 pig kidney cell line. *Biochimica et Biophysica Acta* 1110, 51–58.

Auld, R.C. (1889) Some cases of solid-hoofed hogs and two-toed horses. *American Naturalist* 23, 447–449.

Baker, J.R. (1976) Subaortic stenosis in the pig. *Veterinary Record* 98, 485–486.

Baker, L.N. and Andresen, E. (1964) The Bb blood group factor in pigs. *Vox Sanguinis* 9, 359–362.

Barckhaus, R.H., Harmeyer, J., Kaune, R., Gossrau, R. and Hohling, H.J. (1994) Quantitative X-ray microprobe analysis of the epiphyseal growth plate and histochemistry of various intestinal enzymes in piglets suffering from vitamin D-deficiency rickets, type-I. In: Collery, P., Poirier, L.A., Littlefield, N.A. and Etienne, J.C. (eds) *Metal Ions in Biology and Medicine*, Vol. 3. John Libbey Eurotext Ltd, Montrouge, France, pp. 59–64.

Bendixon, H.C. (1944) Littery occurrence of anophthalmia or microphthalmia together with other malformations in swine – presumably due to vitamin A deficiency of the maternal diet in the first stage of pregnancy and the preceding period. *Acta Pathologica et Microbiologica Scandinavica* 54, 161–179.

Berge, S. (1941) The inheritance of paralysed hind legs, scrotal hernia and atresia ani in pigs. *Journal of Heredity* 32, 271–274.

Berruecos, J.M. and Robison, O.W. (1972) Inheritance of gastric ulcers in swine. *Journal of Animal Science* 35, 20–24.

Bertschinger, H.U., Stamm, M. and Vogeli, P. (1993) Inheritance of resistance to oedema disease in the pig – experiments with an *Escherichia coli* strain expressing fimbriae-107. *Veterinary Microbiology* 35, 79–89.

Blakemore, W.F. and Harding, J.D.J. (1974) Ultrastructural observations on the spinal cords of piglets affected with congenital tremor type A IV. *Research in Veterinary Science* 17, 248–255.

Blakemore, W.F., Harding, J.D. and Done, J.T. (1974) Ultrastructural observations on the spinal cord of a Landrace pig with congenital tremor type AIII. *Research in Veterinary Science* 17, 174–178.

Blangero, J., Tissot, R.G., Beattie, C.W. and Amoss, M.S. (1996) Genetic determinants of cutaneous malignant melanoma in Sinclair swine. *British Journal of Cancer* 73, 667–671.

Blom, E. (1973) Studies on boar semen. 1. A new major defect in the sperm head, the SME-defect. *Acta Veterinaria Scandinavica* 14, 633–635.

Blom, E. and Jensen, P. (1977) Study of the inheritance of the SME seminal defect in the boar. *Nordisk Veterinaermedicin* 29, 194–198.

Blood, D.C. and Studdert, V.P. (1988) *Baillière's comprehensive veterinary dictionary.* Baillière Tindall, London, 1124pp.

Blunn, C.T. and Hughes, E.H. (1938) Hydrocephalus in swine. *Journal of Heredity* 29, 203–208.

Bokori, J. (1956) Congenital goitre in piglets. *Magyar Allatorvosok Lapja* 11, 364–369.

Bosma, A.A., Colenbrander, B. and Wensing, C.J.G. (1975) Studies on phenotypically female pigs with hernia inguinalis and ovarian aplasia. II. Cytogenetical aspects. *Proceedings of the Koninklijke Nederlandse Akademie van Wetenschappen* 78, 43–46.

Briskey, E.J. (1964) Etiological status and associated studies of pale, soft, exudative porcine musculature. *Advances in Food Research* 13, 89–178.

Brody, H. and Bailey, P.L. (1939) Unilateral renal agenesia in a fetal pig. *Anatomical Record* 74, 159–163.

Brooksbank, N.H. (1958) Congenital deformity of the tail in pigs. *British Veterinary Journal* 114, 50–55.

Cargill, C.F. and Riley, M.G.I. (1971) Case report: porcine hyperostosis. *Missouri Veterinarian* 21, 15–16.

Clare, T. and Stephens, E.H. (1944) Congenital porphyria in pigs. *Nature* 153, 252–253.

Clayton, G.A., Powell, J.C. and Hiley, P.G. (1981) Inheritance of teat number and teat inversion in pigs. *Animal Production* 33, 299–304.

Cordes, D.O. and Dodd, D.C. (1965) Bilateral renal hypoplasia of the pig. *Pathologia Veterinaria* 2, 37–48.

Dabczewski, Z. (1949) [Studies on a teratological form of the newly born Pulawska swine]. *Bull. int. Acad. Pol. Sci. Lett., Cl. Sci. math. nat., B II Zool.*, 241–260 (in Polish).

Daniels, T.M., Fass, D.N., White, J.G. and Bowie, E.J.W. (1986) Platelet storage pool deficiency in pigs. *Blood* 67, 1043–1047.

Darcel, C. LeQ., Niilo, L., Avery, R.J. and Bainborough, A.R. (1960) Microphthalmia and macrophthalmia in piglets. *Journal of Pathology and Bacteriology* 80, 281–286.

Dawson, M., Wells, G.A.H., Parker, B.N.J. and Scott, A.C. (1990) Primary parenteral transmission of bovine spongiform encephalopathy to the pig. *Veterinary Record* 127, 338.

Detlefsen, J.A. and Carmichael, W.J. (1921) Inheritance of syndactylism, black and dilution in swine. *Journal of Agricultural Research* 20, 595–604.

Donald, H.P. (1949) The inheritance of a tail abnormality associated with urogenital disorders in a pig. *Journal of Agricultural Science* 39, 164–173.

Done, J.T. (1964) Pityriasis rosea in pigs. *Veterinary Record* 76, 1507–1508.

Done, J.T. (1968) Congenital nervous diseases of pigs – a review. *Laboratory Animals* 2, 207–217

Done, J.T. and Harding, J.D.J. (1967) Kongenitaler Tremor der Schweine (Zitterkrankheit der Ferkel): Veranderungen und Ursachen. *Deutsch Tierartzliche Wohenschrift* 74, 333–334, 336.

Done, J.T., Allen, W.M., Bailey, J., De Gruchy, P.H. and Curran, M.K. (1975) Asymmetric hindquarter syndrome (AHQS) in the pig. *Veterinary Record* 96, 482–488.

Durr, M. (1937) Die Vererbung der Glasaugigkeit beim Schwein. *Zeitschrift Zuchtungsbiologie* B 37, 129–158.

Emerson, J.L. and Delez, A.L. (1965) Cerebellar hypoplasia, hypomyelinogenesis, and congenital tremors of pigs, associated with prenatal hog cholera vaccination of sows. *Journal of the American Veterinary Medical Association* 147, 47–54.

Emsbo, P. (1956) Subaortic stenosis: a comparative study of congenital subvalvular aortic stenosis (left conus stenosis) in swine and man. *Nordisk Veterinaermedicin* 8, 261–270.

Enfield, F.D. and Rempel, W.E. (1961) Inheritance of teat number and relationship of teat number to various maternal traits in swine. *Journal of Animal Science* 20, 876–879.

Erhard, L., Wittman, W., Cwik, S. and Schmid, D.O. (1988) Studies on the blood group and halothane status of progeny of two German Landrace boars. *Bayerisches Landwirtschaftliches Jahrbuch* 65, 221–230.

Evensen, O. (1993) An immunohistochemical study on the cytogenetic origin of pulmonary multinucleate giant cells in porcine dermatosis vegetans. *Veterinary Pathology* 30, 162–170.

Fisher, K.R.S., Partlow, G.D. and Holmes, C.J. (1989) Anatomical observations of holoprosencephaly in swine. *Journal of Craniofacial Genetics and Developmental Biology* 9, 135–146.

Flatla, J.L., Hansen, M.A. and Slagsvold, P. (1961). Dermatosis vegetans in pigs: symptomatology and genetics. *Zentralblatt für Veterinarmedizin* 8, 25–42.

Fredeen, H.T. and Newman, J.A. (1962a) Rib and vertebral numbers in swine. I. Variation observed in a large population. *Canadian Journal of Animal Science* 42, 232–239.

Fredeen, H.T. and Newman, J.A. (1962b) Rib and vertebral numbers in swine. II. Genetic aspects. *Canadian Journal of Animal Science* 42, 240–251.

Friedman, L., Gaines, D.W., Newell, R.F., Sager, A.O., Matthews, R.N. and Braunberg, R.C. (1994) Body and organ growth of the developing Hormel–Hanford strain of male miniature swine. *Laboratory Animals* 28, 376–379.

Fujii, J., Otsu, K., Zorzato, F., Deleon, S., Khanna, V.K., Weiler, J.E., Obrien, P.J. and Maclennan, D.H. (1991) Identification of a mutation in porcine ryanodine receptor associated with malignant hyperthermia. *Science* 253, 448–451.

Gabris, J. (1974) Is anaemia genetically conditioned in piglets? *Veterinarstvi* 24, 411–412.

Gatlin, C.L., Jordan, W.H., Shryock, T.R. and Smith, W.C. (1996) The quantitation of turbinate atrophy in pigs to measure the severity of induced atrophic rhinitis. *Canadian Journal of Veterinary Research – Revue Canadienne de Recherche Veterinaire* 60, 121–126.

Gelati, K.N., Rempel, W.E., Makambera, T.P.E. and Anderson, J.F. (1973) Heterochromia irides in miniature swine. *Journal of Heredity* 64, 343–347.

Gibbons, R.A., Sellwood, R., Burrows, M. and Hunter, P.A. (1977) Inheritance of resistance to neonatal *E. coli* diarrhoea in the pig: examination of the genetic system. *Theoretical and Applied Genetics* 51, 65–70.

Gibson, E.A., Blackmore, R.J., Wijeratne, W.V.S. and Wrathall, A.E. (1976) The 'barker' (neonatal respiratory distress) syndrome in the pig: its occurrence in the field. *Veterinary Record* 98, 476–479.

Gitter, M. and Bowen, P.D.G. (1962) Unusual cerebellar conditions in pigs. II. Cerebellar hypoplasia in pigs. *Veterinary Record* 74, 1152–1154.

Glawischnig, E. (1965) [A contribution to chronic arthritis deformans of the knee and ankle joint in the pig]. *Wiener Tierarztliche Monatsschrift* 52, 17–21.

Gluhovschi, N., Bistriceanu, M., Nafornita, M., Iusco, V. and Bratu, M. (1967) [Hereditary anomaly of pigs characterized by shortening of the vertebral column and reduction in the number of vertebrae]. *Recueil de Medecine Veterinaire* 143, 827–839.

Goedegebuure, S.A., Rothschild, M.F., Christian, L.L. and Ross, R. 1988. Severity of osteochondrosis in genetic lines of Duroc swine divergently selected for front leg weakness. *Livestock Production Science* 19, 487–498.

Gotoh, E. and Ando, M. (1993) The pathogenesis of femoral head deformity in congenital dislocation of the hip – experimental study of the effects of articular interpositions in pigs. *Clinical Orthopaedics and Related Research* 303–309.

Grange, P.A. and Mouricout, M.A. (1996) Transferrin associated with the porcine intestinal mucosa is a receptor specific for K88AB fimbriae of *Escherichia coli*. *Infection and Immunity* 64, 606–610.

Gregg, R.E. (1946) An arterial anomaly in the fetal pig. *Anatomical Record* 95, 53–65.

Gregory, J.F. (1971) Observations on the occurrence of diaphragmatic defects of swine in a commercial production herd. *Veterinary Medicine and Small Animal Clinician* 66, 361–367.

Griffin, R.M. (1965) Congenital diaphragmatic hernia in the piglet. *Veterinary Record* 77, 492–494.

Groth, A.H. and Rosdail, J.R. (1955) Cystic bile ducts and renal tubules in baby pigs. *Journal of the American Veterinary Medical Association* 126, 133–134.

Hala, K. and Hojny, J. (1964) Blood groups of the N system in pigs. *Folia Biologica* 10, 239–244.

Hale, F. (1935) Relation of vitamin A to anophthalmos in pigs. *American Journal of Ophthalmology* 18, 1087–1093.

Hallqvist, C. (1933) Ein Fall von letalfaktoren beim Schwein. *Hereditas* 18, 215–224.

Halnan, C.R.E. (1970) A cephalothoracophagus pig. *Journal of Anatomy* 106, 204.

Hamori, D. (1941) Parrot mouth and hog mouth as inherited deformities. *Allatorvosi Lapok* 64, 57–59.

Hamori, D. (1962) [Predisposition to hernia in pigs]. *Zuchthygiene, Fortpflanzungsstorungen und Besamung der Haustiere* 6, 80–84.

Hamori, D. (1966) Anal proplase in piglets, resulting from inheritance of agenesia of the anal sphincter. *Magyar Allatorvosok Lapja* 21, 41–42.

Hamori, D. (1983) *Constitutional Disorders and Hereditary Diseases in Domestic Animals.* Elsevier, Amsterdam.

Hanset, R. (1991) Gene identification in pigs. In: Maijala, K. (ed.) *Genetic Resources of Pig, Sheep and Goat. World Animal Science,* Vol. B8. Elsevier, Amsterdam, pp. 109–123.

Harding, J.D.J., Done, J.T., Harbourne, J.F., Randall, C.J. and Gilbert, F.R. (1973) Congenital tremor type A III in pigs: an hereditary sex linked cerebrospinal hypomyelinogenesis. *Veterinary Record* 92, 527–529.

Hasler-Rapacz, J., Prescott, M.F., Vonlindenreed, J., Rapacz, J.M., Hu, Z.L. and Rapacz, J. (1995) Elevated concentrations of plasma lipids and apolipoproteins B, CIII, and E are associated with the progression of coronary artery disease in familial hypercholesterolemic swine. *Arteriosclerosis Thrombosis and Vascular Biology* 15, 583–592.

Hayes, H.M. (1974) Congenital umbilical and inguinal hernias in cattle, horses, swine, dogs and cats: risk by breed and sex among hospital patients. *American Journal of Veterinary Research* 35, 839–842.

Head, K.W., Campbell, J.G., Imlah, P., Laing, A.L., Linklater, K.A. and McTaggart, H.S. (1974) Hereditary lymphosarcoma in a herd of pigs. *Veterinary Record* 95, 523–527.

Hendrick, D.A., Smith, A.C., Kratz, J.M., Crawford, F.A. and Spinale, F.G. (1990) The pig as a model of tachycardia and dilated cardiomyopathy. *Laboratory Animal Science* 40, 495–501.

Hill, M.A. (1990a) Causes of degenerative joint disease (osteoarthrosis) and dyschondroplasia (osteochondrosis) in pigs. *Journal of the American Veterinary Medical Association* 197, 107–113.

Hill, M.A. (1990b) Economic relevance, diagnosis, and countermeasures for degenerative joint disease (osteoarthrosis) and dyschondroplasia (osteochondrosis) in pigs. *Journal of the American Veterinary Medical Association* 197, 254–259.

Hjarre, A. (1953) Vegetierende Dermatosen mit Riesenzellenpneumonien beim Schwein. *Deutsche Tierarztliche Wohenschrift* 60, 106–110.

Hogan, A.G., Muehrer, M.E. and Bogart, R. (1941) A hemophilialike disease in swine. *Proceedings of the Society for Experimental Biology and Medicine*, vol. 48, p. 217.

Hojny, J. and Hala, K. (1965) Blood group system O in pigs. In: Matousek, J. (ed.) *Blood Groups of Animals.* Publishing House of the Czechoslovak Academy of Sciences, Prague, pp. 163–168.

Hojny, J. and Nielsen, P.B. (1992) Allele E(bdgjmr) (E17) in the pig E-blood group system. *Animal Genetics* 23, 523–524.

Hojny, J., Schoffel, J., Geldermann, H. and Cepica, S. (1994) The porcine M blood group system: Evidence to suggest assignment of its Ml factor to a new system (P). *Animal Genetics* 25, 99–101.

Hradecky, J., Hruban, V., Hojny, J., Pazdera, J. and Stanek, R. (1985) Development of a semi-inbred line of Landrace pigs. I. Breeding performance and immunogenetic characteristics. *Laboratory Animals* 19, 279–283.

Huang, S.Y., Tsou, H.L., Chiu, Y.T., Shyu, J.J., Wu, J.J., Lin, J.H. and Liu, S.K. (1996) Heritability estimate of hypertrophic cardiomyopathy in pigs (*Sus scrofa domestica*) *Laboratory Animal Science* 46, 310–314.

Hughes, E.H. (1935) Polydactyly in swine. *Journal of Heredity* 26, 415–418.

Hunter, R.H.F. and Greve, T. (1996) Intersexuality in pigs – clinical, physiological and practical considerations. *Acta Veterinaria Scandinavica* 37, 1–12.

Huston, R., Saperstein, G., Schoneweis, D. and Leipold, H.W. (1978) Congenital defects in pigs. *Veterinary Bulletin* 48, 645–675.

Ilancic, D. and Fuks, R. (1962) [Monobrachia – a degenerative abnormality in pigs]. *Veterinaria* 11, 83–86.

Jansen, J.H. (1993) Porcine membranoproliferative glomerulonephritis with intramembranous dense deposits (porcine dense deposit disease) *APMIS* 101(4), 281–289.

Jansen, J.H., Hogasen, K. and Grondahl, A.M. (1995) Porcine membranoproliferative glomerulonephritis type II – an autosomal recessive deficiency of factor H. *Veterinary Record* 137, 240–244.

Jensen, P.T., Nielsen, D.H., Jensen, P. and Bille, N. (1984) Hereditary dwarfism in pigs. *Nordisk Veterinaermedicin* 36, 32–37.

Johansson, I. (1964) Hereditary defects in pigs in Swedish. *Lantbrukshogskolans Mededeelingen Ser. A*, No. 14, 42 pp.

Johansson-Moller, M., Chaudhary, R., Hellmen, E., Hoyheim, B., Chowdhary, B. and Andersson, L. (1996) Pigs with the dominant white coat colour phenotype carry a duplication of the KIT gene encoding the mast/stem cell growth factor receptor. *Mammalian Genome* 7, 822–830.

Johnson, L.E. (1940) 'Streamlined' pigs. A new legless mutation. *Journal of Heredity* 31, 239–242.

Johnson, D.H. and Lush, J.L. (1939) 'Legless', a new lethal in swine. *Genetics* 24, 79.

Johnson, T.B., Fyfe, D.A., Thompson, R.P., Kline, C.H., Swindle, M.M. and Anderson, R.H. (1993) Echocardiographic and anatomic correlation of ventricular septal defect morphology in newborn Yucatan pigs. *American Heart Journal* 125, 1067–1072.

Kaman, J., Drabek, J. and Zert, Z. (1991) Congenital disproportional chondrodysplasia in pigs. *Acta Veterinaria Brno* 60, 237–251.

Kernkamp, H.C.H. and Kanning, H.H. (1955) Primary megacolon (Hirschsprung's disease) in swine. *North American Veterinarian* 36, 642–643.

King, W.A. and Linares, T. (1980) Three cases of segmental aplasia of the uterus in inbred gilts. *Acta Veterinaria Scandinavica* 21, 149–152.

Kingsbury, B.F. (1909) Report of a case of hermaphroditism (hermaphroditismus verus lateralis) in *Sus scrofa*. *Anatomical Record* 3, 278–282.

Koch, P., Fischer, H. and Schumann, H. (1957) *Erbpathologie der Landwirtschaftlichen Haustiere.* Paul Parey, Berlin.

Koch, W., Jochle, W. and Kolb, K.H. (1958) Heredity in porcine atrophic rhinitis. *Deutsche Tierarztliche Wochenschrift* 65, 681–683.

Konig, H., Weber, W. and Kupferschmied, H. (1972) Zur Nebenhodenaplasie beim Stier und Eber. (a) Darstellung von 18 Fallen mit rezessivem Erbgang beim Simmentaler Fleckvieh. (b) Auftreten der Anomalie bei einem Eber und drei Sohnen. [Epididymal aplasia in bull and boar. (a) Description of 18 cases with recessive inheritance in the Simmental breed. (b) Occurrence of anomalies in a boar and three of its male offspring]. *Schweizer Archiv fur Tierheilkunde* 114, 73–82.

Korovetzkaya, N.N. (1938) Inuchenie nekotorkh nasledstrenykh urodstv U Svini [Some inherited deformities in pigs]. *Usp Zootec Nauk* [Moscow] 5, 19–39.

Kosanke, S.D., Pierce, K.R. and Bay, W.W. (1978) Clinical and biochemical abnormalities in porcine GM-2-gangliosidosis. *Veterinary Pathology* 15, 685–699.

Kowalczyk, T., Hoekstra, W.E., Puestow, K.L., Smith, I.D. and Grummer, R.H. (1960) Stomach ulcers in swine. *Journal of the American Veterinary Medical Association* 137, 339–344.

Krajsa, V. (1961) [The pathogenesis of some hereditary abnormalities of the nervous system in pigs]. *Veterinarni Medicina* 6, 435–446.

Kronacher, C. (1924) Verebungsversuche und Beobachtungen an Schweinen. *Zeitschrift für induktive Abstammungs und Vererbungslehre* 34, 1–120.

Kronevi, T., Hansen, H.J. and Jonsson, O.J. (1975) Cerebellar hypoplasia of unknown etiology in pigs. *Veterinary Record* 96, 403–404.

Kruger, G. (1964) [Agnathia inferior in piglets]. *Tierarztliche Umschau* 20, 32.

Lee, D.M., Mok, T., Hasler-Rapacz, J. and Rapacz, J. (1990) Concentrations and compositions of plasma lipoprotein subfractions of Lpb5-Lpu1 homozygous and heterozygous swine with hypercholesterolemia. *Journal of Lipid Research* 31, 839–847.

Leipold, H.W. and Dennis, S.M. (1972) Syndactyly in a pig. *Cornell Veterinarian* 62, 269–273.

Leroy, P., Naveau, J., Elsen, J.M. and Sellier, P. (1990) Evidence for a new major gene influencing meat quality in pigs. *Genetical Research* 55, 33–40.

Lewis, C.J. (1965) Inguinal hernia with attendant vesicocele in the pig. *Veterinary Record* 77, 699.

Lingaas, F. and Ronningen, K. (1991) Epidemiological and genetical studies in Norwegian pig herds. 5. Estimates of heritability and phenotypic correlations of the most common diseases in Norwegian pigs. *Acta Veterinaria Scandinavica* 32, 115–122.

Lojda, L. (1975) The cytogenetic pattern in pigs with hereditary intersexuality similar to the syndrome of testicular feminization in man. *Documenta Veterinaria* 8, 71–82.

Lomo, O.M. (1985) Arthrogryposis and associated defects in pigs: indication of simple recessive inheritance. *Acta Veterinaria Scandinavica* 26, 419–422.

Liu, S.K., Chiu, Y.T., Shyu, J.J., Factor, S.M., Chu, R., Lin, J.H., Hsou, H.L., Fox, P.R. and Yang, P.C. (1994) Hypertrophic cardiomyopathy in pigs: quantitative pathologic features in 55 cases. *Cardiovascular Pathology* 3, 261–268.

Ludvigsen, J., Basse, Clausen, H. and Jonsson, P. (1963) Congenital paralysis of hind quarter in pigs. In: *Landokonomisk Forsogsia boratoriums efteraismode, Arbog 1963*. Danish National Institute of Animal Science, Copenhagen, pp. 414–417.

Lundeheim, N. (1979) Genetic analysis of respiratory diseases in pigs. *Acta Agriculturae Scandinavica* 29, 209–215.

Lundstrom, K., Andersson, A. and Hansson, I. (1996) Effect of the RN gene on technological and sensory meat quality in crossbred pigs with Hampshire as terminal sire. *Meat Science* 42, 145–153.

Lutz, W. (1996) A case of a foreshortened skull among a litter of wild boar (*Sus scrofa* l 1758) [German]. *Zeitschrift für Jagdwissenschaft* 42, 53–60.

McClymont, G.L. (1954) Paresis associated with spinal cord myelin sheath degeneration in new born pigs. *Australian Veterinary Journal* 30, 345–346.

McKay, W.M. (1955) Siamese twin pigs. *Veterinary Record* 67, 734.

McKay, R.M. and Rahnefeld, G.W. (1990) Heritability of teat number in swine. *Canadian Journal of Animal Science* 70, 425–430.

McKusick, V.A. (1997). *Mendelian Inheritance in Man*, 12th edn. Johns Hopkins University Press, Baltimore.

McManus, C.A., Partlow, G.D. and Fisher, K.R.S. (1994) Conjoined twin piglets with duplicated cranial and caudal axes. *Anatomical Record* 239, 224–229.

McPhee, H.C. and Buckley, S.S. (1934) Inheritance of cryptorchidism in swine. *Journal of Heredity* 25, 295–303.

McTaggart, H.S., Head, K.W. and Laing, A.H. (1971) Evidence for a genetic factor in the transmission of spontaneous lymphosarcoma (Leukaemia) of young pigs. *Nature* 232, 557–558.

McTaggart, H.S., Laing, A.H., Imlah, P., Head, K.W. and Brownlie, S.E. (1979) The genetics of hereditary lymphosarcoma of pigs. *Veterinary Record* 105, 36.

Maier, W.M., Nussbaum, G., Domenjoud, L., Klemm, U. and Engel, W. (1990) The lack of protamine-2 (P2) in boar and bull spermatozoa is due to mutations within the P2 gene. *Nucleic Acids Research* 18, 1249–1254.

Malynicz, G.L. (1982) Complete polydactylism in Papua New Guinea village pig, with otocephalic homozygous monsters. *Annales de Genetique et de Selection Animale* 14, 415–420.

Maneely, R.B. (1951) Blindness in new born pigs. *Veterinary Record* 63, 398.

Mariani, P., Moller, M.J., Hoyheim, B., Marklund, L., Davies, W., Ellegren, H. and Andersson, L. (1996) The extension coat color locus and the loci for blood group O and tyrosine aminotransferase are on pig chromosome 6. *Journal of Heredity* 87, 272–276.

Marklund, L., Wintero, A.K., Thomsen, P.D., Johansson, M., Fredholm, M., Gustafsson, U. and Andersson, L. (1993) A linkage group on pig chromosome 4 comprising the loci for blood group L, Gba, Atp1B1 and three microsatellites. *Animal Genetics* 24, 333–338.

Mayer, J. and Pirchner, F. (1995) Asymmetry and inverted nipples in gilts. *Archiv Für Tierzucht – Archives of Animal Breeding* 38, 87–91.

Meyer, H. (1964) [On the dependence of the position of the first phalanx of pigs on the length of the first and second phalanx and body length]. *Deutsche Tierarztliche Wochenschrift* 71, 548–551.

Meyer, H. and Drommer, W. (1968) Inherited hypotrichosis in pigs. *Deutsche Tierarztliche Wochenschrift* 75, 13–18.

Meyer, H. and Plonait, H. (1968) An inherited disorder of calcium metabolism in the pig (hereditary rickets) *Zentralblatt fur Veterinarmedizin* 15A, 481–493.

Mickelson, J.R. and Louis, C.F. (1996) Malignant hyperthermia – excitation–contraction coupling, CA2+ release channel, and cell CA2+ regulation defects [Review]. *Physiological Reviews* 76, 537–592.

Morrill, C.C. (1947) Thick forelegs in baby pigs. *North American Veterinarian* 28, 738.

Muller, S., Wanke, R. and Distl, O. (1995) Segregation of melanocytic lesions in German Landrace crosses with Munich miniature swine Troll [German]. *Deutsche Tierarztliche Wochenschrift* 102, 391–394.

Neeteson, F.A. (1964) Some congenital defects on pigs: congenital myclonia, oedema and atresia ani. *Tijdschrift voor Diergeneeskunde* 89, 1003–1010.

Nes, N. (1958) Hereditary abnormalities of the tongue, cleft palate and harelip in pigs. *Nordisk Veterinaermedicin* 10, 625–643.

Nicholas, F.W. (1996) *Introduction to Veterinary Genetics.* Oxford University Press, Oxford.

Nielsen, P. (1961) The M blood group system of the pig. *Acta Veterinaria Scandinavica* 2, 246–253.

Nielsen, P.B. (1991) Mm, a new factor in the porcine M-blood group system. *Animal Genetics* 22, 183–185.

Nordby, J.E. (1929a) An inherited skull defect in swine. A preliminary report on the inheritance of a frontoparietal skeletal defect involving the meningocele and proencephalus types of cranioshisis in swine. *Journal of Heredity* 20, 228–232.

Nordby, J.E. (1929b) Congenital skin, ear, and skull defects in a pig. *Anatomical Record* 42, 267–280

Nordby, J.E. (1930) Congenital ear and skull defects in swine. *Journal of Heredity* 21, 499–501.

Nordby, J.E. (1932) Inheritance of whorls in the hair of swine. *Journal of Heredity* 23, 397–404.

Nordby, J.E. (1933) Congenital melanotic skin tumors in swine. *Journal of Heredity* 24, 361–364.

Nordby, J.E. (1934a) Kinky tail in swine. *Journal of Heredity* 25, 171–174.

Nordby, J.E. (1934b) Congenital defects in the mammae of swine. *Journal of Heredity* 25, 499–502.

Ollivier, L. (1980) Le determinisme genetique de l'hypertrophie musculaire chez le porc. *Annals de Genetique et de Selection Animale* 12, 383–394.

Ollivier, L. and Lauvergne, J.J. (1967) [A study of the inheritance of the muscular hypertrophy of the Piétrain pig: preliminary results]. *Annales de Medecine Veterinaire* 111, 104–109.

Ollivier, L. and Sellier, P. (1982) Pig genetics: a review. *Annals de Genetique et de la Selection Animale* 14, 481–544.

Olsen, J.H., Hald, B., Thorup, I. and Carstensen, B. (1993) Distribution in Denmark of porcine nephropathy and chronic disorders of the urinary tract in humans. In: Creppy, E.E., Castegnaro, M. and Dirheimer, G. (eds) *Human Ochratoxicosis and its Pathologies*, Inserm/John Libbey Eurotext, Paris, pp. 209–215.

Osborne, J.C., Davis, J.W. and Farley, H. (1968) Hirschsprung's disease: a review and report of the entity in a Virginia swine herd. *Veterinary Medicine and Small Animal Clinician* 63, 451–453.

O'Toole, D., Ingram, J., Welch, V., Bardsley, K., Haven, T., Nunamaker, C. and Wells, G. (1994a) An inherited lower motor neuron disease of pigs – clinical signs in 2 litters and pathology of an affected pig. *Journal of Veterinary Diagnostic Investigation* 6, 62–71.

O'Toole, D., Wells, G., Ingram, J., Cooley, W. and Hawkins, S. (1994b) Ultrastructural pathology of an inherited lower motor neuron disease of pigs. *Journal of Veterinary Diagnostic Investigation* 6, 230–237.

Parish, W.E. and Done, J.T. (1962) Seven apparently congenital non-infectious conditions of the skin of the pig, resembling congenital defects in man. *Journal of Comparative Pathology* 72, 286–298.

Patterson, D.S.P. and Sweasey, D. (1970) Lipid hexose : phosphorus ratio as an aid to the diagnosis of congenital myelin defects in lambs and piglets. *Acta Neuropathologia* 15, 318–326.

Patterson, D.S.P., Sweasey, D., Brush, P.J. and Harding, J.D.J. (1973) Neurochemistry of the spinal cord in British Saddleback piglets affected with congenital tremor, type A-IV, a second form of hereditary cerebrospinal hypomyelinogenesis. *Journal of Neurochemistry* 21, 397–406.

Pearce, F. (1996) BSE may lurk in pigs and chickens. *New Scientist* 150, 5.

Petrov, A. (1974) A little known lethal character in pigs (in Belgian). *Zhivotnovdni Nauki* 11, 75–79.

Ptak, W. (1963) Polydactyly in a wild boar. *Acta Theriologica* 6, 312–314.

Purtell, C., Maeda, N., Ebert, D.L., Kaiser, M., Lundkatz, S., Sturley, S.L., Kodoyianni, V., Grunwald, K., Nevin, D.N., Aiello, R.J. and Attie, A.D. (1993) Nucleotide sequence encoding the carboxyl-terminal half of apolipoprotein-B from spontaneously hypercholesterolemic pigs. *Journal of Lipid Research* 34, 1323–1335.

Rapacz, J. and Hasler-Rapacz, J. (1990) The pig and its plasma lipoprotein polymorphism in studies of atherosclerosis. *Journal of Animal Breeding and Genetics – Zeitschrift fur Tierzuchtung und Zuchtungsbiologie* 131–146.

Rasmusen, B.A. (1964) Gene interaction and the A–O blood group system in pigs. *Genetics* 50, 191–198.

Rasmusen, B.A. (1982) Linkage between genes at the H blood group locus and the loci for C and J blood groups in pigs. *Animal Blood Groups and Biochemical Genetics* 13, 285–289.

Read, W.K. and Bridges, C.H. (1968) Cerebrospinal lipodystrophy in swine. *Pathologia Veterinaria* 5, 67–74.

Reiland, S., OrdellGustafson, N. and Lundeheim, N. (1980) Heredity of osteochondrosis (swine). *Proceedings of the International Pig Veterinary Society Congress*, Copenhagen, p. 328.

Rhoad, A.O. (1934) Woolly hair in swine. *Journal of Heredity* 25, 371–375.

Richter, A., Wissel, J., Harlizius, B., Simon, D., Schelosky, L., Scholz, U., Poewe, W. and Loscher, W. (1995) The campus syndrome in pigs – neurological, neurophysiological, and neuropharmacological characterization of a new genetic animal model of high-frequency tremor. *Experimental Neurology* 134, 205–213.

Riggs, P.K. and Chrisman, C.L. (1991) Identification of aphidicolin-induced fragile sites in domestic pig chromosomes. *Genetics, Selection, Evolution* 23, S187–S190.

Rimaila-Parnanen, E. (1982) Recessive mode of inheritance in progressive ataxia and incoordination in Yorkshire pigs. *Hereditas* 97, 305–306.

Rob, O. (1963) [A primary defect of the acrosome of the spermatozoa of a boar as an inherited form of sterility]. *Veterinarni Medicina* 8, 409–418.

Roberts, E. and Carroll, E. (1931) The inheritance of hairlessness in swine. *Journal of Heredity* 22, 125–132.

Roberts, E. and Morrill, C.C. (1944) Inheritance and histology of wattles in swine. *Journal of Heredity* 35, 149–151.

Roberts, L.D. (1948) Microphthalmia in swine. *Journal of Heredity* 39, 146–148.

Robinson, R. (1991) Genetic defects in the pig. *Journal of Animal Breeding and Genetics* 108, 61–65.

Roels, S., Hassoun, A. and Hoorens, J. (1995) Accumulation of protoporphyrin isomers-I and isomers-III, and multiple decarboxylation products of uroporphyrin in a case of porphyria in a slaughtered pig. *Journal of Veterinary Medicine Series A – Zentralblatt Fur Veterinarmedizin Reihe A – Physiology Pathology Clinical Medicine* 42, 145–151.

Ronne, M. (1995) Localization of fragile sites in the karyotype of *Sus scrofa domesticata* – present status. *Hereditas* 122, 153–162.

Roperto, F., Galati, P. and Rossacco, P. (1993) Immotile cilia syndrome in pigs – a model for human disease. *American Journal of Pathology* 143, 643–647.

Rothschild, M.F. and Christian, L.L. (1988) Genetic control of front leg weakness in Duroc swine. I. Direct response to five generations of divergent selection. *Livestock Production Science* 19, 459–471.

Rothschild, M.F., Christian, L.L. and Blanchard, W. (1988) Evidence for multigene control of cryptorchidism in swine. *Journal of Heredity* 79, 313.

Roussi, J., Samama, M., Vaiman, M., Nichols, T., Pignaud, G., Bonneau, M., Sigman, J., Decastro, H., Griggs, T. and Drouet, L. (1996) An experimental model for testing von Willebrand factor function – successful SLA-matched crossed bone marrow transplantations between normal and von Willebrand pigs. *Experimental Hematology* 24, 585–591.

Runnels, P.L., Vijtiuk, N. and Valpotic, I. (1992) Association of porcine intestinal F4 receptor with *Escherichia coli* induced diarrheal disease and body weight gain in weaned pigs. *Periodicum Biologorum* 94, 141–148.

Sabec, D. (1963) [Arthrosis of the tarsal joints in Swedish Landrace pigs in S.R. Slovenia]. *Veterinaria. Zbornik Radova iz Oblasti Animalne Proizvodnje* 12, 250–256.

Sabec, D., Schilling, E. and Schultz, L.C. (1961) Arthritis deformans of the hock joint in pigs. *Deutsche Tierarztliche Wochenschrift* 68, 231–236.

Sailer, J. (1955) Imperfect epithelium formation – a hereditary defect in pigs. *Tierarztliche Umschau* 10, 215–216.

Seibold, H.R. and Davis, C.L. (1967) Generalized myositis ossificans (familial) in pigs. *Pathologia Veterinaria* 4, 79–88.

Seibold, H.R. and Roberts, C.S. (1957) A microscopic congenital cerebellar anomaly in pigs. *Journal of the American Veterinary Medical Association* 130, 26–27.

Shirota, K., Tanaka, N., Une, Y., Nomura, Y. and Kamei, M. (1995) Congenital renal glomerular fibrosis in a case of swine chondrodysplasia – note. *Journal of Veterinary Medical Science* 57, 151–154.

Simpson, Q.I. and Simpson, J.P. (1908) Genetics in swine hybrids. *Science* 27, 941.

Skold, B.H. and Getty, R. (1961) Spontaneous atherosclerosis of swine. *Journal of the American Veterinary Medical Association* 139, 655–660.

Smith, C. (1966) A note on the heritability of leg weakness scores in pigs. *Animal Production* 8, 345–348.

Smith, A.D.B., Robinson, O.J. and Bryant, D.M. (1936) The genetics of the pig. *Bibliographica Genetica* 12, 1–160.

Smith, H.J. and Stevenson, R.G. (1973) Congenital hydrocephalus in swine. *Canadian Veterinary Journal* 14, 311–312.

Spillman, W.J. (1906) Inheritance of coat colour in swine. *Science* 24, 441–443.

Sprague, L.M. (1958) On the recognition and inheritance of the soluble blood group property 'Oc' of cattle. *Genetics* 43, 906–912.

Stevenson, J.R. and Walker, W.F. (1954) A case of atresia of the ileum with a divided kidney in the foetal pig. *Anatomical Record* 118, 211–214.

Stigler, J., Distl, O., Kruff, B. and Krausslich, H. (1991) Segregation analysis of hereditary defects in pigs. *Zuchtungskunde* 63, 294–305.

Stigler, J., Distl, O., Kruff, B. and Krausslich, H. (1992) Studies of economically important congenital anomalies of pigs. *Tierarztliche Umschau* 47, 883–886.

Stowe, H.D. and Miller, E.R. (1985) Genetic predisposition of pigs to hypo- and hyperselenemia. *Journal of Animal Science* 60, 200–211.

Straw, B.E. and Rothschild, M.F. (1992) Genetic influences on liability to acquired disease. In: Leman, A.D., Straw, B.E., Mengeling, W.L., D'Allaire, S. and Taylor, D.J. (eds) *Diseases of Swine.* Iowa State University Press, Ames, Iowa, pp. 709–717.

Swindle, M.M., Thompson, R.P., Carabello, B.A., Smith, A.C., Hepburn, B.J.S., Bodison, D.R., Corin, W., Fazel, A., Biederman, W.W.R., Spinale, F.G. and Gillette, P.C. (1990) Heritable ventricular septal defect in Yucatan miniature swine. *Laboratory Animal Science* 40, 155–161.

Teneberg, S., Willemsen, P., Degraaf, F.K. and Karlsson, K.A. (1990) Receptor-active glycolipids of epithelial cells of the small intestine of young and adult pigs in relation to susceptibility to infection with *Escherichia coli* K99. *FEBS Letters* 263, 10–14.

Thiele, G.L., Rempel, W.E., Fass, D.N., Bowie, E.J.W., Stewart, M. and Zoecklein, L. (1986) Inheritance of a new bleeding disease in a herd of swine with Willebrand's disease. *Journal of Heredity* 77, 179–182.

Thielscher, H.H. (1994) Cor-bigeminum in a dicephalic pig. *Tierarztliche Umschau* 49, 765–766

Tissot, R.G., Beattie, C.W. and Amoss, M.S. Jr (1987) Inheritance of Sinclair swine cutaneous malignant melanoma. *Cancer Research* 47, 5542–5545.

Titze, K. (1963) Hare lip and cleft jaw and palate in piglets. *Deutsche Tierarztliche Wochenschrift* 70, 654–657.

Toyama, Y. and Itoh, Y. (1993) Ultrastructural features and pathogenesis of knobbed spermatozoa in a boar. *American Journal of Veterinary Research* 54, 743–749.

Uhlhorn, H., Dalin, G., Lunedheim, N. and Ekman, S. (1995) Osteochondrosis in wild boar Swedish Yorkshire crossbred pigs (F2 generation). *Acta Veterinaria Scandinavica* 36, 41–53.

Vaiman, D., Pailhoux, E., Payen, E., Saidimehtar, N. and Cotinot, C. (1995) Evolutionary conservation of a microsatellite in the Wilms Tumour (WT) gene: mapping in sheep and cattle. *Cytogenetics and Cell Genetics* 70, 112–115.

Vandenbroeck, M. and Maghuinrogister, G. (1995) Anatomical and histophysiological particularities of cryptorchidism: Unilateral cryptorchidism in the pig. 2. Particular testicular developments associated with unilateral cryptorchidism in the pig. *Annales de Medecine Veterinaire* 139, 167–174.

Vitums, A. (1972) Persisent truncus arteriosus in a newborn pig. *Anatomischer Anzeiger* 131, 280–285.

Vogt, D.W., Ellersieck, M.R., Deutsch, W.E., Akremi, B. and Islam, M.N. (1986) Congenital meningocele–encephalocele in an experimental swine herd. *American Journal of Veterinary Research* 47, 188–191.

Wellmann, G. (1953) Beobachtungen uber die Bauchflechte (*Pityriasis rosea*) der Ferkel und die Erblichkeit der Disposition dazu. *Tierartzliche Umschau* 8, 292–294.

Wells, G.A.H. and Bradley, R. (1978) Piétrain creeper syndrome: a primary myopathy of the pig? *Neuropathology and Applied Neurobiology* 4, 237–238.

Wells, G.A.H., Pinsent, P.J.N. and Todd, J.N. (1980) A progressive, familial myopathy of the Piétrain pig: the clinical syndrome. *Veterinary Record* 106, 556–558.

Wells, G.A.H., O'Toole, D.T. and Wijeratne, W.V.L. (1987) An inherited neuronal system degeneration of the pig. *Neuropathology and Applied Neurobiology* 13, 233 (abstract).

Wensing, C.J.G. (1976) Testicular descent, cryptorchidism and inguinal hernia. *Proceedings of the Twentieth World Veterinary Congress, Thessalonika* pp. 149–151.

Wiesner, E. and Willer, S. (1974) *Veterinarmedizinische Pathogenetik.* Gustav Fisher Verlag, Jena.

Wijeratne, W.V.S. and Wells, G.A.H. (1980) Inherited renal cysts in pigs. Results of breeding experiments. *Veterinary Record* 107, 484–488.

Wijeratne, W.V.S., Crossman, P.J. and Gould, C.M. (1970) Evidence of a sire effect on piglet mortality. *British Veterinary Journal* 126, 94–99.

Wijeratne, W.V.S., Beaton, D. and Cuthbertson, J.C. (1974) A field occurrence of congenital meningoencephalocoele in pigs. *Veterinary Record* 95, 81–84.

Wilson, R.F., Nalbandov, A.V. and Krider, J.L. (1949) A study of impaired fertility in female swine. *Journal of Animal Science* 8, 558–568.

Wrathall, A.E. (1976) An inherited respiratory distress syndrome in newborn pigs. *Proceedings of the Fourth International Pig Veterinary Society Congress*, Iowa State University, Ames, p. 10.

Wrathall, A.E., Bailey, J., Wells, D.E. and Hebert, C.N. (1977) Studies on the barker neonatal respiratory distress syndrome in the pig. *Cornell Veterinarian* 67, 543–598.

Yang, M.Y. and Long, S.E. (1993) Folate sensitive common fragile sites in chromosomes of the domestic pig (*Sus scrofa*). *Research in Veterinary Science* 55, 231–235.

Young, G.A. (1952) A preliminary report on the etiology of edema of newborn pigs. *Journal of the American Veterinary Medical Association* 121, 394–396.

Yuyama, Y., Yoshimatsu, K., Ono, E., Saito, M. and Naiki, M. (1993) Postnatal change of pig intestinal ganglioside bound by *Escherichia coli* with K99 fimbriae. *Journal of Biochemistry* 113, 488–492.

Zeveren, A. van, Weghe, A. van de, Bouquet, Y. and Varewyck, H. (1988) The porcine stress linkage group. II. The position of the halothane locus and the accuracy of the halothane test diagnosis in Belgian Landrace pigs. *Journal of Animal Breeding and Genetics* 105, 187–194.

5 Biochemical Genetics

R.K. Juneja[1] and P. Vögeli[2]

[1]*Department of Animal Breeding and Genetics, Swedish University of Agricultural Sciences, Box 7023, S-750 07 Uppsala, Sweden;*
[2]*Department of Animal Science, Swiss Federal Institute of Technology, ETH Zentrum, CH-8092, Zurich, Switzerland*

Introduction

During the 1950s, a number of innovative techniques and pioneering works were described that enabled the detection of divers qualitative genetic variations at the biochemical level within a species. These variations, reported first mostly in humans, were mainly of the following categories:

1. Blood groups and immunological variants of plasma proteins (called allotypes), both studied by serological methods (for a review see Irwin, 1976).

2. Inherited variants of proteins, studied by the technique of gel electrophoresis (Smithies, 1955; Raymond and Weintraub, 1959) in combination with the histochemical staining methods used first by Markert and Möller (1959).

The electrophoresis methods, using blood or other crude tissue extracts as the source of proteins, brought about a revolution for revealing and enabling the typing of the different allelic variants of a given protein in a population. In this context, a polymorphic genetic system was defined as one in which there are two or more allelic variants, and in which an allele does not have a frequency of more than 99% in the population. These polymorphisms, in particular those of blood, were obviously easily available genetic markers that could be used for different aspects of genetics and breeding. Similar studies were undertaken in different farm animals. In pigs, detailed reports on the genetic polymorphism of some of the blood plasma proteins like transferrin, ceruloplasmin, haemopexin, prealbumin and amylase were described between 1960 and 1965 (reviewed by Gahne, 1979). Andresen (1962) provided the first extensive study on the porcine blood groups. Studies on the allotypes of pig plasma proteins were also initiated (Rasmusen, 1965). Since the mid 1960s, the blood genetic markers have been used for parentage testing and in various population and linkage studies in pigs. A remarkable breakthrough in the porcine blood typing work was achieved around 1980 when it was established that as many as five blood marker loci were fairly closely linked to the economically important *HAL* gene (the gene determining halothane sensitivity/predisposition to develop malignant hyperthermia in pigs). The use of these blood markers to reduce the frequency of the detrimental HAL^n allele, brought Mendelian genetics back to pig breeding (reviewed by Archibald and Imlah, 1985).

This chapter reviews the different porcine biochemical and serological genetic markers (biochemical polymorphisms, blood groups and allotypes) and their significance in pig genetics and breeding (previous reviews on these topics were given by Gahne, 1979 and Ollivier and Sellier, 1982). The gene frequency data on several blood markers in different pig breeds were given by Franceschi and Ollivier (1981), Tanaka *et al.* (1983), Vögeli (1990), Van Zeveren *et al.* (1990), Oishi *et al.* (1991) and Vackova *et al.* (1996). Throughout this review we have used the term Euro-American breeds which refers to two European breeds: Landrace and Yorkshire/Large White and two American ones: Duroc and Hampshire.

Biochemical Polymorphisms

The biochemical polymorphisms studied in pigs comprise mainly the proteins of blood plasma (serum) and red cells and also to some extent those present in milk, semen and other tissues (a highly detailed compilation of references on the previous studies on blood protein polymorphisms in pigs, was given by Barthelemy and Darre, 1987). It is notable that, unlike in many other animal species, biochemical polymorphism of haemoglobin has as yet not been reported in pigs.

Plasma proteins

The different known plasma protein polymorphisms of pigs are listed in Table 5.1. Many of the porcine plasma proteins show a high degree of polymorphism and are visualized by inexpensive protein staining procedures. Thus the pheno-

Table 5.1. Electrophoretically detectable variants of blood plasma proteins in the pig.

Locus	Protein	Alleles	References
A1BG	α1B-glycoprotein	*F, S, 0*	Juneja and Gahne (1987)
A2M	α2-macroglobulin	*A, B, C*	Schröffel (1965); Hamel (1984)
AMY	Amylase	*A, BF, B, C, Y*	Hesselholt (1969); Kurosawa et al. (1989)
AMY1	Amylase1 (salivary)	*A, B, C,D*	Jumkov and Nikonchik (1977); Rozhkov and Galimov (1990)
C6	Complement component 6	*A, B, C, D, E*	Shibata *et al.* (1993)
CP	Ceruloplasmin	*A, B, C*	Imlah (1964); Oishi *et al.* (1980b)
PON1	Arylesterase/Paraoxonase	*A, 0*	Gahne *et al.* (1972)
ES	Esterase	*E, F, 0*	Grunder and Kristjansson (1974)
GC	Vitamin D binding protein	*F, S*	Ljungqvist and Hyldgaard-Jensen (1983)
HPX	Haemopexin	*0, 1, 1F, 2, 3, 3F, G, 4, 5, X, Y*	Hesselholt and Hristic (1966); Oishi *et al.* (1991)
PLP1	Plasma protein 1	*F, S*	Juneja and Andersson (1994)
PI1	Protease inhibitor 1	*F, S*	Gahne and Juneja (1986)
PO1A	Postalbumin 1A	*A, B, C, D, E, F, G, H, I, R, S, S′, T, U*	Gahne and Juneja (1986)
PO1B	Postalbumin 1B	*b, d, f, r, r′, s, t, u, v, x, y*	Gahne and Juneja (1986)
PI2	Protease inhibitor 2	*F, I, M, N, P, S, Y, Z*	Gahne and Juneja (1986)
PI3	Protease inhibitor 3	*A, B1, B2, C, D*	Stratil *et al.* (1990)
PI4	Protease inhibitor 4	*A1, A2, B1, B2, C1, C2, D*	Cizova *et al.* (1993a)
TF	Transferrin	*A, B, C, D, D′, E, F, I, X*	Hesselholt (1969); Cizova *et al.* (1993b)

Note 1: The porcine proteins PI1, PO1A, PO1B, PI2, PI3 and PI4 belong to the protease inhibitor (*PI*) gene cluster – see text. PI1 is identical to the polymorphic prealbumin (Pr) reported by Kristjansson (1963).

Note 2: Polymorphism of a few other pig plasma proteins, viz. albumin$_1$ (Kristjansson, 1966), alkaline phosphatase (see text), esterase (Kubek, 1970) and plasminogen (Rapacz *et al.*, 1989) have also been reported. These are, however, not included in the above table since either the phenotypes described lacked clarity and reproducibility or the mode of inheritance is not yet clear. Some unpublished data on new variants of the *PI* gene cluster and of *CP* are also not given in the above table but are mentioned in the text.

types of some of these proteins have been studied in several breeds. We have given here some salient features of the different polymorphisms.

α1B-Glycoprotein (A1BG)

A genetic polymorphism of an unidentified pig plasma protein (called post-albumin 2, Po2), was reported by Juneja *et al.* (1983). This protein was later identified as A1BG by conducting immunoblotting with antiserum specific to human plasma proteins and by comparing amino acid composition and N-terminal sequence with that of human A1BG (Juneja *et al.*, 1987; Stratil *et al.*, 1987; Van de Weghe *et al.*, 1988a). Two common alleles are present in most of the breeds. In addition, a rare null allele was observed in one boar line of the Swedish Yorkshire breed (Juneja and Gahne, 1987). By planned mating, a female pig homozygous for the null allele was obtained and this animal was able to produce antibodies against the porcine A1BG. There was no apparent symptom of any disorder in that pig (Gahne *et al.* 1987). The function of A1BG was recently shown to be as a metalloproteinase inhibitor (Catanese and Kress, 1992).

α2-Macroglobulin (A2M) and plasma protein 1 (PLP1)

Three alleles of an unidentified plasma protein called slow α2 globulin (SA2) were reported by Schröffel (1965). Hamel (1984), using immunosubtraction with antiserum to human A2M protein, provided evidence that SA2 is the porcine A2M. It is, in general, difficult to obtain a good resolution of the porcine A2M phenotypes. Hamel (1984) pointed out that the porcine A2M typing was possible only by the analysis of fresh serum samples. Juneja and Andersson (1994) reported polymorphism of an unidentified pig plasma protein called PLP1. On the basis of the comparative gene mapping data, it was discussed that PLP1 may be the monomeric form of the A2M protein. Apart from the tetrameric form, a monomeric form of A2M has been reported to occur in the plasma of pigs and other mammals (see Ryan-Poirier and Kawaoka, 1993).

Alkaline phosphatase (AKP)

Genetic polymorphism of the pig serum alkaline phosphatases has been studied by different authors but the mode of inheritance is not yet clear. Kierek-Jaszczuk and Geldermann (1985) reported four different electrophoretic zones of AKP activity in the pig serum samples. Since the properties and tissue origins for AKPs of A′, A and C zones differed from those of the B zone, they proposed that the serum AKPs must be controlled by at least two different loci.

Amylase (AMY)

Genetic polymorphism of a plasma amylase was reported by Imlah (1965). Five alleles have been reported (Hesselholt, 1969; Kurosawa *et al.*, 1989). The common Euro-American breeds have predominantly AMY^B allele (Vackova *et al.*, 1996) while the East Asian breeds showed considerable frequency each of AMY^A, AMY^B and AMY^C alleles (Oishi *et al.*, 1991). Rozhkov and Galimov (1990) discussed that this amylase is most probably a maltase and provided evidence that the polymorphic plasma amylase 2 reported earlier (Jumkov and Nikonchik, 1977) was the porcine salivary amylase. They proposed the symbols AMY1 and AMY2 for the salivary and the pancreatic amylase, respectively. Four alleles of AMY^1 were reported in pig plasma (Rozhkov and Galimov, 1990). Some other workers, by conducting the electrophoretic analysis of the pancreatic extracts, have given preliminary data on the AMY^2 polymorphism in pigs (Strumeyer *et al.*, 1988; Belanger *et al.*, 1994).

Arylesterase/paraoxonase (PON1)

Gahne *et al.* (1972) reported that a set of at least eight alleles control the wide discontinuous variation of the plasma arylesterase activity in Swedish Landrace pigs. By electrophoresis of the plasma samples, only two phenotypes, presence and absence (homozygous null type) of the enzyme activity, could be distinguished. Both in human and mouse, plasma arylesterase is called paraoxonase with locus symbol PON^1. It is interesting to note that also in human and mouse, the reported genetic polymorphism of PON^1 involves the plasma levels of this enzyme. The allelic variation in human PON^1 was indicated to influence the amount of some of the high density lipoproteins in the plasma (reviewed in Sorensen *et al.*, 1995). Further studies are needed on the plasma arylesterase (PON1) deficient pigs since in human this enzyme is known to have important anti-atherogenic functions (Sorensen *et al.*, 1995).

Ceruloplasmin (CP)

Imlah (1964), using specific staining, described two *CP* alleles in pigs. Most of the breeds are monomorphic and possess the CP^B allele. CP^A is present with a low frequency in the Landrace breeds. A third allele CP^C was reported to occur with a low frequency in the East Asian Ohmini breed (Oishi *et al.*, 1980b). An additional *CP* variant (called CP^D) with mobility slightly faster than that of CP^B was observed in each of the two European wild boars (CP phenotype BD and DD) used in the Scandinavian pig gene mapping project (Juneja, unpublished results). Knyazev *et al.* (1985) reported, in brief, that all the five African wild pigs studied were homozygous for a new *CP* variant (called CP^X) with mobility considerably slower than that of the CP^B variant.

Complement component 6 (C6)

Genetic polymorphism of *C6* in pigs was described by using agarose gel isoelectric focusing followed by immunoblotting with antiserum to rabbit C6 (Shibata *et al.*, 1993). Five common alleles were observed and each of the four Euro-American breeds studied showed a high degree of polymorphism. A rare homozygous null phenotype (complete deficiency of C6) was reported in two pigs (one of 24 Yorkshire and one of 21 Berkshire pigs studied).

Haemopexin (HPX)

A total of 12 *HPX* alleles have been reported in pigs. The porcine *HPX* shows a high degree of polymorphism with two to four common alleles present in each breed (Hesselholt, 1969; Franceschi and Ollivier, 1981; Oishi *et al.*, 1991). A null allele, HPX^G, was reported to occur with a high frequency (about 0.5) in the Göttingen Miniature pigs (Oishi *et al.*, 1980a). The *HPX* bands were visualized by using benzidine or dianisidine staining solutions in different studies. Both of these are carcinogenic substances. This problem is obviated by using any of the two other methods given later. These are by using immunoblotting with antiserum to porcine HPX (Kalab and Stratil, 1989) or by protein staining after two-dimentional (2D) electrophoresis (Juneja and Gahne, 1987).

α-Protease inhibitors (PI)

Genetic polymorphisms of six different pig plasma PI proteins have been described. These are encoded by a gene cluster comprising six tightly linked loci (*PI1, PO1A, PO1B, PI2, PI3* and *PI4)* and that spans a genetic distance of about 2 cM (Gahne and Juneja, 1985a, 1986; Kuehne, 1990; Stratil *et al.*, 1990; Cizova *et al.*, 1993a). Gahne and Juneja (1986) described a high resolution method of 2-D horizontal electrophoresis, followed by protein staining which enabled an excellent separation of the different PO1A, PO1B and PI2 variants. Later a much simpler method of 2-D electrophoresis was reported for the routine typing of the PI variants (Juneja and Gahne, 1987). The frequencies of the *PI* haplotypes (involving *PI1, PO1A, PO1B* and *PI2* loci) have been reported in several breeds. Each breed studied showed five to ten common and several less frequent haplotypes. In each of Swedish Yorkshire and Landrace breeds, about 40 different *PI* haplotypes were observed. The PI variants are extremely useful genetic markers because they show an extensive genetic polymorphism and many of the breeds possess almost entirely different *PI* haplotypes (Gahne and Juneja, 1986; van de Weghe *et al.*, 1987; Kuehne, 1990; Kuryl *et al.*, 1989, 1992). Kuehne (1990) using the method of Gahne and Juneja (1986) indicated further subtypes of some of the PO1A, PO1B and PI2 variants (those data are not included in Table 5.1 since they have not been formally published and the photographs given were not clear). The biochemical data indicate that the porcine *PI* gene cluster is homologous to human plasma α_1-protease inhibitor and α_1-antichymotrypsin gene families (Gahne and Juneja, 1985a; Stratil *et al.*, 1988, 1995).

Transferrin (TF)

Genetic polymorphism of *TF* was first reported independently by Ashton (1960) and by Kristjansson (1960). The TF phenotypes have been studied in a vast number of breeds. Two common and about seven rare alleles are known. The TF^B allele is the predominant allele in the different European and Asian breeds. The TF^A allele is present mainly in Yorkshire, Hampshire and Duroc breeds. A relatively high frequency (0.05–0.35) of TF^C was observed in the East Asian breeds (Oishi *et al.*, 1991). Kurosawa and Tanaka (1988) reported that TF^C was the predominant allele (frequency 0.6–1.0) in most of the wild pig populations of Japan. The rare *TF* alleles are present only in some specific breed, e.g. TF^E only in Hampshire and TF^J in European wild pigs (see Barthelemy and Darre, 1987). The relative mobilities of different *TF* variants and a report on the partially deficient variant TF^F found in European wild pigs, were given by Cizova *et al.* (1993b).

Vitamin D binding protein (GC)

The *GC* polymorphism was reported by using agarose gel isoelectric focusing and immunofixation with antiserum to human GC protein. Two alleles were observed in the Yorkshire and Duroc breeds while the Danish Landrace did not show this variation (Ljungqvist and Hyldgaard-Jensen, 1983).

Red cell enzymes

Genetic polymorphisms have been described for 12 enzymes of the red cells and one of the leucocytes in the pigs (Table 5.2). The phenotypes and allele frequencies of the following four enzymes have been reported in several breeds.

Adenosine deaminase (ADA)

Six different allelic variants of erythrocyte *ADA* have been reported. The different ADA variants are associated with marked differences in enzyme activity. Pigs of phenotype 00 (null type) lack ADA in the red cells but have normal ADA activity in the leucocytes and other tissues. The ADA^0 allele is present with a high frequency (0.8–0.9) in Hampshire and Pitman-Moore miniature pigs. The other common Euro-American breeds show a high degree of *ADA* polymorphism. The six east Asian breeds examined were monomorphic and had only the ADA^A allele (Widar and Ansay, 1975; Bigi *et al.*, 1990; Oishi *et al.*, 1991; Cepica *et al.*, 1991). Hyldgaard-Jensen, in a series of reports, has proposed that the *ADA 00* pigs may be particularly prone to develop inferior meat quality notably of the type DFD (dark, firm and dry) meat (see Hyldgaard-Jensen, 1990).

Table 5.2. Electrophoretically detectable variants of red cell enzymes in the pig.

Locus	Enzyme	Alleles	References
ACP	Acid phosphatase	*A, B*	Meyer and Verhorst (1973)
ADA	Adenosine deaminase	*A, B, 0, Aw, C, D*	Widar and Ansay (1975); Cepica *et al.* (1991)
CA	Carbonic anhydrase	*A, B*	Kloster *et al.* (1970)
CAT	Catalase	*B1, B2*	Baranov (1970)
ESD	Esterase D	*A, B*	Tanaka *et al.* (1980)
G6PD	Glucose-6-phosphate dehydrogenase	*A, B*	Verhorst (1973)
GPI	Glucose phosphate isomerase	*A, B, C*	Saison (1970); Van de Weghe *et al.* (1988b)
GPX	Glutathione peroxidase	*1, 2*	Agar and Board (1984)
PEPC	Peptidase C	*F, S*	Saison (1973)
PEPE	Peptidase E	*A, B*	Randi *et al.* (1986)
PGD	6-phosphogluconate dehydrogenase	*A, B, C*	Saison (1970); Kurosawa and Tanaka (1991)
PGM2	Phosphoglucomutase 2	*A, B*	Safarova *et al.* (1972)
PGM3	Phosphoglucomutase 3 (leucocytes)	*F, S*	Pretorius *et al.* (1977)

Glucose phosphate isomerase (GPI) and 6-phosphogluconate dehydrogenase (PGD)

Two common alleles each of *GPI* and *PGD* were described first by Saison (1970). A third, rare allele at each of these two loci has been reported (Archibald and McTeir, 1988; Van de Weghe *et al.*, 1988b). In many earlier studies on pigs, the term PHI (phosphohexose isomerase), instead of GPI, was used. Kurosawa and Tanaka (1991) described a new *PGD* variant *C'* with a frequency of about 0.2 in a wild pig population of East Asia and indicated that it may be identical to the C variant. The porcine *GPI* and *PGD* loci have proved to be highly valuable genetic markers since both are not only closely linked to the halothane locus but also these enzymes are extremely stable and can be typed in samples stored at −20°C for several years. Furthermore, both show considerable polymorphism in many of the breeds and can be typed easily using large-scale procedures (Gahne and Juneja, 1985b).

Esterase D (ESD)

Two alleles have been reported (Tanaka *et al.*, 1980). The variant allele ESD^B is present with a frequency of about 0.1 in Landrace, Hampshire and Duroc breeds. The Yorkshire and most of the East Asian breeds have only the ESD^A allele (Oishi *et al.*, 1991; Vackova *et al.*, 1996). This is a useful marker since the staining is inexpensive and it is stable as regards transport and storage of samples.

Miscellaneous proteins

Milk proteins

Genetic polymorphisms have been reported for seven different milk proteins of pigs (reviewed by Ng-Kwai-Hang and Grosclaude, 1992; Chung *et al.*, 1992). Two alleles were reported of each of the following: α_{S2} casein (Erhardt, 1989a), β-casein (Glasnak, 1968; Erhardt, 1989a), an unidentified casein Z (Glasnak, 1968), α-lactalbumin (Bell *et al.*, 1981a) and post-lactoglobulin (Chung *et al.*, 1992). Three alleles were described of β-lactoglobulin (Bell *et al.*, 1981b; Erhardt, 1989b) and four alleles of the $whey_2$ protein/X protein (Althen and Gerrits, 1972; Chung *et al.*, 1992). The milk protein gene frequencies in the major pig breeds were given in different studies. These, in general, show considerable polymorphism and all are clearly visualized by protein staining (see Erhardt, 1991; Chung *et al.*, 1992).

Semen and other tissue proteins

Preliminary data on the genetic polymorphism of three unidentified proteins in the boar seminal plasma have been reported (Dostal, 1970; Tsuji *et al.*, 1993). Matousek (1972) described polymorphism of two unidentified proteins in the epididymal fluid of boars. One of the systems of lactate dehydrogenase (LDH-C) is polymorphic in boar spermatozoa (Valenta *et al.*, 1967). Two alleles of sorbitol dehydrogenase were reported using kidney as a source material (Op't Hof *et al.*, 1972).

Blood Groups

The early stages in the development of porcine blood groups were reviewed by Andresen (1962). From a simple beginning of two blood types, A and –, detected by naturally occurring anti-A, the porcine blood groups are now divided into 16 genetic systems and include 79 blood factors and 82 alleles (Table 5.3). Andresen (1962) described ten genetic systems, EAA, EAB, EAE, EAF, EAG, EAH, EAI, EAJ, EAK and EAL (EA denotes erythrocyte antigen). Later, an additional six systems, EAC, EAD, EAM, EAN, EAO and EAP, were discovered (Nielsen, 1961; Andresen and Baker, 1964; Hala and Hojny, 1964; Hojny and Hala, 1965; Saison *et al.*, 1967; Hojny *et al.*, 1994). With the exception of the naturally occurring anti-A and anti-0, all other reagents used to detect the blood group antigens were isolated by absorption from alloimmune pig sera (Vögeli, 1990). The blood groups gene frequency data given below in the Euro-American

Table 5.3. Current status of porcine blood groups.

Systems	Blood factors	Alleles[1]	Test method[2]
EAA	A(A, A_W), 0	*A*, —	1, 4
EAB	a,b	*a*, —	1
EAC	a	*a*, —	4
EAD	a, b	*a, b*	1
EAE	a, b, d, e, f, g, h, i, j, k, l, m, n, o, p, r, s, t	*bdgkmps, deghkmnps, aeglns, defhkmnps, bdfkmps, aeflns, degklnps, aegils, deghjmnt, abgklps, abgkmps, aegmnops,bdgklps, deghjmnr, abgkmops, bdgjmt, bdgjmr*	1
EAF	a, b, c, d	*ac, ad, bc, bd*	1
EAG	a, b	*a, b*	1
EAH	a, b, c, d, e	*a, b, ab, cd, bd, be*, —	4
EAI	a, b	*a, b*	2
EAJ	a, b	*a, b*, —	2
EAK	a, b, c, d, e, f, g	*acf, bf, acef, ade, adeg*, —	4
EAL	a, b, c, d, f, g, h, i, j, k, l, m	*adhi, bcgi, bdfi, agim, adhjk, adhjl*	2, 3
EAM	a, b, c, d, e, f, g, h, i, j, k, m	*ab(e), aem, aejm, ade(m), b, bcd, bcdi, bd, bdg, be(f)m, cd, cdi, (cdk), d, djk, dk, ef, efm, h*, —	2, 4
EAN	a, b, c	*a, b, bc*	2
EAO	a, b	*a, b*	3
EAP	a	*a*, —	2
S	—	*S, s*	—

[1]The genes for which all phenogroups have not yet been detected (open systems) are symbolized by a dash (—).
[2]1, Direct agglutination test; 2, antiglobulin (Coombs) test; 3, dextran test; 4, haemolytic test.

breeds are from the reports of Vögeli (1990), Van Zeveren *et al.* (1990) and Vackova *et al.* (1996) and that in the East Asian breeds are from Oishi *et al.* (1991).

EAA and S systems

Sprague (1958) identified three types, A, 0 and -, with A cells being lysed by cattle anti-J, 0 cells by cattle anti-0 and '-' cells not reacting with either reagent. Genetic studies have shown that the antigenic expression of the phenotypes depends firstly on the EAA system with the two antigenic factors A and 0, and on a second epistatic S system with the two alleles *S* and *s* (Rasmusen, 1964). In adults the four phenotypes are regulated by the following genotypes: 'A', (EAA A/A or A/-) + (S S/S or S/s); '0', EAA -/- + (S S/S or S/s); 'A weak', (EAA A/A or A/-) + S s/s; and 'dash', EAA -/- + S s/s (Vögeli *et al.*, 1983). The porcine A and 0 antigens are not true red cell antigens. These are serum and tissue antigens which are secondarily attached to red cells when the serum concentrations are sufficiently high (Hojny and Hradecky, 1971). The epistatic S system is reported to be undetectable in pigs under 5 weeks of age. Such animals always show type *s/s* implying that the phenotypic expression of factors A and 0 is suppressed in early life. Most of the Euro-American and the East Asian breeds show considerable polymorphism in the EAA and S systems.

EAB, EAD, EAG, EAI and EAO systems

These systems are closed and each consists of two blood factors and two alleles (Andresen, 1962; Hojny and Hala, 1965; Hojny *et al.*, 1967; Saison *et al.*, 1967; Hradecky and Linhart, 1970). In the EAB system, the Landrace and Yorkshire are almost monomorphic while both of the alleles are equally frequent in the Duroc and Hampshire breeds. In the EAD system, all of the four Euro-American breeds have the EAD^b allele with a frequency of more than 0.9. Both alleles of EAG and of EAI systems are present with almost equal frequencies in each of the four Euro-American breeds. In contrast, the East Asian breeds were almost monomorphic in the EAG system. In the EAO system, EAO^b is the predominant allele in different Euro-American breeds while the East Asian breeds have both EAO^a and EAO^b with almost equal frequencies.

EAC, EAJ and EAP systems

The EAC and EAP systems are one-factor, two-allelic open systems (Andresen and Baker, 1964; Hojny *et al.*, 1994). The Ml antigen was later shown to belong to the new blood system P and the factor designation was changed from Ml to Pa (Hojny *et al.*, 1994). The EAJ system is currently a two-factor, three-allelic system which is open (Hojny and Hradecky, 1972). In the Euro-American breeds, *EAC* is almost monomorphic while *EAJ* shows considerable polymorphism.

EAE system

Currently 18 factors have been identified in the EAE system and the genetic control is by 17 alleles (Hojny and Nielsen, 1992). Most of the breeds show a high degree of polymorphism in this system. Four common alleles (EAE^{aeglns}, $EAE^{bdgkmps}$, $EAE^{defhkmnps}$ and $EAE^{deghkmnps}$) are generally present in the Euro-American breeds.

EAF system

This is recognized as a four-allelic closed system (Hojny *et al.*, 1984). EAF^{bd} is the predominant allele in the Euro-American breeds (Vögeli, 1990). Similar results were given by others but the frequency of EAF^{a} was reported to be about 0.5 in the Hampshire breed (Oishi *et al.*, 1991; Vackova *et al.*, 1996).

EAH system

This open system consists of seven alleles (Andresen, 1964; Hojny, 1973). Most of the breeds show a high degree of polymorphism in this system. The alleles EAH^{a}, EAH^{cd} and EAH^{-} are the most frequent ones in the Euro-American breeds.

EAK system

The present status of the EAK system is a seven-factor and six-allelic open system (Nielsen and Vögeli, 1982). In the Hampshire breed, only two alleles (EAK^{acef} and EAK^{bf}) were found. In Durocs, besides the above two alleles, EAK^{-} was also observed with a frequency of about 0.2. System EAK is apparently closed in the Hampshire breed. The EAK^{adeg} allele is reported to occur in the Landrace and Miniature pigs, but is absent in the Yorkshire breed.

EAL system

This system is closed and is composed of 12 factors and 6 alleles (Linhart, 1971). The EAL^{bcgi} is the most frequent allele in the European breeds. The alleles EAL^{adhjl} and EAL^{bdfi} are the most predominant ones in the Duroc and the Hampshire breed, respectively.

EAM system

This is the most complex porcine blood group system. It consists of 12 blood factors and 20 alleles and is still classified as an open system (Hojny *et al.*, 1979). Serological difficulties in the determination of the EAM-blood factors limit the practical use of this system. EAM^{-} is the predominant allele in the different Euro-American breeds.

EAN system

Blood group substances Na, Nb and Nc, in addition to occurring on the red cells, are soluble plasma substances (Hojny and Glasnak, 1970) resembling the A and 0 substances. Three alleles are known in this system (Saison, 1967). The different Euro-American breeds possess only two alleles (EAN^{a} and EAN^{b}) and show considerable polymorphism in this system.

Lipoprotein Allotypes

The lipoprotein allotypes and phenotypes of all pig sera are determined by the double immunodiffusion (DID) test in agar as described by Rapacz *et al.* (1978). Genes controlling the allotypes were assigned to the five lipoprotein loci, *B*, *R*, *S*, *T* and *U*, giving rise to the five lipoprotein systems: LPB, LPR, LPS, LPT and LPU. While LPB, LPS, LPT and LPU encompass lipoprotein antigens of the class low density lipoproteins (LDL), antigens LPR occur on molecules at very low density lipoproteins (VLDL) and very high density lipoproteins (VHDL) (Rapacz, 1982).

LPB system

The LPB allotypes are inherited in 11 complex groups defined as phenogroups or haplotypes (Rapacz, 1982; Rapacz *et al.*, 1994), each determined by one of the 11 APOB (apolipoprotein B) alleles, LPB^{1} to LPB^{10} and LPB^{101} (Table 5.4). Each *LPB* haplotype is composed of an individual (mutant) allotype (LPB1–LPB10 and LPB101) specific for swine only and a set of common epitopes (LPB11–LPB18, LPB20 and LPB111) shared with primates including humans (Rapacz, 1978). An additional 14 semi-common LPB allotypes (LPB21–LPB27 and LPB31–LPB37) forming seven pairs of alternative epitopes, each shared by two or more *LPB* haplotypes have been described and assigned to LPB^{1} to LPB^{8} alleles (Rapacz and Hasler-Rapacz, 1990). The semi-common LPB allotypes are not listed in Table 5.4. The alleles LPB^{5} and $LPB^{5,8}$ are the most frequent ones in the Landrace, Yorkshire and Hampshire breeds while the Duroc breed is almost monomorphic for the LPB^{5} allele. The LPB system, in general, shows a high degree of polymorphism in most of the breeds (Wysshaar, 1988; Full *et al.*, 1990; Vackova

Table 5.4. Representation of 11 allelic genes of 11 LPB haplotypes (phenogroups) with their alloantigenic determinants (allotypic specificity) designated by numerals.

Allele	Specificities[1]									
LPB^{1*}	**1**	12	13	14	15	16	17	18	20	111
LPB^{2}	11	**2**	13	14	15	16	17	18	20	111
LPB^{3}	11	12	**3**	14	15	16	17	18	20	111
LPB^{4}	11	12	13	**4**	15	16	17	18	20	111
LPB^{5}	11	12	13	14	**5**	16	17	18	20	111
LPB^{6}	11	12	13	14	15	**6**	17	18	20	111
LPB^{7}	11	12	13	14	15	16	**7**	18	20	111
LPB^{8}	11	12	13	14	**5**	16	17	**8**	20	111
LPB^{9}	11	12	13	14	15	16	17	18	**9** 20	111
LPB^{10}	11	12	13	14	15	16	17	18	**10**	111
LPB^{101}	11	12	13	14	15 (w)	—	17	18	20	**101**

[1]1–10, 101, individual specificities (given in bold); 11–20, 111, common specificities. The LPB 19 (hypothetical alternative to LPB 9 specificity) is not tested.

*Each allele of the LPB region determines a complex antigenic product bearing a set of distinct allotypic specificities.

et al., 1996). The alleles LPB^{9}, LPB^{10} and LPB^{101} are found in Göttingen Miniature, Clown Miniature and Vietnamese Potbelly pigs (Rapacz *et al.*, 1994). As given later, there are indications that mutations in APOB are correlated with susceptibility to atherosclerosis not only in humans but also in swine (Rapacz *et al.*, 1986a).

LPR system

Rapacz (1982) described two codominant alleles LPR^{1} and LPR^{2}. A third unconfirmed codominant allele $LPR^{1,3}$, which controls the linear subgroup of allotype LPR^{1}, was published recently (Hojny *et al.*, 1993). The *LPR* locus controls the synthesis of allele-specific apolipoprotein R (Rapacz *et al.*, 1986b). The *LPR* locus shows considerable polymorphism in the Landrace and Hampshire breed. The Duroc and Yorkshire breeds have predominantly the LPR^{2} allele with a frequency of about 0.9 (Wysshaar, 1988; Vackova *et al.*, 1996).

LPS, LPT and LPU systems

Only a single marker, *LPS1*, has been identified so far in the open LPS system (Rapacz, 1982). Two allotypes, LPT1 and LPT2, determined by two codominant genes are identified in the open LPT system. There are two allelic genes, LPU^{1} and LPU^{2}, identified in the LPU system (Rapacz, 1982). Since studies on these lipoprotein systems are preliminary, data on gene frequencies are very scarce.

Immunoglobulin and other Allotypes

Allotypes of undefined immunoglobulin (Ig) class were first demonstrated by the antiglobulin inhibition test by Rasmusen (1965) and Nielsen (1972). Rapacz *et al.* (1973) identified the first two immunoglobulin gamma (IgG) allotypes and showed their control by two codominant alleles at an autosomal locus. Later, Rapacz and Hasler-Rapacz (1982) used SDS–PAGE for separation of heavy and light chains of IgG and showed that the IgG A1 and A2 allotypes are located in the heavy chains of the IgG molecule.

Immunoglobulin gamma heavy chain (IGH) allotype system

The method for the preparation of anti-IgG reagents and testing of individual serum samples, using single and double immunodiffusion or immunoelectrophoresis in agar gels, has been described by Rapacz and Hasler-Rapacz (1982). They identified nine different IGH allotypes (IGH1 A1, A2, A3; IGH2 B1, B2; IGH3 C1, C2; IGH4 D1 and D2), controlled by four tightly linked loci designated *IGH1, 2, 3* and *4*. An additional unconfirmed IGH allotype (IGH C3) was published recently (Vackova *et al.*, 1996). The IGH allotypes, in general, show a high degree of polymorphism in most of the breeds (Rapacz and Hasler-Rapacz, 1982; Vackova *et al.*, 1996).

Globulin (GP) allotype system

For this allotype system the designation GP (globulin, pig) is used. The four GP alloantigens are controlled by three complex alleles (GP^{Ab}, GP^{aB} and GP^{ab}) forming a closed system (Janik *et al.*, 1983). The allotypes GP A and GP B are inherited as codominant alleles, whereas GP a and GP b appear as their alternative markers, respectively, and are inherited as recessive alleles. The GP a and GP b alloprecipitins react only with sera of corresponding homozygotes detecting allotypes which appear as recessive markers. Most of the breeds have predominantly the GP^{ab} allele with a frequency of 0.6 or more (Vackova *et al.*, 1996). In electrophoresis, the molecules carrying these antigens migrate in the beta globulin region.

P1–G3 allotype system

Allotypes P1 and G3 are controlled by two mutually excluding alleles at an autosomal locus (Hojny *et al.*, 1992). Allele frequencies in Euro-American breeds indicate considerable predominance of P1.

G9, G16 and other allotypes

G9 and G16 serum protein antigens are hypothesized to be controlled by two codominant alleles at a single autosomal locus (Kuryl *et al.*, 1991). The individuals lacking both G9 and G16 serum protein antigens could be considered as homozygotes of the third (null) allele. In electrophoresis, the molecules carrying these antigens migrate in the alpha globulin region. Polymorphism of three other allotypic specificities (P3, P16 and F1), until now not classified, was reported by Vackova *et al.* (1996). The G9 and F1 variants show considerable polymorphism in the Euro-American breeds.

Significance of the Blood Genetic Markers

The porcine blood genetic markers have been used in different aspects of genetics and breeding as given below. The following markers have been studied most frequently since these are amenable for large-scale typing of the samples. Those studied plasma proteins were A1BG, AMY, CP, HPX, PI1, PO1A, PO1B, PI2 and TF and the red cell enzymes were ADA, GPI, PGD and ESD. The blood group systems have been used only in some laboratories.

Parentage testing

It has been reported from many different countries that the cases of wrongly assigned parentages (generally sires) in the elite breeding herds of pigs were as high as 10–20% before active blood typing was introduced. Our experience shows that this figure is about 5% even in herds where active blood typing is practised. Correct pedigrees are necessary for planned matings, genetic selection, progeny tests and accurate heritability estimates. It is thus important that the cases of wrong parentages are identified. In Sweden and some other countries, only the blood protein markers have been used for this purpose as outlined here (for details see Gahne and Juneja, 1985b; Juneja and Gahne, 1987). The plasma samples are analysed by a simple method of 2-D electrophoresis, followed by protein staining. The first dimensional separation was conducted by agarose gel electrophoresis (pH 5.4) while the second one was by horizontal PAGE (pH 9.0).This allows the phenotyping of eight plasma proteins (A1BG, CP, HPX, PI1, PO1A, PO1B, PI2 and TF) on the same gel. All of these except CP show a high degree of polymorphism. The haemolysate samples are analysed by agarose gel electrophoresis and phenotyping for GPI, PGD and ESD is conducted on the same gel. By this scheme, an exclusion probability (the probability to detect a wrongly assigned parent) of about 90% is achieved in the Landrace and Yorkshire breeds. This is remarkable considering that both the plasma and the haemolysate samples are analysed only one time each. In some other countries (e.g. Germany), the scheme involves the typing of eight blood group systems and three enzyme systems GPI, PGD and ESD which enables an exclusion probability of 95–98% in the common European pig breeds (J.N. Meyer, Göttingen, 1995, personal communication).

Linkage data and marker-assisted selection

The blood typing of pigs in families over the years revealed some linkages among the blood marker loci and also between the blood marker loci and some disease loci as given, in brief, below.

Genetic linkage studies

The following linkages involving the blood marker loci have been reported: *EAK* and *HPX* (Imlah, 1965); *EAI* and *AMY* (Andresen, 1966); *TF* and a lethal factor (Imlah, 1970); *TF* and intestinal *Escherichia coli* receptor locus (*K88R)* (Gibbons *et al.*, 1977); *EAJ, EAC* and *SLA* (Swine leucocyte antigen) (Hradecky *et al.*, 1982); *S, GPI, HAL, H, A1BG* and *PGD* – commonly termed as the halothane linkage group (Andresen, 1971, 1979; Jörgensen *et al.*, 1976; Rasmusen, 1981; Juneja *et al.*, 1983); *PI, IGH* (immunoglobulin gamma heavy chain), *G9* and *G16* (globulin allotypes) (Juneja *et al.*, 1986; Kuryl *et al.*, 1991); *TF* and *CP* (Juneja *et al.*, 1989); and *S* and *ECF18R* (Vögeli *et al.*, 1996, see below).

The polymorphic blood markers were typed in the reference families of the ongoing pig gene mapping projects. The result is that all of the commonly studied blood marker loci have now been assigned to different chromosomes (for references see Chapter 9). The chromosome and the markers are: chromosome 1 (*EAA*), chromosome 3 (*LPB, LPT, LPU*), chromosome 4 (*EAL*), chromosome 5 (*PLP1*), chromosome 6 (*S, EAH, A1BG, GPI, PGD, EAO*), chromosome 7 (*EAC, EAJ, PI, IGH, G9* and *G16*), chromosome 8 (*EAF, PGM1*), chromosome 9 (*EAE, EAK, EAN, HPX, LPR*), chromosome 11 (*EAM, ESD*), chromosome 12

(*EAD*), chromosome 13 (*TF, CP*), chromosome 15 (*EAG*), chromosome 18 (*EAI, AMY*).

Prediction of the halothane genotypes of individual pigs

During 1970–1983, five blood genetic markers were shown to be closely linked to the *HAL* locus (see above). In Sweden and some other countries, the typing of the blood protein loci *GPI, A1BG* and *PGD*, in conjunction with the halothane testing of the offspring, was used to deduce the *HAL*-marker loci haplotypes of each member of a given family. It was shown that the *HAL* genotypes of individual pigs could be determined with an accuracy of about 95% (Gahne and Juneja, 1985b). Thus the *HAL* heterozygotes could be identified, which was previously not possible by using the halothane test alone. This method was used on a country-wide scale and the frequency of the HAL^n allele was brought down almost to zero in nearly all of the nucleus breeding herds. Since not only the *HAL nn* but also the *HAL Nn* pigs tended to have inferior meat quality (Chapter 16) and the *HAL nn* pigs were likely to suffer heavy mortality during transport (see Chapter 10), the use of the *HAL*-linked markers has yielded huge economic benefits to the pig industry and has promoted the cause of animal welfare. For instance, in Sweden during 1984 to 1994, the proportion of pigs dying during transport and mixing was reduced from 2 per 1000 to 2 per 10,000. This approach was subsequently used in several other countries (see e.g. Vögeli *et al.*, 1988). This practical application of the *HAL*-linked blood genetic markers in pigs has been considered as one of the best examples of marker-assisted selection in livestock improvement.

The knowledge on the *HAL*-linked protein marker loci proved to be precious information for some recent works that culminated in the identification of the HAL/malignant hyperthermia gene as the calcium release channel (*CRC,* also called *RYR1*) gene. A direct PCR-based DNA method is now available for the identification of the *HAL* mutation in individual pigs (Fujii *et al.*, 1991; McLennan and Phillips, 1992). The *GPI, A1BG* linkage was reported first in pigs and this linkage has been observed in all the other three mammalian species (human, horse and dog) studied so far (Juneja *et al.*, 1987, 1994; Eiberg *et al.*, 1989). Since the *HAL* locus is located between *GPI* and *A1BG* loci in pigs (see Chapter 9), the linkage between these three loci can be expected to be present in many other mammals.

Blood group S and oedema disease

Oedema disease (*Escherichia coli* enterotoxaemia) and postweaning diarrhoea (enteric colibacillosis) occur in pigs aged 4–12 weeks, and these are responsible for considerable economic losses. Both diseases are caused by toxins produced by *E. coli* that colonize the surface of the small intestine. Resistance to porcine oedema disease and postweaning diarrhoea is associated with the absence of colonization of the intestine with toxigenic *E. coli* strains belonging to certain serotypes. Colonization depends on specific binding between adhesive bacterial fimbriae (e.g. F18) and the intestinal brush border enterocytes. The specific binding is mediated by receptors expressed on the surface of the brush borders. Adhesion of fimbriae F18 has been shown to be genetically controlled by the host (Bertschinger *et al.*, 1993). Inheritance data support the hypothesis that the susceptibility to adhesion by these bacteria is controlled by a dominant allele and resistance by the alternative recessive allele at a genetic locus designated *ECF18R* (*E. coli* F18 receptor). This locus has been recently mapped to porcine chromosome 6 based on its close linkage to the epistatic *S* locus controlling the expression of A–0 blood group antigens (Vögeli *et al.*, 1996). The explanation for the association between blood group *S* phenotype and oedema disease and postweaning diarrhoea may relate to the expression of fucosylated blood group antigens in the intestinal surface mucous cells. It would be of interest to explore the statistical–genetic basis for the association between *S* blood types and susceptibility to adhesion of F18 *E. coli* strains. The possibility of a direct effect of the *S* blood group alleles has to be considered. However, the *ECF18R* locus appears to be situated about 0.5 cM from the *S* locus (Vögeli *et al.*, 1996). Therefore, linkage disequilibrium is a more likely cause of the observed association.

This tight linkage between the blood group locus *S* and the gene determining susceptibility to the oedema disease is of great significance for the basic as well as for the practical aspects of pig genetics and breeding. It is anticipated that a PCR-based DNA method will shortly be available that will enable determination of the genotypes of the individual pigs for the locus of this very important porcine disease.

Lipoprotein alleles and lipid metabolism

Pigs have been used extensively in studies on low density lipoprotein (LDL) metabolism and atherosclerosis. Apolipoprotein B-100 (apoB) is the primary protein component of LDL in mammalian species. ApoB is a ligand for the LDL receptor and mediates the clearance of LDL from the bloodstream. As given above, the apoB gene (LPB system) is highly polymorphic in pigs and 11 *LPB* alleles have been reported. The allele LPB^5 has been found to be associated with elevated levels of plasma LDL, resulting in part from a delayed clearance of LDL (Rapacz *et al.*, 1986a). The LDL from LPB^5 pigs binds to the LDL receptor on pig fibroblasts with one-sixth the affinity of LDL from non-LBP^5 pigs (Lowe *et al.*, 1988). Nichols *et al.* (1992) showed that, on the administration of atherogenic diet, pigs with *LPB1/5* and *5/8* genotypes developed significantly higher serum cholesterol levels than the pigs of other *LPB* genotypes. Thus the apoB genetic variation significantly influences LDL metabolism. It is, however, not yet clear to what extent variation in LDL clearance influences the development of atherosclerotic plaques. Rapacz and Hasler-Rapacz have developed a strain of pigs that exhibits spontaneous elevations of plasma cholesterol, LDL and apoB. This has been designated as familial hypercholesterolaemic (FHC) strain. These pigs provide a unique animal model for studying the genetic basis of coronary artery disease (Prescott *et al.*, 1995).

Blood markers and economic traits

Several studies have been reported on the search for association between blood genetic markers and pig performance traits (see Ollivier and Sellier, 1982). The blood group locus *H* was reported to exert a major effect on litter size in pigs (Jensen *et al.*, 1968; Rasmusen and Hagen, 1973). This was later discussed to be the effect of the *HAL* locus and on the existence of linkage disequilibrium between *HAL* and *H* loci in the pigs studied (Archibald and Imlah, 1985). Relationship between the parental *TF* genotypes and litter size in pigs has been indicated in some studies (Kristjansson, 1964; Zyczko, 1990) and was ruled out in some others (Jensen *et al.*, 1968; Huang and Rasmusen, 1982). Clamp *et al.* (1992), studying one large sire family with 30 litters, indicated a significant effect of the *GPI* locus on daily live weight gain during postweaning to 103 kg live weight in pigs. They discussed that the effect observed was most probably not due to the *HAL* locus and indicated the possibilty of some other gene affecting growth rate in the same chromosomal region. Recently Nystrom *et al.* (1997) reported a significant association between the TF genotypes and body weights at 6–9 weeks of age in one herd of Yorkshire pigs.

Population studies

The porcine blood genetic markers have been used for estimating heterozygosity within populations and also for studying relationships between different breeds. Average heterozygosity (H), to some extent, reflects the state of inbreeding. The H estimates based on a sufficiently large number of polymorphic blood marker loci have been reported in a few studies (Oishi *et al.*, 1990; Van Zeveren *et al.*, 1990; Vögeli, 1990; Glodek *et al.*, 1993; Cepica *et al.*, 1995). In all of the above studies, the estimate of relative H was higher in the Landrace and Yorkshire breeds (ranging from 36 to 43%) as compared with that in the Duroc and Hampshire breeds (30–35%). It should be noted that the above values indicate relative and not absolute values of H since only the known polymorphic systems were examined in those studies; the relative values of H are nevertheless useful for comparing the inbreeding extent of different populations (see Cepica *et al.*, 1995). The lowest value of relative H was observed in the breeds that have passed through a serious bottleneck or that have a smaller population size, e.g. the Piétrain breed. A limited number of Chinese breeds have been studied (e.g. Meishan and Jinhua) and showed less heterozygosity than the common Euro-American breeds (Oishi *et al.*, 1990, 1991). Earlier Widar *et al.* (1975), on the basis of a study on 19 randomly selected red cell enzyme loci, estimated the absolute value of H in Belgian Landrace and Piétrain pigs as about 7% and 3%, respectively. Studies on the extent of allozyme variability have also been reported in the European and American wild pigs; the estimate of the absolute value of H was about 3% in each case (Smith *et al.*, 1980; Hartl and Csaikl, 1987; Randi *et al.*, 1992).

The relationship between different pig breeds, on the basis of blood genetic markers, was given in several reports. Tanaka *et al.* (1983) and Oishi *et al.* (1991) reported that a considerable genetic distance, close to 0.5, separates the two major lines of pig evolution – the East Asian native breeds and the Euro-American breeds. In the above two studies, the relationships among 14 different breeds of China and other East Asian countries were also reported. The relationships among the four major pig breeds using the two European ones (Landrace and Large White) and the two American ones (Duroc and Hampshire) were given in some reports. The most distant of the four breeds was found to be Duroc in some reports (Vögeli, 1990; Oishi *et al.*, 1991) and Hampshire in some others (see Cepica *et al.*, 1995). Oishi *et al.* (1991) showed that another American miniature breed, Pitman-Moore, was closely related to Hampshire. On the basis of blood groups data, the European Landrace pig breeds were classified into two groups: those of Germany, Holland, Switzerland and Hungary in one group; and those of Denmark, Sweden, Czechoslovakia and Russia in the other group (Major *et al.*, 1970; Vögeli, 1990). This is supported, to some extent, by the limited data available on the protease inhibitor (*PI*) haplotype frequencies. For instance, the *PI* haplotype *SE-F* is the most frequent one in Swedish and Norwegian Landrace (ca. 30%) and it is present with a frequency of 5% in Czech Landrace but it is totally absent in the Belgian Landrace (see Kuryl *et al.*, 1989).

The information on the extent of genetic distances between breeds is of considerable significance in pig breeding. Recently Cepica *et al.* (1995) showed that the greater the genetic distance between the two parental populations, the more heterozygosity increases in the crosses (in relation to the mean heterozygosity of the populations). They also discussed the possibility of a positive correlation between the degree of heterozygosity and fitness-associated traits and heterosis in pigs. Previously Dinklage and Hohenbrink (1970) reported that the number of pigs born and reared per litter were more when the matings resulted in heterozygous offspring with regard to blood groups. Recently Christensen *et al.* (1996), in a study based on typing of 21 genetic markers, provided evidence for a highly significant positive correlation between degree of estimated heterozygosity and growth rate in an experimental population of inbred pigs.

Conclusions

We have reviewed the different conventional blood markers in pigs and their use for parentage control, studies on genetic structure of breeds and populations, estimation of heterozygosity of populations, estimation of genetic distance between breeds and for linkage studies and gene mapping. Some of these works resulted in the detection of the economically important halothane linkage group. The recent finding of linkage between the blood group locus *S* and the gene determining susceptibility to oedema disease is also likely to gain much importance in the practical pig breeding.

One of the main applications of the blood markers (polymorphic proteins and blood groups) has been for parentage testing in pigs. For this purpose, the use of protein markers alone (eight plasma proteins and three red cell enzymes analysed by two simple electrophoresis methods given earlier) has proved to be a highly efficient and inexpensive method in pigs. Similarly, the typing of porcine blood groups provided a very powerful tool for parentage testing. It is very likely that the newer molecular markers using the highly polymorphic microsatellites will replace the conventional blood markers for routine parentage testing. However, this will require that genotypings by the molecular markers are also as reproducible and cost effective as that by the conventional blood markers.

Extensive genetic polymorphism was observed at the loci of the α-protease inhibitor (*PI*) gene cluster in pigs; further work is needed at the DNA level to understand the basis of the PI variation. Further work on the detection of new protein polymorphisms in pigs is important and useful since these loci are highly valuable for the purpose of comparative gene mapping. The incentive to search for possible associations between the protein genotypes and the inherited diseases of pigs may also increase as we know more about the functions of these proteins. For instance, α2-macroglobulin was recently shown to be the major inhibitor of the influenza virus in pig serum and transferrin was shown to be a specific receptor for K88ab fimbriae of *E. coli* in pigs (Ryan-Poirier and Kawaoka 1993; Grange and Mouricout 1996).

As mentioned earlier, Rapacz and co-workers in a series of articles have elucidated the association between the lipoprotein genotypes and the lipid metabolism in pigs. The porcine blood groups, due in particular to the pioneering and extensive works of E. Andresen and J. Hojny, have been explored and described in much detail. The porcine blood groups now consist of 16 genetic systems and 82 alleles. Further research on the porcine blood groups may also prove to be of considerable interest since some of the human blood group systems have been found to be of much physiological importance. The recent examples in humans are the identification of the Duffy blood group antigen as receptor for the malaria parasite *Plasmodium vivax* and the Lewis blood group antigen as receptor for *Helicobacter pylori* (Boren *et al.*, 1993; Horuk *et al.*, 1993). Thus further data on the conventional genetic markers will continue to enrich our knowledge of the basic and applied genetics of the pig.

Recently Meijerink *et al.* (1997) reported close linkage of two alpha (1,2) fucosyltransferase genes (FUT^1 and FUT^2) with the *S* and *ECF18R* loci in pigs. A PCR-based method was given for the mutation at nucleotide 307 in FUT^1 that showed a high linkage disequilibrium with the *ECF18R* locus. Whether the FUT^1 or the FUT^2 gene products are involved in the synthesis of carbohydrate structures responsible for bacterial adhesion remains to be determined. A brief study indicating biochemical polymorphism of porcine neutral glycosidase in liver samples was reported by Cygan *et al.* (1997).

References

Agar, N.S. and Board, P.G. (1984) Phenotypic variation of erythrocyte glutathione peroxidase in the pig. *Animal Blood Groups and Biochemical Genetics* 15, 63–66.

Althen, T.G. and Gerrits, R.J. (1972) Polymorphism of whey$_2$ protein of sow's milk. *Journal of Dairy Science* 55, 331–333.

Andresen, E. (1962) Blood groups in pigs. *Annals of the New York Academy of Sciences* 97, 205–225.

Andresen, E. (1964) Further studies on the H blood group system in pigs, with special reference to a new red-cell antigen Hc. *Acta Genetica* 14, 319–326.

Andresen, E. (1966) Blood groups of the I system in pigs: association with variants of serum amylase. *Science* 153, 1660–1661.

Andresen, E. (1971) Linear sequence of the autosomal loci PHI, H and 6 – PGD in pigs. *Animal Blood Groups and Biochemical Genetics* 2, 119–120.

Andresen, E. (1979) Evidence indicating the sequence *Phi, Hal, H* of the three closely linked loci in pigs. *Nordisk Veterinaermedicin* 31, 443–444.

Andresen, E. and Baker, L.N. (1964) The C blood-group system in pigs and the detection and estimation of linkage between the C and J systems. *Genetics* 49, 379–386.

Archibald, A.L. and Imlah, P. (1985) The halothane sensitivity locus and its linkage relationships. *Animal Blood Groups and Biochemical Genetics* 16, 253–263.

Archibald, A.L. and McTeir, B.L. (1988) A new allele at the *Pgd* locus in pigs. *Animal Genetics* 19, 189–191.

Ashton, G.C. (1960) Thread protein and β-globulin polymorphism in the serum proteins of pigs. *Nature* 186, 991–992.

Baranov, O.K. (1970) An immunoelectrophoretic study of protein polmorphism in hemolysates of pig and cattle. *Biochemical Genetics* 4, 549–563.

Barthelemy, G. and Darre, R. (1987) Les variants electrophoretiques de proteines sanguines chez le porc et le sanglier: presentation et applications. *Revue de Medecine Veterinaire* 138, 839–851.

Belanger, B., Benkel, B., LeBel, D. and Pelletier, G. (1994) Polymorphism of α-amylase in a small population of Yorkshire purebred pigs. In: *Proceedings of the 4th World Congress on Genetics Applied to Livestock Production* vol. 21. University of Guelph, Canada, pp. 120–123.

Bell, K., Mckenzie, H.A. and Shaw, D.C. (1981a) Porcine α-lactalbumin A and B. *Molecular and Cellular Biochemistry* 35, 113–119.

Bell, K., Mckenzie, H.A. and Shaw, D.C. (1981b) Porcine β-lactoglobulin A and C: Occurrence, isolation and chemical properties. *Molecular and Cellular Biochemistry* 35, 103–111.

Bertschinger, H.U., Stamm, M. and Vögeli, P. (1993) Inheritance of resistance to oedema disease in the pig: experiments with an *Escherichia coli* strain expressing fimbriae 107. *Veterinary Microbiology* 35, 79–89.

Bigi, D., Gahne, B. and Juneja, R.K. (1990) Further data on adenosine deaminase polymorphism in pigs and evidence for a new allele. *Hereditas* 112, 71–76.

Boren, T., Falk, P., Roth, K.A., Larson, G. and Normark, S. (1993) Attatchment of *Helicobacter pylori* to human gastric epithelium mediated by blood group antigens. *Science* 262, 1892–1895.

Catanese, J.J. and Kress, L.F. (1992) Isolation from opossum serum of a metalloproteinase inhibitor homologous to human α1B-glycoprotein. *Biochemistry* 31, 410–418.

Cepica, S., Brchanova, I., Hradecky, J. and Hojny, J. (1991) Two new common alleles of erythrocyte adenosine deaminase (*ADA*) in pigs. *Animal Genetics* 22, 173–176.

Cepica, S., Wolf, J., Hojny, J., Vackova, I. and Schröffel J., Jr (1995) Relations between genetic distance of parental pig breeds and heterozygosity of their F_1 crosses measured by genetic markers. *Animal Genetics* 26, 135–140.

Christensen, K., Fredholm, M., Wintero, A.K., Jorgensen, J.N. and Andersen, S. (1996) Joint effect of 21 marker loci and effect of realized inbreeding on growth in pigs. *Animal Science* 62, 541–546.

Chung, E.R., Han, S.K., Shin, Y.C., Chung, H.Y. and Kim, J.E. (1992) Studies on biochemical polymorphism of milk protein as genetic markers in pigs. *Asian – Australasian Journal of Animal Sciences* 5, 285–294.

Cizova, D., Gabrisova E., Stratil, A. and Hojny J. (1993a) An additional locus (*PI4*) in the protease inhibitor (*PI*) gene cluster of pigs. *Animal Genetics* 24, 315–318.

Cizova, D., Stratil, A., Muller, E. and Cepica, S. (1993b) A new, partially deficient transferrin variant in the pig. *Animal Genetics* 24, 305–306.

Clamp, P.A., Beever, J.E., Fernando, R.L., McLaren, D.G. and Schook, L.B. (1992) Detection of linkage between genetic markers and genes that affect growth and carcass trait in pigs. *Journal of Animal Science* 70, 2695–2706.

Cygan, J., Neufeld-Kaiser, W., Jara, G. and Daniel, W.L. (1997) Comparative biochemistry of a cytosolic artiodactyl glycosidase. *Comparative Biochemistry and Physiology* 116B, 437–446.

Dinklage, H. and Hohenbrink, H. (1970) Untersuchung uber den Einfluss heterozygoter Blutgruppengenorte auf Merkmale der Zuchtleistung bei der Deutschen Landrasse. *Zuchtungskunde* 4, 284–293.

Dostal, J. (1970) Protein polymorphism in boar seminal plasma. In: *Proceedings of the XIth European Conference on Animal Blood Groups and Biochemical Polymorphisms,* Polish Scientific Publishers, Warsaw, pp. 297–300.

Eiberg, H., Bisgaard, M.L. and Mohr, J. (1989) Linkage between α_1B-glycoprotein (A1BG) and Lutheran (LU) red blood group system: assignment to chromosome 19: new genetic variants of A1BG. *Clinical Genetics* 36, 415–418.

Erhardt, G. (1989a) Isolierung und Charakterisierung von Caseinfraktionen sowie deren genetische Varianten in Schweinemilch. *Milchwissenschaft* 44, 17–20.

Erhardt, G. (1989b) Isolierung und Charakterisierung der Molkenproteine sowie der genetische Varianten in Schweinemilch. *Milchwissenschaft* 44, 145–149.

Erhardt, G.(1991) Anwendungsmöglichkeiten hochauflösender elektrophoretischer Trennverfahren bei tierzuchterischen Fragestellungen. Wissenschaftlicher Fachverlag, Giessen, Germany.

Franceschi, P.F. and Ollivier, L. (1981) Frequences de quelques genes importants dans les populations porcines. *Zeitschrift für Tierzuchtung und Zuchtungsbiologie* 98, 176–186.

Fujii, J., Otsu, K., Zorzato, F., De Leon, S., Khanna, V.K., Weiler, J.E., O'Brien, P.J. and MacLennan, D.H. (1991) Identification of a mutation in porcine ryanodine receptor associated with malignant hyperthermia. *Science* 253, 448–451.

Full, C., Meyer, J.N., Brandt, H. and Glodek, P. (1990) The allotypes of lipoproteins and globulins in the pig. 1. Allotype frequencies in German pig breeds. *Journal of Animal Breeding and Genetics* 107, 221–228 (in German).

Gahne, B. (1979) Immunogenetics and biochemical genetics as a tool in pig breeding programmes. *Acta Agricultura Scandinavica* (Suppl. 21), 185–197.

Gahne, B. and Juneja, R.K. (1985a) Close genetic linkage between four plasma α-protease inhibitor loci (*Pi1, Po1A, Po1B* and *Pi2*) in pigs. In: Peeters, H. (ed.) *Protides of Biological Fluids,* vol. 33. Pergamon Press, New York, pp. 119–122.

Gahne, B. and Juneja, R.K. (1985b) Prediction of the halothane (Hal) genotypes of pigs by deducing *Hal, Phi, Po2, Pgd* haplotypes of parents and offspring: results from a large-scale practice in Swedish breeds. *Animal Blood Groups and Biochemical Genetics* 16, 265–283.

Gahne, B. and Juneja, R.K. (1986) Extensive genetic polymorphism of four plasma α-protease inhibitors in pigs and evidence for tight linkage between the structural loci of these inhibitors. *Animal Genetics* 17, 135–157.

Gahne, B., Bengtsson, S. and Kleppenes, O. (1972) At least eight alleles controlling the arylesterase activity in pig serum. *Proceedings of the XIIth European Conference on Animal Blood and Biochemical Polymorphisms*, Budapest. pp. 379–382.

Gahne, B., Juneja, R. K. and Runsten, B. (1987) Genetic polymorphism of human, pig and dog plasma α1B-glycoprotein: phenotyping by 2-D electrophoresis or immunoblotting and evidence for genetic deficiency in pigs. *Journal of Biochemical and Biophysical Methods* 14 (suppl. 2), 52 (abstract).

Gibbons, R.A., Sellwood, R., Burrows, M. and Hunter, P.A. (1977) Inheritance of resistance to neonatal *E. coli* diarrhoea in the pig: examination of the genetic system. *Theoretical and Applied Genetics* 51, 65–70.

Glasnak, V. (1968) Inter- and intraspecific differences in milk proteins of cattle and swine. *Comparative Biochemistry and Physiology* 25, 355–357.

Glodek, P., Meyer, J.N., Brandt, H. and Pfeiffer, H. (1993) Die genetische Distanz zwischen ost- und westdeutschen Schweinerassen. 1. Mitteilung: die Eigenständigkeit und der Heterozygotiegrad der Rassen. *Archiv für Tierzucht* 36, 621–630.

Grange, P.A. and Mouricout, M.A. (1996) Transferrin associated with the porcine intestinal mucosa is a receptor for K88ab fimbriae of *Escherichia coli. Infection and Immunity* 64, 606–610.

Grunder, A.A. and Kristjansson, F.K. (1974) Genetic control of serum esterases in day-old pigs. *Animal Blood Groups and Biochemical Genetics* 5, 143–151.

Hala, K. and Hojny, J. (1964) Blood groups of the N system in pigs. *Folia Biologica* (Praha) 10, 239–244.

Hamel, R.P. (1984) Verbesserte Darstellung und Identifizierung von Proteinen und deren Polymorphismen im Blutserum von Schweinen mit Hilfe der Polyacrylamidgel-elektrophorese und der isoelektrischen Fokussierung. PhD thesis, Justus-Liebig University Giessen, Germany.

Hartl, G.B. and Csaikl, F. (1987) Genetic variability and differentiation in wild boars (*Sus scrofa ferus* L.): comparison of isolated populations. *Journal of Mammology* 68, 119–125.

Hesselholt, M. (1969) *Serum protein polymorphisms in swine. Electrophoretic identification, genetics and application.* Thesis, Munksgaard, Copenhagen.

Hesselholt, M. and Hristic, V. (1966) Haemopexin polymorphism in pigs. *Acta Veterinaria Scandinavica* 7, 187–188.

Hojny, J. (1973) Further contribution to the H blood group system in pigs. *Animal Blood Groups and Biochemical Genetics* 4, 161–168.

Hojny, J. and Glasnak, V. (1970) A comparison of A, Na and Nb substances in the blood, milk and saliva of sows. *Animal Blood Groups and Biochemical Genetics* 1, 47–51.

Hojny, J. and Hala, K. (1965) Blood group system O in pigs. In: Matousek, J. (ed.) *Proceedings of the 9th European Conference on Animal Blood Groups*, Prague, 1964, pp. 163–168.

Hojny, J. and Hradecky, J. (1971) The time and appearance of blood group antigens in miniature pigs. *Animal Blood Groups and Biochemical Genetics* 2, 105–112.

Hojny, J. and Hradecky, J. (1972) A contribution to the study on H, J and M blood group systems in pigs. In: *Proceedings of the XIIth European Conference on Animal Blood Groups and Biochemical Polymorphisms,* Budapest, 1970, Akademiai Kiado, pp. 299–303.

Hojny, J. and Nielsen, P.B. (1992) Allele E^{bdgjmr} (E^{17}) in the pig E blood group system. *Animal Genetics* 23, 523–524.

Hojny, J., Gavalier, M., Hradecky, J. and Linhart, J. (1967) New blood group factors in pigs. In: *Proceedings of the Xth European Conference of Animal Blood Groups and Biochemical Polymorphisms,* Paris, 1966, pp. 151–158.

Hojny, J., Hradecky, J. and Camacho, A. (1979) Further factors and alleles of the M blood group system in pigs. In: *Proceedings of the XVIth International Conference on Animal Blood Groups and Biochemical Polymorphisms,* Leningrad, 1978, vol. III, pp. 114–119.

Hojny, J., Hradecky, J. and Linhart, J. (1984) New blood group allele (F^{ad}) in the pig. *Animal Blood Groups and Biochemical Genetics* 15, 227–228.

Hojny, J., Janik, A., Schröffel J., Jr and Cizova, D. (1992) New genetically closed allotype system (P1–G3) in pigs. *Animal Genetics* 23 (Suppl. 1), 26 (abstract).

Hojny, J., Janik, A. and Schröffel J., Jr (1993) A new lipoprotein allotype Lpr 3 and allele $Lpr^{1,3}$ in the pig. *Animal Genetics* 24, 445–446.

Hojny, J., Schröffel J., Jr, Geldermann, H. and Cepica, S. (1994) The porcine M blood group system: evidence to suggest assignment of its M1 factor to a new system (P). *Animal Genetics* 25 (Suppl. 1), 99–101.

Horuk, R., Chitnis, C. E., Darbonne, W. C., Colby, T. J., Rybicki, A., Hadley, T. J. and Miller, L. H. (1993) A receptor for the malaria parasite *Plasmodium vivax*: the erythrocyte chemokine receptor. *Science* 261, 1182–1184.

Hradecky, J. and Linhart, J. (1970) Db – next blood group factor of the D system in pigs. *Animal Blood Groups and Biochemical Genetics* 1, 65–66.

Hradecky, J., Hruban, V., Pazdera, J. and Klaudy, J. (1982) Map arrangement of the SLA chromosomal region and the J and C blood group loci in the pig. *Animal Blood Groups and Biochemical Genetics* 13, 223–224.

Huang, M. Y. and Rasmusen, B. A. (1982) Parental transferrin types and litter size in pigs. *Journal of Animal Science* 54, 757–762.

Hyldgaard-Jensen, J.F. (1990) Adenosine deaminase and porcine meat quality. II. Effects of adenosine analogues on plasma free fatty acids, glucose and lactate in pigs representing high and low adenosine deaminase red cell activity. *Acta Veterinaria Scandinavica* 31, 145–152.

Imlah, P. (1964) Inherited variants in serum ceruloplasmins of the pig. *Nature* 203, 658–659.

Imlah, P. (1965) A study of blood groups in pigs. In: Matousek, J. (ed.) *Blood Groups of Animals.* Publishing House of the Czechoslovak Academy of Sciences, Prague, pp. 109–122.

Imlah, P. (1970) Evidence for the Tf locus being associated with an early lethal factor in a strain of pigs. *Animal Blood Groups and Biochemical Genetics* 1, 5–13.

Irwin, M.R. (1976) The beginnings of immunogenetics. *Immunogenetics* 3, 1–13.

Janik, A., Hojny, J. and Duniec, M. (1983) Allotype polymorphism of serum globulin (Gp system) in pigs. *Animal Blood Groups and Biochemical Genetics* 14, 63–70.

Jensen, E. L., Smith, C., Baker, L.N. and Cox, D.F. (1968) Quantitative studies on blood group and serum protein systems in pigs. II. Effects on production and reproduction. *Journal of Animal Science* 27, 856–862.

Jörgensen, P.F., Hyldgaard-Jensen, J., Moustgaard, J. and Eikelenboom, G. (1976) Phosphohexose isomerase (PHI) and porcine halothane sensitivity. *Acta Veterinaria Scandinavica* 17, 370–372.

Jumkov, V.A. and Nikonchik, L.I. (1977) Serum amylase 2 polymorphism in pigs. *Animal Blood Groups and Biochemical Genetics* 8, 247–250.

Juneja, R.K. and Andersson, L. (1994) Genetic variation at pig plasma protein locus *PLP1* and its assignment to chromosome 5. *Animal Genetics* 25, 353–355.

Juneja, R.K. and Gahne, B. (1987) Simultaneous phenotyping of pig plasma α-protease inhibitors (PI1, PO1A, PO1B, PI2) and four other proteins (PO2, TF, CP, HPX) by a simple method of 2D horizontal electrophoresis. *Animal Genetics* 18, 197–211.

Juneja, R.K., Gahne, B., Edfors-Lilja, I. and Andresen, E. (1983) Genetic variation at a pig serum protein locus, *Po-2* and its assignment to the *Phi, Hal, S, H, Pgd* linkage group. *Animal Blood Groups and Biochemical Genetics* 14, 27–36.

Juneja, R.K., Gahne, B., Rapacz, J. and Hasler-Rapacz, J. (1986) Linkage between the porcine genes encoding immunoglobulin heavy-chain allotypes and some serum α-protease inhibitors: a conserved linkage in pig, mouse and human. *Animal Genetics* 17, 225–233.

Juneja, R.K., Gahne, B. and Stratil, A. (1987) Polymorphic plasma postalbumins of some domestic animals (pig PO2, horse Xk and dog Pa proteins) identified as homologous to human plasma α_1B-glycoprotein. *Animal Genetics* 18, 119–124.

Juneja, R.K., Kuryl, J., Gahne, B. and Zurkowski, M. (1989) Linkage between the loci for transferrin and ceruloplasmin in pigs. *Animal Genetics* 20, 307–311.

Juneja, R.K., Johansson, S. and Sandberg, K. (1994) Linkage between the *GPI* and *A1BG* loci in dogs: a conserved linkage in mammals. *Animal Genetics* 25 (Suppl. 2), 49 (abstract).

Kalab, P. and Stratil, A. (1989) Phenotyping of pig α_1B-glycoprotein (PO2) and haemopexin by 1D polyacrylamide gel electrophoresis and immunoblotting. *Animal Genetics* 20, 295–298.

Kierek-Jaszczuk, D. and Geldermann, H. (1985) Serum and tissue alkaline phosphatases in pigs. *Animal Blood Groups and Biochemical Genetics* 16, 205–216.

Kloster, G., Larsen, B. and Nielsen, P.B. (1970) Carbonic anhydrase polymorphism in cattle and swine. *Acta Veterinaria Scandinavica* 11, 318–321.

Knyazev, S.P., Tikhonov, V.N., Suzuki, S., Kurosawa, Y. and Tanaka, K. (1985) Genetic peculiarities of domestic and wild pigs (Suidae) of Eurasia by serum polymorphic systems. 2. Polymorphism by transferrin, ceruloplasmin and amylase systems. *Zoologiceskij Zurnal* 64, 1712–1717 (in Russian).

Kristjansson, F.K. (1960) Genetic control of two blood serum proteins in swine. *Canadian Journal of Genetics and Cytology* 2, 295–300.

Kristjansson, F.K. (1963) Genetic control of two prealbumins in pigs. *Genetics* 48, 1059–1063.

Kristjansson, F.K. (1964) Transferrin types and reproductive performance in the pig. *Journal of Reproduction and Fertility* 8, 311–317.

Kristjansson, F.K. (1966) Fractionation of serum albumin and genetic control of two albumin fractions in pigs. *Genetics* 53, 675–679.

Kubek, A. (1970) Electrophoretical study of the esterases in pig serum. *Proceedings of the XIth European Conference on Animal Blood Groups and Biochemical Polymorphisms*, Warsaw, 1968, Junk, The Hague, pp. 355–358.

Kuehne, R.A. (1990) Serumproteinpolymorphismen der Schweizerischen Schweinrassen dargestellt anhand zweidimensionaler Elektrophorese. PhD thesis, Federal Institute of Technology (ETH), Zurich, Switzerland.

Kurosawa, Y. and Tanaka, K. (1988) Electrophoretic variants of serum transferrin in wild pig populations of Japan. *Animal Genetics* 19, 31–35.

Kurosawa, Y. and Tanaka, K. (1991) PGD variants in several wild pig populations of East Asia. *Animal Genetics* 22, 357–360.

Kurosawa, Y., Tanaka, K., Tomita, T., Katsumata, M., Masangkay, J.S. and Lacuata, A.Q. (1989) Blood groups and biochemical polymorphisms of Warty (or Javan) pigs, Bearded pigs and a hybrid of domestic × Warty pigs in the Philippines. *Japanese Journal of Zootechnical Sciences* 60, 57–69.

Kuryl, J., Zurkowski, M., Hojny, J. and Hradecky, J. (1989) Plasma α-protease inhibitors in Norwegian Landrace and Czech Landrace pigs. *Animal Genetics* 20, 299–305.

Kuryl, J., Janik, A., Nogaj, A. and Wroblewski, T. (1991) Cosegregation of two allotypes G9 and G16 with the α-protease inhibitor variants in pigs. *Animal Genetics* 22, 295–298.

Kuryl, J., Janik, A., Wroblewski, T. and Nogaj, A. (1992) The system of plasma α-protease inhibitors in chosen pig breeds raised in Poland. *Genetica Polonica* 33, 141–145.

Linhart, J. (1971) Lm, a new blood factor of the L system in pigs. *Animal Blood Groups and Biochemical Genetics* 2, 243–245.

Ljungqvist, L. and Hyldgaard-Jensen, J. (1983) Genetic polymorphism of the vitamin D binding protein (Gc protein) in pig plasma determined by agarose isoelectrofocusing. *Animal Blood Groups and Biochemical Genetics* 14, 293–297.

Lowe, S.W., Checovich, W.J., Rapacz, J. and Attie, A.D. (1988) Defective receptor binding of low density lipoproteins from pigs possessing mutant apolipoprotein B alleles. *Journal of Biological Chemistry* 263, 15467–15473.

McLennan, D.H. and Phillips, M.S. (1992) Malignant hyperthermia. *Science* 256, 789–794.

Major, F., Dinklage, H. and Gruhn, R. (1970) Gene frequency of blood groups from different European Landrace pigs. In: *Proceedings of the XIth European Conference on Animal Blood Groups and Biochemical Polymorphisms,* Warsaw, Poland, pp. 271–274.

Markert, C.L. and Möller, F. (1959) Multiple forms of enzymes: tissue, ontogenetic and species specific patterns. *Proceedings of the National Academy of Sciences, USA* 45, 753–763.

Matousek, J.N. (1972) Two protein polymorphic regions in the fluid of the epididymis in boars. In: *Proceedings of the XIIth European Conference on Animal Blood Groups and Biochemical Polymorphisms,* Akademiai Kiado, Budapest, Hungary, pp. 397–400.

Meijerink, E., Fries, R., Vögeli, P., Masabanda, J. Wigger, G., Stricker, C., Neuenschwander, H.U., Bertschinger, H.U. and Stranzinger, G. (1997) Two α(1,2) fucosyltransferase genes on porcine chromosome 6q11 are closely linked to the blood group inhibitor (*S*) and *Escherichia coli* F18 receptor (*ECF18R*) loci. *Mammalian Genome* 8, 736–741.

Meyer, J. and Verhorst, D. (1973) The evidence of erythrocyte acid phosphatase by starch gel electrophoresis in the pig. *Animal Blood Groups and Biochemical Genetics* 4, 129–131.

Ng-Kwai-Hang, K.F. and Grosclaude, F. (1992) Genetic polymorphism of milk proteins. In : Fox, P.F. (ed.) *Advanced Dairy Chemistry– 1:Proteins.* Elsevier Applied Science, London, pp. 405–455.

Nichols, T.C., Bellinger, D.A., Davis, K.E., Koch, G.G., Reddick, R.L., Read, M.S., Rapacz, J., Hasler-Rapacz, J., Brinkhous, K.M. and Griggs, T.R. (1992) Porcine von Willebrand disease and atherosclerosis. Influence of polymorphism in apolipoprotein B100 genotype. *American Journal of Pathology* 140, 403–415.

Nielsen, P.B. (1961) The M blood group system of the pig. *Acta Veterinaria Scandinavica* 2, 246–253.

Nielsen, P.B. (1972) Isoantigens of the immunoglobulins in pigs. *Acta Veterinaria Scandinavica* 13, 143–145.

Nielsen, P.B. and Vögeli, P. (1982) A new Kd subgroup designated Kg in the porcine K blood group system. *Animal Blood Groups and Biochemical Genetics* 13, 65–66.

Nyström, P.E., Juneja, R.K., Johansson, K., Andersson-Eklund, L. and Andersson, K. (1997) Association of the transferrin locus on chromosome 13 with early body weights in pigs. *Journal of Animal Breeding and Genetics* 114, 363–368.

Oishi, T., Esaki, K. and Tomita, T. (1980a) Genetic relationship among Göttingen Miniature, European and East Asian pigs investigated from blood groups and biochemical polymorphism. *Japanese Journal of Zootechnical Science* 51, 226–228.

Oishi, T., Tomita, T. and Komatsu, M. (1980b) New genetic variants detected in the haemopexin and ceruloplasmin systems of Ohmini miniature pigs. *Animal Blood Groups and Biochemical Genetics* 11, 59–62.

Oishi, T., Tanaka, K., Tomita, T., Tsuji, S., Noguchi, H., Tanji, T. and Moki, T. (1990) Genetic variations of blood groups and biochemical polymorphism in Jinhua pigs. *Japanese Journal of Swine Science* 27, 202–208 (in Japanese with summary in English).

Oishi, T., Amano, T. and Tanaka, K. (1991) Phylogenetic relationship among twelve pig breeds analysed by principal component analysis based on blood groups and biochemical polymorphisms. *Animal Science and Technology (Japan)* 62, 750–756 (In Japanese with tables and summary in English).

Ollivier, L. and Sellier, P. (1982) Pig genetics: a review. *Annales de Genetique et de Selection Animale* 14, 481–544.

Op't Hof, J., Osterhoff, D.R. and De Beer, G. (1972) Polymorphism of sorbitol dehydrogenase and 6-phosphogluconate dehydrogenase in swine (*Sus scrofa*). *Animal Blood Groups and Biochemical Genetics* 3, 237–239.

Prescott, M.F., Hasler-Rapacz, J., Von Linden, R.J. and Rapacz, J. (1995) Familial hypercholesterolemia associated with coronary atherosclerosis in swine bearing different alleles of apolipoprotein B. *Annals of New York Academy of Sciences* 748, 283–293.

Pretorius, A.M.G., Schmid, D.O., Cwik, S., Meyer, J. and Albert, E.D. (1977) PGM_3 locus and its genetic polymorphism in lymphocytes of the pig. *Journal of Immunogenetics* 4, 363–365.

Randi, E., Apollonio, M. and Toso, S. (1986) Electrophoretic polymorphism of erythrocyte leucine aminopeptidase in the wild boar, *Sus scrofa*. *Animal Genetics* 17, 359–362.

Randi, E., Massei, G. and Genov, V. (1992) Allozyme variability in Bulgarian wild boar populations. *Acta Theriologica* 37, 271–278.

Rapacz, J. (1978) Lipoprotein immunogenetics and atherosclerosis. *American Journal of Medical Genetics* 1, 377–405.

Rapacz, J. (1982) Current status of lipoprotein genetics applied to livestock production in swine and other domestic species. In: *Proceedings of the 2nd World Congress on Genetics Applied to Livestock Production* (Madrid), Neografis S.L., Madrid, vol. VI, pp. 365–374.

Rapacz, J. and Hasler-Rapacz, J. (1982) Immunogenetic studies on polymorphism, postnatal passive acquisition and development of immunoglobulin gamma (IgG) in swine. In: *Proceedings of the 2nd World Congress on Gentics Applied to Livestock Production* (Madrid), Neografis S.L., Madrid, vol. VIII, pp. 601–606.

Rapacz, J. and Hasler-Rapacz, J. (1990) The pig as a model for studying genetic polymorphisms of apolipoproteins in relation to lipid levels and atherosclerosis. In: Berg, K., Retterstol, N. and Rofsum, S. (eds) *From Phenotype to Gene in Common Disorders,* Munksgaard, Copenhagen, pp. 115–137.

Rapacz, J., Hasler-Rapacz, J. and Li, D. (1973) Two immunoglobulin allotypes in swine. *Genetics* 74, 223–224 (abstract).

Rapacz, J., Hasler-Rapacz, J. and Kuo, W.H. (1978) Immunogenetic polymorphism of lipoproteins in swine. 2. Five allotypic specificities (Lpp6, Lpp11, Lpp12, Lpp13 and Lpp14) in the Lpp system. *Immunogenetics* 6, 405–424.

Rapacz, J., Hasler-Rapacz, J., Taylor, K.M., Checovich, W.J. and Attie, A.D. (1986a) Lipoprotein mutations in pigs are associated with elevated plasma cholesterol and atherosclerosis. *Science* 234, 1573–1577.

Rapacz, J., Hasler-Rapacz, J. and Kuo, W.H. (1986b) Immunogenetic polymorphism of lipoproteins in swine: genetic, immunological and physiochemical characterization of two allotypes Lpr1 and Lpr2. *Genetics* 113, 985–1007.

Rapacz, J. Jr, Reiner, Z., Ye, S.Q., Hasler-Rapacz, J., Rapacz, J. and McConathy, W.J. (1989) Plasminogen polymorphism in swine. *Comparative Biochemistry and Physiology* 93B, 325–331.

Rapacz, J., Hasler-Rapacz, J., Hu, Z.L., Rapacz, J.M., Vögeli, P., Hojny, J. and Janik, A. (1994) Identification of new apolipoprotein B epitopes and haplotypes and their distribution in swine populations. *Animal Genetics* 25 (Suppl. 1), 51–57.

Rasmusen, B.A. (1964) Gene interaction and the A–O blood-group system in pigs. *Genetics* 50, 191–198.

Rasmusen, B.A. (1965) Isoantigens of gamma globulin in pigs. *Science* 148, 1742–1743.

Rasmusen, B.A. (1981) Linkage of genes for PHI, halothane sensitivity, A–O inhibition, H red blood cell antigens and 6-PGD variants in pigs. *Animal Blood Groups and Biochemical Genetics* 12, 207–209.

Rasmusen, B.A. and Hagen, K.L. (1973) The H blood group system and reproduction in pigs. *Journal of Animal Science* 37, 568–573.

Raymond, S. and Weintraub, L. (1959) Acrylamide gel as a supporting medium for zone electrophoresis. *Science* 130, 711.

Rozhkov, Y.I. and Galimov, I.R. (1990) Salivary gland amylase polymorphism in pigs and cattle detected by affinity electrophoresis. *Animal Genetics* 21, 277–283.

Ryan-Poirier, K.A. and Kawaoka, Y. (1993) α_2-Macroglobulin is the major neutralizing inhibitor of influenza A virus in pig serum. *Virology* 193, 974–976.

Safarova, P., Karadjole, I., Hyldgaard-Jensen, J., Nielsen, P.B. and Lycik, G. (1972) Phosphoglucomutase polymorphism in porcine red cells. *Acta Veterinaria Scandinavica* 13, 134–136.

Saison, R. (1967) Two new antibodies, anti-Nb and anti-Nc in the N blood-group system in pigs. *Vox Sanguinis* 12, 215–220.

Saison, R. (1970) Serum and red cell enzyme systems in pigs. In: *Proceedings of the 11th European Conference on Animal Blood Groups and Biochemical Polymorphism,* Warsaw, 1968, Junk, The Hague, pp. 321–328.

Saison, R. (1973) Red cell peptidase polymorphism in pigs, cattle, dogs and mink. *Vox Sanguinis* 25, 173–181.

Saison, R., Rasmusen, B.A. and Hradecky, J. (1967) Da, a factor in a new blood group system in pigs. *Canadian Journal of Genetics and Cytology* 9, 794–798.

Schröffel, J. (1965) Genetic determination of the serum 'thread proteins' and the slow α_2-globulin polymorphism in pigs. In: Matousek, J. (ed.) *Blood Groups of Animals.* Publishing House of the Czechoslovak Academy of Sciences, Prague, pp. 321–329.

Shibata, T., Akita, T. and Abe, T. (1993) Genetic polymorphism of the sixth component of complement (C6) in the pig. *Animal Genetics* 24, 97–100.

Smith, M.W., Smith, M.H. and Brisbin Jr, I.L. (1980) Genetic variability and domestication in swine. *Journal of Mammology* 61, 39–45.

Smithies, O. (1955) Zone electrophoresis in starch gels: group variations in the serum proteins of normal human adults. *Biochemical Journal* 61, 629–641.

Sorenson, R.C., Primo-Parmo, S.L., Camper, S.A. and La Du, B.N. (1995) The genetic mapping and gene structure of mouse paraoxonase/arylesterase. *Genomics* 30, 431–438.

Sprague, L.M. (1958) On the distribution and inheritance of a natural antibody in cattle. *Genetics* 43, 913–918.

Stratil, A., Gahne, B., Juneja, R.K., Hjerten, S. and Spik, G. (1987) Pig plasma postalbumin – 2(α_1B-glycoprotein): isolation, partial characterization and immunological cross-reactivity with other mammalian sera. *Comparative Biochemistry and Physiology* 88B, 953–961.

Stratil, A., Gahne, B., Juneja, R.K., Hjerten, S. and Spik, G. (1988) Pig plasma protease inhibitor gene complex: isolation and partial characterization of three inhibitors. *Comparative Biochemistry and Physiology* 90B, 409–418.

Stratil, A., Cizova, D., Hojny, J. and Hradecky, J. (1990) Polymorphism of pig serum α-protease inhibitor-3 (PI3) and assignment of the locus to the *Pi1, Po1A, Po1B, Pi2, Igh* linkage group. *Animal Genetics* 21, 267–276.

Stratil, A., Cizova-Schröffelova, D., Gabrisova, E., Pavlik, M., Coppieters, W., Peelman, L., Van de Weghe, A. and Bouquet, Y. (1995) Pig plasma α-protease inhibitors PI2, PI3 and PI4 are members of the antichymotrypsin family. *Comparative Biochemistry and Physiology* 111B, 53–60.

Strumeyer, D.H., Kao, W., Roberts, T., Forsyth-Davis, D. (1988) Isozymes of α-amylase in the porcine pancreas: population distribution. *Comparative Biochemistry and Physiology* 91B, 351–357.

Tanaka, K., Kurosawa, Y., Kurokawa, K. and Oishi, T. (1980) Genetic polymorphism of erythrocyte esterase-D in pigs. *Animal Blood Groups and Biochemical Genetics* 11, 193–197.

Tanaka, K., Oishi, T., Kurosawa, Y. and Suzuki, S. (1983) Genetic relationship among several pig populations in East Asia analysed by blood groups and serum protein polymorphisms. *Animal Blood Group and Biochemical Genetics* 14, 191–200.

Tsuji, S., Asao, M., Kusunoki, H. and Oishi, T. (1993) Genetic polymorphism of boar seminal plasma proteins. *Animal Science and Technology* (Japan) 64, 221–227.

Vackova, I., Cepica, S., Wolf, J., Hojny, J. and Schröffel, J., Jr (1996) A genetic blood-marker study on six pig breeds bred in Czechia. *Journal of Animal Breeding and Genetics* 113, 63–76.

Valenta, M., Hyldgaard-Jensen, J. and Moustgaard, J. (1967) Three lactic-dehydrogenase isoenzyme systems in pig spermatozoa and the polymorphisms of the sub-units controlled by a third locus C. *Nature* 216, 506–507.

Van de Weghe, A., Van Zeveren, A., Bouquet, Y. and Varewyck, H. (1987) The serum protease inhibitor linkage group in Belgian pig breeds. *Animal Genetics* 18, 63–69.

Van de Weghe, A., Coppieters, W., Bauw, G., Vandekerckhove, J. and Bouquet, Y. (1988a) The homology between serum proteins Po2 in pig, Xk in horse and α_1B-glycoprotein in human. *Comparative Biochemistry and Physiology* 90B, 751–756.

Van de Weghe, A., Yablanski, T.S., Van Zeveren, A. and Bouquet, Y. (1988b) A third variant of glucose phosphate isomerase in pigs. *Animal Genetics* 19, 55–58.

Van Zeveren, A., Bouquet, Y., Van de Weghe, A. and Coppieters, W. (1990) A genetic blood marker study on 4 pig breeds. 1. Estimation and comparison of within breed variation. *Journal of Animal Breeding and Genetics* 107, 104–112.

Verhorst, D. (1973) Polymorphism in glucose-6-phosphate dehydrogenase in the German Large-White. *Animal Blood Groups and Biochemical Genetics* 4, 65–68.

Vögeli, P. (1990) Blood Groups of Pigs, Serological and Genetic studies (in German). PhD thesis, Federal Institute of Technology (ETH), Zurich, Switzerland.

Vögeli, P., Gerwig, C. and Schneebeli, H. (1983) The A–O and H blood group systems, some enzyme systems and halothane sensitivity of two divergent lines of Landrace pigs using index selction procedures. *Livestock Production Science* 10, 159–169.

Vögeli, P., Kuhne, R., Gerwig, C., Kaufmann, A., Wysshaar, M. and Stranzinger, G. (1988) Bestimmung des Halotangenotyps (HAL) mit Hilfe der S, PHI, HAL, H, PO2, PGD Haplotypen von Eltern und Nachkommen beim Schweizerischen Veredelten Landschwein. *Zuchtungskunde* 60, 24–37.

Vögeli, P., Bertschinger, H.U., Stamm, M., Stricker, C., Hagger, C., Fries, R., Rapacz, J. and Stranzinger, G. (1996) Genes specifying receptors for F18 fimbriated *Escherichia coli,* causing oedema disease and postweaning diarrhoea in pigs, map to chromosome 6. *Animal Genetics* 27, 321–328.

Widar, J. and Ansay, M. (1975) Adenosine deaminase in the pig: tissue-specific patterns and expression of the silent ADA^0 allele in nucleated cells. *Animal Blood Groups and Biochemical Genetics* 6, 109-116

Widar, J., Ansay, M. and Hanset, R. (1975) Allozymic variation as an estimate of heterozygosity in Belgian pig breeds. *Animal Blood Groups and Biochemical Genetics* 6, 221–234.

Wysshaar, M. (1988) Polymorphe Lipoproteinsysteme und ihre Beziehungen zu Leistungseigenschaften beim Schwein. PhD dissertation, Federal Institute of Technology (ETH), Zurich, Switzerland.

Zyczko, K. (1990) Preliminary studies on the fertility of sows differing in the phenotype of transferrin (Tf). *World Review of Animal Production* 25, 37–40.

Molecular Genetics

C. Moran
Department of Animal Science, University of Sydney, Sydney, NSW 2006, Australia

Introduction

The objective of this review is to provide a detailed characterization of the molecular genetics of the pig, *Sus scrofa*. It will be assumed that the reader is familiar with basic molecular biological concepts concerning DNA structure and replication, as well as with the mechanisms of transcription and translation, so these will not be specifically discussed in this review. Supplementary information should be sought in textbooks, such as *Genes VI* by Lewin (1997).

Since the pig has been subjected to far less molecular scrutiny than intensively studied mammalian species, like humans and mice, it will be necessary at times to extrapolate and generalize from other mammals in order to attempt to plug the gaps and deficiencies in the knowledge of the pig. However, it is reasonable to assume that the pig has a typical mammalian genomic structure and molecular biology, given the universality of some features across all mammals studied and also given the concordance with other mammals in those features specifically examined in the pig. Wherever possible this review will cite studies specific to the pig.

It is instructive to begin by placing the genomic structure and molecular biology of the pig in the context of the enormous diversity of chromosome number and structure, and a huge variability in genomic DNA content found among vertebrates. These differences are not related to variation in the developmental complexity of the organisms, but reflect the evolutionary history of chromosomal rearrangement and duplication as well as growth or diminution in the abundance of repetitive and non-coding DNA. At one extreme, the pufferfish, *Fugu rubripes*, and related tetraodontoid fish have particularly small haploid genomes of only 4×10^8 bp, the smallest recorded for any vertebrate (Brenner *et al.*, 1993). The *Fugu* genome, which is 7.5 times smaller than a mammalian genome, contains less than 10% repetitive DNA. At the other extreme, the haploid genome of the lungfish, *Protopterus aethiopicus*, is 350 times larger than that of the pufferfish at 1.4×10^{11} bp. Nevertheless the pufferfish is clearly as complex an organism as a lungfish. Its DNA has been shown to encode as many genes as a mammalian genome since random sequence from its genome is 7.5 times more likely to be coding than mammalian genomic DNA (Brenner *et al.*, 1993). In contrast to the huge range in genome size found in fish, mammals display little variability, with an approximately twofold range in DNA content from 2.1×10^9 bp in muntjac deer, *Muntiacus muntjak*, to 5×10^9 bp in the aardvark, *Orycteropus afer* (John and Miklos, 1988). Nevertheless the same processes of chromosomal rearrangement and growth/diminution of repetitive DNA have occurred during mammalian evolution within the more constrained circumstance of a relatively constant amount of genomic DNA. Although there is relatively little change in total amount, the repetitive components of the mammalian genome are capable of rapid concerted evolution causing differentiation between even closely related species. Other non-coding components of the genome may also diverge relatively rapidly between species.

Nuclear Genome Size and Complexity

Haploid genome size has been estimated quantitatively for numerous mammalian species using spectrophotometry of Feulgen stained sperm nuclei, yielding estimates of about 3 pg of DNA, equivalent to about 3×10^9 base pairs. Schmitz *et al.* (1992) have used flow karyotyping to estimate the size of all porcine chromosomes, yielding a cumulative total of 2.72×10^9 bp for an X chromosome containing haploid nucleus and 2.62×10^9 for one containing a Y chromosome. However, far from all of this DNA specifies unique information or indeed any developmental information at all. Repeated sequences in particular decrease the information content of the genome. Due to the ability of repeated sequences to find more quickly a complementary partner when genomic DNA is denatured and then allowed to re-anneal in solution, the kinetics of renaturation has allowed estimation of the proportion of repeated and unique DNA and the proportions in various classes of repeat. Such reassociation studies of mammals, first performed about 30 years ago, have found that 40–60% of the genome is repeated. Some components of the genome, which reassociate with great rapidity, are repeated several millionfold. The fact that a large proportion of the genome is repeated led to the initial realization that much of the genome is almost certainly non-functional. Subsequently a substantial class of unique DNA, namely introns, was found interleaved throughout the exons or coding DNA. Generally this DNA has no obvious function and intronic sequences can evolve rapidly due to the lack of functional constraint on any mutations which occur. It is now estimated that as little as 10% of the mammalian genome is functional, in the sense that it specifies transcription or its regulation and only 2–3% of the genome consists of exons.

The mammalian genome can be physically partitioned into classes which show an imperfect correspondence to functional and non-functional DNA. Bernardi *et al.* (1985) have shown that the nuclear DNA of warm blooded vertebrates can be separated by density gradient ultracentrifugation on the basis of GC content into four or five classes, called isochores. These classes in mammals, ranging from lightest (lowest GC content) to heaviest (highest GC content), have been labelled L1, L2, L3, H1, H2 and H3. Isochores H2 and H3 comprise only 8% and 4% respectively of the total DNA in humans, but contain about 50% of the genes. Thus coding genes tend to be found in GC rich regions of the mammalian genome. This clustering of coding genes into GC rich isochores correlates with the cytogenetically observable clustering of genes into R (Giemsa negative) bands, which are GC rich.

Repetitive DNA

The level of repetition, size of repeat units and pattern of dispersion of repeated sequences throughout the genome are important criteria for classifying repetitive DNA. In addition, structural and functional characteristics of the repeat units can be used for further categorization. Repetitive DNA can be broadly classified into two categories, namely tandemly repeated and dispersed repeated DNA. Tandemly repeated DNA ranges from highly repeated sequences, which consist of small and non-functional repeat units, to less highly repeated sequences, which may sometimes be functional, for example the ribosomal RNA genes. In the dispersed repeated category are transposable elements, some of which retain the capability of further movement, and pseudogenes.

Highly repetitive – centromeric and interstitial

Cesium chloride gradient ultracentrifugation for many years was the technique of choice for isolating genomic DNA. In many species, it was found that the genomic DNA resolved as a major band in the gradient, corresponding to the bulk of the genome, and a smaller satellite band, which had a different G : C content and therefore density from the bulk of the genome. Analysis of satellite bands from numerous species revealed that they consisted of small sequences, repeated hundreds of thousands of times. Specific repeats can comprise as much as 5% of the total genome and most species have more than one repeat class. Satellite DNA occurs in all species, mainly around the centromeres, where the tandemly repeated units can comprise many megabases of DNA. In well characterized organisms such as humans and mice, certain repeats, such as the alphoid repeats of humans, have been shown to have a chromosome-specific distribution, whereas other repeats are distributed more widely among the chromosomes.

Two distinct centromeric repeat families have been characterized in the pig. Jantsch *et al.* (1990) have described the Mc1 family of repeats restricted to the metacentric chromosomes in the pig karyotype and the Ac2 family, whose distribution is restricted to the acrocentric chromosomes. The repeat unit of the heterogeneous Mc1 family is about 100 bp. Some family members occur in the pericentromeric region of all metacentric chromosomes, whereas others are found only on specific metacentric chromosomes. The Ac2 family is homogeneous and found in the subterminal pericentromeric regions of all acrocentric chromosomes. The Ac2 family is derived from a highly conserved 14 bp repeat unit. Jantsch *et al.* (1990) have speculated that the association of the centromeres of porcine acrocentric, but not metacentric, chromosomes observed during pachytene of meiosis, may allow interchromosomal exchanges which have maintained the homogeneity of the Ac2 family. Miller *et al.* (1993) have independently isolated what appears to be another member of the Mc2 family and showed that it hybridizes to the centromeres of all chromosomes except the Y chromosome under conditions of low stringency, but only to certain metacentric chromosomes under conditions of high stringency.

Apart from the centromeric repeats, McGraw *et al.* (1988) have isolated a 3.8 kb repeat element with 80% sequence identity to an element independently isolated by Mileham *et al.* (1988). This element is male specific, at least at the level of sensitivity of dot blot and Southern hybridization, with more than 200-fold greater copy number in males than females. It contains no internal repetitions and no open reading frames. Akamatsu *et al.* (1989) subsequently isolated and sequenced a 1.25 kb repetitive element visible in *Eco*RI digests of both male and female porcine genomic DNA. It contains imperfect internal repetitions of about 44 bp and has been used as a gender-neutral probe for verifying presence of target DNA in embryo sexing assays. No estimates of copy number were provided for any of these repetitive elements.

Dispersed repetitive

Transposable elements comprise an important group of dispersed repetitive elements, which have important biological features including the capacity for insertional mutagenesis and at least a theoretical capacity for horizontal transmission. They can be classified into two categories, namely DNA mediated elements and RNA mediated elements.

DNA mediated elements

DNA mediated elements, also called transposons, transpose via a DNA intermediate using a cut and paste mechanism. They have short inverted terminal repeats and, except for defective elements, encode their own transposase, the enzyme which catalyses their transposition. Transposons are common in arthropods, plants and bacteria. They were discovered independently in *Drosophila* and maize due to mutational effects of insertion of the elements. Until quite recently, there was no evidence of their occurrence in mammals and it was even suggested that DNA mediated transpositional mechanisms may be somehow inimical to mammalian cells. Morgan (1995) discovered sequences in humans quite closely related to the *mariner* transposon of *Drosophila*, which are known also to be present in a wide range of arthropods and other invertebrates. Mariner-like elements (MLEs) are about 1250 bp in length with inverted repeats of 20–40 bp. Auge-Gouillou *et al.* (1995) have identified MLEs in humans, mouse, rat, chinese hamster, sheep and cattle. Sequence analysis of the human, bovine and ovine MLEs, which were the most closely related from this group, indicated that they were more similar to the MLEs found in *Hyalophora cecropia*, a lepidopteran, than to the *mariner* elements of *Drosophila*. Indeed the mammalian and lepidopteran MLEs were more similar to each other than MLEs from the two insect groups, although it should be pointed out that the mammalian elements were isolated with degenerate primers for *cecropia* MLEs. It has been estimated (Auge-Gouillou *et al.*, 1995) that there are more than one hundred MLEs in the human, ovine and bovine genomes. While many mammalian MLEs are likely to be the decaying and non-functional remnants of ancient horizontal transmission, there is indirect evidence, from the congruence of a mammalian recombination hotspot and a human MLE (Reiter *et al.*, 1996), that MLE transposase is being expressed in mammals.

Given the ubiquitous distribution of MLEs from fungi to vertebrates and the fact that they have been found in all mammalian species examined, it can be predicted confidently that MLEs will occur in the pig. It can also be predicted that other families of DNA mediated transposons will be found in porcine and other mammalian genomes in the future, as intense research effort is focused on mammalian transposons in light of the MLE discoveries.

RNA mediated elements

RNA mediated elements, called retroposons or retrotransposons, comprise the other major category of transposable elements and transpose via an RNA intermediate, usually with the original inserted copy remaining in position. Reverse transcriptase, an enzyme which synthesizes DNA from an RNA template, makes a DNA copy from the RNA intermediate prior to integration of the element into the genome. The categories of RNA mediated elements are retroviruses and LINEs (long interspersed nuclear elements), which fit into the general category of retrotransposons, which have direct long terminal repeats (LTRs). Retrotransposons encode their own reverse transcriptase as well as integrase and other enzymes necessary for transposition. Retroposons comprise a similar category, which encode reverse transcriptase, but lack integrase and LTRs. SINEs (short interspersed nuclear elements) have neither LTRs nor reverse transcriptase and generally have more or less degenerate poly-(A) tails. Like pseudogenes, they must depend on reverse transcriptase and an integration mechanism from other sources.

Retroviruses. Endogenous retroviruses are very common in the genome of the mouse, comprising up to 5% of the genome. Retroviruses encode an *env* gene, which specifies the envelope proteins required for packaging infectious retroviral particles. Although they are generally capable of completing the infectious viral cycle and thus are capable of horizontal transmission, most endogenous viruses are stably inherited and are vertically transmitted only. Todaro *et al.* (1974) showed that the PK-15 type C retroviruses secreted by pig cells in culture are present in multiple copies in the DNA of all pig tissues and cells. Thus, these retroviruses are endogenous and transmitted vertically. Benveniste and Todaro (1975) were able to show that a related virus was found only in close wild relatives within the Suidae, such as the bush pig and the wart hog, and was absent from the peccary and other artiodactyls such as cattle. More intriguingly, a related viral sequence was found in rodents, including a number of species of mice, with a less related sequence in rats. They interpreted this as evidence of the horizontal transfer of the retrovirus from the common ancestor of the genus *Mus* after its divergence from the rat ancestor to the pig lineage. This occurred before the divergence of wart hogs, bush pigs and domestic pigs from a common ancestor but after the peccary had diverged about 5–10 MYA. Clearly, nature has been experimenting with gene transfer in pigs long before it became technically possible as a result of human DNA technology. Interest in porcine retroviruses as potential hazards to xenotransplantation burgeoned after Patience *et al.* (1997) showed that they could replicate in human cells. Subsequently Le Tissier *et al.* (1997) distinguished two categories of porcine retrovirus on *env* sequence.

LINEs. LINEs are a class of retroposon distinguishable from endogenous retroviruses mainly by their lack of *env* sequences. Although LINES encode reverse transcriptase, only a very small proportion of the elements in the genome contain a fully functional reverse transcriptase gene which has escaped mutational inactivation. Most LINE elements are truncated and stranded in their current genomic location, where they are doomed to gradually decay into random sequence due to unconstrained mutation. One LINE family, the mouse L1 family is very old and homologous elements have been found in plants and even protists, leading to the suggestion that it is the ancestor of retroviruses, which acquired its infectious capability secondarily. Only about 10% of mouse L1 elements are the full length of about 7 kb, with the remainder being shorter truncated elements down to 500 bp in size. The L1 family alone is present at more than 80,000 sites within the mouse genome. One could expect a similar number of related elements in the porcine genome, particularly given the almost universal distribution of L1-like elements in eukaryotes. In humans and mouse, LINE elements predominate in the G + C poor G-bands. Miller (1994) has described the use of a porcine LINE (EMBL Accession X15410), which he has called L1Ss, in PCR genotyping of pigs. Sequence has been obtained from the 5′ ends of two L1S elements, one of which is associated with the hox 2.4 gene and the other of which is clearly truncated. Miller did not report any estimate of the copy number of L1Ss. Thomsen and Miller (1996) have subsequently found that this porcine LINE is uniformly distributed throughout the euchromatic part of the porcine genome, with a slight bias towards G-bands.

SINEs. The use of the term SINE (short interspersed nuclear element) will be restricted here to elements which transpose via an RNA intermediate, although it is occasionally used in the literature for any small interspersed element. Because of their small size, SINEs do not encode reverse transcriptase, although many of them are transcribed. SINEs were originally discovered in humans (*Alu* repeats) and mice (B1 and B2 repeats) due to their very high copy number in the genome, with the *Alu* repeat occurring about 500,000 times in the human genome. SINEs are generally less than 200 bp in size and are derived from small cellular RNA species, which have been reverse transcribed and then integrated into the genome. In effect, they are processed pseudogenes which have attained extremely high copy number. Structural RNAs, such as tRNAs and 7SL RNA (a component of signal recognition particle ribonucleoprotein), have independently become the target for reverse transcription and integration in humans and mice to form SINEs. It is believed that certain mutations within structural RNAs predispose them to become templates for reverse transcriptase. In particular, nucleotide changes in the 3′ region of an RNA that cause self complementarity lead to the formation of hairpin loops. This intra-strand priming allows these RNAs to act as templates for synthesis of a DNA copy (cDNA) by reverse transcriptase. In the presumably rare circumstances of expression of reverse transcriptase in a cell, these mutated structural RNAs form effective templates for the synthesis of multiple cDNAs that are able to integrate into the genome.

In the pig, SINEs are ubiquitous. Three families of SINEs have been recognized, with the PRE-1 (porcine repetitive element-1) family (Singer *et al.*, 1987) being the most abundant of the recognized families. The consensus PRE-1 element is 233 bp in length and Singer *et al.* (1987) have estimated that there are 50,000–100,000 copies interspersed throughout the porcine genome. Ellegren (1993a,b) has estimated from database searching that PRE-1 occurs about once every 12 kb in the porcine genome, implying that there are 250,000 copies if the element is randomly distributed. Other estimates (Frengen *et al.*, 1991; Takahashi *et al.*, 1992) are as high as 500,000. Yasue *et al.* (1991), using a *Hinc* fragment containing three PRE-1 elements, discovered evidence for an uneven distribution of PRE-1 along the chromosomes, surprisingly favouring the vicinity of the centromeres. They found no correlation with the distribution of Q bands. In contrast, Sarmiento *et al.* (1993) have shown by *in situ* hybridization that a PRE-1 element, which they called C11, is distributed uniformly throughout the porcine genome, but spares the centromeres of acrocentric chromosomes. The element used by Sarmiento *et al.* (1993) was cloned from within the porcine MHC and showed a pattern of apparent non-random association with other porcine repetitive elements within the MHC but not elsewhere in the genome (Singer *et al.*, 1983). It is difficult to reconcile these observations unless there are subfamilies of the PRE-1 element with different patterns of genomic distribution. Thomsen and Miller (1996) have also examined the distribution of PRE-1 and have found no reproducible clustering of label, but in agreement with Sarmiento *et al.* (1993), signal was absent or reduced in the centromere regions, particularly of acrocentric chromosomes. However, they did find some evidence of heterogeneity of distribution of particular elements within the PRE-1 family with one element apparently preferentially located in strong R-bands.

The 3′ terminus of a PRE-1 element consists of a poly-A tail of variable length. Miller and Archibald (1993) also found that $(CA)_n.(GT)_n$ repeats occur at the 3′ termini of PRE-1 in 12% of cases, whereas Alexander *et al.* (1995) found 24% of their randomly isolated $(CA)_n.(GT)_n$ microsatellite loci to be associated with PRE-1. Wilke *et al.* (1994) found about 36% of their $(CA)_n.(GT)_n$ repeats from small insert clones to be associated with PRE-1 with 80% of the $(CA)_n.(GT)_n$ repeats occurring at the 3′ end of the SINE. As described elsewhere, both the $(A)_n.(T)_n$ and $(CA)_n.(GT)_n$ repeats provide a valuable source of genetic markers. The consensus PRE-1 element contains internal imperfect direct repeats of about 38 nucleotides and an RNA polymerase III split promoter, which is found in all the structural RNAs which are transcribed by RNA polymerase III. Singer *et al.* (1987) found PRE-1 to be transcribed in liver and thymus, but did not test any other tissue. The PRE-1 consensus sequence is homologous to $tRNA^{arg}$ genes (Takahashi *et al.*, 1992), having a similar evolutionary origin to the mouse B2 repeat which is derived from $tRNA^{ser}$ (Daniels and Deininger, 1985).

Recently two new families, ARE-1P (Artiodactyl Repetitive Element-1Porcine) and ARE-2P were discovered (Alexander *et al.*, 1995) during sequencing of clones containing AC_n microsatellite sequences. Related AREs were subsequently found in cattle. Alexander *et al.* (1995) estimated the porcine haploid copy number of ARE-1P as about 3.6×10^4 and about 2.4×10^4 for ARE-2P, with 8×10^3 copies apparently being dimers of ARE-1P and ARE-2P. These dimeric elements generally enclose an $(CA)_n.(GT)_n$ microsatellite repeat. True stretches of polyadenylation were not observed in all ARE-1P and ARE-2P elements, although the high adenine content implies that a poly(A) tail may have been present but has degenerated. Direct repeats within the ARE SINEs were absent, in contrast to PRE-1.

The evolutionary origin of the ARE SINEs is not immediately apparent, although the tRNAs and 7Sl RNA can be ruled out. The ARE SINE consensus sequences lack an RNA polymerase III split promoter implying that they are transcribed by RNA polymerase II (Alexander *et al.*, 1995), which transcribes protein coding genes and the small nuclear ribonucleoprotein genes.

The PRE-1 family is absent from the Bovidae and more distantly related lineages such as mice and humans (Takahashi *et al.*, 1992). Yasue and Wada

(1996) have shown that PRE-1 is present in wart hogs and the collared peccary at about the same level as in domestic pigs. This specificity to the Suidae implies that it originated after their divergence from the ruminants. By examining the differences among 22 PRE-1 sequences, Yasue and Wada (1998) were able to estimate the time of divergence of these elements from an ancestral element as about 4.3 million years before present, setting an upper limit on the age of these three species. By contrast, the presence of ARE SINEs in pigs, cows and possibly sheep implies that these sequences were undergoing retroposition before the radiation of Suidae and Bovidae (Alexander *et al.*, 1995) and thus are much older than PRE-1.

Processed pseudogenes. Pseudogenes are genomic elements which are non-functional but which have recognizable similarity to functional genes. Processed pseudogenes arise from reverse transcription of mRNA followed by integration of the cDNA copy. Processed pseudogenes are smaller than their parental genes since they lack introns, which are spliced out during transcription, but they do possess post-transcriptionally acquired poly(A) tails. They also lack the 5′ and 3′ regulatory sequences necessary for the controlled expression of the parental genomic copy of the gene, including promoter elements. Other than by coincidentally acquiring a new promoter, processed pseudogenes are not transcribed. Freed of adaptive constraints, they decay relatively rapidly due to accumulation of mutations.

To date, Harbitz *et al.* (1993) have described the sole porcine example of a processed pseudogene, which is derived from the glucosephosphate isomerase locus. This is obviously an ancient element, since it contains 181 transitions, 78 transversions, 11 deletions and 1 insertion, and has several stop codons relative to the functional cDNA, with which it retains only 83% nucleotide identity. Assuming a neutral mutation rate of 0.7–1.0% per nucleotide per million years, Harbitz *et al.* (1993) estimated that the pseudogene arose 20 MYA. The pseudogene has been truncated at the 3' end and lacks the poly(A) tail which typifies most processed pseudogenes. Only a single chromosomal site has been found for this *GPI* pseudogene. However given the ubiquity of pseudogenes in other species and the fact that 14 processed pseudogenes of the argininosuccinate synthetase locus have been mapped to many autosomes and both the X and Y chromosomal location in humans (Su *et al.*, 1984), one can predict that more single copy and multicopy processed pseudogenes will be found in the porcine genome.

Minisatellites

Minisatellites are tandemly repeated elements, with the tandem repeat groups dispersed throughout the genome. Together with microsatellites, they contradict the artificial dichotomy of repeats into tandem versus dispersed classes. Minisatellites have repeating units of 9–64 or more nucleotides and were first described in humans (Jeffreys *et al.*, 1985) but have subsequently been discovered in numerous other species. The number of core repeat units varies enormously at particular sites, providing a useful source of genetic markers, whose detection and application are described elsewhere in this volume. Because of this hypervariability in tandem copy number, minisatellites are also called VNTRs (variable number tandem repeats) (Nakamura *et al.*, 1987).

Coppieters *et al.* (1990) discovered a porcine minisatellite, consisting of 15–18 bp repeat units, while attempting to clone the serum protein PO2 locus. The core sequence showed a high degree of similarity to the minisatellite core sequence found by Jeffreys *et al.* (1985) in humans. When used in low stringency hybridization, the clone containing the core repeat units detected a highly variable fingerprint pattern in pigs, proving multiple locations of the element in the genome. It also detected a variable fingerprint pattern in horses and rabbits.

Davies *et al.* (1992) subsequently described a minisatellite located within the porcine glucosephosphate isomerase locus, with a repeating unit of 39 bp (AGGACCCAGGGTCATGTACAGGTAGGCAGAGCTGGTCTG), flanked by divergent repeats of 53–54 bp. The core unit had at least 14 perfect tandem copies. The repeat unit has no homology with other tandem repeat sequences. The minisatellite apparently has but a single genomic location, since a fingerprint pattern was not obtained even under conditions of low hybridization stringency.

Signer *et al.* (1994) reported a porcine minisatellite isolated from a porcine genomic library using the human core sequence of Jeffreys *et al.* (1985) as probe. The porcine minisatellite hybridized to multiple genomic locations under conditions of low to intermediate stringency, but interpretable single locus banding patterns could be obtained by high stringency hybridization and washing. The element was found by linkage analysis and *in situ* hybridization to map to the pseudoautosomal region of the X and Y chromosomes. No estimation was made of the size of the core unit, variation in number of the repeat units or number of genomic sites of related elements, although allelic variants ranged from 3 to 18 kb.

Coppieters *et al.* (1994) isolated a porcine minisatellite using a CAC/CAT triplet probe. Several clones were independently isolated, which contained the same 30 bp repeat unit (consensus GATGAGGATGGGGGATTGGAGATG-GATGGA). The minisatellite maps to the telomeric region of chromosome 14q. It was estimated that there are more than 100 tandemly repeated units.

Despite their hyperpolymorphism, minisatellites have fallen into disfavour as genetic markers in recent years, mainly because they are not amenable to PCR, due to the large repeat size and high repeat number. Their preferential location in telomeric regions of chromosomes (Royle *et al.*, 1988) was another reason for abandoning them as general genetic markers, but they are likely to have a resurgence in popularity as genetic maps become more dense and the search for telomeric markers becomes more directed.

Microsatellites

Like minisatellites, microsatellites are VNTRs. However they have the advantage of small repeat units (one to six nucleotides) and reasonably small copy number which makes them very suitable for PCR amplification. The extremely high number of locations, which appear to be randomly located throughout the genome, makes them ideal genetic markers and the linkage maps of most mammalian species, including the pig, have been built largely of microsatellite markers.

Certain categories of microsatellite repeats predominate in mammalian genomes, with $(AC)_n$ repeats being most widely exploited for mapping in mammals. Moran (1993) has surveyed the incidence of all possible classes of microsatellite repeats in the pig. He systematically searched for repeat clusters in excess of about 20 nucleotides for mononucleotide repeats through to tetranucleotides for all porcine sequences lodged in the Genbank database. The interpretation of the results is complicated by the fact that microsatellites were found in both cDNA and genomic DNA sequences of the approximately 181 nuclear genes represented at that time. Nevertheless, an approximate feel for the range and distribution of types of repeats could be gained. Eight $(A)_n.(T)_n$ mononucleotides and one $(C)_n.(G)_n$ repeat were found. For the dinucleotides, seven $(CA)_n.(GT)_n$ repeats and three $(GA)_n.(CT)_n$ were recognized. Five trinucleotides representing only four of the 20 possible types were found. Finally nine tetranucleotides, representing only six of 60 possible categories were found. The distribution of the elements within each category was 27.3% mononucleotides, 30.3% dinucleotides, 15.1% trinucleotides and 27.3% tetranucleotides. The $(A)_n.(T)_n$ mononucleotides at 24% and the $(CA)_n.(GT)_n$ dinucleotides at 21% were by far the most common categories of specific repeats found. Ellegren (1993b) has also searched the DNA sequence databases for mononucleotide and dinucleotide repeats only, recording the occurrence of runs of ten or more nucleotides. He also found $(A)_n.(T)_n$ and $(CA)_n.(GT)_n$ repeats to predominate, but found the $(A)_n.(T)_n$ repeats to be about five times more common than $(CA)_n.(GT)_n$ dinucleotides. Using data from genomic sequences only, he estimated that $(A)_n.(T)_n$ mononucleotides occurred about once every 7 kb and $(CA)_n.(GT)_n$ repeats occurred about every 30 kb, implying 400,000 copies of $(A)_n.(T)_n$ repeats and 100,000 copies of $(CA)_n.(GT)_n$ in the porcine genome. Ellegren (1993b) found that one quarter of the $(A)_n.(T)_n$ repeats were associated with PRE-1 SINEs, implying perhaps that quite a few porcine SINEs remain to be discovered. Neither Moran (1993) nor Ellegren (1993b) reported any $(AT)_n.(TA)_n$ repeats although Moran later identified an $(AT)_n.(TA)_n$ repeat associated with a PRE-1 SINE in the 3′ UTR of the soluble angiotensin binding protein gene, which enabled Zhang *et al.* (1995) to map this locus.

Wintero *et al.* (1992) have estimated that there are 65,000–100,000 $(CA)_n.(GT)_n$ repeat loci in the porcine genome, from the proportion of positive clones in a genomic library. Using *in situ* hybridization, they showed the chromosomal distribution of these repeats to be more or less uniform across all chromosomes, except that they were under-represented in the centromeric regions, the q arm of the Y chromosome, the telomere, 16q21, which is a C-band-positive interstitial site, and the NOR region on 10p. Thus $(CA)_n.(GT)_n$ microsatellite repeats seem to be less common in regions containing other categories of repeated DNA.

The origin of $(A)_n.(T)_n$ repeats in mammalian genomes can be easily explained in terms of retroposition events, which incorporate poly(A) tails into the chromosomal DNA (Ellegren, 1993a), although it is far from certain that all $(A)_n.(T)_n$ mononucleotides are generated in this way. The gradual decay of poly(A) tails can explain the origin of at least some of the less common related repeats found, and clearly is responsible for the $(AT)_n.(TA)_n$ repeat mapped by Zhang *et al.* (1995). However when it comes to the more common dinucleotide categories, such as $(CA)_n.(GT)_n$ and $(GA)_n.(CT)_n$, it is unclear whether this generalization can be extended further. In support of the evolution of $(CA)_n.(GT)_n$ repeats from degenerate poly(A) tails is the frequent observation by many groups of the contiguity in the porcine genome of the SINE element, PRE-1, and $(CA)_n.(GT)_n$ repeats as mentioned previously. Indeed, Alexander *et al.* (1995) discovered two new elements, ARE-1P and ARE-2P in the process of examining the sequences flanking their other $(CA)_n.(GT)_n$ sequences finding them present in 6% of cases. Assuming these are degenerate poly(A) tails, this means that either there is a very large number of heterogeneous uncharacterized SINE elements, for which no consensus sequence has been derived, or alternatively that the other 70% of $(CA)_n.(GT)_n$ repeats have evolved by some other mechanism. As replication slippage is widely accepted as the mechanism for generation of diversity of microsatellite length, it is conceivable that any non-coding sequence containing two or more dinucleotide repeats is capable, in theory, of growing via this mechanism to the size where it would be recognized as a microsatellite locus. However, the lesser propensity of small repeats to generate new alleles would argue against this hypothesis.

Telomeric repeats

From the early days of studies of chromosome behaviour, it was observed that the ends or telomeres of chromosomes had special properties, which protected them from the degradation and fusion seen with recently broken chromosome ends. It was also realized that conventional DNA polymerases, which require priming with a short RNA fragment, would be incapable of synthesizing DNA right to the ends of discontinuously growing strands, implying that there would be a gradual shortening of the chromosomal ends during chromosome duplication. It is now recognized that mammalian chromosomes are capped by arrays of the hexamer, $(TTAGGG)_n$ (Blackburn, 1991), which can be repeated for up to 10–15 kb in humans and for 100–150 kb in rodents. This repeat is very rare elsewhere in the genome. Interstitial sites generally are points of evolutionary chromosomal rearrangement, such as the telomere to telomere fusion which gave rise to human chromosome 2 (Ijdo *et al.*, 1991) and the interstitial site near the centromere of mouse chromosome 6 (Yen *et al.*, 1995).

In germline cells, the telomere repeats are synthesized by a totally distinct mechanism from normal genomic DNA, using a ribonucleoprotein reverse transcriptase enzyme called telomerase (Morin, 1989). Telomerase contains RNA complementary to the telomeric repeats which it synthesizes and thus is not dependent on the telomeric DNA template. Telomerase is not expressed in somatic cells in mammals. Consequently the duplication of chromosome ends by conventional polymerases will lead to a reduction in the number of copies of the telomeric repeats and it has been observed that the number of copies of telomeric repeats decreases as an animal ages.

Meyne *et al.* (1989) demonstrated the conservation of the $(TTAGGG)_n$ telomeric sequence among 91 species of vertebrates, including fish, amphibians, reptiles, birds and mammals by *in situ* hybridization of the repeat to the telomeres of all chromosomes in all the species examined, although the pig was not among the 70 mammalian species tested. Recently Gu *et al.* (1996) have demonstrated using primed *in situ* synthesis (PRINS) that this sequence resides at the telomeres of all porcine chromosomes. They also found an interstitial site at chromosome 6q21–q22, at a point where there is a disjunction in the syntenic relationship between pig 6 and human 1p and 19q. This site of hybridization gradually disappeared with increasingly stringent conditions of hybridization of the primer, implying that the internal site is non-functional and has undergone gradual mutational decay. Gu *et al.* (1996) interpreted this interstitial position as evidence for a tandem fusion in the origin of porcine chromosome 6.

A future challenge for gene mappers is to find genetic markers very close to the ends of chromosomes. Strategies to exploit variation in the telomere sequence have been developed to map telomere markers in the mouse (Elliott and Yen, 1991) which may also be applicable in the pig. However the gradual decrease in telomere repeat number in somatic cells with increasing number of cell cycles makes this a difficult task.

Expressed repetitive sequences

Mammals, like other organisms, contain some repeated sequences which are expressed and have important cellular functions. In some cases, the repetition of the gene is necessary to ensure adequate levels of expression of the gene product. In other cases, duplication of the gene locus has permitted evolutionary experimentation as the duplicate is able to diverge and evolve new functions.

Ribosomal RNA

Ribosomal RNA is a structural component of ribosomes and thus comprises a vital component of the cellular protein synthesis apparatus. In eukaryotes, the ribosome consists of a small subunit (40S), containing the 18S rRNA and a large subunit (60S) containing 28S, 5.8S and 5S rRNA subunits. The 28S, 18S and 5.8S genes are clustered and expressed coordinately. Cells must express rRNAs at extremely high levels to meet the cellular demand for protein synthesis and have a special enzyme, RNA polymerase I, dedicated specifically to this task. rRNA generally comprises about 80% of the RNA found in a cell. Multiple copies of the rRNA genes expedite transcription. In humans there are 150–300 copies of the 18S/5.8S/28S rRNA cluster distributed between five autosomes, with the number of copies varying between individuals and chromosomal sites. The positions of the tandemly repetitive ribosomal gene clusters, which are separated by untranscribed spacers, are cytologically visible at features called secondary constrictions or nucleolus organizer regions (NORs). Their expression is controlled by a promoter which directs transcription of all members of the cluster as a 45S pre-rRNA transcript, which is processed into 18S, 5.8S and 28S subunits.

In the pig, three sites of ribosomal RNA 18S/5.8S/28S cluster, called RNR1, RNR2 and RNR3, are located on chromosome 8p1.2, 10p1.2–1.3 and 16q2.1 respectively (Czaker and Mayr, 1980; Miyake *et al*., 1988; Popescu *et al*., 1989; Bosma *et al*., 1991).

The 5S (about 120 bp) ribosomal RNA genes are also tandemly repeated and separated by untranscribed spacers. However, they are transcribed by RNA polymerase III and map to quite different locations from the 28S/18S/5.8S cluster in humans, having a major site on human chromosome 1q42.11–42.13 with a minor site nearby at 1q31 (Lomholt *et al*., 1995a). Likewise the 5S rRNA genes in pigs map to 14q23 (Lomholt *et al*., 1995b), nowhere near the other ribosomal RNA genes.

Ling and Arnheim (1994) have characterized the promoter region of the 18S/5.8S/28S cluster in the pig and compared it with equivalent mammalian promoters. This revealed conservation of the regulatory regions found in other mammalian promoters. The only unusual feature was that there was a C nucleotide 16 bp upstream of the transcription start point in the pig, whereas G is conserved in all other eukaryotic promoters.

tRNAs

Like 5S rRNA, tRNAs are transcribed by RNA polymerase III. As for rRNAs, the very high cellular demand for tRNAs can be met only by transcription from multiple copies of the genes. Generally there are ten to several hundred genes for each tRNA. Interestingly the repeat clusters contain genes for several different tRNAs. For example, in the rat, tRNAAsp, tRNAGly and tRNAGlu form a cluster, which is tandemly repeated about ten times. The members of the tandem repeats are not identical and some may be pseudogenes. Very little is known about the tRNA genes of the pig, but it can be safely assumed that they follow a typical mammalian pattern of dispersed distribution and duplication.

Other multigene families

In mammals, there are many protein-encoding gene families in which the members show developmental or tissue specificity of expression. For example, the globin gene family consists of members coordinately expressed at embryonic, fetal and adult stages and includes pseudogenes. The α-actin family consists of members whose expression is specific for skeletal muscle or for cardiac muscle. In some cases, the members of these gene families are located very closely in the genome, whereas in other more heterogeneous gene families, the members may be dispersed. The individual members of these gene families are generally structurally similar to the single copy genes discussed later.

Functional genes. In the pig, a number of multigene families have been recognized. For example, the tumour necrosis factor alpha (*TNF*α) and tumour necrosis factor beta (*TNF*β) loci are located within the major histocompatibility complex and separated by a distance of only 2.5 kb (Chardon *et al.*, 1991). The *ATP1A1* and *ATP1B1*, the alpha 1 and beta 1 polypetides of the Na^+, K^+-ATPase also comprise dispersed members of a putative gene family on chromosome 4 (Lahbib-Mansais *et al.*, 1993) which is conserved in humans and mice.

The casein family of milk proteins has been thoroughly characterized in the pig and retains a similar structure to that found in other mammals. Four members, namely αS1 (Alexander and Beattie, 1992a), αS2 (Alexander *et al.*, 1992), β (Alexander and Beattie, 1992b) and κ (Levine *et al.*, 1992), have been mapped to chromosome 8 and the first three at least have been shown to be very tightly linked (Archibald *et al.*, 1994).

Two very well characterized porcine multigene families, namely the protease inhibitors and the immunoglobulin heavy chain gene families, coincidentally map near each other on porcine chromosome 7. *PI1*, *PO1A*, *PO1B*, *PI2*, *PI3* and *PI4* form a tight linkage group adjacent to *IGH1*, *IGH2*, *IGH3* and *IGH4* (Gahne and Juneja, 1986; Juneja *et al.*, 1986; Van de Weghe *et al.*, 1987; Vogeli *et al.*, 1987; Stratil *et al.*, 1990; Cizova *et al.*, 1993) on 7q24–26 (Gu *et al.*, 1994; Musilova *et al.*, 1995). *PO1A* and *PO1B*, originally called postalbumin 1A and 1B respectively, appear to be cysteine protease inhibitors (Gahne and Juneja, 1985). *PI1*, *PI2*, *PI3* and *PI4* are serine protease inhibitors, with different loci and even alleles at particular loci being identified as trypsin or chymotrypsin inhibitors. Stratil *et al.* (1990) noted that there are many weakly staining alleles and null alleles at the *PI3* locus, which is consistent with the evolutionary expectation within a multigene family containing redundant members.

As more and more cDNA clones from different tissues are being sequenced, members of porcine gene families are being recognized by their sequence similarity to homologues in other mammalian multigene families, particularly in humans. For example, Tuggle and Schmitz (1994) have sequenced fragments of clones from a porcine skeletal muscle cDNA library and have identified the porcine equivalents of human skeletal alpha actin and rat alpha7-integrin. Interestingly in the case of alpha-actinin2, the porcine skeletal clone appeared to be slightly more similar to the human cardiac family member than to its human skeletal homologue.

Pseudogenes. Pseudogenes can arise from duplication of a locus followed by an inactivating mutation such as a frameshift, the incorporation of a stop codon or loss or mutation of promoter elements. Like the processed pseudogenes already described, these duplicate elements are no longer subject to adaptive evolutionary constraint and diverge relatively rapidly from the parental sequences. Such divergent, non-functional loci are frequently recognized as 'degenerate' members of gene families.

Vage *et al.* (1994) have reported a pseudogene within the porcine major histocompatibility complex. The MHC class II *DRB* locus was found to have an expressed version, called *DRB1* and a pseudolocus, called *DRB2*, present in about 60% of chromosomes. *DRB2* has a single nucleotide deletion in codon 54, which would cause a frameshift and total loss of function. Ironically the strong selection pressure for MHC diversity has meant that the functional *DRB1* was found to be much more variable than the pseudogene. Brunsberg *et al.* (1995) have confirmed the findings of Vage *et al.* (1994) and have discovered an additional *DRB* pseudogene. Both were detected using primers which span intron–exon boundaries. Thus the pseudogene is clearly a degenerate member of a gene family. Similarly Mege *et al.* (1991) have described a pseudogene within the interferon family, which they called psi IFN-alpha II-1, which has multiple frameshift mutations. Clearly the evolutionary experimentation following gene duplication can lead, in some cases to, functional novelty but is probably more likely to lead to loss of function and decay of the gene.

Single Copy DNA

Many single copy protein-encoding genes have now been cloned and characterized in the pig and it is well beyond the scope of this review to attempt to cover all of them. Some will be discussed in other chapters. Porcine genes appear in all respects quite typical for mammalian genes. The porcine *calcium release channel (CRC)* locus, which was the subject of an intensive search by several laboratories, has an extremely large message of over 15 kb (Fujii *et al.*, 1991), but so also do the human and rabbit homologues (Zorzato *et al.*, 1990). The size of the entire genomic copy of this locus including introns has not been determined but is likely to be about 120 kb given that 25,780 nucleotides of genomic DNA are required to specify only 3305 nucleotides of the porcine message from nucleotide 4624 to 7929 (Leeb *et al.*, 1993). Protein encoding sequences account for only 13% of the genomic DNA in this region, again emphasizing the large proportion of the genome which is non-coding and of unknown or possibly no function, even in the vicinity of coding genes.

Gene Structure and Function

Conventional genes

The porcine growth hormone gene demonstrates the structural and functional features of a typical protein-encoding gene. The porcine gene retains extensive sequence similarity with the human, rat and bovine homologues, even in the non-coding promoter, intronic and untranslated regions (Vize and Wells, 1987). It is instructive to review the regulatory and structural features of this locus in the sequence of Vize and Wells (1987). Although many of the sequence elements which regulate the expression of this locus have not been identified, regulatory features held in common with many other mammalian genes can be seen. Starting at position −30 from the cap site is the sequence TATAAAA, the Goldberg–Hogness box, required for correct positioning of RNA polymerase II and initiation of transcription. An A nucleotide at position +1, the cap site, marks the start of the messenger RNA, which is capped by addition of 7-methylguanylate in an unusual 5′–5′ linkage by a capping enzyme. This capping protects the RNA from degradation. The initiating methionine codon, ATG, which marks the start of translation of all eukaryotic genes, commences at position +64. Thus the porcine growth hormone message has a 5′-UTR (untranslated region) of 63 nucleotides. Only 10 coding nucleotides occur before the start of the first intron. It is virtually universal for the the splice donor site, that is the first two nucleotides of an intron, to be GT, but here the 242 bp intron commences with GC, which is very unusual. However, intron 1 terminates with a typical splice acceptor dinucleotide, AG. Exon 2 is 162 nucleotides long. Intron 2, which is 210 nucleotides long, has typical splice donor and acceptor sites, namely GT and AG respectively, as do all the other introns in this gene. Exon 3 is 117 nucleotides in length and is followed by intron 3 of 197 nucleotides. Exon 4 is 162 nucleotides in length followed by intron 4 of 278 nucleotides. Exon five consists of 198 nucleotides which specify amino acids followed by the translation STOP codon, TAG. Following this coding sequence is a 3′-UTR of 102 nucleotides and 80 nucleotides after the end of the stop codon is the sequence AATAAA, which is a polyadenylation signal. This directs an enzyme called poly(A) polymerase to add a poly(A) tail to the end of the message at a point 16 nucleotides downstream from the end of the signal. The addition of the poly(A) tail is not template determined and is essential for the stability of the mRNA. Of the 1745 nucleotides from the cap site to the start of the poly(A) tail, there are 1092 nucleotides of non-coding DNA in UTRs and introns, comprising 63% of the sequence. Although some of this non-coding DNA is functional, for example the polyadenylation signal, it is worth emphasizing again that, as for other loci in mammals, only a relatively small proportion of this locus specifies amino acids, and this proportion becomes even smaller if all sequences from the start of the 5′ to the end of the 3′ regulatory sequences are considered.

Of course there are other 5′ and 3′ regulatory sequences which are essential for the tissue specific and temporally regulated expression of any gene like growth hormone. In general, these include positive and negative proximal promoter elements, which are binding sites for transcription factors, which either assist or hinder directly or indirectly the binding of RNA polymerase II. Additionally there are enhancer elements, which are large sequences, often quite distant either upstream or downstream from the gene, which increase the level of transcription, and silencer elements, which are short, often repeated, sequences which down-regulate expression.

Somatically rearranged genes

Unlike conventional genes, the immunoglobulin and T cell receptor loci show extensive patterns of somatic rearrangement as a normal part of the diversity-generating mechanism, which enables the huge range of immune responses found in mammals. In essence, gene families of constituent parts at these loci are assembled by DNA breakage and joining events into more or less unique combinations within committed lineages of immune cells. The control of transcription of these newly assembled functional units and the processing of transcripts occurs as for conventional genes.

The antibody molecules produced by B cells are comprised of light chains, fitting into two major categories, λ and κ, and heavy chains, consisting of five major types, α, δ, ε, γ and μ, which correspond to the five classes of antibodies, IgA, IgD, IgE, IgG and IgM. The light chain families map to two locations in human and mouse, whereas the five heavy chain 'loci' map to a single location which, as mentioned previously, has been identified in the pig. To make a λ or κ light chain, a B cell must undergo a process of combinatorial (V–J) joining in which a unique combination of four exons is assembled by chromosomal breakage and joining events. These exons are called L (leader) and V (variable), of which there are several hundred adjacent but different pairs, J (joining) of which there are four or five variants and C (constant) of which there is normally one copy. Many hundreds of different light chains can be specified in this way. Heavy chain diversity is generated in a similar but slightly more complex way. As for light chains, there are about 250 different L and V exon pairs, followed by about 12 D (diversity) exons, followed by about four J exons, followed by multiexonic constant regions corresponding to all of the major classes and subclasses of antibodies. The construction of a functional heavy chain gene requires breakage and rejoining events to assemble a unique combination of L–V–D–J–C coding regions via two events, namely D–J joining followed by V–DJ joining. Again many hundreds of different heavy chains can be produced in this way, which in combination with the many hundreds of different possible light chains yields great antibody diversity.

In T cells, which are responsible for cell-mediated immune responses, the antigen receptors are composed of α, β, γ and δ subunits, which are encoded at different chromosomal locations. The normal adult receptor is an αβ dimer in humans and mouse, but γδ dimers are more prominent in pigs and are found on up to half the T cells. The functional copies of these genes are assembled by a joining mechanism similar to immunoglobulin chains, with α and γ chains having V and J segments and β and δ having VDJ segments in addition to the constant segments. Thus both for antibodies and for T cell antigen receptors, somatic rearrangements are a normal part of the cycle of expression of these gene products, necessitated by the enormous diversity of products required and the impossibility of encoding all possible products in a conventional fashion.

Although most characterization of these immune gene systems has been performed on mouse and human, many elements of the process have been verified in pigs. Lammers *et al.* (1991) have sequenced most of the V and all of the J and C regions of the porcine κ and λ light chain genes and have found that the J region is more closely conserved across species than the constant regions. The porcine κ light chain is unusual in that the constant region terminates with the dipeptide Glu–Ala after the Cys residue that terminates all other mammalian light chains. Kacskovics *et al.* (1994) have identified five putative subclasses of porcine IgG loci (1, 2a, 2b, 3 and 4) on the basis of C region sequences from a single animal. Brown and Butler (1994) sequenced the constant region of the porcine IgA heavy chain and have shown that unlike human and mouse which have two or more loci, the pig has only a single Cα gene, with the sequence and structure conserved relative to other species particularly towards the carboxy terminus of the gene product. Bosch *et al.* (1992) have similarly characterized a porcine clone for the μ heavy chain. Sun *et al.* (1994) have cloned and sequenced 65 H chain V regions expressed with IgG, IgA or IgM C regions. They found that the porcine V regions belonged to a homogeneous family with greater than 80% nucleotide identity among members and that there were probably fewer than 20 copies of the V regions in the pig. Surprisingly the J region was also very homogeneous, suggesting that a single J region is preferentially used. In other mammals, there are three families of V genes with mice having 100 to 1000 copies, humans 100 to 120 copies and rabbits 100 copies. The relatively depauperate V status of pigs raises questions about how they can generate sufficient antibody diversity.

The cDNA clones of the porcine T cell receptor α, β, γ and δ chains have been characterized (Thome *et al.*, 1993; Grimm and Misfeldt, 1994). Thome *et al.* (1993) found evidence for at least three different constant segments for the porcine γ chain family. There is substantial conservation of constant segment coding sequence within the four subunits across pigs, cattle, mice, sheep and humans.

Base Composition and Methylation

As stated previously, genes in humans are found preferentially in GC rich isochores. Also, the GC content of human exons tends to be about 10% higher than in introns. It would be reasonable to expect that porcine coding regions would be more GC rich than non-coding regions. However, no systematic study has been made of G + C content or other aspects of nucleotide composition of porcine genes or genome. A glimpse of the situation in pigs can be obtained from the following data. If one totals the G + C content of 10 of the microsatellite-containing clones (excluding the repetitive microsatellite component) characterized by Robic *et al.* (1994), none of which is known to encode any function, one finds a G + C content of 43% (1607 from 3768 nucleotides). This is consistent with the expectation that porcine non-coding regions would be GC poor. By contrast, the incomplete genomic sequence of the porcine *CRC* locus has a 53% G + C content, rising even higher to 61% for the full cDNA. In the case of the growth hormone locus, the genomic DNA is 61% G + C and the cDNA is 59%. Again the expectation of high GC content in coding DNA is met. However, not all porcine coding genes have a high GC content. The genomic sequence of the relaxin gene, containing about 1 kb of 5′ flanking non-transcribed sequence, about 300 nucleotides of 3′ untranscribed sequence and a single, very large intron of 5495 nucleotides (Haley *et al.*, 1987) is only 36% G + C, and this figure rises to only 40% for the relaxin cDNA.

Variation in nucleotide composition can have more subtle manifestations than simple variation in nucleotide content. The cytosine of the dinucleotide CpG is normally methylated throughout vertebrate genomes as part of a transcription suppression mechanism, except in clusters of this dinucleotide forming so-called CpG islands, which somehow escape methylation and allow expression of the adjacent gene. The methylated CpG is particularly prone to mutation to TpG. In all vertebrates examined, the CpG dinucleotide is about five times less frequent than would be predicted on base composition, because of this mutational pressure.

Number of Expressed Genes in the Pig

No estimate has been made of gene number in the pig. However the generally accepted figure for vertebrates is about 50,000–100,000, regardless of variation in genome size. Bird (1995) has speculated on a saltatory increase in gene number with the origin of the vertebrate lineage about 600 MYA, perhaps associated with the development of a brain and spinal chord, with the result that vertebrates have much higher gene numbers than other eukaryotes. Antequera and Bird (1993) have estimated that there are 45,000 and 37,000 CpG islands in the human and mouse genome respectively. They have also estimated that 56% of human genes and 47% of mouse genes are preceded by CpG islands, providing an overall estimate of gene number in both species of about 80,000. However, if the proportion of genes preceded by CpG islands has been underestimated, the true gene number may be closer to 67,000.

By determining the proportion of 65,397 human ESTs (expressed sequence tags) which correspond to known genes in the Genbank database and then estimating the level of undetected redundancy among this very large sample of ESTs, Fields *et al.* (1994) have estimated human gene number as 64,000. In the best characterized mammal, namely humans, only about 3500 unique coding regions are recognized on the basis of function. Thus in the pig as for other mammals there remains a very large number of genes still to be identified and characterized.

Mitochondrial Genome

Mitochondria are cellular organelles believed to be derived from ancient intracellular bacterial symbionts. In addition to their unique role in synthesis of ATP (oxidative phosphorylation), these organelles have a very peculiar genetic system. They have an autonomously replicating DNA genome, of about 16.5 kb in mammals, but the mechanism of replication is quite unlike that for nuclear DNA. Replication commences with initiation of heavy (H) strand synthesis at a specific origin and results in formation of a displacement (D) loop with a newly synthesized H-strand of about 680 bases, with very few H-strands achieving full length. Initiation of light (L) strand synthesis occurs from a specific origin but only after this region has been exposed by H-strand synthesis. Transcription is initiated from promoters situated within the D-loop region, with the primary transcripts processed to give 12S and 16S rRNAs, tRNAs and a number of mRNAs which are not capped but are polyadenylated. Virtually all the DNA in the mitochondrial genome is functional, specifying rRNA, tRNA or mRNA with very few and in some cases no non-coding nucleotides between adjacent genes. The D-loop does not encode any known function.

Most proteins in the mitochondria are nuclear encoded but some are mitochondrially encoded and translated by a separate system in which the ribosomal RNAs and transfer RNAs are also mitochondrially encoded. In the mammalian mitochondria, proteins are synthesized using a unique mitochondrial code, which employs different initiation and stop codons from the universal code, and messages are decoded by a different mechanism from that used by nuclear genes. This decoding mechanism uses a minimal set of only 22 tRNAs with different wobble anticodons from those employed in nuclear protein synthesis. Formyl-methionine is used to initiate a peptide as in prokaryotes. However, the mammalian mitochondrial rRNA has only low homology with prokaryotic rRNA.

The complete mitochondrial genome has been sequenced for several mammalian species, including human (Anderson *et al.*, 1981), cattle (Anderson *et al.*, 1982), mouse (Bibb *et al.*, 1981) and rat (Gadaleta *et al.*, 1989). For the pig, detailed analysis of mitochondrial sequences is restricted to the cytochrome b gene (Irwin *et al.*, 1991; Honeycutt *et al.*, 1995) and the cytochrome c oxidase II gene (Honeycutt *et al.*, 1995), where the porcine sequences were generated as part of a large phylogenetic study. Additionally, the D-loop region (Ghivizzani *et al.*, 1993) of the pig has been extensively characterized. The porcine D-loop is much larger than in the cow and surprisingly displays greater organizational similarity to the mouse and human D-loop than to the bovine D-loop. Within the porcine D-loop, there is a ten-nucleotide repeat, with copy number ranging from 14 to 29, giving rise to the highest level of mitochondrial heteroplasmy documented in mammals, apparently as a result of replication slippage in the tandem repeat.

Conclusions

This review of the molecular genetics of the pig has aimed to cover as many studies as possible which are specific to the pig and to put this in the context of genomic structure and function common to all mammals. It is very clear that the pig, like all mammals, has a huge proportion of its genome to which no function can be ascribed. Parts of the genome, such as the pseudogenes, are clearly non-functional, being evolutionary relicts which will eventually decay into unrecognizable random sequence. Other parts, such as the microsatellites, have no obvious function but have revolutionized genetic studies in the pig and other species, as they have provided a tool for accessing parts of the genome previously unrecognizable.

However, even that very small percentage of the genome, estimated to be 2–3% which represents coding sequences or the 10% which is likely to be functional, represents a huge challenge for students of genetics. The number of genes already characterized in the pig can be numbered in the hundreds, yet there are as many as 100,000 to be identified and functionally characterized. Of course each gene can occur in a huge variety of allelic variants, many of which are silent but some of which are responsible for the variation in economically important traits in pigs, so the task of identification and exploitation of useful genes is increased accordingly. Huge strides have been made in the genetic and molecular understanding of the genome of the pig and other mammals, but the job of full genomic characterization has really just begun.

Acknowledgements

I wish to thank Lee Alexander, Allan Lohe, David Grimm, Joan Lunney, Dinah Singer, Feng Gu, Leif Andersson, Esther Signer, Marilyn Mennotti-Raymond and Yizhou Chen for helpful information and advice. I also wish to thank Frank Nicholas and Leif Andersson for valuable comments on the manuscript.

References

Akamatsu, M., Chen, Z., Dziuk, P.J. and McGraw, R. (1989) A highly repeated sequence in the domestic pig: a gender-neutral probe. *Nucleic Acids Research* **17**, 10120.

Alexander, L.J. and Beattie, C.W. (1992a) The sequence of porcine α_{s1}-casein cDNA: evidence for protein variants generated by alternate RNA splicing. *Animal Genetics* 23, 283–288.

Alexander, L.J. and Beattie, C.W. (1992b) The sequence of porcine β-casein cDNA. *Animal Genetics* 23, 369–371.

Alexander, L.J., Das Gupta, N.A. and Beattie, C.W. (1992) The sequence of porcine α_{s2}-casein cDNA. *Animal Genetics* 23, 365–367.

Alexander, L.J., Rohrer, G.A., Stone, R.T. and Beattie, C.W. (1995) Porcine SINE-associated microsatellite markers: evidence for new artidactyl SINEs. *Mammalian Genome* 6, 464–469.

Anderson, S., Bankier, A.T., Barrell, B.G., de Bruijn, M.H.L., Coulson, A.R., Drouin, J., Eperon, I.C., Nierlich, D.P., Roe, B.A., Sanger, F., Schreier, P.H., Smith, A.J.H., Staden, R. and Young, I.G. (1981) Sequence and organisation of the human mitochondrial genome. *Nature* 290, 457–465.

Anderson, S., de Bruijn, M.H.L., Coulson, A.R., Eperon, I.C., Sanger, F. and Young, I.G. (1982) Complete sequence of bovine mitochondrial DNA: conserved features of the mammalian mitochondrial genome. *Journal of Molecular Biology* 156, 683–717.

Antequera, F and Bird, A. (1993) Number of CpG islands and genes in human and mouse. *Proceedings of the National Academy of Sciences USA* 90, 11995–11999.

Archibald, A.L., Couperwhite, S., Haley, C.S., Beattie, C.W. and Alexander, L.J. (1994) RFLP and linkage analysis of the porcine casein loci – CASAS1, CASAS2, CASB and CASK. *Animal Genetics* 25, 349–351.

Auge-Gouillou, C., Bigot, Y., Pollet, N., Hamelin, M.-H., Meunier-Rotival, M. and Periquet, G. (1995) Human and other mammalian genomes contain transposons of the *mariner* family. *FEBS Letters* 368, 541–546.

Benveniste, R.E. and Todaro, G.J. (1975) Evolution of type C viral genes. Preservation of ancestral murine type C viral sequences in pig cellular DNA. *Proceedings of the National Academy of Sciences, USA* 72, 4090–4094.

Bernardi, G., Olofsson, B., Filipski, J., Zerial, M., Salinas, J., Cuny, G., Meunier-Rotival, M. and Rodier, F. (1985) The mosaic genome of warm-blooded vertebrates. *Science* 228, 953–958.

Bibb, M.J., Van Etten, R.A., Wright, C.T., Walberg, M.W. and Clayton, D.A. (1981) Sequence and gene organisation of the mouse mitochondrial DNA. *Cell* 26, 167–180.

Bird, A.P. (1995) Gene number, noise reduction and biological complexity. *Trends in Genetics* 11, 94–100.

Blackburn, E. (1991) Structure and function of telomeres. *Nature* 350, 569–573.

Bosch, B.L., Beaman, K.D. and Kim, Y.B. (1992) Characterisation of a cDNA clone encoding for a porcine immunoglobulin μ chain. *Developmental and Comparative Immunology* 16, 329–337.

Bosma, A.A., de Haan, N.A. and Mellink, C.H. (1991) Ribosomal RNA genes located on chromosome 16 of the domestic pig. *Cytogenetics and Cell Genetics* 58, 2124.

Brenner, S., Elgar, G., Sandford, R., Macrae, A., Venkatesh, B. and Aparicio, S. (1993) Characterisation of the pufferfish (*Fugu*) genome as a compact model vertebrate genome. *Nature* 366, 265–268.

Brown, W.R. and Butler, J.E. (1994) Characterisation of a Cα gene of swine. *Molecular Immunology* 31, 633–642.

Brunsberg, U., Edfors-lilja, I., Andersson, L. and Gustafsson, K. (1996) Structure and organisation of porcine MHC class II DRB genes: evidence for genetic exchange between class II loci. *Immunogenetics* 44, 1–8.

Chardon, P., Nunes, M., Dezeure, F., Andres-Cara, D. and Vaiman, M. (1991) Mapping and genetic organisation of the TNF genes in the swine MHC. *Immunogenetics* 34, 257–260.

Cizova, D., Gabrisova, E. Stratil, A. and Hojny, J. (1993) An additional locus (PI4) in the protease inhibitor (PI) gene cluster of pigs. *Animal Genetics* 24, 315–318.

Coppieters, W., Van de Weghe, A., Depicker, A., Bouquet, Y. and Van Zeveren, A. (1990) A hypervariable pig DNA fragment. *Animal Genetics* 21, 29–38.

Coppieters, W., Zijlstra, C., Van de Weghe, A., Bosma, A.A., Peelman, L., Van Zeveren, A. and Bouquet, Y. (1994) A porcine minisatelite located on chromosome 14q29. *Mammalian Genome* 5, 591–593.

Czaker, R. and Mayr, B. (1980) Detection of nucleolus organizer regions (NOR) in the chromosomes of the domestic pig (*Sus scrofa domestica* L.). *Experientia* 36, 1356–1357.

Daniels, G.R. and Deininger, P.L. (1985) Repeat sequence families derived from mammalian tRNA genes. *Nature* 317, 819–822.

Davies, W., Kran, S., Kristensen, T. and Harbitz, I. (1992) Characterisation of a porcine variable number tandem repeat sequence specific for the glucosephosphate isomerase locus. *Animal Genetics* 23, 437–441.

Ellegren, H. (1993a) Variable SINE 3' poly(A) sequences: an abundant class of genetic markers in the pig genome. *Mammalian Genome* 4, 429–434.

Ellegren, H. (1993b) Abundant $(A)_n.(T)_n$ mononucleotide repeats in the pig genome: linkage mapping of the porcine APOB, FSA, ALOX12, PEPN and RLN loci. *Animal Genetics* 24, 367–372.

Elliott, R.W. and Yen, C.-H. (1991) DNA variants with telomere probe enable genetic mapping of the ends of mouse chromosomes. *Mammalian Genome* 1, 118–122.

Fields, C., Adams, M.D., White, O. and Venter, C. (1994) How many genes in the human genome? *Nature Genetics* 7, 345–346.

Frengen, E., Thomsen, P., Kristensen, T., Kran, S., Miller, R. and Davies, W. (1991) Porcine SINEs: characterization and use in species-specific amplification. *Genomics* 10, 949–956.

Fujii, J., Otsu, K., Zorzato, F., De Leon, S., Khanna, V.K., Weiler, J.E., O'Brien, P.J. and MacLennan, D.H. (1991) Identification of a mutation in the porcine ryanodine receptor identified with malignant hyperthermia. *Science* 253, 448–451.

Gadaleta, G., Pepe, G., De Candia, G., Quagliariello, C., Sbisa, E. and Saccone, C. (1989) The complete nucleotide sequence of the *Rattus norvegicus* mitochondrial genome: cryptic signals revealed by comparative analysis between vertebrates. *Journal of Molecular Evolution* 28, 497–516.

Gahne, B. and Juneja, R.K. (1986) Extensive polymorphism of four plasma alpha-protease inhibitors in pigs evidence for tight linkage between the structural loci of these inhibitors. *Animal Genetics* 17, 135–137.

Ghivizzani, S.C., Mackay, S.L.D., Madsen, C.S., Laipis, P.J. and Hauswirth, H.H. (1993) Transcribed heteroplasmic repeated sequences in the porcine mitochondrial DNA D-loop region. *Journal of Molecular Evolution* 37, 36–47.

Grimm, D.R. and Misfeldt, M.L. (1994) Partial cloning and sequencing of the gene encoding the porcine T-cell receptor δ-chain constant region. *Gene* 144, 271–275.

Gu, F., Chowdhary, B.P., Johansson, M., Andersson, L. and Gustavsson, I. (1994) Localisation of the IGHG, PRKACB and TNP2 genes in pigs by *in situ* hybridisation. *Mammalian Genome* 5, 195–198.

Gu, F., Hindkjaer, J., Gustavsson, I. and Bolund, L. (1996) A signal of telomeric sequences on porcine chromosome 6q21–q22 detected by primed in situ labeling. *Chromosome Research* 4, 251–252.

Haley, J., Crawford, R., Hudson, P., Scanlon, D., Tregear, G., Shine, J. and Niall, H. (1987) Porcine relaxin. Gene structure and expression. *Journal of Biological Chemistry* 262, 11940–11946.

Harbitz, I., Chowdhary, B., Kran, S. and Davies, W. (1993) Characterisation of a porcine glucosephosphate isomerase-processed psuedogene at chromosome 1q1.6–1.7. *Mammalian Genome* 4, 589–592.

Honeycutt, R.L., Nedbal, M.A., Adkins, R.M. and Janecek, L.J. (1995) Mammalian mitochondrial DNA evolution: a comparison of the cytochrome *b* and cytochrome *c* oxidase II genes. *Journal of Molecular Evolution* 40, 260–272.

Ijdo, J.W., Baldini, A., Ward, D.C., Reeder, S.T. and Wells, R.A. (1991) Origin of human chromosome 2: an ancient telomere–telomere fusion. *Proceedings of the National Academy of Sciences, USA* 88, 9051–9055.

Irwin, D.M., Kocher, T.D. and Wilson, A.C. (1991) Evolution of the cytochrome *b* gene of mammals. *Journal of Molecular Evolution* 32, 128–144.

Jantsch, M., Hamilton, B., Mayr, B. and Schweizer, D. (1990) Meiotic chromosome behaviour reflects levels of sequence divergence in *Sus scrofa domestica* satellite DNA. *Chromosoma* 99, 330–335.

Jeffreys, A.J., Wilson, V. and Thein, S.L. (1985) Hypervariable 'minisatellite' regions in human DNA. *Nature* 314, 67–73.

John, B. and Miklos, G.L.G. (1988) *The Eukaryote Genome in Development and Evolution.* Allen and Unwin, London, pp. 416.

Juneja, R.K., Gahne, B., Rapacz, J. and Hasler-Rapacz, J. (1986) Linkage between the porcine genes encoding immunoglobulin heavy-chain allotypes and some serum alpha-protease inhibitors: a conserved linkage in pig, mouse and human. *Animal Genetics* 17, 225–233.

Kacskovics, I., Sun, J. and Butler, J.E. (1994) Five putative subclasses of swine IgG identified from the cDNA sequences of a single animal. *Journal of Immunology* 159, 3565–3573.

Lahbib-Mansais, Y., Yerle, M., Dalens, M., Chevalet, C. and Gellin, J. (1993) Localisation of pig Na^+, K^+-ATPase and subunit genes to chromosome 4 by radioactive *in situ* hybridization. *Genomics* 15, 91–97.

Lammers, B.M., Beaman, K.D. and Kim, Y.B. (1991) Sequence analysis of porcine immunoglobulin light chain cDNAs. *Molecular Immunology* 28, 877–880.

Leeb, T., Schmolzl, S., Brem, G. and Brenig, B. (1993) Genomic organisation of the porcine skeletal muscle ryanodine receptor (RYR1) gene coding region 4624 to 7929. *Genomics* 18, 349–354.

Le Tissier, P., Stoye, J.P., Takeuchi, Y., Patience, C. and Weiss, R.A. (1997) Two sets of human–tropic pig retrovirus. *Nature* 389, 681–682.

Levine, W.B., Alexander, L.J., Hoganson, G.E. and Beattie, C.W. (1992) Cloning and sequencing of the porcine κ-casein cDNA. *Animal Genetics* 23, 361–363.

Lewin, B. (1997) *Genes VI.* Oxford University Press, Oxford and New York, pp. 1260.

Ling, X. and Arnheim, N. (1994) Cloning and identification of the pig ribosomal gene promoter. *Gene* 150, 375–379.

Lomholt, B., Frederiksen, S., Nederby-Neilsen, J. and Hallenberg, C. (1995a) Additional assignment of the genes for 5S RNA in the human chromosome complement. *Cytogenetics and Cell Genetics* 70, 76–79.

Lomholt, B., Christensen, K., Hallenberg, C. and Frederiksen, S. (1995b) Porcine 5S rRNA genes map to 14q23 revealing syntenic relation to human HSA6- and 7. *Mammalian Genome* 6, 439–441.

McGraw, R., Jacobson, R.J. and Akamatsu, M. (1988) A male-specific repeated DNA sequence in the domestic pig. *Nucleic Acids Research* 16, 10389.

Mege, D., Lefevre, F. and Labonnardiere, C. (1991) The porcine family of interferon-omega: cloning, structural analysis and functional analysis of five related genes. *Journal of Interferon Research* 11, 341–350.

Meyne, J., Ratliff, R.L. and Moysis, R.K. (1989) Conservation of the human telomere sequence $(TTAGGG)_n$ among vertebrates. *Proceedings of the National Academy of Sciences, USA* 86, 7049–7053.

Mileham, A.J., Siggens, K.W. and Plastow, G.S. (1988) Isolation of a porcine male specific DNA sequence. *Nucleic Acids Research* 16, 11842.

Miller, J.R. (1994) Use of porcine interspersed repeat sequences in PCR-mediated genotyping. *Mammalian Genome* 5, 629–632.

Miller, J.R. and Archibald, A.L. (1993) 5′ and 3′ SINE-PCR allows genotyping of pig families without cloning and sequencing steps. *Mammalian Genome* 4, 243–246.

Miller, J.R., Hindkjaer, J. and Thomsen, P.D. (1993) A chromosomal basis for the differential organisation of a porcine centromere-specific repeat. *Cytogenetics and Cell Genetics* 62, 37–41.

Miyake, Y.-I., O'Brien, S.J. and Kaneda, Y. (1988) Regionalization of the rDNA gene on pig chromosome 10 by *in situ* hybridisation. *Japanese Journal of Veterinary Science* 50, 341–345.

Moran, C. (1993) Microsatellite repeats in pig (*Sus domestica*) and chicken (*Gallus domesticus*) genomes. *Journal of Heredity* 84, 274–280.

Morgan, G.T. (1995) Identification in the human genome of mobile elements spread by DNA-mediated transposition. *Journal of Molecular Biology* 254, 1–5.

Morin, G.B. (1989) The human telomere terminal transferase enzyme is a ribonucleoprotein that synthesises TTAGGG repeats. *Cell* 59, 521–529.

Musilova, P., Lahbib-Mansais, Y., Yerle, M., Cepica, S., Stratil, A. Copieters, W. and Rubes, J. (1995) Assignment of pig α_1-antichymotrypsin (AACT or PI2) gene to chromosome region 7q23–26. *Mammalian Genome* 6, 445.

Nakamura, Y., Leppert, M., O'Connell, P., Wolff, R., Holm, T., Culver, M., Fujimoto, E., Hoff, M., Kumlin, E. and White, R. (1987) Variable number of tandem repeat (VNTR) markers for human gene mapping. *Science* 235, 1616–1622.

Patience, C., Takeuchi, Y. and Weiss, R.A. (1997) Infection of human cells by an endogenous retrovirus of pigs. *Nature Medicine* 3, 282–286.

Popescu, C.P., Boscher, J. and Malynicz, G.L. (1989) Chromosome R-banding patterns and NOR homologies in the European wild pig and four breeds of domestic pig. *Annals of Genetics* 32, 136–140.

Reiter, L.T., Murakami, T., Koeuth, T., Pentao, L., Muzny, D.M., Gibbs, R.A. and Lupski, J.R. (1996) A recombination hotspot responsible for two inherited preipheral neuropathies is located near a *mariner* transposon-like element. *Nature Genetics* 12, 288–297.

Robic, A., Dalens, M., Woloszyn, N., Milan, D, Riquet, J and Gellin, J. (1994) Isolation of 28 new porcine microsatellites revealing polymorphism. *Mammalian Genome* 5, 580–583.

Royle, N.J., Clarkson, R.E., Wong, Z. and Jeffreys, A.J. (1988) Clustering of hypervariable minisatellites in proterminal regions of human autosomes. *Genomics* 3, 352–360.

Sarmiento, U.M., Sarmiento, J.I., Lunney, J.K. and Rishi, S.S. (1993) Mapping the porcine SLA class I gene (*PD1A*) and the associated repetitive element (*C11*) by fluorescence *in situ* hybridisation. *Mammalian Genome* 4, 64–65.

Schmitz, A., Chaput, B., Fouchet, P., Guilly, M.N., Frelat, G. and Vaiman, M. (1992) Swine chromosomal DNA quantification by bivariate flow karyotyping and karyotype interpretation. *Cytometry* 13, 703–710.

Signer, E., Gu, F., Gustavsson, I., Andersson, L. and Jeffreys, A.J. (1994) A pseudoautosomal minisatellite in the pig. *Mammalian Genome* 5, 48–51.

Singer, D.S., Lifshitz, R., Abelson, L., Nyirjesy, P. and Rudikoff, S. (1983) Specific association of repetitive DNA sequences with major histocompatibility genes. *Molecular and Cellular Biology* 3, 903–913.

Singer, D.S., Parent, L.J. and Ehrlich, R. (1987) Identification and DNA sequence of an interspersed repetitive DNA element in the genome of the miniature swine. *Nucleic Acids Research* 15, 2780.

Stratil, A., Cizova, D., Hojny, J. and Hradecky, J. (1990) Polymorphism of pig serum alpha-protease inhibitor-3 (PI3) and the assignment of the locus to the PI1, PO1A, PO1B, PI2, IGH linkage group. *Animal Genetics* 21, 267–276.

Su, T.-S., Nussbaum, R.L., Airhart, S., Ledbetter, D.H., Mohandas, T., O'Brien, W.F. and Beaudet, A.L. (1984) Human chromosomal assignment for 14 argininosuccinate synthetase pseudogenes: cloned DNAs as reagents for cytogenetic analysis. *American Journal of Human Genetics* 36, 954–964.

Sun, J., Kacskovics, I., Brown, W.R. and Butler, J.E. (1994) Expressed swine V_H genes belong to a small V_H gene family homologous to human V_HIII. *Journal of Immunology* 153, 5618–5627.

Takahashi, H., Awata, T. and Yasue, H. (1992) Characterisation of swine short interspersed repetitive sequences. *Animal Genetics* 23, 443–448.

Thome, M., Saalmuller, A. and Pfaff, E. (1993) Molecular cloning of porcine T cell receptor α, β, γ and δ chains using polymerase chain reaction fragments of the constant regions. *European Journal of Immunology* 23, 1005–1010.

Thomsen, P.D. and Miller, J.R. (1996) Pig genome analysis: differential distribution of SINE and LINE sequences is less pronounced than in human and mouse genomes. *Mammalian Genome* 7, 42–46.

Todaro, G.J., Benveniste, R.E., Lieber, M.M. and Sherr, C.J. (1974) Characterisation of a type C virus released from the porcine cell line PK(15). *Virology* 58, 65–74.

Tuggle, C.K. and Schmitz, C.B. (1994) Cloning and characterisation of pig muscle cDNAs by an expressed sequence tag (EST) approach. *Animal Biotechnology* 5, 1–13.

Vage, D.I., Olsaker, I., Lingaas, F. and Lie, O. (1994) Isolation and sequence determination of porcine class II DRB alleles amplified by PCR. *Animal Genetics* 25, 73–75.

Van de Weghe, A., Van Zeveren, A., Bouquet, Y. and Varewyck, H. (1987) The serum protease inhibitor linkage group in Belgian pig breeds. *Animal Genetics* 18, 63–69.

Vize, P.D. and Wells, J.R.E. (1987) Isolation and characterization of the porcine growth hormone gene. *Gene* 55, 339–344.

Vogeli, P., Kuhne, R., Wysshar, M., Stranzinger, G. and Morel, P. (1987) Recombination rates and gene order for some serum alpha-protease inhibitors and immunoglobulin heavy-chain allotypes in pigs. *Animal Genetics* 18, 351–360.

Wilke, K., Jung, M., Chen, Y. and Geldermann, H. (1994) Porcine $(GT)_n$ sequences: structure and association with dispersed and tandem repeats. *Genomics* 21, 63–70.

Wintero, A.K., Fredholm, M. and Thomsen, P.D. (1992) Variable $(dG\text{-}dT)_n.(dC\text{-}dA)_n$ sequences in the porcine genome. *Genomics* 12, 281–288.

Yasue, H. and Wada, Y. (1996) Swine SINE (PRE-1 sequence) distribution in swine-related animal species and its phylogenetic analysis in swine genome. *Animal Genetics* 27,95–98.

Yasue, H., Takahashi, H., Awata, T. and Popescu, P.C. (1991) Uneven-distribution of short interspersed repetitive sequence, PRE-1, on swine chromosomes. *Cell Structure and Function* 16, 475–479.

Yen, C.-H., Matsuda, Y., Chapman, V.M. and Elliott, R.W. (1995) A genomic clone containing a telomere array maps near the centromere of mouse chromosome 6. *Mammalian Genome* 6, 96–102.

Zhang, W., Haley, C. and Moran, C. (1995) Mapping the soluble angiotensin binding protein (*ABP1*) locus to porcine chromosome 16. *Animal Genetics* 26, 337–339.

Zorzato, F., Fujii, J., Otsu, K., Phillips, M., Green, N.M., Lai, F.A., Meissner, G. and MacLennan, D.H. (1990) Molecular cloning of cDNA encoding human and rabbit forms of the Ca^{2+} release channel (ryanodine receptor) of skeletal muscle sarcoplasmic reticulum. *Journal of Biological Chemistry* 265, 2244–2256.

Immunogenetics

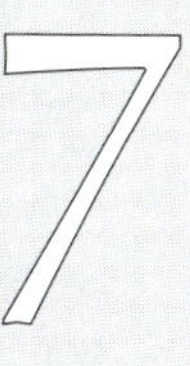

J.K. Lunney[1] and J.E. Butler[2]

[1]Immunology and Disease Resistance Laboratory, LPSI, ARS, USDA, Building 1040, Room 105, Beltsville, MD 20705-2350, USA;
[2]Department of Microbiology and Iowa Interdisciplinary Immunology Program, University of Iowa, 3-550 Bowen Science Building, Iowa City, IA 52242-1109, USA

Introduction

The immune system protects swine against infection and coordinates immune responses to an unknown array of foreign antigens. The immune system adds an additional level of complexity to understanding the genetics of swine as a result of the evolution of lymphocytes, i.e. the T cells and B cells. Unlike other somatic cells, the lymphocyte utilizes a combination of genetic mechanisms, including gene segment rearrangement, untemplated somatic point mutation and templated somatic mutation (gene conversion), to generate a vast repertoire of T-cell receptor (TCR) and B-cell receptor (BCR) phenotypes with which to sample the antigenic environment (Paul, 1993). Without the use of such mechanisms, a TCR and BCR repertoire of such size (approx. 10^{10} for each) would require more DNA than is in the entire typical mammalian genome. The BCR is a membrane immunoglobulin (Ig) that serves as the surface antigen receptor for B cells. The recognition of antigen is through the variable regions of both the heavy and light Ig polypeptide chains.

An equally important genetic element is the major histocompatibility complex (MHC) which is also highly complex, not due to somatic recombination and mutation events, but rather resulting from the large number of allelic variants at many of the >200 loci in the complex (Klein, 1982; Paul, 1993; Campbell and Trowsdale, 1997). The MHC antigens are vital to the overall immune response because of their role in binding and presentation of foreign antigens to the TCR. The allelic variants at these loci serve to increase the range of foreign antigens that can be recognized and presented to the TCR. Other loci in this region serve to process and transport foreign antigens to the MHC molecule for effective presentation to the TCR. The major class I MHC antigens are expressed on all cells whereas the class II MHC antigens are primarily expressed on the surface of so-called professional antigen presenting cells, such as B cells, macrophages, and follicular dendritic cells (Schook and Lamont, 1996).

The genetic complexity provided by loci encoding the MHC, TCR or BCR is the basis for the moniker 'The science of self and non-self recognition' adopted by Jan Klein in his classic immunology textbook (Klein, 1982). This chapter will discuss each of these three important genetic loci in the swine system. The genome map locations of the swine cytokines, i.e. proteins produced by immune effector cells, which regulate these cellular responses, will also be presented.

Chromosome Map Locations of Loci Encoding Elements of the Adaptive Immune System

Loci encoding immunoglobulin polypeptides

The immunoglobulin (Ig) molecules which comprise the BCRs and their secreted form, the antibodies, are composed of two different polypeptide chains, a heavy chain (52–75 kDa depending on Ig heavy chain isotype) and a light chain (22 kDa for both types of light chains, κ and λ). In mammalian species that have been examined, the loci encoding the heavy chain and the two light chain types (κ and λ) are not linked. Similarly, the swine heavy chain locus (*IGHC*) has been mapped to 7q26 while the κ locus (*IGKC*) maps to chromosome 3 and λ locus (*IGL*) to chromosome 14q17–q21 (Gu *et al.*, 1994; Frönicke *et al.*, 1995; Rettenberger *et al.*, 1996b,c). Although both the swine J-chain, which is involved in IgM and IgA structure, and the secreted portion of the poly-Ig receptor have been identified for some time in swine (see below), the linkage groups to which they belong have not yet been identified.

The T-cell receptor gene families

The TCR in swine is of two types: one is composed of a set of α and β heterodimers and the other is a γ/δ heterodimeric pair. Like the immunoglobulins, the loci are complex because they are made up of both variable (Vα, Vβ, Vγ or Vδ) segments and constant segments (Cα, Cβ, Cγ and Cδ). The constant regions of α-, β-, γ- and δ- have been cloned (Thome *et al.*, 1993; Grimm and Misfeldt, 1994). The locus encoding the TCR α-chain (*TCRA*) has been mapped to chromosome 7 and the β-chain (*TCRB*) to chromosome 18 (Rettenberger *et al.*, 1996a).

Major histocompatibility complex

The swine MHC was identified in 1970 simultaneously by Vaiman *et al.* (1970) and Viza *et al.* (1970), and is referred to as the swine leucocyte antigen (SLA) complex. This complex includes a set of linked genes (or haplotype) that encode molecules intimately involved in regulating immune responses. The class I and class II SLA genes encode glycoproteins which bind and present antigenic peptides to T cells resulting in their stimulation. Extensive, molecular research has shown that the SLA complex encodes a large number of genes, including multiple class I and class II SLA genes, peptide transporter genes, and the class III genes, including the complement components, tumour necrosis factor and heat shock proteins (reviewed in Warner and Rothschild, 1991; Lunney, 1994b; Schook *et al.*, 1996). The SLA complex was first linked to the J blood group and to chromosome 7 by SLA serological typing of large numbers of individuals for SLA class I and II specificities (Hruban *et al.*, 1976; Vaiman *et al.*, 1979; Lunney *et al.*, 1986). These studies indicated that SLA class I and class II were separated by as little as 0.3–0.4 cM. This short linkage distance was confirmed by molecular genotyping which suggested SLA distances of 1.1–1.5 cM (Edfors-Lilja *et al.*, 1993; Smith *et al.*, 1995), making the swine MHC the smallest among mammalian MHC so far examined. The SLA complex was first physically mapped to swine chromosome 7 (SSC7) by Rabin *et al.* (1985);

this was confirmed by others (Sarmiento *et al.*, 1993b). Most recently, Smith *et al.* (1995) have confirmed that the SLA complex actually spans the centromere of chromosome 7 with SLA class II (*DQB*) on the q arm (7q1.1), and class III (*TNFB*; 7p1.1) and SLA class I (*SLA-I*; *PD6*; 7p1.1) genes on the p arm of the chromosome. The presence of the centromere within the SLA complex does not appear to alter the function of the associated genes. This physical assignment of the swine MHC spanning the centromere of SSC7 is unique among mammals studied to date (Smith *et al.*, 1995).

Cytokine loci

Significant progress has been made in understanding the molecular basis of immune responses in swine. The complex cellular interactions involved in disease responses have recently been reviewed (Lunney *et al.*, 1996) and international workshops have been held to assign specific CD numbers to the antigens recognized by anti-swine lymphoid cell monoclonal antibodies (Lunney, 1994a; Saalmueller, 1996). It is now clear that a wide range of cytokines regulate swine immune responses (Myers and Murtaugh, 1995; Murtaugh *et al.*, 1996). The genes for many of these cytokines have been mapped in swine, as is noted in Table 7.1.

The Immunoglobulin Genes

Each immunoglobulin protein is composed of heavy and light chains, each chain of which has a highly variable *N*-terminal region combined with a *C*-terminal constant region. To produce these mature protein chains a series of somatic rearrangement events take place prior to transcription of the immunoglobulin loci. These loci are actually composed of major subloci, in particular the 5′ gene segments encoding the variable regions (V_H, $V\kappa$ or $V\lambda$) and the downstream segments encoding the various constant regions (C_H, $C\kappa$ and $C\lambda$) (Paul, 1993). The C_H sublocus is further subdivided into loci encoding the various antibody isotypes, e.g. $C\mu$ for IgM, $C\gamma$ for IgGs, $C\varepsilon$ for IgE and $C\alpha$ for IgA. The variable region sublocus can be further subdivided into regions containing V_H, D_H and J_H segments for heavy chains and V_L and J_L segments for κ- and λ-chains. In all species, the cell surface BCRs are in the form of Ig monomers, composed of two heavy and two light chains, with a transmembrane tail and a very short cytoplasmic 'stub'. Since this portion of the constant region sublocus encodes the BCRs on virgin B cells, it will be discussed separately. The secreted forms of Igs lack the transmembrane tail, contain a short secreted segment and some are even secreted as polymers of the basic Ig monomer, e.g. pentameric IgM and dimeric IgA.

Readers wanting to become more familiar with the somatic rearrangement and mutational events that occur among the variable region segments should consult a recent immunology textbook, e.g. Paul (1993). Briefly, pre-B cells

Table 7.1. Map locations of swine cytokines.

Cytokine	Symbol	Physical map location	Linkage map location	Reference
Interferon-α	*IFNA*	1q2.5–q2.6prox	1: 91 cM	Sarmiento *et al.*, 1993a; Ellegren *et al.*, 1994b; Marklund *et al.*, 1996
Interferon-γ	*IFNG*	5p1.2–q1.1	5: 28 cM	Ellegren *et al.*, 1994a; Marklund *et al.*, 1996
Interleukin-1α	*IL1A*	3q1.2–1.3	3: 66 cM	Mellink *et al.*, 1994; Ellegren *et al.*, 1994a; Marklund *et al.*, 1996
Interleukin-1β	*IL1B*	3q1.1–1.4	3: 22 cM	Mellink *et al.*, 1994; Ellegren *et al.*, 1994a; Marklund *et al.*, 1996
Interleukin-2	*IL2*		8: 74 cM	Ellegren *et al.*, 1994a; Marklund *et al.*, 1996
Interleukin-4	*IL4*	2		Rettenberger *et al.*, 1996a
Interleukin-6	*IL6*	9		Rettenberger *et al.*, 1996a
Transforming growth factor-β2	*TGFB2*	10		Rettenberger *et al.*, 1996a
Transforming growth factor-β3	*TCFB3*	7		Rettenberger *et al.*, 1996a
Tumour necrosis factor-α	*TNFA*	7p1.1–q1.1	7: 54 cM	Chardon *et al.*, 1991; Solinas *et al.*, 1992; Nunes *et al.*, 1993; Smith *et al.*, 1995.
Tumour necrosis factor-β	*TNFB*	7p1.1–q1.1	7: 54 cM	Chardon *et al.*, 1991; Solinas *et al.*, 1992; Nunes *et al.*, 1993; Ellegren *et al.*, 1994a; Smith *et al.*, 1995; Marklund *et al.*, 1996.

acting under the influence of recombinase activation genes (RAGs) rearrange segments of the V_H, D_H and J_H to yield a contiguous V–D–J segment. In mice, with approximately 100 V_H segments, 15 D_H segments and 4 J_H segments, it means that 6,000 combinations are possible by 'combinatorial joining'. The number of variants is further increased due to imprecise joining, i.e. 'junctional diversity'. Finally, the rearranged VDJ region is subjected to somatic point mutation, by an as yet undefined mechanism that is 10^7 greater than the normal rate of somatic mutation in mammalian cells. This translates into approximately one amino acid change per cell division in lymph node and spleen germinal centres where the process occurs. Therefore, it is estimated that an average of 10^8 VDJ variants can be produced. Considering only the 1000 variants of $V\kappa$ generated by a similar process, the estimated total number of V_H–$V\kappa$ variants possible for a complete Ig polypeptide, consisting of Ig_H–$Ig\kappa$, is 10^{11} for mice. The potential significance of these repertoire-generating recombination and mutation events, and variations among species, is discussed below.

Organization of the porcine immunoglobulin loci

The constituency of the heavy chain loci

The only swine Ig locus that has been studied in detail to date is that encoding the variable and constant regions of the heavy chain. So far, genes encoding Cα (IgA), Cε (IgE), Cμ (IgM) and multiple genes encoding Cγ (IgG) have been identified and partially or completely sequenced (Butler *et al.*, 1996a). Unlike rodents and primates, no swine gene encoding Cδ (IgD) has been found (Butler *et al.*, 1996b). The region of the locus encoding the V_H segments is quite unusual in swine; unlike the mouse and human, it contains < 20 V_H genes, only one J_H and an undetermined number of D_H segments. The discussion below will cover three separate regions of the locus: the V_H–D_H–J_H region, the 5′ constant region containing Cμ and finally, the 3′ constant region encoding the remaining C_H genes.

The swine V_H–D_H–J_H sublocus

Figure 7.1 is a map of 3′ end of the V_H region including part or all of the D_H region. The complete genomic sequence of V_H2 and two D_H segments (designated D_HA and D_HB) are given together with the signal sequences of the latter. Sequence analyses of > 85 swine VDJ transcripts indicate that they are more than 85% homologous to each other and belong to the V_H3 family (Sun *et al.*, 1994). V_H3 is the largest V_H family in mice and humans and is considered to be the ancestral V_H family (Schroeder *et al.*, 1990). Analysis of genomic Southern blots using single-strand conformational polymorphisms (SSCP) indicate that swine have ~20 V_H genes, the first mammal for which such a small number was reported and similar to the number originally reported for

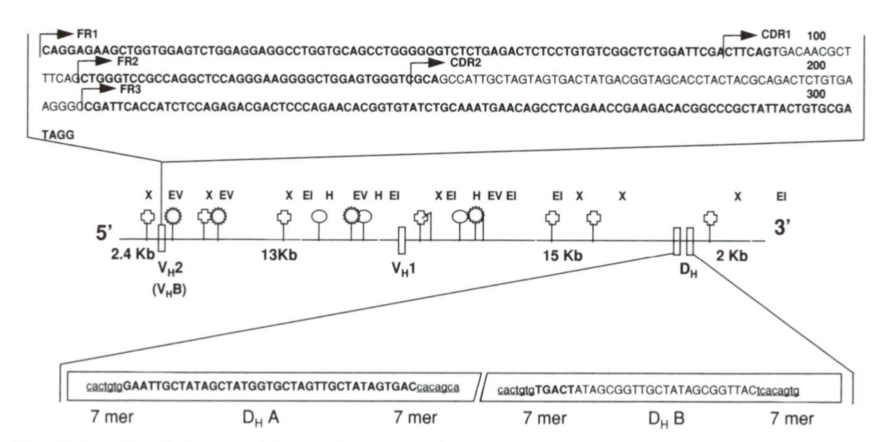

Fig. 7.1. Physical map of the V_H–D_H cosmid from a porcine genomic cosmid library. Restriction sites for *BAM*HI (B), *Hin*dIII (H), *Eco*RI (EI), *Eco*RV (EV) and *Xho*I (X) are indicated. V_H1 and V_H2 have been sequenced and V_H1 appears to be a pseudogene. V_H2 is one of several V_H genes expressed by fetal and neonatal piglets as are the two D_H segments (D_HA and D_HB). The sequences are underlined. (From Sun and Butler, 1996.)

chickens (McCormack and Thompson, 1990). Exactly how many of these may be pseudogenes is currently unknown, although it has already been verified that V_H1, the most 3′ V_H gene, is a pseudogene as a consequence of a faulty reading frame. Studies in a newborn piglet indicate that at least five different V_H genes can be expressed (Sun and Butler, 1996), which differs from the pattern in chickens in which all but the most 3′ V_H genes are pseudogenes (McCormack and Thompson, 1990).

The D_HA and D_HB segments are located 15 kb downstream from V_H1 (Fig. 7.1). These are the only D_H segments that to date have been identified at the genomic level but there is evidence from VDJ transcripts that more than two exist. The distance between the D_H region and J_H in swine is as yet unknown, since no cosmid overlapping these two regions has been cloned. As shown in Fig. 7.2, the J_H segment is located ~5 kb upstream from Cμ and adjacent to the heavy chain enhancer element (Enh_H). Analyses of the > 85 VDJ transcripts, and Southern blot analyses of both genomic and cosmid DNA restriction digests, indicate that only a single J_H is present in swine; this is the pattern observed in chickens but not in other mammals. The swine J_H segment is most homologous to mouse J_H4 and human J_H6 (Butler *et al.*, 1996b).

The 5′ C_H sublocus in swine and the virgin BCR

In mice and humans, newly formed B cells express the membrane forms of two different Ig isotypes on their surface, IgM and IgD. Membrane IgM (mIgM) is not pentameric like the secreted form, but rather monomeric with a 42 amino acid transmembrane segment in place of the octadecapeptide tailpiece of the secreted, pentameric form. Likewise, mIgD contains a *C*-terminal transmembrane segment.

Figure 7.2 is a map of a cosmid clone, which contains the gene segments encoding the various forms of IgM (Sun and Butler, 1997). Also presented in Fig. 7.2 is the complete genomic sequence of porcine Cm. Each domain of the porcine IgM heavy chain is encoded by a separate exon (Cμ1–Cμ4) as are the segments encoding the transmembrane region (μM1 + μM2) and the tailpiece of the secreted form (CμS). Upstream from the exon encoding Cμ1 is switch μ. Switch μ in swine (Sμ), like switch μ in other species and switch regions for other C_H genes, contains multiple repeat sequences predominantly the sequence GAGCT, but also GGGCT, GGGTT and others (Sun and Butler, 1997). Swine Sμ is 3.3 kb in length, and its sequence is slightly more homologous to mouse than human Sμ and contains > 200 GAGCT repeats (Sun and Butler, 1997). Switch regions are involved in the process of switch recombination, a rearrangement event which permits the rearranged VDJ segment expressed with Cμ on virgin B cells to be brought into juxtaposition with other downstream C_H genes so that IgG, IgA or IgE antibodies can be generated which have the same specificity as the original BCR since they express the same V regions. However, concomitant with switch, the rearranged VDJ and VJ encoding the original BCR are somatically mutated, which may result in the expressed mutant BCR having a greater affinity for the antigen than the original BCR. This interesting process

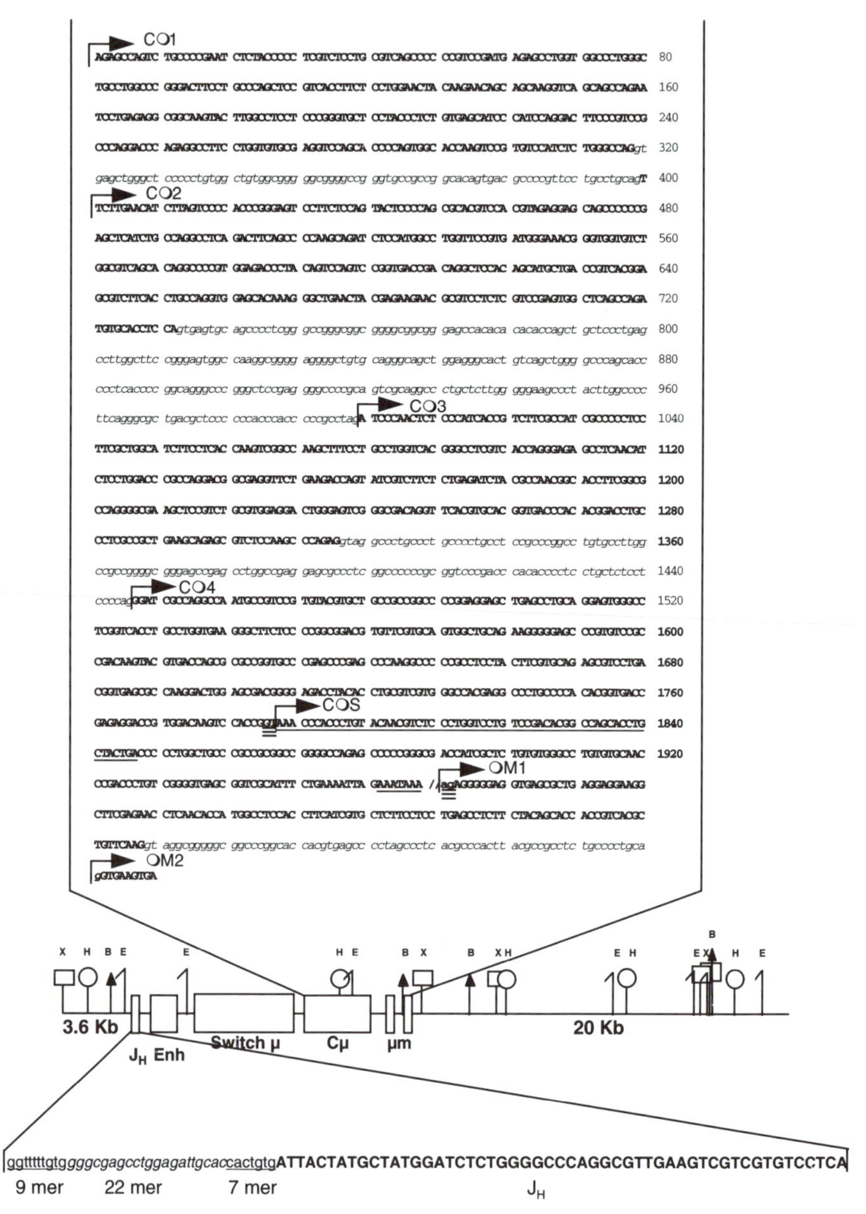

Fig. 7.2. The J_H–Cμ region of the porcine heavy chain. The location of various elements has been mapped using homologous and heterologous DNA probes. Restriction sites for *BAM*HI (B), *Eco*RI (E), *Hin*dIII (H) and *Xba* (X) are indicated. The genomic sequence of the single swine J_H is given below the map and the complete genomic sequence of porcine Cμ and its membrane exons (μm1 and μm2) are given above. Exons are in bold font, introns in italics. The segments encoding the 19 amino acid tailpiece (CμS) is underlined and the donor and acceptor splice sites for the membrane exon are underlined. (From Sun and Butler, 1997.)

of 'affinity maturation' occurs in germinal centres of lymph nodes and spleen under the influence of external antigen.

In rodents and primates, the gene segment located about 5 kb downstream of μM encodes Cδ (IgD). However, using heterologous probes for mouse IgD, or conserved PCR primers for human, mouse and rat IgD, we were unable to clone or amplify a Cδ homologue from swine genomic DNA or from the Cμ cosmid shown in Fig. 7.2 (Sun and Butler, 1997). We were similarly unable to do so from the rabbit and cattle, a result consistent with the studies of Knight and Crane (1994) and Naessens and colleagues (personal communication). Thus it appears that swine, sheep, rabbit and cattle lack IgD. The IgD gene has also not been cloned from chicken DNA (Weill and Reynaud, personal communication). This means that swine differ from rodents and primates in their lack of the second BCR (mIgD).

The 3′ C_H locus of swine

The organization of the swine 3′ C_H locus remains to be determined although considerable information on the C_H genes encoded by this region is available. We have chosen to distinguish between the 3′ segment which encodes Cγ, Cε, Cα and the 5′ segment which encodes the Cμ region (and Cδ in rodents and primates) because of differences in their function (Paul, 1993). As discussed above, the 5′ Cμ region encodes the virgin BCRs and the secreted IgM antibodies that characterize primary immune response and so-called T-cell independent responses to the polysaccharide antigens of microorganisms. In contrast, the 3′ region encodes the constant regions of the antibodies which typify secondary immune responses and responses to T-cell-dependent antigens. The IgGs, IgA and IgE antibodies encoded by the 3′ region have unique or characteristic biological activities that are determined by their Ig protein structure.

Swine are similar to most other species studied (Table 7.2) in that they have only a single Cα gene (Brown and Butler, 1994). However, swine Cα occurs in two allelic forms, *IgA*^{a} and *IgA*^{b}, the latter of which is missing two-thirds of its hinge region (Brown *et al.*, 1995). To the extent that it has been studied, swine have a single gene for Cε and a single Cμ with possibly some allotypic variation in the latter (Sun and Butler, 1997). Perhaps the most interesting region of the 3′ end of the C_H locus is that which encodes the various IgGs. Swine express at least five IgG subclasses (IgG1, IgG2a, IgG2b, IgG3 and IgG4) but genomic blots indicate they have more than five Cγ genes (Kacskovics *et al.*, 1994). Exactly how many Cγ genes exist, whether 8, 10, 12 or more, remains to be determined. In humans, which have four Cγ, two Cε and two Cα genes, the entire 3′ C_H region is clearly the result of gene duplication. As swine have only a single Cα and Cε, the multiple Cγ genes must be the result of gene duplication in only one part of the 3′ C_H region; this would parallel the apparently exclusive duplication of Cα genes in the rabbit (Table 7.2).

Immunoglobulin accessory polypeptides

The two polymeric Igs, secreted IgM (a pentamer) and secretory IgA (a dimer), each contain one molecule of J chain. J chain is a cysteine-rich polypeptide

Table 7.2. Immunoglobulin diversity among animals.[1]

	C_H-region isotypes					C_L-region types[2]		V-gene families			
Species	IgM	IgD	IgG	IgE	IgA	λ	κ	H	λ	κ	Generation of antibody diversity (GOD)[3]
Mouse	1	1	4	1	1	3 (5%)	1 (95%)	14	3	4	CJ, SM
Human	1	1	4	1	2	7 (40%)	1 (60%)	7	7	7	CJ, SM
Bovine	1	0	3	1	1	4 (?) (>95%)	1 (<5%)	1(?)	2(?)	?	CJ, SM; CVS
Sheep	1	0	2(?)	1	1(?)	> 1(?) (>95%)	1(?) (<5%)	1(?)	6	3	CJ, SM,
Rabbit	1	0	1	1	13	8 (20%)	2 (80%)	1	?	?	CJ, SM, CVS
Swine	1	0	> 6(?)	1	1	1(?) (40%)	1(?) (60%)	1	?	?	CJ, SM, CVS
Horse	1	?	> 4(?)	1	1	4 (93%)	1 (7%)	?	1	?	?
Chicken	1	0	1	0	1	1 (100%)	0	1	1	0	CJ, SM (?), CVS

Source: From Butler (1997).

[1]Question marks in parentheses mean that the maximum number (or phenomenon) has not been established but is most likely the number or event indicated. Question marks without parentheses mean data are unavailable.

[2]Per cent values in parenthesis on the second line indicate the proportion of serum Ig bearing λ and κ chains, respectively.

[3]CJ = combinatorial joining; SM = somatic point mutation (untemplated); CVS = somatic gene conversion (templated). It is assumed that combinatorial joining also involves junctional diversity through P- and N-nucleotide additions. N-nucleotide additions, resulting from terminal deoxytransferase activity, are more often found at later stages of fetal and neonatal development.

important in the polymerization and in the maintenance of the polymeric conformation of these Igs (Paul, 1993). A second polypeptide, the poly-Ig receptor, which is expressed on the baso-lateral surface of mucosal epithelial cells, is required for transcytosis of dimeric IgA to the lumenal side of mucosal epithelia. In the process of transcytosis, the *N*-terminal 85 kDa portion of this 115 kDa poly-Ig receptor is cleaved at the lumenal side. This results in the 85 kDa segment, called 'secretory component', becoming an integral part of the structure of the secreted IgA dimer, which is then called SIgA or 'secretory IgA'.

The polymorphism of swine immunoglobulin genes

The V_H-region genes of swine

Polymorphism of V_H genes is the rule, as is clearly demonstrated in extensively studied species like mice and human (Willems van Dijk *et al.*, 1989; Walter and Cox, 1991). Nevertheless, the V_H6 family in humans behaves monomorphically (Sanz *et al.*, 1989). Polymorphism may be expressed as simple allelic variation or as the absence of a gene. In swine, restriction fragment length polymorphism (RFLP) analyses of V_H genes using *Eco*RI showed little or no polymorphism among 11 animals of which six were related animals and five represented different breeds (Fig. 7.3). However, *Eco*RI sites are found only in introns, so that restriction polymorphism within exons would be overlooked in this analysis. Nevertheless, mice and humans show considerable polymorphism when analysed in the same manner. Thus, RFLP analyses extend our sequence analyses of > 85 V_H exons in which we observed highly conserved or nearly identical leader

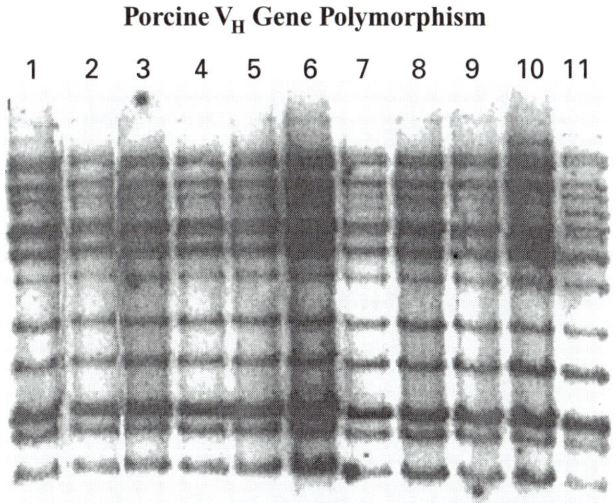

Porcine V_H Gene Polymorphism

1 2 3 4 5 6 7 8 9 10 11

Fig. 7.3. Genomic Southern blot analysis of sperm DNA from 11 different swine using a pan-V_H probe. The six animals on the left (1–6) are from the same breed and are genetically related. The five animals on the right (7–11) represent five different breeds. Sperm was kindly provided by the Pig Improvement Company and the purified DNA was digested with *Eco*RI.

and framework regions, by suggesting that lack of diversity apparently also extends to the population level.

Among the two D_H segments characterized, one appears to occur in the form of two allelic variants (Sun and Butler, 1996). Variations in germline J_H (Fig. 7.2) have not yet been detected. Since there is only a single J_H, it is likely that it is highly conserved as any non-functional mutant would result in total antibody deficiency and certain death for an animal with such a mutation.

The C_H-region genes of swine

Rapacz and Hasler-Rapacz (1982) have published on allotypic variation among serum IgGs in swine. Their studies would predict four to five IgG genes with two allelic variants of each. Since sequence data are not available from their studies, it is impossible to determine if these represent the five IgG subclasses described from > 60 cDNA sequences expressed from a single pig (Kacskovics *et al.*, 1994). In the latter study, putative allelic variants of both IgG1 and IgG2a were identified. However, these require confirmation by appropriate breeding studies and progeny tests.

Insufficient data are available to determine whether polymorphism occurs in the genes for IgM and IgE, whereas the situation for porcine IgA provides a most interesting example of allelic polymorphism. The Cα gene in swine occurs in the form of two codominant allelic variants (IgA^a and IgA^b) which differ by a G to A mutation in the splice acceptor in the first intron (Fig. 7.4; Brown *et al.*, 1995). This mutation results in the use of a cryptic downstream AG splice site in the IgA^b allelic variant so that two-thirds of the hinge region of the expressed IgA

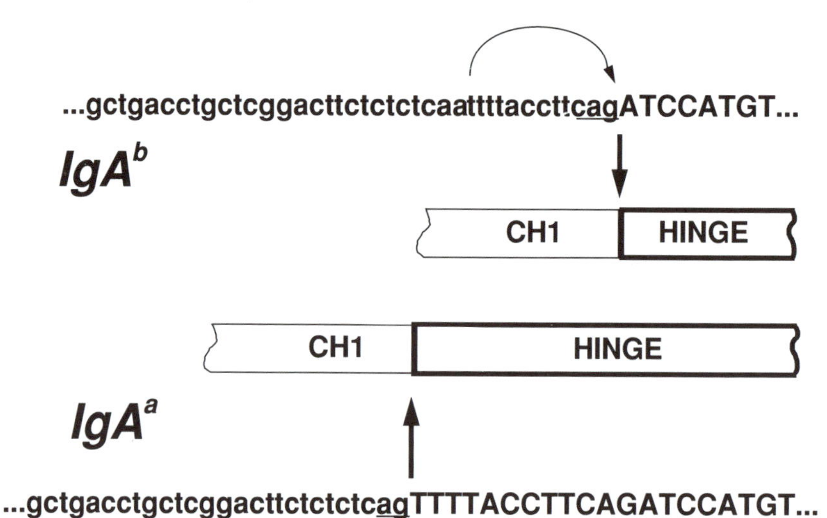

Fig. 7.4. Hinge deletion allelic variants of porcine *IgA*. The normal ag splice acceptor site in IgA^a has been mutated to an aa in IgA^b so that a downstream ag site is used (indicated by curved arrow). This results in the loss of four amino acids in the IgA^b heavy chain hinge region. The actual splice sites are underlined. Intronic sequences are in lower case, exonic sequences in capitals. (From Brown *et al.*, 1995.)

is absent. The Mendelian inheritance of IgA^a and IgA^b has been confirmed by progeny testing (Brown *et al.*, 1995). The biological effect(s) of such a structural change has not yet been determined although it is known from studies on human IgAs, that the hinge of IgA is important in bacterial protease susceptibility (Plaut, 1983).

The genetics of antibody repertoire development in swine

The unique property of lymphocytes to somatically alter their genes by rearrangement and mutation has already been discussed. However, not all vertebrates use the same mechanisms for these changes in their B cells. An extreme variation on the theme is seen in chickens. Chickens have only one functional V_H gene, the remaining 25 being pseudogenes which serve only as donors during templated somatic mutation (gene conversion) (McCormack and Thompson, 1990). This process occurs in the embryo and during the first few weeks of life in the bursa of Fabricius. Segments of upstream pseudogenes are translocated by non-homologous recombination into the single function V_H gene. B cells leaving the bursal follicles with these changes are potentially able to undergo further somatic mutation in secondary lymphoid tissues. Recently it has been shown that rabbits also utilize gene conversion, preferring as recipient their most 3′ V_H gene from the more than 200 genes in their genome (Becker and Knight, 1990). This mutation process occurs in the rabbit appendix and ileal Peyer's patches (Weinstein *et al.*, 1994). Our data suggest that swine also use a conversion mechanism (Sun and Butler, 1996). In contrast to chickens, swine use more than one functional V_H gene and, unlike rabbits, swine do not appear to preferentially utilize their most 3′ functional V_H gene. When Ig transcripts in a newborn piglet were sequenced, six out of 42 were found to be hybrids of five different putative germline V_H genes (Sun and Butler, 1996). Segmental changes (transpositions) mostly involved complementarity determining regions 1 and 2, i.e. CDR1 and CDR2. Although swine differ from chickens in their complement of functional V_H genes, both have only a single J_H, which acts to limit V-region diversification by combinatorial joining. This contrasts with mice, rabbits and humans which have four, five, and six J_H segments, respectively.

Implications for B-cell repertoire and disease resistance

Table 7.2 summarizes the known complexity of Ig genes and the mechanisms used to generate Ig diversity in different species. In all species shown to use gene conversion, i.e. chickens, rabbits, swine and cattle, all have but a single family of V_H gene. Chickens, rabbits and swine have V_H genes that belong to the ancestral V_H3 family whereas V_H2 is apparently the only family used in cattle and sheep (Dufour *et al.*, 1996; Berens *et al.*, 1997; Saini *et al.*, 1997; Sinclair *et al.*, 1997).

The pattern of B-cell repertoire development among different homeothermic vertebrates is shown in Fig. 7.5. This figure highlights differences in the mechanisms used to generate Ig diversity and the anatomical site and timing of the process. In primates and rodents, the P–R group, the primary antibody repertoire is generated through somatic rearrangement (combinatorial joining and junctional diversity) in bone marrow and the secondary antibody repertoire is generated in germinal centres by untemplated somatic (point) mutation. For these species, B-cell development continues throughout life in the bone marrow (Osmond, 1990) and cells of the B lineage predominate (Kantor and Herzenberg, 1993). In chickens, B cells formed in the fetal liver migrate to the bursa and therein diversify their repertoire by gene conversion before leaving for the peripheral (secondary) lymphoid organs (McCormack and Thompson, 1990). The bursa then involutes and the chicken no longer has the ability to make more B cells or to alter its V-region genes by gene conversion. Rabbit B cells migrate from the fetal liver to the appendix and undergo both gene conversion and somatic point mutation before seeding the periphery. The rabbit appendix does not involute but becomes a secondary lymphoid organ (Weinstein *et al.*, 1994). In any case, all rabbit B cells are of the B-1 lineage and their formation ceases 6 weeks after birth.

Thus both chickens and rabbits have a fixed but apparently self-replicating population of B cells from which to generate their antibodies whereas animals in the P–R group can continue to generate new B cells throughout life (Fig. 7.5).

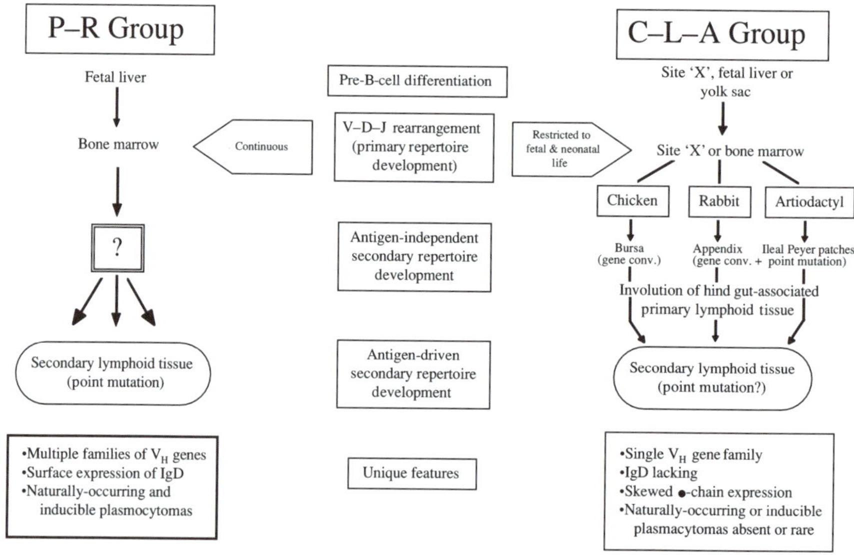

Fig. 7.5. Variations among homeothermic animals in B-cell repertoire development. Classification of animals on the basis of secondary repertoire development. P–R group = primate–rodent group, C–L–A = chicken–lagomorph–artidactyl group. Primary repertoire development is defined as V–D–J or VJ (light chain) combinatorial joining and includes junctional diversity. (From Butler, 1997.)

Swine and all artiodactyls studied have ileal Peyer's patches which involute after one year of life. It is most likely that this organ, like the chicken bursa and rabbit appendix, is the site of Ig gene conversion. Thus the chicken–lagomorph–artiodactyl (C–L–A) group of animals are linked by their use of a single V_H family, their use of gene conversion, their lack of IgD and their discontinuous generation of B cells, although the latter has not been confirmed in artiodactyls. We have speculated that IgD in the P–R group is involved with secondary repertoire development and with affinity maturation in the germinal centres of lymph nodes and the spleen. Furthermore, we suggest IgD may not be needed in animals of the C–L–A group because of a higher surface density of the IgM BCR and the greater role played by intrinsic factors as opposed to environmental antigens, in expanding the B-cell repertoire in these species (Butler *et al*., 1996b; Butler, 1997).

The T-cell Receptor Gene Families

The T-cell receptor gene loci

The TCR in swine is composed of two types, the αβ TCR heterodimers and the γδ TCR heterodimers. Like the Igs, the TCR loci are complex because they are made up of both variable (Vα, Vβ, Vγ or Vδ) segments and constant segments (Cα, Cβ, Cγ and Cδ). The general structure of the TCR is that of a heterodimer which resembles the Fab region of an antibody but with a transmembrane tail similar to the *C*-terminal tail of the BCRs. In mice and humans, less than 5% of the circulating T cells bear the γδ TCR whereas in artiodactyls, including swine, γδ TCR T cells may comprise as much as 40–50% of the circulating T-cell pool (Haas *et al*., 1993; Binns, 1994; Yang and Parkhouse, 1996). Indeed, in young pigs the number of αβ TCR cells is relatively low (Yang and Parkhouse, 1996). The swine T-cell subsets are complex and unique in that the expression of CD4 and CD8 define four populations of extrathymic T cells; $CD4^+CD8^-$, $CD4^-CD8^+$, $CD4^-CD8^-$ and $CD4^+CD8^+$ T cells. The $CD4^-CD8^-$ double negative population contains most of the γδ T cells (Saalmüller *et al*., 1994; Yang and Parkhouse, 1996) similar to the $CD2^-$ $CD4^-$ $CD8^-$ subset in cattle (Mackay *et al*., 1986). The predominance of γδ T cells in the periphery in swine and ruminants, and their association with local tissues, has stimulated interest in studying these cells in these species. Their abundance, even in the circulating pool, facilitated the establishment of porcine γδ T-cell clones (Grimm *et al*., 1993).

Like the BCRs, the *N*-terminal domain of the TCRs is encoded by variable region gene segments which also undergo rearrangement in the thymus under the influence of the recombination activating genes (RAGs). The Vα and Vγ of mouse TCR resemble Ig light chain variable regions, with V and J segments (but lacking D segments) whereas both Vβ and Vδ TCR have V, D and J segments. However, in addition to ~100 Vα gene segments, there are ~50 Jα segments. The mouse Vγ region resembles Ig Vλ in being composed of tandem repeats of Vγ–Jγ–Cγ; Vβ also has tandem repeats but only of Dβ–Jβ–Cβ whereas a series of

5′ Vβ genes resemble the Ig V_H locus. It is interesting to note that the Vδ complex (Vδ–Dδ–Jδ–Cδ) in mice is located within the Cα locus, between the ~100 Vα gene segments and Jα. This would indicate that rearrangement in the Cα locus would effectively inactivate Vδ.

Both αβ and γδ TCRs have been identified serologically in swine and the constant regions of the α, β, γ and δ TCR chains have been cloned (Thome *et al.*, 1993; Grimm and Misfeldt, 1994; Yang *et al.* 1995). The homology of C_β is higher (77%) than that of other mammalian species whereas Cγ and Cδ appear most homologous to their counterparts in ruminants (Thome *et al.*, 1993; Grimm and Misfeldt, 1994). Yang *et al.* (1995) cloned the porcine Vδ region, identifying five putative Vδ families and more than 30 distinct Vδ1 genes. Vδ1 also predominates in sheep and cattle (Hein and Dudler, 1993). This Vδ diversity greatly exceeds that seen in rodents and humans and is the most diverse yet seen in any species. Yang and colleagues (1995) putatively identified three Dδ and four Jδ; this is similar to what has been observed in humans. Especially noteworthy is the 'Ig-like' CDR3 region of the porcine Vδ region, suggesting that perhaps this TCR evolved to recognize confirmational epitopes, like antibodies, but in the context of MHC presentation. This would be distinct from the linear peptides recognized by αβ TCRs (Rock *et al.*, 1994). The pig offers an excellent model in which to investigate TCR diversity and the mechanisms of TCR antigen recognition in local tissues (Itohara *et al.*, 1990; Havran *et al.*, 1991).

Major Histocompatibility Complex or Swine Leucocyte Antigens (SLA)

The major histocompatibility complex is one of the most gene-dense areas of the genome and encodes many genes that regulate immune responses. In humans the class II region contains one gene every 40 kb while the class III region has one gene every 15 kb, and in some areas one gene every 1–2 kb (Campbell and Trowsdale, 1997). Detailed studies of the SLA complex support a similar gene density for swine as will be discussed in detail below. Figure 7.6 is an updated map of the swine MHC, the SLA complex, based on pulse field gel electrophoresis (PFGE) studies and mapping of cosmid, phage and yeast artificial chromosome (YAC) clones (Xu *et al.*, 1992; Wu *et al.*, 1995; Brule *et al.*, 1996; Peelman *et al.*, 1996 a,b; Chardon *et al.*, 1996; Singer, LeGuern, personal communication). Recent reviews of the SLA complex include Warner and Rothschild (1991), Lunney (1994b), and Schook *et al.*(1996).

SLA class I antigens

The classical, expressed class I locus products are designated *SLA-A,B,C* and encode 45 kDa glycoproteins that non-covalently associate with 12 kDa β_2-microglobulin (Chardon *et al.*, 1978; Lunney and Sachs, 1978; Vaiman *et al.*,

1979). Class I antigens are expressed on the surface of most cells; most mature swine lymphocytes express high levels of these antigens. These molecules serve to present antigens to CD8+ T cells to stimulate immune responses. SLA-typing alloantisera have been developed in laboratories worldwide; the last international comparative test was reported by Renard *et al.* (1988). Serologic evidence clearly proves the existence of two expressed SLA class I gene products and indicates a third in some SLA haplotypes (Lunney and Sachs, 1978; Vaiman *et al.*, 1979; Ivanoska *et al.*, 1991). More recently monoclonal antibodies (mAb) have been produced and tested for SLA class I antigens and allelic reactivity as has been discussed in detail elsewhere (Lunney *et al.*, 1988; Warner and Rothschild, 1991; Schook *et al.*, 1996). Most anti-SLA class I mAb, when tested on outbred swine populations, seem to detect determinants common to a number of SLA class I antigens and/or alleles (Ivanoska *et al.*, 1991; Kristensen *et al.*, 1992) and, thus, these mAb are designated as being broadly polymorphic or as reacting with 'public' SLA class I specificities (Lunney, 1994b; Schook *et al.*, 1996).

Singer *et al.* (1982) were the first to clone and express a swine class I gene. Using RFLP analyses of genomic DNA, her group proceeded to show that the SLA^d haplotype encoded only six to eight class I genes (Singer *et al.*, 1982, 1988). Chardon, Vaiman and their colleagues confirmed that swine had a limited number of class I genes when they analysed DNA from a wide range of SLA haplotyped swine, although they suggested seven to ten class I genes (Chardon *et al.*, 1985a,b). In their studies, considerable polymorphism was observed between distinct haplotypes; polymorphic RFLP fragments in Southern blots appeared to be inherited in accord with serological typing results. Most recently Gaycken *et al.* (1994) identified nine class I genes in their screening of a porcine cosmid library with human cDNA probes. Singer and her colleagues determined that only two swine class I genes (*PD1*, *PD14*) could be expressed at the protein level while a third (*PD6*) was only detected at the mRNA level (Ehrlich *et al.*, 1987; Singer *et al.*, 1988). By sequence analyses, they confirmed that the class I genes *PD1*, *PD7* and *PD14* are 80–85% homologous to each other but only 49% and 55% homologous to *PD6* and *PD15*; *PD6* shares only 45% homology with *PD15*.

Class I proteins are very heterogeneous between individuals, thus it is difficult to characterize and maintain panels of typing antisera to determine the SLA class I alleles of outbred swine. Therefore, scientists worldwide have been testing molecular methods as alternatives. First, RFLP patterns were tested to establish SLA specificities, however, the complexity of the patterns and the presence of bands for the classical and non-classical class I genes caused difficulties in class I specificity assignment, although allele inheritance could be followed within families (Flanagan *et al.*, 1988; Jung *et al.*, 1989; Smith *et al.*, 1995). Now, with the availability of many swine class I gene sequences, scientists have developed PCR–RFLP and PCR based sequencing methods that have proven to be more predictable for typing specific SLA class I alleles (Shia *et al.*, 1991; D. Smith, personal communication).

To estimate the size of the SLA class I region, Singer and her colleagues (unpublished data) used PFGE to show that the three class I SLA genes, *PD1*,

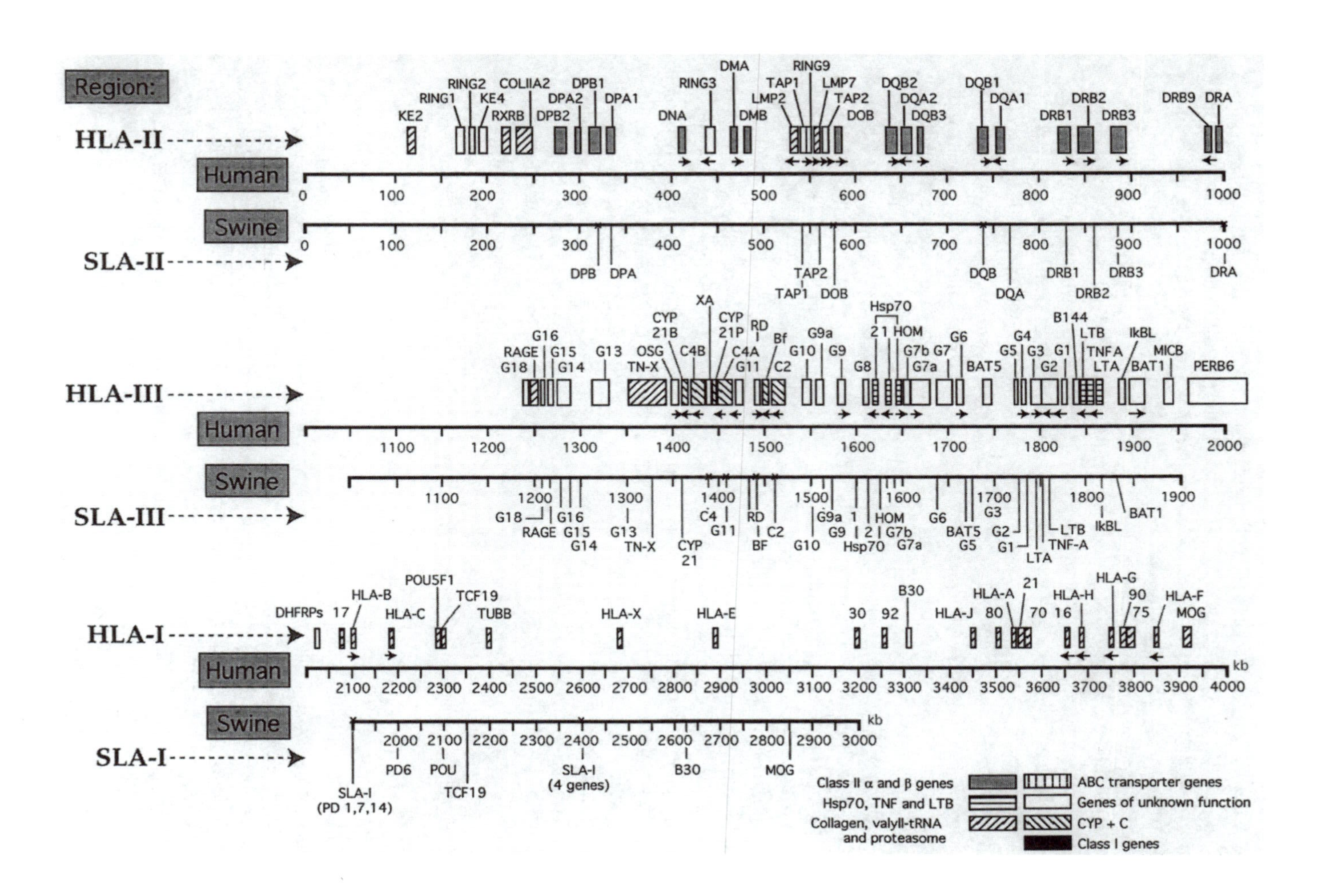

Region:
HLA-II
Human
SLA-II
Swine
HLA-III
Human
SLA-III
Swine
HLA-I
Human
SLA-I
Swine
KE2
RING1
RING2
KE4
RXRB
COLIIA2
DPB2
DPA2
DPB1
DPA1
DNA
RING3
DMA
DMB
LMP2
TAP1
RING9
LMP7
TAP2
DOB
DQB2
DQA2
DQB3
DQB1
DQA1
DRB1
DRB2
DRB3
DRB9
DRA
DPB
DPA
TAP1
TAP2
DOB
DQB
DQA
DRB1
DRB2
DRB3
DRA
G18
RAGE
G16
G15
G14
G13
TN-X
OSG
CYP 21B
C4B
XA
CYP 21P
C4A
RD
Bf
G11
C2
G10
G9a
G9
G8
Hsp70 2 1 HOM
G7b
G7a
G7
G6
BAT5
G5
G4
G3
G2
G1
B144
LTB
TNFA
LTA
IkBL
BAT1
MICB
PERB6
G18
RAGE
G16
G15
G14
G13
TN-X
CYP 21
C4
G11
RD
BF
C2
G10
G9a
G9
1 2
Hsp70
HOM
G7b
G7a
G6
BAT5
G5
G3
G2
G1
LTA
TNF-A
LTB
IkBL
BAT1
DHFRPs
17
HLA-B
HLA-C
POU5F1
TCF19
TUBB
HLA-X
HLA-E
30
92
B30
HLA-J
80
HLA-A
21
70
16
HLA-H
HLA-G
90
75
HLA-F
MOG
SLA-I (PD 1,7,14)
PD6
POU
TCF19
SLA-I (4 genes)
B30
MOG
kb
Class II α and β genes
Hsp70, TNF and LTB
Collagen, valyll-tRNA and proteasome
ABC transporter genes
Genes of unknown function
CYP + C
Class I genes

PD7 and *PD14*, all reside on the same DNA fragment (<320 kb) and that they are separated by no more than 500 kb from the fragment that contains *PD6* and *PD15*, as noted in the comparative map (Fig. 7.6). This restricted distance for the SLA class I region is supported by the recombination data, collected using traditional serologic procedures of large numbers (>3000) of piglets in pedigreed families, wherein a distance of 0.03–0.05 cM for the SLA class I region (*SLA-I*) was reported (Vaiman *et al.*, 1979; Lunney *et al.*, 1986). Xu *et al.* (1992) established that the class I genes were linked to the class III complement *C2* gene on a genomic fragment of 420 kb. Most recently, Chardon *et al.* (1996) have identified a single YAC clone which contains the class III *BAT1* gene, and four class I genes, including *PD6*, within a distance of 260 kb. Additional YAC clones link *PD6*, *POU*, *TCF19*, and four other class I genes to *B30* within a maximal distance of 650 kb, and to *MOG* with an additional 225 kb (Chardon *et al.*, 1996). The limited number of SLA class I genes has been confirmed by RFLP analyses using appropriate non-classical class I MHC DNA probes (Singer *et al.*, personal communication), although mapping of expressed sequence tags (ESTs) may reveal other expressed swine class I region genes. Overall, the detailed mapping data support the conclusion that SLA class I genes are fewer in number, and the SLA class I region significantly more compressed, than those for humans or mice.

SLA class II antigens

The SLA class II region (*SLA-II*) encodes a variety of class II genes (Fig. 7.6), of which only the *SLA-DR* and *SLA-DQ* appear to be expressed at the protein level. The expressed proteins consist of 33–35 kDa α (A) chains associated with 27–28 kDa β (B) chains (Lunney and Sach, 1978; Lunney *et al.*, 1983; Chardon

Fig. 7.6. Detailed map of the swine major histocompatibility or swine leucocyte antigen (SLA) complex as compared to the human leucocyte antigen (HLA) complex. This SLA map was developed by Lunney (1994) based on data from colleagues worldwide and combined with the HLA map published by Campbell and Trowsdale (1993, 1997). It is based on pulse field gel electrophoretic analyses by Xu *et al.* (1992), Wu *et al.* (1995) and unpublished results from several laboratories (Singer, LeGuern, personal communications) as well as mapping of overlapping cosmid and YAC clones by Peelman *et al.* (1996a,b), Brule *et al.* (1996) and Chardon *et al.* (1996). Maximal distances (from pulse field studies and overlapping clones) are noted by **x** on the SLA map; unknown distances are assumed to be homologous to the human distance. Mapping of the outer limits of the SLA class II region indicates that the *SLA-DP* gene loci are ~300 kb from *SLA-DO* and another 150 kb from the *SLA-DQ* loci (LeGuern, personal communication). From the pulse field gel electrophoresis data, it is clear that the distance from the SLA class I loci to the SLA class III loci is at least 100 kb shorter than for the homologous human genes; similarly, the distance between SLA class III and class II loci is 50 kb shorter. Most importantly, the SLA class I genes are many fewer in number and appear to be more compressed than in humans or mice, even when probed with appropriate non-classical class I MHC DNA (Singer *et al.*, personal communication), reducing the SLA class I region by ~1000 kb as compared to the human class I region.

et al., 1981). These proteins present antigen to CD4+ T cells and thus stimulate a wide array of immune responses. There is a differential expression of class II antigens. As expected from other species, swine B cells and macrophages express both SLA-DR and SLA-DQ antigens but, unexpectedly, T cells express higher levels of SLA-DR than SLA-DQ antigens. Moreover, there is preferential expression of class II antigens on the CD8+ T-cell subset (Pescovitz *et al.*, 1985; Lunney and Pescovitz, 1987). The importance/relevance of this unusual class II T cell expression has yet to be fully explained.

Molecular methods have been used to probe the complexity of genes encoded within the SLA class II region. Using both human and swine class II probes, RFLP analysis of genomic DNA has revealed a high degree of polymorphism for SLA class II genes. Analyses using class II porcine probes revealed a greater number of restriction fragments, as compared with human class II probes, and indicated that at least three α chain genes and five to eight β chain genes are encoded by each chromosome bearing the SLA complex (Chardon *et al.*, 1985a,b; Vaiman *et al.*, 1986; Warner, 1986; Sachs *et al.*, 1988). This region includes loci, which encode proteins that transport (*TAP*) or chaperone (*DM*) foreign antigen peptides as they become associated with the class II antigens for presentation to the TCR. RFLP mapping studies, performed with transporter-associated (*TAP*)*1* and *TAP2* probes (Vaske *et al.*, 1994a,b) indicated that these genes map between the *DM* and *DQ* genes but due to limited informative meioses the exact chromosomal location is assigned on this map by homology to the HLA region map. Extensive evidence from LeGuern, Sachs and colleagues on genomic DNA from NIH minipigs indicates that, for each SLA haplotype, there is one monomorphic *DRA* and two polymorphic *DRB* genes; one (or more) polymorphic *DQA* and *DQB* gene(s); no more than one *DPA* and one *DPB* gene, but no evidence for an expressed SLA-DP protein; evidence only for a *DNB* and *DZA* gene, and none for *DQZ* genes (Fig. 7.6; Sachs *et al.*, 1988; Hirsch *et al.*, 1990, 1992; Pratt *et al.*, 1990; LeGuern, personal communication). Detailed PCR-based sequencing of genomic DNA from wild boars and domestic pigs revealed the existence of a third *DRB* gene and indicated that two out of three *DRB* genes are pseudogenes (Brunsberg *et al.*, 1996). Using Swiss Large White and American Hampshire breeds the porcine *DRA* probe revealed two to five polymorphic fragments indicating that there may be polymorphism at the *DRA* locus (Shia *et al.*, 1991). These authors confirmed only one polymorphic *DQA* locus, and multiple loci for *DQB* and *DRB*, with the latter exhibiting high complexity in these outbred pigs, which the authors concluded as indicative of multiple *DRB* genes. Their evidence revealed six allelic RFLP patterns for *DQA* and *DQB*, five for *DRA* and seven for *DRB* (Shia *et al.*, 1995).

Detailed sequencing efforts with DQA, DQB, DRA, and DRB cDNA clones, designated $SLADQA^{C}$, $SLADQB^{C}$, $SLADRA^{C}$ and $SLADRB^{C}$, showed 75–87% identity to the corresponding HLA genes (Gustafsson *et al.*, 1990a,b; Pratt *et al.*, 1990). Locus-specific PCR primers have helped to assess SLA class II expression as well as to assess allelic frequencies using PCR–RFLP analyses (Gustafsson *et al.*, 1990a,b; Vage *et al.*, 1994; Shia *et al.*, 1995; Komatsu *et al.*, 1996). Based on analyses of exon 2, which encodes the β1 domain of the class II protein, 16

DQB and nine *DRB* alleles were detected (Vage *et al*., 1994; Shia *et al*., 1995; Brunsberg *et al*., 1996; Schook *et al*., 1996). Sequencing of the PCR product for the 94 amino acids of the β1 domain of *DRB* revealed 13 *DRB* alleles with an average of 13.1 ± 3.5 polymorphic residues compared with the *SLADRB*^{C} allele (Gustafsson *et al*., 1990a; Shia *et al*., 1995; Schook *et al*., 1996). Similarly, sequencing of the PCR product for the 93 amino acids of the β1 domain of *DQB* revealed seven *DQB* alleles with an average of 12.0 ± 2.1 polymorphic residues compared with the *SLADQB*^{C} allele (Gustafsson *et al*., 1990b; Vage *et al*., 1994; Shia *et al*., 1995; Schook *et al*., 1996). Komatsu *et al*. (personal communication) have found numerous new alleles of *DQB* in Ifugao (Philippine) native pigs and of *DRB1* in Xiang (Chinese) miniature pigs, and of *DQA* in both breeds. The β1 domain of the class II antigen contains the putative antigen recognition site (ARS) (Brown *et al*., 1988) which binds the peptide antigen and presents it to T cells. Of the 16 amino acids that compose the putative swine ARS, ten residues (62.5%) in the *DQB* ARS and 15 residues (93.7%) in *DRB* ARS were found to be variable as compared with 13% for *DQB* and 20% for *DRB* non-ARS residue substitutions in the ARS (Schook *et al*., 1996).

Class II specific alloantisera are available for certain SLA haplotypes but there has never been an international comparison workshop to assign specificities. MAb reactive with swine class II MHC antigens can be split into groups based on their class II locus reactivity, i.e. those reactive with SLA-DR or SLA-DQ antigens and those that are not yet specified (Lunney *et al*., 1988). The first mAb to be assigned a swine locus reactivity were the anti-mouse-I-E mAb that showed strong cross-reactions with both human and swine class II DR antigens (Lunney *et al*., 1983) as confirmed by Hirsch *et al*. (1992) using cell lines transfected by SLA-DRB genes. MAb recognizing SLA-DQ antigens are also available (Lunney, 1994b). Only two mAb that recognize unique, polymorphic determinants of swine class II DQ antigens have recently been reported (Sachs *et al*., personal communication).

SLA class III antigens

The class III region (*SLA-III*) of the SLA complex includes a broad array of genes, including those for serum complement products (*C2, C4, Bf*), the steroid 21-hydroxylase (*CYP21*) enzyme, the cytokine tumour necrosis factor (*TNF*), heat shock protein (*HSP70*), and the lymphotoxins (*LTA, LTB*). Early SLA mapping studies showed that the levels of serum complement components were regulated by one or more genes within the SLA complex (Kirszenbaum *et al*., 1985; Lie *et al*., 1987). Since that time Chardon, Peelman and their colleagues have extensively analysed the SLA class III region. The *CYP21* and *TNF* genes have been cloned and they and the complement *Bf, C2* and *C4* genes have each been mapped to the SLA complex (Fig. 7.6; Lie *et al*., 1987; Chardon *et al*., 1988, 1991; Geffrotin *et al*., 1990, 1991; Kuhnert *et al*., 1991; Ruohonen-Lehto *et al*., 1993). Later, non-immune factors such as the opposite strand gene were mapped to this region (*OSG*, Brughelle-Mayeur *et al*., 1992; now termed *TN-X*, Peelman *et al.*, 1996a) and the RNA helicase, *BAT1*, has also been mapped to this region (Peelman *et al.*, 1995). After extensive investigations using overlapping cosmid and YAC clones, the evidence clearly shows that there is only one gene each for *CYP21, C2* and *C4*, even though other species have two or more *CYP21* and *C4* genes (Geffrotin *et al.*, 1990, 1991; Xu *et al.*, 1992; Chardon *et al.*, 1996; Peelman *et al.*, 1996a,b). The high gene density in the class III region of humans (Campbell and Trowsdale, 1997) is paralleled in swine as confirmed by the detailed comparative mapping studies of Peelman and his colleagues (Fig. 7.6; Brule *et al.*, 1996; Peelman *et al.*, 1996a,b). Indeed, isolation of overlapping clones has verified that the swine MHC is even more compact, using 250 kb for this central region (*OSG* to *G7*) as compared with 300 kb for humans (Brule *et al.*, 1996; Campbell and Trowsdale, 1997).

SLA Complex map

Mapping of the outer limits of the SLA class II region indicates that the *DP* gene loci are ~300 kb from *DO* and another 150 kb from the *DQ* loci (Fig. 7.6; LeGuern, personal communication). Data from these groups have verified that the SLA complex map is quite homologous to the HLA complex; yet, from the detailed mapping data of Peelman, Chardon and their colleagues, it is now clear that regions within the SLA complex are shorter than the corresponding region of the HLA complex (Campbell and Trowsdale, 1993, 1997). Specifically the distance between SLA class II and class III genes, i.e. between *DRA* and *G18*, is at least 50 kb shorter than HLA distance (Fig. 7.6; Brule *et al.*, 1996). As noted above, the core class III is 50 kb smaller than humans, and Chardon *et al.* (1996) used YAC clones to verify that the distance between the end of the SLA class III genes (*BAT1*) and the class I genes is at least 100 kb shorter. Most importantly, the very limited number of swine class I genes (7–8 genes as compared with 17–40 for humans and mice, respectively), appears to shorten the SLA class I region by >1000 kb (Chardon *et al.*, 1996), resulting in a total distance of <2900 kb for the full SLA complex as compared with ~4000 kb for the human HLA complex (Fig. 7.6). This shortened SLA genomic length is in agreement with its short linkage distance (<1.1 cM, Smith *et al.*, 1995).

SLA alleles associated with immune and disease responses

The hallmark of the MHC is its association with genes involved in controlling an array of immune responses from complement genes to the major molecules involved in antigen presentation (Klein, 1982). In Table 7.3 is summarized the research produced over the last 12 years that assessed the influence of SLA-encoded genes on immune and disease traits. Many of these studies were only possible because of the availability of the MHC inbred lines of NIH minipigs produced by Sachs *et al.* (1976) and the well characterized panels of SLA typing reagents produced by Vaiman and his colleagues for outbred pigs (Renard *et al.*, 1988; Vaiman *et al.*, 1988).

As expected from the now classic immune response gene control studies in rodents, swine with different SLA haplotypes developed very different titres of antibodies to defined antigens (lysozyme, (T,G)-A—L) as well as for responses to sheep red blood cells and an atrophic rhinitis vaccine (Table 7.3). Similarly, cellular responses to defined antigens showed weak associations with specific SLA haplotypes. Because of the difficulty and expense of performing controlled disease challenge studies only limited numbers of such studies have been performed (Lewin *et al.*, 1991; Lunney and Grimm, 1994). Lunney and colleagues have established that both primary and secondary responses to the foodborne helminth parasite *Trichinella spiralis* are regulated by SLA associated genes (Lunney and Murrell, 1988; Madden *et al.*, 1990, 1993; Bugarski *et al.*, 1996). Preliminary studies with the foodborne protozoan parasite *Toxoplasma gondii* show no such SLA association (Dubey *et al.*, 1996). Earlier *in vitro* studies of SLA control of anti-bacterial responses (Lacey *et al.*, 1989; Lumsden *et al.*, 1993) need to be confirmed *in vivo* by actual challenges of SLA-defined mini-pigs. As the SLA genes of Chinese pig breeds are identified it seems likely that novel associations of SLA genes with skin diseases will arise, given the unusual skin structure of Meishan pigs and their increased tendency for skin diseases, such as hog mange and *Actinomyces pyogenes* infection (Komatsu *et al.*, personal communication).

SLA alleles have been documented to be associated with an array of production traits, as has been reviewed by Vaiman *et al.* (1988), Warner and Rothschild (1991) and Schook *et al.* (1996) and in chapters within this book. Just recently, Zavazava and Eggert (1997) reviewed data that reinforced earlier research indicating MHC association with behavioural responses. Several independent factors, such as MHC determination of commensal flora and associated odours, seem to be involved with MHC-determined behavioural responses in animals. Warner and Rothschild (1991) reviewed the importance of the MHC in reproductive responses, especially for genes involved in early embryogenesis. Rothschild *et al.* (1996) cautioned that, when several independent production traits are being evaluated, care must be taken in interpreting apparent SLA associations with QTL effects. Thus, research results indicating SLA associations with QTL effects or disease responses must be confirmed in larger populations and in repeated studies in order to be validated. Once proven these results could help guide breeders in selectively increasing the frequency of certain SLA alleles, i.e. those which are known to be associated with enhanced disease resistance or QTL effects.

Importance of Immunogenetics in Pig Health and Disease Resistance

There are many physiological similarities between swine and human immunogenetics, such as the similarity in their MHC structure and function. However,

Table 7.3. Immune parameters associated with specific SLA haplotypes.

Immune parameter	Breed	SLA association	Reference
Serum immunoglobulin level			
IgG	NIH minipigs	Higher in SLA class *II*^{d}	Mallard *et al.*, 1989a
IgM	NIH minipigs	None significant	Mallard *et al.*, 1989a
IgA	German breeds	Higher in *H4*	Buschmann *et al.*, 1985
Specific antibody production			
Anti-(T,G)-A—L	NIH minipigs	Higher in SLA class *II*^{d} [H10], lower in class *II*^{c} [H*4*]	Lunney *et al.*, 1986
	NIH minipigs	Higher in SLA class *II*^{d}	Mallard *et al.*, 1989b
Anti-lysozyme	Large Whites	Higher in *H12*, lower in *H10*	Vaiman *et al.*, 1978
	NIH minipigs	Higher in SLA class *II*^{c} [*H4*]; lower in SLA class *II*^{d} [H*10*]	Lunney *et al.*, 1986
Anti-sheep RBC	NIH minipigs	Higher in SLA class *II*^{d}	Mallard *et al.*, 1989b
Anti-*B. bronchiseptica*	Various breeds	Higher in *H10*	Rothschild *et al.*, 1984
Antibody avidity	NIH minipigs	None significant	Appleyard *et al.*, 1992
Serum heamolytic complement	Large Whites	Higher in *H10*, lower in *H12*	Vaiman *et al.*, 1978
	NIH minipigs	None significant	Mallard *et al.*, 1989b
Cellular responses			
DTH response	NIH minipigs	Higher in SLA class *II*^{d}	Mallard et al, 1989c
Bacterial phagocytosis anti-*Salmonella*	NIH minipigs	Higher in *SLA*^{aa}, *SLA*^{ac}	Lumsden *et al.*, 1993
Parasite antigen proliferation	NIH minipigs	Higher in *SLA*^{c}	Lunney and Murrell, 1988
Interferon induction	NIH minipigs	None significant	Jordan *et al.*, 1995
Bactericidal activity	NIH minipigs	Lower in *SLA*^{aa}, *SLA*^{ad}	Lacey *et al.*, 1989
Macrophage O_2 production	Inbred Yorkshires	None with class *II* haplotype	Groves *et al.*, 1993
Disease responses			
Melanoma appearance	Sinclair Swine	Initiator penetrance associated with *SLA B* haplotype	Tissot *et al.*, 1989, 1993; Blangero *et al.*, 1996
	Hybrid minipigs	Associated with SLA class *I* L9.10	Hruban et al, 1994
	Munich miniature swine	None significant	Mueller *et al.*, 1995
Response to primary *Trichinella spiralis* infection	NIH minipigs	Lower parasite burden in *SLA*^{cc}	Lunney and Murrell, 1988
Response to secondary *Trichinella spiralis* infection	NIH minipigs	Faster anti-parasite response in *SLA*^{aa}	Madden *et al.*, 1990, 1993
Response to primary *Toxoplasma gondii* infection	NIH minipigs	None significant	Dubey *et al.*, 1996

Source: Adapted from Schook *et al.* (1996).

swine B-cell repertoire development may prove to be remarkably different. Thus, the implications of swine immunogenetics research for swine health and productivity are extensive. First, if swine Ig production is shown to follow the same patterns as found in chickens and rabbits, then their B-cell development is not continuous and thus all antibody-producing cells are derived from a population developed *in utero* or within the first year of life. This would imply that vaccination protocols should be adjusted to maximize B-cell responses to protect swine health and productivity. In order to fully appreciate this process, it is necessary to understand how environmental or intrinsic factors influence this process. While intestinal flora apparently do not influence bursal activity in chickens, it does appear, at least, to cause proliferation and an increase in the size of the appendix in rabbits (Knight and Crane, 1994). Recently, so-called B-cell superantigens, of both bacterial and intrinsic origin, have been described (Silverman, 1994). These are speculated to cause the expansion of certain B cells including the preferential usage of the most 3′ V_H gene in rabbits (Pospisil *et al.*, 1996). It would seem that there is yet a great deal to be learned about the immunogenetics of B-cell development in swine.

For MHC controlled responses, the last five years have seen a tremendous expansion of our understanding of the complexity of the SLA complex, its genetic structure and the heterogeneity of its gene products. This knowledge will enable researchers to expand their studies to assess the effects of specific SLA alleles on QTL and disease responses and to identify exactly which genes enable pigs to resist infection by specific pathogens. The clear evidence that SLA antigens are modulated during viral disease responses is just one indication of the role of these molecules in controlling infectious diseases (Gonzalez-Juarrero et al., 1992a,b).

Several authors have reviewed the potential of using genetic approaches to improve animal disease resistance (Warner *et al.*, 1987; Rothschild, 1990). Mallard, Wilkie and their colleagues have performed a series of experiments to establish populations of pigs that they predict will be more immunologically active and thus more resistant to infectious diseases (Mallard *et al.*, 1989c; Lacey *et al.*, 1989). Such general immune resistance properties may indeed be incorporated into commercial breeding operations (McLaren, 1992). International swine genome mapping efforts (Archibald *et al.*, 1995; Yerle *et al.*, 1995; Rohrer *et al.*, 1996) will facilitate efforts to identify loci that control disease resistance. Such studies are in progress for specific candidate genes, such as the intestinal *Escherichia coli* K88 receptor (Edfors-Lilja *et al.*, 1993) and the bacterial resistance associated gene, *NRAMP* (Feng *et al.*, 1996; Tuggle *et al.*, 1996). As management changes in the pig industry alter the range of pathogens to which pigs are exposed, and as consumers demand pork products free of antibiotic contamination, it becomes increasingly more important that disease resistant breeding stock be available. Disease-resistant pigs, in well-managed facilities, will help decrease drug usage by producers and increase the health of the nation's food supply.

References

Appleyard, G.D., Mallard, B.A., Kennedy, B.W. and Wilkie, B.N. (1992) Antibody avidity in swine lymphocyte antigen-defined miniature pigs. *Canadian Journal of Veterinary Research* 56, 303–307.

Archibald, A.L., Haley, C.S., Brown, J.F., Couperwhite, S., McQueen, H.A., Nicholson, D., Coppieters, W., Van de Weghe, A., Stratil, A., Wintero, A.K., Fredholm, M., Larsen, N.J., Nielsen, V.H., Milan, D., Woloszyn, N., Robic, A., Dalens, M., Riquet, J., Gellin, J., Caritez, J.-C., Burgaud, G., Ollivier, L., Bidanel, J.-P., Vaiman, M., Rendard, C., Geldermann, H., Davoli, R., Ruyter, D., Verstege, E.J.M., Groenen, M.A.M., Davies, W., Heyheim, B., Keiserud, A., Andersson, L., Ellegren, H., Johansson, M., Marklund, L., Miller, J.R., Anderson, D.V., Signer, E., Jeffreys, A.J., Moran, C., Le Tissier, P., Rothschild, M.F., Tuggle, C.K., Vaske, D., Helm, J., Liu, C.-C., Rahman, A., Yu, T.-P., Larson, R.G. and Schmitz, C.B. (1995) The PiGMaP Consortium linkage map of the pig (*Sus scrofa*). *Mammalian Genome* 6, 157–175.

Becker, R.S. and Knight, K.L. (1990) Somatic diversification of immunoglobulin heavy chain VDJ genes: Evidence for somatic gene conversion in rabbit. *Cell* 63, 987–997.

Berens, S.J., Wylie, D.E. and Lopez, O.J. (1997) Use of a single Vh family and long CDR3s in the variable region of cattle heavy chains. *International Immunology* 9, 189–197.

Binns, R.M. (1994) The null γ/δ-TCR$^+$T cell family in the pig. *Veterinary Immunology and Immunopathology* 43, 69–77.

Blangero, J., Tissot, R.G., Beattie, C.W. and Amoss, M.S., Jr (1996) Genetic determinants of cutaneous malignant melanoma in Sinclair swine. *British Journal of Cancer* 73, 667–671.

Brown, J.H., Jardetzky, T., Saper, M.A., Samroui, B., Bjorkman, P.J. and Wiley, D.C. (1988) A hypothetical model of the foreign antigen-binding site of class II histocompatibility molecules. *Nature* 332, 845–853.

Brown, W.R. and Butler, J.E. (1994) Characterization of the single Cα gene of swine. *Molecular Immunology* 31, 633–642.

Brown, W.R., Kacskovics, I., Amendt, B., Shinde, R., Blackmore, N., Rothschild, M. and Butler, J.E. (1995) The hinge deletion variant of porcine IgA results from a mutation at the splice acceptor site in the first intron. *Journal of Immunology* 154, 3836–3842.

Brughelle-Mayeur, C., Geffrotin, C. and Vaiman, M. (1992) Sequences of the swine 21-hydroxylase gene (*CYP21*) and a portion of the opposite strand overlapping gene of unknown function previously described in human. *Biochemica Biophysica Acta* 1171, 153–161.

Brule, A., Chardon, P., Rogel-Gaillard, C., Mattheeuws, M. and Peelman, L.J. (1996) Cloning of the *G18-C2* porcine MHC class III subregion. *Animal Genetics* 27 (Suppl. 2), 76.

Brunsberg, U., Edfors-Lilja, I., Andersson, L. and Gustafsson, K. (1996) Structure and organization of pig MHC class II *DRB* genes: evidence for genetic exchange between loci. *Immunogenetics* 44, 1–8.

Bugarski, D., Cuperlovic, K. and Lunney, J.K. (1996) MHC (SLA) class I antigen phenotype and resistance to *Trichinella spiralis* infection in swine: a potential relationship. *Acta Veterinaria Beograd* 46, 115–126.

Buschmann, H., Krausslich, H., Hermann, H., Meyer, J. and Kleinschmidt, A. (1985) Quantitative immunological parameters in pigs – experiences with the evaluation of an immunocompetence profile. *Zeitschrift fur Tierzuchtung und Zuchtungsbiologie* 102, 189–195.

Butler, J.E. (1997) Immunoglobulin gene organization and the mechanism of repertoire development. *Scandinavian Journal of Immunology* 45, 455–462.

Butler, J.E., Sun, J., Navarro, P., Kacskovics, I. and Brown, W.R. (1996a) The C_H and V_H immunoglobulin genes of swine: implications for repertoire development. *Veterinary Immunology and Immunopathology* 54, 7–17.

Butler, J.E., Sun, J. and Navarro, P. (1996b) The swine heavy chain locus has a single J_H and no identifiable IgD. *International Immunology* 8, 1897–1904.

Campbell, R.D. and Trowsdale, J. (1993) Map of the human Mhc. *Immunology Today* 14, 349–352.

Campbell, R.D. and Trowsdale, J. (1997) A map of the human major histocompatibility complex. *Immunology Today* 18, 43 (poster).

Chardon, P., Vaiman, M., Renard, C. and Arnoux, B. (1978) Pig histocompatibility antigens and β2-microglobulin. *Transplantation* 26, 107–112.

Chardon, P., Renard, C. and Vaiman, M. (1981) Characterization of class II histocompatibility antigens in pigs. *Animal Blood Groups and Biochemical Genetics* 12, 59–65.

Chardon, P., Renard, C., Kirszenbaum, M., Geffrotin, C., Cohen, D. and Vaiman, M. (1985a) Molecular genetic analyzes of the major histocompatibility complex in pig families and recombinants. *Journal of Immunogenetics* 12, 139–149.

Chardon, P., Vaiman, M., Kirszenbaum, M., Geffrotin, C., Renard, C. and Cohen, D. (1985b) Restriction fragment length polymorphism of the major histocompatibility complex of the pig. *Immunogenetics* 21, 161–171.

Chardon, P., Geffrotin, C. and Vaiman, M. (1988) Genetic organization of the SLA complex. In: Warner, C., Rothschild, M., and Lamont, S. (eds) *The Molecular Biology of the Major Histocompatibility Complex of Domestic Animal Species.* ISU Press, Ames, Iowa, pp. 63–78.

Chardon, P., Nunes, M., Dezeure, F., Andres-Cara, D. and Vaiman, M. (1991) Mapping and genetic organizatiion of the *TNF* genes in the swine MHC. *Immunogenetics* 34, 257–260.

Chardon, P., Rogel-Gaillard, C., Save, J-C., Bourgeaux, N.,-Renard, C., Velten, F. and Vaiman, M. (1996) Establishment of the physical continuity between the pig SLA class I and class III regions. *Animal Genetics* 27 (Suppl. 2), 76.

Dubey, J.P., Lunney, J.K., Shen, S.K., Kwok, O.C.H., Ashford, D.A. and Thulliez, P. (1996) Infectivity of low numbers of *Toxoplasma gondii* oocysts to pigs. *Journal of Parasitology* 82, 438–443.

Dufour, V., Malinge, S. and Nan, F. (1996) The sheep Ig variable region repertoire consists of a single V_H family. *Journal of Immunology* 156, 2163–2170.

Edfors-Lilja, I., Ellegren, H., Wintero, A.K., Ruohonen-Lehto, M., Fredholm, M., Gustafsson, U., Juneja, R.K. and Andersson, L. (1993) A large linkage group on pig chromosome 7 including the Mhc class I, class II (*DBQ*), and class III (*TNFB*) genes. *Immunogenetics* 38, 363–372.

Edfors-Lilja, I., Gustafsson, U., Duval-Ilflah, Y. and Andersson, L.A. (1995) The porcine intestinal receptor for *Escherichia coli* K88*ab*, K88*ac*: regional localization on chromosome 13 and influence of IgG response to the K88 antigen. *Animal Genetics* 26, 237–242.

Ehrlich, R., Lifshitz, R., Pescovitz, M.D., Rudikoff, S. and Singer, D.S. (1987) Tissue specific expression and structure of a divergent member of a class I Mhc gene family. *Journal of Immunology* 139, 593–602.

Ellegren, H., Chowdhary, B.P., Johansson, M. and Andersson, L. (1994a) Integrating the porcine physical and genetic linkage map using cosmid derived markers. *Animal Genetics* 25, 155–164.

Ellegren, H., Chowdhary, B.P., Johansson, M., Marklund, L., Fredholm, M., Gustavsson, I. and Andersson, L. (1994b) A primary linkage map of the porcine genome reveals a low rate of genetic recombination. *Genetics* 137, 1089–1100.

Feng, J., Li, Y., Hashad, M., Schurr, E., Gros, P., Adams, L.G. and Templeton, J.W. (1996) Bovine natural resistance associated macrophage protein (N*rampl*) gene. *Genome Research* 6, 956–964.

Flanagan, M., Jung, Y.C., Rothschild, M.F. and Warner, C.M. (1988) RFLP analysis of SLA class I genotypes in Duroc swine. *Immunogenetics* 27, 465–469.

Fronicke, L., Chowdhary, B.P., Scherthan, H. and Gustavsson, I. (1995) A comparative map of the porcine and human genomes demonstrates ZOO-FISH and gene mapping-based chromsomal homologies. *Mammalian Genome* 6, 176–186.

Gaycken, U., Shabahang, M., Meyer, J.N. and Glodek, P. (1994) Molecular characterization of the porcine Mhc class I region. *Animal Genetics* 25, 357–363.

Geffrotin, C., Chardon, P., de Andres-Cara, D.F., Feil, R., Renard, C. and Vaiman, M. (1990) The swine steroid 21-hydroxylase gene (*CYP21*): cloning and mapping within the swine leucocyte antigen complex. *Animal Genetics* 21, 1–13.

Geffrotin, C., Renard, C., Chardon, P. and Vaiman, M. (1991) Marked genetic polymorphism of the swine steroid 21-hydroxylase gene, and its location between the SLA class I and class II regions. *Animal Genetics* 22, 311–322.

Gonzalez-Juarrero, M., Lunney, J.K., Sanchez-Vizcaino, J.M. and Mebus, C. (1992a) Modulation of splenic swine leukocyte antigen (SLA) and viral antigen expression following ASFV inoculation. *Archives of Virology* 123, 145–156.

Gonzalez-Juarrero, M., Mebus, C., Pan, R., Revilla, Y., Alonso, J.M. and Lunney, J.K. (1992b) Swine leukocyte antigen (SLA) and macrophage marker expression on both African swine fever virus (ASFV) infected and non-infected primary porcine macrophage cultures. *Veterinary Immunology and Immunopathology* 32, 243–259.

Grimm, D.R. and Misfeldt, M.L. (1994) Partial cloning and sequencing of the gene encoding the porcine T cell receptor d-chain constant region. *Gene* 144, 271–275.

Grimm, D.R., Richerson, J.T., Theiss, P.M., LeGrand, R.D. and Misfeldt, M.L. (1993) Isolation and characterization of gamma delta T lymphocyte cell lines from Sinclair swine peripheral blood. *Veterinary Immunology and Immunopathology* 38, 1–20.

Groves, T.C., Wilkie, B.N., Kennedy, B.W. and Mallard, B.A. (1993) Effect of selection of swine for high and low immune responsiveness on monocyte superoxide anion production and class II Mhc antigen expression. *Veterinary Immunology and Immunopathology* 36, 347–353.

Gu, F., Chowdhary, B.P., Johansson, M., Andersson, I. and Gustavsson. (1994) Localisation of the immunoglobulin gamma heavy chain (*IGHG*), cAMP dependent protein kinase catalytic beta subunit (*PRKACB*) and transition protein 2 (*TNP2*) genes in pigs by *in situ* hybridisation. *Mammalian Genome* 5, 1985–1988.

Gustafsson, K., Germana, S., Hirsch, F., Pratt, K., LeGuern, C. and Sachs, D.H. (1990a) Structure of miniature swine class II *DRB* genes: conservation of hypervariable amino acid residues between distantly related mammalian species. *Proceedings of the National Academy of Sciences, USA* 87, 9798–9801.

Gustafsson, K., LeGuern, C., Hirsch, F., Germana, S., Pratt, K. and Sachs, D.H. (1990b) Class II genes of miniature swine. IV. Characterization and expression of two allelic class II *DQB* cDNA clones. *Journal of Immunology* 145, 1946–1953.

Haas, W., Pereiva, A. and Tonegawa, S. (1993) Gamma/delta cells. *Annual Review of Immunology* 11, 637–685.

Havran, W.L., Chien, Y.H. and Allison, J.P. (1991) Recognition of γ/δ antigens by skin-derived T-cells with invariant γ/δ antigen receptors. *Science* 252, 1430–1432.

Hein, W.R. and Dudler, L. (1993) Divergent evolution of the T cell repertoire: extensive diversity and developmentally regulated expression of the sheep γ/δ T cell receptor. *EMBO Journal* 12, 715–722.

Hirsch, F., Sachs, D.H., Gustafsson, K., Pratt, K., Germana, S. and LeGuern, C. (1990) Class II genes of miniature swine. III. Characterization of an expressed pig class II gene homologous to HLA-DQA. *Immunogenetics* 31, 52–56.

Hirsch, F., Germana, S., Gustafsson, K., Pratt, K., Sachs, D.H. and LeGuern, C. (1992) Structure and expression of class II alpha genes in miniature swine. *Journal of Immunology* 149, 841–846.

Hruban, V., Simon, N., Hradecky, J. and Jilek, F. (1976) Linkage of the pig main histocompatibility complex and the J blood group system. *Tissue Antigens* 7, 267–271.

Hruban, V., Horak, V. and Fortyn, K. (1994) Presence of specific MHC haplotypes in melanoblastoma-bearing minipigs. *Animal Genetics* 25 (Suppl. 2), C10.

Itohara, S., Farr, A.G., Lafaille, J.J., Bonneville, M., Takagaki, A., Haas, W. and Tonegawa, S. (1990) Homing of a γ/δ thymocyte subset with homogenous T-cell receptors to mucosal epithelia. *Nature* 343, 754–757.

Ivanoska, D., Sun, D.C. and Lunney, J.K. (1991) Production of monoclonal antibodies reactive with polymorphic and monomorphic determinants of SLA class I gene products. *Immunogenetics* 33, 220–223.

Jordan, L.T., Derbyshire, J.B. and Mallard, B.A. (1995) Interferon induction in swine leukocyte antigen-defined miniature pigs. *Research in Veterinary Science* 58, 282–283.

Jung, Y.C., Rothschild, M.F., Flanagan, M.P., Pollak, E. and Warner, C.M. (1989) Restriction fragment length polymorphisms in the swine major histocompatibility complex. *Theoretical and Applied Genetics* 77, 271–274.

Kacskovics, I., Sun, J. and Butler, J.E. (1994) Five subclasses of swine IgG identified from the cDNA sequences of a single animal. *Journal of Immunology* 153, 3565–3573.

Kantor, A.B. and Herzenberg, L.A. (1993) Origin of murine B cell lineages. *Annual Review of Immunology* 11, 501–508.

Kirszenbaum, M., Renard, C., Geffrotin, C., Chardon, P. and Vaiman, M. (1985) Evidence for mapping pig *C4* gene(s) within the pig major histocompatibility complex (*SLA*). *Animal Blood Groups and Biochemical Genetics* 16, 65–72.

Klein, J. (1982) *Immunology: the Science of Self–Non-self Discrimination.* Wiley-Interscience, New York, 687 pp.

Knight, K.L. and Crane, M.A. (1994) Generating the antibody repertoire in rabbit. *Advances in Immunology* 56, 179–218.

Komatsu, M., Kawakami, K., Maruno, H., Onishi, A. and Takena, K. (1996) RT-PCR-based genotyping for swine major histocompatibility complex (SLA) class II genes. *Animal Technology Japan* 67, 211–217.

Kristensen, B., Renard, C., Ostergard, H. and Lunney, J.K. (1992) Reactivity of murine monoclonal anti-SLA class I antibodies on cells from outbred swine. *Animal Genetics* 23, 41.

Kuhnert, P., Wuthrich, C., Peterhans, E. and Pauli, U. (1991) The porcine tumor necrosis factor-encoding genes: sequence and comparative analysis. *Gene* 102, 171–177.

Lacey, C., Wilkie, B.N., Kennedy, B.W. and Mallard, B.A. (1989) Genetic and other effects on bacterial phagocytosis and killing by cultured peripheral blood monocytes of SLA-defined miniature pigs. *Animal Genetics* 20, 371–378.

Lewin, H.A., Clamp, P.A., Beever, J.E. and Schook, L.B. (1991) Mapping genes for resistance to infectious diseases in animals. In: Schook, L.B., Lewin, H.A. and

McLaren, D.G. (eds) *Gene Mapping Techniques and Applications.* Marcel Dekker, New York, pp. 283–299.

Lie, W.R., Rothschild, M.F. and Warner, C.M. (1987) Mapping of *C2, Bf* and *C4* genes to the swine major histocompatibility complex (swine leukocyte antigen). *Journal of Immunology* 139, 3388–3394.

Lumsden, J.S., Kennedy, B.W., Mallard, B.A. and Wilkie, B.N. (1993) The influence of the swine major histocompatibility genes on antibody and cell-mediated immune responses to immunization with an aromatic-dependent mutant of *Salmonella typhimurium. Canadian Journal of Veterinary Research* 57, 14–18.

Lunney, J.K. (1994a) Characterization of swine leukocyte differentiation antigens. *Immunology Today* 14, 147–148.

Lunney, J.K. (1994b) The swine leukocyte antigen (SLA) complex. *Veterinary Immunology and Immunopathology* 43, 19–28.

Lunney, J.K. and Grimm, D.R. (1994) The major histocompatibility complex: current state of knowledge and its use in and impact on livestock improvement. *Proceedings of the 5th World Congress on Genetics Applied to Livestock Production,* Guelph, Canada, vol. 20, pp. 230–237.

Lunney, J.K. and Murrell, K.D. (1988) Immunogenetic analysis of *Trichinella spiralis* infection in swine. *Veterinary Parasitology* 29, 179–199.

Lunney, J.K. and Pescovitz, M.D. (1987) Phenotypic and functional characterization of pig lymphocyte populations. *Veterinary Immunology and Immunopathology* 17, 135–144.

Lunney, J. and Sachs, D.H. (1978) Transplantation in miniature swine. IV. Chemical characterization of MSLA and Ia-like antigens. *Journal of Immunology* 120, 607–612.

Lunney, J.K., Osborne, B.A., Sharrow, S.O., Devaux, C., Pierres, M. and Sachs, D.H. (1983) Sharing of Ia antigens between species. IV. Interspecies cross reactivity of monoclonal antibodies directed against polymorphic, mouse Ia determinants. *Journal of Immunology* 130, 2786–2793.

Lunney, J.K., Pescovitz, M.P. and Sachs, D.H. (1986) The swine major histocompatibility complex: its structure and function. In: Tumbleson, M.E. (ed.) *Swine in Biomedical Research.* Plenum Press, New York, vol. 3, pp. 1821–1836.

Lunney, J.K., Sun, D.C., Ivanoska, D., Pescovitz, M.D. and Davis, W.C. (1988) SLA monoclonal antibodies. In: Warner, C., Rothschild, M. and Lamont, S. (eds) *The Molecular Biology of the Major Histocompatibility Complex of Domestic Animal Species.* ISU Press, Ames, Iowa, pp. 97–119.

Lunney, J.K., Saalmueller, A., Pauly, T., Boyd, P., Hyatt, S., Strom, D., Martin, S. and Zuckermann, F. (1996) Cellular immune responses controlling infectious diseases. In: Tumbleson. M. and Schook, L. (eds) *Advances in Swine in Biomedical Research.* Plenum Press, New York, pp. 307–317.

McCormack, W.T. and Thompson, C.B. (1990) Somatic diversification of the chicken light-chain gene. *Advances in Immunology* 48, 41–67.

McLaren, D.G. (1992) Biotechnology transfer: A pig breeding company perspective. *Animal Biotechnology* 3, 37–54.

Mackay, C.R., Maddox, J.F. and Brandon, M.R. (1986) Three distinct subpopulations of sheep T lymphocytes. *European Journal of Immunology* 16, 19–24.

Madden, K.B., Murrell, K.D. and Lunney, J.K. (1990) *Trichinella spiralis*: MHC associated elimination of encysted muscle larvae in swine. *Experimental Parasitology* 70, 443–451.

Madden, K.B., Moeller, R.F., Jr, Goldman, T. and Lunney, J.K. (1993) *Trichinella spiralis*: Genetic basis and kinetics of the anti-encysted muscle larval response in miniature swine. *Experimental Parasitology* 77, 23–35.

Mallard, B.A., Wilkie, B.N. and Kennedy, B.W. (1989a) The influence of the swine major histocompatibility genes (SLA) on variation in serum immunoglobulin (Ig) concentration. *Veterinary Immunology and Immunopathology* 21, 139–145.

Mallard, B.A., Wilkie, B.N. and Kennedy, B.W. (1989b) Influence of major histocompatibility genes on serum hemolytic complement activity in miniature swine. *American Journal of Veterinary Research* 50, 389–397.

Mallard, B.A., Wilkie, B.N. and Kennedy, B.W. (1989c) Genetic and other effects on antibody and cell mediated immune response in swine leucocyte antigen (SLA)-defined miniature pigs. *Animal Genetics* 20, 167–175.

Marklund, L., Johansson, M., Hoyheim, B., Davies, W., Fredholm, M., Juneja, R.K., Mariani, P., Coppieters, W., Ellegren, H. and Andersson, L. (1996) A comprehensive linkage map of the pig based on a wild pig – Large White intercross. *Animal Genetics* 27, 255–269.

Mellink, C.H.M., Lanbib-Mansais, Y., Yerle, M. and Gellin, J. (1994) Mapping of the regulatory type I alpha and catalytic beta subunits of cAMP-dependent protein kinase and interleukin I alpha and I beta in the pig. *Mammalian Genome* 5, 298–302.

Mueller, S., Wanke, R. and Distl, O. (1995) Segregation of melanocytic lesions in German Landrace crosses with Munich miniature swine Troll. *Deutsche Tierarztliche Wochenschrift* 102, 391–394.

Murtaugh, M., Butler, J.E. and Charley, B. (1996) Soluble effector and regulatory functions in the porcine immune system. In Tumbleson, M., and Schook, L. (eds) *Advances in Swine in Biomedical Research*. Plenum Publishing, New York, pp. 277–290.

Myers, M.J. and Murtaugh, M.P. (1995) *Cytokines in Animal Health and Disease*. Marcel Dekker, New York, 457 pp.

Nunes, M., Yerle, M., Dezeure, F., Gellin, J., Chardon, P. and Vaiman, M. (1993) Isolation of four *HSP70* genes in the pig and localization on chromosomes 7 and 14. *Mammalian Genome* 4, 247–254.

Osmond, D.G. (1990) B cell development in the bone marrow. *Seminars in Immunology* 2, 173–195.

Paul, W. (1993) *Fundamental Immunology*, 3rd edn. Marcel Dekker, New York, 1490 pp.

Peelman, L.J., Chardon, P., Nunes, M., Renard, C., Geffrotin, C., Vaiman, M., Van Zeveren, A., Coppieters, W., van de Weghe, A., Bouquet, Y., Choy, W.W., Strominger, J.L. and Spies, T. (1995) The BAT1 gene in the MHC encodes an evolutionarily conserved putative nuclear RNA helicase of the DEAD family. *Genomics* 26, 210–220.

Peelman, L.J., Chardon, P., Vaiman, M., Mattheeuws, M., Van Zeveren, A., Van de Weghe, A., Bouquet, Y. and R.D. Campbell (1996a) A detailed physical map of the porcine major histocompatibility complex (MHC) class III region: comparison with human and mouse MHC class III regions. *Mammalian Genome* 7, 363–367.

Peelman, L.J., Mattheeuws, M., Van Zeveren, A., Van de Weghe, A. and Bouquet, Y. (1996b) Conservation of the *RD-BF-C2* organization in the pig MHC class-III region: mapping and cloning of the pig *RD* gene. *Animal Genetics* 27, 35–43.

Pescovitz, M.D., Popitz, F., Sachs, D.H. and Lunney, J.K. (1985) Expression of Ia antigens on resting porcine T cells. A marker of functional T cells. In: Streilein, J.W. (ed.)

Advances in Gene Technology: Molecular Biology of the Immune System. ICSU Press, New York, pp. 271–275.

Plaut, A.G. (1983) The IgA1 proteases of pathogenic bacteria. *Annual Review of Microbiology* 37, 603–622.

Pospisil, R., Young-Cooper, G.O. and Mage, R.G. (1996) Preferential expansion and survival of B lymphocytes based on V_H.framework 1 and framework 3 expression. 'Positive' selection in appendix of normal and V_H-mutant rabbits. *Proceedings of the National Academy of Science, USA* 92, 6961–6965.

Pratt, K., Sachs, D.H., Germana, S., El-Gamil, M., Hirsch, F., Gustafsson, K. and LeGuern, C. (1990) Class II genes of miniature swine. II. Molecular identification and characterization of *DRB* (β) genes from the SLA^c haplotype. *Immunogenetics* 31, 1–13.

Rabin, M., Fries, R., Singer, D. and Ruddle, F.H. (1985) Assignment of the porcine major histocompatibility complex to chromosome 7 by in situ hybridization. *Cytogenetics and Cell Genetics* 39, 206–209.

Rapacz, J. and Hasler-Rapacz, J. (1982) Immunogenetic studies on polymorphism, postnatal passive acquisition and development of immunoglobulin gamma (IgG) in swine. In: *Proceedings of the 2nd World Congress on Genetics Applied to Livestock Production*, vol. III. Editorial Garsi, Madrid.

Renard, C., Kristensen, B., Gautschi, C., Hruban, V., Fredholm, M. and Vaiman, M. (1988) Joint report of the first international comparison test on swine lymphocyte alloantigens (SLA). *Animal Genetics* 19, 63–72.

Rettenberger, G., Bruch, J., Fries, R., Archibald, A.L. and Hameister, H. (1996a) Assignment of 19 porcine type I loci by somatic cell hybrid analysis detects new regions of conserved synteny between human and pig. *Mammalian Genome* 7, 275–279.

Rettenberger, G., Bruch, J., Leeb, T., Brenig, B., Klett, C. and Hameister, H. (1996b) Assignment of pig immunoglobulin kappa gene, *IGKC*, to chromosome 3q12–q14 by fluorescence *in situ* hybridization (FISH). *Mammalian Genome* 7, 324.

Rettenberger, G., Bruch, J., Leeb, T., Klett, C., Brenig, B. and Hameister, H. (1996c) Mapping of the porcine immunoglobulin lambda gene, *IGL*, by fluorescence in situ hybridization (FISH) to Chromosome 14q17–q21. *Mammalian Genome* 7, 326.

Rock, E.P., Sibbald, P.R., Davis, M.M. and Chien, Y-H. (1994) CDR length in antigen-specific immune receptors. *Journal of Experimental Medicine* 179, 323–331.

Rohrer, G.A., Alexander, L.J., Hu, Z., Smith, T.P.L., Keele, J.W. and Beattie, W. (1996) A comprehensive map of the porcine genome. *Genome Research* 6, 371–391.

Rothschild, M.F. (1990) Selection under challenging environments. In: Owen, J.B. and Axford, R.F.E. (eds) *Breeding for Disease Resistance in Farm Animals.* CAB International, Wallingford, Oxon, pp. 73–85.

Rothschild, M.F., Chen, H.L., Christian, L.L., Lie, W.R., Venier, L., Cooper, M., Briggs, C. and Warner, C.M. (1984) Breed and swine lymphocyte antigen haplotype differences in agglutination titers following vaccination with *B. bronchiseptica. Journal of Animal Science* 59, 643–649.

Rothschild, M.F., Liu, H.C., Tuggle, C.K., Yu, T.P. and Wang, L. (1996) Analysis of pig chromosome 7 genetic markers for growth and carcass performance traits. *Journal of Animal Breeding and Genetics* 112, 341–348.

Ruohonen-Lehto, M.K., Rothschild, M.F. and Larson, R.G. (1993) Restriction fragment length polymorphisms at the heat shock protein (*HSP70*) gene(s) in pigs. *Animal Genetics* 24, 67–68.

Saalmüller, A. (1996) Characterization of swine leukocyte differentiation antigens. *Immunology Today* 17, 352–354.

Saalmüller, A., Hirst, W., Maurer, S. and Weiland, E. (1994) Discrimination between two subsets of porcine $CD8^+$ cytolytic T lymphocytes by the expression of CD5 antigen. *Immunology* 81, 578–583.

Sachs, D.H., Leight, G., Cone, J., Schwarz, S., Stuart, L. and Rosenberg, S. (1976) Transplantation in miniature swine. I. Fixation of the major histocompatibility complex. *Transplantation* 22, 559–567.

Sachs, D.H., Germana, S., El-Gamil, M., Gustafsson, K., Hirsch, F. and Pratt, K. (1988) Class II genes of miniature swine. *Immunogenetics* 28, 22–32.

Saini, S.S., Hein, W.R. and Kaushik, A. (1997) A single predominantly expressed polymorphic immunoglobulin in Vn gene family, related to mammalian group I, Clan II, is identified in cattle. *Molecular Immunology* 34, 641–651.

Sanz, I., Kelly, P., Williams, C., Scholl, S., Tucker, P. and Capra, J.D. (1989) The smaller human V_H gene families display remarkably little polymorphism. *EMBO Journal* 8, 3741–3749.

Sarmiento, U.M., Sarmiento, J.I., Lunney, J.K. and Rishi, S.S. (1993a) Mapping of the porcine alpha interferon (*IFNA*) gene to chromosome 1 by fluorescence *in situ* hybridization. *Mammalian Genome* 4, 62–63.

Sarmiento, U.M., Sarmiento, J.I., Lunney, J.K. and Rishi, S.S. (1993b) Mapping of the porcine SLA class I gene (PDIA) and the associated repetitive element (C11) by fluorescence *in situ* hybridization. *Mammalian Genome* 4, 64–65.

Schook, L.B. and Lamont, S.J. (eds) (1996) *The Major Histocompatibility Complex of Domestic Animal Species.* CRC Press, New York, 319 pp.

Schook, L.B., Rutherford, M.S., Lee, J-K., Shia, Y-C., Bradshaw, M. and Lunney, J.K. (1996) The swine major histocompatibility complex. In: Schook, L.B. and Lamont, S.J. (eds) *The Major Histocompatibility Complex of Domestic Animal Species.* CRC Press, New York, pp. 212–244.

Schroeder, H.W., Jr, Hillson, J.L. and Perlmutter, R.M. (1990) Structure and evolution of mammalian V_H families. *International Immunology* 2, 41–50.

Shia, Y.-C., Gautschi, C., Ling, M.-S., Beever, J.E., McLaren, D.G., Lewin, H.A. and Schook, L.B. (1991) RFLP analysis of SLA halotypes in Swiss Large White and American Hampshire pigs using SLA class I and class II probes. *Animal Biotechnology* 2, 75–91.

Shia, Y.-C., Bradshaw, M., Rutherford, M.S., Lewin, H.A. and Schook, L.B. (1995) PCR-based genotyping for characterization of *SLA-DQB* and *SLA-DRB* alleles in domestic pigs. *Animal Genetics* 26, 91–99.

Silverman, G.J. (1994) Superantigens and the spectrum of unconventional B-cell antigens. *The Immunologist* 2, 51–55.

Sinclair, M.C., Gilchrist, J. and Aitken, R. (1997). The bovine immunoglobulin repertoire is dominated by a single diversified Vn family. *Journal of Immunology* 159, 3883–3889.

Singer, D.S., Camerine-Otero, R.D., Satz, M.L., Osborne, B., Sachs, D. and Rudikoff, S. (1982) Characterization of a porcine genomic clone encoding a major histocompatibility antigen: expression in mouse L cells. *Proceedings of the National Academy of Science, USA* 79, 1403–1407.

Singer, D.S., Ehrlich, R., Golding, H., Satz, L., Parent, L. and Rudikoff, S. (1988) In: Warner, C., Rothschild, M. and Lamont, S. (eds.) *The Molecular Biology of the MHC of Domestic Animal Species*, ISU Press, Ames, Iowa, pp. 53–62.

Smith, T.P.L., Rohrer, G.A., Alexander, L.J., Troyer, D.L., Kirby-Dobbels, K.R., Janzen, M.A., Cornwell, D.L., Louis, C.F., Schook, L.B. and Beattie, C.W. (1995) Directed integration of the physical and genetic linkage maps of swine chromosome 7 reveals that the SLA spans the centromere. *Genome Research* 5, 259–271.

Solinas, S., Pauli, U., Kuhnert, P., Peterhans, E. and Fries, R. (1992) Assignment of the porcine tumor necrosis factor alpha and beta genes to the chromosome region 7p11–q11 by in situ hybridization. *Animal Genetics* 23, 267–271.

Sun, J. and Butler, J.E. (1996) The molecular characteristic of VDJ transcripts from a newborn piglet. *Immunology* 88, 331–339.

Sun, J. and Butler, J.E. (1997) Sequence analysis of swine switch μ, Cμ and Cμm. *Immunogenetics* 46, 452–460.

Sun, J., Kacskovics, I., Brown, W.R. and Butler, J.E. (1994) Expressed swine V_H genes belong to a small V_H gene family homologous to human V_HIII. *Journal of Immunology* 153, 5618–5627.

Thome, M., Saalmüller, A. and Pfaff, E. (1993) Molecular cloning of porcine T cell receptor, and chains using polymerase chain reaction fragments of the constant regions. *European Journal of Immunology* 23, 1005–1010.

Tissot, R.G., Beattie, C.W. and Amoss, M.S., Jr (1989) The swine leucocyte antigen (SLA) complex and Sinclair swine cutaneous malignant melanoma. *Animal Genetics* 20, 51–57.

Tissot, R.G., Beattie, C.W. Amoss, M.S., Jr, Williams, J.D., and Schumacher, J. (1993) Common swine leucocyte antigen (SLA) haplotypes in NIH and Sinclair miniature swine have similar effects on the expression of an inherited melanoma. *Animal Genetics* 24, 191–193.

Tuggle, C., Schmitz, C.B., Wahls, S., Gingerich Feil, D. and Rothschild, M.F. (1996) Pig *NRAMP* cDNA cloning and analysis of other genes for disease resistance and comparative genome mapping in pigs. *Animal Genetics* 27 (Suppl. 2), 98.

Vage, D.I., Olsaker, I., Lingaas, F. and Lie, O. (1994) Isolation and sequence determination of porcine class II *DRB* alleles amplified by PCR. *Animal Genetics* 25, 73–77.

Vaiman, M. and Renard, C. (1980) Deficit of piglets homozygous for the SLA histocompatibility complex in families. *Animal Blood Groups and Biochemical Genetics* 11, 57–66.

Vaiman, M., Renard, C., Lafage, P., Ameteau, J. and Nizza, P. (1970) Evidence of a histocompatibility system in swine (SLA). *Transplantation* 10, 155–164.

Vaiman, M., Metzger, J., Renard, C. and Vila, J. (1978) Immune response gene(s) controlling the humoral anti-lysozyme response (Ir-Lys) linked to the major histocompatibility complex SLA in the pig. *Immunogenetics* 7, 231–243.

Vaiman, M., Chardon, P. and Renard, C. (1979) Genetic organization of the pig SLA complex. Studies on nine recombinants and biochemical and lyostirp analysis. *Immunogenetics* 9, 353–362.

Vaiman, M., Chardon, P. and Cohen, D. (1986) DNA polymorphism in the major histocompatibility complex of man and various farm animals. *Animal Genetics* 17, 113–121.

Vaiman, M., Renard, C. and Bourgeaux, N. (1988) SLA, the major histocompatibility complex of swine: its influence on physiological and pathological traits. In: Warner, C., Rothschild, M. and Lamont, S. (eds) *The Molecular Biology of the MHC of Domestic Animal Species.* ISU Press, Ames, Iowa, pp. 23–38.

Vaske, D.A., Liu, H.C., Larson, R.G., Warner, C.M. and Rothschild, M.F. (1994a) Rapid communication: TaqI restriction fragment length polymorphism at the porcine transporter associated with antigen processing 2 (*TAP2*) locus. *Journal of Animal Science* 72, 798.

Vaske, D.A., Ruohonen-Lehto, M.K., Larson, R.G., Warner, C.M. and Rothschild, M.F. (1994b) Rapid communication: restriction fragment length polymorphisms at the porcine transporter associated with antigen processing 1 (*TAP1*) locus. *Journal of Animal Science* 72, 255.

Viza, D., Sugar, J.R. and Binns, R.M. (1970) Lymphocyte stimulation in pigs. Evidence for the existence of a single major histocompatibility locus: *PL-A*. *Nature* 227, 949–951.

Walter, M.A. and Cox, D.W. (1991) Nonuniform linkage disequilibriuim within a 1500 kb region of the human immunoglobulin heavy chain complex. *American Journal of Human Genetics* 49, 917–923.

Warner, C.M. (1986) Genetic manipulation of the major histocompatibility complex. *Journal of Animal Science* 63, 279–287.

Warner, C.M. and Rothschild, M.F. (1991) The swine major histocompatibility complex (SLA). In: Srivastava, R., Ram, B. and Tyle, P. (eds) *Immunogenetics of the Major Histocompatibility Complex*. VCH Publishers, New York, pp. 368–397.

Warner, C.M., Meeker, D.L. and Rothschild, M.F. (1987) Genetic control of immune responsiveness: a review of its use as a tool for selection for disease resistance. *Journal of Animal Science* 64, 394–406.

Weinstein, P.D., Anderson, A.O. and Mage, R.G. (1994) Rabbit IgA sequences in appendix germinal centers: V_H diversification by gene conversion-like and hypermutation mechanisms. *Immunity* 1, 647–659.

Willems van Dijk, K., Schroeder, H.W., Jr, Perlmutter, R.M. and Milner, E.C.B. (1989) Heterogeneity in the human immunoglobulin V_H locus. *Journal of Immunology* 142, 2547–2553.

Wu, L., Rothschild, M.F. and Warner, C.M. (1995) Mapping of the SLA complex class III region by pulsed field gel electrophoresis. *Mammalian Genome* 6, 607–613.

Xu, Y., Rothschild, M.F. and Warner, C.M. (1992) Mapping of the SLA complex of miniature swine: mapping of the SLA gene complex by pulsed field gel electrophoresis. *Mammalian Genome* 2, 2–12.

Yang, H. and Parkhouse, M.E. (1996) Phenotypic classification of porcine lymphocyte subpopulations in blood and lymphoid tissues. *Immunology* 89, 76–83.

Yang, Y-G., Ohta, S., Yamada, S., Shimizu, M. and Takagaki, Y. (1995) Diversity of T cell receptor delta-chain cDNA in the thymus of a one month old pig. *Journal of Immunology* 155, 1981–1993.

Yerle, M., Lahbib-Mansais, Y., Mellink, C., Goureau, A., Pinton, P., Echard, G., Gellin, J., Zijlstra, C., DeHaan, N., Bosma, A.A., Chowdhary, B., Gu, F., Gustavsson, I., Thomsen, P.D., Christensen, K., Rettenberger, G., Hameister, H., Schmitz, A., Chaput, B. and Frelat, G. (1995) The PiGMaP consortium cytogenetic map of the domestic pig (*Sus scrofa domestica*). *Mammalian Genome* 6, 176–186.

Zavazava, N. and Eggert, F. (1997) MHC and behavior. *Immunology Today* 18, 8–10.

Cytogenetics and Physical Chromosome Maps

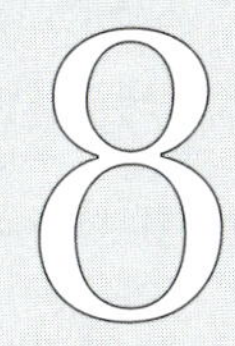

B.P. Chowdhary

Department of Animal Breeding and Genetics, Swedish University of Agricultural Sciences, Box 7023, 750 07, Uppsala, Sweden

Introduction

Pig chromosomes generate special interest among farm animal cytogeneticists. This is evident from the fact that, presently, the domestic pig (*Sus scrofa*) is cytogenetically the most widely studied farm animal. One of the main reasons for this interest has been the relatively fewer chromosomes in the domestic pig (2n = 38) as compared with other farm/domesticated animal species like cattle (2n = 60), sheep (2n = 54), goats (2n = 60), buffaloes (2n = 50), dog (2n = 78), horse (2n = 64) and chicken (2n = 78). This is further supplemented by the fact that pig chromosomes are readily distinguishable from each other, and it is relatively easy to identify most of the chromosomes even without the use of special staining methods. Ever since identification techniques became available, porcine chromosomes have largely been studied for the analysis of karyotypic anomalies in somatic and germ cells. Cytogenetic studies of embryos have also been of great interest during recent years.

The past decade has witnessed an increasing interest in gene mapping in farm animals and, as in other farm animals, this has also led to the construction of gene maps in pigs. The cause has been enhanced by *a priori* knowledge of the porcine chromosomes. Presently, the porcine genome is among the most well studied farm animal genomes and ranks high among the available gene maps of other mammalian species. The physical and the genetic linkage maps (Chapter 9) developed during the past 5 years, have played a significant role in developing a broader understanding of this genome. With the basic or fundamental maps showing an almost uniform coverage of the porcine genome with polymorphic markers, attention is now shifting towards the utilization/further development of the gene maps, in order to tackle problems of economic significance in pigs.

The present chapter aims to address the porcine genome through two broad sections. The first section primarily deals with *cytogenetics*, wherein a brief background to pig chromosomes, different identification methods applied, the standard karyotype, deviant/aberrant conditions, etc., are given. The aim of this section is to introduce the reader to the basic concept of normal and abnormal pig chromosomes and provide a summary of evolutionary aspects of the pig karyotype. Additional/detailed reading is recommended in other textbooks and reviews (Hare and Singh, 1979; Gustavsson, 1980, 1990; Eldridge, 1985; Popescu, 1989). Details on taxonomy and evolution are dealt with by Rothschild and Ruvinsky in Chapter 1 of this book. The second section, which forms a major part of this chapter, deals with the present status of the *physical gene map* in pigs and highlights various approaches hitherto used to demonstrate how the genome is physically orientated. This information, in combination with the genetic linkage maps (discussed in Chapter 9), helps in developing a better understanding of the porcine genome.

The Pig Chromosomes

Cytogenetic studies in pigs were first initiated in the early 1900s when the diploid number of chromosomes was reported to be 18 in males and 20 in females (Wødsedalek, 1913). Subsequent studies almost doubled this number to 40 in both sexes (Hance, 1917). The correct diploid chromosome number of 38 was, however, first reported by Krallinger (1931) and was supported by a number of later studies (Bryden, 1933; Hillebrand, 1936; Crew and Koller, 1939). Controversy arose when Makino (1944) disagreed with these observations (on grounds that the latter authors had misinterpreted two small satellite bearing chromosomes as four acrocentric chromosomes). The diploid chromosome number thus again rose to 40 and was confirmed by various studies during the coming years (Sachs, 1954; Spalding and Berry, 1956; Aparicio, 1960). S. Makino, almost 18 years later, realized the mistake and accepted the diploid chromosome number to be 38 (Makino *et al.*, 1962). Confirmations to this end were forthcoming, and Gimenez-Martin *et al.* (1962), McConnell *et al.* (1963) and Stone (1963) finally demonstrated that the domestic pig has 38 chromosomes.

Evolution of the pig karyotype generates interest because of differences noticed within the family Suidae, and due to polymorphism observed in the chromosome number of the wild pigs. Table 8.1 summarizes these variations and provides an overview of the status of diploid number for different members of family Suidae along with salient characteristics observed within each genus/species/subspecies.

Banding Techniques applied to Pig Chromosomes

Prior to the introduction of banding techniques, a number of variations of the porcine karyotype were presented. However, since the inception of identification methods (e.g. QFQ – Caspersson *et al.*, 1970; RBA – Dutrillaux *et al.*, 1973) using different fluorescent dyes like Quinacrine mustard, Acridine orange, Hoechst 33258, 4′, 6-diamidino-2-phenylindole (DAPI) and enzyme/chemical/heat treatments (e.g. GTG – Seabright 1971; RBG – Dutrillaux and Lejeune 1971), the pig chromosomes have been arranged almost in the same karyotypic format as seen today. The first banded karyotype reported in pig was prepared following Q-banding (Gustavsson *et al.*, 1972). Later, G-banding patterns, which resemble the former, were described (Berger, 1972). Based on these and several contemporary reports (see Gustavsson, 1980), a standard karyotype for the domestic pig was agreed upon at the Reading Conference, 1976 (see Ford *et al.*, 1980). Only GTG-banded chromosomes were discussed and published in the standard. Later, with the reporting of R-banded karyotypes in pigs (Lin *et al.*, 1980; Rønne *et al.*, 1987), which gives almost reverse banding patterns as compared with Q- and G-banding patterns, need for presenting both G- and R-banded karyotypes was felt. Hence an RBA-banded, and a revised GTG-banded karyotype with a much better resolution and more distinct bands as compared with the previous standard karyotype, were presented (Committee for the Standardized Karyotype of the Domestic Pig, 1988). The new karyotype followed the same arrangement of the chromosomes as earlier. In addition, the Committee agreed upon schematic drawings of individual G- and R-banded chromosomes, wherein, for the first time, a landmark system following the internationally agreed band designation method was decided also for the pig chromosomes. It was the first comprehensive description of a karyotype among farm/domesticated animals which enabled the description of normal karyotypes as well as the aberrant patterns with the help of a landmark system. A cursory look at the schematic drawings of G-banded porcine chromosomes arranged according to the standard karyotype (see Fig. 8.1) shows four rows of chromosomes (2n = 38). The top three rows have the meta-/submetacentric

Table 8.1. A summary of the current status of cytogenetics in the Suidae. The five genera of Suidae and the different species of *Sus* are presented with respective chromosome numbers, karyotypic characteristics and references.

Family	Genus / species	2n	Comments	References
SUIDAE	***Babyrousa*** *Babyrousa babyrousa*	38	11 autosomes and X identical to dom. pig. 5 pairs with no direct equivalents. Y is acro. with distinct p arm	Bosma (1980) Bosma and de Hann (1981)
	Phacochoerus *Phacochoerus aethiopus* (Wart hog)	34	14 autosomal pairs and sex chromosomes similar to dom. pig. Others are rob. transl. of chr. no. 13–16, 15–17	Bosma (1978); Melander and Hansen-Melander (1980)
	Potamochoerus *Potamochoerus porcus* (Bush pig)	34	12 pairs of meta/submeta. autosomes; 4 pairs of acro. X-largest meta/submeta. Y-smallest submeta.	Melander and Hansen-Melander (1980); Bosma *et al.* (1991b)
	Hylochoerus *Hylochoerus meinertzhageni* (Giant forest hog)	32	all autosomes and X are meta/submeta. Y still not analysed	Melander and Hansen-Melander (1980)
	Sus			
	Sus scrofa			
	Sus srofa domestica (Domestic pig)	38	Chromosome number same in all breeds of domestic pig hitherto studied	see Bosma *et al.* (1991b)
	Wild pig	36 or 38	Polymorphic system present	
	Asian wild pig	36	16/17 centric fusion translocation[1]	Tikhonov and Troshina (1975)
	West (former USSR, Europe, USA, etc.)	36	15/17 centric fusion translocation[1]	McFee *et al.* (1966); Gropp *et al.* (1969); Rittmannsperger (1971); Gustavsson *et al.* (1973), Tikhonov and Troshina (1975); Bosma (1976); Popescu *et al.* (1980)
	Sus vittatus leucomystax (Japanese wild pig)	38	Karyotype identical to domestic pig	Muramoto *et al.* (191965); Okamoto *et al.* (1982); Bosma *et al.* (1991a)
	Sus porsula salvanius (Pigmy hog)	38	Karyotype similar to dom. pig centromeric region of acrocen. chromosomes have extra band	Bosma (1983)
	Sus verrucosus Javan warty pig	38	Karyotype similar to dom. pig except chromosome 10 and Y	Bosma *et al.* (1991a)
	Sus barbatus The bearded pig	38	Karyotype shows MANY similarities to dom. pig	Bosma *et al.* (1991b)
	Sus celebensis Sulawesi warty pig	38	Karyotype similar to dom. pig except the Y chromosome	Bosma *et al.* (1991a)

[1]Involvement of chromosome 17 in both polymorphic systems of wild pigs indicates that centric fusion translocation has been involved in the evolution of the domestic pig (Gustavsson, 1990) and that such translocations have significantly influenced karyotype evolution within the family Suidae.

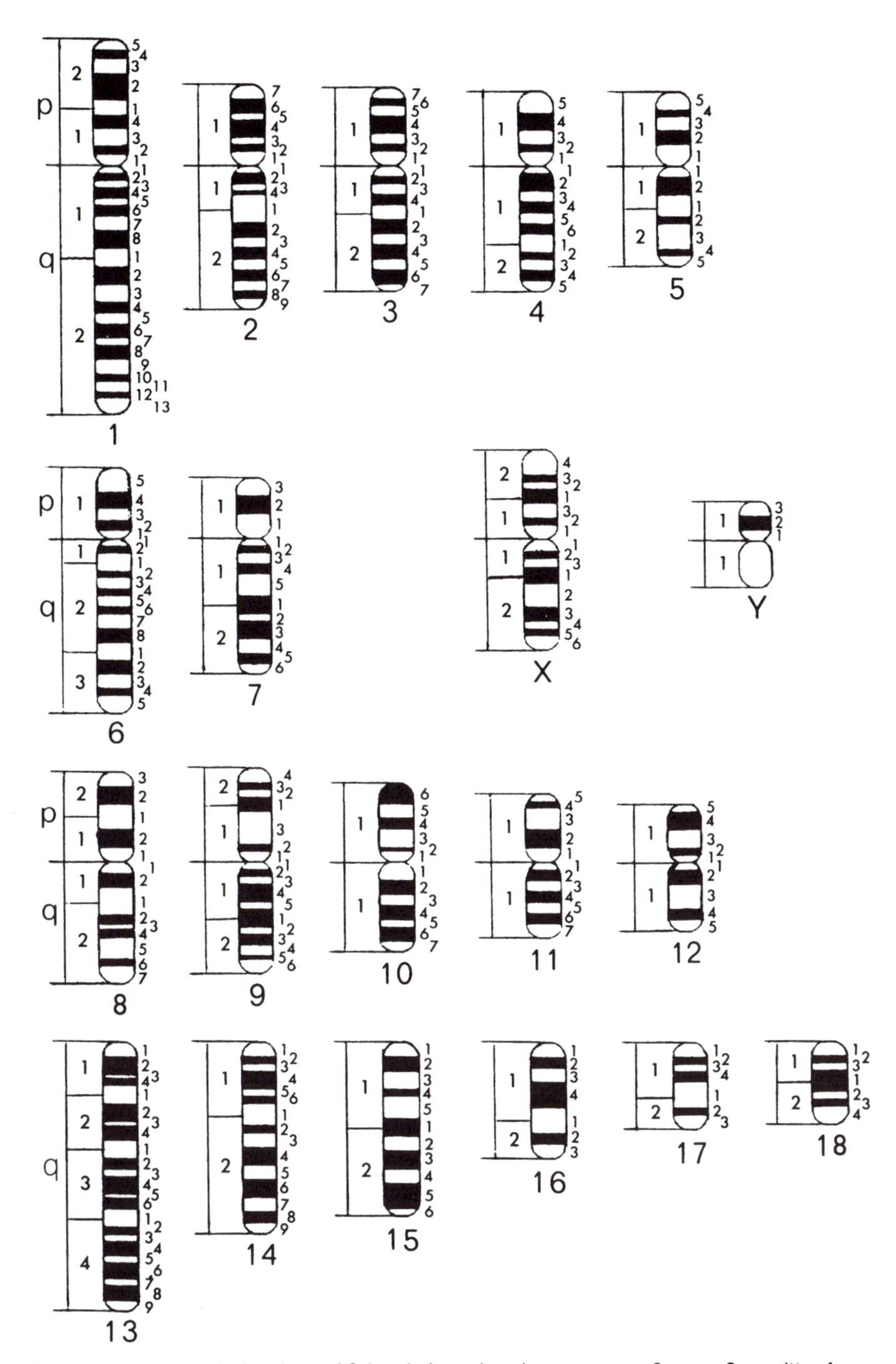

Fig. 8.1. Schematic drawings of G-banded porcine chromosomes. Source: Committee for the Standardized karyotype of the Domestic Pig, 1998.

chromosomes (12 pairs of autosomes and the sex chromosomes) while the last row has the acrocentric chromosomes (6 pairs of autosomes).

In addition to the G-, Q- and R-banding methods, which are basic for the identification of individual chromosomes, porcine chromosomes have also been studied using other staining techniques. Among these, the staining of the nucleolus organizer regions (NORs; also referred to as secondary constriction sites) is significant because it detects active NOR sites on the chromosomes. This has primarily been carried out by conventional silver staining methods using ammoniacal silver solutions (Ag-I technique; Goodpasture and Bloom, 1975; Bloom and Goodpasture, 1976). During recent years, these regions have also been studied using the *in situ* hybridization (ISH) technique. NOR regions comprising genes encoding 18S and 28S RNA are detected close to the centromeric regions of chromosomes 8 and 10 (Lin *et al.*, 1980; Toga-Piquet *et al.*, 1984; Vagner-Capodano *et al.*, 1984; Mellink *et al.*, 1991). Furthermore, a narrow intercalary site suggesting a small rRNA gene cluster was also detected on chromosome 16 (Bosma *et al.*, 1991c) using fluorescent *in situ* hybridization (FISH). Numerical variations and polymorphism (size variations) of Ag-NORs has been demonstrated in pigs at the cellular, individual and population levels (reviewed by Mellink *et al.*, 1992, 1994a). Recently, another class of rRNA genes (5S) were also ISH mapped in pig to 14q22 (Lomholt *et al.*, 1995; Mellink *et al.*, 1996).

Constitutive heterochromatin, which is most often present in the centromeric regions (therefore also referred to as centromeric heterochromatin) of pig chromosomes is primarily detected using the technique of Sumner (1972). These regions, in general, stain negatively with the G-, Q- and R-banding techniques. Although strong C-positive bands are detected on all porcine chromosomes (Hansen-Melander and Melander, 1974; Lin *et al.*, 1982), the uniarmed chromosomes have been shown to be AT (adenine and thymidine) rich while the biarmed chromosomes are GC rich (see Gustavsson, 1990). Further, on the Y chromosome, the centromeric region is C-negative, while the whole long arm is C-positive. Polymorphism in the amount of centromeric heterochromatin has been described both in uniarmed (13–18) and biarmed (1, 2, 8, 11 and 12) chromosomes (reviewed by Eldridge, 1985; Gustavsson, 1990). Of particular interest is the polymorphism associated with the interstitial heterochromatic region observed in chromosome 16 (Glahn-Luft *et al.*, 1982).

The centromeric and telomeric segments of the chromosomes can also be highlighted using the T-banding technique (THA; Dutrillaux, 1973). The technique was first applied to porcine chromosomes by Gustavsson (1983) who showed that in normal individuals, in addition to the centromeric regions, most of the telomeric regions (except 4q, 5q, 8q and 10p) stain bright. Furthermore, bright interstitial bands were also observed on 2q, 6q and 10p. The technique has been found to be useful in delineating reciprocal translocations in domestic pigs (Gustavsson and Settergren, 1983). Transfer of THA-positive structures was observed in two reciprocal translocations involving chromosomes 1 and 7 and 15 and 16 (Gustavsson *et al.*, 1988). Recently, a human telomere repeat probe $(TTAGGG)_n$ was used for FISH localization of telomeric sites in pigs (de La Seña, 1995). In addition to distinct fluorescent signals on the terminal ends of the chromosomes, an interstitial telomeric signal was also observed on chromosome band 6q22 of all the individuals examined. Interestingly, this site coincides with one of the bright bands on 6q detected by Gustavsson (1983) and is at the junction of porcine segments corresponding to human chromosomes 1 and 19 (see Frönicke *et al.*, 1996).

Chromosome Abnormalities – An Overview

As observed in other species, chromosome aberrations have also been found in pigs. These aberrations can be classified into two major categories – *numerical* and *structural.*

Numerical aberrations may either be of the *euploid* or *aneuploid* heteroploidy type. The former types of heteroploidies, where variation of the whole haploid complement is involved, have been studied in porcine sperms, ovum and zygotes (Hancock, 1959; Thibault, 1959; Bomsel-Helmreich, 1961; Hunter, 1967; McFeely, 1966, 1967; Moon *et al.*, 1975; Dolch and Chrisman, 1981). Some of the conditions demonstrated in these studies are:

1. Presence of more than two pronuclei in ovum recovered 30–48 hours after the onset of heat.
2. Occurrence of heteroploid, triploid or mosaic embryos in sows impregnated about 44 hours after the onset of heat.
3. Presence of four X or three X and one Y chromosome in tetraploid zygotes.
4. Detection of tetraploid, triploid or mixed diploid/triploid condition in some of the 10-day-old blastocysts.

Factors ranging from polyspermy, polygyny, suppression of first cleavage division, etc., are among the reasons attributed to these abnormal chromosome numbers.

In contrast to euploidy, where the whole haploid complement is involved, the aneuploid-related heteroploidy could either be associated with autosomes or sex chromosomes. Relatively few cases of aneuploidy of the porcine sex chromosomes have been described. Cases showing 37,XO (Nes, 1968; Lojda, 1975), 39,XXY, 39,XXY/40,XXXY (Breeuwsma, 1970; Hancock and Daker, 1981) have been described. A 40,XXXY case was also incidentally found during cytogenetic analysis of a litter from a translocation carrier (Gustavsson, 1984). In general, aneuploidy of the sex chromosomes is associated with abnormalities of the gonads, and hence sterility. The individuals may phenotypically appear male, female or sometimes intersex (depending on the chromosome constitution) but always demonstrate underdeveloped or deviant external or internal genitalia.

True aneuploidy of the autosomes appears to be a non-existent phenomenon, at least in liveborn piglets. However, examination of early embryos shows that the condition does exist, but is limited to the embryonic stage. Monosomy of chromosome 11 (Smith and Marlowe, 1971) and presumptive double trisomy of chromosomes 17 and 18 (Ruzicska, 1968) indicate the prenatal presence of autosomal aneuploidy in pigs. Contrary to this, mixoploid condition showing

Table 8.2. Summary of different translocations detected in pigs.

Translocation	Breakpoints	Breed	References*
1q–;11q+	1p11;6q35	LR Swe	Hansen-Melander and Melander (1970)
1p+;6q–	1q11;16q11	LW	Lockniskar (1974)
1q–;16q+	1p13;8q27	LR	Förster *et al.* (1981)
1p–;8q+	1q23;14q21	Yo Swe	Gustavsson *et al.* (1982)
1q+;14q–	1p25;14q15	SML	Golish *et al.* (1982)
1p+;14q–	1q21;17q11	Yo Swe	Gustavsson (1984)
1q–;17q+		Yo Swe	Gustavsson (1984)
1p–;3q+	1q17;14q21	LW Ru	Konovalov *et al.* (1987)
1q–;14q+	1p23;11q15	LW	Tarocco *et al.* (1987)
1p–;11q+	1p25;15q13	LR Fin	Kuokkanen and Mäkinen (1988)
1p+;15q–	1q27;15q26	LR Fin	Kuokkanen and Mäkinen (1988)
1q–;15q+	1q2.13;7q24	LW	Blaise *et al.* (1988)
1q+;7q–	1p25;11p15	LR Swe	Gustavsson *et al.* (1988)
1p–;11p+		BL	Tzocheva (1990)
1q+;18q–	1p11;6q11	Ha	Villagomez *et al.* (1990)
1p+;6q–	1q2.11;10p15	LW Swe	Yang *et al.* (1992)
1q;10p	1q2.12;14q22		Ravaoarimanana *et al.* (1992)
1q+;14q–			Zhang *et al.* (1992)
1p–;9p+	1q12;6q22	LW	Ducos *et al.* (1997)
1q–;6q+	2p13;15q24	Ga	Ducos *et al.* (1997)
2p;15q	2p17;4q11	LR Fin	Mäkinen *et al.* (1987)
2p+;4q–	2p14;14q23		Gustavsson, quoted by Kuokkanen and Mäkinen (1988)
2?;14?	3p13;7q21		Villagomez *et al.* (1993)
3p+;7q–		LW	Popescu *et al.* (1983)
3p+;7q–	3p15;13q31	Indian	Sharma *et al.* (1991)
3p+;13q–	3p14;5q23	LW	Ducos *et al.* (1997)
3p–;5q+	4p11;14q11	LF	Ducos *et al.* (1997)
4p+;14q–	4q25;13q41	LW × LF	Popescu and Legault (1979)
4q+;13q–		LR Fin	Mäkinen and Remes (1986)
4q–;15p+		Pi	Blaise *et al.* (1988)
4q–;14p+	4q16;6q28	Indian	Sharma *et al.* (1991)
4q+;6p–	5q12;8q27	LW	Ducos *et al.* (1997)
5q–;8q+	5q11;14q–	Yo Swe	Gustavsson (1984)
5q+;14q–	5q25;15q25	HA × Pi	Popescu *et al.* (1984)
5q+;15q–		LR	Parkanyi *et al.* (1992)
6p+;15q–	6p11;14q11	LR Fin	Bouters *et al.* (1974)
6p+;14q–	6p15;15q13	LW × Essex	Madan *et al.* (1978)
6p+;15q–	6q33;8q26	Pi × LW	Bonneau *et al.* (1991)
6q–;8q+	6q11;16q11	Ga × Ms	Bonneau *et al.* (1991)
6q–;16p+	6q27;14q21	SML	Ducos *et al.* (1997)
6q+;14q–	6p15;13q41	LW × Pi	Ducos *et al.* (1997)
6p+;13q–	7q21;11q11	SML	Ducos *et al.* (1997)
7q–;11q+	7q24;12q15	Yo Swe	Gustavsson *et al.* (1982)
7q–;12q+	7p13;13q21	Yo Fin	Kuokkanen and Mäkinen (1987)
7p+;13q–	7q26;17q11	Ha	Gustavsson *et al.* (1988)
7q+;17q–	7q13;8q27	Ha	Villagomez *et al.* (1995)

Table 8.2. *continued*

Translocation	Breakpoints	Breed	References*
7q;8q	7q24;15q12		Ravaoarimanana *et al.* (1992)
7q+;15q–	8p23;14q27	LW	Konfortova *et al.* (1995)
8p;14q	8q27;13q36		Ravaoarimanana *et al.* (1992)
8q;13q	9p24;11q11		Ravaoarimanana *et al.* (1992)
9p+;11q–	9p24;15q13	Yo Swe	Gustavsson *et al.* (1982)
9p+;15q–	11p1,5;15q1,3	LW × LF	Ducos *et al.* (1997)
11p+;15q–		LR Fin	Henricson and Bäckström (1964)
	11p14;16q14		Gustavsson, quoted by Kuokkanen and Mäkinen (1988)
11p+;16q–		LW × Pi	Ducos *et al.* (1997)
11q+;13q–		LW	Ducos *et al.* (1997)
12q+;15q–	12q13;13q11	LW Ru	Konovalov *et al.* (1987)
12q+;13q–	13q21;14q27	Minisib	Astachova *et al.* (1990)
13q–;14q+		Yo Swe	Hageltorn *et al.* (1976)
	14q29;15q24		Golish and Ritter (1990)
14q+;15q–	15q26;16q21	Ha	Gustavsson and Jönsson (1992)
15q+;16q–	15q13;17q21	Yo	Gustavsson *et al.* (1988)
15q–;17q+	16q23;17q21	LW	Ducos *et al.* (1997)
16q+;17q–		LR × Du	Popescu and Boscher (1986)
	Xq24;13q21	LR × Viet	Astachova *et al.* (1991)
Xq+;13q–		Ha	Gustavsson *et al.* (1989)
Xp+;14q–			Villagomez *et al.* (1990)
rep(X:14)(p+;1–)			Singh *et al.* (1994)

Abbreviations: LR, Landrace; LF, French Landrace; LW, Large White; Yo, Yorkshire; Pi, Piétrain; Ha, Hampshire; Ga, Gascon; Du, Duroc; MS, Meishan; BL, Belgian Landrace; SLM, Syntetic Male Line; Swe, Swedish; Fin, Finnish; Ru, Russian; Viet, local breed from Vietnam. It is expected that approximately another 10–15 new translocations will be added to this list in the very near future. (Courtesy, Dr Alain Ducos.)
*Individual references cited in this table are not listed in the reference list but may be found in Ducos *et al.*, 1997.

extra chromosomes and chromosome fragments (37,XY,-18/38,XY/39,XY+18 and 37,XY,-18/38,XY) have been described in liveborn individuals (Vogt *et al.*, 1974). Furthermore, Bösch *et al.* (1985) incidentally discovered a pregnant true hermaphrodite 38,XY/39,XX,+14 individual, which was mosaic in 29% of the cells examined.

Structural aberrations of the porcine chromosomes may be broadly grouped into *translocations* and *duplications/deletions*. During recent years, *inversions* have also been detected in pigs. Among *translocations*, reciprocal (rcp) translocations are commonly observed in pigs. Of the more than 50 different translocations described in pigs (Table 8.2), approximately 50% have been detected in our laboratory by I. Gustavsson and co-workers during the past 20 years. Comparative analysis of various rcp translocations in pigs (Table 8.2) shows that there are certain chromosomes which are preferentially involved in reciprocal rearrangements, indicating that the distribution of participating chromosomes in translocations is non-random. Among the chromosomes most often involved are 1, 7, 14 and 15. However, among the chromosomes found to be least often involved are 9, 10, 12 18 and Y. Furthermore, within the chromosomes, there are certain sites/bands which are more involved in breakages as compared with others. For example, band 1q21 is involved in reciprocal translocations with five different chromosomes (5, 7, 14, 17 and 18). Similarly, of the five different rearrangement/breakage sites on chromosome 7, band q21 is the most involved. Furthermore, on chromosome 14, bands q11 and q21 appear to be more susceptible to translocation-related breakages. Whether these breakage sites have evolutionary significance or always correspond to fragile sites on the chromosomes is still a matter of open discussion. At least the human–pig comparative painting data (Zoo-FISH; Rettenberger *et al.*, 1995b; Frönicke *et al.*, 1996) do not show a clear-cut correspondence between the translocation breakpoint regions and bands separating conserved syntenic regions. Attempts have been made in the recent past to correlate the translocation breakpoints with the presence of fragile sites (Rønne, 1995). In several instances, a clear correlation between the two was reported. However, the question which still requires clarification is: to which type of fragile sites (these can be caused/produced by several different factors) do translocation breakpoints truly correspond?

Irrespective of the chromosomes involved in reciprocal translocations, it has been unambiguously shown that translocations result in the production of a variety of both balanced and unbalanced gametes (McClintock, 1945; Ford and Clegg, 1969; King, 1980). The latter are considered to be the main reason of reduced fertility in translocation carriers (reviewed by Gustavsson, 1990). As a result of a comprehensive survey of the Swedish pig population, it was found that at least 50% of all the breeding boars removed from the population due to lower than average litter size carried reciprocal translocation. The decrease in the litter size typically ranged from 25 to 50%, and was attributed mainly to the early elimination of chromosomally unbalanced translocation zygotes – sometimes even before implantation (Åkesson and Henricson, 1972; Hageltorn *et al.*, 1976; King *et al.*, 1981; Popescu and Boscher, 1982; Gustavsson *et al.*, 1983). Several other reports, mostly describing individual cases of reciprocal translocations in pigs, also indicate similar trends (reviewed by Gustavsson, 1990). It is worth mentioning that, in general, the semen characteristics of translocation carriers appear normal. Further, most often, the balanced carriers have a normal body confirmation, although serious malformations have been described in stillborn piglets carrying a presumptive balanced rcp(1q–;11q+) translocation (Hansen-Melander and Melander, 1970). Chromosomal analysis of stillborn piglets from a translocation carrier boar rcp(14;15)(q29;q24) showed partial monosomy and partial trisomy of the chromosomes involved along with anatomical deformities (Gustavsson and Jönsson, 1992). Early postnatal death of 9% of the piglets from a translocation carrier rcp(7;17)(q26;q21) has also been reported (Villagomez *et al.*, 1995a). Cytogenetically, the piglets had unbalanced karyotypic constitution. It has been proposed that reciprocal translocations in males might sometimes induce degenerative changes in the testicles (Chandley *et al.*, 1972). This has to some extent been supported in a few investigations carried out during recent years (Gustavsson *et al.*, 1989; Villagomez, 1993).

During the last decade, synaptonemal complex (SC) analysis of the translocation carriers in pigs has been instrumental in understanding the underlying mechanism of pairing of the involved chromosomes during meiosis 1, and the effect it has on the production of unbalanced gametes (Gustavsson *et al.*, 1988, 1989; Villagomez, 1993). An interesting dimension to this work has also been the preparation of a SC karyotype of boar spermatocytes (Villagomez, 1993) wherein the bivalents are arranged according to the autosomal somatic karyotype, based on their relative lengths and arm ratios (Fig. 8.2).

In addition to the reciprocal translocations, *centric-fusion translocations* (also traditionally referred to as Robertsonian translocations – rob), though rare, have also been reported in pigs. Robertsonian translocations between chromosomes 13 and 17 have been described in normal (Miyake *et al.*, 1977; Alonso and Cantu, 1982), intersex (Masuda *et al.*, 1975) and even malformed piglets (Miyake *et al.*, 1977). The rob carriers, like the rcp carriers, demonstrate varying degrees of decrease in litter size (Schwerin *et al.*, 1986; McFeely *et al.*, 1988).

Duplications and deletions of whole chromosome/arm, which are primarily caused by reciprocal translocations and inversions, are rarely observed in live pigs/piglets, primarily because gametes carrying such a condition when fertilized, result in early death of the embryos. McFeely (1966) reported spontaneous deletion of a whole arm of a medium sized biarmed chromosome in 10-day-old embryos of cytogenetically normal parents. Cases of tertiary trisomy and monosomy of chromosome 17 have recently been described from a balanced rcp carrier, as discussed above (Villagomez *et al.*, 1995a), although the piglets died within a day after birth. One of the piglets with the tertiary trisomy of chromosome 17, however, survived and demonstrated normal body confirmation and testis size (Villagomez *et al.*, 1995b). In spite of histologically normal testes, 17.5% of the spermatozoa of this boar demonstrated acrosomal defects.

Chromosomal aberrations, numerical or structural, are overwhelmingly considered to arise spontaneously, and are therefore considered as new mutations. However, as is observed by the follow-up of the litters from chromosomally aberrant individuals, some of these may be considered as inherited. In addition to above described aberrations, some unspecified conditions, e.g. presence of an extra undefined centric fragment (39,XX,+cen) and mixoploid condition (38,XX/39,XX,+cen) from male offspring of an autosomal aneuploid individual (see above) have also been described (Vogt *et al.*, 1974). Unspecified structural chromosomal aberrations have also been reported in 29% of the cells of a phenotypically normal boar which produced stillborn/malformed piglets at a high rate (Oprescu *et al.*, 1976).

Hermaphroditism, which is associated with the presence of genitalia representing both sexes, is common in pigs. The animals may be classified as classical, true or unilateral hermaphrodites, or even as male and female pseudohermaphrodites (reviewed by Gustavsson, 1990). It is extremely rare for true intersex pigs to become pregnant (Bösch *et al.*, 1985). The majority of hermaphrodite pigs have been cytogenetically known to be 'females' (see

Miyake, 1973). However, in some instances, the chromosomal constitution could be that of a male. A genetic background to pig hermaphroditism, indicating it to be inherited either as a dominant gene or a recessive sex-linked gene was also suggested (Hulot, 1970; Lojda, 1975). However, other more established factors like deletion in the X or Y chromosome, or translocation of the Y

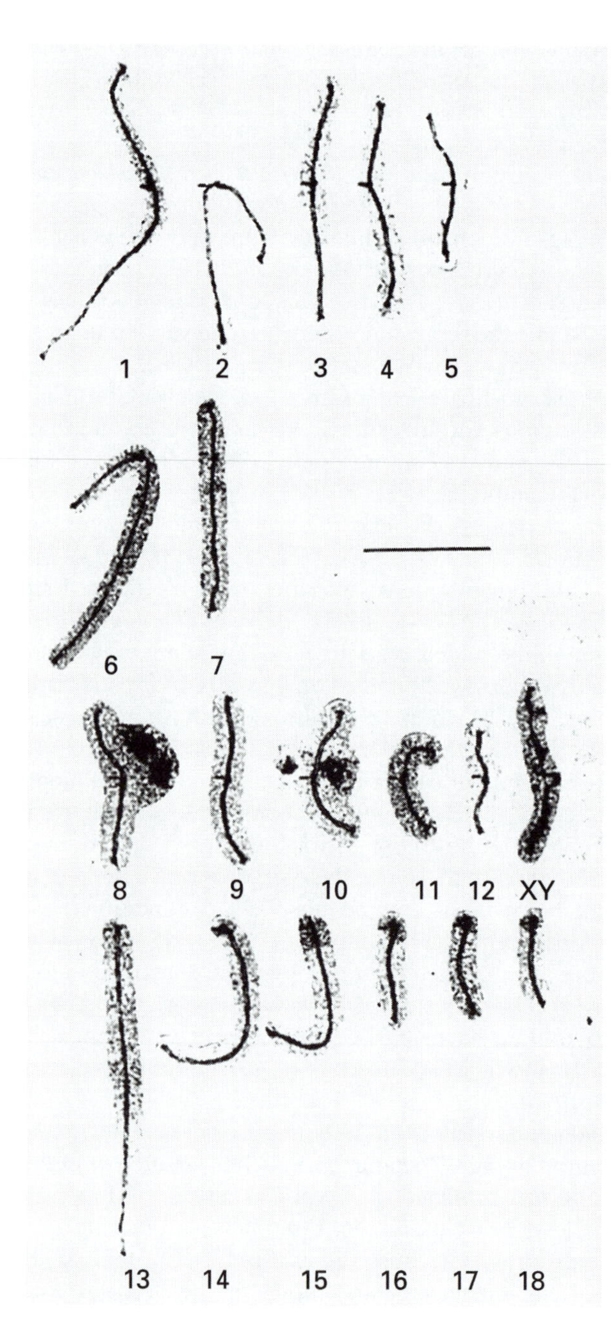

Fig. 8.2. Synaptonemal complex (SC) karyotype of the domestic pig. The bivalents have been arranged according to the system used for mitotic chromosomes. Note that kinetochores (–) in the sex chromosomes are not identified. Bar indicates 5 μm. (Courtesy D.A.F. Villagomez, PhD thesis.)

chromosome (specifically the male determining factor) to an autosome or even the X chromosome, may result in hermaphrodites. Other reasons such as overcrowding of embryos resulting in diffusion of adrenal hormones, etc., have also been suggested (Breeuwsma, 1970).

Chimerism in pigs, like in other species, is the result of the intrauterine fusion of the umbilical cord, resulting in vascular exchange between female and male pigs. The resulting individuals most often show an admixture of XX/XY type of cells in blood. Cases of whole body XX/XY (Basrur and Kanagawa, 1971) and XX/XXY (Toyama, 1974) chimerism have also been seen. Christensen and Bräuner Nielsen (1980) have shown that XX/XY chimerism in females does not necessarily result in intersexuality. However, Bosma *et al.*, (1975) and Colenbrander and Wensing (1975) observed XX/XY chimerism in blood and XX chromosome constitution in other body cells of female pigs with hernia inguinalis and ovarian aplasia. Except for the prepuce development, the animals had normal external female genitalia. However, internally, gross variations were observed.

Lastly, it might be worth mentioning that chromosomal abnormalities in pigs have also been studied with the help of methods other than banding, viz. flow cytometry. Matsson *et al.* (1986) were among the first to try this approach to study and sort a translocation involving chromosomes 13 and 14. The peak corresponding to the very small part of 13q− could not be distinguished in the flow karyotype although the peak for extra large 14q+ was fairly evident. Further, Hausmann *et al.* (1993) demonstrated a 6/15 translocation in pigs using slit scan flow cytometry. Correlation of the peak with the two chromosomes was carried out by post-sorting GTG banding. Recently, Jensen *et al.* (1994) applied the technique to semen of boars carrying different translocations. Although this approach has been proposed as a cost effective method to detect translocations which are otherwise difficult to analyse using conventional banding techniques, the technique has limitations (e.g. lack of internal standard) in clearly differentiating between normal and imbalanced gametes.

Gene Mapping in Pigs – Physical Gene Maps

Gene mapping studies in pigs date back to almost three decades when Andresen (1963) initiated the basic study of pig blood groups. This initial era of immunogenetic and biochemical genetics, which made significant contributions for almost two decades (for details see Chapters 5, 6 and 7), can be termed as the beginning of the construction of gene maps in pigs. By the end of the 1970s, six linkage groups were known in pigs (see Gahne, 1979). The next decade made an addition of only three new linkage/syntenic groups. However, assigning these syntenic/linkage groups to specific chromosomes (except the X chromosome) still eluded researchers, because: (i) detailed study of the porcine karyotype was still in progress; and (ii) no appropriate methods were available to associate a gene or group of genes to a specific chromosome. Nonetheless, the early 1980s marked the formal beginning of physical gene mapping in pigs because of the introduction of a new technique, viz. somatic hybridization. This decade witnessed a slow but steady progress in identification of new syntenic/linkage groups and their assignment to specific chromosomes or chromosomal regions (see below). Towards the end of the 1980s, development and application of new techniques enhanced the power of physical gene mapping in pigs (reviewed by Chowdhary, 1991). Further, during the past 6–7 years, methodological advancements and coordinated efforts of a number of research groups from different countries have significantly increased the pace of the development of porcine physical gene maps. The following section of the chapter deals with individual physical mapping techniques hitherto applied and summarizes the results obtained through them.

Somatic Hybridization

Analysis of somatic cell hybrid (SCH) panels has played a significant role in the development of physical gene maps in humans. The technique basically involves fusion of two cell lines originating from different species (e.g. A and B) resulting in hybrid cells which unilaterally lose chromosomes of the species under investigation (e.g. A) and 'stabilize' with one or a few chromosomes of that species plus the chromosomal complement of the other species (B). Each such hybrid cell with a specific combination of the chromosomes of the two species is referred to as a 'clone'. The 'clones' thus obtained are then analysed to ascertain synteny between two loci (reviewed by Kao, 1983). In some cases, the clones may have only one complete chromosome or an arm/segment of species 'A'. Such hybrid clones are more useful than the above described clones because they increase the precision of mapping of the DNA fragments. By the mid-1980s, when more than 600 loci were assigned to human chromosomes/chromosomal subregions using the SCH technique, only 10–15 loci were mapped in pigs using this method. Among the first studies related to the use of SCH panels for gene mapping in pigs include those reporting the mapping of *HPRT*, *G6PD*, *PGK* and the *GLA* enzyme loci to the X chromosome (Förster 1980; Förster *et al.*, 1980; Leong *et al.*, 1983b). Later, 10–12 more loci were added to this tally, all of them to autosomes. Some of these loci (e.g. *LDHB* and *PEPB*) were even assigned to specific chromosomal regions because the hybrids were known to contain only a specific part of the chromosome concerned (Ryttman *et al.*, 1986, 1988).

As in other species, the initial era of SCH genetics in pigs was restricted to the mapping of enzyme genes only because the available methodology required analysis of gene products in SCH clone lysates. Although in humans, filter hybridization using Southern blots (Southern, 1975) greatly improved the specificity and sensitivity of detecting the presence of anonymous as well as specific gene sequences in the hybrid cell genomic DNA (see Ruddle, 1981), its application in pig somatic cell genetics, though very limited, began much later (e.g. Harbitz *et al.*, 1990b; Poulsen *et al.*, 1991; Thomsen *et al.*, 1991). Although other farm animal gene maps, specifically cattle, benefited much from this advancement (see Womack, 1990), pig gene mappers drifted towards other rather attractive and comparatively more accurate physical mapping methods, viz. *in situ* hybridization (ISH). Due to this, there was a big upsurge in chromosomal localization of loci and SCH analysis was considered as a tiresome and time-consuming mapping method with very little output.

During the mid-1990s, the SCH analysis once again gained momentum in pigs (Rettenberger *et al.*, 1994d; Zijlstra *et al.*, 1994; Yerle *et al.*, 1996a), primarily due to the substitution of PCR-based analysis in place of Southern blot analysis. The new technique not only averted the use of radioactive material; but also enabled the mapping of any DNA segment with known sequence/primer pair through a simple PCR reaction. The approach has almost revolutionized physical assignment of DNA segments, and as far as is known, over 150 loci (microsatellite, coding sequences and anonymous DNA segments) have been mapped in pigs within the past 3 years (Rettenberger *et al.*, 1994b,c,d, 1995a, 1996a,b; Bruch *et al.*, 1996; Robic *et al.*, 1996, Zijlstra *et al.*, 1996). A summary of different SCH panels used, methodology applied and loci mapped is presented in Table 8.3. The table shows that a total of 236 loci have so far been mapped using SCH analysis, of which 18 have been mapped using cell lysate enzyme analysis, 3 by Southern blot analysis, and more than 175 by PCR analysis. Further, of the total loci mapped, 121 are microsatellite markers, 102 specific genes and 13 anonymous sequences. The recent upsurge of SCH analysis based mapping data is encouraging, and at least two mapping panels (Rettenberger *et al.*, 1994b,c,d; Robic *et al.*, 1996) are accurate enough to assign a marker either to a chromosomal arm or in some cases even to specific bands. The third panel developed in Utrecht (see Zijlstra *et al.*, 1996) is also fairly useful, although it is not completely informative for some chromosome pairs.

Cytogenetic analysis, or more specifically identification of chromosomes in hybrid cell lines, used to be difficult in the early days of SCH analysis. Initially, the technique of Bobrow and Cross (1974) was used to differentiate between the human and rodent chromosomes. Later improved chromosome identification methods, e.g. the QFQ- (Caspersson *et al.*, 1970) and THA-banding (Dutrillaux, 1973), were applied. Modifications in these techniques proposed by Gustavsson (1983) were used to differentiate between the pig and mouse chromosomes (see Ryttman *et al.*, 1986). Radioactive *in situ* hybridization with total porcine genomic DNA was also carried out on metaphase preparations of individual cell lines to judge the number of porcine chromosomes present (Fries *et al.*, 1990). During recent years, two separate approaches, which are easy to perform and provide more accurate information as compared with other available methods, have been used for chromosome identification. The most direct and efficient of these involves extraction of DNA from the hybrid cell line, labelling it non-radioactively and then hybridizing it to normal porcine metaphase preparations. The results highlight those porcine chromosomes or chromosomal segments which are represented in the hybrid cell line (Frengen *et al.*, 1994; Yerle *et al.*, 1996a; Zijlstra *et al.*, 1996). Chromosome numbers and fairly accurate band descriptions of the segments can then be used to define each hybrid cell line. Alternatively, whole chromosome paints (WCPs) representing each porcine chromosome may be used to 'paint' metaphase spreads from the hybrid cell

Table 8.3. Somatic cell hybrid mapping in pigs.

Type of hybrid line	No. of clones	Loci mapped	Method	References
Pig–rodent	?	*TBP10*	P	Brenig and Leeb (1996)
Pig–mouse	21	*4 (MS)*	P	Brown et al. (1994)
Pig–mouse	21	*EPO, MDH2, ZP3*	P	Bruch *et al.* (1996)
Pig–mouse	27	*LEPR*	P	Ernst *et al.* (1997)
Pig–mouse Pig–hamster	27	*APOC3, C3, CACNP1, CANX, CAP2A2, CAPN1, CDC42, CKB, COX8, CST3, DNCL, EIF4E, FTL, GBP1, HARS, HMG1, KICT, LDHA, PST1, RFC5, SCP2, TPM4, TUBAL1, U2AF1, UGT1, VIL1, SLC3A1, SSC1G2, SSC20B10, SSC2C10, SSC9C8 (loci underlined are cDNA clone numbers)*	P	Fridolfsson *et al.* (1996); Fridolfsson *et al.* (unpublished observations)
Pig–mouse	27	*G6PD, HPRT, PGK1*	E	Förster 1980; Förster *et al.* (1980)
Pig–mouse	27	*PGM1, PEPB, LDHA, LDHB, ME1, MDH1*	E	Förster and Hecht (1984)
Pig–mouse	27	*PMI, NP, SOD1*	E	Förster and Hecht (1985)
Pig–hamster Pig–mouse	29	*MPI, NP, PKM2*	E	Gellin *et al.* (1981)
Pig–hamster Pig–mouse	18 19	*CRC (RYR), GPI*	S	Harbitz *et al.* (1990b)
Pig–mouse Pig–hamster	29	*2(MS)* *ACTB, ALDOB, ATPS, BCMA, C1QA, dystroglucan, WARS, ZFP*	P	Jørgensen *et al.* (1996c)
Pig–hamster Pig–mouse	29	*DAX1, SOX2, SOX9, CYP19*	P	Lahbib-Mansais *et al.* (1996)
Pig–mouse	21	*RYR2*	P	Leeb *et al.* (1995)
Pig–mouse	11	*SOD-1*	E	Leong *et al.* (1983a)
Pig–mouse	7	*GLA G6PD, HPRT*	E	Leong *et al.* (1983b)
Pig–mouse	21	*OBS*	P	Neuenschwander *et al.* (1996)
Pig–mouse	26	*ANPEP*	S	Poulsen *et al.* (1991)
Pig–mouse	21	*INSL3*	P	Rettenberger *et al.* (1994b)
Pig–mouse	21	*3 (MS)*	P	Rettenberger *et al.* (1994c)
Pig–mouse	21	*PRM1, TYP2, UBC*	P	Rettenberger *et al.* (1994d)
Pig–mouse	21	*7 (MS)* *ANPEP, ATP2, BNP1, CGA, DAGK, FSHB, IFNG, IGF1, IL1B, SPP1*	P	Rettenberger *et al.* (1995a)
Pig–mouse	21	*ACO2, ADRA2, CAST, CCK, CHAT, IGKC, IGLV, IL4, IL6, INHA, LIF, MX1, PTH, RBP2, TCRA, TCRB, TGFB2, TGFB3, UOX*	P	Rettenberger *et al.* (1996a)
Pig–mouse	21	*APOC3*	P	Rettenberger *et al.* (1996b)
Pig–mouse Pig–hamster	8 19	*100 (MS)*	P	Robic *et al.* (1996)

Table 8.3. *Continued*

Type of hybrid line	No. of clones	Loci mapped	Method	References
Pig–mouse	34	*LDHA, LDHB, MPI, PEPB, PGM1*	E	Ryttman et al. (1986)
Pig–mouse	18	*MPI, NP*	E	Ryttman *et al.* (1988)
Pig–hamster	39	*PGD*	E	Thomsen *et al.* (1991)
Pig–mouse				
Pig–hamster		*HOXB, TK1*	S	Thomsen and Zhdanova (1995)
Pig–mink	17			
Pig–mouse	27	*6 (MS)*	P	Yerle *et al.* (1996a)
Pig–hamster		*IFNA, TNFB*		
Pig–mink	17	*TK1, UMPH2*	E	Zhdanova *et al.* (1994)
Pig–hamster	15	*12 (MS)*	P	Zijlstra *et al.* (1996)
Pig–mouse	6			

The table provides a list of all somatic cell hybrid panels hitherto generated, the number of clones present in each, the genes/microsatellites (MS) mapped, and the approch used (E = enzyme analysis; P = PCR analysis; S = Southern blot analysis) to map these loci.

lines. These may be generated from flow-sorted porcine chromosomes (Dixon *et al.*, 1992; Bouvet *et al.*, 1993; Yerle *et al.*, 1993b), or through chromosomal microdissection (see below, page 234–235). Compared to the above described method, these approaches are more tedious and provide less accurate information. One can only find the presence or absence of genetic material corresponding to a certain porcine chromosome and it is difficult to define accurately the origin in terms of arm or band numbers.

Recently, somatic hybrid panels have also been used to map cDNA sequences derived from a porcine small intestine cDNA library. Fridolfsson *et al.* (1996) designed primers from a set of sequenced unique clones (Winterø *et al.*, 1996) and used a PCR-based approach on the Toulouse somatic cell hybrid panel (Yerle *et al.*, 1996a), to map a total of 20 (plus 11 more; personal communication) genes to specific chromosomes/chromosomal regions in pigs.

Radiation Hybrid Mapping

Radiation hybrid (RH) mapping is basically a somatic cell genetic technique which was developed as a general approach for mapping mammalian genomes. Suggested first by Pontecorvo (1971), and utilized for mapping of the human chromosomes by Goss and Harris (1975), the technique had a rebirth during the 1990s, when it was used for the construction of high resolution maps in humans (see Cox *et al.*, 1990; Gyapay *et al.*, 1996). The technique primarily involves breaking the chromosome of interest or the whole genome of a species (e.g. human or pig) into several fragments by a high dose of X-ray irradiation. The broken fragments are recovered in recipient cells (e.g. rodent) and approximately a hundred such hybrid clones are analysed for the presence or absence of DNA markers. The underlying principle for mapping rests on the fact that the further apart two markers are on a chromosome, the greater are the chances that they will be separated by X-ray breakage and vice versa. The frequency of breakage helps in estimating the distance statistically, and this in turn assists in ordering a set of markers. As far as known, at least two porcine whole genome radiation hybrid panels have been established (Yerle *et al.*, 1996b; A. Archibald, personal communication). Attempts are underway to characterize the two panels on a cytogenetic and molecular level. It is expected that a total of 100–150 independent hybrid clones will comprise each final informative panel. The panel will then be used for the construction of a high resolution gene map of the pig.

Chromosomal *In Situ* Hybridization – Radioactive and Non-radioactive

In situ hybridization (ISH) is a technique that has generated a great deal of interest among researchers in a variety of biological fields ranging from genetics, pathology, virology, neurobiology, etc., to developmental biology/embryology, biochemistry and other related disciplines. It allows the direct detection/visualization, and hence localization, of specific nucleic acid sequences (both DNA and RNA) within a tissue, cell or chromosome and yields both molecular and morphological information about each cell (Lawrence and Singer, 1986). With specific reference to physical gene mapping, ISH still remains one of the most important techniques to localize genes/anonymous DNA segments to specific chromosomal regions. The technique involves hybridization *in situ* of a modified (labelled) DNA probe to complementary site(s) on the chromatin fibre. The modifications could be radioactive or non-radioactive while the chromatin fibre could be metaphase or interphase chromosome or even stretched DNA fibre. Application of the ISH technique to localize DNA fragments on porcine chromosomes dates back almost ten years. The time period up to now may broadly be divided into the radioactive and non-radioactive ISH eras, each of which is separately discussed below.

Radioactive ISH

This approach involves labelling of DNA probes using radioactively modified nucleotides. Although a number of radioactive labels, viz. ^{32}P, ^{35}S, and ^{125}I have been used for localization of a few DNA sequences, tritium (^{3}H) has been the label of choice primarily because of lower background irradiation coupled with high sensitivity. The labelled nucleotides are introduced in the probe by nick translation or random primer labelling, and are detected by autoradiography (see Chowdhary *et al.*, 1989). Statistical analysis of radioactive signals on the chromosomes is carried out on at least 50 to 100 metaphase cells to detect hybridization signal at a particular site on a specific chromosome. Chromosome identification, which is an integral part of any ISH localization enabling demarcation of the chromosomal site for a locus, can be carried out prior to or after hybridization. Although pre-hybridization banding (Q-, R- and G-banding; e.g. Ansari *et al.*, 1988; Chowdhary *et al.*, 1989, Yerle *et al.*, 1990a,b) was most prevalent during the early days of farm animal gene mapping, post-hybridization banding (mostly G-banding; e.g. Chowdhary *et al.*, 1989) became more popular due to the associated practical ease.

The pig was the first farm animal in which the radioactive ISH technique was applied to localize a DNA sequence chromosomally. Using this technique, Geffrotin *et al.* (1984), Rabin *et al.* (1985) and Echard *et al.* (1986) mapped the porcine major histocompatibility (*MHC*) locus to the pericentromeric region of chromosome 7. This initial success did not result in a sudden increase in ISH data in pigs. Nevertheless, a few groups initiated organized ISH experiments to expand the porcine physical gene map, resulting in the localization of the *GPI* (Davies *et al.*, 1988; Chowdhary *et al.*, 1989) and *IFNA* (Yerle and Gellin, 1989) loci. Thereafter, a rapid increase in the ISH mapping data was observed, and most of the loci mapped until 1993 were localized using radioactive ISH. This chromosomally assigned a number of already known and newly identified linkage/syntenic groups, resulting in a rapid expansion of the pig gene map. To date, using this technique, a total of 68 loci (all representing coding sequences) have been mapped to specific chromosomal bands in pigs (Table 8.4a). One of the major advantages of this technique has been the ability to use cross-species small genomic as well as cDNA sequences as probes for chromosomal localization. This has been specifically important considering that at the onset of organized genome analysis in pigs, very few cloned porcine sequences were available for mapping.

Although radioactive ISH has been instrumental in the construction of physical gene maps in pigs (as well as in other farm animals) during the initial mapping phase, the technique suffered from major drawbacks, viz. safety measures for handling radioactive material, long (minimum 3 weeks) autoradiographic exposure times, limited spatial resolution on the chromosomes and statistical analysis of grain distribution over many metaphase cells (sometimes >100). These considerations led to a major shift towards the use of non-radioactive approaches (Pinkel *et al.*, 1986; Lawrence *et al.*, 1988) to localize DNA sequences. Presently, the majority of ISH localizations are carried out using non-radioactive methods although some work is still being done using the radioactive approach.

Non-radioactive ISH

The early 1990s witnessed the introduction of non-radioactive techniques in farm animals. Although technically the underlying principles of this approach are the same as radioactive ISH, three major differences are worth mentioning with respect to non-radioactive ISH. Firstly, the probe DNA is modified using biotin, digoxigenin (DIG), di-/tri-nitrophenol, 2-acetylaminofluorene or other

Table 8.4a. Genes mapped by *in situ* hybridization.

Gene symbol	Gene name	Location	Reference
Radioactive *in situ* hybridization mapped genes			
AACT	Alpha-1-antichymotrypsin	7q2.3–q2.6	Musilova *et al.* (1995)
ALB	Albumin	8q1.2	Chowdhary *et al.* (1993a)
AMH	Antimüllerian hormone	2	Lahbib-Mansais *et al.* (1996)
ANPEP	Alanyl (membrane) aminopeptidase N, CD13	7cen–q2.1	Poulsen *et al.* (1991)
*APOB**	Apoliopoprotein B	3q2.4–qter	Solinas *et al.* (1992a)
APOE	Apoliopoprotein E	6q1.1–q2.1	Yerle *et al.* (1990b); Chowdhary *et al.* (1994c)
ATP1A1	Na$^+$,K$^+$ ATPase alpha subunit	4q1.6–q2.3	Mellink *et al.* (1992); Lahbib-Mansais *et al.* (1993)
ATP1B1	NA$^+$,K$^+$ ATPase beta subunit	4q1.3–q2.1	Mellink *et al.* (1992); Lahbib-Mansais *et al.* (1993)
CALCR	Calcitonin receptor	9q1.1–q1.2	Zolnierowicz *et al.* (1994)
CP	Ceruloplasmin	13q3.1–q3.3	Chaudhary *et al.* (1993)
*CRC** *(RYR)*	Calcium release channel	6q1.2	Harbitz *et al.* (1990b); Yang *et al.* (1992)
CS1	Citrate synthetase	5p1.2–p1.3	Chaudhary *et al.* (1992)
CYP19	Cytochrome P450, subfamily XIX	1	Lahbib-Mansais *et al.* (1996)
DAO	D-amino acid oxidase	14q2.1–q2.3	Mellink *et al.* (1992); Mellink *et al.* (1993)
*EDN1**	Edothelin-1	7p1.2–p1.3	Lahbib-Mansais *et al.* (1995)
ENO1	Enolase 1	6q2.2–q2.4	Yerle *et al.* (1990b)
ESR	Oestrogen receptor	1p2.4–2.5	Chowdhary *et al.* (1994c)
FSHB	Follicle stimulating hormone, beta polypeptide	2p1.2–p1.6	Mellink *et al.* (1995)
FSHR	Follicle stimulating hormone receptor	3q2.2–q2.3	Remy *et al.* (1995)
*GGTA1**	Alpha 1,3 galactosyl-transferase	1q2.10–q2.11	Strahan *et al.* (1995)
*GH**	Growth hormone, somatotropin	12p1.4	Thomsen *et al.* (1990); Yerle *et al.* (1993a)
GHR	Growth hormone receptor	16q1.3–q1.4	Chowdhary *et al.* (1994a)
*GPI**	Glucose phosphate isomerase	6q1.2	Chowdhary *et al.* (1989); Yerle *et al.* (1990b); Yang *et al.* (1992)
GRP78	Glucose-regulated protein 78kD	1q2.10–q2.13	Mellink *et al.* (1992); Mellink *et al.* (1993)
*HXB**	Tenascin C; hexabrachion	1q2.11–q2.13	Garrido *et al.* (1995)
*IFNA**	Leucocyte interferon alpha	1q2.5	Yerle *et al.* (1986); Yerle and Gellin (1989)
IFNB1	Interferon beta I	2p1.5 (Chr. 1 by linkage)	Graphodatsky *et al.* (1991)
IFNG	Interferon gamma	5p1.1–q1.1	Johansson *et al.* (1993)
IGF1R	Insulin-like growth factor-1 receptor	1q1.7–q2.1	Lahbib-Mansais *et al.* (1995)
IGHG	Immunoglobulin gamma heavy chain	7q2.1–q2.6	Gu *et al.* (1993a); Gu *et al.* (1994)
IL1A	Interleukin 1 alpha	3q1.2–q1.3	Mellink *et al.* (1994b)
IL1B	Interleukin 1 beta	3q1.1–q1.4	Mellink *et al.* (1994b)
INSR	Insulin receptor	2q1.1–q2.1	Gu *et al.* (1993b)
LHB	Luteinizing hormone	6q2.1	Mellink *et al.* (1995)

Table 8.4a. *continued*

Gene symbol	Gene name	Location	Reference
*LHCGR**	Luteinizing hormone/ choriogonadotropin receptor	3q2.2–q2.3	Yerle *et al.* (1992)
*LIPE**	Hormone sensitive lipase	6q1.2	Mellink *et al.* (1992); Gu *et al.* (1993b); Mellink *et al.* (1993)
LPL	Lipoprotein lipase	14q1.2–q1.4	Gu *et al.* (1992)
LRP2	Calcium sensing protein	15q2.2–q2.4	Chowdhary *et al.* (1994c); Chowdhary *et al.* (1995b)
ME1	Cytozolic malic enzyme	1p1.2	Nunes *et al.* (1996)
NADPH?	NADPH oxidase-light chain subunit	6p1.4–p1.5	Cepica *et al.* (1996)
NGFB	Nerve growth factor beta	4q1.6–q2.3	Lahbib-Mansais *et al.* (1994)
NP	Nucleoside phosphorylase	7q2.1–q2.2	Gellin *et al.* (1990)
PGD	6-Phosphogluconate dehydrogenase	6q2.2–q2.5	Harbitz *et al.* (1990a); Yerle *et al.* (1990a)
PI	Alpha-1-antitrypsin; protease inhibitor	7q2.4–q2.6	Mellink *et al.* (1992); Archibald *et al.* (1996)
PLAU	Plasminogen activator urokinase type	14q2.4–q2.6	Mellink *et al.* (1992); Mellink *et al.* (1993)
*POLR2**	Large polypeptide RNA polymerase II	12q	Rettenberger *et al.* (1992)
PRKACB	cAMP-dependent protein kinase catalytic beta subunit	6q3.1–q3.3	Gu *et al.* (1993a); Gu *et al.* (1994); Mellink *et al.* (1994b)
PRKARIA	cAMP-dependent protein kinase regulatory type 1 alpha	12p1.3–p1.4	Mellink *et al.* (1992); Mellink *et al.* (1994b)
RLN	Relaxin	1q2.8–q2.9	Chowdhary *et al.* (1994c); Ellegren *et al.* (1994a)
*RN5S@**	RNA 5S cluster	14q2.3	Lomholt *et al.* (1995)
SFI	Steroidogenic factor 1	1	Lahbib-Mansais *et al.* (1996)
*SLA-1**	Major histocompatibility complex class I genes	7p1.2–q1.2	Geffrotin *et al.* (1984); Rabin *et al.* (1985); Rettenberger *et al.* (1992); Sarimento *et al.* (1993d)
*SLA-2**	Major histocompatibility complex class II genes	7p1.2–q1.2	Rettenberger *et al.* (1992)
*SPPI**	Secreted phosphoprotein 1	8q2.5–q2.7	Chowdhary *et al.* (1994c)
SRY	Sex-determining region Y	Yp1.2–p1.3	Yang *et al.* (1993)
*SUDO1**	RFLPs sudo class I	7p1.2–p1.3	PigBase (unclear)
TF	Transferrin	13q3.1	Chowdhary *et al.* (1993a)
TGFB1	Transforming growth factor, beta I	6qcen–q2.1	Yerle *et al.* (1990a)
TNFA	Tumour necrosis factor alpha	7p1,1–q1.1	Solinas *et al.* (1992b)
TNFB	Tumour necrosis factor beta	7p1.1–q1.1	Solinas *et al.* (1992b)
*TNP2**	Transition protein 2	3p1.1–p1.2	Gu *et al.* (1993a); Gu *et al.* (1994); Rettenberger *et al.* (1994d)
*TP53**	Tumour protein p53	12q1.2–q1.4	Rettenberger *et al.* (1992); Rettenberger *et al.* (1993)
TPP2	Tripeptidyl peptidase 2	11q1.7	Chowdhary *et al.* (1993b)
TYR	Tryosinase	9p1.2–p1.3	Gu *et al.* (1994a); Chowdhary *et al.* (1994c)
*UBB**	Ubiquitin	14q1.2–q1.4	Rettenberger *et al.* (1992)
UBC	Polyubiquitin	14q1.2–q1.5	Rettenberger *et al.* (1994d)
WARS	Tryptophanyl-tRNA synthetase	7q3.2–q3.3	Graphodatsky *et al.* (1991)
WT1	Wilms' tumour gene 1	2	Lahbib-Mansais *et al.* (1996)

Table 8.4a. *continued*

Gene symbol	Gene name	Location	Reference
Fluorescent *in situ* hybridization (FISH) mapped genes			
ACADM	Acyl-coenzyme A dehydrogenase	6q3.2	Hamasima *et al.* (1996)
ACO2	Aconitase 2, mitochondrial	5p1.4–p1.5	Kusumoto *et al.* (1996); Yasue *et al.* (1996)
ACP5	Uteroferrin	2q1.2–q2.1	Yasue *et al.* (1995)
ACR	Acrosin	5p1.5	Yasue *et al.* (1996)
ADCY5	Adenylate cyclase 5	13q4.1	A. Sjöberg (pers. comm.)
APOA1	Apolipoprotein A1	9p1.1–p1.2	R. Chaudhary (pers. comm.)
*APOB**	Apolipoprotein B	3q2.4–qter	Sarimento and Kadavil (1993a); Murakami *et al.* (1994)
APOC3	Apolipoprotein C3	9p13	R. Chaudhary (pers. comm.)
ATP2B2	ATPase, Ca^{2+} transporting, plasma membrane	13q1.3–3.2	A. Sjöberg (pers. comm.)
CAPN	Skeletal muscle calpain	1q1.5–q1.7	Briley *et al.* (1993); Briley *et al.* (1996)
CHAT	Choline acetyltransferase	14q2.8–q2.9	Murakami *et al.* (1994); Murakami *et al.* (1996)
CHGA	Chromogranin A	7q2.4–q2.6	Kojima *et al.* (1996)
COL8A1	Collagen, type VIII, alpha 1	13q4.4–4.6	A. Sjöberg (pers. comm.)
*CRC** (*RYR*)	Calcium release channel	6q1.2	Chowdhary *et al.* (1992); Chowdhary *et al.* (1994b,c); Chowdhary *et al.* (1995a)
DPP4	Dipeptidylpeptidase IV	15q2.1	Thomsen *et al.* (1993)
*EDN1**	Endothelin-1	7p1.3–pter	Kojima *et al.* (1996)
*GGTA1**	Alpha 1,3 galactosyl-transferase	1q2.10–q2.11	Strahan *et al.* (1995)
*GH**	Growth hormone	12p1.4	Chowdhary *et al.* (1992); Chowdhary *et al.* (1994b)
*GP**	Glucose phosphate isomerase	6q1.2	Chowdhary *et al.* (1992); Chowdhary *et al.* (1994b,c); Chowdhary *et al.* (1995a)
GPI1	Glucose phosphate isomerase pseudogene	1q1.6–q1.7	Chowdhary *et al.* (1992); Harbitz *et al.* (1993)
GST	Glutathion S-transferase	14q2.1	R. Chaudhary (pers. comm.)
GSTM1	Glutathione S-transferase, MU	5q2.4	R. Chaudhary (pers. comm.)
HARS	Histidyl-tRNA synthetase	2q2.8–q2.9	R. Chaudhary (pers. comm.)
HSPA1	Heat shock 70 kDa protein 1	7p1.1	Nunes *et al.* (1993)
HSPA2	Heat shock 70 kDa protein 2	7p1.1	Nunes *et al.* (1993)
HSPA3	Heat shock 70 kDa protein 3	7p1.1	Nunes *et al.* (1993)
HSPA6	Heat shock 70 kDa protein 6	14q2.4–q2.5	Nunes *et al.* (1993)
*HXB**	Tenascin; hexabrachion	1q2.11–q2.13	Awata *et al.* (1995)
*IFNA**	Interferon alpha	1q2.5	Sarimento *et al.* (1993c)
IGA	Immunoglobulin alpha	7q2.6	R. Chaudhary (pers. comm.)
IGKC	Immunoglobulin kappa	3q1.2–q1.4	Rettenberger *et al.* (1996d)
IGL	Immunoglobulin lambda	14q2.1 14q2.2–q2.3	Rettenberger *et al.* (1996c); R. Chaudhary (pers. comm.)
IL6	Interleukin 6	9q1.4–q1.5	Bruch *et al.* (1996)
INSL3	Leydig insulin-like hormone	2q1.2–q1.3	Rettenberger *et al.* (1994b)
IVL	Involucrin	4q2.2–q2.3	Sarimento and Kadavil (1993b)

Table 8.4a. *continued*

Gene symbol	Gene name	Location	Reference
KIT	Mast/stem growth factor receptor	8p1.2	Johansson-Moller et al. (1996); Sakurai *et al.* (1996)
LAMA	Laminin alpha	1p2.4–p2.5	R. Chaudhary (pers. comm.)
LCAT	Lechitin cholesterol acyl transferase	6p13	Frengen *et al.* (1997)
LCT	Lactase phlorizin hydrolase	15q1.3	Thomsen *et al.* (1995)
*LHCGR**	Luteinizing hormone/ choriogonadotrophin receptor	3q2.2–q2.3	Yerle *et al.* (1992)
*LIPE**	Hormone-sensitive lipase	6q1.2	Chowdhary *et al.* (1994b,c); Chowdhary *et al.* (1995a)
LYZ	Lysozyme	5p1.1	R. Chaudhary (pers. comm.)
MUC	Submaxillary gland mucin	5q2.3	Johansson *et al.* (1993)
NFIC	Nuclear factor I/C	2q1.2–q1.3	Frönicke and Scherthan (1997)
NFKBI	Nuclear factor of kappa light polypeptide gene enhancer in B-cells inhibitor	7q1.5–q2.1	Musilová *et al.* (1996)
NPY5R	Neuropeptide Y receptor Y5	8p1.1–p1.2	A. Sjöberg (pers. comm.)
ORM1	Alpha 1-acid glycoprotein	1q2.10–q2.12	Murakami *et al.* (1994); Murakami *et al.* (1995)
PCCB	Propionyl coenzyme A carboxylase, beta polypeptide	13q3.1–3.2	A. Sjöberg (pers. comm.)
PD1A	Porcine SLA class I	7p1.2–p1.4	Sarimento *et al.* (1993d)
PDGFRA	Platelet-derived growth factor receptor alpha	8p1.2	Johansson-Moller *et al.* (1996)
PGA	Pepsinogen A	2p1.7	Zijlstra *et al.* (1992)
*POLR2**	Large polypeptide RNA polymerase II	12q	Rettenberger *et al.* (1992)
RLNCE	Pro-relaxin converting enzyme	2q2.1	Chowdhary *et al.* (1994c)
RN	Rendement Napole ('acid meat' gene)	15q2.1–q2.2	Milan *et al.* (1996b)
		15q2.5	A. Sjöberg (pers. comm.)
*RN5S@**	RNA 5S cluster	14q2.3	Mellink *et al.* (1996)
RNR1	Ribosomal RNA (NOR)	8p1.2	Mellink *et al.* (1994a)
RNR2	Ribosomal RNA (NOR)	10p1.2–p1.3	Miyake *et al.* (1988); Mellink *et al.* (1994a)
RNR3	Ribosomal RNA (NOR)	16q2.1	Bosma *et al.* (1991c)
*SLA-I**	Major histocompatibility complex class I genes	7p1.2–q1.2	Rettenberger *et al.* (1992)
*SLA-II**	Major histocompatibility complex class II genes	7p1.2–q1.2	Rettenberger *et al.* (1992)
SLC2A1	Solute carrier family 2	6q3.4–q3.5	Kusumoto and Yasue (1995)
*SPP1**	Secreted phosphoprotein 1	8q2.7–qter	Alexander *et al.* (1996c)
*SUDO1**	RFLPs sudo class I	7p1.2–p1.3	PigBase (unclear)
SYK	Spleen tyrosine kinase	14q1.4	Kojima *et al.* (1996)
TBP10	Tat-binding protein/26S protease subunit family	12	Brenig and Leeb (1996)
TGFB2	Transforming growth factor beta 2	10p1.6	Rettenberger *et al.* (1996e)
TNP1	Transition protein 1	15q2.4–q2.5	Rettenberger *et al.* (1994a)
*TNP2**	Transition protein 2	3p1.1–p1.2	Rettenberger *et al.* (1994d)
*TP53**	Tumour protein p53	12q1.2–q1.4	Rettenberger *et al.* (1992)
TSHB	Thyrotorophin b subunit	4q2.2–q2.4	Kojima *et al.* (1996)
*UBB**	Ubiquitin	14q1.2–q1.5	Rettenberger *et al.* (1992)

Table 8.4a. *continued*

Gene symbol	Gene name	Location	Reference
UOX	Urate oxidase	6q26–q32	Rettenberger et al. (1996e)
vWF	von Willebrand factor	5q2.1–q2.2	Sjöberg *et al.* (1996)
ZP3	Zona pellucida glycoprotein	3pter–p1.5	Bruch *et al.* (1996)

*Genes mapped by RISH and FISH.

labels. Secondly, detection is carried out with specific antibodies, either by fluorescence or enzyme reaction; and thirdly, relatively larger species-specific probes (generally >5 kb) are used, due to which pre-hybridization annealing of repeat sequences is to be carried out to obtain distinct hybridization signals. As far as is known, the non-radioactive ISH experiments carried out in pigs have *mainly* used the fluorescent detection system (hence the term FISH – fluorescent *in situ* hybridization), with biotin as the most commonly used labelling agent. In some instances, DIG labelled probes have also been used (Chowdhary *et al.*, 1994c, 1995a).

The initial applications of FISH in pigs included the use of cosmid probes for localizing microsatellite loci (Ellegren *et al.*, 1993, 1994a,b). Thereafter, a number of specific genes were also mapped, but all the probes used were of lambda or cosmid origin. Except for the Y-specific repeat probe, the only small sized clone (700 bp) which gave very strong hybridization signals in pig was for the submaxillary gland mucin (*MUC*) gene (Johansson *et al.*, 1993). This was most likely because the probe sequence was several times tandemly repeated within the genome, resulting in an enhanced signal. Marked increase in the physical mapping data during a period of 4–5 years, coupled with the available genetic linkage mapping information, enabled for the first time the amalgamation of the two maps (128 genetic markers of which 47 were ISH mapped) to provide the first estimate of the likely average genome length in pigs (1873 ± 139 cM; Ellegren *et al.*, 1994b). Similar observations were concurrently reported by Rohrer *et al.* (1994). Presently, approximately 69 coding sequences and over 154 anonymous/microsatellite containing sequences have been FISH mapped in pigs (Table 8.4b).

The FISH technique proved a major advancement in localization of DNA sequences on chromosomes not only because of greater precision as compared with radioactive ISH but also due to a considerably reduced experimental time to map a locus (2–3 days vs. 2–3 months). Using fairly elongated chromosomes, the mapping resolution of FISH extends up to 1 Mb (Trask *et al.*, 1989), which is considerably better than other chromosomal localization methods. A further advancement in the application of this technique was the evolvement of the multicolour FISH approach, which enabled physical ordering of closely located loci both on metaphase and interphase chromosomes. The porcine *GPI*, *CRC* and *LIPE* genes, which map to the same chromosomal band (6q12; Chowdhary *et al.*, 1994b), were thus ordered in the hitherto only metaphase/interphase multicolour FISH study carried out in farm animals (Chowdhary *et al.*, 1995a).

This helped in resolving the long-standing confusion on the relative order of the *GPI* and *CRC* (halothane locus) genes with respect to the centromere. Presently, another set of experiments is underway in our laboratory to order eight yeast artificial chromosome (YAC) clones in the subterminal region of chromosome 15 in pigs, which is now known to harbour the porcine *RN* gene (Rendement Napoli locus; responsible for its effect on meat quality in pigs).

Probes in the size range of 5–40 kb are conventionally suited for FISH experiments, while probes within the 3–4 kb range are considered to be on the borderline of use in FISH. Probes shorter than this limit are usually difficult to localize primarily due to technical limitations of detecting the fluorescent signal. Although several developments showing increased sensitivity in the FISH technique have been reported during recent years (Nilsson *et al.*, 1994; Raap *et al.*, 1995), the methodology has still to evolve for its routine use in the localization of small single copy sequences (<2 kb). Nonetheless, during the past year, attempts have been made to localize cDNA sequences in pigs. Of the 40–45 clones used, as yet only 18 could be localized to specific chromosomal regions after repeated experiments (R. Chaudhary and P. Thomsen, personal communication). For the remaining probes it was extremely difficult to differentiate the specific and the background signal. This initial success has at least demonstrated the possibilities of mapping 1–2 kb cDNA sequences in pigs using FISH. The approach can be further refined with future advancements in the methodology.

During recent years, large insert libraries, viz. P1 bacteriophage library (A. Archibald, personal communication) and yeast artificial chromosome (YAC; Frengen, 1993; Leeb *et al.*, 1995; Rogel-Gaillard *et al.*, 1996; Alexander *et al.*, 1997) libraries have been constructed. Mapping of such large fragment clones can be of significance in the integration of physical and genetic linkage maps because:

1. The clones can be readily mapped by FISH;
2. The clones have a high probability of containing simple tandem repeat sequences which may be readily converted into markers;
3. The clones facilitate the construction of expressed sequence tagged (EST) maps.

During the recent past, some clones from the above mentioned libraries have been mapped to specific chromosomal locations using FISH (McQueen *et al.*, 1994; Bruch *et al.*, 1996; Milan *et al.*, 1996a; Rettenberger *et al.*, 1996e; A. Sjöberg, personal communication). Although chimerism is one of the problems associated with these probes, preliminary experiments with their use in physical mapping have been successful. As yet, approximately 25 PAC/YAC clones, some of them containing specific genes, have been localized in pigs. Recently, cross-species hybridization of PAC clones was also tried in pigs. A pooled DNA probe from PACs containing the human *LCAT* gene cluster was used in FISH experiments on porcine metaphase chromosomes, assigning it to chromosome 6p13 (Frengen *et al.*, 1997). This corresponds to recent success obtained by using bovine cosmid clones on reindeer (Prakash *et al.*, 1996) and two muntjac species (Frönicke *et al.*, 1997) chromosomes. The findings point towards the

Table 8.4b. Anonymous DNA sequences and microsatellites mapped using ISH.

Symbol	Location	References
Fluorescent *in situ* hybridization		
A47	14q1.4	Looft *et al.* (1994)
BHT34	6p1.1–p1.1	Høyheim *et al.* (1992)
BHT49	1q2.9–q2.11	Høyheim *et al.* (1992)
DYZ1	Yq	Thomsen *et al.* (1992)
P1-44M18	1p2.4	McQueen *et al.* (1994)
S0000 (T20)	Yq (14q2.4–q2.5)	Chowdhary *et al.* (1992)
S0001	4p1.2–p1.3	Marklund *et al.* (1993)
S0002	3q2.5–q2.6	Fredholm *et al.* (1992)
S0015	14q2.9	Coppieters *et al.* (1994)
S0038	10p1.6	McQueen *et al.* (1994)
S0039	10q1.3–q1.4	McQueen *et al.* (1994)
S0041	1q1.7–q1.8	Yerle *et al.* (1994a)
S0042	8cen	Yerle *et al.* (1994a)
S0043	2q1.1–q2.1	Yerle *et al.* (1994a)
S0044	6p1.1–p1.3	Yerle *et al.* (1994a)
S0045	10p1.1	Yerle *et al.* (1994a)
S0046	14q2.1–q2.2	Yerle *et al.* (1994a)
S0047	7q1.2–q1.4	Yerle *et al.* (1994a)
S0048	11cent	Yerle *et al.* (1994a)
S0049	2q2.1	Yerle *et al.* (1994a)
S0050	2p1.6	Yerle *et al.* (1994a)
S0051	6q2.3–q2.4	Yerle *et al.* (1994a)
S0052	6q2.3	Yerle *et al.* (1994a)
S0053	7q1.5	Yerle *et al.* (1994a)
S0054	13q3.1–q3.2	Yerle *et al.* (1994a)
S0055	14q2.6–q2.7	Yerle *et al.* (1994a)
S0056	1q2.12–q2.13	Yerle *et al.* (1994b); Robic *et al.* (1995)
S0058	14q2.1	Yerle *et al.* (1994b); Robic *et al.* (1995)
S0059	6q2.5–q2.6	Yerle *et al.* (1994b); Robic *et al.* (1995)
S0076	13q1.2	Winterø *et al.* (1994a)
S0077	16q1.4	Winterø *et al.* (1994b)
S0095	9p1.1–q1.1	Chowdhary *et al.* (1992); Ellegren *et al.* (1993)
S0096	12p1.1–p1.2	Chowdhary *et al.* (1992); Ellegren *et al.* (1993)
S0102	7q1.1–q1.2	Ellegren *et al.* (1994b)
S0105	16qter	Ellegren *et al.* (1994b)
S0106	12q1.3	Ellegren *et al.* (1994b)
S0107	4q1.2	Ellegren *et al.* (1994b)
S0108	6q2.4–q2.5	Ellegren *et al.* (1994b)
S0109	9p1.2	Ellegren *et al.* (1994b)
S0110	2p1.5–p1.6	Ellegren *et al.* (1994b)
S0162	14q2.1–q2.2	Ellegren *et al.* (1994b)
S0163	9p1.1–p1.2	Ellegren *et al.* (1994b)
S0165	3q2.7	Ellegren *et al.* (1994b)
S0166	14q2.1	Chowdhary *et al.* (1992); Ellegren *et al.* (1994b)
S0167	3q2.4–q2.5	Ellegren *et al.* (1994b)
S0169	14q2.5	Ellegren *et al.* (1994b)

Table 8.4b. *continued*

Symbol	Location	References
S0170	2p1.1	Ellegren *et al.* (1994b)
S0171	15q1.3	Ellegren *et al.* (1994b)
S0172	3q2.4–q2.5	Ellegren *et al.* (1994b)
S0173	10p1.2–q1.1	Ellegren *et al.* (1994b)
S0201	10p1.3–p1.4	Yerle *et al.* (1994a)
S0202	6p1.4–p1.5	Yerle *et al.* (1994a)
S0203	2p1.5–p1.6	Yerle *et al.* (1994a)
S0206	3p1.4	Yerle *et al.* (1994b); Robic *et al.* (1995)
S0207	7q2.4–q2.5	Yerle *et al.* (1994b); Robic *et al.* (1995)
S0208	13q4.7–q4.8	Yerle *et al.* (1994a)
S0214	4q1.6–q2.1	Yerle *et al.* (1994b); Robic *et al.* (1995)
S0216	3q2.3–q2.5	Yerle *et al.* (1994b); Robic *et al.* (1995)
S0294	6p1.1–q1.1	Høyheim *et al.* (1994a)
S0295	9q2.1–q2.3	Høyheim *et al.* (1995a)
S0296	17q1.3–q1.4	Høyheim *et al.* (1995b)
S0297	6p1.4–p1.5	Høyheim *et al.* (1994b)
S0298	16q1.4	Høyheim *et al.* (1995c)
S0301	4p1.5	Høyheim *et al.* (1994c)
S0302	1q2.9–q2.11	Ellegren *et al.* (1994a)
S0306	18q2.4	Thomsen and Høyheim (1994)
*S0323***	14q1.2–q1.3	Signer *et al.* (1996)
*S0324***	Yp1.3;Xp2.1–p2.4	Signer *et al.* (1994)
S0325	6cen	Signer *et al.* (1996)
*S0326***	16q2.3	Signer *et al.* (1996)
S0359	17q2.1	Yerle *et al.* (1994b)
S0360	14q2.8–q2.9	Yerle *et al.* (1994b); Yerle *et al.* (1996c)
S0361	9q2.1–q2.2	Yerle *et al.* (1994b); Yerle *et al.* (1996c)
S0362	10q1.2	Yerle *et al.* (1994b)
S0364	15q1.3–q1.4	Yerle *et al.* (1994b); Yerle *et al.* (1996c)
S0365	14q1.3–q1.4	Yerle *et al.* (1994b); Yerle *et al.* (1996c)
S0366	10p1.1–q1.1	Yerle *et al.* (1994b)
S0367	1q1.8–q2.1	Yerle *et al.* (1994b)
S0400	14q1.3–q1.4	Yerle *et al.* (1994b); Yerle *et al.* (1996c)
S0531	1p2.2	Jørgensen *et al.* (1996a)
S0532	13q4.8	Jørgensen *et al.* (1996b)
SW1065	15q2.2	Alexander *et al.* (1996b)
SW1200	5q2.2–q2.3	Alexander *et al.* (1996c)
SW1400	13q2.1–q2.2	Alexander *et al.* (1996c)
SW1416	15q1.1–q1.2	Alexander *et al.* (1996b)
SW1647	6q3.1–q3.3	Alexander *et al.* (1996c)
SW1983	15q2.5	Alexander *et al.* (1996b)
SW2401	9p2.1	Alexander *et al.* (1996c)
SW2404	4p1.5	Alexander *et al.* (1996c)
SW2406	6p1.4	Alexander *et al.* (1996c)
SW2407	9p1.3–p2.1	Alexander *et al.* (1996c)
SW2408	3q2.5–p2.6	Alexander *et al.* (1996c)
SW2409	4p1.5	Alexander *et al.* (1996c)

Table 8.4b. *continued*

Symbol	Location	References
SW2410	8p2.2	Alexander *et al.* (1996c)
SW2412	13q2.3–q2.4	Alexander *et al.* (1996c)
SW2413	11q1.6	Alexander *et al.* (1996c)
SW2416	1q2.1	Alexander *et al.* (1996c)
SW2425	5p1.2	Alexander *et al.* (1996c)
SW2427	17q2.3	Alexander *et al.* (1996c)
SW2431	17q2.3	Alexander *et al.* (1996c)
SW2435	4q2.2–q2.3	Alexander *et al.* (1996c)
SW2439	14q2.1	Alexander *et al.* (1996c)
SW2440	13q4.8–q4.9	Alexander *et al.* (1996c)
SW2441	17q1.2–q1.4	Alexander *et al.* (1996c)
SW2442	2q2.1	Alexander *et al.* (1996c)
SW2443	2p1.7	Alexander *et al.* (1996c)
SW2445	2p1.6	Alexander *et al.* (1996c)
SW2448	13q3.1	Alexander *et al.* (1996c)
SW2453	Xq2.5	Alexander *et al.* (1996c)
SW2456	Xp1.3	Alexander *et al.* (1996c)
SW2458	13q2.1–q2.2	Alexander *et al.* (1996c)
SW2459	13q3.1	Alexander *et al.* (1996c)
SW2470	Xp2.1–p2.2	Alexander *et al.* (1996c)
SW2472	13q3.2	Alexander *et al.* (1996c)
SW2490	12p1.5	Alexander *et al.* (1996c)
SW2491	10p1.4	Alexander *et al.* (1996c)
SW2494	12p1.4–p1.5	Alexander *et al.* (1996c)
SW2495	9p1.1–q1.1	Alexander *et al.* (1996c)
SW2505	6q2.5–q2.6	Alexander *et al.* (1996c)
SW2507	14q2.6–q2.7	Alexander *et al.* (1996c)
SW2508	14q2.1	Alexander *et al.* (1996c)
SW2509	4p1.5	Alexander *et al.* (1996c)
SW2513	2p1.6	Alexander *et al.* (1996c)
SW2514	2q2.8	Alexander *et al.* (1996c)
SW2515	14q2.9	Alexander *et al.* (1996c)
SW2517	16q2.2	Alexander *et al.* (1996c)
SW2520	14q2.6–q2.7	Alexander *et al.* (1996c)
SW2525	6p1.4–p1.5	Alexander *et al.* (1996c)
SW2527	3p1.4	Alexander *et al.* (1996c)
SW2535	6p1.5	Alexander *et al.* (1996c)
SW2537	7q2.5	Alexander *et al.* (1996c)
SW2540	18q1.2–q1.3	Alexander *et al.* (1996c)
SW2551	1q2.8	Alexander *et al.* (1996c)
SW2557	6q2.3	Alexander *et al.* (1996c)
SW2571	9p1.1–p1.2	Alexander *et al.* (1996c)
SW2586	2q2.1	Alexander *et al.* (1996c)
SW2593	14q2.5–q2.6	Alexander *et al.* (1996c)
SW2597	3q1.1–q1.2	Alexander *et al.* (1996c)
SW2604	14q2.4	Alexander *et al.* (1996c)
SW2611	8p2.3	Alexander *et al.* (1996c)

Table 8.4b. *continued*

Symbol	Location	References
SW2612	14q2.1–q2.2	Alexander *et al.* (1996c)
SW2618	3p1.1–p1.2	Alexander *et al.* (1996c)
vSW2623	2p1.7	Alexander *et al.* (1996c)
SW920	10q	Johansson *et al.* (1995)
SW973	6p1.5	Alexander *et al.* (1996c)
SWC19	10p1.1	Alexander *et al.* (1996c)
SWC22	13q3.2	Alexander *et al.* (1996c)
SWC23	12q1.3	Alexander *et al.* (1996c)
SWC27	14q2.8–q2.9	Alexander *et al.* (1996c)
SWC9	2p1.7	Alexander *et al.* (1996c)
SWR2417	17q2.1	Alexander *et al.* (1996c)
SWR2516	2p1.7	Alexander *et al.* (1996c)
SW973	6p1.5-pter	Rohrer *et al.* (1994)
SW1200	5q2.2–q2.3	Rohrer *et al.* (1994)
SW1400	13q2.1–q2.2	Alexander *et al.* (1996a)
SW 1647	6q3.1–q3.2	Alexander *et al.* (1996a)
OPN	8q2.7-qter	Rohrer *et al.* (1994)
DISC-PCR		
PGHAS	1q1.8–q2.2	Troyer *et al.* (1994)
S0088	15q2.1	Milan *et al.* (1996b)
SW307	1q2.3–q2.4	Troyer *et al.* (1994)
SW373	1q2.6–q2.7	Troyer *et al.* (1994)
SW64	1q1.6–q1.8	Troyer *et al.* (1994)
SW936	15q2.1–q2.2	Milan *et al.* (1996b)
SW974	1q2.6	Troyer *et al.* (1994)

**Mapped by radioactive *in situ* hybridization.

possibility of using large sized probes (or even pools of them) for comparative mapping and detection of corresponding chromosomal sites of genes of interest. Additional methodological modifications will, however, be necessary to increase the success rate of such FISH experiments.

The FISH technique has also been applied to map precisely the location of nucleolus organizer regions (NORs) and the telomeric repeat sequences in pigs. Using a 5.2 kb human rDNA probe, Bosma *et al.* (1991c) detected a secondary constriction site on chromosome 16, in addition to the two usually detected on chromosomes 8 and 10 by the traditional silver staining method (Goodpasture and Bloom, 1975). It was proposed that the site is transcriptionally inactive due to which there was no signal with the latter technique, but using FISH it was possible to detect the presence of these sequences on the chromosome. The application of FISH to detect the telomeric sequences in pigs (de la Seña *et al.*, 1995) not only showed distinct hybridization sites on the terminal region of the chromosomes, but also demonstrated for the first time the presence of an interstitial site on the long arm of chromosome 6. Whether it correlates to the bright T-band seen on 6q (Gustavsson *et al*., 1983) is unclear, because other similar bright interstitial T-bands could not be detected in this FISH experiment. Nonetheless, the FISH technique has proved extremely useful in localizing those rDNA and telomeric sites which could not be mapped/ detected using routine methods.

Fibre-FISH and DNA-combing

Fibre-FISH

Fluorescent *in situ* hybridization to prometaphase chromosomes enables assignment of DNA clones to within 1–3 Mb resolution on chromosomal bands (Lawrence *et al*., 1990). Hybridization to extended chromatin in interphase cells further improves this resolution to within 50–750 kb, and provides excellent means to physically order DNA probes which, even on prometaphase chromosomes, are visible at the same site (Trask *et al*., 1989; Lichter *et al*., 1991; Chowdhary *et al*., 1995a; A. Sjöberg, unpublished results, personal communication). A further refinement in this resolution has been achieved during recent years, and has proven very useful in developing high resolution physical maps for DNA stretches of up to 500 kb. Although a variety of approaches have been used to decondense (Heng *et al*., 1992; Wiegant *et al*., 1992; Haaf and Ward, 1994) or extend (Parra and Windle 1993; Fidlerová *et al*., 1994; Houseal *et al*., 1994) the chromatin fibres, the improved method described by Heiskanen *et al*. (1994) provides greater optical mapping resolution.

Briefly, pulse field gel electrophoresis (PFGE) blocks are prepared by embedding peripheral blood lymphocytes in low melting point agarose. After appropriate lysis of the cells within the blocks, protein free extended DNA fibres are obtained on microscopic slides (Plate 1a). This target DNA is conveniently hybridized with closely located/overlapping probes using routine FISH conditions. Hybridization sites are visualized as stretches of fluorescent signals (see Plate 1b). Order of the clones can be determined by using differential labelling (biotin or digoxigenin) and detection (FITC, rhodamine or Texas red) conditions. Overlapping clones can be determined by the detection of a yellow signal caused by an admixture of green and red signals of individual clones. One of the other advantages of this technique is to predict physical (kilobase; kb) distances between clones. The size of the probe can, for example, be used to calibrate the size of the hybridization signal, which in turn can be used to roughly estimate the distance in kilobase, between the hybridized probes.

Study of the linear arrangement of four lambda clones located within about 200 kb of the porcine major histocompatibility (SLA) region on chromosome 7 (Sjöberg *et al*., 1997) was among the first such experiments carried out in pigs. The result also estimates the size of the four clones, and the gap distances between them. On this basis, an estimate of the total stretch of DNA studied is also proposed (Plate 1c). The study also carried out a quantitative analysis using three different approaches (the Watson–Crick, Probe Size and Relative Length standards) to compare the estimates obtained using different conversion parameters for the measured micrometers (mm) values. It was observed that although relative length-based estimates, in general, do not differ much from those derived using the Watson–Crick standard and size of one of the clones used, a certain degree of bias could be involved in them. The latter two types of estimates were in fair agreement with each other. Observations that average condensation of DNA-fibres produced by the fibre-FISH technique corresponds to that expected from a totally linearized DNA molecule (Weier *et al.*, 1995), shows that the Watson–Crick-based conversions are more reliable.

Recently, the fibre-FISH technique was also applied to physically orientate the plasmid subclones derived from a lambda clone containing the porcine erythropoietin (*EPO*) gene (Liu *et al.*, 1998). This is the first study in pigs (as well as farm animals) where small sized clones, ranging between 2.7–3.5 kb, were used for fibre-FISH. The study pointed out that although small sized probes may be successfully used for high resolution visual mapping on stretched DNA fibres, the hybridization signals of some of the probes may be extremely small (sometimes a single dot) due to the presence of only a small unique DNA stretch (exon/codon) within the probe. Such signals (and hence the probes) may not be best suited for use as a 'standard' in converting micrometer measurements into kilobases. This was obvious when a disarray was observed between the hybridization signal-based converted physical estimates and the restriction analysis-based estimates, for at least one of the probe used.

The fibre-FISH technique is useful also in finding transcriptional orientation of clones, and in detection of duplications, deletions or inversions in a particular genomic fragment (see Palotie *et al.*, 1996). As mentioned above, the detection of the latter three is easier if a sizeable segment of DNA (2–3 kb) is involved. Currently, attempts are underway to quantify the extent of duplication in the *KIT* gene in pigs (Johansson-Moller *et al.*, 1996), using the fibre-FISH technique. The hybridization signals of the porcine *KIT* probe observed using routine FISH on metaphase chromosomes shows clear distinction in the intensity of hybridization signal between the purebred white (e.g. Large White Yorkshire) and coloured pigs (e.g. Hampshire and Duroc). Crosses between these types of pig have a heterozygous condition and show stronger signal on one of the homologues. Preliminary findings indicate that these differences can be visualized and quantified with the help of measurements of hybridization signals on DNA fibres originating from these pigs. Recently, fibre-FISH has also been utilized for positional cloning of a disease gene (*CLN5*) in humans (Laan *et al.*, 1996). The approach, and the associated success, implies that the technique can find significant use in cloning genes of interest in pigs and other farm animal species.

DNA-combing

This is an approach which technically resembles fibre-FISH to a large extent (see Bensimon *et al.*, 1994), but differs in: (i) the source of target DNA; and (ii) how it is stretched on glass slides for *in situ* hybridization. The target may include any cloned DNA which is applied in solution, on a silanated glass slide. One of the ends of the DNA molecule binds to the glass surface because such treated surfaces have a high pH-dependent binding specificity for DNA ends. Gradual evaporation of the solution under a glass coverslip stretches, straightens and fixes the DNA molecules, thus making it available as a target for hybridization (Plate 2a). The stretched DNA follows similar condensation patterns to those observed by mechanical stretching during fibre-FISH. One of the biggest advantages of this approach is that the stretching is more uniform as compared with the latter, and a set of subclones originating from the stretched clone can be used as hybridization probes, resulting in their rapid optical ordering. The technique is also best suited to construct a contig map of a set of subclones originating from a BAC or YAC clone.

The only study hitherto carried out in pigs (and also in farm animals) relates to ordering a set of plasmid subclones derived from a lambda clone containing the porcine *EPO* gene. These subclones were also used for the agarose-based fibre-FISH method (see above). Distinct hybridization signals on combed DNA fibres (Plate 2b) enabled the ordering of the three subclones (E4-S1, E4-S2 and E4-S3; Liu *et al.*, 1998). The results were in complete agreement with those obtained by fibre-FISH on whole genomic DNA embedded in agarose.

In Situ PCR

Among the various techniques for the localization of DNA sequences to specific chromosomal regions, two related techniques are worth mentioning, viz. the PRimed *IN Situ* PCR (PRINS) and the Direct *In situ* Single-Copy PCR (DISC-PCR). Theoretically, in both techniques the PCR is run directly on the slide, with the chromosomal DNA as the template. The PRINS technique has found good use in humans and other species for the localization of repeat sequences, e.g. the centromeric and telomeric repeats (Koch *et al.*, 1989; Hindkjaer *et al.*, 1995). However, for mapping single copy small sequences, the technique has still to be developed further. As far as is known, this approach has been applied in pigs only by Gu *et al.* (1996) to confirm the observations of de la Seña *et al.* (1995) concerning the detection of interstitial telomeric site on the long arm of chromosome 6 (see page 228).

DISC-PCR is technically the same approach as PRINS except that until now only the enzyme based system has been used to detect the signal on the chromosome, as compared with the fluorescence detection in the former. One of the major breakthroughs associated with this technique was the ability to localize small microsatellite sequences (averaging 200 bp) directly on metaphase chromosomes. Troyer *et al.* (1994), who coined the term DISC-PCR, used the technique for localizing five microsatellite sequences (*PGHAS*, *SW64*, *SW307*, *SW974* and *SW373*) on chromosome 1 in pigs. Since then, the technique has been used to map two more loci, viz. *SW19* and *Sw920*, on chromosome 10. Recently, this technique has also been used to map another microsatellite locus (*SW936*) in order to localize the porcine *RN* gene to 15q 21–q22 (Milan *et al.*,

1996a,b). Contrary to these observations, YACs containing *SW936* and some adjacent microsatellite loci were very recently mapped to the subterminal region of this chromosome (15q25; our unpublished results). During the past year, DISC-PCR has also been used to localize the *IFNA*, *P450arom*, *CAPN* and *S0082* loci to chromosome 1q, orientate a linkage group by mapping two microsatellite loci on chromosome 10, map three microsatellite loci to chromosome 13, and localize several microsatellite markers to chromosome 15 (D. Troyer, personal communication).

Although DISC-PCR is an important step forward for developing concordance between the physical and linkage maps, it suffers from some drawbacks. The most important among these are:

1. Several metaphase cells (50 or above) have to be analysed for individual localization, and statistical analysis is needed to support the observations.
2. Chromosome morphology is so much disturbed that their identification is difficult after the reaction.
3. Precise mapping cannot be done because the grains scored for each PCR reaction follow a radioactive ISH-like distribution covering several bands.
4. Due to unknown reasons, there is a tendency of shift (by a few bands) in the signal from the real location of the sequences. The differences observed between the findings of Troyer *et al.* (1994) and those reported by other groups (e.g. Robic *et al.*, 1996; see Chapter 9) may partly be attributed to this effect.

Thus, although the technique appears to be very promising, it still has to evolve through methodological modifications, before being used on a routine basis. In the recent past, some changes in the washing steps have been tried to prevent the signal shift. Furthermore, fluorescent labelled nucleotides have been used to skip the enzymatic detection. To overcome the difficulties of chromosome identification, attempts have also been made to conduct G-banding prior to carrying out DISC-PCR (Milan *et al.*, 1996b). Although these modifications have partly contributed to increasing the efficiency of the technique, its routine use as a mapping method has still to undergo the test of time.

Pulsed Field Gel Electrophoresis

Long-range physical gene mapping using pulsed field gel electrophoresis (PFGE) has contributed significantly in improving the mapping resolution of various genomes within a range of several kilobases up to 10 Mb (see review by Anand, 1986). Because of the extent of coverage obtained using this method, the PFGE maps overlap with the cytogenetic banding techniques (Bickmore and Sumner, 1989), and provide direct measurements of chromosomal distances. The approach thus bridges the resolution gap between molecular and genetic mapping methods. The technique involves slicing of mammalian genomic DNA with rare cutting enzymes that generate several hundred kilobase fragments, which after transfer to membrane filters, are hybridized with DNA probes. This results in a banding pattern which can be used to construct long-range restriction maps defining the position of the enzyme cleavage site. In pigs, the application of this technique has been restricted mainly to the construction of detailed physical maps of the major histocompatibility complex (SLA) region. Xu *et al.* (1992) observed that the order of the SLA region is similar to that of human MHC (class II–class III–class I). Later, Wu *et al.* (1995) determined the fine order of the genes within the class III region, and showed the order to be class II–class III (*CYP21/C4–Bf/C2/HSP70–TNFα*)–class I. The size of the class III region located between the *CYP21* and *TNFα* loci was estimated to be 500 kb. Recently, Peelman *et al.* (1996) proposed a precise organization of the 700 kb segment of DNA between *G18* and *BAT1*, with more than 30 genes mapped to it. A comparison of the order of these loci in pig, human and mouse showed that, as compared with the class I and class II regions, the class III region is more conserved in the three species. The physical continuity of the SLA class I and class III regions was also studied using YAC clones (Chardon *et al.*, 1996) and it was found that a maximum distance of 260 kb exists between the class I sequences and *BAT1* (the most terminal locus of *SLA* class III region).

In addition to fine mapping of the SLA region, the PFGE technique has also been applied to study the *GPI-CRC* region located on chromosome 6 (Frengen and Davies, 1995). The two genes were limited to a region of less than 50 and 180 kb, respectively, in a 310 kb and 1.1 Mb restriction map. No overlap of the maps defining the two probes was observed. Furthermore, the limit of the *GPI* locus was estimated to be at least 500 kb from the *CRC* locus. As far as is known, compared with human and mouse, PFGE has been used to a limited extent in pigs. This can be attributed partly to the late introduction of this technique to pig genome analysis, and also to the introduction of other physical mapping methods (multicolour ISH to elongated chromosomes, interphase chromatin or extended DNA-fibres), whereby visual long range physical maps can be constructed relatively easily.

Flow Sorting of Chromosomes

Among the various livestock/domesticated species, the pig is the most attractive animal for flow sorting individual chromosomes, primarily due to small chromosomal number, distinct shape and size, and variation in base pair ratio. Flow sorting of porcine chromosomes was first carried out by Grunwald *et al.* (1986), Matsson *et al.* (1986), Geffrotin *et al.* (1987) and Grunwald *et al.* (1989), who used both single and dual laser systems, obtaining fairly distinct separation of the chromosomes from normal as well as translocated cell lines. Grunwald *et al.* (1986) and Geffrotin *et al.* (1987) used the flow sorted chromosomes for spot blot hybridization, thus demonstrating that the technique can have applications in pig gene mapping. Using a dual laser fluorescence activated cell sorter (FACS), Dixon *et al.* (1992) and Schmitz *et al.* (1992) demonstrated for the first time that pig chromosomes can be resolved into 19–20 peaks. It was observed that most, if not all, of the pig chromosomes could be separated into definite peaks. Later, Schmitz *et al.* (1992), with the help of PCR analysis on flow sorted chromosomes from a translocation carrier, were able to identify the peaks for chromosomes 3 and 7. Furthermore, peaks for X and Y chromosomes were also confirmed.

One of the first pieces of direct evidence to correlate the flow karyotype peaks with specific pig chromosomes came when Langford *et al.* (1992), using a novel approach of chromosome painting, demonstrated that 5 of the 20 clusters observed in the bivariate flow karyotype of a male pig corresponded to chromosomes 1, 13, 18, X and Y. The authors used the then newly devised degenerate oligonucleotide primed (DOP) PCR technique (Telenius *et al.*, 1992) to amplify and non-radioactively label DNA from individual peaks. The labelled DNA, when used as a probe on porcine metaphase chromosomes in a FISH experiment, clearly showed that each studied cluster represented (painted) a specific pig chromosome. This breakthrough was significant for unambiguous correlation of individual peaks to particular chromosomes. Later, two independent studies (Langford *et al.*, 1993; Yerle *et al.*, 1993b) established complete correspondence of each of the clusters in the porcine bivariate flow karyotype with a specific chromosome, thus providing a 'standard' flow karyotype for the pig. While Langford *et al.* (1993) applied the earlier used DOP-PCR technique for DNA amplification and labelling, Yerle *et al.* (1993b) used the Priming Authorizing Random Mismatching (PARM-) PCR technique proposed by Milan *et al.* (1993). Identification of pig chromosomes in a bivariate flow karyotype has also been carried out by GTG banding of mouse, pig and seven pig–mouse hybrid cell lines (Bouvet *et al.*, 1993), followed by tentative assignment of 12 porcine chromosomes (1, 2, 5, 6, 10, 11, 13, 14, 16, 18, X and Y) to individual peaks, on the basis of DNA content versus relative chromosome length. Further, in the seven lines, seven of the eight peaks distinct from those of the mouse cell line could be correlated with the presence of chromosomes 5, 9, 10, 11 or 16, 14, 15 and 18.

The success associated with flow cytometry in pigs increased expectations that this technique, as in humans, will open new avenues of research in porcine genome analysis. It was anticipated that chromosome specific libraries will soon be constructed for individual porcine chromosomes. However, contrary to expectations, very few chromosome specific libraries are known to have been made during the past 3 years (chromosome 1 – Miller *et al.*, 1992; chromosome 13 – Davies *et al.*, 1994, chromosome 18 – Ellegren and Basu, 1995; chromosome 6 – Lunney *et al.*, 1995). These libraries have partially contributed in providing markers for the porcine genetic linkage map. Of these, the chromosome 18 enriched library was of significance because it enabled construction and assignment of the first linkage map (comprising markers *S0062*, *S0177*, *S0179* and *SW787*) to this chromosome. However, the drawback associated with the library (as in other cases too) was that, of the 11 unique microsatellites obtained, only two could be mapped to chromosome 18, and the remaining to five different porcine chromosomes (chromosomes 3, 4, 7, 9 and 11). Relatively better accuracy was reported by Davies *et al.* (1994), who mapped 10 of the 13 microsatellites isolated, to the target chromosome (chromosome 13). Contamination of individual flow karyotype clusters with DNA/segments from other chromosomes is an inherent problem of the flow cytometry technique. In humans, this was largely overcome by the use of specific cell lines (e.g. somatic cell hybrids containing a single human chromosome; see Gray and Cram, 1990) for flow sorting chromosomes. However, similar resources are not abundantly available in pigs. Those available through recent construction of somatic hybrid panels (see page 213), have not been exploited for this purpose. Under these circumstances, flow cytometry does not appear to be the best alternative for library construction and search of polymorphic markers in the porcine genome. Novel microdissection and microcloning approaches are now available (see below), wherein chromosome-specific DNA can be obtained and cloned with greater ease, rapidity and accuracy.

Whole chromosome paints (WCPs), obtained through flow cytometry, have to some extent found use in cytogenetic analysis of translocations. Konfortova *et al.* (1995) flow sorted a 7/15 translocation chromosome and used the amplified DNA as probe to paint and demonstrate the chromosomes involved. One of the significant contributions of the WCPs has been their use in partly confirming homeologies (Goureau *et al.*, 1996) previously delineated between the human and porcine karyotypes (Rettenberger *et al.*, 1995b; Frönicke *et al.*, 1996). Recently, the porcine WCPs were also used on cattle (Schmitz *et al.*, 1996), babirusa (*Babyrousa babyrousa*; Bosma *et al.*, 1996a) and white lipped peccary (*Tayassu pecari*; Bosma *et al.*, 1996a,b), in an attempt to elucidate the comparative status of their karyotypes with respect to that of the domestic pig (see more in section on comparative painting, pages 239–240).

Microdissection of Chromosomes

Development of skeletal physical and genetic linkage maps in pigs has basically been carried out using random mapping approaches. However, in the present perspectives, these approaches will be of relatively lesser significance because there is an increasing need now to utilize the knowledge gained through the basic maps, for the search of genes governing traits of economic and biological importance. This will require detailed analysis of specific genomic regions. Chromosomal microdissection provides one of the most convenient alternatives to meet this end. Briefly, the technique involves microscopic scraping of specific chromosomal regions, followed by PCR-based amplification of the scraped DNA, and cloning of the resultant product into a desired vector (see Meltzer *et al.*, 1992; Guan *et al.*, 1994). The library thus obtained consists of several thousand probes representative of the region scraped. One of the biggest advantages of this technique over chromosome specific libraries constructed using flow sorted individual chromosomes is that scraping and library construction can be carried out not only for a whole chromosome but also for an individual arm, segment or even a single band. Further, the libraries thus obtained are devoid of impurities/background commonly associated with flow sorted chromosomes.

Microdissection of porcine chromosomes is still in its initial stages of development. As yet, limited numbers of chromosomes/chromosomal segments have been dissected in pigs, and only some of these have been used for the construction of libraries and selection of markers. Chaudhary *et al.* (unpublished observations) recently microdissected pig chromosomes 1p, 1q24–qter, 2qcen–2q12, 4q, 8, 13q11–q31, 13q11–q49, 13q32–q43, 13q32–qter 15 and 16q21–q23, and constructed whole/partial chromosome painting probes. Use of these painting probes on porcine chromosomes confirmed their origin and demonstrated that they were adequately representative of the region scraped. As a follow-up to the above work, a chromosome-1p and -15-specific library were constructed, and work is in progress for screening the library for microsatellite markers. Library construction of other scrapings is currently being carried out. Recently, Mendiola *et al.* (1995) and Louis *et al.* (1996) scraped chromosome 6 and prepared a small insert genomic library which was screened with (GT)10 probe. The search provided nine markers which were developed and genotyped in a Yorkshire × Piétrain pig reference family and genetically mapped to chromosome 6. The WCP thus created was also used on human metaphase chromosomes (Zoo-FISH), and as expected, correspondence of this porcine chromosome with regions of human chromosome 1p and 19q was observed. Applying similar approaches, Troyer *et al.* (1996) scraped porcine chromosomes 13q33–q36 and 1p21–pter. The subgenomic libraries thus obtained were hybridized with a (CA)10 probe providing positive clones presently being used for marker development in the chromosomal regions of origin.

There are several potential applications of DNA libraries generated from scraped chromosomes or chromosomal regions. These can be used as follows:

1. For direct sequencing of the clones because of their convenient size, thereby overcoming the need for subcloning, etc.
2. To identify coding sequences within the clones, which could then serve as candidate genes for a specific condition mapped to a particular region.
3. To screen genomic libraries for isolating region-specific lambda, cosmid, P1/BAC or YAC clones, which can then be used for high resolution mapping in that region.
4. For cross-species painting, in an attempt to develop fine scale comparative maps.

In addition to these applications, chromosomal microdissection can be of significant use in cytogenetic analysis of unidentifiable chromosomes, e.g. minute unbalanced or balanced translocations involving small chromosomal regions, which may normally escape detection through routine banding analysis. Further, minute deletions (terminal or intercalary), which may otherwise pass unnoticed, can readily be traced using this technique. Thus, intractable chromosomal rearrangements can be readily analysed by microdissecting the abnormal/suspected chromosome and reverse painting it on normal metaphase preparations, whereby the originating chromosomes are highlighted. This technique is also finding extensive use in cancer cytogenetics.

Chromosome Painting

The term chromosome painting derives from the use of whole or partial chromosome paints (WCPs and PCPs) as probes for *in situ* hybridization to chromosomal preparations (interphase or metaphase chromosomes), such that the representative chromosome is highlighted (or painted) in the preparation (Lichter *et al.*, 1988; Pinkel *et al.*, 1988). Because the probe does not represent a single genomic site but is a cocktail of numerous sites of the originating chromosome (such that the whole chromosome or part thereof is almost completely represented), the observed signal is in fact an aggregation of several hybridization sites uniformly covering (or 'painting') the chromosome. Broadly speaking, the probes used for chromosome painting (CP) may originate from two different sources: (i) from flow sorted chromosomes (see pages 232–234); and (ii) from microdissected chromosomes (see pages 234–235). The extent of representation of the chromosomal DNA in both types of probes is, as judged from the paints, fairly high. The only advantage of the latter type is that they can be rapidly and conveniently prepared with minimal or no contamination of other chromosomal material, and that one can easily decide about the size of chromosomal segment which is to be used in the preparation of the probe. Paints for individual porcine chromosomes generated using the two techniques have been described in respective sections of this chapter, wherein likely uses of these paints have also been highlighted. It may, however, be mentioned that porcine WCPs have not yet found significant use in cytogenetic analysis.

Comparative chromosome painting

During recent years, comparative genome analysis has become an integral part of studying the genomes of various mammalian species. The techniques hitherto applied for studying homeology between genomes have hinged primarily on gross cytogenetics (mainly banding patterns) and available gene mapping data. Similarities in banding patterns, for example, within the Bovidae (cattle, sheep, goat and buffalo), are a good indication of homology at the DNA level. This has been demonstrated by several studies (see Prakash *et al.*, 1997). However, sometimes banding patterns can be of little consequence even in evolutionarily closely related species (e.g. species belonging to the family Equidae). Further, in evolutionarily diverged species, although one may find partial analogy between the banding and gene mapping data (see human–mouse comparison in Sawyer and Hozier, 1986), comparative banding patterns provide negligible information. In such cases, one utilizes the gene mapping data to arrive at a conclusion. The limitation with the latter is that it is useful only in identifying conserved linkage/syntenic groups between species. However, it does not precisely define chromosome segment homeology. Comparative painting provides an ideal solution to this problem by highlighting (painting) corresponding chromosomal segments.

Chromosome painting demonstrates the use of WCPs on chromosomes of the species of origin, e.g. use of individual porcine WCPs on porcine metaphase chromosomes. However, in an attempt to develop a better understanding of gross organization of diverged/less related genomes, recent years have witnessed the use of WCPs across species. The majority of attempts hitherto made concentrate on comparison of a specific mammalian genome with that of the human. This restriction/limitation was primarily because WCPs developed from individual chromosome-specific libraries (CSLs) were first, and most readily, available (commercially as well as from research organizations) only in humans (Deaven *et al.*, 1986; Fuscoe *et al.*, 1986). The use of WCPs across species, specially evolutionarily diverged species, was not as straightforward as their use in the species of origin. Methodological modifications proposed by Scherthan *et al.* (1994), viz. use of 8–10 times higher probe concentration than that used for species specific painting, 3–7 times prolonged hybridization time, efficient blocking of repeat sequences using Cot-1 DNA, and less stringent post-hybridization washing conditions, enabled their use across evolutionarily diverged species. Following this breakthrough, the porcine karyotype was the first to be completely studied using individual human CSLs. The work carried out by Rettenberger *et al.* (1995b) and Frönicke *et al.* (1996) delineated homeologous chromosomal segments between the pig and human karyotypes. Forty-seven porcine chromosomal segments corresponding to all human chromosomes except the Y were detected (Table 8.5; Frönicke *et al.*, 1996), resulting in a nearly complete coverage of the porcine karyotype. The comparative gene mapping information available between the two species further strongly supported the observed segmental homeologies. Frönicke *et al.* (1996) conducted a detail survey of the available physical and genetic linkage data in pigs and humans, and provided a comprehensive and updated comparative status of the two genomes (Plate 3).

The human–pig comparative painting results (Frönicke *et al.*, 1996) also enabled comparison of genome fractions (GF) of individual human chromosomes (expressed as per cent length of the haploid human chromosomal complement and referred to as relative length) with the sum of per cent fraction(s) of the homologous porcine segment(s) painted by each human chromosome (Table 8.6). Comparative porcine values for each of the human chromosomes were estimated based on the per cent of the pig genome painted by each human CSL. Of the 23 comparisons made, a fairly close agreement in the corresponding values was observed for 14 human chromosomes (see Table 8.6). In the remaining cases (human chromosomes 4, 7, 8, 9, 10, 15, 16, 19 and 20) the values did not differ by greater than 0.5–0.7% of either genome. The comparative values for human chromosomes 1, 4, 7 and 9 were, however, particularly different (approximately 0.8–1.2%). The observed partial differences may be attributed to polymorphic heterochromatic blocks (e.g. human chromosomes 1, 9 and 16; see Francke, 1994) and dispersed, small segmental homologies beyond the detection limits of Zoo-FISH (animal fluorescent *in situ* hybridization). Nonetheless, the findings show that in spite of diverged karyotypes, the two species, in general, demonstrate close confirmation between the GF values of their homologous segments. The observations agree with similar comparisons made between cattle and humans (Chowdhary *et al.*, 1996).

Recently, in an attempt to refine further the known human–pig chromosomal homeologies, Goureau *et al.* (1996) conducted bidirectional chromosome painting. The experiment involved the use of individual human CSLs on porcine metaphase chromosomes, as well as the use of WCPs generated from flow sorted individual porcine chromosomes on human metaphase chromosomes. As in other experiments (Rettenberger *et al.*, 1995b; Frönicke *et al.*, 1996), the

Table 8.5. Comparative status of human and porcine chromosomes as identified by painting and gene mapping (GM). The porcine fragment(s) painted by individual human chromosomes are presented.

Human	Expected homologies on pig chromosomes		Homologous porcine segments
chrom. no.	GM-basis	Painting basis	observed through Zoo-FISH
1	4,6,9,14	4,6,9,10	4q16–qter(S);6q23–q28(S); 6q32–qter(S);9q24–q26;10p(S)
2	3,12,15	3,15	3cen–qter(S);15q11–q14(S); 15q21–qter(S)
3	13	13	13cen–q46(S)
4	8	8	8(S)
5	2,14,16	2,16	2q21(middle)–qter(S);16(S)
6	1,7	1,7	1p(S);7pter–q13(MS)
7	9,12,18	9,18	9cen–q23(S);18(S)
8	14	4,14,15	4pter–q15(S);14q12–q13(M); 15q15(M)
9	1	1,10?	1q24–qter(S);10cent–q11(MS)
10	14	10,14	10q12–qter(S);14q23–qter(S)
11	2,9	2,9	2pter–p11(WM);9pter–p11(M)
12	5,14	5,14	5p14–qter(S);14q14(MS)
13	11	11	11
14	7	1,7	1q22–q23(S);7q15–q22(S); 7q24–qter(S)
15	7	1,7	1q14–q18(S);7q14(M); 7q23–q24(MS)
16	3	3,6	3p(M);6p(S)
17	12,14	12	12(S)
18	—	1,6	1q11–q13(S);1q18(half)–q21(S); 6q28–q31(MS)
19	2,6	2,6	2cent–q21(MS);6cent-q21(MS)
20	—	17	17(S)
21	7,9,13	13	13q47–qter(S)
22	5,14	5,14	5pter–p14(S);14q16–q22(MS)
X	X	X	X(S)

The intensity of signal on each of the porcine segments painted is presented in brackets as follows: (S) = strong; (M) = medium; (W) = weak. In a few cases, intermediate signal intensities (MS = medium strong, WM = weak medium) were also observed. (?) questions validity of this result because hybridization was on the peri-centromeric region.

human painting probes painted 95% of the porcine chromosomes, the reverse painting of porcine WCPs on human chromosomes was only 60% successful. The latter detected 39 homologous segments in humans. The approach has been useful in discerning some of the details for homologies known only on a gross level by the human–pig paints.

The comparative karyotypes of pigs and cattle with respect to that of humans enable indirect comparison between the genomes of the two former species. However, more direct evidence of segmental homology between the two species was recently provided by Schmitz *et al.* (1996). The authors used porcine WCPs (except the Y chromosome), generated from flow sorted individual porcine chromosomes, on cattle metaphase spreads, and proposed a pig–cattle comparative karyotype. Probes for chromosomes 9, 4 and X were used as a single pool because of inadequate resolution from flow cytometry. Furthermore, as mentioned earlier, porcine WCPs have also been used on babirusa (*Babyrousa babyrousa*; Bosma *et al.*, 1996a) and white lipped peccary (*Tayassu pecari*; Bosma *et al.*, 1996b). Using WCPs for pig chromosomes 1, 3 and 6 on babirusa chromosomes, and conducting comparative GTG banding studies, homology between the karyotypes of the two species was fully elucidated. Further, each of the three WCPs when applied to white lipped peccary chromosomes, showed correspondence to parts of two different chromosomes of the peccary karyotype (Bosma *et al.*, 1996b).

Table 8.6. Comparison of genome fractions (GF) of individual human chromosomes (expressed as per cent length of the haploid human chromosomal complement) with the sum of per cent fraction(s) of the homologous porcine segment(s) painted by each. A fairly close agreement in the corresponding values for the homologous fractions in the two genomes was observed for a majority of the chromosomes. The human values were obtained from Harris *et al.* (1986), while the pig values were estimated from Lin *et al.* (1980). Figures in brackets indicate values adjusted for one centromere, while summing up the values of the porcine segments homologous to each human chromosome. A correction factor of 0.4 was therefore added or subtracted from the porcine values with no or two centromeres, respectively. The value of 0.4 was an approximate average for individual centromore derived from the banded images of the porcine standard karyotype.

Chrom. segm. no.	Human genome fraction	Pig S GF of segm. painted	Chrom. segm. no.	Human genome fraction	Pig S GF of segm. painted
1	8.17	7.39	12	4.51	4.46
2	7.97	8.13 (7.73)	13	3.62	3.30
3	6.67	7.02	14	3.46	3.11 (3.56)
4	6.38	5.50	15	3.30	2.75 (3.15)
5	6.11	6.30	16	3.08	3.87 (3.47)
6	5.71	6.05 (5.65)	17	2.81	3.10
7	5.37	4.92 (4.52)	18	2.67	2.25
8	4.87	4.82	19	2.07	1.94 (1.54)
9	4.59	3.39	20	2.21	2.90
10	4.51	5.01	21	1.48	0.99 (1.39)
11	4.51	4.59 (4.19)	22	1.70	1.42 (1.82)
			X	5.15	5.20

A further advancement to the human–pig comparisons has very recently been carried out by using arm-specific paints (ASPs) from five human chromosomes, viz. 2, 3, 5, 16 and 19 (Chaudhary *et al.*, 1998). The authors observed that human chromosomes 6 and 16 ASPs showed complete conservation of individual arms as single blocks/arms both in pig and horse (Fig. 8.3). A similar trend was, in general, also observed for human chromosome 19 ASPs. However,

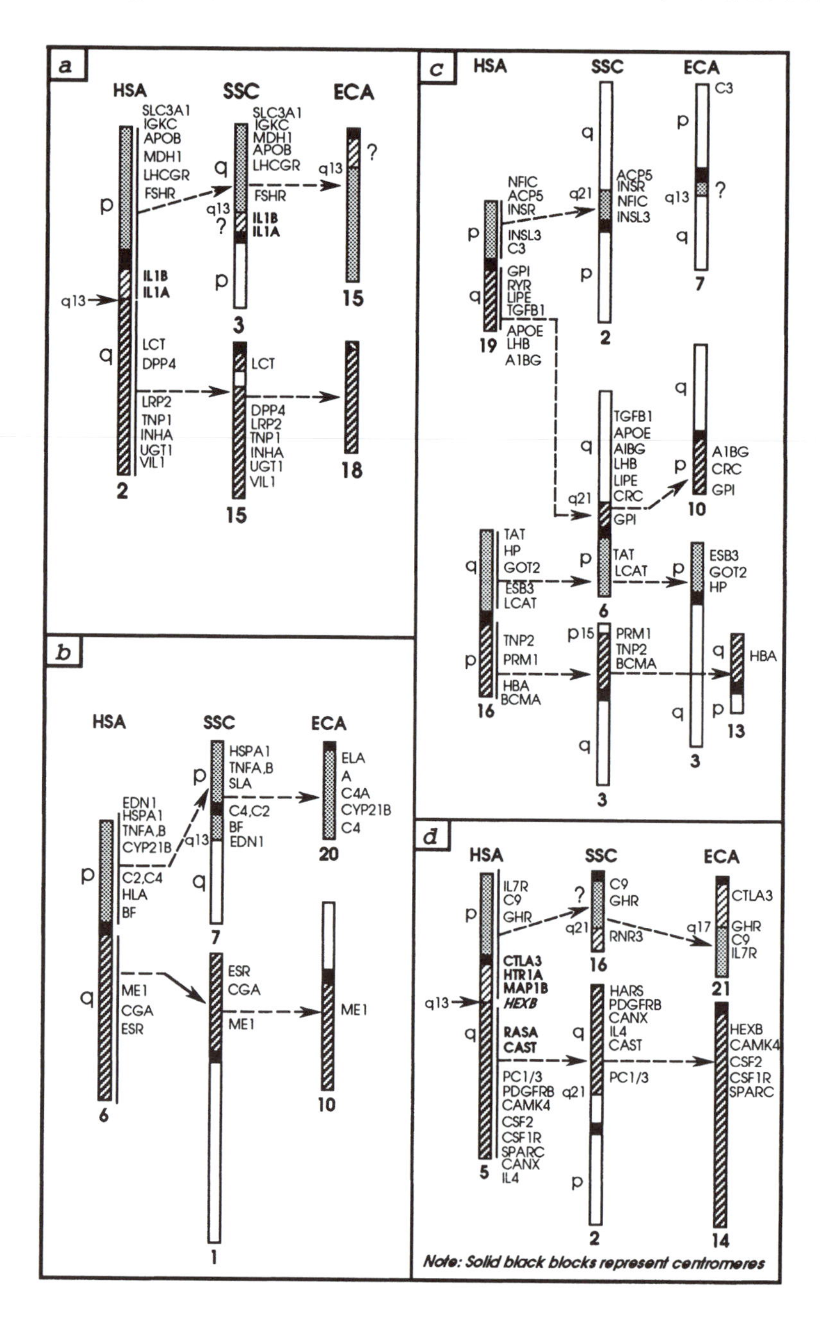

contrary to these observations, human chromosome 2 and human chromosome 5 ASPs demonstrated that homology in pig (and horse) is not restricted to individual arms of the two human chromosomes. Band q13 of both human chromosomes is the most likely point which bifurcates the chromosomes into two evolutionarily conserved segments. The study provides more accurate comparative information between human and pig than hitherto known with the help of whole chromosome paints for individual human chromosomes.

Future Prospects

Physical gene mapping in pigs has made unprecedented progress in the past 3–4 years. Pig *cytogenetics*, which has always been an integral part of these developments (by providing a chromosomal framework for exhibiting the gene/marker locations), has not undergone a molecular genetics type multifaceted development during the same period. In fact, pig cytogenetics, like that of other farm animals, had enjoyed a very fruitful evolution period during the 1970s and 1980s. The next decade mainly centred on the extension of the earlier work (e.g. detecting new chromosomal aberrations). However, application of electron microscopic methods (SC analysis) to provide a basis to fertility problems caused by chromosomal aberrations, was a new dimension to cytogenetic analysis. With all the achievements hitherto made, it is vital to explore where pig cytogenetics can and should lead to from this point onwards. With the advent of several novel techniques which can be applied directly to the chromosomes, there are new possibilities for the cytogeneticists to address those problems which were earlier difficult or impossible to approach using traditional methods. The magic words 'going molecular', with FISH as a major tool at hand, can prove a boom to pig cytogeneticists as well. A shift from traditional cytogenetics to molecular cytogenetics is thus the primary need of the hour. Availibility of WCPs (from flow sorted or microdissected chromosomes) to scrutinize translocations more closely, can be one of the beginnings. Possibilities of microdissecting

Fig. 8.3. Schematic drawings showing homologous chromosomal regions detected in pig (SSC) and horse (ECA) karyotypes, using arm-specific paints (ASPs) from five human (HSA) chromosomes, viz. HSA2 (a), HSA6 (b), HSA16 and 19 (c) and HSA5 (d). Each arm of the five human chromosomes is denoted with a specific pattern, and the same pattern is used for those regions of pig and horse chromosomes which were detected to be homologous with individual paints. Striped segments on (a) and (d) denote parts of the human q arms that share synteny homology with the p arms. For each chromosome, the species abbreviation is stated above while the chromosome number is stated below. Besides the chromosomes, a list of genes hitherto mapped have been presented; in humans, only the comparative data are presented. Vertical lines besides the human chromosomes show the region most likely involved in homology with specified (follow arrows) porcine and equine chromosomes. Band (q13) shown on the left of HSA2 (a) and HSA5 (d) indicates the most likely site where two evolutionarily conserved blocks on these chromosomes meet. Gene symbols in bold, adjacent to the above band indicate those genes which help in deducing these sites.

abnormal chromosomes and painting them back to normal metaphases is another way of approaching the same problem. The latter can be of significant use when banding techniques are of little or no consequence in chromosome identification. Origin of very small chromosome(s) (or sometimes extra chromosomal material) can very conveniently be detected within a week through microdissection-mediated analysis. Deletions, which normally are difficult to detect if they are not of noticeable size, can now be studied using the same approach. PRINS labelling can also be applied as a viable alternative for rapid detection of numerical and structural chromosome abnormalities. The biggest advantage associated with all the FISH-based analysis methods mentioned above is that the degree of accuracy is much higher than that obtained through traditional banding methods.

Pig cytogenetics has evolved as a branch with practical significance to pig breeders. To keep on enjoying this status, it would be necessary to resolve and address some of the hitherto unanswered questions concerning, e.g. association between chromosomal abnormalities and congenital defects. The large amount of work done in humans can be a blueprint to plan such a study. Little is known about the reasons for polymorphism, e.g. of the heterochromatic regions, and inheritance/segregation in the progenies. Fragile sites, although reported in pigs, have not yet been analysed with an aim to study their effect on the population in general. It would be interesting to discover if, as in humans, some fragile sites are associated with congenital deformities/defects. Very little is known about those fragile sites which might have evolutionary significance. Further, cytogenetic studies of pig embryos/zygotes at different stages of the life cycle can be further expanded particularly to study the fate of unbalanced zygotes, and in an attempt to understand why and how long they survive. This could also help in providing some explanations to malformed and stillborn piglets. The above described novel chromosome analysis techniques can also be applied to investigate apparently normal karyotype bearing individuals which show deformities, defects and malformations.

In spite of the fact that, in practical terms, linkage maps are considered as the 'useful' maps, and physical maps primarily as the 'reference' maps, *physical gene mapping* has provided vital information about the organization of the genome at the individual chromosome level. With fresh impetus from a variety of new developments in the recent past, physical gene mapping has enabled studying those finer details, which it has not been possible to examine with linkage mapping alone. The fact that physical gene mapping plays an equally significant role as linkage mapping in humans shows that, in farm animals too, this mapping approach will be needed to complement the linkage maps, in search of genes of interest.

Genome analysis in pigs, as in some other farm animals, has passed over the stage of random mapping. Utilizing the basic knowledge gained, and that constantly generated, efforts are now being directed partly towards increasing the resolution of the maps, and partly to initiate site-specific genome analysis. Chromosome microdissection is one of the techniques which is anticipated to make important contributions in generating markers specific to a small segment, or a band. Recent success with microdissection-mediated marker generation for some porcine chromosomes has paved the way for the application of this technique to produce an array of probes within very restricted chromosomal regions. It is expected that within the next year or two, partial or whole chromosome libraries will be available for all the porcine chromosomes. The technique will also find use in targeting those areas of the genome which are relatively deficient in informative markers. Microdissected band-specific libraries, in combination with available lambda/cosmid or even large insert libraries (mainly YACs), will be vital in constructing contigs within a specified region of the genome. This in turn can help in closing in on genes of interest. The large insert libraries (YACs and P1 bacteriophage), which are overcoming the initial phase of testing the extent of representation and chimerism, will not only be useful in identification and characterization of markers, but will also greatly facilitate their rapid chromosomal localization. The possibilities of finding expressed sequence tagged sites (ESTs) within YACs will make localization of such clones even more significant.

Fine physical mapping techniques like interphase mapping, fibre-FISH and DNA combing will be of significance in instances where closely located clones will require physical ordering. The latter two techniques will prove ideal for detecting deletions and duplications of DNA segments, and also for constructing a contig map within a range of approximately 750 kb. Prospects of positional cloning using the fibre-FISH approach is yet another possibility which might be explored by pig gene mappers in the near future. The radiation hybrid mapping technique, which has very recently been introduced to farm animals, appears promising not only in terms of increasing the resolution of the map, but also in acting as a bridge between physical and genetic linkage mapping. The technique will considerably increase accurate ordering of the markers mapped. It is expected that the pig gene map, like other farm animal gene maps, will benefit from these developments.

During recent years comparative gene mapping has gained vital importance in farm animals. Advances made through techniques like Zoo-FISH have enabled the development of a better insight to the comparative organization of some diverged mammalian genomes including pig, cattle, human and mouse (Chaudhary *et al*., 1998). As in other 'map poor' species, development of comparative gene maps in pigs will be essential in retriving information from the 'map rich' genomes, e.g. human and mouse. Further, recent success with fine comparative mapping using arm-specific human paints on pig chromosomes (Chaudhary *et al*., 1998) provides possibilities to explore the homeologies at a finer level. Additional refinement could be obtained by using segment-specific paints for a chromosome (e.g. five segments of human chromosome 3 painted to pig chromosomes) as probes. Initial success with the use of human YAC pool across pig opens yet another possibility to explore homeology at an even more refined level. All this comparative work will have to be constantly supplemented with precise mapping of coding sequences/ESTs. A combination of the two new pieces of information will, in true terms, enable the utilization of the human map for searching for candidate loci for genes of interest in pigs.

Acknowledgements

I am extremely grateful to Ms Terje Raudsepp for her untiring efforts in literature search and the enormous contributions she made in preparation of some of the tables. The spontaneous helping hand extended by Ms Anna Tömten in preparing final versions of some of the figures is also gratefully acknowledged. I am thankful to Prof. Ingemar Gustavsson for his comments/discussions on the cytogenetics part of the chapter. The work was supported by funds provided by the Swedish Council for Forestry and Agricultural Research.

References

Åkesson, A. and Henricson, B. (1972) Embryonic death in pigs caused by unbalanced karyotype. *Acta Veterinaria Scandinavica* 13, 151–160.

Alexander, L.J., Rohrer, G.A. and Beattie, C.W. (1996a) Cloning and characterization of 414 polymorphic porcine microsatellites. *Animal Genetics* 27, 137–148.

Alexander, L.J., Smith, T.P.L., Rohrer, G.A., Beattie, C.W. and Broom, M.F. (1996b) Construction of a porcine yeast artificial chromosome (YAC) library. *Proceedings XXVth International Conference of Animal Genetics* 21–25 July, Tours, France, p. 118.

Alexander, L.J., Troyer, D.L., Rohrer, G.A., Smith, T.P.L., Schook, L.B. and Beattie, C.W. (1996c) Physical assignments of 68 porcine cosmids and lambda clones containing microsatellites. *Mammalian Genome* 7, 368–372.

Alexander, L.J., Smith, T.P.L., Beattie, C.W. and Broom, M.F. (1997) Construction and characterization of a large insert porcine YAC library. *Mammalian Genome* 8, 50–51.

Alonso, R.A. and Cantu, J.M. (1982) A Robertsonian translocation in the domestic pig (*Sus scrofa*). 37,XX-13-17, t.rob. (13;17). *Annales de Genetique* 25, 50–52.

Anand, R. (1986) Pulsed field gel electrophoresis: a technique for fractionating large DNA molecules. *Trends in Genetics* 2, 278–283.

Andresen, K. (1963) A study of blood groups of the pig. Thesis, Munksgaard, Copenhagen, 299 pp.

Ansari, H.A., Hediger, R., Fries, R. and Stranzinger, G. (1988) Chromosomal localization of the major histocompatibility complex of the horse (ELA) by in situ hybridization. *Immunogenetics* 28, 362–364.

Aparicio, R.D. (1960) Cytogenetic study of spermatogenesis in the pig. *Archivos de Zootecnia.* (Cordoba) 9, 103.

Archibald, A.L., Couperwhite, S., Mellink, C.H.M., Lahbib-Mansais, Y. and Gellin, J. (1996) Porcine alpha-1-antitrypsin (PI): cDNA sequence, polymorphism and assignment to chromosome 7q2.4–q2.6. *Animal Genetics* 27, 85–89.

Awata, T., Yamakuchi, H., Kumagi, M. and Yasue, H. (1995) Assignment of the tenascin gene (HXB) to swine chromosome 1q21.1–q21.3 by fluorescence *in situ* hybridization. *Cytogenetics and Cell Genetics* 69, 33–34.

Basrur, P.K. and Kanagawa, H. (1971) Sex anomalies in pigs. *Journal of Reproduction and Fertility* 26, 369–372.

Bensimon, A., Simon, A., Chiffaudel, A., Croquette, V., Heslot, F. and Bensimon, D. (1994) Alignment and sensitive detection of DNA by a moving interface. *Science* 265, 2096–2098.

Berger, R. (1972) Étude du caryotype du porc avec une nouvelle technique. *Experimental Cell Research* 75, 298–300.

Bickmore, W.A. and Sumner, A.T. (1989) Mammalian chromosome banding – an expression of genome organization. *Trends in Genetics* 5, 144–148.

Bloom, S.E. and Goodpasture, C. (1976) An improved technique for selective silver staining of nucleolar organizer regions in human chromosomes. *Human Genetics* 34, 199–206.

Bobrow, M. and Cross, J. (1974) Differential staining of human and mouse chromosomes in interspecific cell hybrids. *Nature* 251, 77–79.

Bomsel-Helmreich, O. (1961) Hétéroploidie expérimentale chez la truie. *Proceedings 4th International Congress Animal Reproduction*, The Hague, vol. 3, pp. 578–581.

Bösch, B., Höhn, H. and Rieck, G.W. (1985) Hermaphroditismus verus bei einem gravidem Mutterschwein mit einem 39, XX,14+-Mosaik. *Zuchthygiene* 20, 161–168.

Bosma, A.A. (1976) Chromosomal polymorphism and G-banding patterns in the wild boar (*Sus scrofa* L.) from the Netherlands. *Genetica* 46, 391–399.

Bosma, A.A. (1978) The chromosomal G-banding pattern in the wart hog, *Phacochoerus aethiopicus* (Suidae: Mammalia) and its implications for the systematic position of the species. *Genetica* 49, 15–19.

Bosma, A.A. (1980) The karyotype of the babirusa (*Babyrousa babyrussa*): karyotype evolution in the Suidae. *4th European Colloquium on Cytogenetics of Domestic Animals* pp. 238–241.

Bosma, A.A. and de Haan, N.A. (1981) The karyotype of *Babyrousa babyrussa* (Suidae, Mammalia). *Acta Zoologica et Pathologica Anverpiensia* 76, 17–27.

Bosma, A.A., Colenbrander, B. and Wensing, C.J.G. (1975) Studies of phenotypically female pigs with hernia inguinalis and ovarian aplasia. II Cytogenetical aspects. *Proceedings Konilijke Nederlanse Akademie van Wetenschap*, Ser. C 78, 43–46.

Bosma, A.A., Oliver, W.L.R. and Macdonald, A.A. (1983) The karyotype, including G- and C-banding patterns, of the pigmy hog *Sus (Porcula) salvanius* (Suidae, Mammalia). *Genetica* 61, 99–106.

Bosma, A.A., de Haan, N.A., Blouch, R.A. and Macdonald, A.A. (1991a) Comparative cytogenetic studies in *Sus verrucosus, Sus celebensis* and *Sus scrofa vittatus* (Suidae: Mammalia). *Genetica* 83, 189–194.

Bosma, A.A., de Haan, N.A. and Macdonald, A.A. (1991b) The current status of cytogenetics of the Suidae: A Review. *Frädrich-Jubiläumsband.* Bongo, Berlin 18, 258–272.

Bosma, A.A., de Haan, N.A. and Mellink, C.H. (1991c) Ribosomal RNA genes located on chromosome 16 of the domestic pig. *Cytogenetics and Cell Genetics* 58, 2124.

Bosma, A.A., de Haan, N.A., Mellink, C.H.M., Yerle, M. and Zijlstra, C. (1996a) Chromosome homology between the domestic pig and the babirusa (family Suidae) elucidated with the use of porcine painting probes. *Cytogenetics and Cell Genetics* 75, 32–35.

Bosma, A.A., de Haan, N.A., Mellink, C.H.M., Yerle, M. and Zijlstra, C. (1996b) Chromosome homology between pig, the babirusa, and the white-lipped peccary studied by heterologous painting. *Proceedings of 12th European Colloquium on Cytogenetics of Domestic Animals*, 25–28 June, Zaragoza, Spain. Published in: *Archivos de Zootecnia* 45, 243–244.

Bouvet, A., Konfortov, B.A., Miller, N.G.A., Brown, D. and Tucker, E.M. (1993) Identification of pig chromosomes in pig–mouse somatic cell hybrid bivariate flow karyotypes. *Cytometry* 14, 369–376.

Breeuwsma, A.J. (1970) Studies on intersexuality in pigs. PhD thesis, Research Institute for Animal Husbandry, 'Schoonord', Ziest, The Netherlands.

Brenig, B. and Leeb, T. (1996) Isolation and chromosomal assignment of the porcine *TBP10* gene, a member of the Tat-binding protein/26S protease subunit family. *Proceedings XXVth International Conference on Animal Genetics,* 21–25 July, Tours, France, p. 119.

Briley, G.P., Riggs, P.K., Hancock, D.L., Womack, J.E. and Bidwell, C.A. (1993) Localization of porcine sceletal muscle calpain by FISH and PCR amplification of flow sorted chromosomes. *8th North American Colloquium on Domestic Animal Cytogenetics and Gene Mapping,* 13–16 July, Guelph, Canada, p. 79.

Briley, G.P., Riggs, P.K., Womack, J.E., Hancock, D.L. and Bidwell, C.A. (1996) Chromosomal localization of porcine skeletal muscle calpain gene. *Mammalian Genome* 7, 226–228.

Brown, J.F., Hardge, T., Rettenberger, G. and Archibald, A.L. (1994) Four new porcine polymorphic microsatellite loci (S0032, S0034, S0036, S0037). *Animal Genetics* 25, 365.

Bruch, J., Rettenberger, G., Leeb, T., Meier-Ewert, S., Klett, C., Brenig, B. and Hameister, H. (1996) Mapping of type I loci from human chromosome 7 reveals segments of conserved synteny on pig chromosomes 3, 9, and 18. *Cytogenetics and Cell Genetics* 73, 164–167.

Bryden, W. (1933) The chromosomes of the pig. *Cytologia* 5, 149–153.

Caspersson, T., Zech, L., Johansson, C. and Modest, E.J. (1970) Identification of human chromosomes by DNA-binding fluorescent agents. *Chromosoma* 30, 215–227.

Cepica, S., Musilová, P., Stratil, A., Juráková, M. and Rubes, J. (1996) Assignment of porcine NADPH oxidase-light chain subunit (p22-phox) to chromosome 6p14-p15 by radioactive *in situ* hybridization. *Proceedings XXVth International Conference on Animal Genetics,* 21–25 July, Tours, France, p. 120.

Chandley, A., Christie, S., Fletcher, J. Frackiewicz, A. and Jacobs, P.A. (1972) Translocation heterozygosity and associated subfertility in man. *Cytogenetics* 11, 516–533.

Chardon, P., Rogel-Gaillard, C., Save, J.-C., Bourgeaux, N., Renard, C., Velten, F. and Vaiman, M. (1996) Establishment of the physical continuity between the pig SLA class I and class III regions. *Proceedings XXVth International Conference on Animal Genetics,* 21–25 July, Tours, France, p. 121.

Chaudhary, R., Chowdhary, B.P., Harbitz, I. and Gustavsson, I. (1992) Localization of the citrate synthetase (CS) gene to the p12–p13 bands of chromosome 5 in pigs by *in situ* hybridization. *Hereditas* 117, 39–44.

Chaudhary, R., Chowdhary, B.P., Harbitz, I. and Gustavsson, I. (1993) Localization of the ceruloplasmin (CP) gene to the q32–q33 bands of chromosome 13 in pigs by *in situ* hybridization. *Hereditas* 119, 7–10.

Chaudhary, R., Raudsepp, T., Guan, X-Y., Zhang, H. and Chowdhary, B.P. (1998) Zoo-FISH with microdissected arm specific paints for HSA2, 5, 6, 16 and 19 refines known homology with pig and horse chromosomes. *Mammalian Genome* (in press).

Chaudhary, R., Winterø, A-K., Fredholm, M. and Chowdhary, B.P. (1998a) FISH mapping of seven cDNA clones in pigs. *Chromosome Research* (in press).

Chowdhary, B.P. (1991) Gene mapping in pigs: present status and future prospects. *Proceedings of 7th North American Colloquium on Domestic Animals Cytogenetics and Gene Mapping,* pp. 75–88.

Chowdhary, B.P., Harbitz, I., Mäkinen, A., Davies, W. and Gustavsson, I. (1989) Localization of the glucose phosphate isomerase gene to the p12–q21 segment of chromosome 6 in pig by *in situ* hybridization. *Hereditas* 111, 73–78.

Chowdhary, B.P., Thomsen, P.D., Ellegren, H., Harbitz, I., Andersson, L. and Gustavsson, I. (1992) Precise localization of some genetic markers in pigs using nonradioactive *in situ* hybridization. *Animal Genetics* 23. (Suppl 1), p. 92.

Chowdhary, B.P., Johansson, M., Chaudhary, R., Ellegren, H., Gu, F., Andersson, L. and Gustavsson, I. (1993a) *In situ* hybridization mapping and restriction fragment length polymorphism analysis of the porcine albumin (ALB) and transferrin (TF) genes. *Animal Genetics* 24, 85–90.

Chowdhary, B.P., Johansson, M., Gu, F., Bräuner-Nielsen, P., Tomkinson, B., Andersson, L. and Gustavsson, I. (1993b) Assignment of the linkage group EAM-TYRP2-TPP2 to chromosome 11 in pigs by *in situ* hybridization mapping of the TPP2 gene. *Chromosome Research* 1, 175–179.

Chowdhary, B.P., Ellegren, H., Johansson, M. Andersson, L. and Gustavsson, I. (1994a) *In situ* hybridization mapping of the growth hormone receptor (GHR) gene assigns a linkage group (C9, FSA, GHR, and S0105) to Chromosome 16 in pigs. *Mammalian Genome* 5, 160–162.

Chowdhary, B.P., Thomsen, P.D., Harbitz, I. and Gustavsson, I. (1994b) Precise localization of the genes encoding for glucosephosphate isomerase (GPI), calcium release channel (CRC), hormone sensitive lipase (LIPE) and growth hormone (GH) in pigs, using nonradioactive *in situ* hybridization. *Cytogenetics and Cell Genetics* 67, 211–214.

Chowdhary, B.P., de la Seña, C.A and Gustavsson, I. (1994c) *In situ* hybridization mapping in pigs: a summary of results from Uppsala. *Proceedings 11th European Colloquium on Cytogenetics of Domestic Animals*, 2–5 August, Copehagen, Denmark, pp. 17–19.

Chowdhary, B.P., de la Seña, C., Harbitz, I., Eriksson, L. and Gustavsson, I. (1995a) FISH on metaphase and interphase chromosomes demonstrates the physical order of the genes for GPI, CRC, and LIPE in pigs. *Cytogenetics and Cell Genetics* 71, 175–178.

Chowdhary, B.P., Lundgren, S., Johansson, M., Hjalm, G., Akerstrom, G., Gustavsson, I. and Rask, L. (1995b) *In situ* hybridization mapping of a 500-kDa calcium sensing protein gene (LRP2) to human chromosome region 2q31–q32.1 and porcine chromosome region 15q22–q24. *Cytogenetics and Cell Genetics* 71, 120–123.

Chowdhary, B.P., Frönicke, L., Gustavsson, I. and Scherthan, H. (1996). Comparative analysis of the cattle and human genomes: detection of ZOO-FISH and gene mapping-based chromosomal homologies. *Mammalian Genome* 7, 297–302.

Colenbrander, B. and Wensing, C.J.G. (1975) Studies on phenotypically female pigs with hernia inguinalis and ovarian aplasia. I. Morphological aspects. *Proceedings XXVth International Conference on Animal Genetics*, Ser. C 78, 33–42.

Committee for the Standardized Karyotype of The Domestic Pig. (1988) Standard karyotype of the domestic pig. *Hereditas* 109, 151–157.

Coppieters, W., Zijlstra, C., Van de Weghe, A., Bosma, A.A., Peelman, I., Van Zeveren, A. and Bouquet, Y. (1994) A porcine minisatellite located on chromosome 14q29. *Mammalian Genome* 5, 591–593.

Cox, D.R., Burmeister, M., Price, E.R., Kim, S. and Myers, R.M. (1990) Radiation hybrid mapping: a somatic cell genetic method for contructing high-resolution maps of mammalian chromosomes. *Science* 250, 245–250.

Crew, F.A.E. and Koller, P.C. (1939) Cytogenetical analysis of the chromosomes in the pig. *Proceedings of Royal Society Edinburgh* 59, 163.

Christensen, K. and Bräuner Nielsen, P.B. (1980) A case of blood chimerism (XX,XY) in pigs. *Animal Blood Groups and Biochemical Genetics* 11, 55–57.

Davies, W., Harbitz, I., Fries, R., Stranzinger, G. and Hauge, J.G. (1988) Porcine malignant hyperthermia carrier detection and chromosomal assignment using a linked probe. *Animal Genetics* 19, 203–212.

Davies, W., Hoyheim, B., Chaput, B., Archibald, A.L. and Frelat, G. (1994) Characterization of microsatellites from flow-sorted porcine chromosome 13. *Mammalian Genome* 5, 707–711.

Deaven, L.L., Van Dilla, M.A., Bartholdi, M.F., Carrano, A.V., Cram, L.S. and Fuscoe, J.C. (1986) Construction of human chromosome-specific DNA libraries from flow-sorted chromosomes. *Cold Spring Harbor Symposium on Quantitative Biology,* 51, pt 1, 159–167.

de la Seña, C., Chowdhary, B.P. and Gustavsson, I. (1995) Localization of the telomeric $(TTAGGG)_n$ sequences in chromosomes of some domestic animals by fluorescence in situ hybridization. *Hereditas* 123, 269–274.

Dixon, S.C., Miller, N.G.A., Carter, N.P. and Tucker, E.M. (1992) Bivariate flow cytometry of farm animal chromosomes: a potential tool for gene mapping. *Animal Genetics* 23, 203–210.

Dolch, K.M. and Chrisman, C.L. (1981) Cytogenetic analysis of preimplantation blastocysts from prepuberal gilts treated with gonadotropins. *American Journal of Veterinary Research* 42, 344–346.

Ducos, A., Berland, H., Pinton, A., Séguéla, A., Blanc, M.F., Darré, A., Sans, P. and Darré, R. (1997) Les translocations réciproques chez le porc: état des lieux et perspectives. *Journées Recherche Porcine en France* 29, 375–382.

Dutrillaux, B. (1973) Nouveau systeme de marquage chromosomique: Les bandes T. *Chromosoma* 41, 395–402.

Dutrillaux, B. and Lejeune, J. (1971) Cytogénétique humaine. Sur une nouvelle technique d'analyse du caryotype humain. *Comptes Rendus de l'Académie des Sciences (Paris)* 272, 2638–2640.

Dutrillaux, B., Laurent, C., Couturier,J. and Lejeune, J. (1973) Coloration des chromosomes humains par l'acridine orange après traitement par le 5-bromodeoxyuridine. *Comptes Rendus Hebdomadaires Seances de l'Académie des Sciences Série D,* 276, 3179–3181.

Echard,G., Yerle, M., Gellin, J., Dalens, M. and Gillois, M. (1986) Assignment of the major histocompatibility complex to the p1.4–q1.2 region of chromosome 7 in the pig (*Sus scrofa domestica* L.) by *in situ* hybridization. *Cytogenetics and Cell Genetics* 41, 126–128.

Eldridge, F.E. (1985) *Cytogenetics of Livestock.* Avi Publishing Company, Westport, Connecticut. 298 pp.

Ellegren, H. and Basu, T. (1995) Filling the gaps in the porcine linkage map: isolation of microsatellites from chromosome 18 using flow sorting and SINE-PCR. *Cytogenetics and Cell Genetics* 71, 370–373.

Ellegren, H., Johansson, M., Chowdhary, B.P., Marklund, S., Ruyter, D., Marklund, L., Nielsen, P.B., Edfors-Lilja, I., Gustavsson, I., Juneja, R.K. and Andersson, L. (1993) Assignment of 20 microsatellite markers to the porcine linkage map. *Genomics* 16, 431–439.

Ellegren, H., Chowdhary, B.P., Fredholm, M., Hoyheim, B., Johansson, M., Nielsen, P.B., Thomsen, P.D. and Andersson, L. (1994a) A physically anchored linkage map of pig chromosome 1 uncovers sex- and position-specific recombination rates. *Genomics* 24, 342–350.

Ellegren, H., Chowdhary, B.P., Johansson, M. and Andersson, L. (1994b) Integrating the porcine physical and genetic linkage map using cosmid derived markers. *Animal Genetics* 25, 155–164.

Fidlerová, H., Senger, G., Kost, M., Sanseau, P. and Sheer, D. (1994) Two simple procedures for releasing chromatin from routinely fixed cells for fluorescence *in situ* hybridization. *Cytogenetics and Cell Genetics* 65, 203–205.

Ford, C.E. and Clegg, H.M. (1969) Reciprocal translocation. *British Medical Bulletin* 25, 110–114.

Ford, C.E., Pollock, D.L. and Gustavsson, I. (eds) (1980) Proceedings of the First International Conference for the Standardization of Banded Karyotype of Domestic Animals, University of Reading, Reading, UK, 2–6 August 1976. *Hereditas* 92, 145–162.

Förster, M. (1980) Localization of X-linked genes in cattle and swine by somatic hybrid cells. *Proceedings 4th European Colloqium on Cytogenetics of Domestic Animals,* Uppsala, Sweden, pp. 322–331.

Förster, M. and Hecht, W. (1984) Some provisional gene assignments in pig. *Proceedings. 6th European Colloqium on Cytogenetics of Domestic Animals,* pp. 351–355.

Förster, M. and Hecht, W. (1985) Genlokalisierung fur die Superoxid Dismutase (SOD1), Nukleosid Phosphorylase (NP) und Mannose Phospho Isomerase (MPI) beim Schwein. *Zuchtungskunde* 57, 249–255.

Förster, M., Stranzinger, G. and Hellkuhl, B. (1980) X-chromosome gene assignment of swine and cattle. *Naturwissenschaften* 67, 48–49.

Francke, U. (1994) Digitalized and differentially shaded human chromosome idiograms for genome applications. *Cytogenetics and Cell Genetics* 65, 206–219.

Fredholm, M., Winterö, A.K., Christensen, K., Thomsen, P.D., Nielsen, P.B., Kristensen, B., Juneja, R.K., Archibald, A.L. and Davies, W. (1992) Mapping of genetic markers in porcine backcross families. *Animal Genetics* 23 (Suppl. 1), 84–85.

Frengen, E. (1993) Detailed physical mapping in the porcine Halothane linkage group. PhD Thesis, Oslo, Norway. pp. 36.

Frengen, E. and Davies, W. (1995) Long-range mapping of the calcium release channel and glucosephosphate isomerase loci using pulsed-field gel electrophoresis. *Animal Genetics* 26, 181–184.

Frengen, E., Thomsen, P.D., Schmitz, A., Frelat, G. and Davies, W. (1994) Isolation of region-specific probes from pig chromosome 6 by coincidence cloning. *Mammalian Genome* 5, 497–502.

Frengen, E., Thomsen, P.D., Brede, G., Solheim, J., de Jong, P.J. and Prydz, H. (1997) The gene cluster containing the LCAT gene is conserved between human and pig. *Cytogenetics and Cell Genetics* 76, 53–57.

Fridolfsson, A.-K., Hori, T., Winterö, A.K., Fredholm, M., Yerle, M., Robic, A., Andersson, L. and Ellegren, H. (1996) Assignment of transcripts to the porcine genome map by SSCP and somatic cell hybrid mapping. *Proceeding of XXVth Internatational Conference on Animal Genetics,* 21–25 July, Tours, France, p. 122.

Fries, R., Vögeli, P. and Stranzinger, G. (1990) Gene mapping in the pig. In: McFeely, R.A. (ed.) *Domestic Animal Cytogenetics.* Academic Press, San Diego, California, pp. 273–304.

Frönicke, L. and Scherthan, H. (1997) Assignment of the porcine nuclear factor I/CTF (*NFI/CTF*) gene to Chromosome 2q12–13 by FISH. *Chromosome Research* 5, 254–261.

Frönicke, L., Chowdhary, B.P., Scherthan, H. and Gustavsson, I. (1996) A comparative map of the porcine and human genomes demonstrates Zoo-FISH and gene mapping-based chromosomal homologies. *Mammalian Genome* 7, 285–290.

Frönicke, L., Chowdhary, B.P. and Scherthan, H. (1997) Segmental homology between cattle (*Bos taurus*), Indian- *(Muntiacus muntjak vaginalis)* and Chinese-muntjac *(M. Reevesi)* karyotypes. *Chromosoma* 106, 108–113.

Fuscoe, J.C., Van Dilla, M.A. and Deaven, L.L. (1986) Construction and availability of human chromosome-specific gene libraries. *Progress in Clinical Biological Research* 209A, 465–472.

Gahne, B. (1979) Immunogenetics and biochemical genetics as a tool in pig breeding programmes. *Acta Agricultura Scandinavica* (Suppl. 21), 185–197.

Garrido, J.J., Lahbib-Mansais, Y., Geffrotin, C., Yerle, M. and Vaiman, M. (1995) Localization of the tenascin-C gene to pig chromosome 1. *Mammalian Genome* 6, 221.

Geffrotin, C., Popescu, C.P., Cribiu, E.P., Boscher, J., Renard, C., Chardon, P. and Vaiman, M. (1984) Assignment of MHC in swine to chromosome 7 by *in situ* hybridization and serological typing. *Annales de Genetique* 27, 213–219.

Geffrotin, C., Grunwald, D., Chardon, P. and Vaiman, M. (1987) Swine MHC: mapping by chromosome flow sorting and spot hybridization. *Animal Genetics* 18 (Suppl. 1), 118.

Gellin, J., Echard, G., Benne, F. and Gillois, M. (1981) Pig gene mapping: PKM2-MPI-NP synteny. *Cytogenetics and Cell Genetics* 30, 59–62.

Gellin, J., Yerle, M. and Gillois, M. (1990) Assignment of nucleoside phosphorylase gene to q2.1–q2.2 region of chromosome 7 in the pig, *Sus scrofa domestica* L. *Animal Genetics* 21, 207–210.

Gimenez-Martin, G., Lopez-Saez, J.F. and Monge, E.G. (1962) Somatic chromosomes of the pig. *Journal of Heredity* 53, 281, 290.

Glahn-Luft, B., Dzapo, V. and Wassmuth, R. (1982) Polymorphism of C-banding in swine. *Proceedings of 5th European Colloquium on Cytogenetics of Domestic Animals*, pp. 312–313.

Goodpasture, C. and Bloom, S. (1975) Visualization of nucleolar organizer regions in mammalian chromosomes using sliver staining. *Chromosoma* 53, 37–50.

Goss, S.J. and Harris, H. (1975) New methods for mapping genes in human chromosomes. *Nature (London)* 255, 680–684.

Goureau, A., Yerle, M., Schmitz, A., Riquet, J., Milan, D., Pinton, P., Frelat, G. and Gellin, J. (1996) Human and porcine correspondence of chromosome segments using bidirectional chromosome painting. *Genomics* 36, 252–262.

Graphodatsky, A., Lushnikova, T., Biltueva, L., Eremina, V., Rubtsov, N., Filippov, V., Astakhova, N.M., Rivkin, M., Shumny, T., Ermolaev, V. and Ruvinsky, A. (1991) Localization of some human genes to mammalian chromosomes by *in situ* hybridization. *Cytogenetics and Cell Genetics* 58, 1983.

Graphodatsky, A., Filippov, V., Biltueva, L., Astachova, N., Violetta, E., Lushnikova, T. and Ruvinsky, A.D. (1992) Localization of the pig gene ESD to Chromosome 13 by *in situ* hybridization. *Mammalian Genome* 3, 52–53.

Gray, J.W. and Cram, L.S. (1990) Flow karyotyping and chromosome sorting. In: Melamed, M.R., Lindmo, T., Mendelsohn, M.J. (eds) *Flow Cytometry and Sorting.* Wiley-Liss, New York, pp. 503–529.

Gropp, A., Giers, D. and Tettenborn, U. (1969) Das Chromosomenkomplement des Wildschweins (*Sus scrofa*). *Experientia* 25, 778.

Grunwald, D., Geffrotin, C., Chardon, P., Frelat, G. and Vaiman, M. (1986) Swine chromosomes: flow sorting and spot blot hybridization. *Cytometry* 7, 582–588.

Grunwald, D., Frelat, G. and vaiman, M. (1989) Animal flow cytogenetics. In: Yen, A. (ed.) *Flow Cytometry: Advanced Research and Clinical Applications*, vol. 1. CRC Press, Boca Raton, Florida, pp. 131–142.

Gu, F. (1994) *In situ* hybridization mapping of genetic markers in the porcine genome. PhD thesis, Swedish University of Agricultural Sciences, Uppsala, Sweden.

Gu, F., Harbitz, I., Chowdhary, B.P., Davies, W. and Gustavsson, I. (1992) Mapping of the porcine lipoprotein lipase (LPL) gene to chromosome 14q12–q14 by *in situ* hybridization. *Cytogenetics and Cell Genetics* 59, 63–64.

Gu, F., Chowdhary, B.P., Johansson, M., Andersson, L. and Gustavsson, I. (1993a) Mapping of the immunoglobulin gamma heavy chain (IGHG), cAMP-dependent protein kinase catalytic beta subunit (PRKACB) and transition protein 2 (TNP2) genes in pigs by *in situ* hubridization. *8th North American Colloquium on Domestic Animal Cytogenetics and Gene Mapping,* 13–16 July, Guelph, Canada, p. 69.

Gu, F., Harbitz, I., Chowdhary, B.P., Mosnes, M. and Gustavsson, I. (1993b) Chromosomal localization of the hormone sensitive lipase (LIPE) and insulin receptor (INSR) genes in pigs. *Hereditas* 117, 231–236.

Gu, F., Chowdhary, B.P., Johansson, M., Andersson, L. and Gustavsson, I. (1994) Localization of the IGHG, PRKACB, and TNP2 genes in pigs by *in situ* hybridization. *Mammalian Genome* 5, 195–198.

Gu, F., Hindkjær, J. Gustavsson, I. and Bolund, L. (1996) A signal of telomeric sequences on porcine chromosome 6q21–q22 detected by primed in situ labelling. *Chromosome Research* 4, 251–252.

Guan, X.-Y., Meltzer, P.S. and Trent, J.M. (1994) Rapid generation of whole chromosome painting probes (WCPs) by chromosome microdissection. *Genomics* 22, 101–107.

Gustavsson, I. (1980) Banding techniques in chromosome analysis of domestic animals. *Advances in Veterinary and Comparative Medicine* 24, 245–289.

Gustavsson, I. (1983) The THA technique as applied to porcine chromosomes. *Hereditas* 99, 311–313.

Gustavsson, I. (1984) Reciprocal translocations in the domestic pig (an interim report of a Swedish survey). *Proceedings of 6th European Colloquium on Cytogenetics of Domestic Animals*, pp. 80–86.

Gustavsson, I. (1990) Chromosomes of the pig. In: McFeely, R.A. (ed.) *Domestic Animal Cytogenetics.* Academic Press, Harcourt Brace Jovanovich, San Diego, California, pp. 73–107.

Gustavsson, I. and Jönsson, L. (1992) Stillborns, partially monosomic and partially trisomic, in the offspring of a boar carrying a translocation: rcp(14;15)(q29;q24). *Hereditas* 117, 31–37.

Gustavsson, I. and Settergren, I. (1983) Reciprocal chromosome translocations and decreased litter size in the domestic pig. *Journal of Dairy Science* 66 (Suppl. 1), 248.

Gustavsson, I., Hageltorn, M., Johansson,C. and Zech, L. (1972) Identification of the pig chromosomes by quinacrine mustard fluorescence technique. *Experimental Cell Research* 70, 471–474.

Gustavsson, I., Hageltorn, M., Zech, L. and Reiland, S. (1973) Identification of the chromosomes in centric fusion/fission polymorphic system of the pig (*Sus scrofa* L.). *Hereditas* 75, 153–155.

Gustavsson, I., Settergren, I. and King W.A. (1983) Occurrence of two different reciprocal translocations in the same litter of domestic pigs. *Hereditas* 99, 257–267.

Gustavsson, I., Switonski, M., Larsson, K., Plöen, L. and Höjer, K. (1988) Chromosome banding studies and synaptonemal complex analyses of four reciprocal translocations in the domestic pig. *Hereditas* 109, 169–184.

Gustavsson, I., Switonski, M., Ianuzzi, L. Plöen, L. and Larsson, K. (1989) Banding studies and synaptonemal complex analyses of an X-autosome translocation in the domestic pig. *Cytogenetics and Cell Genetics* 50, 188–194.

Gyapay, G., Schmitt, K., Fizames, C., Jones, H., Vega-Czarny, N., Spillett, D., Muselet, D., Prud'Homme, J.-F., Dib, C., Auffray, C., Morissette, J., Weissenbach, J. and Goodfellow, P.N. (1996) A radiation hybrid map of the human genome. *Human Molecular Genetics* 5, 339–346.

Haaf, T. and Ward, D.C. (1994) High resolution ordering of YAC contigs using extended chromatin and chromosomes. *Human Molecular Genetics* 4, 629–633.

Hageltorn, M., Gustavsson, I. and Zech, L. (1976) Detailed analysis of a reciprocal translocation (13q–;14q+) in the domestic pig by G- and Q-staining techniques. *Hereditas* 109, 169–184.

Hamasima, N., Suzuki, H., Kimura, M., Ito, T., Murakami, Y. and Yasue, H. (1996) Isolation of the pig medium-chain acyl-CoA dehydrogenase gene and assignment to chromosome 6q3.2. *Proceedings of XXVth International Conference on Animal Genetics*, 21–25 July, Tours, France, p. 124.

Hance, R.T. (1917) The diploid chromosome complexes of the pig (*Sus scrofa*) and their variations. *Journal of Morphology* 30, 155–222.

Hancock, J.L. (1959) Polyspermy of pig ova. *Animal Production* 1, 103–106.

Hancock, J.L. and Daker, M.G. (1981) Testicular hypoplasia in a boar with abnormal sex chromosome constitution (39 XXY). *Journal of Reproduction and Fertility* 61, 395–397.

Hansen-Melander, F. and Melander,Y. (1970) Mosaicism for translocation heterozygosity in a malformed pig. *Hereditas* 64, 199–202.

Hansen-Melander, F. and Melander,Y. (1974) The karyotype of the pig. *Hereditas* 77, 149–158.

Harbitz, I., Chowdhary, B.P., Chaudhary, R., Kran, S., Frengen, E., Gustavsson, I., Christensen, K. and Hauge, J. (1990a) Isolation, characterization and chromosomal assignment of a partial cDNA for porcine 6-phosphogluconate dehydrogenase. *Hereditas* 112, 83–88.

Harbitz, I., Chowdhary, B.P., Thomsen, P.D., Davies, W., Kaufmann, U., Kran, S., Gustavsson, I., Christensen, K. and Hauge, J.G. (1990b) Assignment of the porcine calcium release channel gene, a candidate for the malignant hyperthermia locus, to the 6p11–q21 segment of chromosome 6. *Genomics* 8, 243–248.

Harbitz, I., Chowdhary, B.P., Kran, S. and Davies, W. (1993) Characterization of a porcine glucosephosphate isomerase-processed pseudogene at chromosome 1q1.6–q1.7. *Mammalian Genome* 4, 589–592.

Hare, W.C.D. and Singh, E.L. (1979) *Cytogenetics in Animal Reproduction*. Commonwealth. Agriculture Bureau, Farnham Royal, England. 96 pp.

Hausmann, M., Popescu, C.P., Boscher, J., Kerbæuf, D., Dölle, J. and Cremer, C. (1993) Identification and cytogenetic analysis of an abnormal pig chromosome for flow cytometry and sorting. *Zeitschrift für Naturforschung* 48c, 645–653.

Heiskanen, M., Karhu, R., Hellsten, E., Peltonen, L., Kallioniemi, O.P. and Palotie, A. (1994) High resolution mapping using fluorescence *in situ* hybridization to extended DNA fibers prepared from agarose-embedded cells. *BioTechniques* 5, 928–933.

Heng, H.H.Q., Squire, J. and Tsui, L.-C. (1992) High resolution mapping of mammalian genes by *in situ* hybridization to free chromatin. *Proceedings of the National Academy of Sciences of the USA* 89, 9509–9513.

Hillebrand, P. (1936) Chromosomal investigation of three different breeds of domestic swine (Deutsches weisses Edelschwein, veredeltes Landschwein und Berkshire). Diss. Phil. Breslau.

Hindkjær, J., Brandt, C.A., Koch, J., Lund, T.B., Kølvraa, S. and Bolund, L. (1995) Simultaneous detection of centromere-specific probes and chromosome painting libraries by a combination of primed *in situ* labelling and chromosome painting (PRINS-painting). *Chromosome Research* 3, 41–44.

Houseal, T.W., Dackowski, W.R., Landes, G.M. and Klinger, K.W. (1994) High resolution mapping of overlapping cosmids by fluorescence *in situ* hybridization. *Cytometry* 15, 193–198.

Høyheim, B., Keiserud, A., Chaput, B., Schmitz, A., Frelat, G. and Davies, W. (1992) Rapid isolation of porcine chromosome 6-specific microsatellites. *Animal Genetics* 23 (Suppl. 1), 91.

Høyheim, B., Keiserud, A. and Thomsen, P.D. (1994a) A polymorphic porcine dinucleotide repeat SO294 (BHT 34) at chromosome 6p11–q11. *Animal Genetics* 25, 431.

Høyheim, B., Keiserud, A. and Thomsen, P.D. (1994b) A polymorphic porcine dinucleotide repeat SO297 (BHT 164) at chromosome 6p14–q15. *Animal Genetics* 25, 431.

Høyheim, B., Keiserud, A. and Thomsen, P.D. (1994c) A polymorphic porcine dinucleotide repeat SO301 (BHT 12) at chromosome 4p15. *Animal Genetics* 25, 432.

Høyheim, B., Keiserud, A. and Thomsen, P.D. (1995a) A highly polymorphic porcine dinucleotide repeat SO295 (BHT 40) at chromosome 9q21–q23. *Animal Genetics* 26, 57.

Høyheim, B., Keiserud, A. and Thomsen, P.D. (1995b) A highly polymorphic porcine dinucleotide repeat SO296 (BHT 137) at chromosome 17q13. *Animal Genetics* 26, 58.

Høyheim, B., Keiserud, A. and Thomsen, P.D. (1995c) A polymorphic porcine dinucleotide repeat SO298 (BHT 287) at chromosome 16q14. *Animal Genetics* 26, 56–57.

Hulot, F. (1970) Analyse chromosomique de deux porcs intersexués (*Sus scrofa domesticus*). *Annales de Genetique et Selection Animale* 2, 355–361.

Hunter, R.H.F. (1967) The effects of delayed insemination on fertilization and early cleavage in the pig. *Journal of Reproduction and Fertility* 13, 355–361.

Jensen, P.Ø., Larsen, J.K., Christensen, I.J., Christensen, K., Pedersen, K.M. and Gustavsson, I. (1994) Non-disjunction events and Y chromosome size variation evaluated by flow cytometry of sperm cells from cattle and swine. *Proceedings 11th European Colloquium on Cytogenetics of Domestic Animals,* pp. 66–70.

Johansson, M., Chowdhary, B., Gu, F., Ellegren, H., Gustavsson, I. and Andersson, L. (1993) Genetic analysis of the gene for porcine submaxillary gland mucin: physical assignment of the MUC and interferon gamma genes to chromosome 5. *Journal of Heredity* 84, 259–262.

Johansson, M., Ellegren, H. and Andersson, L. (1995) Comparative mapping reveals extensive linkage conservation – but with gene order rearrangements – between the pig and the human genomes. *Genomics* 25, 682–690.

Johansson-Moller, M., Chaudhary, R., Hellmen, E., Hoyheim, B., Chowdhary, B.P. and Andersson, L. (1996) Pigs with the dominant coat colour phenotype carry a duplication of the KIT gene encoding the mast/stem cell growth factor receptor. *Mammalian Genome* 7, 822–830.

Jørgensen, C.B., Hoyheim, B. and Thomsen, P.D. (1996a) A polymorphic porcine dinucleotide repeat S0531 (BHT10) at chromosome 1p22. *Animal Genetics* 27, 433–442.

Jørgensen, C.B., Hoyheim, B. and Thomsen, P.D. (1996b) A polymorphic porcine dinucleotide repeat S0532 (BHT487) at chromosome 13q48. *Animal Genetics* 27, 433–442.

Jørgensen, C.B., Winter¿, A.K., Yerle, M. and Fredholm, M. (1996c) Mapping of genes isolated from a porcine small intestine cDNA library. *Proceedings of XXVth International Conference on Animal Genetics,* 21–25 July, Tours, France, p. 125.

Kao, F.-T. (1983) Somatic cell genetics and gene mapping. *International Review of Cytology,* vol. 85. Academic Press, pp. 109–146.

King, W.A. (1980) Spontaneous chromosome translocation in cattle and pigs: A study of the causes of their fertility reducing effects. PhD thesis, Swedish University of Agricultural Science, Uppsala.

King, W.A., Gustavsson, I., Popescu, C.P. and Linares, T. (1981) Gametic products transmitted by rcp(13q–;14q+) translocation heterozygous pigs, and resulting embryonic loss. *Hereditas* 95, 239–246.

Koch, J.E., Kølvraa, S., Petersen, K.B., Gregersen, N. and Bolund, L. (1989) Oligonucleotide-priming methods for the chromosome-specific labeling of alpha satellite DNA in situ. *Chromosoma* 98, 259–265.

Kojima, M., Ohata, K., Harumi, T., Murakami, Y. and Yasue, H. (1996) Isolation and chromosomal assignment of four genes in the pig by fluorescence *in situ* hybridization. *Proceedings of XXVth International Conference on Animal Genetics,* 21–25 July, Tours, France, p. 126.

Konfortova, G.D., Miller, N.G. and Tucker, E.M. (1995) A new reciprocal translocation (7q+;15q–) in the domestic pig. *Cytogenetics and Cell Genetics* 71, 285–288.

Krallinger, H.F. (1931) Cytological studies on some domestic animals. *Arch. Tierernaehr. Tierz. Abt. B,* 5, 127–187.

Kusumoto, H. and Yasue, H. (1995) Assignment of the glucose transporter 1 gene (SLC2A1) to swine chromosome 6q34-qter. *Cytogenetics and Cell Genetics* 71, 377–379.

Kusumoto, H., Muladno and Yasue, H. (1996) Molecular cloning of a swine genomic fragment containing the heart aconitase (ACO) gene and its assignment on swine chromosome 5p14–p15. *Procedeengs of XXVth International Conference on Animal Genetics,* 21–25 July, Tours, France, p. 127.

Laan, M., Isosomppi, J., Klockars, T., Peltonen, L. and Palotie, A. (1996) Utilization of FISH in positional cloning: an example on 13q22. *Genome Research* 6, 1002–1012.

Lahbib-Mansais, Y., Yerle, M., Dalens, M., Chevalet, C. and Gellin, J. (1993) Localization of pig Na+, K+-ATPase and subunit genes to chromosome 4 by radioactive *in situ* hybridization. *Genomics* 15, 91–97.

Lahbib-Mansais, Y., Mellink, C., Yerle, M. and Gellin, J. (1994) A new marker (NGFB) on pig chromosome 4, isolated by using a concensus sequence conserved among species. *Cytogenetics and Cell Genetics* 67, 120–125.

Lahbib-Mansais, Y., Yerle, M. and Gellin, J. (1995) Localization of IGF1R and EDN genes to pig chromosomes 1 and 7 by *in situ* hybridization. *Cytogenetics and Cell Genetics* 71, 225–227.

Lahbib-Mansais, Y., Barbosa, A., Yerle, M., Parma, P., Milan, D., Pailhoux, E., Gellin, J. and Cotinot, C. (1996) Mapping in pig of genes involved in sexual differentiation: AMH, CYP19, DAX1, SF1, SOX2, SOX9 and WT. *Proceedings of XXVth International Conference on Animal Genetics,* 21–25 July, Tours, France, pp. 127–128.

Langford, C.F., Telenius, H., Carter, N.P., Miller, N.G.A. and Tucker, E.M. (1992) Chromosome painting using chromosome-specific probes from flow-sorted pig chromosomes. *Cytogenetics and Cell Genetics* 61, 221–223.

Langford, C.F., Telenius, H., Miller, N.G.A., Thomsen, P.D. and Tucker, E.M. (1993) Preparation of chromosome-specific paints and complete assignment of chromosomes in the pig flow karyotype. *Animal Genetics* 24, 261–267.

Lawrence, J.B. and Singer, R.H. (1986) Intracellular localization of messenger RNAs for cytoskeletal proteins. *Cell* 45, 407–415.

Lawrence, J.B., Villnave, C.A. and Singer, R.H. (1988) Sensitive, high-resolution chromatin and chromosome mapping in situ: presence and orientation of two closely integrated copies of EBV in lymphoma line. *Cell* 52, 51–61.

Lawrence, J.B., Singer, R.H. and McNeil, J.A. (1990) Interphase and metaphase resolution of different distances within the human dystrophin gene. *Science* 249, 928–932.

Leeb, T., Rettenberger, G., Hameister, H., Brem, G. and Brenig, B. (1995) Construction of a porcine YAC library and mapping of the cardiac muscle ryanodine receptor gene to chromosome 14q22–q23. *Mammalian Genome* 6, 37–41.

Leong, M.M., Lin, C.C. and Ruth, R.F. (1983a) Assignment of superoxide dismutase (SOD-1) gene to chromosome No. 9 of domestic pig. *Canadian Journal of Genetics and Cytology* 25, 233–238.

Leong, M.M., Lin, C.C. and Ruth, R.F. (1983b) The localization of genes for HPRT, G6PD, and alpha-GAL onto the X-chromosome of domestic pig (*Sus scrofa domesticus*). *Canadian Journal of Genetics and Cytology* 25, 239–245.

Lichter, P., Cremer, T., Borden, J., Manuelidis, L. and Ward, D.C. (1988) Delineation of individual human chromosomes in metaphase and interphase cells by in situ suppression hybridization using recombinant DNA libraries. *Human Genetics* 80, 224–234.

Lichter, P., Boyle, A.L., Cremer, T. and Ward, D.C. (1991) Analysis of genes and chromosomes by non-isotopic in situ hybridization. *Genetic Analysis – Techniques and Applications* 8, 24–35.

Lin, C.C., Biederman, B.M., Jamro, H.K., Hawthorne, A.B. and Church, R.B. (1980) Porcine (*Sus scrofa domestica*) chromosome identification and suggested nomenclature. *Canadian Journal of Genetics and Cytology* 22, 103–116.

Lin, C.C., Joyce, E., Biederman, B.M. and Gerhart, S. (1982) The constitutive heterochromatin of porcine chromosomes. *Journal of Heredity* 73, 231–233.

Liu, W-S., Harbitz, I., Gustavsson, I. and Chowdhary, B.P. (1998) Mapping of the porcine erythropoietin gene to chromosome 3p16-p15 and ordering of four related subclones by fiber-FISH and DNA-combing. *Hereditas* 128 (1).

Lojda, L. (1975) The cytogenetic pattern in pigs with hereditary intersexuality similar to the syndrome of testicular feminization in man. *Doc. Vet. (Brno)* 8, 71–82.

Lomholt, B., Christensen, K., Hallenberg, C. and Frederiksen, S. (1995) Porcine 5S rRNA gene map to 14q23 revealing syntenic relation to human HSPA6- and 7. *Mammalian Genome* 6, 439–441.

Looft, C., Nagel, M., Renard, C., Chardon, P., Yerle, M., Reinisch, N. and Kalm, E. (1994) Isolation and mapping of porcine microsatellites on chromosome 14. *Animal Genetics* 25 (Suppl. 2).

Louis, C.F., Ambady, S., Ponce de Leon, F.A, Buoen, L.C. and Mendiola, J.R. (1996) Production and use of a microdissected swine chromosome 6 genomic library. *Proceedings of XXVth International Conference on Animal Genetics*, 21–25 July, Tours, France, p. 148.

Lunney, J.K., Grimm, D.R., Goldman, T., Holley, R., Mendiola, J. and Louis, C.F. (1995) Characterization of a porcine chromosome 6 specific library. In: *First International Workshop on Porcine Chromosome 6, Animal Genetics* 26, 377–401.

McClintock, B. (1945) *Neurospora*. I. Preliminary observations of the chromosomes of *Neurospora crassa*. *American Journal of Botany* 32, 671–678.

McConnell, J., Fechheimer, N.S. and Gilmore, L.O. (1963) Somatic chromosomes of the domestic pig. *Journal of Animal Science* 22, 374–379.

McFee, A.F., Banner, M.W. and Rary, J.M. (1966) Variation in chromosome number among European wild pigs. *Cytogenetics* 5, 75–81.

McFeely, R.A. (1966) A direct method for the display of chromosomes from early pig embryos. *Journal of Reproduction and Fertility* 11, 161–163.

McFeely, R.A. (1967) Chromosome abnormalities in early embryos of the pig. *Journal of Reproduction and Fertility* 13, 579–581.

McFeely, R.A., Klunder, L.R. and Goldman, J.B. (1988) A Robertsonian translocation in a sow with reduced litter size. *Proceedings of 8th European Colloquium on Cytogenetics of Domestic Animals*, pp. 35–37.

McQueen, H.A., Yerle, M. and Archibald, A.L. (1994) Anchorage of an unassigned linkage group to pig chromosome 10 using Pi clones. *Mammalian Genome* 5, 646–648.

Makino, S. (1944) The chromosome complex of the pig (*Sus scrofa*). (Chromosome studies in domestic mammals, III). *Cytologia* 13, 170–178.

Makino, S., Sasaki, M.S., Sofuni, T. and Ishikawa,T. (1962) Chromosome condition of an intersex swine. *Proceedings of Japanese Academy of Sciences* 38, 686–689.

Marklund, L., Vinterö, A.K., Thomsen, P.D., Johansson, M., Fredholm, M., Gustafsson, U. and Andersson, L. (1993) A linkage group on pig chromosome 4 comprising the loci for blood group L, GBA, ATP1B1 and three microsatellites. *Animal Genetics* 24, 333–338.

Masuda, H., Okamoto, A. and Waide, Y. (1975) Autosomal abnormality in a swine. *Japanese Journal of Zootechnology Science* 46, 671–676.

Matsson, P., Annerén, G. and Gustavsson, I. (1986) Flow cytometric karyotyping of mammals, using blood lymphocytes: Detection and analysis of chromosomal abnormalities. *Hereditas* 104, 49–54.

Melander, Y. and Hansen-Melander, E. (1980) Chromosome studies in African wild pigs (Suidae, Mammalia). *Hereditas* 92, 283–289.

Mellink, C.H.M., Bosma, A.A., De Haan, N.A. and Wiegant, J. (1991) Distribution of rRNA genes in breeds of domestic pig studied by non-radioactive in situ hybridization and selective silver staining. *Genetics, Selection and Evolution* 23, (Suppl. 1), 168s–172s.

Mellink, C.H.M., Lahbib-Mansais, Y., Yerle, M. and Gellin, J. (1992) Chromosomal localization of eight markers in the pig by radioactive *in situ* hybridization. *Animal Genetics* 23 (Suppl. 1), 86–87.

Mellink, C.H.M., Lahbib-Mansais, Y., Yerle, M. and Gellin, J. (1993) Localization of four new markers to pig chromosomes 1, 6 and 14 by radioactive *in situ* hybridization. *Cytogenetics and Cell Genetics* 64, 256–260.

Mellink, C.H.M., Bosma, A.A. and de Haan, N.A. (1994a) Variation in size of Ag-NORs and fluorescent rDNA *in situ* hybridization signals in six breeds of domestic pig. *Hereditas* 120, 141–149.

Mellink, C., Lahbib-Mansais, Y., Yerle, M. and Gellin, J. (1994b) Mapping of the regulatory type I alpha and catalytic beta subunits of cAMP-dependent protein kinase and interleukin 1 alpha and 1 beta in the pig. *Mammalian Genome* 5, 298–302.

Mellink, C., Lahbib-Mansais, Y., Yerle, M. and Gellin, J. (1995) PCR amplification and physical localization of the genes for pig FSHB and LHB. *Cytogenetics and Cell Genetics* 70, 224–227.

Mellink, C.H., Bosma, A.A., de Haan, N.A. and Zijlstra, C. (1996) Physical localization of 5S rRNA genes in the pig by fluorescence *in situ* hybridization. *Hereditas* 124, 95–97.

Meltzer, P.S., Guan, X.-Y., Burgess, A. and Trent, J.M. (1992) Rapid generation of region specific probes by chromosome microdissection and their application. *Nature Genetics* 1, 24–28.

Mendiola, J.R., Ambady, S., Buoen, L.C., Paszek, A.A., Schook, L.B., Ponce de Leon, F.A. and Louis, C.F. (1995) Generation of a swine chromosome 6 specific library by microisolation. *Proceedings 9th North American Colloquium on Domestic Animals. Cytogenetics and Gene Mapping,* A & M University, Texas.

Milan, D., Yerle, M., Schmitz, A., Chaput, B., Vaiman, M., Frelat, G. and Gellin, J. (1993) A PCR-based method to amplify DNA with random primers: determining the chromosomal content of porcine flow-karyotype peaks by chromosome painting. *Cytogenetics and Cell Genetics* 62, 139–141.

Milan, D., Woloszyn, N., Giteau, M., Navas, A., Yerle, M., Rogel-Gaillard, C., Chardon, P., Gellin, J., Elsen, J.M. and Le Roy, P. (1996a) Toward the identification of RN gene involved in meat quality in pigs. *Proceedings of XXVth International Conference on Animal Genetics,* 21–25 July, Tours, France, p. 182.

Milan, D., Woloszyn, N., Yerle, M., Le Roy, P., Bonnet, M., Riquet, J., Lahbib-Mansais, Y., Caritez, J.-C., Robic, A., Sellier, P., Elsen, J.-M. and Gellin, J. (1996b) Accurate mapping of the 'acid meat' RN gene on genetic and physical maps of pig chromosome 15. *Mammalian Genome* 7, 47–51.

Miller, J.R., Dixon, S.C., Miller, N.G.A., Tucker, E.M., Hindkjær, J. and Thomsen, P.D. (1992) A chromosome 1 specific DNA library from the domestic pig *(Sus scrofa domestica). Cytogenetics and Cell Genetics* 61, 128–131.

Miyake, Y.-I. (1973) Cytogenetical studies on swine intersexes. *Japanese Journal of Veterinary Research* 21, 41–49.

Miyake, Y.-I., Kawata, K., Ishikawa, T. and Umezu, M. (1977) Translocation heterozygosity in a malformed piglet and its normal littermates. *Teratology* 16, 163–168.

Miyake, Y.-I., O'Brien, S.J. and Kaneda, Y. (1988) Regional localization of rDNA gene on pig chromosome 10 by *in situ* hybridization. *Japanese Journal of Veterinary Science* 50, 341–345.

Moon, R.G., Rashad, M.N. and Mi, M.P. (1975) An example of polyploidy in pig blastocysts. *Journal of Reproduction and Fertility* 45, 147–149.

Murakami, Y., Itoh, T., Ohata, K. and Yasue, H. (1994) Chromosomal assignment of porcine genes by fluorescence *in situ* hybridization. *Proceedings XXIVth International Conference on Animal Genetics,* 23–29 July, Prague, Czech Republic, p. 122.

Murakami, Y., Itoh, T., Ohata, K. and Yasue, H. (1995) Assignment of alpha 1-acid glycoprotein gene (ORM1) to swine chromosome region 1q2.10–q2.12 by fluorescence *in situ* hybridization. *Cytogenetics and Cell Genetics* 70, 186–187.

Murakami, Y., Itoh, T. and Yasue, H. (1996) Assignment of the choline acetyltransferase gene to porcine chromosome 14q25–q27 by fluorescence *in situ* hybridization. *Mammalian Genome* 7, 325.

Muramoto, J., Makino, S., Ishikawa, T. and Kanagawa, H. (1965) On the chromosomes of the wild boar and the boar-pig hybrids. *Proceedings of the Japanese Academy* 41, 236–239.

Musilová, P., Lahbib-Mansais, Y., Yerle, M., Cepica, S., Stratil, A., Coppieters, W. and Rubes, J. (1995) Assignment of pig alpha-1-antichymotrypsin (AACT or P12) gene to chromosome region 7q23–q26. *Mammalian Genome* 6, 445.

Musilová, P., Lee, D.A., Stratil, A., Cepica, S., Rubes, J., Lowe, X. and Wyrobek, A. (1996) Assignment of the porcine *IKBA* gene (*IkBa*) encoding a cytoplasmic inhibitor of the NF-kB to chromosome 7q15–q21 by FISH. *Mammalian Genome* 7, 323–324.

Nes, N. (1968) Betydningen av kromosomaberrasjoner hos dyr. *Forskning og Forsoek i Landbruket* 19, 393–410.

Neuenschwander, S., Rettenberger, G., Meijerink, E., Jörg, H. and Stranzinger, G. (1996) Partial characterization of porcine obesity gene (*OBS*) and its localization to chromosome 18 by somatic cell hybrids. *Animal Genetics* 27, 275–278.

Nilsson, M., Malmgren, H., Samiotaki, M., Kwiatkowski, M., Chowdhary, B.P. and Landegren, U. (1994) Padlock probes: circularizing oligonucleotides for localized DNA detection. *Science* 265, 2085–2087.

Nunes, M., Yerle, M., Dezeure, F., Gellin, J., Chardon, P. and Vaiman, M. (1993) Isolation of four *HSP70* genes in the pig and localization on Chromosomes 7 and 14. *Mammalian Genome* 4, 247–251.

Nunes, M., Lahbib-Mansais, Y., Geffrotin, C., Yerle, M., Vaiman, M. and Renard, C. (1996) Swine cytosolic malic enzyme: cDNA cloning, sequencing and localization. *Mammalian Genome* 7, 815–821.

Okamoto, A., Shiobara, T., Tomita, T. and Ohmi, H. (1980) Chromosome studies of the wild boar and Ohmini pigs. *Bulletin of College of Agriculture, Utsunomiya University* 11, 1–8.

Oprescu, S., Voiculescu, I., Oprescu, S. and Gessner, E. (1976) Cytogenetic investigations on a phenotypically normal boar responsible for repeated abortions and on his progeny. *Proceedings 8th International Congress on Animal Reproduction and Artificial Insemination,* Cracow, vol. 4, pp. 738–741.

Palotie, A., Heiskanen, M., Laan, M., and Horelli-Kuitunen, N. (1996) High-resolution fluorescence in situ hybridization: a new approach in genome mapping. *Annales Medicinae* 28, 101–106.

Parra, I. and Windle, B. (1993) High resolution visual mapping of stretched DNA by fluorescent hybridization. *Nature Genetics* 5, 17–21.

Peelman, L.J., Chardon, P., Vaiman, M., Mattheeuws, A., Van Zeveren, A., Van de Weghe, A., Bouquet, Y., Campbell, R.D. (1996) A detailed physical map of the porcine major histocompatibility complex (MHC) class III region: comparison with human and mouse MHC class III region. *Mammalian Genome* 7, 363–367.

Pinkel, D., Gray, J.W., Trask, B., van den Engh, G., Fuscoe, J. and van Dekken. (1986) Cytogenetic analysis by in situ hybridization with fluorescently labeled nucleic acid probes. *Cold Spring Harbor Symposia on Quantitative Biology,* vol LI, pp. 151–157.

Pinkel, D., Landegent, J., Collins, C., Fuscoe, J., Segraves, R., Lucas, J. and Gray, J. (1988) Fluorescence in situ hybridization with human chromosome-specific libraries: Detection of trisomy 21 and translocations on chromosome 4. *Proceedings of the National Academy of Sciences, USA* 85, 9138–9142.

Pontecorvo, G. (1971) Induction of directional chromosome elimination in somatic cell hybrids. *Nature* (London) 230, 267–269.

Popescu, C.P. (1989) Cytogénétique des mammifères d'élevage. *Institut National de la Recherche Agronomique,* Paris.

Popescu, C.P. and Boscher, J. (1982) Cytogenetics of preimplantation embryos produced by pigs heterozygous for the reciprocal translocation (4q+;14q–). *Cytogenetics and Cell Genetics* 34, 119–123.

Popescu, C.P., Quéré, J.P. and Franceschi, P. (1980) Observations chromosomiques chez le sanglier français (*Sus scrofa scrofa*). *Annales de Génétique Sélection Animale* 12, 395–400.

Poulsen, P.H., Thomsen, P.D. and Olsen, J. (1991) Assignment of the porcine aminopeptidase N (PEPN) gene to chromosome 7cen-q21. *Cytogenetics and Cell Genetics* 57, 44–46.

Prakash, B., Kuosku, V., Olsaker, I., Gustavsson, I. and Chowdhary, B.P. (1996) Comparative FISH mapping of bovine cosmids to reindeer chromosomes demonstrates conservation of the X-chromosome. *Chromosome Research* 4, 214–217.

Prakash, B., de la Seña, C., Olsaker, I., Gustavsson, I. and Chowdhary, B.P. (1998) Comparative FISH mapping of 28 bovine cosmids to homologous cattle, goat, and river buffalo chromosomes. *Cytogenetics and Cell Genetics*, (in press).

Raap, A.K., van de Corput, M.P.C., Vervenne, R.A.W., van Gijlswijk, R.P.M., Tanke, H.J. and Wiegant, J. (1995) Ultra-sensitive FISH using peroxidase-mediated deposition of biotin or fluorochrome tyramides. *Human Molecular Genetics* 4, 529–534.

Rabin, M., Fries, R., Singer, D. and Ruddle, F.H. (1985) Assignment of the porcine major histocompatibility complex to chromosome 7 by *in situ* hybridization. *Cytogenetics and Cell Genetics* 39, 206–209.

Remy, J.J., Lahbib-Mansais, Y., Yerle, M., Bozon, V., Couture, L., Pajot, E., Greber, D. and Salesse, R. (1995) The porcine follitropin receptor: cDNA cloning, functional expression and chromosomal localization of the gene. *Gene* 163, 257–261.

Rettenberger, G., Klett, C. and Hameister, H. (1992) The application of comparative gene mapping for the rapid construction of a porcine gene map. *Animal Genetics* 23 (Suppl. 1), 86.

Rettenberger, G., Klett, C. and Hameister, H. (1993) Localization of the tumour protein, TP53, on porcine chromosome 12q12–q14. *Animal Genetics* 24, 307–309.

Rettenberger, G., Adham, I.M., Engel, W., Klett, C. and Hameister, H. (1994a) Assignment of the porcine transition protein TNP1 to chromosome 15q24–q25 by fluorescence *in situ* hybridization (FISH). *Mammalian Genome* 5, 249–250.

Rettenberger, G., Burkhardt, E., Adham, I.M., Engel, W., Fries, R., Klett, C. and Hameister, H. (1994b) Assignment of the Leydig insulin-like hormone to porcine chromosome 2q12–q13 by somatic cell hybrid analysis and fluorescence *in situ* hybridization. *Mammalian Genome* 5, 307–309.

Rettenberger, G., Fredholm, M. and Fries, R. (1994c) Chromosomal assignment of porcine microsatellites by use of a somatic cell hybrid mapping panel. *Animal Genetics* 25, 343–345.

Rettenberger, G., Fries, R., Engel, W., Scheit, K.L.H., Dolf, G. and Hameister, H. (1994d) Establishment of a partially informative porcine somatic cell hybrid panel and assignment of the loci for transition protein 2 (TNP2) and protamine 1 (PRM1) to chromosome 3 and polyubiquitin (UBC) to chromosome 14. *Genomics* 21, 558–566.

Rettenberger, G., Bruch, J., Beattie, C.W., Moran, C., Fries, R. and Hameister, H. (1995a) Chromosomal assignment of seventeen porcine microsatellites and genes by use of somatic cell hybrid mapping panel. *Animal Genetics* 26, 269–273.

Rettenberger, G., Klett, C., Zechner, U., Kunz, J., Vogl, W. and Hameister, H. (1995b) Visualization of the conservation of synteny between humans and pigs by heterologous chromosomal painting. *Genomics* 26, 372–378.

Rettenberger, G., Bruch, J., Fries, R., Archibald, A.L. and Hameister, H. (1996a) Assignment of 19 porcine type I loci by somatic cell hybrid analysis detects new regions of conserved synteny between human and pig. *Mammalian Genome* 7, 275–279.

Rettenberger, G., Bruch, J. and Hameister, H. (1996b) Chromosomal assignment of the porcine gene for apolipoprotein C3 (APOC3) to chromosome 9 by somatic cell hybrids. *Mammalian Genome* 7, 322–323.

Rettenberger, G., Bruch, J., Leeb, T., Klett, C., Brenig, B. and Hameister, H. (1996c) Mapping of the porcine immunoglobulin lambda gene, IGL, by fluorescence *in situ* hybridization (FISH) to chromosome 14q17–q21. *Mammalian Genome* 7, 326.

Rettenberger, G., Bruch, J., Leeb, T., Brenig, B., Klett, C. and Hameister, H. (1996d) Assignment of pig immunoglobulin kappa IGKC, to chromosome 3q12–q14 by fluorescence *in situ* hybridization (FISH). *Mammalian Genome* 7, 324–325.

Rettenberger, G., Leeb, T., Meier-Ewert, S., Bruch, J., Klett, C., Brenig, B. and Hameister, H. (1996e) Mapping of the porcine urate oxidase and transforming growth factor beta 2 genes by fluorescence *in situ* hybridization. *Chromosome Research* 4, 147–150.

Rittmannsperger, C. (1971) Chromosomenuntersuchungen bei Wild- und Hausschweinen. *Annales de Génétique Sélection Animale* 3, 105–107.

Robic, A., Parrou, J.-L., Yerle, M., Goureau, A., Dalens, M., Milan, D. and Gellin, J. (1995) Pig microsatellites isolated from cosmids revealing polymorphism and localized on chromosomes. *Animal Genetics* 26, 1–6.

Robic, A., Riquet, J., Yerle, M., Milan, D., Lahbib-Mansais, Y., Dubut-Fontana, C. and Gellin, J. (1996) Porcine linkage and cytogenetic maps integrated by regional mapping of 100 microsatellites on somatic cell hybrid panel. *Mammalian Genome* 7, 438–445.

Rogel-Gaillard, C., Bourgeaux, N., Save, J.-C., Chaput, B. and Renard, C. (1996) A three genome-equivalent pig YAC library allowing an efficient recovery of unique and repeated sequences. *Proceedings of XXVth International Conference on Animal Genetics*, 21–25 July, Tours, France, p. 153.

Rohrer, G.A., Alexander, L.J., Keele, J.W., Smith, T.P. and Beattie, C.W. (1994) A microsatellite linkage map of the porcine genome. *Genetics* 136, 231–245.

Rønne, M. (1995) Localization of fragile sites in the karyotype of *Sus scrofa domestica*: present status. *Hereditas* 122, 153–162.

Rønne, M., Stefanova, V., Di Berardino, D. and Strandby, B. (1987) The R-banded karyotype of the domestic pig (*Sus scrofa*). *Hereditas* 106, 219–231.

Ruddle, F.H. (1981) A new era in mammalian gene mapping: gene mapping by somatic cell genetics and recombinant DNA technologies. *Nature* 294, 115–120.

Ruzicska, P. (1968) Double trisomy in a pig embryo. *Mammalian Chromosome Newsletter* 9, 240–241.

Ryttman, H., Thebo, P., Gustavsson, I., Gahne, B. and Juneja, R.K. (1986) Further data on chromosomal assignments of pig enzyme loci *LDHA, LDHB, MPI, PEPB* and *PGM1*, using somatic cell hybrids. *Animal Genetics* 17, 323–333.

Ryttman, H., Thebo, P. and Gustavsson, I. (1988) Regional assignment of NP and MPI on chromosome 7 in pig, Sus scrofa. *Animal Genetics* 19, 197–200.

Sachs, L. (1954) Chromosome numbers and experimental polyploidy in the pig. *Journal of Heredity* 45, 21–24.

Sakurai, M., Zhou, J.H., Ohtaki, M., Itoh, T., Murakami, Y. and Yasue, H. (1996) Assignment of c-KIT to swine chromosome 8p12–p21 by fluorescence *in situ* hybridization. *Mammalian Genome* 7, 397.

Sarmiento, U.M. and Kadavil, K. (1993a) Mapping of the porcine apolipoprotein B (APOB) gene to chromosome 3 by fluorescence *in situ* hybridization. *Mammalian Genome* 4, 66–67.

Sarmiento, U.M. and Kadavil, K. (1993b) Mapping of the porcine involucrin (IVL) gene to Chromosome 4 by fluorescence *in situ* hybridization. *Mammalian Genome* 4, 60–61.

Sarmiento, U.M., Sarmiento, J.I., Lunney, J.K. and Rishi, S. (1993c) Mapping of the porcine alpha interferon (IFNA) gene to Chromosome 1 by fluorescence *in situ* hybridization. *Mammalian Genome* 4, 62–63.

Sarmiento, U.M., Sarmiento, J.I., Lunney, J.K. and Rishi, S. (1993d) Mapping of the porcine SLA class I gene (PD1A) and the associated repetitive element (C11) by fluorescence *in situ* hybridization. *Mammalian Genome* 4, 64–65.

Sawyer, J.R. and Hozier, J.C. (1986) High resolution of mouse chromosomes: banding conservation between man and mouse. *Science* 232, 1632–1635.

Scherthan, H., Cremer, T., Arnason, U., Weier, H.-U., Lima-de-Faria, A. and Frönicke, L. (1994) Comparative chromosome painting discloses homologous segments in distantly related mammals. *Nature Genetics* 6, 342–347.

Schmitz, A., Chardon, P., Gainche, I., Chaput, B., Guilly, M.-N., Frelat, G. and Vaiman, M. (1992) Pig standard bivariate flow karyotype and peak assignment for chromosomes X, Y, 3, and 7. *Genomics* 14, 357–362.

Schmitz, A., Oustry, A., Vaiman, D., Chaput, B., Frelat, G. and Cribiu, E.P. (1996) Comparative karyotype of pig and cattle using whole chromosome painting probes. *Proceedings of XXVth International Conference on Animal Genetics*, 21–25 July, Tours, France, p. 102.

Schwerin, M., Golisch, D. and Ritter, E. (1986) A Robertsonian translocation in swine. *Genetics, Selection and Evolution* 18, 367–374.

Seabright, M. (1971) A rapid banding technique for human chromosomes. *Lancet* ii, 971–972.

Signer, E.N., Gu, F., Gustavsson, I., Andersson, L. and Jeffreys, A.J. (1994) A pseudoautosomal minisatellite in the pig. *Mammalian Genome* 5, 48–51.

Signer, E.N., Gu, F. and Jeffreys, A.J. (1996) A panel of VNTR markers in pigs. *Mammalian Genome* 7, 433–437.

Sjöberg, A., Seaman, W.T., Bellinger, D.A., Griggs, T.R., Nichols, T.C. and Chowdhary, B.P. (1996) FISH mapping of the porcine vWF gene to chromosome 5q21 extends synteny homology with human Chromosome 12. *Hereditas* 124, 199–202.

Sjöberg, A., Peelman, L.J. and Chowdhary, B.P. (1997) Application of three different methods to analyze fiber-FISH results obtained using four lambda clones from the porcine MHC III region. *Chromosome Research* 5, 247–253.

Smith, J.H. and Marlowe, T.J. (1971) A chromosomal analysis of 25-day-old pig embryos. *Cytogenetics* 10, 385–391.

Solinas, S., Hasler-Rapacz, J., Maeda, N., Rapacz, J. and Fries, R. (1992a) Assignment of the pig apolipoprotein B locus (APOB) to chromosome region 3q24–qter. *Animal Genetics* 23, 71–75.

Solinas, S., Pauli, U., Kuhnert, P., Peterhans, E. and Fries, R. (1992b) Assignment of the porcine tumour necrosis factor alpha and beta genes to the chromosome region 7p11–q11 by *in situ* hybridization. *Animal Genetics* 23, 267–271.

Southern, E.M. (1975) Long range periodicities in mouse satellite DNA. *Journal of Molecular Biology* 94, 51–69.

Spalding, J.F. and Berry, R.O. (1956) A chromosome study of the wild pig (*Pecari angulatus*) and the domestic pig (*Sus scrofa*). *Cytologia* 21, 81–84.

Stone, L.E. (1963) A chromosome analysis of the domestic pig (*Sus scrofa*) utilizing a peripheral-blood culture technique. *Canadian Journal of Genetics and Cytology* 5, 38–42.

Strahan, K.M., Gu, F., Preece, A.F., Gustavsson, I., Andersson, L. and Gustafsson, K. (1995) cDNA sequence and chromosome location of pig 1,3 galactosyltransferase. *Immunogenetics* 41, 101–105.

Sumner, A.T. (1972) A simple technique for demonstrating centromeric heterochromatin. *Experimental Cell Research* 75, 304–306.

Telenius, H., Carter, N.P., Bebb, C.E. Nordenskjöld, M., Ponder, B.A.J. and Tunnacliffe, A. (1992) Degenerate oligonucleotide-primed PCR: general amplification of target DNA by a single degenerate primer. *Genomics* 13, 718–725.

Thibault, C. (1959) Analyse de la fécondation del'oeuf de la truie après accouplement ou insémination artificielle. *Colloq. Reprod. Artif. Insemin. Pig, Inst. Natl. Rech. Agron.*, Paris pp. 165–188.

Thomsen, P.D. and Hoyheim, B. (1994) S0306 – a molecular marker at pig chromosome 18q24. *Animal Genetics* 25, 377.

Thomsen, P.D. and Zhdanova, N.S. (1995) Reverse painting for identification of pig chromosomes in hybrid cell lines: assignment of the *HOXB* and the *TK1* gene to pig Chromosome 12p. *Mammalian Genome* 6, 670–672.

Thomsen, P.D., Fredholm, M., Christensen, K. and Schwerin, M. (1990) Assignment of the porcine growth hormone gene to chromosome 12. *Cytogenetics and Cell Genetics* 54, 92–94.

Thomsen, P.D., Bosma, A.A., Kaufmann, U. and Harbitz, I. (1991) The porcine *PGD* gene is preferentially lost from chromosome 6 in pig × rodent somatic cell hybrids. *Hereditas* 115, 63–67.

Thomsen, P.D., Hindkjaer, J. and Christensen, K. (1992) Assignment of a male-specific DNA repeat to Y-chromosomal heterochromatin. *Cytogenetics and Cell Genetics* 61, 152–154.

Thomsen, P.D., Qvist, H., Marklund, L., Andersson, L., Sjostrom, H. and Noren, O. (1993) Assignment of the dipeptidase IV (DPP4) gene to pig chromosome 15q21. *Mammalian Genome* 4, 604–607.

Thomsen, P.D., Johansson, M., Troelsen, J.T. and Andersson, L. (1995) The lactase phlorizin hydrolase (LCT) gene maps to pig chromosome 15q13. *Animal Genetics* 26, 49–52.

Tikhonvov, V.N. and Troshina, A.I. (1975) Chromosome translocations in the karyotype of wild boars *Sus scrofa* L. of the European and the Asian areas of USSR. *Theoretical and Applied Genetics* 45, 304–308.

Toga-Piquet, C., Henderson, A.S., Grillo, J.M., Vagner-Capodano, A.M. and Stahl, A. (1984) Localisation des gènes ribosomique et activité nucléolaire dans les lymphocytes du porc (*Sus scrofa domestica*) stimulés par la phytohémagglutinine. *Comptes Rendus Hebdomadaires des Séances de l'Academie des Sciences* (III) 298, 383–386.

Toyama, Y. (1974) Sex chromosome mosaicisms in five swine intersexes. *Japanese Journal of Zootechnological Sciences* 45, 304–308.

Trask, B.J., Pinkel, D. and van den Engh, G. (1989) The proximity of DNA sequences in interphase cell nuclei is correlated to genomic distance and permits ordering of cosmids spanning 250 kilobase pairs. *Genomics* 5, 710–717.

Troyer, D.L., Goad, D.W., Xie, H., Rohrer, G.A., Alexander, L.J. and Beattie, C.W. (1994) Use of direct *in situ* single-copy (DISC) PCR to physically map five porcine microsatellites. *Cytogenetics and Cell Genetics* 67, 199–204.

Troyer, D., Rohrer, G., Stone, R., Hu, J., Alexander, L. and Beattie, C. (1996) Nonoverlapping subgenomic libraries from porcine chromosomes. *Proceedings of XXVth International Conference on Animal Genetics*, 21–25 July, Tours, France, p. 132.

Vagner-Capodano, A.M., Henderson, A.S., Lissitzky, S. and Stahl, A. (1984) The relationships between ribosomal genes and fibrillar centres in thyroid cells cultivated in vitro. *Biology of the Cell* 51, 11–22.

Villagomez, D.A.F. (1993) Zygotene–pachytene substaging and synaptonemal complex karyotyping of boar spermatocytes. *Hereditas* 118, 87–99.

Villagomez, D.A.F., Gustavsson, I., Jönsson, L. and Plöen, L. (1995a) Reciprocal translocation, rcp(7;17)(q26;q11), in a boar giving reduced litter size and increased rate of piglets dying in the early life. *Hereditas* 122, 257–267.

Villagomez, D.A.F., Gustavsson, I. and Plöen, L. (1995b) Synaptonemal complex analysis in a boar with tertiary trisomy, product of a rcp(7;17)(q26;q11) translocation. *Hereditas* 122, 269–277.
Vogt, D.W., Arakaki, D.T. and Brooks, C.C. (1974) Reduced litter size associated with aneuploid cell lines in a pair of full-brother Duroc boars. *American Journal of Veterinary Research* 35, 1127–1130.
Weier, H.-U.G., Wang, M., Mullikin, J.C., Zhu, Y., Chen, J.-F., Greulich, K.M., Bensimon, A. and Grey, J.W. (1995) Quantitative DNA fiber mapping. *Human Molecular Genetics* 4, 1903–1910.
Wiegant, J., Kalle, W., Mullenders, L., Brookes, S., Hoovers, J.M.N., Dauwerse, J.G., Van Ommen, G.J.B. and Raap, A.R. (1992) High resolution *in situ* hybridization using DNA halo preparations. *Human Molecular Genetics* 1, 587–591.
Winterø, A.K., Chowdhary, B. and Fredholm, M. (1994a) A porcine polymorphic microsatellite locus (S0076) at chromosome 13q12. *Animal Genetics* 25, 430.
Winterø, A.K., Fredholm, M. and Thomsen, P. (1994b) A porcine polymorphic microsatellite locus (S0077) at chromosome 16q14. *Animal Genetics* 25, 122.
Winterø, A.K., Fredholm, M. and Davies, W. (1996) Evaluation and characterization of a porcine small intestine cDNA library: analysis of 839 clones. *Mammalian Genome* 7, 509–517.
Wødsedalek, J.E. (1913) Spermatogenesis of the pig with special reference to the accessory chromosomes. *Biological Bulletin* 25, 8.
Womack, J.E. (1990) Gene mapping in the cow. In: McFeely, R.A. (ed.) *Domestic Animal Cytogenetics.* Academic Press, Harcourt Brace Jovanovich, San Diego, California, pp. 73–107.
Wu, L., Rothschild, M.F. and Warner, C.M. (1995) Mapping of the SLA complex class III region by pulsed field gel electrophoresis. *Mammalian Genome* 6, 607–610.
Xu, Y., Rothschild, M.F. and Warner, C.M. (1992) Mapping of the SLA complex of miniature swine: Mapping of the SLA gene complex by pulsed field gel electrophoresis. *Mammalian Genome* 2, 2–10.
Yang, H., Jung, H.-R., Solinas-Toldo, S., Lang, E., Bolt, R., Fries, R. and Stranzinger, G. (1992) A reciprocal whole-arm translocation, rcp(1;6) (1p6p;1q6q) in a boar, localization of the breakpoints, and reassignment of the genes for glucose phosphate isomerase (GPI) and calcium release channel (CRC). *Cytogenetics and Cell Genetics* 61, 67–74.
Yang, H., Fries, R. and Stranzinger, G. (1993) The sex-determining region Y (SRY) gene is mapped to p12–p13 of the Y chromosome in pig (*Sus scrofa domestica*) by *in situ* hybridization. *Animal Genetics* 24, 297–300.
Yasue, H., Kusumoto, H. and Mikami, H. (1995) Assignment of the uteroferrin gene (ACP5) to swine chromosome 2q12–q21 by fluorescence *in situ* hybridization. *Cytogenetics and Cell Genetics* 71, 249–252.
Yasue, H., Kusumoto, H., Nisamatu, N., Muladno and Awata, T. (1996) Molecular cloning of swine genomic fragments containing acrosin (ACR) and heart aconitase (ACO) genes, isolation of linkage markers from the fragments, and their assignment to swine chromosomes. *Proceedings of XXVth International Conference on Animal Genetics*, 21–25 July, Tours, France, p. 133.
Yerle, M. and Gellin, J. (1989) Localization of leucocyte interferon gene in the q2.5 region of pig chromosome 1 by *in situ* hybridization. *Genetics, Selection and Evolution* 249–252.

Yerle, M., Gellin, J., Echard, G., Lefevre, F. and Gillois, M. (1986) Chromosomal localization of leucocyte interferon gene in the pig (*Sus scrofa domestica* L.) by *in situ* hybridization. *Cytogenetics and Cell Genetics* 42, 129–132.

Yerle, M., Archibald, A.L., Dalens, M. and Gellin, J. (1990a) Localization of the *PGD* and *TGFb*-1 loci to pig chromosome 6q. *Animal Genetics* 21, 411–417.

Yerle, M., Gellin, J., Dalens, M. and Galman, O. (1990b) Localization on pig chromosome 6 of markers GPI, APOE, and ENO1, carried by human chromosomes 1 and 19, using *in situ* hybridization. *Cytogenetics and Cell Genetics* 54, 86–91.

Yerle, M., Galman, O., Lahbib-Mansais, Y. and Gellin, J. (1992) Localization of the pig luteinizing hormone/choriogonadotropin receptor gene (LHCGR) by radioactive and nonradioactive *in situ* hybridization. *Cytogenetics and Cell Genetics* 59, 48–51.

Yerle, M., Lahbib-Mansais, Y., Thomsen, P.D. and Gellin, J. (1993a) Localization of the porcine growth hormone gene to chromosome 12p1.2–p1.5. *Animal Genetics* 24, 129–131.

Yerle, M., Schmitz, A., Milan, D., Chaput, B., Monteagudo, L., Vaiman, M., Frelat, G. and Gellin, J. (1993b) Accurate characterization of porcine bivariate flow karyotype by PCR and fluorescence in situ hybridization. *Genomics* 16, 97–103.

Yerle, M., Goureau, A., Gellin, J., Le Tissier, P. and Moran, C. (1994a) Rapid mapping of cosmid clones on pig chromosomes by fluorescent *in situ* hybridization. *Mammalian Genome* 5, 34–37.

Yerle, M., Robic, A., Goureau, A., Milan, D., Pinton, P., Woloszyn, N. and Gellin, J. (1994b) Contribution to the establishment of connections between genetic and cytogenetic maps of the porcine genome. *Animal Genetics* 25 (Suppl. 2), 48.

Yerle, M., Echard, G., Robic, A., Mairal, A., Dubut-Fontana, C., Riquet, J., Pinton, P., Milan, D., Lahbib-Mansais, Y. and Gellin, J. (1996a) A somatic cell hybrid panel for pig regional gene mapping characterized by molecular cytogenetics. *Cytogenetics and Cell Genetics* 73, 194–202.

Yerle, M., Pinton, P., Robic, A., Delcros, C. and Gellin, J. (1996b) Generation of a porcine radiation hybrid panel: a new tool for mapping the pig genome. *Proceedings of XXVth International Conference on Animal Genetics*, 21–25 July, Tours, France, p. 133.

Yerle, M., Robic, A. and Renard, C. (1996c) Five molecular markers localized by FISH on pig chromosomes. *Animal Genetics* 27, 217.

Yerle, M., Lahbib-Mansais, Y., Pinton, P., Robic, A., Goureau, A., Milan, D. and Gellin, J. (1997) The Cytogenetic map of the pig. *Mammalian Genome* 8, 592–607.

Zhdanova, N.S., Astakhova, N.M., Kuznetsov, S.B., Schuler, L. and Serov, O.L. (1994) Characterization of pig–mink cell hybrids: assignment of the *TK1* and *UMPH2* genes to pig chromosome 12. *Mammalian Genome* 5, 781–784.

Zijlstra, C., Bosma, A.A. and Meijerink, P.H.S. (1992) Chromosomal assignment of the pepsinogen A gene in the pig by in situ hybridization. *Animal Genetics* 23 (Suppl. 1), 84.

Zijlstra, C., Bosma, A.A. and de Haan, N.A. (1994) Comparative study of pig–rodent somatic cell hybrids. *Animal Genetics* 25, 319–327.

Zijlstra, C., Bosma, A.A., de Haan, N.A. and Mellink, C. (1996) Construction of a cytogenetically characterized porcine somatic cell hybrid panel and its use as a mapping tool. *Mammalian Genome* 7, 280–284.

Zolnierowicz, S., Cron, P., Solinas-Toldo, S., Fries, R., Lin, H.Y. and Hemmings, B.A. (1994) Isolation, characterization, and chromosomal localization of the porcine calcitonin receptor gene. *Journal of Biological Chemistry* 269, 19530–19538.

Genetic Linkage Maps

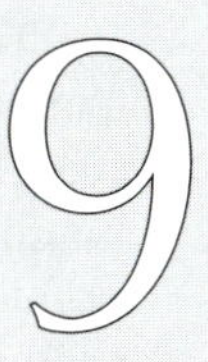

A.L. Archibald and C.S. Haley
Roslin Institute (Edinburgh), Roslin, Midlothian EH25 9PS, UK

Introduction

The genetic information that makes a pig what it is and controls variation between and within breeds is encoded in DNA. Each DNA strand contains a series of functional gene loci interspersed with long tracts which seemingly have little function. The DNA is packed into the 18 pairs of autosomes and the sex chromosomes found in practically every cell of the animal. The total complement of genetic information is often referred to as the genome.

The two copies of a gene carried on alternative members of the homologous chromosome pair may differ from each other in their exact DNA sequence, and these two alleles may or may not encode functionally different products, depending upon the exact nature of the change at the DNA level. Across a population there may be a number of different alleles present, all differing from each other at the level of the DNA. Such variation found in a population is referred to as polymorphism and genes or loci displaying variation are said to be polymorphic. During reproduction, a zygote receives one of each of the pairs of chromosomes from each parental gamete. Hence progeny receive half their genome, and a sample of the genetic variation, from each parent.

Linkage represents a deviation from Mendel's laws of inheritance

Polymorphic genetic loci are a prerequisite for linkage mapping. In essence it is necessary to be able to distinguish the copy of a gene (the allele) that has been received from each parent. Mendel's basic laws of inheritance explain the transmission of the two different alleles to subsequent offspring. First, given enough offspring it can be demonstrated that each allele has an equal chance of being passed on to next generation. This phenomenon is known as 'equal segregation'. Second, the allele transmitted for one gene has no influence on the allele(s) which are passed on for other gene(s). This second law is termed 'independent assortment'. Both equal segregation and independent assortment can be explained by the organization of genes on pairs of homologous chromosomes. For each homologous pair an individual receives one copy from each parent and passes one copy to each offspring. Thus, equal segregation is explained by the separation of pairs of homologous chromosomes during the formation of gametes. Independent assortment is a consequence of the independent behaviour of separate pairs of homologous chromosomes. For example, an individual might pass on a copy of chromosomes 1, 2, 4 and 5 that it received from its mother and its father's copy of chromosomes 3, 6, 7 and 8 to a given offspring.

Genetic linkage was first revealed as a deviation from Mendel's predictions on independent assortment. Just as equal segregation and independent assortment can be explained by the chromosomal theory of inheritance, so too linkage is a consequence of the chromosomal organization of the genome. Genes that are located close to each other on the same chromosome do not assort independently at meiosis. Genes that are far apart on the *same* chromosome do appear to assort independently. The DNA that constitutes the backbone of each chromosome is known to be a single continuous molecule. How then is it possible for alleles at two different loci on the same chromosome to segregate independently as if they were located on two independent chromosomes? The explanation lies in the exchanges or crossovers between a pair of homologous chromosomes that occur at meiosis during the formation of the gametes. Thus, an individual can produce a gamete containing a recombinant chromosome carrying parts of the chromosomes it inherited from its parents. As already noted, interchromosomal recombination arises from the independent assortment of chromosomes and creates equal numbers of parental and non-parental (recombinant) genotypes. Intrachromosomal recombination is the result of crossing-over between a pair of homologous chromosomes. The proportion of intrachromosomal recombinant genotypes is a measure of the crossing-over frequency. In general, the further apart two loci are on a chromosome the greater is the chance that a crossover event will have taken place between them and so the greater will be the proportion of recombinants (the recombination rate). Hence the recombination rate can be used to provide a measure of the distance between two loci on the same chromosome. If the recombination rate between two loci is equal to 50%, then in effect, the loci segregate independently and appear unlinked.

The requirements for linkage mapping

There are two requirements for genetic (or linkage) mapping – pedigrees in which the relationships are known and polymorphic genetic loci. When we can identify alternative alleles at a locus and use them to follow the inheritance of a section of chromosome in a mapping study, the loci are often referred to as genetic markers. Genetic (or linkage) mapping involves following the segregation of polymorphic genetic markers in families to establish whether or not alleles at one marker locus co-segregate with alleles at other marker loci, in other words determining whether the marker loci are linked and if so, measuring the distance between them.

Polymorphic genetic markers

Polymorphic genetic markers are measurable characteristics that vary between individuals. We now readily accept that polymorphic genetic markers are synonymous with variations in DNA sequences that can be revealed with a range of molecular techniques. However, linkage mapping and analysis predates the development of DNA technology. Underlying DNA variation can also be revealed in gene products. For example, discrete visible traits, including coat colour, can be used as polymorphic genetic markers.

Although a polymorphism can refer to any level of variation, clearly a genetic marker that only exhibits variation in 1 in 100 individuals is unlikely to be useful for linkage studies. Rather there is a requirement for informative genetic markers, i.e. highly polymorphic genetic markers such that most individuals are heterozygous and we can trace the inheritance of their alleles in their progeny. The concept of a marker's polymorphic information content (PIC) has been defined by Botstein *et al.* (1980) in a paper that considered the possibility of constructing restriction fragment length polymorphism (RFLP)-based linkage maps of entire genomes, as a measure of the frequency with which marker alleles can be traced from parents to their progeny. The informativeness of a marker in a given population is also often expressed in terms of the degree of heterozygosity, i.e. the proportion of individuals, usually in the parental generation, that carry two different alleles at the locus of interest. The heterozygosity is closely related to PIC and both measures have a maximum value of unity and a minimum value of zero.

The markers that were developed as genetic analysis moved from the farm to the laboratory were so-called biochemical genetic and immunogenetic markers. This class of marker is described in greater detail in Chapters 5, 6 and 7. Briefly, variation in protein sequences were revealed as differences in mobility in electric fields (electrophoretic variants), as differences in antigenicity or as differences in function (e.g. enzymatic activity). The first two types of marker were preferred as they are discrete. As noted below, the initial linkage analyses and maps were developed with such biochemical and immunogenetic markers. These markers, however, represent no more than 10% of the markers on current linkage maps, which are dominated by DNA (or molecular genetic) markers.

Molecular genetic markers have several advantages over those types used in earlier studies. Firstly, with modern technologies they are abundant and relatively easy to identify. They can have very high heterozygosity making them useful for linkage studies. Finally, molecular genetic markers can also be typed from a small sample of DNA extracted from any tissue and taken from animals of any age. The DNA can be stored indefinitely, while many biochemical markers can only be tested on fresh tissue.

Three main categories of molecular genetic markers are being employed in genetic mapping studies in pigs. First, RFLPs have been established by Southern blot analyses (e.g. Archibald *et al.*, 1994, 1996a; Couperwhite *et al.*, 1992a,b). Most of the RFLP markers developed for linkage mapping in the pig have used expressed sequences as probes. RFLP markers are generally only diallelic, corresponding to the presence or absence of a recognition site for the restriction endonuclease employed, and hence tend to have a low heterozygosity or PIC. The highest heterozygosity expected with two alleles is 0.5 when the two alleles are at equal frequencies.

The second main category of molecular genetic marker employed in linkage studies in the pig is based on variable number of tandem repeats of a particular DNA sequence, different alleles having different numbers of the repeating motif. This category can be subdivided into two classes depending upon the length of the repeat unit. If the repeat unit is one to five nucleotides then these markers are termed *microsatellites*(Litt and Luty, 1989; Weber and May, 1989). Where the repeat unit is greater than five nucleotides the loci are generally classed as *minisatellites* (Jeffreys *et al.*, 1985). Confusingly, minisatellites are also often referred to as VNTR (variable number of tandem repeat) loci in a manner that tends to exclude microsatellites. When the repetitive core sequence from a minisatellite locus is used to probe a genomic Southern blot many polymorphic bands are revealed. These multiple banded patterns, often referred to as 'DNA fingerprints' arise from the multiple loci that share a common or related tandem repeat motif.

Although both minisatellite and microsatellite loci are almost intrinsically highly polymorphic, the latter have some key advantages for linkage mapping. Microsatellite loci are abundant (~65–100,000 loci in the pig) and evenly distributed throughout the genome (Winterø *et al.*, 1992). In contrast minisatellite loci in humans exhibit a skewed distribution with clustering in telomeric regions (Royle *et al.*, 1988). Recent studies by Gilles Vergnaud and colleagues indicate that a similar clustering of minisatellite loci is also found in the pig genome (Amarger *et al.*, 1996). The genotyping of microsatellite loci is also easier. As these markers are always genotyped using the polymerase chain reaction (PCR), primer sequence information facilitates the sharing of markers and a common technology is used for the markers. With the development of fluorescent DNA fragment analysers it is possible to capture the data electronically as the PCR products are being fractionated and to automate the sizing of the microsatellite alleles (Ziegle *et al.*, 1992). In contrast, genotyping minisatellites generally relies upon Southern blot technology, which is laborious, time-consuming, requires DNA in greater quantity and quality and the distribution and storage of specific cloned probes if markers are to be used by several groups.

The third main category of molecular genetic marker employed in pig linkage mapping studies is single stranded conformation polymorphisms (SSCPs) (Orita *et al.*, 1989; Archibald *et al.*, 1996b). Briefly, short sequences (100–400 nucleotides) are amplified using the polymerase chain reaction, the PCR products are denatured to yield single stranded DNA molecules that are resolved by native polyacrylamide gel electrophoresis. The rate of migration is determined in part by the influence of their DNA sequence on the conformation of these single stranded molecules. Thus, subtle differences between individuals including point mutations can be revealed with this technique. Several groups are currently developing SSCPs from short expressed sequence tags (ESTs) as a means of increasing the gene content of linkage maps.

Most of the molecular genetic markers mapped by linkage analysis in the pig fall into the three categories outlined above. Of the other types of molecular markers, one merits attention – amplified fragment length polymorphisms (AFLPs) (Vos *et al.*, 1995). The advantages of AFLP technology include the capability to develop thousands of polymorphic markers without any cloning or sequencing steps and the capability of acquiring thousands of genotypes each day. A potential limitation on the informativeness of AFLPs is the extent to which both alleles can be detected at any given locus. If each fragment observed on an AFLP gel is only scored as present or absent then the markers are treated as dominant/recessive and are less informative for linkage analysis than codominant markers in which both alleles can be scored. However, scientists at Keygene, where AFLP technology was developed, claim that by quantifying the relative amounts of each fragment in AFLP patterns from different individuals it is possible to determine whether there are one or two copies of each band (allele). Zabeau and his colleagues at Keygene initially developed the AFLP system for genome analysis in plants. However, this technology has been recently transferred to animals, including pigs, where it is being evaluated. The future integration of maps built using AFLPs with those based on other markers will help the worth of the technology to be evaluated.

Pedigrees for linkage mapping

The pedigrees used for textbook illustrations of linkage analyses rely upon pure inbred lines in which every individual within the line is genetically identical, and homozygous for all loci. Such lines of inbred mice are available for experimental work. There are two principal types of pedigrees or crosses employed in such linkage studies – backcrosses and intercrosses. Initially, two pure (but distinct) inbred lines are crossed to generate F_1 individuals. Although it is assumed that the F_1 animals are heterozygous at all loci, they are in fact only heterozygous for those loci that are fixed for different alleles in the pure inbred parental lines. In a backcross design the F_1 animals are crossed to either (or both) of the pure inbred parental lines. In an intercross design the F_1 animals are crossed to each other or *inter se* to generate F_2 individuals.

These types of pedigree structures are often cited as being employed in livestock studies, including pig linkage experiments. The use of these terms is somewhat misleading as the animals used in such livestock linkage studies are not inbred or genetically uniform. Crosses can be established in which for the (few) loci of interest it is known that the parental animals are multiple heterozygotes or multiple homozygotes (e.g. Andresen, 1970; Otsu *et al.*, 1991). However, as these pigs are not pure inbred lines, the cross may be, in effect, an intercross for some loci and a backcross for others.

As will be discussed later, the pedigrees being employed in the major linkage- and QTL-mapping studies in pigs are based on an intercross design with purebred founders of diverse genetic origins. This intercross design of pedigree is more powerful for linkage analysis than the more conventional backcross, at least when markers are codominant. In a backcross, individuals which are heterozygous for the loci of interest are crossed to individuals which are homozygous for these loci. Segregation of markers can only be detected from the heterozygous parent. In the intercross, both parents are heterozygous and thus provide twice as much information on segregation.

Genetic or recombination distances

Measurement of the distance between two loci involves the estimation of the recombination fraction (r). This is the probability that the two loci recombine – i.e. that there is an odd number of crossovers between the loci. The simplest case to consider is the backcross performed with true inbred lines (Fig. 9.1). In this case with codominant marker loci we can have four classes of progeny which correspond to two types of gametes from the F_1, recombinant and non-recombinant. With both the parental and grandparental (from the inbred lines) genotypes known, it is possible simply to count the number of recombinant gametes. This is because the linkage phase in the heterozygous F_1 parent can be deduced from its parents (i.e. we know that A1 and B1 are on one gamete and

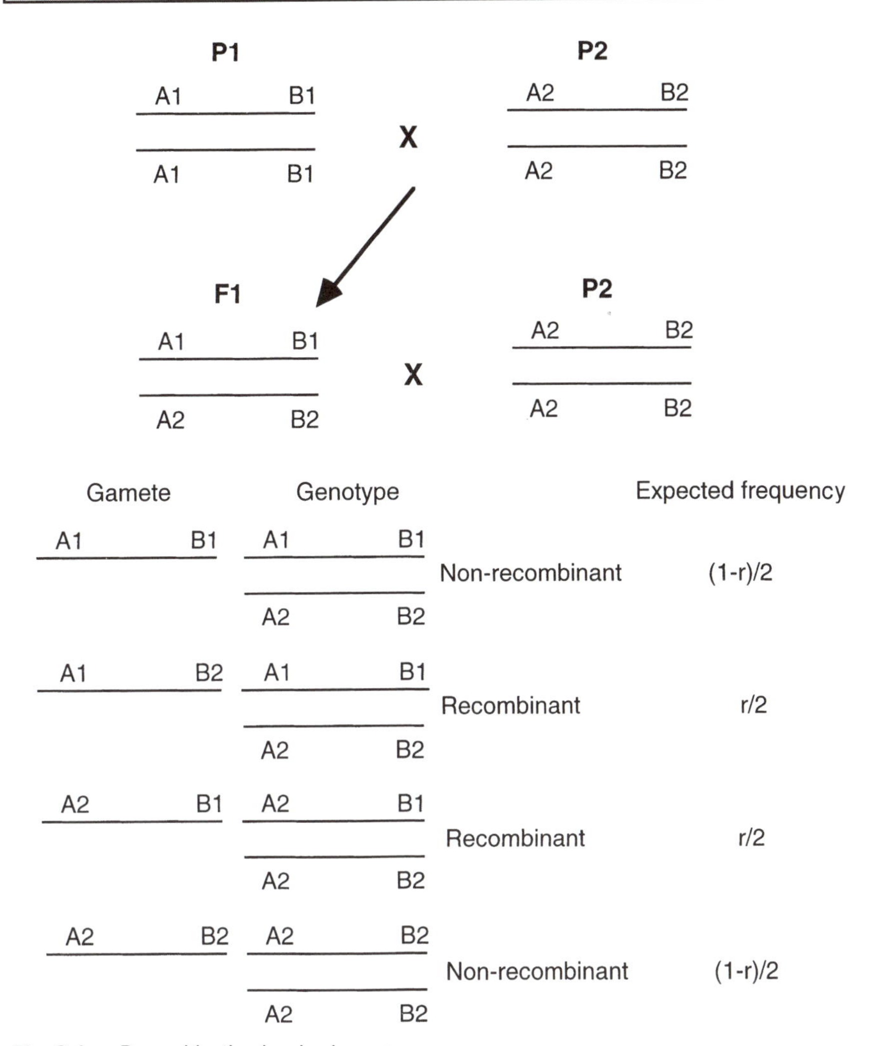

Fig. 9.1. Recombination in a backcross.

A2 and B2 are on the other). The number of recombinants divided by the total number of gametes provides our estimate of the recombination fraction. If the markers are unlinked, then by random segregation of the markers approximately half the gametes are expected to be recombinant and half non-recombinant, i.e. the recombination fraction is expected to be 0.5. A simple statistical test of whether the loci are linked is whether the number of recombinant to non-recombinant gametes deviates from a one to one ratio using a χ^2 (chi-square) test. Often recombination fractions are converted into percentage terms, i.e. $r \times 100$.

In fact generally in farm animal studies the cross is more complex than this because, even in three generation pedigrees, the linkage phase may be unknown and so it is not possible to count recombinant gametes. Thus, it is necessary to use more sophisticated methods based upon maximum likelihood to estimate the recombination fraction. The principle can be illustrated by reference to the backcross situation. Imagine in a family of 10 backcross individuals that we did not observe any recombinant individuals. If the loci were actually unlinked, the probability of one individual being non-recombinant is 0.5 and the probability of 10 being so is $0.5^{10} \approx 0.001$ (i.e. not very likely). On the other hand, if the true recombination rate between the two loci is 0.1, then the probability of one non-recombinant individual is 0.9 and the probability of 10 is $0.9^{10} \approx 0.35$ (i.e. much more likely). The ratio of these two figures is 350, i.e. a recombination rate of 0.1 is 350 times more likely than one of 0.5. Similarly, if the true recombination rate was 0, then the observed result has a probability of 1 and the ratio of the two probabilities is 1000. The basic principle is to find a value of the recombination rate that maximizes this likelihood ratio (in this example the value is 0).

When using the maximum likelihood method the test for the significance of a linkage is not based upon the χ^2 but upon the $\log_{10}$ of the likelihood ratio or log odds, with the value of the test statistic often abbreviated to 'lod score'. In the example above the lod score for $r = 0$ is 3 ($\log_{10}$ (1000)). The significance threshold for this test, based on the assumption that a sequential test is being performed (i.e. new data are being added to old) and on arguments concerning the prior probability of linkage, is usually taken to be 3, with values greater than this being taken as evidence of linkage (Morton, 1955). Negative lod scores suggest an absence of linkage (i.e. linkage at the recombination fraction at which the lod was calculated is *less* likely than no linkage) and a lod score of <-2.0 is taken to be good evidence of lack of linkage at the level of the recombination fraction used in the calculation. These chosen values have stood the test of time, giving a reasonable level of false positive results. It might be thought that as the number of loci being tested increases so would the chance of finding a false positive result. But the chance of finding the true linkage increases also and the two effects balance until more than around 100 loci are being tested. In this case the lod threshold should be increased slightly, but in practice, with the dense maps now available it is only very uninformative markers (or those with many errors) that do not give very high lod scores in a reasonable study.

A useful feature of lod scores is that if they have been calculated for the same pair of loci and the same recombination fraction in two independent families they are additive, i.e. the joint lod score is the sum of the individual ones. However, different families will always produce slightly different recombination fractions that maximize the likelihood in that family, if only by chance, and it is not possible simply to sum maximum lods from different families to derive the overall maximum lod.

In analysing data where the phase is unknown, it is necessary to calculate the likelihood ratio taking into account alternative phases. This can get rather complex, particularly with pedigrees over several generations. Fortunately there are several widely available computer programs that can be used for this task, including LINKAGE (Lathrop and Lalouel, 1988) and its variants and CRI-MAP (Green *et al.*, 1990).

Multi-locus genetic maps

In building a map with a number of linked loci we could perform all the pairwise analyses for linkage and then try and build a map based on these distances. In fact it is more efficient to perform multi-locus analyses. With more than two loci the distances measured in recombination fractions are not additive, that is, with three loci, A, B and C, the distance between loci A and C will not be the sum of the distances between A and B and between B and C. This is because of the occurrence of double recombinants which may lead to observed recombination between A and B and B and C but not between A and C (Fig. 9.2). In fact if there is no interference in recombination (i.e. the probability of a recombination between B and C is not affected by the occurrence of one between A and B), the recombination fraction between A and C (r_{AC}) is expected to be:

$$r_{AC} = r_{AB} + r_{BC} - 2r_{AB}r_{BC}$$

where r_{AB} and r_{BC} are the recombination fractions between A and B and between B and C, respectively. The computation of maps based upon multiple loci is much more complicated than the pairwise analyses and it is not possible to compare all maps if a number of loci are involved (because the number of possible orders becomes vary large). Again, LINKAGE and CRI-MAP provide efficient computational tools to perform these analyses and enable calculation and maximization of the likelihood for a map including several or many loci.

If we know or can assume the amount of interference that occurs between recombination events, it is possible to convert the distances in recombination fractions into a measure of distance between loci which is additive. The measure used is the morgan, named after the famous geneticist, which is divided into 100 centimorgans (cM). The formulae used to convert recombination fractions into linear map distances are known as mapping functions and a number have been derived depending upon the degree of interference assumed (Kosambi, 1944). If we assume that no interference occurs, then the 'Haldane' mapping function applies:

$$M = -0.5 \log_e(1 - 2r)$$

with the inverse relation:

$$r = 0.5[1 - e^{(-2M)}]$$

where M is the distance in morgans, r the recombination fraction and e the mathematical constant. When r is small it is approximately equal to the distance M, but at larger values the two measures differ somewhat.

Thus it is possible to build a linkage map of a chromosome incorporating several or many marker loci. The distances between the loci are estimated and usually expressed in centimorgans. The ultimate aim of a multi-locus map is to provide the correct order of all markers on each linkage group. Unfortunately this is usually a vain hope because most mapping populations do not contain enough individuals, and hence meioses to order all markers. There may be no individual that contains a crossover between a pair of closely linked markers and therefore either order of the markers may be equally likely. Because of this, it has become the practice in human genetic studies to define two types of multi-locus map; 'framework maps' and 'comprehensive' maps (Keats *et al.*, 1991). In a framework map the markers are ordered with a high degree of certainty. Usually such maps are chosen to be 1000 times more likely (equivalent to a lod score of 3) than the next best order that can be found and the markers tend to be chosen to be those which are more informative and evenly spaced (see Fig. 9.3). Other good criteria for choice of framework markers may be those

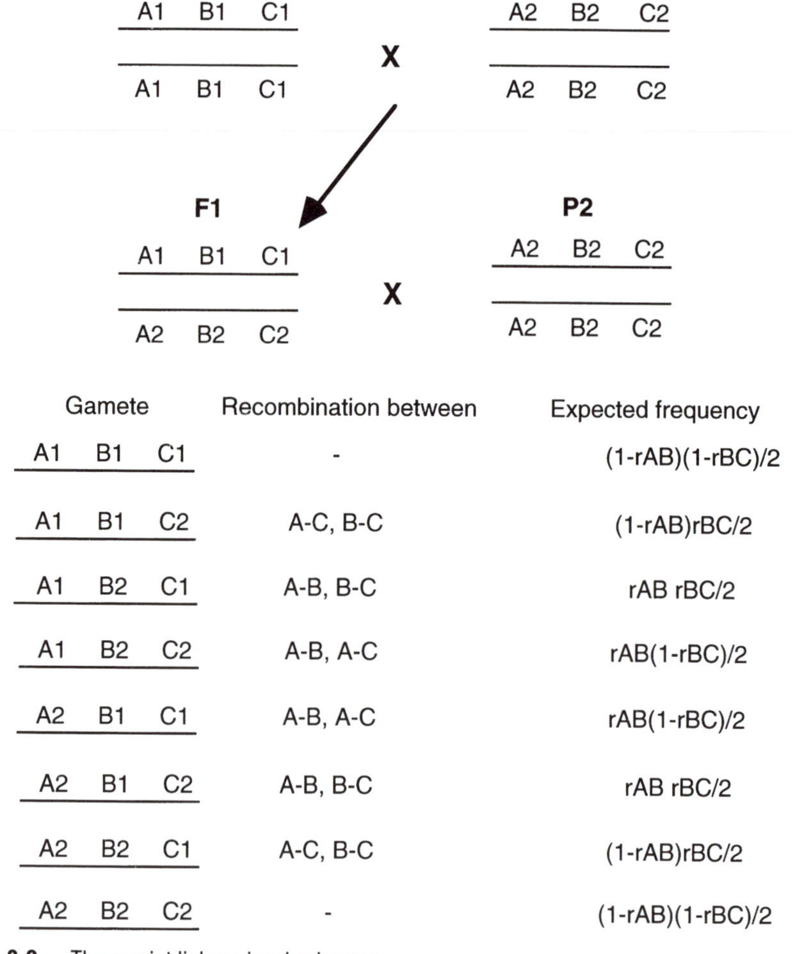

Gamete	Recombination between	Expected frequency
A1 B1 C1	-	(1-rAB)(1-rBC)/2
A1 B1 C2	A-C, B-C	(1-rAB)rBC/2
A1 B2 C1	A-B, B-C	rAB rBC/2
A1 B2 C2	A-B, A-C	rAB(1-rBC)/2
A2 B1 C1	A-B, A-C	rAB(1-rBC)/2
A2 B1 C2	A-B, B-C	rAB rBC/2
A2 B2 C1	A-C, B-C	(1-rAB)rBC/2
A2 B2 C2	-	(1-rAB)(1-rBC)/2

Fig. 9.2. Three point linkage in a backcross.

which are physically located and those which are likely to be useful in several populations, such as microsatellites. (It should be noted that there is not just one possible framework map, as replacing one marker with another which is completely linked to it can produce an equally good framework map.) In a comprehensive map all loci are inserted in their best possible positions (see Figs 9.4 and 9.5). There are likely to be many comprehensive maps which have very similar likelihoods and one thing that can be stated with some confidence is that any comprehensive map is likely to be incorrect in some respect!

In building a multi-locus map it is only rarely that it is possible to calculate the likelihood of all possible orders of markers and choose the highest. As the number of loci increases the number of possible orders mushrooms making it computationally infeasible to look at all possible orders. This is also rarely necessary, as earlier studies will generally have built a strongly supported map (or at least a map where the order of selected 'framework' markers is known with some certainty). Thus, in practice, map building becomes an exercise of adding new markers to established maps and finding their best location. This can be supported by flipping the order of groups of adjacent markers to see if the insertion of a new marker has produced a new global order. Furthermore, a map built starting from a single pair of markers can be checked by rebuilding the map starting from a second or third pair of loci. This simple heuristics, although not infallible, gives some confidence in the final derived map. Again, a program such as CRI-MAP provides the tools to perform such analyses.

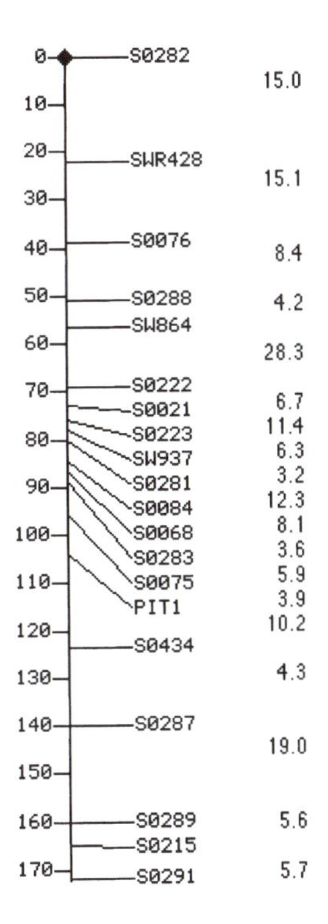

Fig. 9.3. A 'framework' linkage map of porcine chromosome 13. The sex-averaged map is shown. The order of the markers on this 'framework' map is supported by odds of at least 1000:1. The support for the order of adjacent markers is shown to the right of the map, e.g. the order of markers *S0282* and *SWR428* is supported by odds of 10^{15}:1.

Data errors

A potential major problem in map building is the occurrence of data errors. Errors are inevitable and can arise at many points (e.g. from misclassification of animals and samples, misloading or reading of gels, misrecording of data or

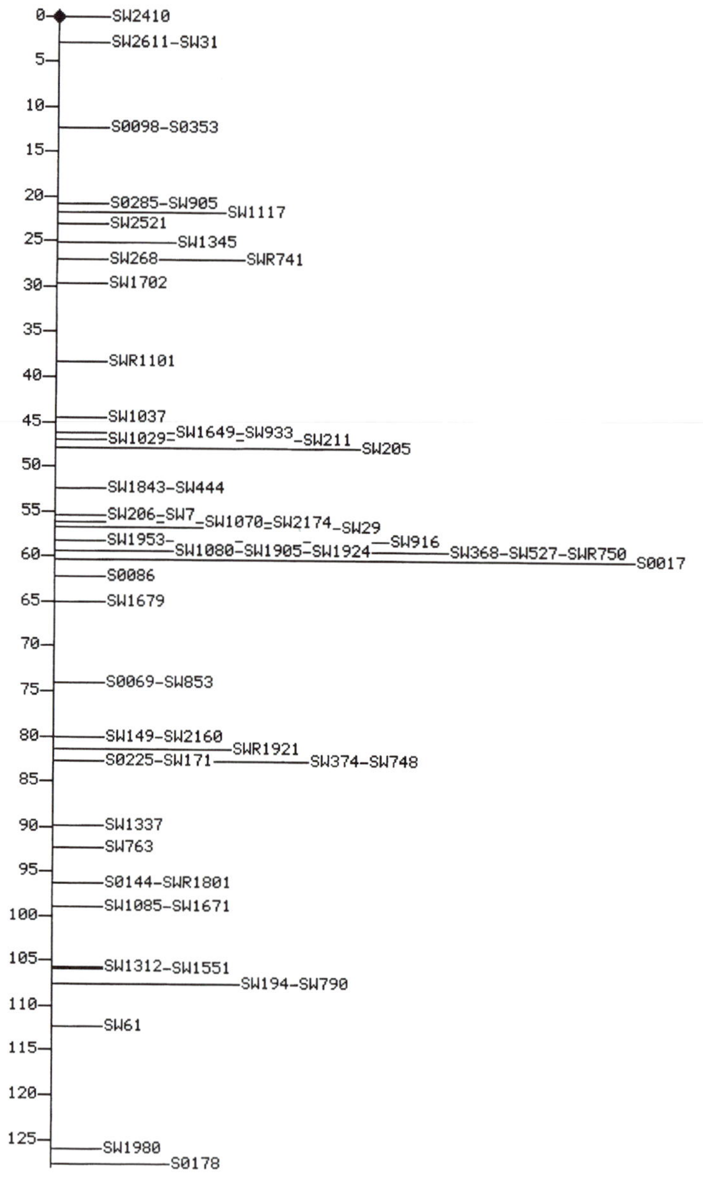

Fig. 9.4. A 'comprehensive' linkage map of porcine chromosome 8 as developed by the USDA-MARC group (Rohrer *et al.*, 1996; http://sol.marc.usda.gov/). In this linkage map all the markers are inserted in their best possible location.

errors on data entry into databases, etc.). Every effort should be made in mapping to put in place procedures that minimize the possibility that errors get through the system undetected. Such systems include double recording of gels, double entry of data, repetition of some measurements to check on error rates, etc.

Some errors will be detected in the analysis process because they create Mendelian segregation errors (i.e. progeny genotypes that are impossible given the parents' genotypes). Early detection of such errors in the data management

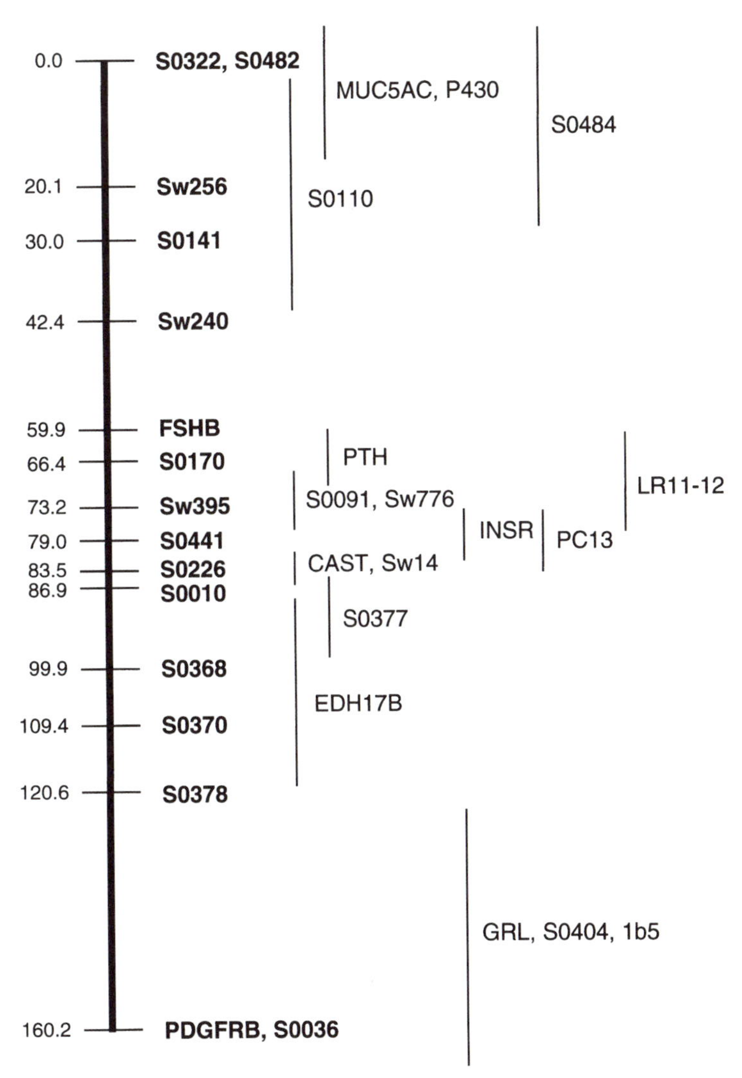

Fig. 9.5. A 'comprehensive' map of porcine chromosome 2 (from Archibald *et al.*, in preparation). In this alternative comprehensive map display a framework map of the chromosome is shown, with the range of possible locations of non-framework markers illustrated by the bars to the right of the framework map.

software/database is a great advantage as it can allow them to be rectified before the original gels are archived and the data added to the project database. The remaining errors are more insidious and usually create apparent recombinant events. The effect of such undetected errors can be more or less serious depending on their frequency. Buetow (1991) has looked at the influence of errors both on the ability to infer the correct order of markers and on map distances. He looked at rates of errors from 0.5% to 1.5% in maps where the markers were approximately 1 to 3 cM apart. As errors usually generate false recombinants, it is not surprising that they lead to an inflation of map distances. Buetow (1991) found a consistent effect, however, no matter which actual marker spacing was used, map distances between markers were inflated by twice the error rate (e.g. a true distance of 1 cM is estimated to be 3 cM if there is a 1% error rate). Thus a low error rate might not have a major effect in a low resolution map, but in a map with 1 cM spaced markers, an error rate as low as 0.5% could double the total map length! This effect is to some extent responsible for the increase in total map lengths observed as maps become denser. In addition, errors tend to reduce the support for the correct order of markers and increase the likelihood that an incorrect order would be found, even some that were well supported (i.e. 1000 times more likely than the correct order). Buetow (1991) found that building maps by combining pairwise analyses (using the MAP90 program of Morton and Collins, 1990) was generally more robust in the presence of errors than a multipoint analysis using CRI-MAP (Green *et al.*, 1990). The errors studied by Buetow (1991) were in individual progeny. Genotyping errors of key parental animals can lead to apparently large numbers of errors in their progeny and to the inference of very incorrect marker order.

There are several means of detecting errors that do not lead to non-Mendelian segregation, but no completely infallible means. If the order of markers has been correctly inferred, an error will often lead to an apparent double recombinant between the marker and markers on either side. In a dense map this is very unlikely. CRI-MAP provides a means of detecting such events and they can then be double checked. However, this method is dependant on the correct order and errors leading to an incorrect order could lead to apparent double recombinants for genotypes that are actually correct. If the insertion of a marker in the map causes a significant change in the distance between flanking markers this can also indicate errors (and from Buetow, 1991, the error rate can be estimated as half the size of the expansion caused). Marked differences in recombination rate between pedigrees or between families within a pedigree can also be an indication of errors. None of these methods is infallible and should not replace scrupulous quality control of the data collection process in the first place.

Pedigree structure

Most of the map development that has been undertaken in pigs has made use of pedigrees derived from crosses between divergent breeds. This decision is due to the increased heterozygosity found in F_1 animals from such a cross and was critical when the emphasis was on more lowly polymorphic markers such as RFLPs, but also has an impact for markers such as microsatellites (Rohrer *et al.*, 1994). Where genotypic data from all three generations (pure-bred grandparents, F_1 parents and progeny) are available, it is often, but not always, possible to know the linkage phase of markers. If only two generations of genotypic data are collected it is necessary to infer linkage phase from the progeny. Theoretical investigations have been made of the relative powers of alternatives for detecting linkage (van der Beek and van Arendonk, 1993). For a recombination rate of < 0.20, between two and six progeny are required to replace information on phase lost if the grandparents are not typed. These authors conclude that, with large numbers of progeny per family, data from grandparents are not really necessary. However, most investigators seem happier with the extra security in the ability to detect errors provided by information from grandparents.

Variation in recombination frequency

Although true marker order will generally remain constant between different individuals (apart from relatively rare rearrangements), the same is not true of recombination distance. Differences in the occurrence of recombination, and hence of map length, between the sexes in various species has been long documented (Haldane, 1922). It has also now become clear that the ratio of male to female meioses varies through the genome. It is not clear that ignoring this effect has a major influence on the correct construction of linkage maps, but nonetheless, the available programs allow separate estimation of distances in the two sexes. Thus maps can be constructed making use of this option, or at the very least, a map derived assuming equal recombination can be checked for alternative orders allowing for differences between the sexes.

It should also be noted that differences in recombination rates between populations and between families are well documented in various species and are to some extent under genetic control. In principle these effects should also be taken into account when merging data from different families and populations. In practice this is rarely done and does not seem to have been a major cause of mapping errors.

Early Linkage Studies in the Pig

The linkage map of the pig as summarized by Echard (1990) comprised seven linkage groups. The two best characterized linkage groups centred on the *HAL* and *SLA* loci have been assigned to chromosomes 6 and 7 respectively through *in situ* hybridization localizations of one or more members of the group.

Genome-wide Linkage Mapping Studies

During the 1990s concerted efforts have been made to develop comprehensive linkage maps of the porcine genome. The major contributors to this effort have been international collaborative projects based in Europe, the PiGMaP consortium (Archibald *et al.*, 1995; Haley and Archibald, 1992) and the related Nordic collaboration (Ellegren *et al.*, 1994a; Marklund *et al.*, 1996), and the largely single centre work of the USDA Meat Animal Research Center (Rohrer *et al.*, 1994, 1996). The combined efforts of these groups have placed more than 1500 polymorphic genetic markers on the porcine linkage map.

PiGMaP Linkage Consortium

Among the objectives of the *Pig Gene Mapping Project* (or PiGMaP), initiated with support from the EC BRIDGE programme, was a collaborative effort to develop a 20 cM linkage map of the pig genome. Development of this genetic (linkage) map was initiated by a group of ten laboratories within the PiGMaP collaboration. Subsequently, other laboratories in Europe, Australia, Japan and the United States have joined the PiGMaP Linkage Consortium which currently includes over 20 laboratories.

Reference pedigrees were established in five centres (Roslin, France, Germany, The Netherlands, Sweden) (see Archibald *et al.*, 1995 for details). These pedigrees take the form of three-generation families in which grandparents from genetically divergent breeds have been crossed to produce the parental (F_1) generation which have subsequently been intercrossed. In the Scottish, French and Dutch pedigrees the founder grandparental breeds are the Chinese Meishan and the European Large White (Yorkshire). The Swedish and German pedigrees have European Wild Boar and European improved breeds as their grandparents. DNA from 118 F_2 pigs plus their respective parents and grandparents has been distributed to participating laboratories for genotyping. The 20 Swedish pigs represent a subset of the 200 pigs that make up the F_2 generation of the Nordic mapping pedigrees (Ellegren *et al.*, 1994a).

Genotyping data from the participating laboratories are entered into the project database (ResPig) at the Roslin Institute. A rapid linkage analysis is offered through which genotypes for new markers (loci) are subjected to two-point linkage analysis with the CRI-MAP program versus all the loci in the database. Estimates of two-point recombination distances and lod scores are returned to the submitting laboratory within a few days. As collaborating laboratories have on-line access to ResPig across the Internet, participants can also perform their own analyses. Thus the PiGMaP Linkage Consortium offers a common resource for mapping new polymorphic markers.

The first release of the PiGMaP Linkage Consortium linkage maps of the porcine genome was developed by segregation analysis of 239 genetic markers – 81 of these markers correspond to known genes. Linkage groups were assigned to all 18 autosomes plus the X chromosome. As 69 of the markers on the linkage map had also been mapped physically (by others), there is significant integration of linkage and physical map data. Six informative markers failed to show linkage to these maps. As in other species, the genetic map of the heterogametic sex (male) was significantly shorter (~16.5 morgans) than the genetic map of the homogametic sex (female) (~21.5 morgans). The sex averaged genetic map of the pig was estimated to be ~18 morgans in length. Mapping information for 61 Type I loci (genes) enhances the contribution of the pig gene map to comparative gene mapping. Because these linkage maps incorporate both highly polymorphic Type II loci, predominantly microsatellites, and Type I loci they will be useful both for large experiments to map quantitative trait loci and for the subsequent isolation of trait genes following a comparative and candidate gene approach (see O'Brien, 1991 for a definition of Type I and Type II loci).

Nordic Pig Gene Mapping Collaboration

Three Nordic laboratories (in Uppsala, Copenhagen and Oslo), established comprehensive linkage maps of the porcine genome. Briefly, a three-generation pedigree was established in which two European Wild Boars were crossed with eight Large White sows to generate F_1 pigs which were crossed *inter se* to yield 200 F_2 offspring. Initially, these pigs were genotyped for 105 polymorphic genetic markers (Ellegren *et al.*, 1994a) and maps elaborated using the LINKAGE package (Lathrop and Lalouel, 1988). More recently, further genetic markers have been genotyped, including markers selected from the USDA-MARC and PiGMaP maps. The revised maps from the Nordic collaboration (Marklund *et al.*, 1996) contain 236 mapped markers with a total (sex-averaged) map length of ~ 2300 cM. Again the linkage maps developed by the Nordic collaborators include a relatively high proportion of genes (89 Type I loci) and 72 loci which have been physically mapped by *in situ* hybridization.

USDA-MARC Linkage Maps

The most detailed linkage maps of the porcine genome have been developed by the USDA Meat Animal Research Center at Clay Center, Nebraska, USA. The linkage mapping efforts of the USDA-MARC pig gene mapping group differ from those of the two international collaborative groups. First, the USDA-MARC maps are dominated by microsatellite markers. Second, only two generations of pigs have been genotyped. Although the design of the mapping pedigrees is presented as being a backcross in concept, it is not a true backcross. Rohrer and his colleagues argue that as their maps are based upon a backcross design they cannot estimate sex-specific recombination distances. In a true backcross F_1 (heterozygous) indidivuals are crossed to one of the purebred (homozygous) grandparental lines (see Fig. 9.1). The heterozygosities of many of the microsatellite markers employed in the USDA-MARC mapping pedigrees exceed 50% in the parental generation. Thus, for many pairs of markers both parents (sexes) must be heterozygous and informative for linkage.

The first linkage maps released by the USDA-MARC group were predominantly microsatellite-based (Rohrer *et al.*, 1994): 366 anonymous microsatellites and 15 Type I markers (microsatellites associated with genes and RFLP markers) were assigned to 24 linkage groups. The assignment of linkage groups to chromosomes was largely effected through markers developed and mapped (physically and genetically) by the European collaborating laboratories. The second release of linkage maps from the USDA-MARC group featured 1042 linked loci in 19 linkage groups corresponding to the 18 autosomes and the X chromosome (Rohrer *et al.*, 1996). The total map length was estimated as 2286.6 cM (sex-averaged). Through the efforts of Alexander *et al.* (1996) a substantial number of new markers were added that had been mapped both physically and genetically to give a total of 124 physically anchored loci on the linkage maps.

Current Status of the Pig Genetic (Linkage) Maps

The status of the linkage maps of the porcine genome is constantly changing. Moreover, it is not practical to combine the information from different linkage studies into a common set of maps. Thus, not only is not feasible to provide a consensus linkage map of the pig genome but there is not sufficient space in this chapter to summarize all the published maps. The public genome databases provide the most effective and timely summary of the status of the pig genetic (linkage) maps.

All the published linkage maps and two-point linkage data are stored in the pig genome database (PiGBASE). This database is freely accessible via the World Wide Web (URL = http://www.ri.bbsrc.ac.uk/pigmap/pig_genome_mapping.html with mirror sites at http://tetra.gig.usda.gov:8400/pigbase/manager.html and http://dirk.invermay.cri.nz/). Additional access can be obtained through http://www.public.iastate.edu/~pigmap. The Anubis map viewer software developed by our colleague Chris Mungall (Mungall, 1996) allows users to compare the various linkage maps side by side (see Fig. 9.6). In the revised database design for PiGBASE (Archibald *et al.*, 1996c) users also have access to information on the individual polymorphic markers *e.g.* PCR conditions and primer sequences, restriction enzymes and probes.

The USDA-MARC group have also made their linkage maps and the underlying information about the markers accessible on the World Wide Web (http://sol.marc.usda.gov/). A summary of the polymorphic markers in published linkage maps is given in Tables 9.1 and 9.2. See Appendices 1 and 2 for further details.

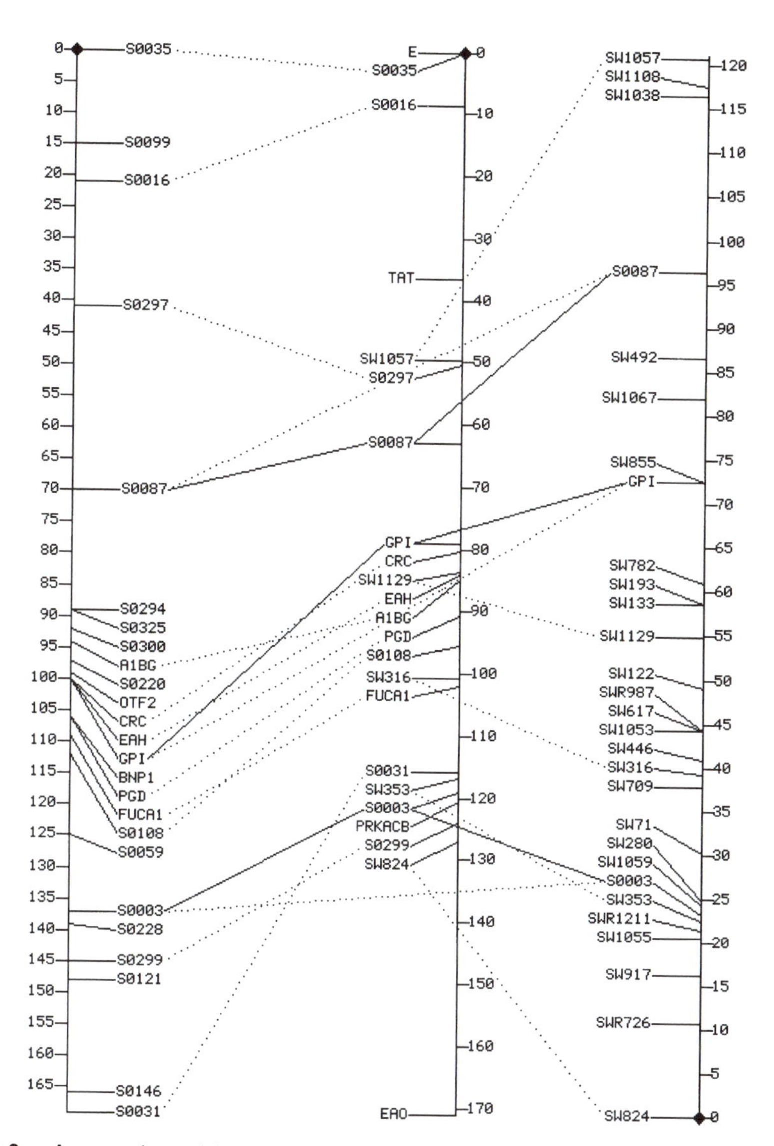

Fig. 9.6. A comparison of three linkage maps of porcine chromosome 6. The maps are, from left to right: PiGMaP release 1 (Archibald *et al.*, 1995); Nordic release 2 (Marklund *et al.*, 1996); USDA-MARC release 1 (Rohrer *et al.*, 1994). This comparative map display was developed with the Anubis Map Manager software (version 2) (Mungall, 1996). Markers that appear on all three maps are joined by solid lines and those that appear on two maps are joined by dotted lines. The older USDA-MARC map has been used for this figure in order to make it more legible – the higher marker density on the second release USDA-MARC map is more readily viewed using Anubis and the on-line databases (http://www.ri.bbsrc.ac.uk/genome_mapping.html).

The development of consensus maps

With physical mapping methods such as *in situ* hybridization the locus of interest is mapped with respect to physical landmarks such as chromosome bands. By reference to these common landmarks, mapping data from a variety of experiments can be brought together in single maps. In contrast, in linkage mapping objects or markers are mapped with respect to each other. Moreover, recombination distances between markers are not constant but can vary depending upon the sex, breed or even identity of the individual. Although linkage data from disparate experiments can be brought together by considering all the two-point distances and lod scores (Morton and Collins, 1990), truly integrated

Table 9.1. Summary of genetic markers mapped on the major genome-wide linkage maps.

Map name	PiGMaP release 1	Nordic release 1	Nordic release 2	USDA-MARC release 1	USDA-MARC release 2
Reference	Archibald *et al.*, 1995	Ellegren *et al.*, 1994	Marklund *et al.*, 1996	Rohrer *et al.*, 1994	Rohrer *et al.*, 1996
Type I (genes)					
RFLP or PCR–RFLP	59	34	47	7	18
Microsatellites within genes	14	17	16	8	28
SSCP	3		1		
Phenotypic, including coat colour	–		3		
Protein variants	7	11	11		
Blood groups	6	10	11		
Total (Type I)	89	72	89	15	46
Type II					
Microsatellites	143	1	135	369	996
Minisatellites	6	51	8		
Other	3	4	7		
Total (Type II)	152	56	150	369	996
Total mapped markers	241	128	239	384	1042

Table 9.2. Number of genetic markers mapped in more than one of the major genome-wide linkage maps.

Map name[1]	Also on USDA-MARC release 2	Also on Nordic release 2	Also on PiGMaP release 1
MARC release 2	n/a	149	152
Nordic release 2[2]	149	n/a	121
PiGMaP release 1	152	121	n/a

[1]For references, see Table 9.1.
[2]The data set used to develop the second release of the Nordic linkage maps has been added to the data set being used to build the second release of the PiGMaP Consortium maps.

linkage maps can only be developed by joint analysis of the original genotyping and segregation data. The more markers that are scored in all individuals the more robust will be the joint analysis and the resulting maps. If joint analyses cannot be arranged then it would be desirable at least to align the major linkage maps. Map alignment can be effected by reference to markers typed in more than one experiment (see Fig. 9.6).

All three major linkage mapping groups have contributed to map alignment by genotyping markers developed by others as summarized in Table 9.2 (see also McQueen *et al.*, 1995; Zhang *et al.*, 1995; Kapke *et al.*, 1996; Marklund *et al.*, 1996; Rohrer *et al.*, 1996). Full map integration has been achieved for chromosomes 6 and 7 through international workshops (Paszek *et al.*, 1995; Rohrer *et al.*, 1997). For both these chromosomes the data sets from the USDA-MARC, PiGMaP and Nordic groups were merged. Data from the Minnesota mapping pedigrees were also added. A number of common markers were selected for both these chromosomes. The selected markers were genotyped in all four mapping pedigrees prior to each chromosome workshop. The resulting consensus framework maps are very robust and the workshop comprehensive maps provide a valuable catalogue of mapped polymorphic markers for QTL-mapping studies.

A second release of linkage maps has been submitted for publication by the PiGMaP Linkage Consortium. There are more than 650 polymorphic markers currently in the project database. Moreover the second release maps have been developed by merging the genotypic data held on the Nordic mapping pedigrees and the original PiGMaP reference mapping pedigrees. Thus, the second release PiGMaP maps represent a very significant effort in pig linkage map integration.

Lessons from the Published Maps

In addition to providing a resource for further studies, some genetic messages can be derived from the maps developed so far. In common with other mammalian species, in pigs the genetic map derived from male meioses is significantly less than that from female (Ellegren *et al.*, 1994a,b; Archibald *et al.*, 1995). Individual chromosomes and chromosomal regions deviate, however, from this overall pattern, a remarkable example being on chromosome 1 (see Fig. 9.7). Here one region showing only 3 cM between adjacent markers in females displays a distance of around 50 cM in males, a phenomenon surely worthy of further study. Ollivier (1995) has compared maps from Ellegren *et al.* (1994a), Rohrer *et al.* (1994) and Archibald *et al.* (1995) and concluded that there are real differences in length between the maps, with the wild boar based map being the shortest.

Of regions that are relatively marker poor, the most obvious example is the X chromosome, a feature common across a number of mammalian species. It has been suggested that this represents the relatively reduced recombination

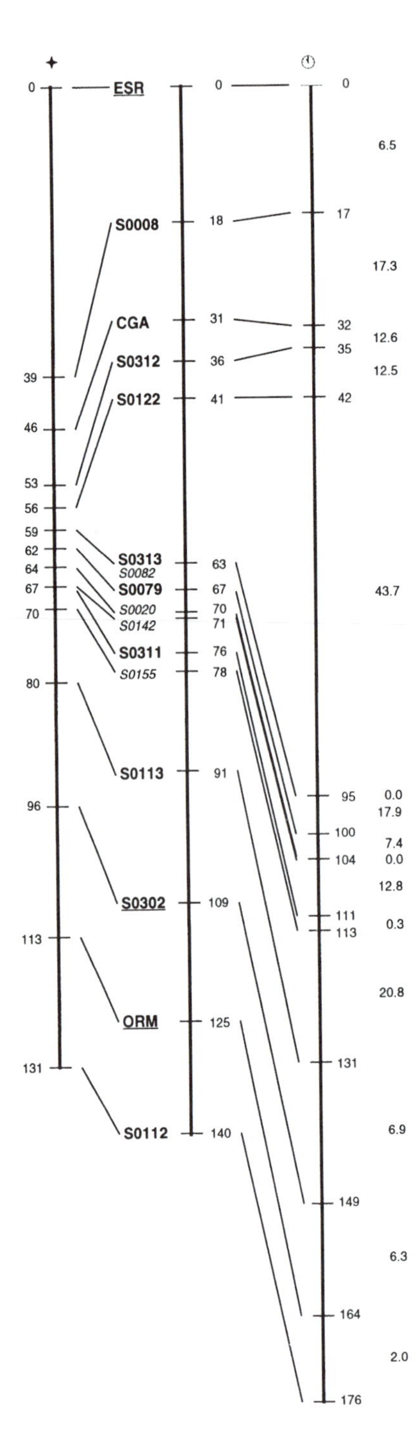

Fig. 9.7. A comparison of the sex-specific and sex-average linkage maps of chromosome 1 (after Archibald *et al.*, 1995). The three maps shown are, from left to right, the female-specific, sex-averaged and male-specific recombination maps, with the estimated Kosambi map distances. To the right of the maps is given the support for order of adjacent markers based upon the sex-separate maps (see Fig. 9.3). The most remarkable example of a sex difference in recombination distances in pigs can be seen in the interval between markers *S0122* and *S0313*.

(i.e. none in males) and greater selective pressure on recessive loci in the homogametic sex leading to purges of neutral variation via hitch-hiking effects.

Estimates of the sex-averaged map length of the pig (making some allowance for telomeric regions which may have no markers) is around 25 M (Marklund *et al.*, 1996; Rohrer *et al.*, 1996). This is considerably less of that of 38 M estimated for man, despite the two genomes containing similar amounts of DNA (Morton, 1991). Although we must bear in mind our previous comments about the increasing effect that errors have on map lengths as the maps become more dense (and correspondingly expect the pig map to grow as it becomes more dense), this may well be a real difference. We should note that mice and cattle, with similar amounts of DNA to pig, have shorter and longer map lengths, respectively (Dietrich, *et al.*, 1994; Barendse *et al.*, 1997). As there are substantially more markers on the mouse linkage map than either of the pig or cattle maps, it is evident that the length of genetic maps is not solely determined by the physical size of the genome.

The Need for Linkage Maps – Mapping Trait Genes

An animal's phenotype is the result of complex actions of the blueprint defined in the genes inherited from its parents and environmental factors. In agricultural species humans have sought to modify both these components. Changes to environmental factors can be immediate but are expensive and require continuous inputs – improved feedstuffs, housing and vaccines. In contrast, genetic improvements are slower but are permanent and cumulative and have minimal maintenance costs.

The age-old philosophy of breeding from the best has proved to be an effective approach to the genetic improvement of livestock over the 10,000 years since animals were domesticated. During the past fifty years more sophisticated approaches to animal breeding have been developed founded on quantitative genetics. Animal geneticists and breeders are concerned to understand the genetic control of traits critical to agricultural performance such as growth, litter size and disease resistance. However, apart from an assumption that the control of many of these traits is polygenic, their genetic architecture remains unknown. Most of the genes of agricultural importance (and many controlling susceptibility to human diseases) do not have a large enough effect on their own to produce qualitative differences between individuals, rather variation in several or many genes combines to produce continuous or quantitative variation between animals. From studying quantitative variation alone it is not possible to study the effects of individual loci affecting quantitative traits (so-called quantitative trait loci or QTLs). Tracking the inheritance of markers in populations whose performance is recorded should allow some of the loci to be identified and the architecture of the genetic control of production traits to be at least partially determined. The general principle is simple: if it is possible to find significant associations between the inheritance of a particular chromosomal region (as determined by following marker inheritance) and trait variation in a sufficiently large population, this suggests it contains a gene or genes affecting the traits in question.

QTL-Mapping

A number of major programmes are underway in Europe and the United States to map QTLs in pigs. The first results confirmed the possibility of identifying chromosomal regions with major effects on performance traits (Andersson *et al.*, 1994). Briefly, a three-generation pedigree was established in which two European Wild Boars were crossed with eight Large White sows to generate F_1 pigs which were crossed *inter se* to yield 200 F_2 offspring. These pigs were genotyped for 105 polymorphic genetic markers and growth rates were recorded from birth to 70 kg. After slaughter, fat levels and intestinal length were noted. The new statistical methods developed (Haley *et al.*, 1994) to map QTLs in this type of outbred population were tested extensively with simulated data in order to validate the methods and to determine the appropriate significance thresholds. The largest effects revealed in the subsequent analyses were attributable to a region of chromosome 4. Fatness, whether measured as the percentage of fat in the abdominal cavity or backfat thickness, appeared to be under the control of QTLs that mapped to the proximal end of chromosome 4. QTLs for intestinal length and growth rate (from birth to 70 kg) were located distal to the 'fatness' QTLs on chromosome 4.

One encouraging feature of this study is the evidence that QTLs can be revealed in relatively small experiments – 200 pigs typed for ~100 markers of which only a proportion were highly polymorphic microsatellites. Over a thousand microsatellite markers have been mapped in pigs with the potential when integrated into a single linkage map to give 90% coverage of the genome, with markers assigned to all 18 autosomes and spaced at intervals of 5 cM or less. By employing both more markers and more animals it may be possible to map QTLs to smaller intervals of 5–10 cM. We should note that precision mapping of QTLs in F_2 and backcross populations is not an easy task. For example, Knott and Haley (1992) found that a QTL explaining 11% of the variance (a very large effect) was only mapped to an accuracy of ±4 cM using a sample of 1000 F_2 animals. Approaches that select individuals with appropriate recombination events as parents of subsequent generations, and hence over several generations, dissect regions containing QTLs into smaller and smaller sections, may be able to map QTL more precisely with fewer individuals, but at the expense of taking a number of years to generate the required precision.

The approach outlined above is used for scanning of the whole genome for 'anonymous' QTLs, i.e. for use when one wants to screen the whole genome and identify any region with an effect on the trait of interest. The problem with the approach is that it demands the typing of the target pig population for markers covering the entire genome, i.e. several hundred individuals with 100 or more markers. In addition to the genotyping load that this entails, the large number of statistical tests carried out in the course of a genome scan requires a stringent significance threshold to avoid too many false positive results being reported (Lander and Kruglyak, 1995).

An alternative approach to performing an entire scan of the genome is to focus on candidate genes for the trait under study and to look for associations between polymorphisms within these genes and performance. This approach requires less genotyping, but assumes that one has sufficient knowledge to identify good candidate genes whereas an approach using a whole genome scan will identify associations whether with known candidate genes or not. Furthermore, even if a chosen candidate gene is causal of variation, it is unlikely that a polymorphism identified within the gene will directly identify the causative lesion, unless one is very lucky. Polymorphisms are likely to exist every few hundred base pairs in the pig genome, thus there may be quite a number of different polymorphisms within or close to the intron/exon structure of a typical candidate gene of several or many kilobases in length. Thus the observed polymorphism must be in linkage disequilibrium with the causative lesion in order for an association between the polymorphism and a trait to be identified. An example of an apparent notable success of the candidate gene approach is the identification of associations between the oestrogen receptor locus (*ESR*) and litter size in pigs (Rothschild *et al.*, 1996). The original association has been confirmed in larger samples and several different breeds. The effect associated with ESR differs between populations, so either the extent of linkage disequilibrium differs between populations or the gene interacts with the genetic background (i.e. displays epistasis). In fact it is still an open question as to whether the causative lesion is actually within *ESR* or within some other closely linked gene.

From Mapping to Isolating QTLs

The experience of isolating disease genes in humans from a knowledge of their map locations proves the enormity of such exercises. The so-called candidate gene approach has been noticeably more successful for the isolation of disease genes than positional cloning (Collins, 1995). This latter approach also proved effective in the identification of the gene responsible for susceptibility to malignant hyperthermia in pigs (Davies *et al.*, 1988; Fujii *et al.*, 1991). This genetic disease, often known as the porcine stress syndrome, causes significant economic losses from sudden stress deaths and poor meat quality that are compensated, in part, by reductions in body fat associated with the disease allele. The response of the pig breeding industry, at least in the UK, to the availability of a diagnostic DNA test for the mutation responsible for this disease is instructive. Polymorphic markers linked to the disease locus have been known for about fifteen years and could have been used to eliminate the disease gene by marker assisted selection. However, it is only since the advent of the direct DNA test that programmes to eliminate the gene have been widely implemented. Thus, while the mapping of QTLs to intervals between flanking polymorphic genetic markers facilitates selection of desirable alleles at the QTL by marker assisted selection, there will be desire to isolate the genes corresponding to the QTL. The isolation of the trait gene will not only allow the implementation of more effective selection but also the exploitation of desirable natural alleles or genetically engineered superior alleles through gene transfer.

Once a quantitative trait locus has been mapped to an interval between two microsatellite or other Type II markers (as defined by O'Brien, 1991), there will be a need to identify markers closer to the QTL and ultimately to isolate or identify the nature of the QTL. The candidate gene approach requires that the relevant genetic (linkage) maps are well populated with known genes. Genetic markers within or adjacent to known genes have been classed as Type I markers (O'Brien, 1991). The mapping of Type I markers or genes provides a means to access the abundance of mapping information in the 'map-rich' species – humans and mice. Syntenic or linkage relationships over short distances (e.g. 5–10 cM between humans and mice) are often conserved across species (Ellegren *et al.*, 1993). Indeed, heterologous chromosome painting studies have demonstrated that there is extensive conservation of synteny between pigs and humans (Rettenberger *et al.*, 1995; Goureau *et al.*, 1996). Therefore, if a QTL is mapped to an interval between two genes or Type I loci that are also syntenic in other species, genes that map to the corresponding interval in those species can be added to the list of potential candidate genes for the QTL. This 'comparative positional candidate gene' approach is likely to be the route of preference in livestock for the foreseeable future.

Conclusions

The ideal genetic (linkage) map of a livestock species should include both Type I and Type II loci. Specific populations have been established for QTL-mapping in pigs. Although such populations can be genotyped for a battery of microsatellite markers, it would be too time consuming to type these large populations for a wide range of Type I markers, many of which are based upon Southern blot/RFLP analyses. Rather it would be valuable to establish a linkage map to which Type I and Type II markers are mapped. Such a map can be seen as a community resource from which individual laboratories can select markers appropriate to their needs – additional microsatellite markers, with which to explore a particular 10–20 cM region, or Type I markers, from which to launch a comparative or candidate gene approach to identifying a QTL. Therefore the published genome-wide linkage maps (Archibald *et al.*, 1995; Marklund *et al.*, 1996; Rohrer *et al.*, 1996) fulfil this requirement of a community resource. The distribution of DNA from these reference families and its continued typing with new markers will continue to expand this resource for the foreseeable future and to offer smaller groups the opportunity to integrate their markers into the genome-wide maps.

Acknowledgements

The pig genome research programme at the Roslin Institute is supported by funds from the Biotechnology and Biological Sciences Research Council, the Ministry of Agriculture, Fisheries and Food, and the European Commission.

References

Alexander, L.J., Troyer, D.L., Rohrer, G.A., Smith, T.P.L., Schook, L.B. and Beattie, C.W. (1996) Physical assignments of 68 porcine cosmid and lambda clones containing polymorphic microsatellites. *Mammalian Genome* 7, 368–372.

Amarger, V., Gauguier, D., Yerle, M., Monfouilloux, S., Giraudeau, F., Pinton, P., Schott, J.J., Apiou, F., Lepetit, D., Aubert, D., Lauthier, V., Leneel, T., Ritter, O., Lathrop, M., Dutrillaux, B., Buard, J. and Vergnaud, G. (1996) The distribution of minisatellites in the human, pig, and rat genome reflects the history of rearrangements involving chromosome-ends. *Proceedings of XXVth International Conference on Animal Genetics*, 21–25 July 1996, Tours, France, *Animal Genetics* 27 (Suppl. 2), 75 (abstract C076).

Andersson, L., Haley, C.S., Ellegren, H., Knott, S.A., Johansson, M., Andersson, K., Andersson-Eklund, L., Edfors-Lilja, I., Fredholm, M., Hansson, I., Håkansson, J. and Lundström, K. (1994) Genetic mapping of quantitative trait loci for growth and fatness in pigs. *Science* 263, 1771–1774.

Andresen, E. (1970) Close linkage between the locus for phosphohexose isomerase (PHI) and the H blood group locus in pigs. *Animal Blood Groups and Biochemical Genetics* 1, 171–172.

Archibald, A.L., Couperwhite, S., Haley, C.S., Beattie, C.W. and Alexander, L.J. (1994) RFLP and linkage analysis of the porcine casein loci – *CASAS1*, *CASAS2*, *CASB* and *CASK*. *Animal Genetics* 25, 349–351.

Archibald, A.L., Haley, C.S., Brown, J.F., Couperwhite, S., McQueen, H.A., Nicholson, D., Coppieters, W., Van de Weghe, A., Stratil, A., Winterø. A.K., Fredholm, M., Larsen, N.J., Nielsen, V.H., Milan, D., Woloszyn, N., Robic, A., Dalens, M., Riquet, J., Gellin, J., Caritez, J.-C, Burgaud, G., Ollivier, L., Bidanel, J-P., Vaiman, M., Renard, C., Geldermann, H., Davoli, R., Ruyter, D., Verstege, E.J.M., Groenen, M.A.M., Davies, W, Høyheim, B., Keiserud, A., Andersson, L., Ellegren, H., Johansson, M., Marklund, L., Miller, J.R., Anderson Dear, D.V., Signer, E., Jeffreys, A.J., Moran, C., Le Tissier, P., Muladno, Rothschild, M.F., Tuggle, C.K., Vaske, D., Helm, J., Liu, H.-C., Rahman, A., Yu, T.-P., Larson, R.G. and Schmitz, C.B. (1995) The PiGMaP consortium linkage map of the pig (*Sus scrofa*). *Mammalian Genome* 6, 157–175.

Archibald, A.L., Couperwhite, S., Mellink, C.H.M., Lahbib-Mansais, Y. and Gellin, J. (1996a) Porcine alpha-1-antitrypsin (*PI*): cDNA cloning and gene mapping. *Animal Genetics* 27, 85–89.

Archibald, A.L., Couperwhite, S. and Jiang, Z.H. (1996b) The porcine *TTR* locus maps to chromosome 6q. *Animal Genetics* 27, 351–353.

Archibald, A.L., Hu, J., Mungall, C., Hillyard, A.L., Burt, D.W., Law, A.S. and Nicholson, D. (1996c) A generic single species genome database. XXVth International Conference on Animal Genetics, 21–25 July 1996, Tours, France, *Animal Genetics* 27 (Suppl. 2), 55 (abstract C001).

Barendse, W., Vaiman, D., Kemp, S.J., Sugimoto, Y., Armitage, S.M., Williams, J.L., Sun, H.S., Eggen, A., Agaba, M., Aleyasin, S.A., Band, M., Bishop, M.D., Byrne, K., Collins, F., Cooper, L., Coppieters, W., Denys, B., Drinkwater, R.D., Easterday, K., Elduque, C., Ennis, S., Erhardt, G., Ferretti, L., Flavin, N., Gao, Q., Georges, M., Gurung, R., Harlizius, B., Hawkins, G., Hetzel, J., Hirano, T., Hulme, D., Jørgensen, C., Kessler, M., Kirkpatrick, B.W., Konfortov, B., Kostia, S., Kuhn, C., Lenstra, J.A., Leveziel, H., Lewin, H.A., Leyhe, B., Lil, L., Martin Burrile, I., McGraw, R.A., Miller, J.R., Moody, D.E., Moore, S.S., Nakane, S., Nijman, I.J., Olsaker, I., Pomp, D., Rando, A., Ron, M., Shalon, A., Teale, A.J., Thieven, U., Urquhart, B.G.D., Våge, D.-I., Van de Weghe, A., Varvio, S., Velmala, R., Vilkki, J., Weikard, R., Woodside, C., Womack, J.E., Zanotti, M. and Zaragoza, P. (1997) A medium-density genetic linkage map of the bovine genome. *Mammalian Genome* 8, 21–28.

Botstein, D., White, R.L., Skolnick, M. and Davis, R.W. (1980) Construction of a genetic linkage map in man using restriction fragment length polymorphisms. *American Journal of Human Genetics* 32, 314–331.

Buetow, K.H. (1991) Influence of aberrant observations on high-resolution linkage analysis outcomes. *American Journal of Human Genetics* 49, 985–994.

Collins, F.S. (1995) Positional cloning moves from perditional to traditional. *Nature Genetics* 9, 347–350.

Couperwhite, S., Kato, Y. and Archibald, A.L. (1992a) A *Taq*I RFLP at the porcine thyroid stimulating hormone β-subunit locus (*TSHB*). *Animal Genetics* 23, 567.

Couperwhite, S., Hemmings, B.A. and Archibald, A.L. (1992b) A *Bam*HI RFLP at the locus encoding the 65 kDa regulatory subunit of porcine protein phosphatase 2A (PPP2ARB). *Animal Genetics* 23, 568.

Davies, W., Harbitz, I., Fries, R., Stranzinger, G. and Hauge, J. (1988) Porcine malignant hyperthermia detection and chromosomal assignment using a linked probe. *Animal Genetics* 19, 203–212.

Dietrich, W.F., Miller, J.C., Steen, R.G., Merchant, M., Damron, D., Nahf, R. Gross, A., Joyce, D.C., Wessel, M., Dredge, R.D., Marquis, A., Stein, L.D., Goodman, N., Page, D.C. and Lander, E.S. (1994) A genetic map of the mouse with 4006 simple sequence length polymorphisms. *Nature Genetics* 7, 220–245.

Echard, G. (1990) *Sus scrofa domestica* L. In: O'Brien, S.J. (ed.) *Genetic Maps.* Cold Spring Harbor Laboratory Press, Cold Spring Harbor, New York, p. 4.110.

Ellegren, H., Fredholm, M., Edfors-Lilja, I., Winterø, A.K. and Andersson, L. (1993) Conserved synteny between pig chromosome 8 and human chromosome 4 but rearranged and distorted linkage maps. *Genomics* 17, 599–603.

Ellegren, H., Chowdhary, B.P., Johansson, M., Marklund, L., Fredholm, M., Gustavsson, I. and Andersson, L. (1994a) A primary linkage map of the porcine genome reveals a low rate of recombination. *Genetics* 137, 1089–1100.

Ellegren, H., Chowdhary, B.P., Fredholm, M., Høyheim, B., Johansson, M., Nielsen, P.B., Thomsen, P.D. and Andersson, L. (1994b) A physically anchored linkage map of pig chromosome 1 uncovers sex- and position-specific recombination rates. *Genomics* 24, 342–350.

Fujii, J., Otsu, K., Zorzato, F., de Leon, S., Khanna, V.K., Weiler, J., O'Brien, P.J. and MacLennan, D.H. (1991) Identification of a mutation in the porcine ryanodine receptor that is associated with malignant hyperthermia. *Science* 253, 448–451.

Goureau, A., Yerle, M., Schmitz, A., Riquet, J., Milan, D., Pinton, P., Frelat, G. and Gellin, J. (1996) Human and porcine correspondence of chromosome segments using bidirectional chromosome painting. *Genomics* 36, 252–262.

Green, P., Falls, K. and Crooks, S. (1990) Documentation for CRI-MAP, version 2.4. Washington University School of Medicine, St Louis.

Haldane, J.B.S. (1922) Sex ratio and unisexual sterility in hybrid animals. *Genetics Journal* 12, 101–109.

Haley, C.S. and Archibald, A.L. (1992) Porcine genome analysis. In: Davies, K.S. and Tilghman, S.M. (eds) *Genome Analysis, Volume 4: Strategies for Physical Mapping.* Cold Spring Harbor Laboratory Press, Cold Spring Harbor, New York, pp. 99–129.

Haley, C.S., Knott, S.A. and Elsen, J.-M. (1994) Mapping quantitative trait loci in crosses between outbred lines using least squares. *Genetics* 136, 1195–1207.

Jeffreys, A.J., Wilson, V. and Thein, S.L. (1985) Hypervariable 'minisatellite' regions in human DNA. *Nature* 314, 67–73.

Kapke, P., Wang, L., Helm, J. and Rothschild, M. F. (1996) Integration of the PiGMaP and USDA maps for chromosome 14. *Animal Genetics* 27,187–190.

Keats, B.J.B., Sherman, S.L., Morton, N.E., Robson, E.B., Buetow, K.H., Cartwright, P.E., Chakravarti, A., Francke, U., Green, P.P. and Ott, J. (1991) Guidelines for human linkage maps; an international system for human linkage maps (ISLM, 1990). *Genomics* 9, 557–560.

Knott, S.A. and Haley, C.S. (1992) Aspects of maximum likelihood interval mapping in an F_2 population. *Genetical Research* 60, 139–151.

Kosambi, D.D. (1944) The estimation of map distances from recombination values. *Annals of Eugenics* 12, 172–175.

Lander, E.S. and Kruglyak, L. (1995) Genetic dissection of complex traits: guidelines for interpreting and reporting linkage results. *Nature Genetics* 11, 241–247.

Lathrop, G.M. and Lalouel, J.M. (1988) Efficient computations in multi-locus linkage analysis. *American Journal of Human Genetics* 42, 498–505.

Litt, M. and Luty, J.A. (1989) A hypervariable microsatellite revealed by *in vitro* amplification of a dinucleotide repeat within the cardiac muscle actin gene. *American Journal of Human Genetics* 44, 397–401.

McQueen, H.A., Haley, C.S., Couperwhite, S. and Archibald, A.L. (1995) A linkage map of pig chromosome 10. *Journal of Heredity* 86, 228–230.

Marklund, L., Johansson Møller, M., Høyheim, B., Davies, W., Fredholm, M., Juneja, R.K., Mariani, P., Coppieters, W., Ellegren, H. and Andersson, L. (1996) A comprehensive linkage map of the pig based on a wild pig–Large White intercross. *Animal Genetics* 27, 255–269.

Morton, N.E. (1955) Sequential test for the detection of linkage. *American Journal of Human Genetics* 7, 277–318.

Morton, N.E. (1991) Parameters of the human genome *Proceedings of the National Academy of Sciences, USA* 88, 7474–7476.

Morton, N.E. and Collins, A. (1990) Standard maps of chromosome 10. *Annals of Human Genetics* 54, 235–251.

Mungall, C. (1996) Visualization tools for genome mapping – the Anubis map manager. XXVth International Conference on Animal Genetics, 21–25 July 1996, Tours, France, *Animal Genetics* 27, (Suppl. 2), 56 (abstract C006).

O'Brien, S.J. (1991) Mammalian genome mapping: lessons and prospects. *Current Opinions in Genetics and Development* 1, 105–111.

Ollivier, L. (1995) Genetic differences in recombination frequency in the pig (*Sus scrofa*) *Genome* 38, 1048–1051.

Orita, M., Iwahana, H., Kanazawa, H., Hayashi, K. and Sekiya, T. (1989) Detection of polymorphisms of human DNA by gel electrophoresis as single-strand conformation polymorphisms. *Proceedings of the National Academy of Sciences, USA* 86, 2766–2770.

Otsu, K., Khanna, V.K., Archibald, A.L. and MacLennan, D. H. (1991) Co-segregation in F1 generation backcrosses of porcine malignant hyperthermia and a probable causal mutation in the skeletal muscle ryanodine receptor gene. *Genomics* 11, 744–750.

Paszek, A.A., Schook, L.B., Louis, C.F., Mickelson, J.R., Flickinger, G.H., Murtaugh, J., Mendiola, J.R., Janzen, M.A., Beattie, C.W., Rohrer, G.A., Alexander, L.J., Andersson, L., Ellegren, H., Johansson, M., Mariani, P., Marklund, L., Høyheim, B., Davies, W., Fredholm, M., Archibald, A.L. and Haley, C.S. (1995) First International Workshop on Porcine Chromosome 6. *Animal Genetics* 26, 377–401.

Rettenberger, G., Klett, C., Zechner, U., Kunz, J., Vogel, W. and Hameister, H. (1995) Visualization of the conservation of synteny between humans and pigs by heterologous chromosome painting. *Genomics* 26, 372–378.

Rohrer, G.A., Alexander, L.J., Keele, J.W., Smith, T.P. and Beattie, C.W. (1994). A microsatellite linkage map of the porcine genome. *Genetics* 136, 231–245.

Rohrer, G.A., Alexander, L.J., Hu, Z., Smith, T.P.L., Keel, J.W. and Beattie, C.W. (1996) A comprehensive map of the porcine genome. *Genome Research* 6, 371–391.

Rohrer, G.A., Alexander, L.J., Beattie, C.W., Wilke, P., Flickinger, G.H., Schook, L.B., Paszek, A.A., Andersson, L., Mariani, P., Marklund, L., Fredholm, M., Høyheim, B., Archibald, A.L., Nielsen, V.H., Milan, D. and Groenen, M.A.M. (1997) A consensus linkage map for swine chromosome 7. *Animal Genetics* 28, 223–229.

Rothschild, M.F., Jacobson, C., Vaske, D., Tuggle, C., Wang, L., Short, T., Eckardt, G., Sasaki, S., Vincent, A., McLaren, D., Southwood, O., Vander Steen, H., Milham, A. and Plastow, G. (1996) The estrogen receptor locus is associated with a major gene influencing litter size in pigs. *Proceedings of the National Academy of Sciences, USA* 93, 201–205.

Royle, N., Clarkson, R., Wong, Z. and Jeffreys, A.J. (1988) Clustering of hypervariable minisatellites in pro-terminal regions of human autosomes. *Genomics* 3, 352–360.

Van der Beek, S. and van Arendonk, J.A.M. (1993) Criteria to optimize designs for detection and estimation of linkage between marker loci from segregating populations containing several families. *Theoretical and Applied Genetics* 86, 269–280.

Vos, P., Hogers, R., Bleeker, M., Reijans, M., Van de Lee, T., Hornes, M., Frijters, A., Pot, J. Peleman, J., Kuiper, M. and Zabeau, M. (1995) AFLP: a new technique for DNA fingerprinting. *Nucleic Acids Research* 23, 4407–4414.

Weber, J.L. and May, P.E. (1989) Abundant class of human DNA polymorphisms which can be typed using the polymerase chain reaction. *American Journal of Human Genetics* 44, 388–396.

Winterø, A.-K., Fredholm, M. and Thomsen, P.D. (1992) Variable (dG-dT)n- (dC-dA)n sequences in the porcine genome. *Genomics* 12, 281–288.

Zhang, W., Haley, C. and Moran, C. (1995) Alignment of the PiGMaP and USDA linkage maps of porcine chromosomes 2 and 5. *Animal Genetics* 26, 361–364.

Ziegle, J.S., Su, Y., Corcoran, K.P., Nie, L., Mayrand, P.E., Hoff, L.B., McBride, L.J., Kronick, M.N. and Diehl, S.R. (1992) Application of automated DNA sizing technology for genotyping microsatellite loci. *Genomics* 14, 1026–1031.

Genetics of Behaviour

J.J. McGlone,[1] C. Désaultés,[2] P. Mormède[2] and M. Heup[1]

[1]*Pork Industry Institute, Department of Animal Science and Food Technology, Texas Tech University, Lubbock, TX 79409-2141 USA;* [2]*Laboratories for Stress Genetics, INRA - INSERM, Université de Bordeaux II, Bordeaux, France*

Introduction

The wild pigs of Asia and Europe were selected for domestication in large part because of their behaviour. While there was considerable variation in pig behaviour within the species, the behavioural traits that favoured domestication included their omnivorous dietary needs, their medium body size, their typically (though not always) docile nature, their relatively weak maternal–neonatal bonds, their precocial nature and their general adaptability (Hale, 1969; Ratner and Boice, 1976; Signoret *et al.*, 1976). The very plasticity of the domestic pig allows them to grow and reproduce in less-than-ideal and often variable environments. To survive and prosper requires a plastic and forgiving genome.

We know that the domestic pig was selected from native strains of wild pigs in Europe and Asia. The wild ancestors, the oldest breeds of domesticated European pigs and common modern-day domesticated pigs share many behavioural characteristics (Mitichashvili *et al.*, 1991). Wild pigs showed similar social, ingestive and exploratory behaviours as their domesticated counterparts (Robert *et al.*, 1987). Different from wild ancestors, the domestic pig has been selected to be more calm, quiet and less active than wild pigs (Robert *et al.*, 1987).

The study of behaviour genetics explores the very nature of a species. We know that species are distinguished from each other by both their morphology and their behaviour. Thus, while sheep are flocking animals, pigs tend to travel in small herds. Ruminants forage, while pigs scavenge as they forage. Intense selection may not make the scavenger a grazer, but it might increase its rate of feed intake and thus its rate of scavenging or feeding. What makes a pig a pig? Its behaviour is certainly a large part of the answer to that question. And its behaviour, like other traits, is a function of both its genes and its environment. Within a species and within the behavioural types of that species, selected breeds or strains of animals often have unique behavioural traits that breed true (like the sheep dog vs. the lap dog). Pig behaviours vary within the species. Some are lethargic, some hyperactive and many are in the middle.

The field of behaviour genetics has been around for several decades (Siegel, 1976) but relatively few investigations have been reported that seek to understand better the genetic basis of pig behaviours. In recent years, the average rate of publication in this field has been less than one paper per year (McGlone, 1991).

The subject of behaviour genetics of farm animals has been reviewed from time to time to synthesize the handful of literature (Siegel, 1976; Hohenboken, 1986, 1987). We believe behaviour genetics is ripe for major advances and it is our objective to review what is known today on this subject in an attempt to provide fruitful direction for further investigations.

Feeding Behaviours

The pig, as a species, is known for its relatively copious feed intake. Pigs can reach feed intakes of 5% of their body weight per day, which exceeds the level of feed intake of most farm animal species (NRC, 1988). All other factors held constant, if we increase feed intake in young animals, we will not only improve the rate of weight gain but we do so at greater feed efficiency (because most of the maintenance needs are already met). Thus, we clearly have great economic reason to increase feed intake in present, lean lines of pigs.

The genetic basis for control of average daily feed intake (ADFI) has been less well studied than the closely correlated trait, average daily gain (ADG). But because ADG and ADFI are highly correlated, selection would be successful if one were to select for feed intake alone. A complete review of feed efficiency is presented in Chapter 15.

Estimates of heritability for ADFI are similar to those for ADG (0.105 ± 0.214 and 0.098 ± 0.070 for ADFI and ADG, respectively; Bereskin, 1986). Bereskin (1986) also reported a phenotypic correlation between ADFI and ADG of 0.656 and a genetic correlation of 0.176 ± 0.226. ADFI was largely ignored in the early literature because it is just one factor that impacts the more direct economic factor of ADG and because feed intake is much more difficult to measure in an individual animal than are body weight changes over time.

Among some modern lines selected for lean meat, multiple selection goals may not result in increased feed intake especially if ADG is emphasized. If one selects intensively for both less body fat and increased weight gain, feed conversion tends to improve, and interestingly, feed intake goes down. Vogeli (1978) reported a 2% decline in feed intake per generation while selecting for reduced fat and increase weight gain.

Quite different problems of feed intake on commercial farms may show genetic variation and thus are amenable to change. For example, adult sow feed intake is too high, which wastes resources and, if left unchecked, causes excessive body fat and body size. The solution is physically to limit feed intake which requires resources and alters sow behaviour.

A second problem of feed intake is the postweaning feeding period in young pigs. Pigs usually do not take to dry feed with much enthusiasm in the hours shortly after weaning. Gradually, over the first few days and weeks, feed intake increases to the species-typical level. A quicker start on dry feed would mean faster and probably more efficient weight gains.

The problem of poor postweaning gains was addressed by Akins (1989). She measured feeding behaviour in the postweaning period for pigs weaned at 28 days of age for half- and full-siblings. She measured feeding durations, number of feeding bouts and time from weaning to first feeding experiences. In addition she calculated a slope of onset of feed intake with time after weaning as X and feeding time as Y. In this model, pigs with a larger slope of feeding duration over time after weaning would have a quicker onset of feed intake.

All measures of postweaning feeding behaviour showed heritability (h^2) estimates that were significantly greater than zero (Table 10.1). Postweaning feeding behaviours showed very high h^2 (0.87). In the same work, the h^2 for postweaning ADG was 0.84 ± 0.14. The comparison with postweaning ADG suggests that the genetic variation for postweaning feeding behaviour was similar to the genetic variation for ADG.

Onset of feeding after weaning also showed significant h^2 for several measures. With standard errors of about 0.13, the estimates of heritability ranged from 0.31 to 0.89 for onset of dry feed intake. These data suggest that selection for early onset of intake of dry feed would have as great a chance of success as selection for body weight gain.

De Haer (1992) examined the genetics of feeding behaviour in group-housed growing pigs. Landrace males and females were compared with Yorkshires of both sexes for feed intake and feeding patterns. Landrace pigs spent more time in each meal, they ate fewer meals with larger intake per meal but had similar overall feed intakes compared with Yorkshire pigs.

Table 10.1. Heritability estimates (on diagonal) and genetic (above) and phenotypic correlations (below the diagonal) for selected behavioural traits in weaning pigs. Data were collected over 24 hours after weaning and are from full-sibling analyses using Harvey's (1987) procedures (Akins, 1989).

	Behaviour (measured in min)				
Behaviour	Feeding	Drinking	Attack	Submit	FB[1]
Feeding	0.87 ± 0.13	0	0	0	0.25
Drinking	0.34†	0.58 ± 0.13	0	0	0.14
Attack	–0.05	0.13	0 ± 0	0	0
Submit	–0.09	–0.09	0.03	0 ± 0	0
FB[1]	0.25*	0.34†	–0.02	0	0.36 ± 0.13

Source: Atkins, 1989.
[1]Number of feeding bouts during the first 3 days after weaning at 28 days of age.
*Differs from zero, $P < 0.05$.
†Differs from zero, $P < 0.01$.

De Haer (1992) used several methods and models to estimate heritability of several measures of feeding behaviour. Feeding behaviours were lowly to moderately heritable. Daily feed intake had a heritability estimate of 0.16 ± 0.16. However, number of meals, number of large meals per day, number of feeder visits per day and feed intake per visit had higher estimates (0.35–0.49, ± 0.20 to ± 0.22).

Sexual Behaviours

Sexual behaviours include courtship behaviours by both sexes, male and female reproductive techniques, and female expression of oestrus. Each of these categories of reproductive behaviour is possibly under genetic control in part or whole. Goy and Jakway (1969) espoused the merits of gaining a better understanding of the genetic basis of sexual behaviours, but little direct work has been reported in pigs. Even reviews (for example, Bichard and David, 1985) of the genetics of prolificacy have not appreciated the role of behaviour in its significant contribution to prolificacy. A complete review of the genetics of reproduction is found in Chapter 11.

We do know that certain breeds of pigs express oestrus, including oestrous behaviour, at a very young age. The Meishan breed of pigs from China shows much earlier onset of oestrus than traditional European breeds of pigs. Reasons for this are not known for the most part.

Onset of oestrus after weaning seems to have a genetic component. Napel *et al.* (1995) reported realized heritabilities of 0.17 to 0.36 ± 0.05 for weaning to oestrous interval. This trait might be amenable to selection or identification of major genes.

Sexual techniques, generally thought to be innate in part, have not been studied for contributions from genes. Male libido, at least, is an economically important trait, probably closely tied to gamete production. Male libido is another area that should be investigated for genetic variation.

Social Behaviours

A variety of species from laboratory animals to farm animals to humans have been investigated to determine genetic underpinnings for control of aggression and violence. This literature is extensive, but not for the pig. The most prominent social behaviours are agonistic behaviours which include aggression, submission and threats.

Geneticists have studied the genetic variation for aggressive behaviour for mice and chickens and a few other species for several decades (reviewed in Siegal, 1976). Generally, the h^2 for aggression is approximately 0.20 (Siegal, 1976). Early work by Beilharz and Cox (1967) showed significant sire and breed effects for expression of social dominance in pigs. Our estimate for aggression in early weaned pigs was zero (see Table 10.1) and we know from experience that certain lines and certain individual parents have tendencies for high or low aggression. So, at least intuitively, we expect significant genetic variation for agonistic behaviours in the pig.

In recent years, major genes have been identified that are related to higher and lower levels of aggression in mice (Hen, 1996). One gene that causes an increase in aggression is a point mutation of the monoamine oxidase A (*MAOA*) enzyme gene. When mice were engineered to knock out the *MAOA* gene, they displayed an increase in brain serotonin and norepinephrine and a very large increase in aggressive behaviour (Cases *et al.*, 1995). A Dutch family with extensive familial increases in aggressive and violent behaviours likewise was shown to have a point-source mutation of the *MAOA* gene (Brunner *et al.*, 1993a,b). The issue is a bit more complicated (see review by Hen, 1996), but one could easily conclude that certain families or strains of pigs which show excessive aggressive behaviours may have similar mutated genetic variants that are attributed to brain neurochemistry.

A second major gene, which might have some role in modulation of aggressive behaviour, was identified in mice. Neuronal nitric oxide synthase (*NOS*) is expressed in developing and mature animals' nervous systems (to our knowledge, it has not been studied in the pig). Mice lacking *NOS* show more offensive attack and have other symptoms of affected reproduction and gastric problems (Nelson *et al.*, 1995).

Social behaviours interact with most other behaviours. For example, socially dominant pigs will inhibit the feed intake of submissive pen mates. Thus, determination of the genetic basis of feeding behaviour has significant environmental variation due to the dominance status of individual members of the group. To get a clean estimate of non-social behaviours, one needs to consider the behaviour of pigs in both the social and non-social conditions. For example, within a group of growing boars, one could be genetically superior in growth rate, but because it was socially submissive, its phenotype indicates it has low feed intake and weight gain. When freed from its environmental constraint, it would express its genetic potential.

Stress-related, Alarm, Fear and Other Emotional Behaviours

Porcine stress syndrome

Genetics of stress in pigs has been dominated by the acute stress syndrome, also known as porcine stress syndrome (PSS) or malignant hyperthermia. PSS is characterized in susceptible animals by intense skeletal muscle contractures with metabolic acidosis and elevation in body temperature, with a rapid fatal outcome (Britt, 1991). This syndrome is triggered by various stressful stimulations, such as handling and transport, by inhalation anaesthetics like halothane and by skeletal muscle relaxants. Major economic losses result not only from sudden deaths, especially in stress situations, but also from the development of pale, soft and exudative meat that arises from postmortem manifestation of the disease in malignant susceptible animals. The frequency of susceptible animals differs widely between porcine breeds, being highest in the Piétrain pig, and frequent in Yorkshire, Poland China, Duroc and Landrace breeds. The defect was shown to be transmitted as an autosomal recessive trait, and was related to the movement of calcium ions through the sarcoplasmic reticulum membrane. The condition results from a single point mutation in a gene on chromosome 6 for the skeletal muscle ryanodine receptor (a calcium channel; Fuji *et al.*, 1991). Selection of susceptible homozygote animals was based on their response to halothane challenge test. The availability of molecular diagnostic assays for the causal mutation (Otsu *et al.*, 1992), which allows the detection of heterozygotes as well as homozygotes, makes it possible to eliminate the mutation. However, many producers continue to use boars which carry the stress gene because the offspring from these boars produce more kilograms of pork at the expense of meat quality (Monin *et al.*, 1981).

From stress to psychobiology of adaptation

In general, the concept of stress derives from the study of the physiological adjustments necessary to maintain homeostasis in a fluctuating environment. Cannon (1935) noticed that the specificity of the sympathetic nervous system responses usually seen with most stimulations was lost when the intensity of the stimulus overflowed the normal regulatory mechanisms or in case of an intense emotional activation, such as in the cat exposed to a barking dog. The cardioascular and metabolic actions of the sympathetic nervous system were considered as necessary adjustments allowing an efficient behavioural adaptation, the 'fight or flight' response. Selye (1936) described the activation of the adrenocortical gland, releasing glucocorticoid hormones under the control of the anterior pituitary gland and of the paraventricular nucleus of the hypothalamus (the hypothalamic–pituitary–adrenal axis or HPA), as a non-specific response to a number of stimulations. Mason (1971) demonstrated that this non-specificity of response was primarily the result of the emotional activation by the environmental stimulus. Thus stress research moved from the field of pure physiology to psychobiology, with physiological responses being intimately interconnected to behavioural adjustments. Both response outputs are the coordinated expression of a central emotional state induced by environmental stimulations. However, these responses are not stereotyped, but depend upon the specific individual pattern of reactivity and the efficiency of the behavioural control over the stimulus (Dantzer and Mormède, 1983).

Measuring interindividual variation of stress responses

Individual features of behavioural reactivity are studied in different experimental settings. The most widely used test is the exposure of individual animals to a novel environment. This paradigm is called an 'open field' test. The ambulation score, exploration time, vocalizations, faecal boli and urinations are the classical measures of the emotional output, involving both behavioural and autonomic responses (Fraser, 1974). The test can be further complicated by the presence of food in the arena, to create a motivational conflict (neophobia test, Mormède *et al.*, 1984), or by adding new objects during the test (Lawrence *et al.*, 1991; Jensen *et al.*, 1995a). The response of the adrenocortical axis to this challenge can also be measured by comparing plasma levels of ACTH and cortisol before and after the test. The open field test has been widely used in laboratory rodents to measure emotionality, the animals displaying a low motor activity and a high elimination being considered as more emotional (Hall, 1934; Archer, 1973).

More recently, Hessing *et al.* (1993) have used a 'backtest' to measure tonic immobility in piglets. In this test, each animal is put on its back and restrained in this supine position for 60 seconds while one hand is placed loosely over the head of the pig. Escape attempts and vocalizations are scored. Previous data obtained in poultry and quail indicated that the duration of tonic immobility could be a measure of fearfulness. Hessing *et al.* (1993, 1994) described the distribution of backtest scores as bimodal, suggesting the existence of two distinct phenotypes (active vs. passive). However, Forkman *et al.* (1995) could not confirm this bimodal distribution (see also the discussion by Jensen *et al.*, 1995b) and several other groups found a unimodal distribution of reactivity traits in the population (Lawrence *et al.*, 1991; Mormède *et al.*, 1994).

Forkman *et al.* (1995) compared the behaviour of piglets in a number of social (rank order, social dependence, aggression) and non-social tests (novelty, extinction, backtest) and used a principal component analysis to study the relationships between these different measures. This analysis suggested three independent personality traits: aggression, sociability and exploration. This approach is quite recent in pigs but confirms the multidimensional features of behavioural responses to social and non-social challenges which have been clearly demonstrated in laboratory rodents and other species. These data show that in non-social situations, general motor activity and emotional reactivity are two independent dimensions (Trullas and Skolnick, 1993), and that in social situations, aggressive behaviour and the search for social contacts are also independent traits, which are not linked to either activity or emotionality. A careful distinction between these traits is important when looking for genetic influences, the basic assumption being that each trait is subserved by different biological mechanisms. A few studies have also used conditioned behaviours to study differences in reactivity between individual pigs (Dantzer *et al.*, 1980) or between genotypes (Dantzer and Mormède, 1983).

Genetics of emotional behaviours

Several divergent selection experiments in laboratory rodents have shown that the traits measured in the open field have a genetic component (Broadhurst, 1962; DeFries *et al.*, 1966). In pigs, Von Borell and Ladewig (1992) showed that the variability of open field scores was larger between litters than within litters. Comparison of different breeds demonstrated a wide range of variation (Mormède *et al.*, 1984), and the study of an F_2 intercross between Large White sires (with a high score of activity in the open field) and Meishan dams (with a low score) showed that the character was normally distributed, suggesting that the trait is under multigenic control (Mormède *et al.*, 1994). In Fig 10.1 defecation and locomotion scores are shown in a novel environment test in Meishan and Large White pigs and in crossbred offspring (see figure legend for explanation). Large White pigs displayed higher defecation and locomotion scores than Meishan pigs and F_1 animals were intermediate for locomotion scores, suggesting that there was no dominance effect for this behavioural trait. For defecation scores, Meishan pigs were dominant over Large Whites. The heritability estimate for the locomotion score, calculated as four times the paternal half-sib correlations, was rather low and was approximately 0.16 (calculated by Beilharz and Cox, 1967). Furthermore, Barnett *et al.* (1988) showed that genetic selection on the basis of growth performance resulted in an increase in activity, but no change in the exploratory activity in a novel environment, together with a moderately increased time of feeding in this novel environment, and a reduced motivation to interact socially.

Social behaviours of pigs towards humans show large individual differences (Lawrence *et al.*, 1991) and also might be under significant genetic control (Hemsworth *et al.*, 1990). Pig fear of humans is an interspecific behavioural interaction. Pigs vary in their behaviour towards humans in ways that are likely to be under some genetic control. Hemsworth *et al.* (1990) measured pigs' fear of humans in a standardized test of pig approach to a stationary human. Pigs that approached more quickly were said to have less fear of humans. The heritability estimate for 'fear of humans' was 0.376 ± 0.19 by REML methods. Thus, just as early domesticated pigs were selected for less fear of humans, there seems to be room for continued selection for ease of pig handling.

Genetics and neuroendocrine emotional responses

In pigs, most available data deal with the hypothalamic pituitary axis (HPA), although large breed/strain variations have also been shown for circulating

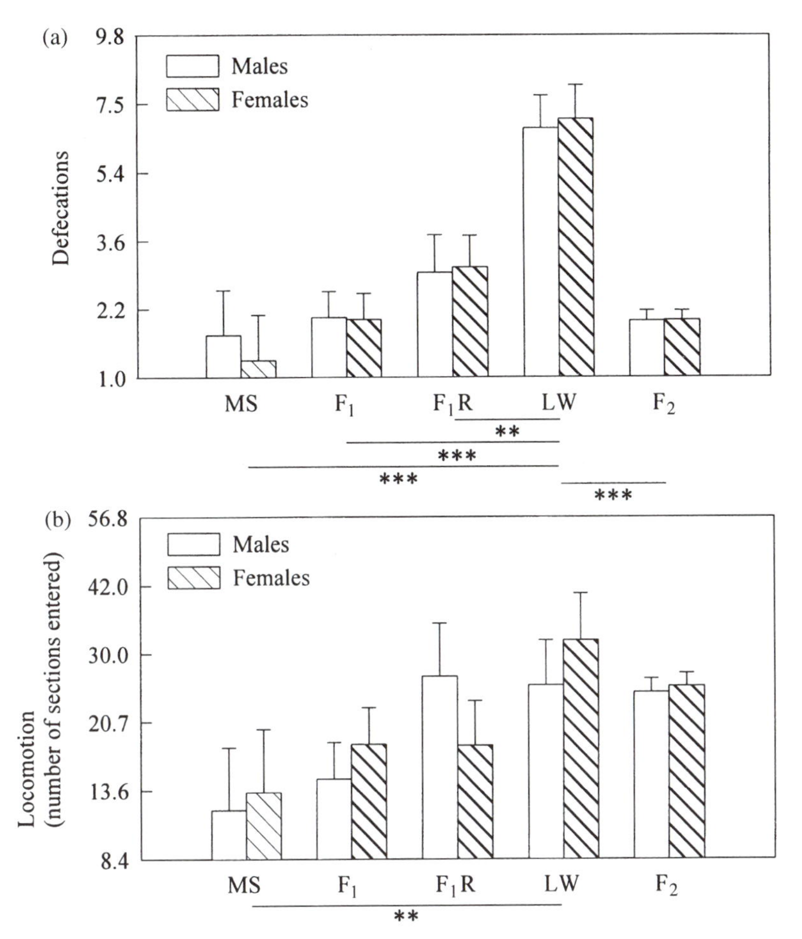

Fig. 10.1. Defecations (a) and locomotion scores (b) of 6-week-old piglets in a novel environment exposure according to sex and genetic type. MS, Meishan, $n = 12$ males and $n = 12$ females; F_1, MS sows × LW boars, $n = 20$ males and $n = 18$ females; F_1R, LW sows × MS boars, $n = 14$ males and $n = 20$ females; LW, Large White, $n = 18$ males and $n = 18$ females; F_2, (MS × LW) × (MS × LW), $n = 244$ males and $n = 250$ females. Values are the least square means and vertical lines are the standard errors. $^{**}P < 0.01$, $^{***}P < 0.001$. Data have been normalized by Box-Cox transformation $(y = x^t - 1)/t$ and scales are not arithmetic.

catecholamines in pigs (Mormède *et al.*, 1984) as well as in laboratory rodents (McCarty and Kopin, 1978). Individual differences in circulating cortisol levels were shown by Hennessy and co-workers to be related to changes in the adrenal response to ACTH (Hennessy *et al.*, 1988; Zhang *et al.*, 1990, 1992), a result obtained also in humans (Bertagna *et al.*, 1994). Indeed, the cortisol response to ACTH shows a very large range of inter-individual variation but is a stable trait in individuals (Von Borell and Ladewig, 1992, in pigs; Bertagna *et al.*, 1994, in humans). Cortisol response to ACTH can be influenced by chronic environmental stress, such as tight restraint (Janssens *et al.*, 1994, 1995) and by genetic factors, as clearly demonstrated by several divergent selection experiments in poultry (Brown and Nestor, 1973; Edens and Siegel, 1975; Satterlee and Johnson, 1988). We have also described large differences in circulating cortisol levels between European and Chinese pigs (Mormède *et al.*, 1984), and we have obtained evidence for multigene control of the adrenocortical activity from the study of an F_2 population from Large White and Meishan pigs (Mormède *et al.*, 1994). Figure 10.2 shows basal plasma cortisol levels in Meishan and Large White pigs and in crossbred animals. Meishan, F_1 and F_2 pigs had higher plasma cortisol levels than Large White pigs. These results showed that the 'high cortisol' trait in Meishan pigs was transmitted by autosomal dominant inheritance. However, we have recently shown that although the difference originated from the adrenal gland, it was independent from its ACTH drive (Désautés *et al.*,

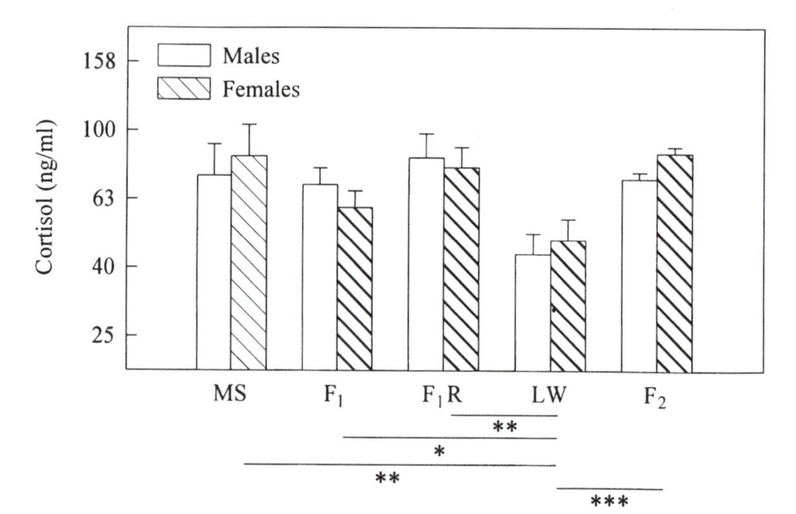

Fig. 10.2. Basal cortisol levels according to sex and genetic type in 6-week-old piglets. MS, Meishan, $n = 12$ males and $n = 12$ females; F_1, MS sows × LW boars, $n = 20$ males and $n = 18$ females; F_1R, LW sows × MS boars, $n = 14$ males and $n = 20$ females; LW, Large White, $n = 18$ males and $n = 18$ females; F_2, (MS × LW) × (MS × LW), $n = 244$ males and $n = 250$ females. Values are the least square means and vertical lines are the standard errors. $^*P < 0.05$, $^{**}P < 0.01$, $^{***}P < 0.001$. Data have been normalized by log transformation and this figure was drawn on a log scale.

submitted). Taken together, these experiments show that multiple mechanisms may explain the genetic control of adrenocortical function.

Several experimental data have demonstrated a link between the HPA axis activity and production traits. Hennessy and Jackson (1987) and Barnett *et al.* (1988) showed that the selection on the basis of growth performance was followed by a reduction of plasma cortisol levels and maximum corticosteroid binding capacity. Further investigations should concentrate on the genetic relationships between different traits of emotional reactivity and production traits.

The modulation by genetic factors of behavioural and neuroendocrine responses to stress in pigs is now well established. The tools to investigate the molecular basis of multigene influence on the different stress responses are available, as shown in mice by Flint *et al.* (1995). This approach is now possible in the pig, as a result of the development of knowledge of the pig genome (Archibald, 1994; Archibald *et al.*, 1995; Rohrer *et al.*, 1994).

'Abnormal' Behaviours

The word 'abnormal' is in quotes because we only label behaviours as abnormal when we do not understand why animals express such behaviours. To the animal, it might be quite natural and normal to bite tails and ears, chew bars and penning materials, drink excessive amounts of water (polydipsia), sit excessively, suck on navels, or other 'abnormal' behaviours. But to us, these are rare, unwanted behaviours with unknown causes or mechanisms. One possible mechanism could be that the genome of these individuals predisposes them to express these 'abnormal' behaviours in given environments. One should not assume that presence of 'abnormal' behaviours is evidence either of a genetic basis or of environmental effects.

We surveyed the behaviour of sows who reside in gestation crates (also called stalls). In our search we were looking for ways that we might group the sows into categories of expression of stereotyped bar biting. As can be seen from Fig. 10.3, the distribution of sow stereotyped bar chewing is very wide. But what should be clear is that there is enough variation in this behaviour to warrant investigation into potential genes that might be in part responsible for the wide variation in this 'abnormal' behaviour.

Domestic pigs show cannibalistic behaviours in two general forms: growing pigs chewing/eating appendages and killing and or eating piglets among periparturient sows (sow savaging). The aggressive behaviour of sows towards their piglets was investigated for a possible genetic basis by van der Steen *et al.* (1988). These Dutch researchers examined sows and their female offspring for maternal cannibalism. The estimated h^2 for cannibalism based on half-sibling data were 0.12 to 0.25 and for daughter–mother regression was 0.49 to 0.87. The higher estimates for the daughter–dam analyses were thought to be due to strong maternal effects. For herds with a moderately high level of sow cannibalism, selection against this behaviour has a good chance of success. Because we see an apparently higher rate of cannibalism among our outdoor gilts where

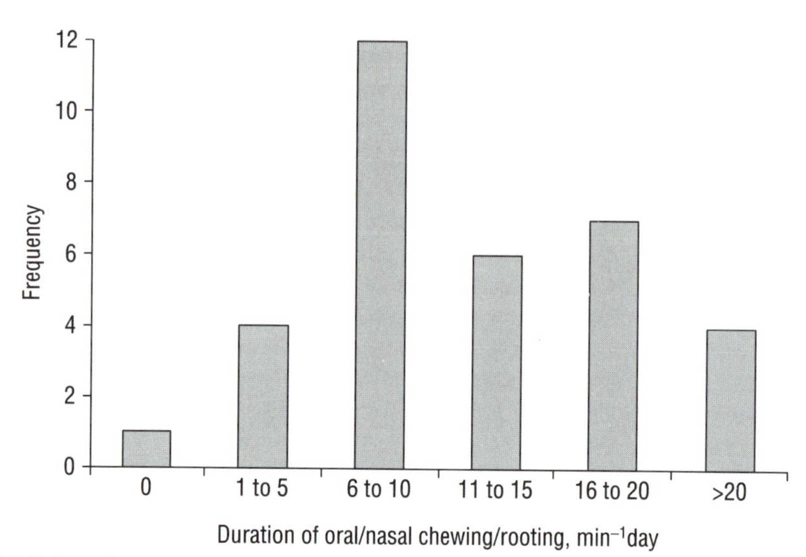

Fig. 10.3. Histogram of oral/nasal chewing and rooting for confined sows. $n = 34$ sows and data represent the average frequency per 2-hour time block over a 24-hour observation period.

piglet access is easier than in an indoor crate (McGlone *et al.*, unpublished observation), we cannot attribute this behaviour to indoor, confined housing.

Growing and adult pigs sit for varying amounts of times. Indeed, some individual pigs never sit while others sit for extended amounts of time. The time spent sitting might be important for a number of reasons. First, sows expend more energy sitting than lying down. Less active pigs might utilize feed more efficiently, as long as they keep eating. Of the crushed piglets found on a farm, about one-half the cases occur when sows move from sitting to lying down (rather than rolling from sternum to the side or lying down from a standing position or stepping on a piglet). If sows did not sit, the crushing rate would be lower (in fact, non-sitting sows have one-third the rate of piglet crushing as sows that sit; Morrow-Tesch and McGlone, 1990).

The heritability estimate of sitting was determined by examining the frequency and duration of sitting among growing full-sibling pigs. The h^2 estimate for sitting frequency and duration were 0.41 ± 0.14 and 0.43 ± 0.14, respectively (McGlone *et al.*, 1991). The results were striking in both the high heritability estimate and the low variation of the estimate.

The entire issue of animal welfare has entered the realm of behaviour genetics (Beilharz, 1982, 1987). If we can identify genes that control 'abnormal' behaviours, we can then discover if animals are more comfortable in the state of showing high or low levels of these 'abnormal' behaviours. Until the genetic bases of 'abnormal' behaviours are known, we cannot know if animals either are not coping with production environments, or have received a copy of one or more genes that cause an expression of these behaviours. For example, what if we could select sows that express high or low levels of stereotyped bar biting?

Which would be best? If this were possible, we could develop the lines and possibly determine which is less stressed or more comfortable. Secondly, what if we could find a gene that controlled tail biting? Would not everyone agree we should select against this behaviour? Final decisions about levels of animal comfort in production systems are likely to be inappropriate in the absence of behavioural genetic information.

Swedish Animal Protection laws preclude their citizens from genetically selecting livestock that suit production environments. Rather, the law states, the environments must be designed to meet the behavioural needs of the pigs. This is a rather strong edict that might in fact not be in the best interest of the pigs' welfare. Presumably, selecting sows who do not sit to save piglets from a death due to crushing would not be allowed because selection would change behaviour. Likewise, selecting against sows who savage their litters would not be allowed under Swedish law. The ancient farmers who selected pigs to be less active are, by Swedish law, said to have made a mistake. Both genetic selection and environmental modifications must be used to create production systems that are more comfortable and conducive to sound pig welfare.

Summary

Domestication of pigs resulted in changes in pig behaviour due to genetic selection and environmental control. Feeding, sexual, social, emotional and stress-related behaviours are thought to have low to moderate heritability estimates. Genes have been identified in mice that modulate agonistic behaviours, but other major genes affecting behaviour have not been identified. Behaviour genetics is a well-established discipline but relatively little work has been done in this field with the domestic pig in spite of great economic opportunity and society concerns. We expect more work in pig behaviour genetics, particularly identification of genes that might influence feeding, sexual, social, emotional and abnormal behaviours.

References

Akins, C.K. (1989) Genetics of behavior and performance of growing pigs. MSc thesis. Texas Tech University, Lubbock, Texas.

Archer, J. (1973) Tests of emotionality in rats and mice: a review. *Animal Behaviour* 21, 205–235.

Archibald, A.L. (1994) Mapping of the pig genome. *Current Opinion in Genetics and Development* 4, 395–400.

Archibald, A.L. *et al.* (1995) The PiGMaP consortium linkage map of the pig (*Sus scrofa*). *Mammalian Genome* 6, 157–175.

Barnett, J.L., Hemsworth, P.H., Cronin, G.M., Winfield, C.G., McCallum, T.H. and Newman, E.A. (1988) The effects of genotype on physiological and behavioural responses related to the welfare of pregnant pigs. *Applied Animal Behaviour Science* 20, 287–296.

Beilharz, R.G. (1982) Genetic adaptation in relation to animal welfare. *International Journal for the Study of Animal Problems* 3, 117–124.

Beilharz, R.G. (1987) The behaviour component of intensive animal welfare. In: Henry, P. and Chenoweth, (eds) *Intensive Animal Welfare*. Australian Veterinary Association (Qld Divsion).

Beilharz, R.G. and Cox, F. (1967) Genetic analysis of openfield behavior in swine. *Journal of Animal Science* 26, 988–990.

Bereskin, B. (1986) A genetic analysis of feed conversion efficiency and associated traits in swine. *Journal of Animal Science* 62(4), 910–917.

Bertagna, X., Coste, J., Raux-Demay, M.C., Letrait, M. and Strauch, G. (1994) The combined corticotropin-releasing hormone/lysine vasopressin test discloses a corticotroph phenotype. *Journal of Clinical Endocrinology and Metabolism* 79, 390–394.

Bichard, M. and David, P.J. (1985) Effectiveness of genetic selection for prolificacy in pigs. *Journal of Reproduction and Fertility* (Suppl.), 33, 127–138.

Britt, B.A. (1991) Malignant hyperthermia: a review. In: Schobaum, E. and Lomax, P. (eds) *Thermoregulation: Pathology, Pharmacology and Therapy*. Pergamon, New York, pp. 179–292.

Broadhurst, P.L. (1962) A note on further progress in psychogenetic selection experiment. *Psychology Reports* 10, 65–66.

Brown, K.I. and Nestor, K.E. (1973) Some physiological responses of turkeys selected for high and low adrenal response to cold stress. *Poultry Science* 52, 1948–1954.

Brunner, H.G., Nelen, M., Breakefield, X.O., Ropers, H.H. and Van Oost, B.A. (1993a) Abnormal behavior associated with a point mutation in the structural gene for monoamine oxidase A. *Science* 262, 578–580.

Brunner, H.G., Nelen, M.R., Van Zandvoort, P., Abeling, N.G.G.M., Van Gennip, A.H., Wolters, E.C., Kulper, M.A., Ropers, H.H. and Van Oost, B.A. (1993b) X-linked borderline mental retardation with prominent behavioral disturbance in monoamine metabolism. *American Journal of Human Genetics* 52, 1032–1039.

Cannon, W.B. (1935) Stresses and strains of homeostasis. *American Journal of Medical Science* 189, 1–14.

Cases, O., Vitalis, T., Seif, I., De Maeyer, E., Sotelo, C. and Gasper, P. (1996) Lack of barrels in the somatosensory cortex of monoamine oxidase A-deficient mice: role of a serotonin excess during the critical period. *Neuron* 16, 297–307.

Dantzer, R., Mormède, P.,(1978) Behavioural and pituitary–adrenal characteristics of pigs differing by their susceptibility to the malignant hyperthermia syndrome induced by halothane anesthesia. 1. Behavioural measures. *Annals de Recherches Veterinaires*, 9, 559–567.

Dantzer, R. and Mormède, P. (1983) Stress in farms animals: a need for reevaluation. *Journal of Animal Science* 57, 6–18.

Dantzer, R., Arnone, M. and Mormède, P. (1980) Effects of frustration on behaviour and plasma corticosteroid levels in pigs. *Physiology of Behaviour* 24, 1–4.

DeFries, J.C., Hegman, J.P. and Weir, M.W. (1966) Openfield behavior in mice: evidence for a major gene effect mediated by the visual system. *Science* 154, 1577–1579.

De Haer, L.C.M. (1992) Relevance of eating pattern for selection of growing pigs. Thesis, Wageningen, the Netherlands.

Edens, W.E. and Siegel, H.S. (1975) Adrenal responses in high and low ACTH response lines of chickens during acute heat stress. *General Comparative Endocrinology* 25, 64–73.

Flint, J., Corley, R., DeFries, J.C., Fulker, D.W., Gray, J., Miller, S. and Collins, A.C. (1995) A simple genetic basis for a complex psychological trait in laboratory mice. *Science* 269, 1432–1435.

Forkman, B., Fuhuhaug, I.L. and Jensen, P. (1995) Personality, coping patterns, and aggression in piglets. *Applied Animal Behaviour Science* 45, 31–42.

Fraser, D. (1974) The vocalizations and other behaviours of growing pigs in an open-field. *Applied Animal Ethology* 1, 3–16.

Fuji, J., Otsu, K., Zorzato, F., De Leon, S., Khana, V.K., Weiler, J.E., O'Brien, P.J. and McLennan, D.H. (1991) Identification of a mutation in porcine ryanodine receptor associated with malignant hyperthermia. *Science* 253, 448–451.

Goy, R.W. and Jakway, J.S. (1969) Role of inheritance in determination of sexual behavior patterns. In: Bliss, E.L. (ed.) *Roots of Behavior: Genetics, Instinct, and Socialization in Animal Behavior/by Thirty-One Authors 1969.* Hafner Pub. Co., New York, pp. 96–112.

Hale, E.B. (1969) Domestication and the evolution of behavior. In: Hafez, E.S.E. (ed.) *The Behaviour of Domestic Animals*, 2nd edn. Baillière Tindall & Cassel, London.

Hall, C.S. (1934) Emotional behavior in the rat. I: defecation and urination as measures of individual differences in emotionality. *Journal of Comparative Psychology* 18, 385–403.

Harvey, W.R. (1987) User's guide for LSMLMW. p.c.l version. Mixed model least-squares and maximum likelihood computer program.

Hemsworth, P.H., Barnett, J.L., Treacy, D. and Madgwick, P. (1990) The heritability of the trait fear of humans and the association between this trait and subsequent reproductive performance of gilts. *Applied Animal Behavior Science* 25, 85–95.

Hen, R. (1996) Mean genes. *Neuron* 16, 17–21.

Hennessy, D.P. and Jackson, P.N. (1987) Relationship between adrenal responsiveness and growth rate. In: *Manipulating Pig Production.* Australian Pig Science Association Committee.

Hennessy, D.P., Stelmasiak, T., Johnston, N.E., Jackson, P.N. and Outh, K.H. (1988) Consistent capacity for adrenocortical response to ACTH administration in pigs. *American Journal of Veterinary Research* 8, 1276–1283.

Hessing, M.J.C., Hagelso, A.M., Van Beek, J.A.M., Wiepkema, P.R., Schouten, W.G.P. and Krukow, R. (1993) Individual behavioural characteristics in pigs. *Applied Animal Behaviour Science* 37, 285–295.

Hessing, M.J.C., Hagelso, A.M., Schouten, W.G.P., Wiepkema, P.R. and Van Beek, J.A.M. (1994) Individual behavioral and physiological strategies in pigs. *Physiology of Behavior* 55, 39–46.

Hohenboken, W.D. (1986) Inheritance of behavioural characteristics in livestock. A Review. *Animal Breeding Abstracts* 54, 623–639.

Hohenboken, W.D. (1987) Behavioral genetics. *Veterinary Clinics of North America, Food Animal Practice* 3(2), 217–229.

Janssens, C.J.J.G., Helmond, F.A. and Wiegant, V.M. (1994) Increased cortisol response to exogenous adrenocorticotropic hormone in chronically stressed pigs : influence of housing conditions. *Journal of Animal Science* 72, 1771–1777.

Janssens, C.J.J.G., Helmond, F.A. and Wiegant, V.M. (1995) Chronic stress and pituitary–adrenocortical responses to corticotropin-releasing hormone and vasopressin in female pigs. *European Journal of Endocrinology* 132, 479–486.

Jensen, P., Forkman, B., Thodberg, K. and Köster, E. (1995a) Individual variation and consistency in piglet behaviour. *Applied Animal Behaviour Science* 45, 43–52.

Jensen, P., Rushen, J. and Forkman, B. (1995b) Behavioural strategies or just individual variation in behaviour? A lack of evidence for active and passive piglets. *Applied Animal Behaviour Science* 43, 135–139.

Lawrence, A.B., Terlouw, E.M.C. and Illius, A.W. (1991) Individual differences in behavioural responses of pigs exposed to non-social and social challenges. *Applied Animal Behaviour Science* 30, 73–86.

McCarty, R. and Kopin, I.J. (1978) Sympatho-adrenal medullary activity and behavior during exposure to footshock stress: a comparison of seven rat strains. *Physiology of Behaviour* 21, 567–572.

McGlone, J.J. (1991) Techniques for evaluation and quantification of pig reproductive, ingestive, and social behaviors. *Journal of Animal Science* 69(10), 4146–4154.

McGlone, J.J., Akins, C.K. and Green, R.D. (1991) Genetic variation of sitting frequency and duration in pigs. *Applied Animal Behaviour Science* 30(3/4), 319–322.

Mason, J.W. (1971) A re-evaluation of the concept of non-specificity in stress theory. *Journal of Psychiatric Research* 8, 323–333.

Mitichashvili, R.S., Tikhonov, V.N. and Bobovich, V.E. (1991) Immunogenetic characteristics of the gene pool Kakhetian pigs and dynamics of population parameters during several generations. *Soviet Genetics* 26(7), 855–863.

Monin, G., Sellier, P., Ollivier, L., Goutefongea, R. and Girard, J.P. (1981) Carcass characteristics and meat quality of halothane positive Piétrain pigs. *Meat Science* 5, 413–423.

Mormède, P. and Dantzer, R. (1978) Behavioural and pituitary–adrenal characteristics of pigs differing by their susceptibility to the malignant hyperthermia syndrome induced by halothane anesthesia. 2. Pituitary–adrenal function. *Annals de Recherches Veterinaires* 9, 569–576.

Mormède, P., Dantzer, R., Bluthe, R.M. and Caritez, J.C. (1984) Differences in adaptative abilities of three breeds of Chinese pigs. Behavioural and neuroendocrine studies. *Genetique, Selection, Evolution* 16, 85–102.

Mormède, P., Garcia-Bellenguer, S., Dulluc, J. and Oliver, C. (1994) Independent segregation of a hyperactive hypothalamic-hypophyso-adrenal axis and a reduced behavioral reactivity in pigs. *Psychoneuroendocrinology* 19, 305–311.

Morrow-Tesch, J. and McGlone, J.J. (1990) Productivity and behavior of sows in level vs sloped farrowing pens and crates. *Journal of Animal Science* 68, 82–87.

National Research Council (1988) *Nutrient Requirements of Swine.* National Academy Press, Washington DC, USA.

Nelson, R.J., Demas, G.E., Huang, P.L., Fishman, M.C., Dawson, V.L., Dawson, T.M. and Synder, S.H. (1995) Behavioural abnormalities in male mice lacking neuronal nirtic oxide synthase. *Nature* 378, 383–386.

Otsu, K., Phillips, M.S., Khanna, V.K., De Leon, S. and MacLennan, D.H. (1992) Refinement of the diagnostic assays for a probable causal mutation for porcine and human malignant hyperthermia. *Genomics* 13, 835–837.

Ratner, S.C. and Boice, R. (1976) Effects of domestication on behaviour. In: Hafez, E.S.E. (ed.) *The Behaviour of Domestic Animals,* 3rd edn. Baillière Tindall & Cassel, London, pp. 3–19.

Robert, S., Dancosse, J. and Dallaire, A. (1987) Some observations on the role of environment and genetics in behaviour of wild and domestic forms of *Sus scrofa. Applied Animal Behaviour Science* 17 (3/4), 253–262.

Rohrer, G.A., Alexander, L.J., Keele, J.W., Smith, T.P. and Beattie, C.W. (1994) A microsatellite linkage map of the porcine genome. *Genetics* 136, 231–245.

Satterlee, D.G. and Johnson, W.A. (1988) Selection of Japenese quail for contrasting blood corticosterone response to immobilization. *Poultry Science* 67, 25–32.

Selye, H. (1936) A syndrome produced by diverse noxious agents. *Nature* 138, 32–33.

Siegel, P.B. (1976) Behavioural genetics. In: Hafez, E.S.E. (ed.) *The Behaviour of Domestic Animals*, 3rd edn. Baillière Tindall & Cassel, London, pp. 20–42.

Signoret, J.P., Baldwin, B.A., Fraser, D. and Hafez, E.S.E. (1976) The behaviour of swine. In: Hafez, E.S.E. (ed.) *The Behaviour of Domestic Animals*, 3rd edn., Baillière Tindall & Cassel, London, pp. 295–329.

ten Napel, J., de Vries, A.G. and Buiting, G.A.J. (1995) Genetics of the interval from weaning to estrus in first letter sows: distribution of data, direct response of selection, and heritability. *Journal of Animal Science* 73, 2193–2203.

Trullas, R. and Skolnick, P. (1993) Differences in fear motivated behaviors among inbred mouse strains. *Psychopharmacology* 111, 323–331.

van der Steen, H.A., Schaeffer, L.R., de Jong, H. and de Groot, P.N. (1988) Aggressive behavior of sows at parturition. *Journal of Animal Science* 66(2), 271–279.

Vogeli, P. (1978) The behavior of fattening performance parameters and quantitative and qualitative carcass characteristics of two divergent lines of pigs using index selection procedures. Doctoral Thesis, Swiss Federal Institute of Technology, Zurich, Switzerland.

Von Borell, E. and Ladewig, J. (1992) Relationship between behaviour and adrenocortical response pattern in domestic pigs. *Applied Animal Behavior Science* 24, 195–206.

Zhang, S.H., Hennessy, D.P. and Cranwell, P.D. (1990) Pituitary and adrenocortical responses to corticotropin-releasing factor in pigs. *American Journal of Veterinary Research* 51, 1021–1025.

Zhang, S.H., Hennessy, D.P., Cranwell, P.D., Noonan, D.E. and Francis, H.J. (1992) Physiological responses to exercice and hypoglycemia stress in pigs of differing adrenal responsiveness. *Comparative Biochemistry and Physiology* 103, 195–703.

Biology and Genetics of Reproduction

M.F. Rothschild[1] and J.P. Bidanel[2]

[1]Iowa State University, Department of Animal Science, 225 Kildee Hall, Ames, IA 50011-3150 USA; [2]Institut National de la Recherche Agronomique, Station de Génétique Quantitative et Appliquée 78352 Jouy-en-Josas Cedex, France

Introduction

Reproduction, a process essential to maintenance of a species, must be under relatively strict genetic control. This genetic control must ensure that the steps in the reproductive process are repeated with great certainty and precision. Natural selection influenced these steps well before the pig was domesticated over 5000 years ago and humans applied artificial selection. However, nature did not remove genetic variation entirely and humankind has effectively altered the pig to fit social, food and environmental needs.

A simple survey across breeds demonstrates that considerable genetic variability exists for several reproductive measures. Average litter size varies among breeds from 4 to 16 pigs per litter for mature sows. Mean age at puberty varies from 3 to 7 months of age. Clearly these breed differences, combined with evidence for genetic variability within breeds, suggest that substantial genetic improvement of reproductive performance in the pig is possible.

Substantial gains in the efficiency of pig production systems can be expected from genetically improving reproductive traits (Tess *et al.*, 1983; de Vries, 1989). Incremental costs related to the production of additional pigs are minimal so that substantial gains can be achieved by improving the number of piglets weaned per breeding animal per unit of time. There is some evidence that genetic improvement of numerical productivity can be enhanced by genetically acting on its component traits, i.e. age at sexual maturity, fertility, prolificacy and piglet viability, or their underlying physiological processes.

Genetic differences for reproductive traits have been observed both among and within breeds and lines. Differences among breeds and lines can be most effectively exploited through the use of crossbreeding. Within breed or line genetic variability is usually characterized by heritability and genetic correlation estimates which quantify the additive genetic variation that can be manipulated via selection of superior animals. The genes responsible for these genetic differences are usually not known. Yet recent advances in the pig genetic map (see Chapters 8 and 9) have made it possible to identify individual genes with large effects on reproductive traits. The purpose of this review is to provide and discuss recent evidence for the underlying genetic control of reproductive traits and methods of genetic improvement of these traits.

Some Aspects of Pig Reproductive Biology

Herd reproductive performance depends on complex physiological pathways which determine male and female reproductive performance such as age at sexual maturity, gamete production, libido, fertilization, embryo, fetal and piglet survival. This section only provides some basic features of sow and boar reproductive biology which may be useful for understanding the rest of the chapter. More detailed elements on pig reproductive biology can be found in books such as Pond and Houpt (1978) and Cole and Foxcroft (1982).

Male

Spermatogenesis in the boar starts at 4 to 6 months of age in most pig breeds, but may begin before 100 days of age in some early maturing breeds such as the Chinese Meishan breed. Sperm quality and quantity then steadily increase with testicular development, testosterone production and libido until sexual maturity at 6 to 8 months of age and then at a much lower rate until boars reach their adult body size. A parallel rise in male accessory glands (seminal vesicle, prostate and bulbo-urethral glands), which produce 95% of the seminal plasma, results in a correlated increase in the volume of the ejaculate. Sexual activity is controlled by gonadotrophic hormones. Follicle stimulating hormone (FSH) stimulates spermatogenesis, whereas luteinizing hormone (LH) stimulates steroid hormones (testosterone, but also other steroids such as androstenone) synthesis and secretion by the interstitial Leydig cells. The action of LH is dependent on the FSH induction of LH receptors on the Leydig cells. Boar ejaculate is characterized by its large volume (around 300 ml on average) and spermatozoa number (80 to 120 billion when semen is collected once a week), which corresponds to total sperm reserves and widely exceeds daily sperm production (10 to 20 billion spermatozoa/day^{-1}). As a consequence, spermatozoa number per ejaculate steadily decreases when the boar is used or collected more than once a week, in spite of a slight increase in sperm production with ejaculation frequency. Large amounts of spermatozoa and semen are necessary to ensure normal conception rate and prolificacy (50 ml of semen and 3 billion sperm are usually considered as minimum requirements for artificial insemination). Frozen boar semen can successfully be employed, but leads to much lower conception rate and litter size than fresh semen, so that its commercial use is currently very limited. These peculiarities limit the use of either natural service (sow to boar ratio cannot exceed 12 to 15) or AI boars (which produce about 1000 semen doses per year) and consequently the dissemination capacity of favourable genes in breeding programmes.

Female

Puberty in gilts, which is usually defined as the moment of first ovulation, occurs at 3–4 months of age in the most early maturing breeds (Chinese) and at an average of 6–7 months of age in the most widely used Western pig breeds. It generally coincides with the first oestrus, though ovulation without external manifestation of oestrus (silent heat) occurs occasionally in pigs, and generates a steroid-secreting activity of corpora lutea. Ovulations then occur every 3 weeks during the second half of a 2–3 day oestrous period in the absence of gestation and have a mean duration of 2–3 hours. The oestrous cycle is controlled by gonadotrophic hormones. FSH stimulates recruitment and development of ovarian follicles. Ovulation and corpora lutea formation are stimulated by LH. Ovulation rate increases with oestrus and parity number until the fourth or fifth parity. Conception rate in the pig is high (80–90%) and has increased with the generalization of double mating (two services 12 or 24 hours apart during oestrus). Ova fertilization begins a few hours after mating and lasts approximately 8 hours. Implantation occurs at about 18 days after fertilization, and gestation length averages about 114 days. The rate of prenatal mortality in pigs is 30 to 40% on average. The largest part of the loss (20 to 30%) occurs before or during the implantation period (Wrathall, 1971). As fertilization rate in pigs is generally close to 100% (Perry and Rowlands, 1962; Wrathall, 1971), most of the ova wastage is due to embryo mortality. An additional 10 to 15% loss occurs at parturition and during lactation, mainly during the first 3 or 4 days of life (Svendsen, 1992). With very few exceptions, the lactating sow has a very limited follicular development, does not ovulate nor show any oestrous symptom. The total removal of the sow from her litter at weaning normally results in an acceleration of follicular growth and in ovulation within 4–10 days.

Traits of interest

A list of reproductive traits of current or potential interest for pig breeding is given in Table 11.1. The reproductive efficiency of natural service boars may be characterized by their age at sexual maturity, their mating ability, the conception rate, the size of the litters resulting from their matings and their longevity. Important component traits are libido, aggressiveness, semen and sperm quantity and quality. Traits of interest are rather similar for AI boars, but with a greater emphasis on semen and sperm quantity and quality and on the ease and frequency of semen collection. Components of litter size at weaning, i.e. ovulation rate, prenatal survival and piglet preweaning survival, are the most important contributors to sow numerical productivity (Tess *et al.*, 1983 ; de Vries, 1989; Ducos, 1994). Other important component traits are uterine capacity (Bennett and Leymaster, 1989), maternal behaviour, teat number and milk production. Decreased age at puberty, which is expected to reduce generation interval and increase the sexual maturity of females at a given age, and reduced intervals from weaning to conception, which depend on both fecundity and sexual behaviour, are also associated with a better numerical productivity, but their economic value is lower than that of litter size.

Chromosomal Abnormalities

Though most genetic control occurs at the individual gene level, gross genetic abnormalities can also affect reproduction. The primary gross genetic abnormality affecting reproduction is when a chromosomal break happens and a reciprocal translocation occurs (see Chapter 8). Evidence of detrimental effects of reciprocal translocations on fertility and prolificacy have been reported (reviewed by Popescu, 1989). Reduction in litter size ranges from 5 to 100% and is primarily due to an increased embryo mortality (Popescu, 1989). A total of 68 reciprocal translocations involving all pig chromosomes except the Y chromosome have been reported so far (reviewed in Ducos *et al.*, 1997). Abnormalities of chromosome number have been found at relatively high frequencies (5–10%) in early pig embryos (McFeely, 1967; Fechheimer and Beatty, 1974; Long and Williams, 1982). These abnormalities generally lead to the death of embryos and may explain a significant part of prenatal mortality. As a consequence, such abnormalities have very rarely been found after birth. Exceptions include

Klinefelter (XXY) syndrome, Turner syndrome (X0) and occasionally trisomy. Other chromosomal abnormalities include intersexuality, which is one of the most frequent genetic defects in pig breeds (0.1 to 0.4% according to Backström and Henricson, 1971). Such abnormalities are rare and have generally a limited impact on overall reproductive efficiency. A notable exception concerns boars carrying translocations, which reduce litter size of their matings and transmit

Table 11.1. Main reproductive traits of interest for animal breeding.

Reproductive function	Sex	Main components or predictive traits
Sexual maturity	Male	Age at first mating or sperm collection
		Testes size
		Size of Cowper's gland
	Female	Age at first progesterone rise
		Age at first oestrus
Sexual behaviour	Male	Ability to mount
	Female	Visible symptoms of oestrus
		Mating stance
Gonadocyte production	Male	Sperm quantity
		Semen volume (per ejaculate, per day)
		Sperm concentration
		Testes size (length, width, circumference, volume and weight)
		Sperm quality
		Sperm motility
		Proportion of:
		dead spermatozoa
		abnormal spermatozoa
	Female	Ovulation rate
Fertility	Male	Conception rate of mates
	Female	Conception rate
		Weaning to oestrus interval
		Weaning to conception interval
		Number of service per conception
Prolificacy	Male	Litter size of mates
	Female	Fertilization rate
		Number of embryos or fetuses
		Embryo, fetus or prenatal survival rates
		Uterus size (length or weight of uterine horns)
		Uterine capacity
		Number of mummified piglets
		Number of piglets born or born alive
Nursing abilities	Female	Number of piglets weaned
		Litter preweaning survival rate
		Teat number
		Milk production
Hormone levels	Male	Testosterone level
	Female	Progesterone level

their abnormality to half of their progeny and may have important economic consequences at a herd or a breeding scheme level (Ducos *et al.*, 1997).

Between-breed Variation and Crossbreeding

Breed differences

Differences between pig breeds in reproductive traits have been reported by a number of authors, often in the context of crossbreeding studies. Though breed differences vary between experiments because of sampling, time-dependent or location-dependent variations, pig breeds can be classified into four main groups which differ in production and reproduction performance levels (Legault, 1985). Dual-purpose breeds, such as Large White, Yorkshire, Landrace and some original lines, exhibit a satisfactory level for both reproduction and production traits. Specialized 'paternal' breeds, such as Piétrain, Belgian Landrace, Hampshire and Poland China and an increasing number of original strains, show medium reproduction and high production performance levels. Specialized 'maternal' breeds essentially include a limited number of native breeds from China, such as the group of the Taihu breeds (e.g. Meishan), which exhibit exceptional reproductive abilities, but poor production performance. Finally, there is a large group of 'native' breeds which generally have poor production and reproduction performance levels, but are well adapted to their particular environment. The reproductive performance of the first three groups of breeds only is considered here.

Several surveys on the performance of prolific breeds of China have been published over the last 15 years (e.g. Zhang *et al.*, 1983; Xu, 1985a,b) and have evidenced the early maturity, high prolificacy and good mothering abilities of the Meishan, Jiaxing, Fengjing, Erhualian and Minzhu breeds. The exceptional reproductive ability of the Meishan breed has been confirmed under intensive management conditions of several European and American countries (see Table 11.2). Meishan gilts reach puberty about 100 days earlier than Large White gilts. Ovulation rate of Meishan gilts at first oestrus is rather low and inferior or similar to that of Large White pigs (Bolet *et al.*, 1986; Christenson, 1993). It then increases with oestrus number, so that Meishan gilts produce a larger number of ova than Large White gilts at the same chronological age. Meishan females also have a significantly higher conception rate than their Large White counterparts (Després *et al.*, 1992) and exhibit an average superiority for litter size at birth which ranges from 2.4 to 5.2 piglets. Larger litters of Meishan females come from a lower prenatal mortality (Bidanel *et al.*, 1990a) or from the combination of a higher ovulation rate and a higher prenatal survival for a given ovulation rate (Haley *et al.*, 1995). Meishan females have a similar or slightly lower proportion of stillbirths than Large White sows, but a higher preweaning mortality rate. However, as shown by Lee and Haley (1995) and Blasco *et al.* (1995), this larger mortality is essentially due to the higher litter size of Meishan sows at birth, as survival rate adjusted for the number of piglets born is clearly in favour of

Table 11.2. Comparative reproductive performance of Chinese Meishan and European or American pig breeds.

Trait	Meishan[1]	Western breed[1]	Western breed used[2]	Parity	Reference
Age at puberty (d)	92 (541)	197 (444)	LW	—	Després *et al.* (1992)
	96	201	Y	—	White *et al.* (1991)
Ovulation rate	9.2 (13)	13.7 (15)	LW	1 (1st oestrus)	Bolet *et al.* (1986)
	11.7 (15)	14.4 (12)	LW	1 (3rd oestrus)	Bolet *et al.* (1986)
	21.7 (30)	17.1 (32)	C	1	Ashworth *et al.* (1990)
	20.3 (38)	19.1 (19)	LW	1–6	Bidanel *et al.* (1990a)
	14.1 (18)	14.1 (19)	LW	1 (4th oestrus)	Hunter *et al.* (1991)
	14.5 (20)	17.3 (17)	C	1	Anderson *et al.* (1992)
	24.9 (14)	15.2 (15)	C	2–3	Anderson *et al.* (1992)
	20.2 (19)	15.4 (14)	C	1	Ashworth *et al.* (1992)
	26.2 (23)	18.3 (20)	LW	3	Wilmut *et al.* (1992)
	16.8 (24)	14.7 (17)	C	1	Christenson (1993)
	27.8 (14)	20.7 (20)	LW	3	Galvin *et al.* (1993)
	22.7 (9)	16.3 (12)	C	2	White *et al.* (1991)
	19.2 (63)	15.1 (56)	LW	1	Haley *et al.* (1995)
	23.0 (33)	17.2 (28)	LW	2	Haley *et al.* (1995)
Conception rate (%)	93.0 (115)	85.2 (210)	LW	1–4	Després *et al.* (1992)
Number born alive	14.0 (19)	10.7 (50)	LW	1–3	Legault and Caritez (1983)
	13.6 (93)	10.3 (44)	LW	1–3	Bidanel *et al.* (1989)
	11.9 (21)	9.5 (61)	WS	1	Young (1990)
	14.3 (124)	10.3 (44)	LW	1–4	Després et al (1992)
	11.8 (21)	7.2 (20)	Y	1	White *et al.* (1991)
	12.6 (148)	9.6 (84)	LW	1–5	Bidanel (1993)
	13.2 (63)	10.0 (56)	LW	1	Haley *et al.* (1995)
Prenatal survival	15.0 (33)	9.8 (28)	LW	2	Haley *et al.* (1995)
(%)	77.9 (38)	67.1 (19)	LW		Bidanel *et al.* (1990a)
	65.0	68.8	LW	1	Haley *et al.* (1995)
Perinatal survival	58.3	55.9	LW	2	Haley *et al.* (1995)
(%)	96.0 (97)	90.8 (44)	LW	1–3	Bidanel *et al.* (1989)
	96.9 (148)	87.3 (84)	LW	1–5	Bidanel (1993)
Birth to weaning	91.6	94.1	LW	1–2	Haley *et al.* (1995)
survival (%)	88.7 (93)	93.9 (44)	LW	1–3	Bidanel *et al.* (1989)
	88.1 (148)	93.8 (84)	LW	1–5	Bidanel (1993)
Weaning to oestrus	81.6	84.6	LW	1–2	Lee and Haley (1995)
interval (days)	5.2 (87)	7.5 (141)	LW	2–4	Després *et al.* (1992)

[1]Average performance (number of records).
[2]LW = Large White; Y = Yorkshire; C = Crossbred; WS = White synthetic.

Meishan sows. Finally, Meishan sows have shorter weaning to oestrus intervals than Large White sows (Després *et al.*, 1992). More detailed results on the reproductive characteristics of the Meishan breed can be found in several reviews (e.g. Bidanel *et al.*, 1990b; Haley and Lee, 1993; Ashworth *et al.*, 1996).

Differences in reproductive performance between dual-purpose breeds are generally limited. However, Landrace gilts tend to reach puberty earlier (Christenson, 1981; Hutchens *et al.*, 1982; Allrich *et al.*, 1985; Bidanel *et al.*, 1996a) while having a slighly lower ovulation rate and a higher prenatal survival rate than Large White gilts (e.g. Bidanel *et al.*, 1996a). Paternal breeds generally have lower reproductive performance than dual-purpose breeds. Hampshire and Belgian Landrace sows show lower ovulations rates (–1.5 to –2 corpora lutea) and farrow about two piglets less than Large White sows, Duroc and Piétrain sows being intermediate (see the review of Blasco *et al.*, 1993). Paternal breeds also tend to have lower maternal abilities, as shown by higher preweaning mortality rates as compared with Large White or Landrace breeds (reviewed by Blasco *et al.*, 1995).

Crossbreeding

Pig producers have long known that crossbreeding is an effective means of improving reproductive performance. This improvement, called heterosis or hybrid vigour, comes from an increase in heterozygosity, which leads to better average genotypic values at dominant loci. As already mentioned, litter traits are controlled by the genes of both piglets and sows, and enhanced performance may come from crossed piglets (i.e. direct or individual heterosis effects) or crossed dams (i.e. sow or maternal heterosis effects). The plethora of crossbreeding experiments makes it difficult to describe all of them but several reviews have been published (Sellier, 1976; Johnson, 1980, 1981; Gunsett and Robison, 1990). An attempt to summarize available data excluding crosses with prolific Chinese breeds is presented in Table 11.3. In terms of sow heterosis, there is an average reduction in age at puberty of 11.3 days for crossbred sows when compared with purebreds. In addition, crossbred females have 2–4% higher conception rates, slightly larger ovulation rates (+0.5 ova) and 0.6 to 0.7 more piglets per litter at birth and 0.80 more piglets at weaning than purebreds. Postfarrowing survival of piglets is higher for crossbred sows (5%) and litter weights are greater (+1 kg at birth and +4.2 kg at 21 days). Litter heterosis effects lead to slightly larger litter size at birth (+0.24 piglet per litter) and to higher piglet survival (+5.8%) and litter weights.

Crossbred sires have been compared with purebred boars in several experiments (Wilson *et al.*, 1977; Neely *et al.*, 1980; reviewed in Buchanan, 1987). At a constant age, testis size and weight and total sperm are greater in crossbred than in purebred boars. Conception rates for first service or during extended breeding periods are higher (5–9%) for crossbred boars, and crossbred sires average 1.22 services per conception as compared with 1.41 services for purebreds (Johnson, 1981). Available results also suggest that crossbred boars have more libido and

Table 11.3. Average heterosis effects for reproductive traits in crosses between Western pig breeds.[1]

Trait	Heterosis value	Number of estimates
Dam heterosis		
Age at puberty (days)	−11.3	13
Ovulation rate	0.52	7
Conception rate (%)	3.0	9
Litter size		
at 30 days of gestation	0.73	3
at birth	0.66	11
at 21 days	0.66	9
at weaning	0.84	9
Embryonic or prenatal survival rate (%)	6.7	3
Birth to weaning survival rate (%)	5.0	3
Litter weight (kg)		
at birth	0.93	9
at 21 days	5.04	7
at 42 days	15.0	3
Litter heterosis		
Litter size		
at 30 days of gestation	0.39	4
at birth	0.24	47
at 21 days	0.30	31
at weaning	0.49	16
Embryonic or prenatal survival rate (%)	−1.1	5
Birth to weaning survival rate (%)	5.8	15
Litter weight (kg)		
at birth	0.59	33
at 21 days	2.47	29
at 42 days	13.35	12

[1]Updated from Sellier (1976), Johnson (1981) and Gunsett and Robison (1990).

are more aggressive than purebred boars (Wilson *et al.*, 1977; Neely and Robison, 1983). Theoretically, crossbred boars should have more variable progeny than purebred boars but experimental results do not confirm this hypothesis.

It should be noted that heterosis values may differ according to breed combinations. For instance, Large White × Landrace crosses generally exhibit lower heterosis values than other crosses between European or American breeds. Conversely, heterosis values in crosses between Large White and Meishan breeds are two- or threefold higher than in Large White × Landrace crosses. Heterosis for age at puberty is around 40–50 days (Legault and Caritez, 1983). Sow heterosis effects on litter size at birth and at weaning exceed two piglets per litter, so that Meishan × Large White sows farrow larger litters than purebred Meishan (Bidanel *et al.*, 1989; Bidanel, 1993; Haley *et al.*, 1995; Lee and Haley, 1995). Similar results have been obtained with Chinese Fengjing and Minzhu breeds (Young, 1995).

These crossbreeding results have been incorporated into crossbreeding programmes practised at the producer level. Comparisons of crossbreeding systems between European or American pig breeds have been conducted by several authors (e.g. Bennett *et al.*, 1983; McLaren *et al.*, 1987). More recently, breeding organizations or companies have developed synthetic lines by crossing breeds or lines known for high maternal performance such as the Meishan breed. These lines are then used as dam or more often grandam lines and crossed with dual-purpose breeds to produce F_1 sows which are sold for crossing with specific paternal genetic types (Bidanel, 1990; McLaren, 1990). Most individual pork producers practise crossbreeding through the use of specialized paternal and maternal genotypes. In some cases, they may also practise rotational or partial rotational crossbreeding programmes (reviewed in McLaren and Bovey, 1992).

Within-breed Genetic Variability

Components of genetic variation

Additive genetic variation is generally assessed by heritabilities. Estimates of heritability for several reproductive traits are summarized in Table 11.4. Regarding male traits, testes and accessory gland measurements have moderate to high heritabilities and are expected to respond easily to selection, while sperm characteristics, testosterone level and libido traits are slightly less heritable.

Female reproductive traits have low to moderate heritabilities. The most heritable traits are those depending solely on the genotype of the female, i.e. age at puberty, ovulation rate and weaning to oestrus interval. Conversely, litter size, conception and survival rates and, to a lesser extent, litter weight, which result from complex interactions between sow, boar and embryo or piglet genotypes, have low heritabilities and are therefore difficult to improve through selection. Several authors have quantified the relative importance of embryo and parental effects on genetic variation in litter traits. They have confirmed the prominent part of sow genotype, but have also shown that both progeny and boar genotypes significantly influence litter traits. The service sire has a rather limited effect on litter size (1–5% of phenotypic variance) according to Ollivier and Legault (1967), See *et al.* (1993), Beauvois *et al.* (1997), but taking into account this effect has been shown to improve genetic evaluation models for litter size (Woodward *et al.*, 1993; Beauvois, 1996). Similarly, low but non-negligible additive direct genetic effects on embryonic survival (4% of phenotypic variance) and litter weight at 21 days (6% of phenotypic variance) were obtained by Gama *et al.* (1991) and Rodriguez *et al.* (1994), respectively.

Some authors hypothesized that the preweaning environment provided by the female's dam, such as the size of the birth litter of the female, may have a significant effect on variation in litter traits and lead to underestimated heritability values (e.g. Vangen, 1980) and lower than expected responses to selection (Van der Steen, 1985; Roehe and Kennedy, 1993). As reviewed by Haley *et al.*

Table 11.4. Heritability estimates (h^2) for male and female reproductive traits in the pig.[1]

Trait	Number of estimates	Mean h^2	Range
Male traits			
Testis width	8	0.37	0.02–0.61
Testis length	6	0.33	0.30–0.39
Testis weight	5	0.44	0.24–0.73
Epididymis weight	4	0.33	0.15–0.55
Size of Cowper's gland	2	0.61	0.56–0.66
Sperm quantity	3	0.37	0.31–0.42
Sperm motility	3	0.17	0.13–0.20
Basal testosterone level	2	0.25	0.14–0.37
Libido	13	0.15	0.03–0.47
Female traits			
Age at puberty	13	0.33	0–0.64
Standing reflex	1	0.29	—
Intensity of vulvar symptoms	1	0.24	—
Ovulation rate	15	0.32	0.10–0.59
Prenatal survival rate	9	0.15	0–0.23
Total number born	85	0.11	0–0.76
Number born alive	96	0.09	0–0.66
Number weaned	42	0.07	0–1.0
Piglet survival to weaning	16	0.05	0–0.97
Litter weight at birth	10	0.29	0–0.54
Litter weight at 21 days	15	0.17	0.07–0.38
Weaning to oestrus interval	4	0.25	0.17–0.36
Rebreeding interval	3	0.23	0.03–0.36

[1]Updated from Lamberson (1990), McLaren and Bovey (1992) and Blasco *et al.* (1993, 1995).

(1988), the impact of birth fraternity size on heritability estimates for litter traits is generally very small. Recent estimations of the heritability of this maternal effect have provided controversial results. Mercer and Crump (1990), Haley and Lee (1992) and Pérez Enciso and Gianola (1992) found few or no maternal effects, while Southwood and Kennedy (1990), Ferraz and Johnson (1993), See *et al.* (1993) and Irgang *et al.* (1994) found significant estimates of maternal heritability ranging from 0.01 to 0.13. Differences between estimates may reflect sampling errors and genetic or management differences (e.g. crossfostering) between populations. However, as shown by Roehe and Kennedy (1993), ignoring maternal genetic effects which are negatively correlated with direct effects leads to reduced selection response as a result of negative maternal response and reduced direct response even when maternal heritability is low.

Suggestions that the genetic correlations between successive litters might be substantially less than one have also been put forward to explain lower than expected response to selection for litter size. In their review of early work on this topic, Haley *et al.* (1988) showed that genetic correlations between adjacent parities are high and considered that the lower estimates obtained between non-adjacent litters are likely to be biased downward due to culling. Recent studies using statistical methods accounting for selection bias have given controversial results. In the Yorkshire breed Irgang *et al.* (1994) and Roehe and Kennedy (1995) obtained genetic correlations between first and second parities ranging from 0.17 to 0.59 and recommend the use of a multiple-trait model. Conversely, estimates of genetic correlations between parities reported by Alfonso (1995), Roehe and Kennedy (1995) and Beauvois (1996) in the Landrace breed were close to unity.

Finally, crossbreeding studies indicate that non-additive genetic effects may be important for most reproductive traits. Though ignoring this variability may substantially bias estimates of additive genetic effects (Johansson *et al.*, 1994), there is, to our knowledge, no estimate of non-additive genetic parameters of reproductive traits in pigs.

Genetic correlations

Phenotypic and genetic correlations between male genital tract measurements are generally large (Legault *et al.*, 1979; Toelle *et al.*, 1984; Bonneau and Sellier, 1986). Testes measurements are also favourably related to total sperm or per cent spermatogenesis (Wilson *et al.*, 1977; Toelle *et al.*, 1984; Young *et al.*, 1986), as well as to basal or induced LH and testosterone levels (Bates *et al.*, 1986; Lubritz *et al.*, 1991). A number of researchers have examined the interest of male traits as indirect selection criteria to improve female reproductive performance. Estimates of genetic correlations between testes measurements and age at first oestrus, ovulation rate or litter size are generally low and do not show any consistent trend (Schinckel *et al.*, 1983; Toelle and Robison, 1985; Bates *et al.*, 1986; Benoit, 1986; Young *et al.*, 1986; Sellier and Bonneau, 1988; Johnson *et al.*, 1994).

Phenotypic and genetic correlations between several female traits are shown in Table 11.5. Age at puberty exhibits negative, i.e. favourable, genetic correlations with ovulation rate and number of embryos (Young *et al.*, 1978; Bidanel *et al.*, 1996a). Conversely, both negative and positive estimates of genetic correlations with litter size have been reported (Young *et al.*, 1978; Rydhmer *et al.*, 1992). Genetic parameters for litter size at birth and its components, i.e. ovulation rate and embryo or fetal survival, have recently been reviewed by Blasco *et al.* (1993). Ovulation rate and prenatal survival show a moderate negative correlation. Litter size at birth appears to be more closely related to prenatal survival than to ovulation rate. Measurements of litter size at birth (total number and number born alive) and at weaning (number weaned) and litter weight exhibit large positive genetic correlations (reviewed by Blasco *et al.*, 1995). However, they appear to be unfavourably correlated with stillbirth or preweaning mortality rates (Blasco *et al.*, 1995).

Numerous authors have estimated genetic correlations of male or female reproductive traits with growth and carcass traits. Testes measurements show favourable genetic relationships with growth traits when measured at a constant

Table 11.5. Means[1] of literature estimates of genetic and phenotypic correlations[2] among reproductive traits.

	AP[3]	OR	NB	NBA	S	NW	LBW	L21W
AP		–0.08	0.04	0.04	0.08	0.10	–0.15	–0.11
OR	0.06		0.08	0.25	–0.38	0.10	0.17	0.09
NB	–0.02	0.12		0.91	–0.11	0.73	0.62	0.45
NBA	–0.03	0.13	0.92		0.16	0.81	0.65	0.61
S	0.11	–0.11	–0.15	–0.05		0.53	0.13	0.80
NW	0.01	0.05	0.71	0.79	0.59		0.70	0.87
LBW	–0.02	0.09	0.79	0.82	0.09	0.71		0.68
L21W	–0.03	0.02	0.44	0.53	0.60	0.80	0.61	

[1]Updated from Lamberson (1990) and Blasco *et al.* (1993, 1995).
[2]Genetic correlations above the diagonal, phenotypic correlation below.
[3]AP = age at puberty; OR = ovulation rate; NB = number born; NBA = number born alive; S = preweaning survival rate; NW = number weaned; LBW = litter weight at birth; L21W = 21-day litter weight.

age (Toelle *et al.*, 1984; Young *et al.*, 1986; Lubritz *et al.*, 1991; Johnson *et al.*, 1994), but relationships are less clear when measurements occur at a constant weight (Benoit, 1986; Young *et al.*, 1986). Estimates of genetic correlations with backfat thickness are generally low and have a varying sign (Toelle *et al.*, 1984; Young *et al.*, 1986; Johnson *et al.*, 1994). Growth traits also appear to be favourably associated with testosterone levels in the study of Lubritz *et al.* (1991).

Similarly, age of gilts at puberty exhibits negative, i.e. favourable, genetic correlations with growth rate (Reutzel and Sumption, 1968; Young *et al.*, 1978; Hutchens *et al.*, 1981; Rydhmer *et al.*, 1992; Bidanel *et al.*, 1996a), while both negative (Rydhmer *et al.*, 1992; Bidanel *et al.*, 1996a) and null or positive (Young *et al.*, 1978; Hutchens *et al.*, 1981; Hixon *et al.*, 1987) relationships with backfat thickness have been reported. Litter size or weights and growth or carcass traits are weakly correlated (reviewed by Brien, 1986 and Haley *et al.*, 1988). Recent estimates of genetic correlations between growth/carcass and litter traits (Short *et al.*, 1994; Rydhmer *et al.*, 1995; Ducos and Bidanel, 1996; Kerr and Cameron, 1996) generally agree with previous estimates, although some significant estimates were obtained by Ducos and Bidanel (1996) and Kerr and Cameron (1996). Growth rate and to some extent carcass lean content might also be unfavourably correlated with intensity of oestrous symptoms, as reported by Rydhmer *et al.* (1994).

Very few estimates of genetic correlations between reproductive and meat quality traits are available in the literature. Most studies concern the genetic relationships between male sexual development and fat androstenone level, which is a major compound responsible for boar taint. A large genetic correlation (0.68 ± 0.05) between the size of bulbo-urethral glands and fat androstenone level in a Large White × Landrace population was reported by Fouilloux *et al.* (1997). Results reported by Willeke *et al.* (1987) and Sellier and

Bonneau (1988) suggest that low fat androstenone level in young boars and sexual precocity in gilts and young boars are genetically antagonistic. The few available estimates of the genetic relationships between reproduction and the meat technological quality are still inconclusive. Litter size at birth was found by Hermesch *et al.* (1995) to have insignificant relationships with pH or drip loss, but negative genetic correlations with meat colour (-0.50 ± 0.17 and -0.53 ± 0.24, respectively for second and third parity litter size). Similarly, Larzul (1997) reported non-significant genetic correlations between muscle glycolytic potential (GP) and litter size or weight, but a negative genetic relationship between GP and age at puberty.

Selection experiments

Several selection experiments dealing with various reproductive traits have been conducted over the last 30 years in pigs. Most experiments attempted, directly or indirectly, to improve litter size. There have been several selection experiments for directly increasing litter size (Ollivier and Bolet, 1981; Bolet *et al.*, 1989; Lamberson *et al.*, 1991). Most of these experiments produced little or no significant response. Ollivier and Bolet (1981) conducted a selection experiment based on average litter size in the first two parities. After 11 generations of selection in a closed line, total response was only 0.26 piglets and not significant (Bolet *et al.*, 1989). However, five subsequent generations of selection within the same line – with 12.5% immigration from the hyperprolific stock described below – yielded a significant genetic gain (Bolet *et al.*, 1987). Lamberson *et al.* (1991) conducted a selection for litter size for eight generations following previous selection for high ovulation rates. Realized heritability was 0.15 ± 0.05 and response after eight generations was estimated to be 0.48 to 1.06 pigs, depending on the method of analysis. Rutledge (1980) also attempted to improve litter size by correcting for fraternity size but response was not significant. McLaren and Bovey (1992), in reviewing these experiments, suggested that the failure or limitations of these experiments was due to several reasons including population size, management problems, maternal effects, inbreeding depression and within-family selection.

Hyperprolific selection is another way of increasing litter size. Such a scheme, which implies extremely intense selection of sows on several litters, combined with backcrossing of their sons to sows of similar high prolificacy, was initiated in France 20 years ago (Legault and Gruand, 1976). Results of the French hyperprolific Large White strain after 20 years showed a genetic superiority of 1.4 pigs/litter (born alive) compared with normal contemporary Large White sows (Bidanel *et al.*, 1994). Hyperprolific sows also had larger ovulation rates, lower age at puberty and FSH concentration at 150 days of age and increased follicle oestradiol concentrations during the follicular phase of the oestrous cycle (Després *et al.*, 1992; Driancourt *et al.*, 1992; Driancourt and Terqui, 1996). Other countries and pig breeding companies have practised hyperprolific selection with positive results (e.g. Sorensen and Vernesen, 1991).

Using total pigs born in five litters, the Pig Improvement Company selected the top 1.7% of their sows and responses to hyperprolific selection experiments have been nearly as high as in the French experience (D.G. McLaren, PIC USA, personal communication).

While heritabilities for litter size are relatively low (around 0.10), heritabilities for ovulation rate and embryo survival appear to be higher (Table 11.4). Such estimates have encouraged researchers to consider selection based on components of litter size (ovulation rate, embryo survival and uterine capacity). Initial experiments at the University of Nebraska dealing with ovulation rate gave a direct response of 3.7 ova (Cunningham *et al.*, 1979) and an indirect response of 0.8 pigs per litter (Lamberson *et al.*, 1991) after nine generations of selection. Following 11 generations of relaxed selection an advantage of 0.74 pigs per litter was maintained (Lamberson *et al.*, 1991). Increased ovulation rate was associated with a faster increase in FSH concentration and a higher FSH peak (Kelly *et al.*, 1988). Bidanel *et al.* (1996b) selected two Large White lines for either increased ovulation rate or increased prenatal survival. After four generations of selection, ovulation rate had increased by 0.6 ova/generation in the high ovulation rate line, but without any correlated response on litter size. Conversely, no significant genetic trends were obtained in the line selected for prenatal survival.

Johnson *et al.* (1984) proposed that index selection for both ovulation rate and embryo survival would be more effective in increasing litter size. Early response over five generations of selection was 0.19 pigs/litter/generation (Neal *et al.*, 1989) and more recent results at generation ten show a difference between the control and select lines of 6.6 ova, 3.3 fetuses at day 50 of gestation and 1.5 pigs born alive (Casey *et al.*, 1994). Selection for ovulation rate and embryo survival resulted in differences in the pattern of oestradiol secretion in young males before puberty and in enhanced FSH secretion in mature boars (Mariscal *et al.*, 1996).

A model in which litter size is the minimum number of viable embryos (a function of ovulation rate) or the minimum allowed by uterine space has been proposed (Bennett and Leymaster, 1989). This model was based on results dealing with use of unilateral hysterectomy–ovariectomy by means of a surgical method to measure uterine capacity. Simulation has been used to address the potential of this approach (Bennett and Leymaster, 1990; Perez-Enciso *et al.*, 1996) and recent experimental results suggest its merit (Leymaster and Bennett, 1994).

In an effort to directly measure the relationship between female and male reproductive measures, selection for increased testis weight at 150 days of age, as predicted by *in vivo* size measurements, was practised for ten generations in a composite Large White/Landrace line (Johnson *et al.*, 1994). Direct response to selection for testes weight was an increase of 19 grams ($P < 0.01$) and realized heritability was 0.35 ± 0.02. Boars from the selected line had larger epididymis weights (Harder *et al.*, 1995) as well as a higher sperm concentration in the semen and a larger sperm production per gram of parenchymal tissue (Huang and Johnson, 1996). Daily sperm production and sperm epididymal storage also increased more rapidly at younger ages in the selected than in the control line (Rathje *et al.*, 1995). Age at puberty decreased by 6 days, but not significantly, in the line selected for higher testicular development when compared with the control line. Ovulation rate also increased by 0.76 ± 0.43 ova in the females of the testicular selection line (Johnson *et al.*, 1994). Johnson and co-workers concluded that testis weight might be used as a selection criterion for improving semen characteristics of AI boars, but should not be used as an indicator trait for genetically improving female reproductive performance.

A significant direct response was obtained by Hixon *et al.* (1987) after one generation of divergent selection for age at puberty. Results from another selection experiment on age at puberty were reported by Lamberson *et al.* (1991). The line involved was selected for decreased age at puberty for eight generations following selection for increased ovulation rate. Age at puberty decreased by about 2 days/generation and realized heritability was 0.25 ± 0.05. Age at puberty was not associated with increased litter size in this selection line.

More recently, selection on circulating levels of testosterone was considered. Robison *et al.* (1994) initiated a divergent selection experiment on gonadotrophin-releasing hormone challenge. Pre-challenge and post-challenge levels of testosterone in the high line were three times those of the low line after ten generations of selection. Heritabilities for pre- and post-challenge testosterone levels were moderate. Prolificacy of the high line females was significantly larger than that of the low line females.

A selection experiment to reduce the interval from weaning to oestrus (IWE) has been practised for eight generations in The Netherlands. Realized heritability was estimated to be 0.17 (ten Napel *et al.*, 1995a). However, the experiment also provided some indication of a genetic antagonism between IWE and litter traits (ten Napel, The Netherlands, personal communication). Ten Napel *et al.* (1995b) divided IWE into the interval between weaning and the start of cyclic activity, the interval between the start of cyclic activity and oestrus, the incidence of silent oestrus and the cycle length. They concluded that genetic variation in IWE is mainly due to genetic variation in the interval of weaning to the start of cyclic activity.

Inbreeding

Inbreeding occurs when related animals are mated and is quantified by a coefficient which measures the probability that the two genes at any locus in an individual are identical by descent (i.e. descend from the same allele carried by a particular ancestor). The inbreeding coefficient F ranges from 0 (completely outbred) to 100% (completely inbred). Inbreeding was first used by breeders to help fix specific genetic characteristics in an effort to help develop breeds. In the United States, inbred lines of pigs were created in the early 1930s for further use in crossbreeding (Craft, 1958). These lines mirrored the extremely successful results obtained in the hybrid corn business. In pigs, inbred lines suffered from much lower fertility, lower piglet survival rates and some reduction in general performance (Craft, 1958), so that this method of improvement has been abandoned. The genetic effect of inbreeding is to increase homozygosity, which is undesirable for three reasons. First, it causes a loss in genetic variation and hence reduces the potential rate of genetic progress. Then it increases the frequency of genetic abnormalities by increasing the number of animals homozygous for recessive deleterious alleles which had been previously hidden in the population. Finally, the reduced proportion of heterozygous individuals will result in lower average genotypic value at dominant loci and hence will cause a decrease of the performance levels. This decrease is called inbreeding depression and is generally larger for the least heritable traits and increases with additional amounts of inbreeding. Inbreeding occurs in any population of finite size and accumulates more rapidly in smaller and in selected populations.

Rates of inbreeding depression for some reproductive traits are presented in Table 11.6. As litter traits depend on dam and offspring genotypes, the effects of inbreeding should be considered at both levels too. Though estimates are not numerous, dam inbreeding seems to strongly reduce ovulation rate and prenatal survival. Litter inbreeding causes an additional reduction of 0.60 embryos per 10% increase in *F* at 25 days of gestation. Estimated decrease in total number of piglets born from a 10% increase in dam inbreeding coefficient is 0.40. The greatest effects of inbreeding are seen in reduced survivability of piglets. Estimates range from a decrease of 0.30 to 0.50 piglets for each 10% increase in litter inbreeding with an additional decrease of 0.20 to 0.40 piglets for each 10% increase in dam inbreeding. In the male, the effects of inbreeding are reduction in sperm numbers and sexual aggressiveness or libido.

Table 11.6. Effect of 10% increase in inbreeding coefficient of litter and dam on reproductive performance.

	Litter			Dam		
Trait	N[1]	Mean	Range	N	Mean	Range
Ovulation rate				2	−1.13	−1.13 to −0.55
Embryo number at 25 days	1	−0.60		2	−1.75	−2.7 to −0.8
Embryo survival to 25 days (%)	1	−3.30		2	−5.62	−10.9 to −0.33
Litter size						
Total number born	15	−0.29	−2.53 to 0.13	12	−0.40	−1.29 to 0
Number of stillborn	2	−0.27	−0.37 to −0.16	2	−0.19	−0.28 to −0.10
Number born alive	6	0.01	−0.20 to 0.17	4	−0.30	−0.63 to 0
Number at 21 days	8	−0.53	−2.78 to 0.20	7	−0.22	−0.43 to 0.03
Litter weight (kg)						
at birth	10	−0.15	−0.64 to 0.20	7	−0.46	−1.12 to −0.07
at 21 days	6	−1.72	−3.35 to 0.10	5	−2.18	−4.33 to −0.91

Source: Johnson, 1990.
[1]N = number of estimates.

Effects of Individual Genes

The limited genetic improvement made by selection and crossbreeding for female and male reproductive traits has encouraged a search for single genes affecting reproduction. Recent developments in segregation analysis and in the area of gene mapping and molecular genetics have now made it possible to search for major genes and quantitative trait loci (QTL) and to study candidate genes which may control reproductive traits.

Early research in this area centred on blood groups and protein polymorphisms and their association primarily with litter size and on the estimation of potential pleiotropic effects of known major genes. First evidence for such associations was provided by Jensen *et al.* (1968) and Rasmusen and Hagen (1973), who reported an association between the *H* locus, located on pig chromosome 6, and litter size, with an unfavourable apparent effect of the H^a allele and by Kristjansson (1964) and Imlah (1970) who reported an apparent effect of alleles at the transferrin locus, located on pig chromosome 13, on pig fertility and prolificacy. However, this transferrin locus effect was not confirmed by other authors (e.g. Jensen *et al.*, 1968; Huang and Rasmusen, 1982). Several other associations between blood group and protein loci and reproductive traits have been investigated, but they have often led to contradictory results (reviewed by Ollivier and Sellier, 1982).

During the 1980s, considerable efforts were made to investigate the role of the pig major histocompatibility complex (*MHC*), called the swine leucocyte antigen (*SLA*) complex on male and female reproductive traits (reviewed in Warner and Rothschild, 1991). The pig *MHC* is a large set of genes located on chromosome 7 (Warner and Rothschild, 1991). Certain *MHC* genotypes have been associated with increased or decreased testicular size and hormone differences and 1.5–5% of the phenotypic variation in these traits was explained by the pig *MHC* (Rothschild *et al.*, 1986b). In the female, several traits appear to be associated with the *SLA* polymorphism. Several reports link the *MHC* to ovulation rate (Rothschild *et al.*, 1984; Conley *et al.*, 1988), litter size, number born alive and number weaned (reviewed in Vaiman *et al.*, 1988; Warner and Rothschild, 1991). There is some evidence suggesting that *MHC* homozygosity of the embryo may be a disadvantage (reviewed in Vaiman *et al.*, 1988). Researchers have also examined whether the pig has a *MHC* gene associated with embryo development and its relationship to litter size. Results from miniature pigs (Ford *et al.*, 1988) suggest that such a gene may exist within the pig MHC. Other reports (Rothschild *et al.*, 1986a; Vaiman *et al.*, 1988; Warner and Rothschild, 1991) indicate that the *MHC* is associated with birth and weaning weights. Whether these effects are direct effects of genes within the *MHC* such as 21-hydroxylase, or are due to linkages with other genes outside the complex is unknown. Further cloning and identification of individual genes on chromosome 7 should help to answer that question.

Recent discoveries such as the *FEC* (fecundity) gene marker in sheep have encouraged the search of individual genes affecting pig reproductive traits. Of initial interest was the investigation of why some Chinese breeds of pigs, like the

Meishan, are so prolific. In 1991, Rothschild and colleagues began a candidate gene investigation of the role of the oestrogen receptor (*ESR*) gene in controlling litter size. Initial results showed that one *ESR* polymorphism found initially in the Meishan and later in the Large White breed (Rothschild *et al*., 1994) was associated with improved litter size in a Meishan × Large White composite line. More recent results (Rothschild *et al*., 1995, 1996) demonstrate that *ESR* is either a major gene or very closely linked to a major gene for litter size. In the above mentioned line, the favourable *B* allele is associated with a first-parity additive effect of +1.15 pigs/litter for each copy of the allele. In second and later parities, the effect of the *B* allele is about +0.5 pigs/litter and appears to act in a dominant manner. The *B* allele is also segregating in several Large White populations (Rothschild *et al*., 1995, 1996; Legault *et al*., 1996), and is approximately +0.4 pigs per litter in first parity and +0.3 pigs in later parities (Short *et al*., 1997) but the effect of the *B* allele seems to differ between populations. These differences may indicate either that the mutation used in the *ESR* test is only a linked marker gene or that differences in the genetic background have an impact on the expression of the *ESR* locus. In any case, the underlying mechanisms of the favourable *ESR* allele are still unknown but it has been hypothesized that *ESR* may affect embryo survival.

Progress achieved in the pig genetic map during the last 5 years (see Chapters 8 and 9) now gives the opportunity to begin the systematic search for loci affecting quantitative traits of economic importance. First results dealing with marker gene effects on reproductive traits using this systematic approach indicate associations between the microsatellite marker *Sw444* region on chromosome 8 and ovulation rate or uterine length in a cross between Meishan and Large White breeds (Wilkie *et al*., 1996). Other associations involving one chromosome 6 region, which seems to differ from the *H* blood group locus region, and number born per litter, as well as regions of chromosomes 4 and 7 and number of stillborns, were suggested by Wilkie *et al*. (1996). A QTL for ovulation rate was also found on chromosome 8 in a cross between a line selected for ovulation rate and a control line (Rathje *et al*., 1996).

Several genes with major effects on economically important traits have been evidenced in pigs. The most widely studied gene is the skeletal muscle ryanodine receptor or halothane sensitivity (*HAL*) locus, which has major effects on several carcass and meat quality traits in pigs (see Chapter 14). Various but fairly inconsistent effects of the *HAL* locus on male and female reproductive performance have been reported. Schlenker *et al*. (1984) found a smaller ejaculate volume and a lower number of sperm for halothane-negative (HN) as compared with halothane-positive (HP) boars while an opposite conclusion was reached by Hillbrand and Glodek (1984). Pfeiffer *et al*. (1986) reported a better semen quality in HN than in HP boars. Schneider *et al*. (1980) in the Swiss Landrace, Baulain and Glodek (1987) in the German Landrace and Sellier *et al*.(1987) in the Piétrain found a favourable sire effect of HP boars on litter size at birth, while Lampo *et al*. (1985) in the Belgian Landrace breed and Sellier *et al*. (1987) in the Piétrain × Large White cross did not find any noticeable difference. With regard to fertility traits, HP sows appear to be similar or even slightly superior to HN sows in the ability to become pregnant (Van der Steen, 1983; Simpson *et al.*, 1986; Baulain and Glodek, 1987; Sellier *et al.*, 1987). A significant advantage of HN over HP sows for number born per litter was reported by Schneider *et al.* (1980) in the Swiss Landrace (+0.55 piglet born alive), by Van der Steen (1983) in the Dutch Landrace (+1.3 piglet born/litter) and by Carden *et al.* (1985) in Piétrain–Hampshire composite lines (+1.20 ± 0.4 piglet born/litter). Several studies on the German Landrace breed (e.g. Willeke *et al.*, 1984; Grosse-Lembeck and Kalm, 1985; Baulain and Glodek, 1987) also showed a slightly, but non-significantly better prolificacy of HN sows as compared with HP sows. Conversely, Simpson *et al.* (1986) and Sellier *et al.* (1987) did not find any difference in litter size between halothane phenotypes. As suggested by Sellier *et al.* (1987), the differences observed between studies tend to indicate that the halothane locus has no direct effect on reproductive performance, but may be in linkage desequilibrium with the H blood group chromosomal region in some populations. Other major genes such as the *RN* gene (Le Roy *et al.*, 1990a), the *MU* gene (Le Roy *et al.*, 1990b), the *IMF* gene (Janss *et al.*, 1997) or a gene with a major effect on the size of bulbo-urethral glands (Fouilloux *et al.*, 1997) have recently been evidenced in pigs, but their effect on reproductive traits has not been investigated so far.

Conclusions and Implications

Large genetic differences for reproductive traits exist both among and within pig breeds. Between-breed variations have been widely exploited over the last decades through breed specialization and the generalization of crossbreeding. Conversely, little had been done until recently to profit from the within-breed variability. Selection plans have mainly been aimed at improving production traits and generally have neglected the least heritable reproductive traits. Things have begun to change over the last 10 years due to the combination of several factors. The economic interest in reducing backfat thickness is now limited in many countries, whereas much can be gained from improved sow and boar reproductive performance. Experimental results have shown that the least heritable traits such as litter size could be successfully selected for in certain circumstances. The use of powerful across-herd genetic evaluation techniques based on Best Linear Unbiased Prediction methodology (e.g. Henderson, 1984) have given geneticists the opportunity to substantially increase the efficiency of selection of these least heritable traits. As a consequence, litter size has become a major component of selection goals in maternal lines of pigs and annual genetic trends of +0.1–0.3 piglet/litter have been obtained in some pig populations over the last few years.

Further gains in prolificacy, but also in sexual maturity and mothering abilities, can be expected in the near future from the increasing use of prolific Chinese breeds or of synthetic lines developed using these prolific breeds in crossbreeding plans. The use of genetic markers should also contribute to more efficient genetic improvement of reproductive traits, which are sex limited and often have a late expression in life. Genes like ESR and genetic markers can be used for marker-assisted introgresssion of favourable genes affecting reproduction from Chinese prolific breeds into commonly used maternal geneoypes, for removing unfavourable alleles for fatness in Chinese × Western synthetic lines or for marker assisted selection within populations. However, as discussed by Visscher and Haley (1995), the use of genetic markers is associated with potential extra gains, but also with extra costs and risks due to poor estimates of QTL position and effects or detection of spurious QTLs. Further research to develop high density maps, to identify the genes responsible for the observed variations and to study gene effects on economically important traits and their underlying physiological processes will be very useful to solve these problems.

Other reproductive traits are likely to have an increasing importance in future genetic improvement programmes. The increasing number of piglets per litter and the regulations against early weaning of piglets should enhance the importance of mothering abilities. They can be measured through number of piglets weaned or preweaning survival rate, although some bias may arise from piglet exchange across litters. They can also be characterized through component traits such as behavioural traits, milk production and associated traits such as sow feed consumption during lactation. Other reproductive traits, such as boar semen quality and quantity, boar and sow longevity, might also be worth considering.

Acknowledgements

The authors would like to thank Louis Ollivier, Pierre Sellier and the anonymous referees for their useful comments and suggestions.

References

Alfonso, L.A. (1995) [Genetic variability and selection of prolificacy in pigs (in Spanish)]. Doctoral thesis, University of Lleida, Spain.

Allrich, R.D., Christenson, R.K. and Ford, J.J. (1985) Age at puberty and estrous activity of straightbred and reciprocal crossed gilts. *Animal Reproduction Science* 8, 281–286.

Anderson, L.H., Christenson, L.K., Christenson, R.K. and Ford, S.P. (1992) Investigations into the control of litter size: comparisons of Chinese (Meishan) and American pigs. In: Rusheng, C. (ed.) *Proceedings of the International Symposium on Chinese Pig Breeds.* Northeast Forestry University Press, Harbin, China, pp. 86–90.

Ashworth, C.J., Haley, C.S., Aitken, R.P. and Wilmut, I. (1990) Embryo survival and fetal growth following reciprocal embryo transfer between Chinese Meishan and Large White gilts. *Journal of Reproduction and Fertility* 90, 595–603.

Ashworth, C.J., Haley, C.S. and Wilmut, I. (1992) Effect of regumate on ovulation rate, embryo survival and conceptus growth in Meishan and Landrace × Large White gilts. *Theriogenology* 37, 433–443.

Ashworth, C.J., Pickard, A.R. and Haley, C.S. (1996) Comparative aspects of embryo survival between European and Chinese breeds. *Pig News and Information* 17, 69N–73N.

Backström, L. and Henricson, B. (1971) Intersexuality in the pig. *Acta Veterinaria Scandinavica* 12, 257–273.

Bates, R.O., Buchanan D.S., Johnson, R.K., Wettemann, R.P., Fent, R.W. and Hutchens, L.K. (1986) Genetic parameter estimates for reproductive traits of male and female littermate swine. *Journal of Animal Science* 63, 377–385.

Baulain, U. and Glodek, P. (1987) Beziehungen zwischen Halothanereaktion and Zuchtleistung bei Sauen verschiedener Populationen. *Züchtungskunde* 59, 122–134.

Beauvois, E. (1996) [Studies on genetic and non genetic factors of variation of litter size in pigs (in French)]. Mémoire de fin d'études, Ecole Nationale d'Ingénieurs des Travaux Agricoles de Bordeaux, France, 64 pp.

Beauvois, E., Labroue, F. and Bidanel, J.P. (1997) [Studies on factors of variation of litter size at birth in Large White and French Landrace pig breeds. Consequences on genetic evaluation for prolificacy (in French)]. *Journées de la Recherche Porcine en France* 29, 353–360.

Bennett, G.L. and Leymaster, K.A. (1989) Integration of ovulation rate, potential embryonic survival and uterine capacity into a model of litter size in swine. *Journal of Animal Science* 67, 1230–1241.

Bennett, G.L. and Leymaster, K.A. (1990) Genetic implications of a simulation model of litter size based on ovulation rate, potential embryonic viability and uterine capacity: simulated selection. *Journal of Animal Science* 68, 980–986.

Bennett, G.L., Tess, W.M., Dickerson, G.E. and Johnson, R.K. (1983) Simulation of breed and crossbreeding effects on costs of pork production. *Journal of Animal Science* 56, 801–813.

Benoit, C. (1986) [Testicular measurements as a selection criterion for genetic improvement of reproduction in farm animals, with an application to pigs (in French)]. Mémoire de Diplôme d'études Approfondies – Université Paris VI, Orsay, France.

Bidanel, J.P. (1990) Potential use of prolific Chinese breeds in maternal lines of pigs. In: Hill, W.G., Thompson, R. and Wooliams, J.A. (eds), *Proceedings of the 4th World Congress on Genetics Applied to Livestock Production*, vol. XV, pp. 481–484.

Bidanel, J.P. (1993) Estimation of crossbreeding parameters between Large White and Meishan porcine breeds. III. Dominance and epistatic components of heterosis on reproductive traits. *Genetics, Selection, Evolution* 25, 263–281.

Bidanel, J.P., Caritez, J.C. and Legault, C. (1989) Estimation of crossbreeding parameters between Large White and Meishan porcine breeds. I. Reproductive performance. *Genetics, Selection, Evolution* 21, 507–526.

Bidanel, J.P., Caritez, J.C. and Lagant, H. (1990a) Ovulation rate and embryo survival in gilts and sows with variable proportions of Meishan and Large White genes. In: Molénat, M. and Legault, C. (eds), *Chinese Pig Symposium.* INRA, Jouy-en-Josas, France, pp. 109–110 (abstract).

Bidanel, J.P., Caritez, J.C. and Legault, C. (1990b) Ten years of experiments with Chinese pigs in France. 1. Breed evaluation. *Pig News and Information* 11, 345–348.

Bidanel, J.P., Gruand, J. and Legault, C. (1994) An overview of twenty years of selection for litter size in pigs using 'Hyperprolific' schemes. In: Smith, C., Gavora, J.S., Benkel, B., Chesnais, J., Fairfull, W., Gibson, J.P., Kennedy, B.W. and Burnside, E.B. (eds) *Proceedings of the 5th World Congress on Genetics Applied to Livestock Production*, vol. 17, pp. 512–515.

Bidanel, J.P., Gruand, J. and Legault, C. (1996a) Genetic variability of age and weight at puberty, ovulation rate and embryo survival in gilts and relations with production traits. *Genetics, Selection, Evolution* 28, 103–115.

Bidanel, J.P., Blasco, A., Dando, P., Gogué, J. and Lagant, H. (1996b) [Results of four generation of selection for ovulation rate and prenatal survival in Large White pigs (in French)] *Journées de la Recherche Porcine en France* 28, 1–8.

Blasco, A., Bidanel, J.P., Bolet, G., Haley, C.S. and Santacreu, M.A. (1993) The genetics of prenatal survival of pigs and rabbits: a review. *Livestock Production Science* 37, 1–21.

Blasco, A., Bidanel, J.P. and Haley, C.S. (1995) Genetics and neonatal survival. In: Varley, M.A. (ed.) *The Neonatal Pig. Development and Survival.* CAB International, Wallingford, Oxon, UK, pp. 17–38.

Bolet, G., Martinat-Botté, F., Locatelli, P., Gruand, J., Terqui, M. and Berthelot, F. (1986) Components of prolificacy of hyperprolific Large-White sows. Comparison with Meishan and control Large-White sows. *Génétique Sélection Evolution* 18, 333–342.

Bolet, G., Renard, C., Ollivier, L. and Dando, P. (1987) [Selection for prolificacy in pigs: response to selection in an open line (in French)]. *Journées de la Recherche Porcine en France* 19, 47–54.

Bolet, G., Ollivier, L. and Dando, P. (1989) [Selection for prolificacy in pigs: response to an eleven generation selection experiment (in French)]. *Genetics, Selection, Evolution* 21, 93–106.

Bonneau, M. and Sellier, P. (1986) Fat androstenone content and development of genital system in young Large White boars: genetic aspects. *World Review of Animal Production* 22, 27–30.

Brien, F.D. (1986) A review of the genetic and physiological relationships between growth and reproduction in mammals. *Animal Breeding Abstracts* 54, 975–997.

Buchanan, D.S. (1987) The crossbred sire: experimental results for swine. *Journal of Animal Science* 65, 117–127.

Carden, A.E., Hill, W.G. and Webb, A.J. (1985) The effects of halothane susceptibility on some economically important traits in pigs. 1. Litter productivity. *Animal Production* 40, 351–358.

Casey, D., Rathje, T.A. and Johnson, R.K. (1994) Second thoughts on selection for components of reproduction in swine. In: Smith, C., Gavora, J.S., Benkel, B., Chesnais, J., Fairfull, W., Gibson, J.P., Kennedy, B.W. and Burnside, E.B. (eds) *Proceedings of the 5th World Congress on Genetics Applied to Livestock Production*, vol. 17, pp. 315–318.

Christenson, R.K. (1981) Influence of confinement and season of the year on puberty and estrous activity of gilts. *Journal of Animal Science* 52, 821–830.

Christenson, R.K. (1993) Ovulation rate and embryonic survival in Chinese Meishan and White crossbred pigs. *Journal of Animal Science* 71, 3060–3066.

Cole, D.J.A. and Foxcroft, G.R. (1982) *Control of Pig Reproduction.* Butterworths, London.

Conley, A.J., Jung, Y.C., Schwartz, N.K., Warner, C.M., Rothschild, M.F. and Ford, S.P. (1988) Influence of SLA haplotype on ovulation rate and litter size in miniature pigs. *Journal of Reproduction and Fertility* 22, 595–601.

Craft, W.A. (1958) Fifty years of progress in swine breeding. *Journal of Animal Science* 17, 960–980.

Cunningham, P.J., England, M.E., Young, L.D. and Zimmerman, D.R. (1979) Selection for ovulation rate in swine: Correlated response in litter size and weight. *Journal of Animal Science* 48, 509–516.

Després, P., Martinat-Botté, F., Lagant, H., Terqui, M. and Legault, C. (1992) Comparison of reproductive performance of three genetic types of sows: Large White (LW), 'hyperprolific Large White' (LWH), Meishan (MS) (in French). *Journées de la Recherche Porcine en France* 24, 25–30.

De Vries, A.G. (1989) A model to estimate economic value of traits in pig breeding. *Livestock Production Science* 21, 49–66.

Driancourt, M.A. and Terqui, M. (1996) Follicular growth and maturation in hyperprolific and Large White sows. *Journal of Animal Science* 74, 2231–2238.

Driancourt, M.A., Prunier, A., Huyghe, J.M., Bidanel, J.P. and Martinat-Botté, F. (1992) hCG induced oestrus and ovulation rate and FSH concentrations in prepuberal gilts from lines differing by their adult ovulation rate. *Animal Reproduction Science* 29, 297–305.

Ducos, A. (1994) [Genetic evaluation of pigs tested in central stations using a multiple trait animal model (in French)]. Doctoral thesis, Institut National Agronomique Paris-Grignon, France.

Ducos, A. and Bidanel, J.P. (1996) Genetic correlations between production and reproductive traits measured on-farm, in the Large-White and French Landrace breeds. *Journal of Animal Breeding and Genetics* 113, 493–504.

Ducos, A., Berland, H., Pinton, A., Séguéla, A., Blanc, A., Darré, A. and Darré, R. (1997) [Reciprocal translocations in pigs: current state of knowledge and prospects (in French)]. *Journées de la Recherche Porcine en France* 29, 375–382.

Fechheimer, N.S. and Beatty, R.A. (1974) Chromosomal abnormalities and sex ratio in rabbit blastocysts. *Journal of Reproduction and Fertility* 37, 331–341.

Ferraz, J.B.S. and Johnson, R.K. (1993) Animal model estimation of genetic parameters and response to selection for litter size and weight, growth, and backfat in closed seedstock populations of Large White and Landrace swine. *Journal of Animal Science* 71, 850–858.

Ford, S.P., Schwartz, N.K., Rothschild, M.F., Conley, A.J. and Warner, C.M. (1988) Influence of SLA haplotype on preimplantation embryonic cell number in miniature pigs. *Journal of Reproduction and Fertility* 84, 99–104.

Fouilloux, M.N., Le Roy, P., Gruand, J., Sellier, P. and Bonneau, M. (1997) [Evidence of a major gene controlling fat androstenone level and sexual maturity in entire male pigs (in French)]. *Journées de la Recherche Porcine en France* 29, 369–374.

Galvin, J.M., Wilmut, I., Day, B.N., Ritchie, M., Thompson, M. and Haley, C.S. (1993) Reproductive performance in relation to uterine and embryonic traits during early gestation in Meishan, Large White and crossbred sows. *Journal of Reproduction and Fertility* 98, 377–384.

Gama, L.T., Boldman, K.G. and Johnson, R.K. (1991) Estimates of genetic parameters for direct and maternal effects on embryonic survival in swine. *Journal of Animal Science* 69, 4801–4809.

Grosse-Lembeck, K. and Kalm, E. (1985) Eignung verschiedener Stressempfindlichkeitstests zur Entwicklung von Mutterlinien, dargestellt an der Deutschen Landrasse in Schleswig-Holstein. *Züchtungskunde* 57, 99–112.

Gunsett, F.C. and Robison, O.W. (1990) Crossbreeding effects on reproduction, growth and carcass traits. In: Young, L.D. (ed.) *Genetics of Swine.* NC-103 Publication.

Haley, C.S. and Lee, G.J. (1992) Genetic factors contributing to variation in litter size in British Large White gilts. *Livestock Production Science* 30, 99–113.

Haley, C.S. and Lee, G.J. (1993) Genetic basis of prolificacy in Meishan pigs. *Journal of Reproduction and Fertility* 48 (Suppl.), 247–259.

Haley, C.S., Avalos, E. and Smith, C. (1988) Selection for litter size in the pig. *Animal Breeding Abstract* 56, 317–332.

Haley, C.S., Lee, G.J. and Ritchie, M. (1995) Comparative reproductive performance in Meishan and Large White pigs and their crosses. *Animal Science* 60, 259–267.

Harder, R.R., Lundstra, D.D. and Johnson, R.K. (1995) Growth of testes and testicular morphology after eight generations of selection for increased predicted weight of testes at 150 days of age in boars. *Journal of Animal Science* 73, 2186–2192.

Henderson, C.R. (1984) *Applications of Linear Models in Animal Breeding.* University of Guelph, Guelph, Ontario, Canada.

Hermesch, S., Luxford, B.G. and Graser, H.U. (1995) Genetic relationships between litter size and meat quality traits in Australian pigs. In: Van Arendonk, J.A.M. (ed.) *Book of Abstracts of the European Association for Animal Production,* G1.18 (abstract).

Hillbrand, F.W. and Glodek, P. (1984) Halothane sensitivity and semen quality of boars in 3 lines of pigs. *35th Annual Meeting of the European Association for Animal Production,* The Hague, The Netherlands, paper P5.11.

Hixon, A.L., Mabry, J.W., Benyshek, L.L., Weaver, W.M. and Marks, M.A. (1987) Estimates of genetic parameters for sexual and compositional maturity in gilts. *Journal of Animal Science* 64, 977–982.

Huang, M.Y. and Rasmusen, B.A. (1982) Parental transferrin types and litter size in pigs. *Journal of Animal Science* 54, 757–762.

Huang, Y.T. and Johnson, R.K. (1996) Effect of selection for size of testes in boars on semen and testes traits. *Journal of Animal Science* 74, 750–760.

Hunter, M.G., Biggs, C., Ashworth, C.J. and Haley, C.S. (1991) Ovarian function in Chinese Meishan pigs. *Biology of Reproduction* 44 (Suppl. 1), 145 (abstract).

Hutchens, L.K., Hintz, R.L. and Johnson, R.K. (1981) Genetic and phenotypic relationships between puberal and growth characteristics of gilts. *Journal of Animal Science* 53, 946–951.

Hutchens, L.K., Hintz, R.L. and Johnson, R.K. (1982) Breed comparisons for age and weight at puberty in gilts. *Journal of Animal Science* 55, 60–66.

Imlah, P. (1970) Evidence for the Tf locus being associated with an early lethal factor in a strain of pig. *Animal Groups and Biochemical Genetics* 1, 5–13.

Irgang, R., Favero, J.A. and Kennedy, B.W. (1994) Genetic parameters for litter size of different parities in Duroc, Landrace and Large White sows. *Journal of Animal Science* 72, 2237–2246.

Janss, L.L.G., van Arendonk, J.A.M. and Brascamp, E.W. (1997) Bayesian statistical analyses for presence of single genes affecting meat quality traits in a crossed pig population. *Genetics* 145, 395–408.

Jensen, E.L., Smith, C., Baker, L.N. and Cox DF (1968) Quantitative studies on blood group and serum protein systems in pigs. II. Effects on production and reproduction. *Journal of Animal Science* 27, 856–862.

Johansson, K., Kennedy, B.W. and Wilhelmson, M. (1994) Precision and bias of estimated genetic parameters in the presence of dominance and inbreeding. In: Smith, C., Gavora, J.S., Benkel, B., Chesnais, J., Fairfull, W., Gibson, J.P., Kennedy, B.W. and Burnside, E.B. (eds), *Proceedings of the 5th World Congress on Genetics Applied to Livestock Production,* vol. 18, pp. 386–389.

Johnson, R.K. (1980) Heterosis and breed effects in swine. North Central Regional Publication No. 262.

Johnson, R.K. (1981) Crossbreeding in swine: experimental results. *Journal of Animal Science* 52, 906–923.

Johnson, R.K. (1990) Inbreeding effects on reproduction, growth and carcass traits. In: Young, L.D. (ed.) *Genetics of Swine.* NC-103 Publication.

Johnson, R.K., Zimmerman, D.R. and Kittock, R.J. (1984) Selection for components of reproduction in swine. *Livestock Production Science* 11, 541–558.

Johnson, R.K., Eckardt, G.R., Rathje, T.A. and Drudik, D.K. (1994) Ten generations of selection for predicted weight of testis in swine: Direct response and correlated response in body weight, backfat, age at puberty and ovulation rate. *Journal of Animal Science* 72, 1978–1988.

Kelly, C.R., Socha, T.E. and Zimmerman, D.R. (1988) Characterization of gonadotropic and ovarian steroid hormones during the periovulatory period in high ovulating select and control line gilts. *Journal of Animal Science* 66, 1462–1474.

Kerr, J.C. and Cameron, N.D. (1996) Genetic and phenotypic relationships between performance test and reproduction traits in Large White pigs. *Animal Science* 62, 531–540.

Kristjansson, F.K. (1964) Transferrin types and reproductive performance in the pig. *Journal of Reproduction and Fertility* 8, 311–317.

Lamberson, W.R. (1990) Genetic parameters for reproductive traits. In: Young, L.D. (ed.) *Genetics of Swine.* NC-103 Publication.

Lamberson, W.R., Johnson, R.K., Zimmerman, D.R. and Long, T.E. (1991) Direct responses to selection for increased litter size, decreased age at puberty or random selection following selection for ovulation rate in swine. *Journal of Animal Science* 69, 3129–3143.

Lampo, P., Nauwynck, W., Bouquet, Y. and Van Zeveren, A. (1985) Effect of stress susceptibility on some reproductive traits in Belgian Landrace pigs. *Livestock Production Science* 13, 279–287.

Larzul, C. (1997) [Genetic variability of an in vivo measurement of muscle glycolytic potential in pigs: relationships with performance, muscle characteristics and meat technological quality (in French)]. Doctoral thesis, Institut National Agronomique Paris-Grignon, France.

Lee, G.J. and Haley, C.S. (1995) Comparative farrowing to weaning performance in Meishan and Large White pigs and their crosses. *Animal Science* 60, 269–280.

Legault, C. (1985) Selection of breeds, strains and individual pigs for prolificacy. *Journal of Reproduction and Fertility* 33 (Suppl.), 151–166.

Legault, C. and Caritez, J.C. (1983) [Experiments on Chinese pigs in France. I. reproductive performance in purebreeding and crossbreeding (in French)]. *Genetics, Selection, Evolution* 15, 225–240.

Legault, C. and Gruand, J. (1976) [Genetic improvement of sow prolificacy through the creation of an 'hyperprolific' line and the use of artificial insemination. Principles and preliminary results)]. *Journées de la Recherche Porcine en France* 8, 201–212.

Legault, C., Gruand, J. and Oulion, F. (1979) [Elaboration and genetic interest of an in vivo measurement of testicular weight in young boars (in French)]. *Journées de la Recherche Porcine en France* 11, 313–322.

Legault, C., Gruand, J., Lebost, J., Garreau, H., Ollivier, L., Messer, L.A. and Rothschild, M.F. (1996) [Frequency and effect of the ESR gene in two Large White lines in France (in French)]. *Journées de la Recherche Porcine en France* 28, 9–14.

Le Roy, P., Naveau, J., Elsen, J.M. and Sellier, P. (1990a) Evidence for a new major gene influencing meat quality in pigs. *Genetical Research* 55, 33–40.

Le Roy, P., Elsen, J.M. and Naveau, J. (1990b) [Studies on the genetic variability of carcass fat content in the Laconie line (in French)]. *Journées de la Recherche Porcine en France* 22, 11–16.

Leymaster, K.A. and Bennett, G.L. (1994) An approach to select for litter size in swine: conceptual, theoretical and applied aspects. In: *Proceedings of the 49th National Breeding Roundtable*, pp. 54–64.

Long, S.E. and Williams, C.V. (1982) A comparison of the chromosome complement of inner cell mass and trophoblast cells in day 10 pig embryos. *Journal of Reproduction and Fertility* 66, 645–648.

Lubritz, D., Johnson, B. and Robison, O.W. (1991) Genetic parameters for testosterone production in boars. *Journal of Animal Science* 69, 3220–3224.

McFeely, R.A. (1967) Chromosome abnormalities in early embryos of the pig. *Journal of Reproduction and Fertility* 13, 579–581.

McLaren, D.G. (1990). Potential of Chinese breeds to improve pork production efficiency in the USA. *Animal Breeding Abstract* 58, 347–369.

McLaren, D.G. and Bovey, M. (1992) Genetic influences on reproductive performance. *The Veterinary Clinics of North America: Food Animal Practice Swine Reproduction*, 8, 435–459.

McLaren, D.G., Buchanan, D.S. and Williams, J.E. (1987) Economic evaluation of alternative crossbreeding systems involving four breeds of swine. II. System efficiency. *Journal of Animal Science* 65, 919–928.

Mariscal, D.V., Wolfe, P.L., Bergfeld, E.G., Cupp, A.S., Kojima, F.N., Fike, K.E., Sanchez, T., Wehrman, M.E., Johnson, R.K., Kittok, R.J., Ford, J.J. and Kinder, J.E. (1996) Comparison of circulating concentrations of reproductive hormones in boars of lines selected for size of testes or number of ovulations and embryonic survival to concentrations in respective control lines. *Journal of Animal Science* 74, 1905–1914.

Mercer, J.T. and Crump, R.E. (1990) Genetic parameter estimates for reproduction traits in purebred Landrace pigs. In: Hill, W.G., Thompson, R. and Wooliams, J.A. (eds), *Proceedings of the 4th World Congress on Genetics Applied to Livestock Production*, vol. XV, pp. 489–492.

Neal, S.M., Johnson, R.K. and Kittok, R.J. (1989) Index selection for components of litter size in swine: Response to five generations of selection. *Journal of Animal Science* 67, 1933–1945.

Neely, J.D. and Robison, O.W. (1983). Estimates of heterosis for sexual activity in boars. *Journal of Animal Science* 56, 1033–1038.

Neely, J.D., Johnson, B.H. and Robison, O.W. (1980) Heterosis estimates for measures of reproductive traits in crossbred boars. *Journal of Animal Science* 51, 1070–1077.

Ollivier, L. and Bolet, G. (1981) [Selection for prolificacy in pigs: results of a ten generation selection experiment (in French)]. *Journées de la Recherche Porcine en France* 13, 261–267.

Ollivier, L. and Legault, L. (1967) [Direct effect of the boar on the size and weight of litters produced by artificial insemination (in French)]. *Annales de Zootechnie* 16, 247–254.

Ollivier, L. and Sellier, P. (1982) Pig genetics: a review. *Annales de Génétique et de Sélection Animale* 14, 481–544.

Pérez-Enciso, M. and Gianola, D. (1992) Estimates of genetic parameters for litter size in six strains of Iberian pigs. *Livestock Production Science* 32, 283–293.

Pérez-Enciso, M., Bidanel, J.P., Baquedano, I. and Noguera, J.L. (1996) A comparison of alternative genetic models for litter size in pigs. *Animal Science* 63, 255–264.

Perry, J.S. and Rowlands, I.W. (1962) Early pregnancy in the pig. *Journal of Reproduction and Fertility* 4, 175–188.

Pfeiffer, H., Lengerken, G.V., Schwalbe, M., Horn, P. and Kovac, G. (1986) Relationships between the stress susceptibility and the fertility of pigs. *37th Annual Meeting of the*

European Association for Animal Production, Budapest, Hungary, paper GP3-20 (6pp).
Pond, W.G. and Houpt, K.A. (1978) *The Biology of the Pig*. Cornell University Press, Ithaca, New York.
Popescu, C.P. (1989) Cytogenetics of farm animals (in French)]. INRA, Paris, France, 114 pp.
Rasmusen, B.A. and Hagen, K.L. (1973) The H blood-group system and reproduction in pigs. *Journal of Animal Science* 37, 568–573.
Rathje, T.A., Johnson, R.K. and Lundstra, D.D. (1995) Sperm production in boars after nine generations of selection for increased weight of testes. *Journal of Animal Science* 73, 2177–2185.
Rathje, T.A., Rohrer, G.A. and Johnson, R.K. (1996) Quantitative trait loci affecting reproductive traits in swine. *Journal of Animal Science* 74 (Suppl. 1), 122 (abstract).
Reutzel, L.F. and Sumption, L.J. (1968) Genetic and phenotypic relationships involving age at puberty and growth rate of gilts. *Journal of Animal Science* 27, 27–30.
Robison, O.W., Lubritz, D. and Johnson, B. (1994) Realized heritability estimates in boars divergently selected for testosterone levels. *Journal of Animal Breeding and Genetics* 111, 35–42.
Rodriguez, M.C., Rodriganez, J. and Silio, L. (1994) Genetic analysis of maternal ability in Iberian pigs. *Journal of Animal Breeding and Genetics* 111, 220–227.
Roehe, R. and Kennedy, B.W. (1993) Effect of selection for maternal and direct genetic effects on genetic improvement of litter size in swine. *Journal of Animal Science* 71, 2891–2904.
Roehe, R. and Kennedy, B.W. (1995) Estimation of genetic parameters for litter size in Canadian Yorkshire and Landrace swine with each parity of farrowing treated as a different trait. *Journal of Animal Science* 73, 2959–2970.
Rothschild, M.F., Zimmerman, D.R., Johnson, R.K., Venier, L. and Warner, C.M. (1984) SLA haplotype differences in lines of pigs which differ in ovulation rate. *Animal Blood Groups and Biochemical Genetics* 15, 155–158.
Rothschild, M.F., Renard, C., Bolet, G., Dando, P. and Vaiman, M. (1986a) Effect of swine lymphocyte antigen haplotypes on birth and weaning weights in pigs. *Animal Genetics* 17, 267–272.
Rothschild, M.F., Renard, C., Sellier, P., Bonneau, M. and Vaiman, M. (1986b) Swine lymphocyte antigen (SLA) haplotype effects on male genital trait development and androsterone level. In: Dickerson, G. E. and Johnson, R.K. (eds) *Proceedings of the 3rd World Congress on Genetics Applied to Livestock Production*, vol. XI, pp. 197–202.
Rothschild, M.F., Jacobson, C., Vaske, D.A., Tuggle, C.K., Short, T.H., Sasaki, S., Eckardt, G.R. and McLaren, D.G. (1994) A major gene for litter size in pigs. In: Smith, C., Gavora, J.S., Benkel, B., Chesnais, J., Fairfull, W., Gibson, J.P., Kennedy, B.W. and Burnside, E.B. (eds) *Proceedings of the 5th World Congress on Genetics Applied to Livestock Production*, vol. 21, pp. 225–228.
Rothschild, M.F., Vaske, D.A., Tuggle, C.K., Messer, L.A., McLaren, D.G., Short, T.H., Eckardt, G.R., Mileham, A.J. and Plastow, G.S. (1995) Estrogen receptor locus is a major gene for litter size in the pig. In: Van Arendonk, J.A.M. (ed.) *Book of Abstracts of the European Association for Animal Production*, p. 53 (abstract).
Rothschild, M.F., Jacobson, C., Vaske, D.A., Tuggle, C., Wang, L., Short, T., Eckardt, G., Sasaki, S., Vincent, A., McLaren, D.G., Sowthwood, O., van der Steen, H., Mileham, A. and Plastow, G. (1996) The Estrogen receptor locus is associated with a major

gene influencing litter size in pigs. *Proceedings of National Academy of Sciences USA* 93, 201–205.

Rutledge, J.J. (1980) Fraternity size and swine reproduction. I. Effect on fecundity of gilts. *Journal of Animal Science* 51, 868–870.

Rydhmer, L., Johansson, K., Stern, S. and Eliasson-Selling, L. (1992) A genetic study of pubertal age, litter traits, weight loss during lactation and relations to growth and leanness in gilts. *Acta Agriculturae Scandinavica* 42, 211–219.

Rydhmer, L., Eliasson-Selling, L., Johansson, K., Stern, S. and Andersson, K. (1994) A genetic study of estrus symptoms at puberty and their relationship to growth and leanness in gilts. *Journal of Animal Science* 72, 1964–1970.

Rydhmer, L., Lundeheim, N. and Johansson, K. (1995) Genetic parameters for reproduction traits in sows and relations to performance-test measurements. *Journal of Animal Breeding and Genetics* 112, 33–42.

Schinckel, A.P., Johnson, R.K., Pumpfrey, R.A. and Zimmerman, D.R. (1983) Testicular growth in boars of different genetic lines and its relationship to reproductive performance. *Journal of Animal Science* 56, 1065–1076.

Schlenker, G., Jugert, L., Mudra, K., Pohle, M. and Heinze, C. (1984) Spermqualität von Ebern unterschieldlicher Halothanempfindlichkeit. *Monatshefte für Veterinärmedizin* 39, 760–763.

Schneider, A., Schwörer, D. and Blum, J. (1980) Beziehung des Halothan-genotyps zu den Produktions- und Reproduktionsmerkmalen der Schweizerischen Landrasse. *Annales de Génétique et de Sélection Animales* 12, 417 (abstract).

See, M.T., Mabry, J.W. and Bertrand, J.K. (1993) Restricted maximum likelihood estimation of variance components from field data for number of pigs born alive. *Journal of Animal Science* 71, 2905–2909.

Sellier, P. (1976) The basis of crossbreeding in pig: A review. *Livestock Production Science*, 3, 203–226.

Sellier, P. and Bonneau, M. (1988) Genetic relationships between fat androstenone level in males and development of male and female genital tract in pigs. *Journal of Animal Breeding and Genetics* 105, 11–20.

Sellier, P., Cousin, V. and Dando, P. (1987) Effects of halothane sensitivity on male and female reproductive performance in Piétrain lines. *Annales de Zootechnie* 36, 249–264.

Short, T.H., Wilson, E.R. and McLaren, D.G. (1994) Relationships between growth and litter traits in pig dam lines. In: Smith, C., Gavora, J.S., Benkel, B., Chesnais, J., Fairfull, W., Gibson, J.P., Kennedy, B.W. and Burnside, E.B. (eds) *Proceedings of the 5th World Congress on Genetics Applied to Livestock Production*, vol. 17, pp. 413–416.

Short, T.H., Rothschild, M.F., Southwood, O.I., McLaren, D.G., DeVries, A., van der Steen, H., Eckardt, G.R., Tuggle, C.K., Helm, J., Vaske, D.A., Mileham, A.J. and Plastow, G.S. (1997). The effect of the estrogen receptor locus on reproduction and production traits in four commercial lines of pigs. *Journal of Animal Science* (in press).

Simpson, S.P., Webb. A.J. and Wilmut, I. (1986) Performance of British Landrace pigs selected for high and low incidence of halothane sensitivity. 1 – Reproduction. *Animal Production* 43, 485–492.

Sorensen, D.A. and Vernesen, A.H. (1991) Large scale selection for number of born piglets using an animal model. *42th Annual Meeting of the European Association for Animal Production*, Berlin, Germany, paper G3.16 (8 pp.).

Southwood, O.I. and Kennedy, B.W. (1990) Estimation of direct and maternal genetic variance for litter size in Canadian Yorkshire and Landrace swine using an animal model. *Journal of Animal Science* 68, 1841–1847.

Svendsen, J. (1992) Perinatal mortality in pigs. *Animal Reproduction Science* 28, 59–67.

ten Napel, J., de Vries, A.G., Buiting, G.A.J., Luiting, P., Merks, J.W.M. and Brascamp, E.W. (1995a) Genetics of the interval from weaning to estrus in first-litter sows: Distribution of the data, direct response to selection and heritability. *Journal of Animal Science* 73, 2193–2203.

ten Napel, J., Kemp, B., Luiting, P. and de Vries, A.G. (1995b) A biological approach to examine genetic variation in weaning to oestrus interval in first-litter sows: a review. *Livestock Production Science* 41, 81–93.

Tess, M.W., Bennett, G.L. and Dickerson, G.E. (1983) Simulation of genetic changes in life cycle efficiency of pork production. II – Effects of components on efficiency. *Journal of Animal Science* 56, 354–368.

Toelle, V.D. and Robison, O.W. (1985) Estimates of genetic relationship between testes measurements and female reproductive traits in swine. *Zeitschrift für Tierzüchtung und Züchtungsbiologie* 102, 125–132.

Toelle, V.D., Johnson, B.H. and Robison, O.W. (1984) Genetic parameters for testes trait in swine. *Journal of Animal Science* 59, 967–973.

Vaiman, M., Renard, Ch. and Bourgeaux, N. (1988) SLA, the major histocompatibility complex in swine: its influence on physiological and pathological traits. In: Warner, C.M., Rothschild, M.F. and Lamont, S.J. (eds) *The Molecular Biology of the Major Histocompatibility Complex of Domestic Animal Species.* Iowa State Press, Ames, Iowa.

Van der Steen, H.A.M. (1983) Maternal and genetic influences on production and reproduction traits in pigs. Doctoral thesis, Agricultural University of Wageningen, The Netherlands.

Van der Steen, H.A.M. (1985) The implication of maternel effects for genetic improvement of litter size in pigs. *Livestock Production Science* 13, 159–168.

Vangen, O.P. (1980) Studies on a two trait selection experiment in pigs. VI. Heritability estimates of reproductive traits. Influence of maternal effects. *Acta Agriculturae Scandinavica* 30, 320–326.

Visscher, P.M. and Haley, C.S. (1995) Utilizing genetic markers in pig breeding programmes. Animal Breeding Abstract 63, 1–8.

Warner, C.M. and Rothschild, M.F. (1991). The swine major histocompatibility complex (SLA). In: Srivastava, R., Ram, B. and Tyle, P. (eds) *Immunogenetics of the Major Histocompatibility Complex.* VCH Publishers, New York, pp. 368–397.

White, B.R., McLaren, D.G., Dziuk, P.J. and Wheeler, M.B. (1991) Attainment of puberty and the mechanism of large litter size in Chinese Meishan females versus Yorkshire females. *Biology of Reproduction* 44 (Suppl. 1), 160 (abstract).

Wilkie, P.J., Paszek, A.A., Flickinger, G.H., Rohrer, G.A., Alexander, L.J., Beattie, C.W. and Shook, L.B. (1996) Scan of 8 porcine chromosomes for growth, carcass and reproductive traits reveals two likely quantitative trait loci. In: *Proceedings of the XXVth International Conference on Animal Genetics*, Tours, France, 187 (abstract).

Willeke, H., Amler, K. and Fisher, K. (1984) Der Einfluss des Halothanstatus der Sau auf deren Wurfgrösse. *Züchtungskunde* 56, 20–26.

Willeke, H., Claus, R., Müller, E., Pirchner, F. and Karg, H. (1987) Selection for high and low level of 5α-androst-16-en-3-one in boars. I. Direct and correlated response of endocrinological traits. *Journal of Animal Breeding and Genetics* 104, 64–73.

Wilmut, I., Ritchie, W., Haley, C.S., Ashworth, C.J. and Aitken, R.P. (1992) A comparison of rate and uniformity of embryo development in Meishan and Large White pigs. *Journal of Reproduction and Fertility* 95, 45–56.

Wilson, E.R., Johnson, R.K. and Wettemann, R.P. (1977) Reproductive and testicular characteristics of purebred and crossbreed boars. *Journal of Animal Science* 44, 939–947.

Woodward, B.W., Mabry, J.W., See, M.T., Bertrand, J.K. and Benyshek, L.L. (1993). Development of an animal model for across-herd genetic evaluation of number born alive in swine. *Journal of Animal Science* 71, 2040–2046.

Wrathall, A.E. (1971) *Prenatal Survival in Pigs. Part I. Ovulation Rate and its Influence on Prenatal Survival and Litter Size in Pigs.* Commonwealth Agricultural Bureau, Farnham Royal, England, 108 pp.

Xu, Z.Y. (1985a) On the biological and economical traits of ten Chinese indigenous breeds of pigs. Part 1. *Pig News and Information* 6, 301–309.

Xu, Z.Y. (1985b) On the biological and economical traits of ten Chinese indigenous breeds of pigs. Part 2. *Pig News and Information* 6, 425–431.

Young, L.D. (1990) Evaluation of Chinese breeds: research plans and early results from the US Meat Animal Research Center. In: Molénat, M. and Legault, C. (eds) *Chinese Pig Symposium.* INRA, Jouy-en-Josas, France, pp. 119–120 (abstract).

Young, L.D. (1995) Reproduction of F1 Meishan, Fengjing, Minzhu and Duroc gilts and sows. *Journal of Animal Science* 73, 711–721.

Young, L.D., Pumfrey, R.A., Cunningham, P.J. and Zimmerman, D.R. (1978) Heritabilities and genetic and phenotypic correlations for prebreeding traits, reproductive traits and principal components. *Journal of Animal Science* 46, 937–949.

Young, L.D., Leymaster, K.A. and Lundstra, D.D. (1986) Genetic regulation of testicular growth and its relationship to female reproductive traits in swine. *Journal of Animal Science* 63, 17–26.

Zhang, W. C., Wu, J.S. and Rempel, W.E. (1983) Some performance characteristics of prolific breeds of pigs in China. *Livestock Production Science* 10, 59–68.

Transgenics and Modern Reproductive Technologies

12

U. Besenfelder,[1] M. Müller[1] and G. Brem[2]
[1]Institut für Tierzucht und Genetik, Veterinärmedizinische Universität Wien (VUW), Josef Baumann Gasse 1, A-1210 Vienna, Austria; [2]Department of Biotechnology in Animal Production, IFA Tulln, Konrad Lorenz Str. 20, A-3430 Tulln, Austria

Modern Reproductive Technologies

Introduction

The ability to copy the *in vivo* development of pre-implantation embryos to an extracorporal *in vitro* system provides powerful possibilities for modern reproductive technologies in research and animal production. The development and improvement of such technologies are concentrating on gamete and embryo collection and preservation, *in vitro* production of embryos, culturing and manipulation of embryos (splitting, nuclear transfer, production of chimeras, establishment of ES cells, gene transfer) and embryo transfer (ET) (see Fig. 12.1).

Basic understanding of the reproductive physiology and the development of novel techniques is facilitated by modern equipment for ultrasonography, endoscopy and cryopreservation. Hormonal synchronization, (super-) ovulation and artificial insemination of donor and recipient animals represent major points which have to be managed prior to embryo manipulation and transfer experiments. Precise data on ovulatory response following gonadotrophin application in swine are not available although efforts have been made to determine the optimal time for synchronization, superovulation and insemination. Ultrasonography has been used to monitor the interval between insemination and ovulation and on the basis of these data to determine the optimal time for insemination (24 to 0 h before ovulation) (Waberski *et al.*, 1994; Soede *et al.*, 1995).

Surveillance of the reproductive organs by ultrasonography depends on the physical composition of the different tissues due to their compactness and liquidity, i.e. it reflects more the look into ('inspection') rather than on to ('adspection') the organs. Endoscopy, however, allows direct observation of the surface of the organs analogous to surgery. In contrast to surgery, endoscopy is minimally invasive, thus permitting repeated use due to significant reduction of operative incriminations and postoperative consequences. Endoscopy has already been used in excellent studies monitoring the peri-ovulatory time in sows, including follicular development until ovulation, numbers of corpora lutea and the remainder of the reproductive tract (Brüssow *et al.*, 1990; Kramer and Lamberson, 1991; Huff and Esbenshade, 1992; Lee *et al.*, 1995a). In addition, endoscopy has been successfully used for artificial (oviductal) insemination, suggesting that this technique may be important for the maintenance of single animals, lines and breeds in cases where small volumes of sperm suspension are available (Morcom and Dukelow, 1980).

This review will concentrate mainly on the 'oocyte and embryo side' of modern reproductive technology, i.e. the hormonal treatment of sows, embryo collection, production, manipulation, preservation and transfer (see Fig. 12.1). Attempts aiming at the preservation and sexing of porcine sperm are described briefly below.

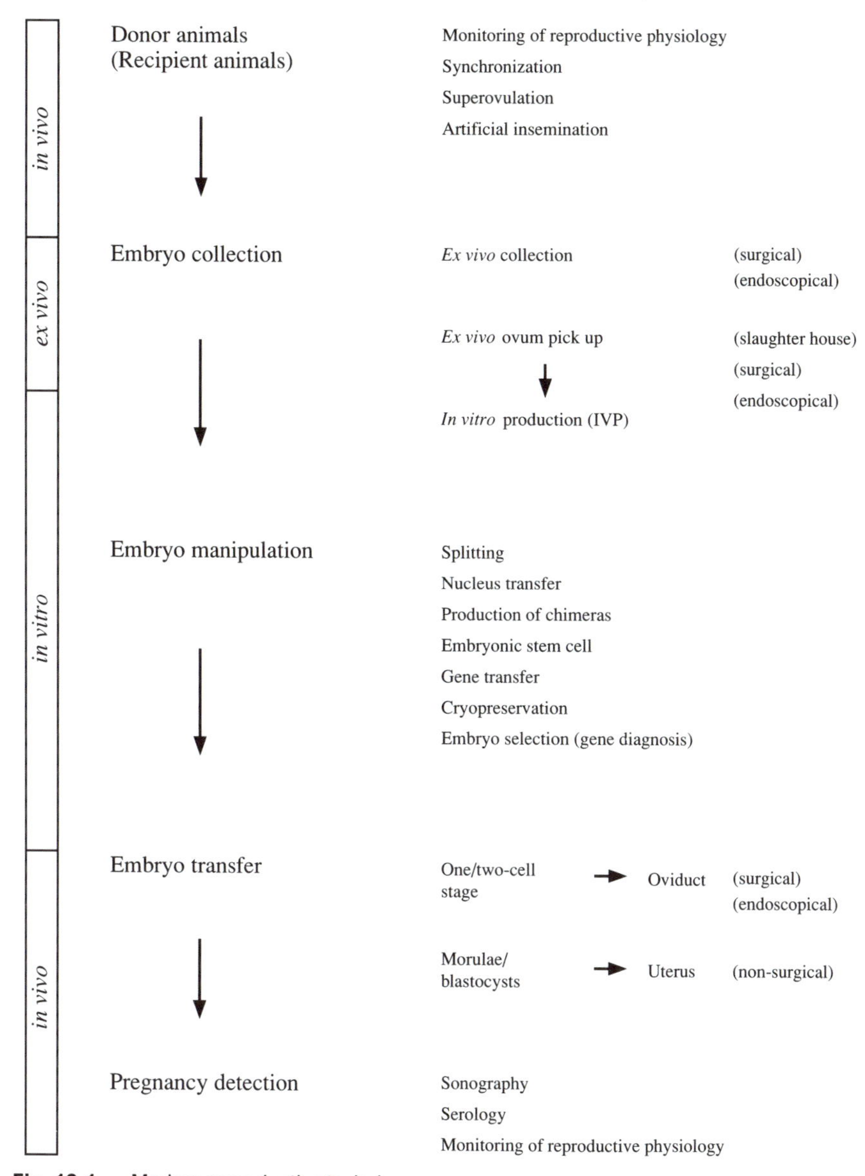

Fig. 12.1. Modern reproductive techniques.

Semen sorting and cryopreservation

Sex determination of semen is of major interest for all species used for modern reproduction technologies. Welch *et al.* (1995) separated intact bovine male and female sperm cells by flow cytometry into purities of 90%. Sperm sorting resulted in a few thousand living cells. Hence, so far this technology can be applied for *in vitro* purposes only. First attempts to establish semen sorting in swine were recently published. Rath *et al.* (1997) transferred embryos into the uterine horns of two recipients after *in vitro* fertilization with x-sorted spermatozoa. Additionally they deposited sorted semen intratubally in 2 gilts. The recipients delivered four and six healthy female piglets and the gilts farrowed 13 of 15 (85%) piglets of the predicted gender, respectively.

More work has been done in the field of cryopreservation. Concentration of cryoprotectant, cooling rate, size of straw and warming velocity are main factors influencing the post-thawed viability of embryos. Freezing protocols show that glycerol could be used as an effective cryoprotectant despite its toxicity (Bwanga *et al.*, 1991a). High concentrations (4–6%) of glycerol have been shown to increase sperm motility while the number of sperm with normal apical ridge decreased (Fiser and Fairfull, 1990; Fiser *et al.*, 1993). Glycerol, at a concentration of 3%, is recommended (Bwanga *et al.*, 1991b; Fiser et al., 1993; Ortman and Rodriguez Martinez, 1994). Main damage of the spermatozoal membrane was found to occur during cooling to 5°C, whereas supercooling to –6°C did not cause further damage (Ortman and Rodriguez Martinez, 1994).

Ice crystal formation depends on the cooling rate. Best spermatozoal quality is expected using a freezing rate of 30°C min^{-1} and 1200°C min^{-1} for thawing (Fiser *et al.*, 1993). Frozen (3% glycerol) thawed ejaculates showed a post-thawed motility rate of 54.5% and 75.5 % of the sperms showed normal apical ridges. After insemination 75% of the oocytes were fertilized (Bwanga *et al.*, 1991b).

Embryo collection and production

Porcine embryos can be collected *ex vivo* from living donors by either surgical or endoscopical flushing of oviducts (or uterine horns) or from slaughtered animals. Another possibility is the *in vitro* production of embryos from collected oocytes. Both techniques are described below.

Embryo recovery

Superovulation of sows usually results in about 30 collectable oocytes or embryos per donor. While both the quantity and quality of embryos isolated from prepuberal gilts are higher than from older animals, multiparous sows offer notable advantages as embryo recipients (see below) in terms of pregnancy rates, embryo and offspring survival. Table 12.1 shows a typical scheme for hormonal treatment of donors and embryo collection used in several laboratories (for review and additional references see Brem and Müller, 1994). Prepuberal gilts or synchronized donors (postweaning sows or treated sows) are usually superovulated with a single pregnant mare's serum gonadotrophin (PMSG) application. Embryo recovery is routinely done by flushing of the Fallopian tube or the uterine horns surgically or after slaughter. Recently, we were able to demonstrate that the principle of creating a functionally closed oviduct and uterus for flushing can be performed not only by surgery but also by the use of endoscopy (Besenfelder *et al.*, 1997). The oviduct and the top of the

Table 12.1. Treatment of donors and embryo recovery from swine (German Landrace, Pietrain, crossbreeds).

Age/body weight	Synchronization	Superovulation (PMSG)	Induction of ovulation (hCG) (h after PMSG)	Insemination (h after hCG)*	References
Prepuberal gilts (75–90 kg)	Not necessary	1250 IU	750 IU i.m. (72 h)	2 × [24 h + 36 h]	Brem *et al.*, 1985
Mature gilts	15 mg Altrenogest for 5–9 d starting at d 12–6 of cyclus	1500–2000 IU (24–30 h after last admin. of Altrenogest)	500 IU (72 h)	2 × [18 h + 36 h]	Hammer *et al.*, 1985
Mature gilts	120 mg Methallibure over a period of 6 d, starting at d 10–15 of cyclus	1500 IU s.c. (24 h after last admin. of Methallibure)	500 IU i.m. (72 h)	2 × [16 h + 36 h]	Pursel *et al.*, 1988

*The fertilized oocytes are collected 24 h after the second insemination.

uterine horns of prepuberal donors were bilaterally flushed by this technique. Endoscopic embryo collection enables complete embryo recovery, minimizes manipulations and efforts and guarantees the repeated use of the donors in reproductive biology. We expect that the use of endoscopy in swine reproduction will represent a powerful tool for embryo recovery and embryo transfer (see below), because it provides free access to the ovaries, oviducts and uteri, allows visual control without surgery and avoids unnecessary movements or distortion of the organs.

Ovum pick up and in vitro production of embryos

Since the first reports of *in vitro* fertilization of pig (Harms and Schmidt, 1970) great attempts have been made to produce porcine embryos *in vitro*. This would provide a pool of pre-implantation stages for the use of modern reproductive applications (see Fig. 12.1). Moreover, *in vitro* production (IVP) of embryos can utilize follicular oocytes which normally undergo atresia. The IVP includes numerous *in vitro* steps which try to mimic the *in vivo* development of embryos: oocyte maturation, fertilization and culture until transfer or cryopreservation. Oocyte recovery is performed by flushing of ovulated oocytes from oviducts or by puncturing of follicles and aspiration of the oocytes. The latter can be done from untreated donors (immature oocytes) or after gonadotrophin treatment of the animals in which follicle maturation already has been started. As has been demonstrated for the collection of embryos (see above), endoscopy can also be successfully used for follicle puncture and oocyte aspiration in pigs (Brüssow and Ratky, 1994). After stimulation of follicular growth with PMSG (1000 IU) follicle aspiration resulted in a recovery rate of 55% (collected oocytes/follicles punctured). The discrepancy observed between the number of counted and punctured follicles was explained by technical difficulties. Nevertheless, this technique offers the possibility of repeated *ex vivo* ovum pick up in pigs and thus represents an alternative source of oocytes in addition to those collected from slaughtered animals. Experience gained from routine bovine IVF programmes indicate that hormonal treatment of the donors is unnecessary. Our own results have demonstrated that repeated use of the endoscopic equipment can significantly increase the efficiency and success rates in oocyte or embryo recovery and embryo transfer programmes.

The different steps required for IVF in pigs have been developed separately. Motlik and Fulka (1974) succeeded in the *in vitro* maturation of porcine oocytes. Baker and Polge (1976) summarized experiments on *in vitro* fertilization of swine oocytes. The capacitation of boar spermatozoa *in vitro* has been demonstrated for the first time by Nagai *et al.* (1984). Finally, IVP of porcine embryos resulted in normal pregnancies with the birth of living piglets (Mattioli *et al.*, 1988; Wu *et al.*, 1992). However, there are differences between the developmental capacity of *in vitro* and *in vivo* produced embryos (Rath *et al.*, 1995). Since the reasons for this are unclear, research aims at the improvement of the crucial IVP steps, i.e. maturation (Funahashi and Day, 1993; Funahashi *et al.*, 1994a; Rath et al., 1995) and fertilization. In terms of fertilization, the main focus is on the activity status of follicular oocytes during sperm penetration (Kikuchi *et al.*, 1995), the problem of polyspermic penetration during the *in vitro* fertilization (Funahashi *et al.*, 1994b) and *in vitro* fertilization by frozen epididymal spermatozoa (Nagai *et al.*, 1988) or by frozen ejaculated spermatozoa (Wang *et al.*, 1991).

The knowledge accumulated so far by many laboratories has enhanced the pig IVP system and succeeded in the production of IVP piglets. Intensified studies of cellular physiology of gametes and zygotes are expected to further improve the efficiency rates (reviewed by Nagai, 1994).

Embryo splitting and nuclear transplantation

There is a great interest in the production of two or more genetically identical individuals (clones). In basic scientific research cloned animals are of high value because they show no (or very limited) genetic variation. In animal production and breeding they guarantee the maintenance of single animals or lines with a high genetic value for commercial use and of rare breeds. In the future, the ability to clone embryos should compensate losses occurring during the procedure of embryo manipulation and transfer including pre- and postnatal mortality.

Embryo splitting is performed by microsurgical (bi-)section of early tubal stage embryos, morulae and blastocyst stages. After blastulation the inner cell mass has to be divided into two equal parts and the resulting zona pellucida-free demi-embryos are transferred to recipients. The split embryos can be either cultured *in vitro* or transferred into foster animals immediately after micromanipulation. Generally, the method needs to be improved because so far only a very low efficiency in twin production has been published (Nagashima *et al.*, 1988b, 1989; Reichelt and Niemann, 1994). Pig embryos seem to be much more sensitive to the splitting procedure than bovine embryos.

In principle the procedure of nuclear transfer is performed by dislocating pronuclei or blastomeres from an embryo into enucleated oocytes. Pronuclear exchange embryos were produced by positioning pronuclei containing karyoplasm in the perivitelline space of enucleated zygotes and subsequent fusion in an electric field. After *in vivo* culture up to 1 week, a high percentage (93%) of embryos showed cleavage to ≥ four cell stage and 38% reached the expanded blastocyst stage. Up to now only one of the manipulated embryos survived until birth and resulted in a piglet derived from nuclear transfer (Prather *et al.*, 1989). The transplantation of blastomeres into recipient oocytes was performed by electrofusion. Nagashima *et al.* (1992) obtained an enucleation rate of 88%, activation rate of 73%, fusion rate of 83% and cleavage rate to the two–four cell stage of 46%. In this study, embryonic development until birth was not reported. The reasons for the poor development of reconstituted porcine embryos are unknown. Studies by several laboratories suggest that nuclear transfer leads to a lack of RNA synthesis which is normally detectable in pig embryos from the four-cell stage (reviewed by Niemann and Reichelt, 1993). Further understanding of the cell cycle and basic molecular and ultrastructural events in pig embryos should improve the success rates of nuclear transfer.

Production of chimeras and embryonic stem cells

A chimera is, by definition, an individual which is generated from cells developing out of two genetically different zygotes. Chimeras are usually produced by aggregation of two early stage embryos (mainly morulae) or the injection of blastomeres or embryonic cells into the cavity of blastocysts. The blastocyst injection method succeeded in the production of one live born chimeric piglet (Kashiwazaki *et al.*, 1992). In this study, day 6 blastocysts were collected from Landrace (white hair), the zona pellucida was removed, and the inner cell mass isolated and dissociated by immunosurgery. About 20 inner cell mass cells were injected into the cavity of a blastocyst derived from a Duroc (brownish hair). Among the 11 piglets produced, one animal showed an obvious chimerism (mixed hair colour). The production of germline chimeras was reported in two independent studies (Anderson *et al.*, 1994; Onishi *et al.*, 1994) by injecting fresh inner cell mass (ICM) into blastocysts.

The handling and manipulation of blastomeres *per se* is required for molecular genetic analyses of pre-implantation embryos. Such genetic diagnosis may aim at inherited disorders (see Chapter 4) or important performance traits (see Chapters 11, 15, 16 and 17). In addition, blastomere handling is an inseparable instrumentation for the production of embryonic stem (ES) cells. Embryonic stem cells in swine are a tool desired for the study of cell differentiation and development, gene regulation but more importantly production of transgenic ('knock-out') pigs. The term 'knock-out', the generation of null mutations of a gene, is explained by the inactivation of an endogenous gene by insertion of cloned sequences. This process is called homologous recombination or targeted disruption. The ES cells have been isolated from the inner cell mass of mouse blastocysts (Evans and Kaufmann, 1981) and EG cells with similar characteristics from cultured primordial germ cells (reviewed by Donovan, 1994). The problems of establishing ES cells of livestock may be related to the limited availability of embryos with a defined genetic background (i.e. there are no comparable inbred strains as there are in mice), and furthermore, the embryonic development of large mammals is not as well understood as that of rodents. Clearly defined criteria for the establishment of ES cells are their maintenance in culture through multiple passages, a reasonably quick cell division rate, and most important their ability to participate in all tissues types, including the germline, indicating their pluri/totipotency.

At present, it is not well established which developmental stage is optimal for the isolation of porcine stem cells. Moreover, the culture conditions, growth factors required and the phenotypic and biochemical criteria for pig ES cells remain to be rigorously established (for review see Pedersen, 1994). The approaches to isolate pluripotential cells from swine are mainly based on morphology, *in vitro* differentiation and biochemical characteristics. ES-like porcine cells have been first described by Piedrahita *et al.* (1988). In 1994, Wheeler showed the generation of live piglets using embryos derived from ES-like cells (Wheeler, 1994). The ES-like cells were injected into the blastocoel cavity of a host blastocyst – 96% of these embryos survived micromanipulation. After embryo transfer 40% developed to term and delivered live offspring. The use of two different coat colour markers of ES cells and recipient blastocysts allowed the detection of chimeric piglets. A total of 72% of piglets consisted of colours of both genotypes. However, so far attempts to establish ES cells in pigs have not fulfilled all of the criteria of pluripotency, i.e. germline chimerism (Pedersen, 1994; Wheeler, 1994). Moreover, it has to be emphasized that the experiments in swine are not a routine method as is the case in rodents.

Cryopreservation – deep freezing and vitrification of embryos

By definition deep freezing methods include the formation of ice crystals, whereas vitrification avoids any crystallization and is a state of high viscosity. Cryopreservation of pig embryos has important applications in animal biotechnology. Embryos of individual animals, lines and strains of a high genetic value would be available at any time and in any place, thus minimizing hygienic barriers. Furthermore, the technique greatly facilitates the performance of embryo transfer programmes (see below) permitting the transfer of frozen–thawed embryos at any time into appropriate recipients.

In contrast to ruminants, porcine embryos are difficult to adapt to the cryopreservation conditions. Deep freezing protocols have been developed stepwise trying to optimize the cooling rates, temperatures, cyroprotectants (including nutrients and supplements) and the embryonic stages used.

Porcine embryos show a drastically reduced viability when exposed to temperatures below 15°C (Polge, 1977). However, Nagashima and colleagues (Nagashima *et al.*, 1988a) demonstrated that expanded and hatched blastocysts could be cultured at 6°C for a given time at higher rates than early blastocysts and morulae. Based on this, different freezing protocols (slow, rapid or stepwise freezing) in combination with different cryoprotectants (glycerol, dimethylsulphoxide (DMSO), etc.) have been tested. The first successful transfer of porcine embryos frozen at −35°C was reported by Hayashi *et al.* (1989). Subsequent studies used different freezing temperatures (−20°C, −196°C) and also reported live-born piglets after embryo transfer (Feng *et al.*, 1990; Kashiwazaki *et al.*, 1991; Mödl *et al.*, 1996). In a recent study pregnancies and live-born piglets were generated from early stage embryos (two–four cell) which had been deep frozen (−196°C) after centrifugation and removal of cytoplasmic lipids (Nagashima *et al.*, 1995). This study provides clear evidence for a critical role of the quantity and quality of cytoplasmic lipids for the freezability of swine embryos. However, the removal of cytoplasmic lipids by micromanipulation is laborious and cannot be applied for routine purposes. Future experiments of deep freezing should aim at the establishment of culture conditions for porcine embryos which improve their ability to tolerate freezing by reducing their lipid content and/or increasing the fluidity of the cell membranes.

An alternative to avoid the problems connected with porcine embryo freezing is provided by vitrification procedures. The method is well established for mice (Rall and Fahy, 1985), rabbits (Smorag *et al.*, 1989) and cattle (Ishimori *et al.*, 1993). Since it provides different cryophysical properties and thus other 'stress' components for the preserved embryos, vitrification might increase the tolerance of porcine embryos against very low temperatures. The survival of pig embryos (expanded and hatched blastocysts) were assessed in various vitrification solutions (ethylene glycol, ficoll, trehalose/DMSO, acetamide, propylene glycol/trehalose/ethylene glycol, propylene glycol, trehalose). So far the optimal equilibration medium has not yet been found, but freeze–thaw experiments indicate that embryos will survive this procedure (Yoshino *et al.*, 1993). The next set of experiments will obviously aim at the development of a less toxic vitrification solution.

Embryo transfer

The term embryo transfer (ET) includes the collection or production of embryos (*ex vivo* or *in vitro*) from donor animals, the temporary culture and/or manipulation and reintroduction into the physiological system (recipient animal). ET depends on the developmental stage of the embryos either into the oviduct or into the uterine horns of the foster mothers. Routinely, the recipients are treated like the donors (see Table 12.1), except with a 12 h delay and a reduction in the dosage of PMSG to 750 IU. The early stage embryos are transferred to the recipients 60–63 h after induction of ovulation. Available transfer techniques are:

1. The surgical procedure;
2. Transfer *per viam naturalem* through the cervix;
3. Endoscopic access to the reproductive organs.

The surgical transfer of porcine embryos into the oviduct and uterine horns was the first procedure established. This technique was mainly developed to introduce new genetic material into specific-pathogen-free pig herds (Holtz *et al.*, 1987; Cameron *et al.*, 1989). Embryos are transferred under the highest possible standard of surgical sterility. The abdominal cavity is entered by ventral midline laparotomy and ovaries, oviducts and uterine horns are moved out of the abdomen. The synchronization of the recipient is visually controlled. The tubal stage embryos are deposited into the ampulla (via infundibulum) (Springman and Brem, 1989) or into the uterine horns, which are reached by puncturing the oviduct with a needle and insertion via oviduct (Cameron et al., 1989) or by direct transmural introduction into the lumen (Blum-Reckow and Holtz, 1991). Migration of the embryos ('spacing') results in their equal distribution in both uterine horns. The reported pregnancy rates after surgical transfer range from 60 to 85%. In addition, significant stress of the animals results from anaesthesia, surgery, manipulation of the genital tract and too much time and effort is required. Therefore, it was attempted to establish non-surgical or minimal invasive methods.

Under appropriate conditions, porcine zygotes can be cultured *in vitro* to the developmental stage (8–16 cell stage) that enables transfer directly into the uterus. It has to be mentioned, however, that prolonged culture of porcine embryos drastically reduces the implantation rate (Blum-Reckow and Holtz, 1991). The *in vitro* culture is a prerequisite for non-surgical ET, first applied by transcervical implantation (Polge and Day, 1968; Sims and First, 1987). A double-walled catheter is placed *per vaginam* through the cervix and the embryos are deposited into the uterine body or horns. This first successful non-surgical ET in swine was monitored by implantation sites rather than farrowing. Transcervical ET performed by several other groups resulted in pregnancy rates of 20–60% (Reichenbach and Nieman, 1994; Glavin *et al.*, 1994; Hazeleger and Kemp, 1994). Despite the obvious benefits of non-surgical ET, there are also some limitations. The success rates are low in comparison to surgical ET, transfer of early stage embryos is impossible and the synchronization of the recipients has to be monitored by additional analyses.

The use of laparoscopy offers direct insight into the abdominal cavity by minimal invasion. During ET the reproductive organs are kept *in situ*. The embryos can be deposited into the oviduct or into the uterine horns under endoscopical view. Previous studies clearly demonstrated that the best conditions for embryo development from the one cell stage to blastocysts are provided by the Fallopian tube (Blum-Reckow and Holtz, 1991; Hagen *et al.*, 1991). The first trial of ET by endoscopy into the uterine horns was reported by Stein-Stefani and Holtz. These transfers resulted in a pregnancy rate of 14%, diagnosed on day 30 (Stein-Stefani and Holtz, 1987). Recently, we described the first successful endoscopic ET programme (embyro recovery and transfer) resulting in the birth of living piglets (Besenfelder *et al.*, 1997). In order to ensure the welfare of the animals as well as to obtain optimal results, the endoscopical technique was developed by the following steps:

1. ET into the Fallopian tube;
2. ET into the uterine horn;
3. Both optimized procedures were combined for unilateral embryo flushing;
4. Bilateral recovery of embryos.

When endoscopic ET was routinely used, 90% of the pigs became pregnant. The manipulation is minimally invasive, which provides the 'bifunctional' use of superovulated sows as embryo donors and piglet producers. In two cases of unilateral endoscopic recovery, the donors remained pregnant and delivered six and nine normal piglets, respectively (Besenfelder *et al.*, 1997).

Conclusions

Modern animal biotechnology facilitates the international exchange of genetic material and guarantees the maintenance of genetic resources by embryo banking. Embryo transfer reduces logistic and hygienic difficulties in pig breeding and is a prerequisite for renovation programmes of swine stocks. Furthermore, the handling of embryos is required for all transgene approaches aiming at the germline alteration of pigs (see below). The use of endoscopy in pig reproduction represents in many aspects the method of choice. The operation is quick and minimally invasive and causes therefore a minimum of stress and complications for the animals. In addition, the equipment is very mobile and can be used anywhere.

Transgenics

Introduction

The term transgenic (Gordon and Ruddle, 1981; Palmiter and Brinster, 1985) refers to organisms which carry in their genome and/or express *in vitro* manipulated gene constructs. The first mammals containing experimentally introduced foreign DNA were generated by microinjection of SV40 DNA into the blastocoel of mice and by infection of mouse embryos with Moloney leukaemia retrovirus (reviewed by Jaenisch, 1988). For animal breeding the three crucial aspects of transgenesis are integration, expression and transmission of the gene construct(s), i.e. gene transfer into the germline. Somatic gene transfer approaches result in (mostly transient) gene expression with the longest duration being a life span. The main use of this technology is human gene therapy, however, in some cases it might be beneficial in animal production.

Gene transfer techniques and applications

The techniques for generating transgenic vertebrates (Table 12.2) have been established in mice. The generation of transgenic farm animals was first reported over a decade ago (Brem *et al.*, 1985; Hammer *et al.*, 1985). The gene transfers were carried out by microinjection of DNA constructs into the pro nuclei of zygotes. At present this method is still the only proven route for the germline

Table 12.2. Methods for germline and somatic gene transfer.

Germline gene transfer (requires early embryonic stages)
Established techniques: DNA microinjection; infection with (retro-)viral vectors; embryonic stem cells
Alternative techniques (efficiency remains to be established): sperm-mediated gene transfer; transfection methods
Somatic gene transfer
Injection of free ('naked') or carrier-bound DNA; particle bombardment ('gene-gun'); aerosols
Infection with viral vector systems; transfection methods; cellular carriers

manipulation of livestock. The other listed gene transfer methods are facing some limitations (cloning capacity and expression efficiency of retroviral vectors), are not (yet) available (embryonic stem cells), are too inefficient (tissue culture transfection methods) or have not been proven to be satisfactory in their reproducibility (sperm-mediated gene transfer) (Brem, 1993; Brem and Müller, 1994). An intriguing simple method would be the sperm-mediated gene transfer, i.e. the mixing of spermatozoa with DNA constructs prior to (*in vitro*) fertilization (Lavitrano *et al.*, 1989). Notwithstanding the controversy this approach has generated (Brinster *et al.*, 1989) and the feasibility in routine application still to be proven, this method should be followed up (reviewed by Schellander and Brem, 1997). Another potential sperm-based gene delivery is proposed by Brinster and Avarbock (1994). Transplanted spermatogonial cells were shown to generate sperm capable of fertilizing oocytes and producing offspring. Provided that culture, transfection and selection conditions for spermatogonial cells with gene constructs are found, this method could be used for the generation of transgenic animals.

Germline gene transfer into farm animals offers the exciting prospect of completely new breeding strategies and novel applications. These fall into three groups:

1. The improvement of production efficiency and quality of animal products.
2. The production of recombinant proteins of high value.
3. The creation of animal models for human diseases and organs for xenotransplantation.

However, despite intensive efforts to transfer genes into livestock, only a few applications have been successful (reviewed Brem and Müller, 1994). Gene transfer into farm animals is facing several obstacles:

1. There is an obvious difference in scale (e.g. time schedules) and in reproductive biology between experiments in mice and livestock. On average only about 1–3% of embryos microinjected and transferred result in transgenic newborns (see Table 12.3).
2. DNA microinjection results in random integration of the gene constructs in the host genome. Thus the transgene expression often underlies 'chromosomal position effects', which results in uncertainty about the expected tissue specificity

Table 12.3. Efficiencies of gene transfer into mammals by DNA microinjection.

	Pig	Mouse	Rat	Rabbit	Sheep	Goat	Cattle
Animals born/injected embryos transferred	5–10%	10–20%	15–25%	10–15%	10–15%	15%	10–15%
Transgenic animals/offspring (integration frequency)	10–15%	15–20%	18%	10–12%	5–15%	7%	2–5%
Transgenic/injected embryos transferred (efficiency)	0.5–1 (–3)%	2–3 (–5)%	4–5%	1–2%	1–2%	1%	0.2%

[1]For references see Brem, 1993; Brem and Müller, 1994; Wall, 1996.

and in variations of transgene expression levels in different transgenic lines carrying identical DNA constructs (reviewed in Brem, 1993). One approach to achieve strict spatio-temporal pattern of expression from genes of interest is the use of yeast artificial chromosome (YAC) gene constructs providing extensive sequences flanking the coding unit of the gene in order to avoid unwanted side effects of transgene expression. Recently, reports of the first YAC DNA transgenic farm animals (rabbits) have been published (Brem *et al.*, 1996).

3. Totipotential cell lines derived from farm animals are currently not available for routine gene transfer experiments. Recently, sheep were cloned by nuclear transfer from an established cell line, which is another important step towards the routine availability of cultured embryonic cells (Campbell *et al.*, 1996; Solter, 1996). Without doubt, the most exciting development in pig embryology will be the *in vitro* establishment of ES cells from this species. The availability of this technique will give new impetus to gene transfer, because it will not only provide the possibility of additive gene transfer and homologous recombination but will notably reduce problems such as low efficiency, non-expression of transgenes or insertional mutations. The establishment of porcine early embryonic stage cell lines has been described (Wheeler, 1994). However, the final proof of pluripotency, i.e. germline chimerism remains to be demonstrated (see page 352).

Somatic gene transfer experiments do not aim at the integration of the gene construct into all cell types. Thus there is no requirement of transferring the DNA in early embryonic stages and 'transient' transgenesis can be achieved by all passive or active transfection methods developed in tissue culture or animal models (Table 12.1). In animal production somatic gene transfer might be used for the improvement of disease resistance (see also Fig. 12.2) or the production of proteins of high value (gene farming, see below) (e.g. Archer *et al.*, 1994). However, except for genetic immunization approaches (see Fig. 12.2; also Müller and Brem, 1994, 1996), the expression levels achieved with somatic gene transfer are currently too low for possible applications in farm animals.

Gene transfer into pigs by DNA microinjection

Gene transfer into pigs by direct microinjection of DNA into the pronuclei of zygotes has become a more or less established method in a number of laboratories. However, it is not yet a routine breeding method, especially in terms of its overall efficiency (for reviews see Wall *et al.*, 1992; Brem and Müller, 1994; Wall, 1996). The different steps required for the production of transgenic swine are depicted in Fig. 12.1. Additional intensive research is required in order to optimize the generation of transgenics (e.g. Martin *et al.*, 1996). The reproductive biology techniques, i.e. the collection, culture and transfer of porcine embryos, are described above. The laboratory equipment required and its use for DNA microinjection is listed in Table 12.4.

A typical gene construct should ideally provide all elements controlling temporal–spatial and tissue-specific expression. The coding portion (structural

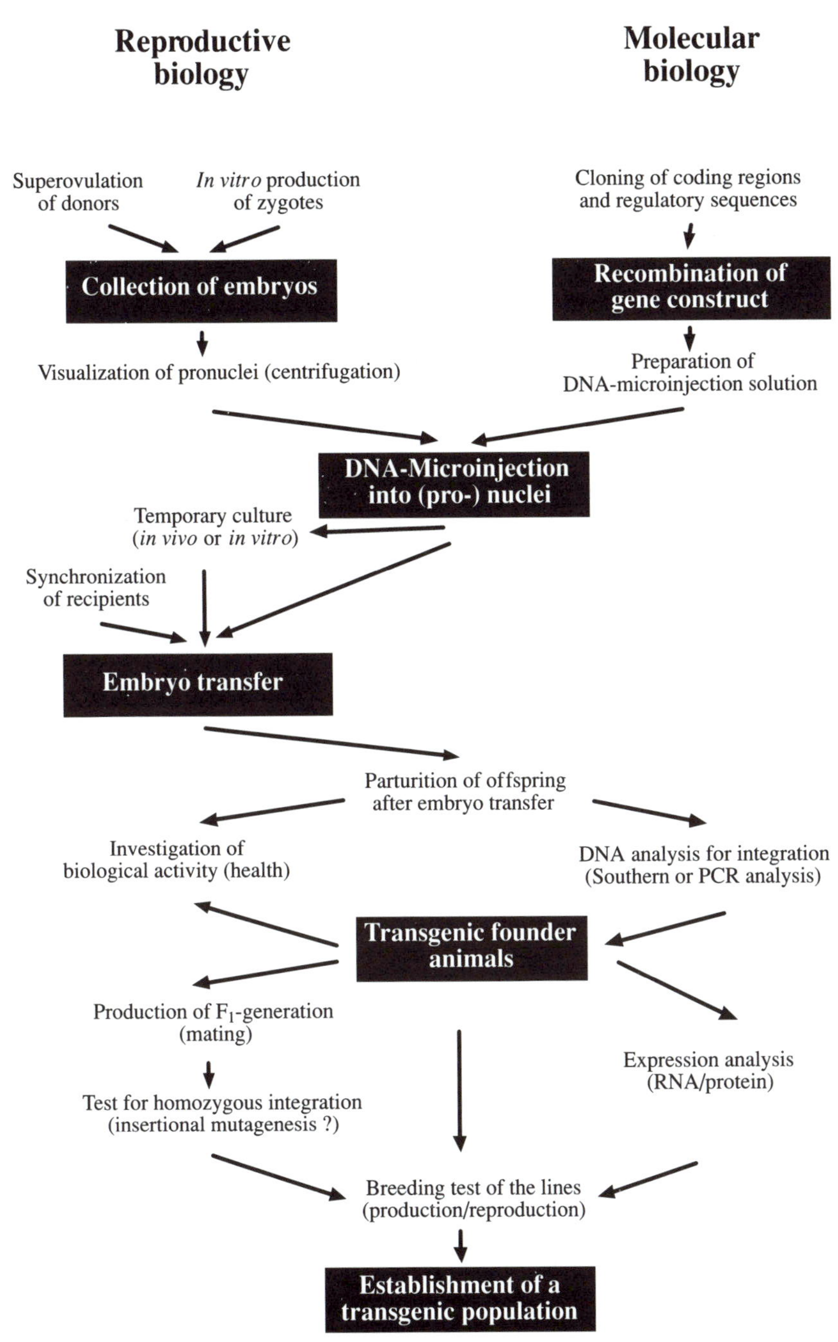

Fig. 12.2. Steps and techniques in reproductive and molecular biology required for the production of transgenics.

Table 12.4. Equipment required for microinjection of DNA.

Equipment	Use
Stereomicroscope with transmitted illumination	Embryo collection and preparation for transfer
Stable (vibration free) working bench	Construction of micromanipulation unit
Inverse microscope (× 20, × 400)	Visualization of (pro-)nuclei and microinjection
Micromanipulators	Moving of holding pipette and of injection pipette
Microinjection device	DNA flow in injection pipette
Injection chamber and temperature controlled criss-cross table	Handling of embryos
Microforge	Production of pipettes
Mechanical pipette puller	Production of injection pipettes (∅ 1 μm)
Centrifuge (× 15,000 g)	Removal of particles from the DNA microinjection solution; removal of cytoplasmic granula for visualization of porcine pronuclei

gene or cDNA) is linked to 5′ *cis*-control elements, i.e. promoter and enhancer/silencer regions, which specify tissue, time, quantity, accuracy and polarity of transcription. 3′ Control regions provide correct RNA processing signals and are thought to be implicated in RNA stability. Although there are no 'golden rules' for an optimal transgene design, it is currently envisaged to combine the coding sequences with 5′ and 3′ flanking regions as big as possible in order to shield the inserted DNA constructs from influences of the surrounding chromatin (see above). In the following, we briefly describe the transgenic approaches applied to swine so far.

Improvement of production efficiency and quality of animal products by transgenic means

Transgenic pigs carrying gene constructs altering growth related functions

Growth is a very complex process which is influenced by the interaction of hormones and autocrine/paracrine factors, nutritional conditions and environmental factors set against a discrete genetic background. Among the genetically determined factors, the genes encoding polypeptides of the growth hormone cascade are of particular interest. In this cascade the hypothalamus is responsible for the circadian production of the positive-acting growth hormone releasing hormone (GHRH, somatoliberin) and its antagonist, somatotrophin release-inhibiting factor (SRIF, somatostatin). Both hormones are also synthesized in the pancreas and gut cells. They control the production of growth hormone (GH, somatotrophic hormone, STH, somatotrophin) in the anterior lobe of the pituitary gland. The GH action is highly dependent on the metabolic status of the organism: low blood glucose levels result in catabolic effects (lipolysis) and a positive energy balance causes anabolic effects which are mainly governed by insulin-like growth factor 1 (IGF-1, somatomedin C). The

IGF-1 is synthesized in the liver but also in other organs and tissues and induces mitogenesis and protein synthesis. The release of GH is tightly controlled by positive and negative feedback involving the hormones of the cascade and metabolic components. In addition it shows marked fluctuations with several peaks occurring over a single day.

GH transgenic mice were produced for the first time in 1982 (Palmiter *et al.*, 1982). They showed an enhanced growth performance with a fourfold increase in growth rates and a twofold increase in final body weight. Subsequently a variety of transgenic mice were generated harbouring *GH* gene constructs encoding *GH* variants or *GH* of other species controlled by various promoters (reviewed by Wanke *et al.*, 1992). The *GHRH* and *IGF-1* gene constructs were also transferred first into mice.

The generation of the first transgenic pigs was based on the reasonable assumption that insertion of additional copies of the *GH* gene controlled by a heterologous (i.e. not controlled by the feedback mechanisms of the *GH* cascade) promoter would result in an enhanced growth performance of the transgenics (Pursel *et al.*, 1990). However, in contrast to the findings in transgenic mice, *GH*-transgenic pigs did not *a priori* show increased growth performance (e.g. Ebert *et al.*, 1988; Vize *et al.*, 1988). This was due to reduction in the food uptake combined with an increased utilization of nutrients in these animals. Only after feeding of a protein-enriched diet supplemented with essential amino acids, minerals and vitamins did *GH*-transgenic pigs attain a higher (15%) daily weight gain than control animals (reviewed by Pursel *et al.*, 1989). In terms of carcass composition, the transgenic pigs showed a massive reduction of the back fat (Pursel *et al.*, 1989; Solomon *et al.*, 1994). It turned out that constitutive and/or high level expression of GH in pigs caused a variety of pathological side effects, including gastric ulcers, severe synovitis, dermatitis, nephritis, cardiomegaly, pneumonia, insulin resistance, osteochondrosis dissecans and reduced fertility (Ebert *et al.*, 1988, 1990; Pursel *et al.*, 1989; Pinkert *et al.*, 1994). In additional experiments promoter elements were used which provided lower constitutive or inducible expression of *GH* constructs (Polge *et al.*, 1989; Wieghart *et al.*, 1990). This resulted in the desired increase in carcass leanness and a reduction of the detrimental side effects. It has to be mentioned, however, that again the desired effects were compromised by negative phenotypic features (Wieghart *et al.*, 1990; Pinkert *et al.*, 1994).

An alternative approach of altering the growth performance involves the differentiation process of muscle cells themselves. The chicken proto-oncogene *c-ski*, was shown to cause large increases in skeletal muscle when expressed in transgenic mice (Ebert and Schindler, 1993). Following this idea, a *c-ski* gene construct was transferred into swine (Pursel *et al.*, 1992). The observable effects were inconsistent among the transgenic pigs generated, probably caused by differences in spatial–temporal expression of the transgene and in levels of *c-ski*. Some pigs showed muscle hypertrophy at the age of 3–7 months. Others experienced weakness in the legs due to muscle atonia. Histological examination demonstrated a high degree of vacuolic degeneration of the muscle tissue.

The pathology associated with growth promotion by transgenesis is due to the inability to tightly and coordinately regulate the expression of genes involved in the growth hormone cascade or muscle development. It will be necessary to mimic the pulsatile release of the endogenous hormones or to direct and restrict the gene expression to the target cells. An attempt to increase muscle mass in pigs by directing IGF-1 expression into the skeletal muscle was only partially successful (Pursel *et al.*, 1996). Transgenic offspring expressed high levels of IGF-1 in a tissue-specific manner, but differences in phenotype of transgenic and control pigs have not been encountered.

It is questionable whether the efforts required for optimizing transgenic approaches to alter growth performance and carcass composition are justified by its potential benefits. In pig breeds that have been already extensively selected for growth performance and leanness, the transgene effects in terms of cost efficiency cannot be expected to be tremendous without the threat of detrimental consequences. Moreover, the production of transgenic meat depends on the consumer acceptance (McCracken, 1993; Pursel and Solomon, 1993).

Transgenic pigs carrying disease altering genes

Approaches to reduce disease susceptibility of livestock will be a benefit in terms of animal welfare and will also be of economic importance. The costs of disease and health control have been estimated to account for 10–20% of total production costs. Attempts to select for improved disease resistance by conventional breeding programmes are hampered by several problems (for review see Müller and Brem, 1991). Novel immunization strategies based on nucleic acid technologies focus on two main issues: additive gene transfer and the development of nucleic acid vaccines (see Fig. 12.3). The aim is to stabilize or transiently express components known to provide or influence non-specific or specific host defence mechanisms against infectious pathogens. In pigs, two approaches aimed at the improvement of health have been tested: the transfer of a specific disease resistance gene and of antibody encoding genes.

A well-examined specific disease resistance gene is the *Mx1* gene product of certain mouse strains. The mouse Mx1 protein belongs to a family of polypeptides with GTPase activity synthesized in interferon- (IFN-) treated vertebrate cells. Some Mx proteins have been shown to block the replication of certain negative-stranded RNA viruses (reviewed by Staeheli *et al.*, 1993). Synthesis of mouse Mx1 protein in various cell lines and transgenic mice demonstrated that it is both necessary and sufficient to promote resistance to influenza A viruses in previously susceptible cells and animals (for review see Müller and Brem, 1994; Pavlovic *et al.*, 1995). The cloning and functional characterization of this specific disease resistance gene enabled a gene transfer programme to study whether *Mx1* transgenic pigs would show reduced susceptibility to influenza infections (Müller *et al.*, 1992). Two transgenic pig lines were established showing IFN-inducible expression of transgene-specific mRNA. Despite extensive protein analysis no mouse Mx1 protein could be detected in the transgenic animals. Although the gene construct was shown to be tightly regulated in mouse cell

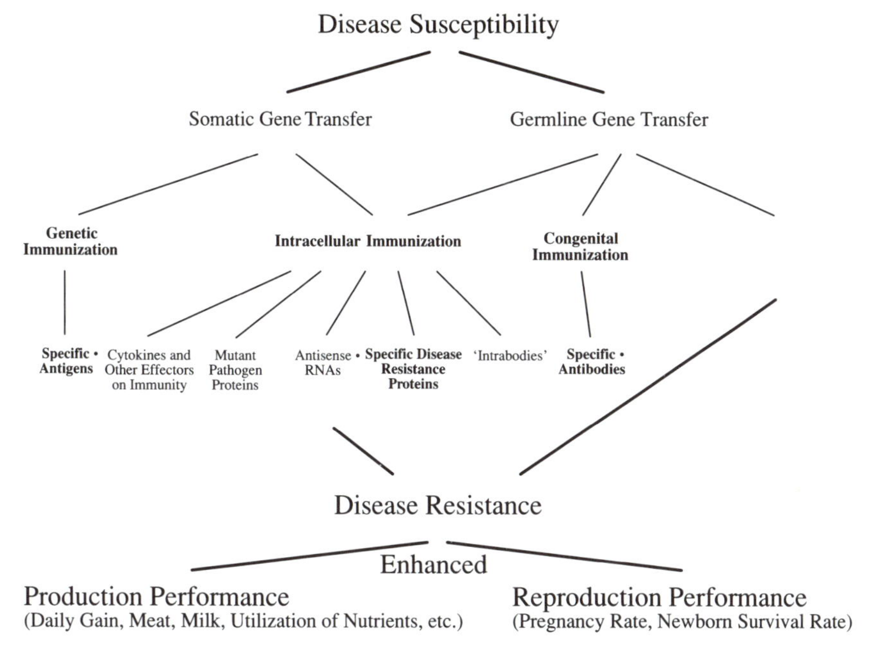

Fig. 12.3. Enhanced disease resistance resulting in enhanced production and reproduction performance: strategies to improve disease resistance by transgenic means (adapted from Müller and Brem, 1996). The disruption of specific genes by homologous recombination ('gene knockout') requires totipotential embryonic cells, which remain to be established for pigs. The strategies applied to swine are indicated by bold letters and explained in greater detail in the text. For the somatic gene transfer, i.e. nucleic acid ('genetic') immunization tested in swine see Haynes *et al.* 1996. The other concepts are described elsewhere in greater detail (Müller and Brem, 1994, 1996; Müller *et al.* 1996). A prerequisite for the practical use of the possible applicable methods in veterinary medicine and animal genetics is an evaluation of the benefits and costs and of the safety aspects.

lines it showed a basal low-level transcriptional activity in tissues of the transgenic pigs. Permanent Mx synthesis was shown to be deleterious during embryonic development. Thus, in our transgenic pigs, the observed leakiness of the promoter used may only have allowed the embryonic development of animals with abolished translation of the gene construct. The gene transfer experiments in pigs demonstrated that the choice of the regulatory elements controlling *Mx1* transgene expression is crucial. Therefore, future gene constructs should guarantee tight transgene regulation during embryonic development and highly inducible Mx1 synthesis after exogenous stimuli (Müller *et al.*, 1992).

'Congenital immunization' is defined as transgenic expression and germline transmission of a gene encoding an immunoglobulin specific for a pathogen and therefore providing congenital immunity without prior exposure to that pathogen (Müller *et al.*, 1996). As shown by many investigations, cloned genes coding for monoclonal antibodies can be expressed in large amounts in transgenic mice. The approach was tested in farm animals by expressing the gene constructs encoding mouse monoclonal antibodies in transgenic rabbits, pigs and sheep (Lo *et al.*, 1991; Weidle *et al.*, 1991). Both experiments resulted in transgene expression but revealed also some unexpected findings, e.g. aberrant sizes of the transgenic antibody or little antigen binding capacity. These results were explained by heterologous immunoglobulin chain associations and/or deviant post-translational modifications. Nevertheless, both experiments illustrate the potential of introducing beneficial traits such as germline-encoded immunity into farm animals. It remains to be investigated, however, whether the efforts required for optimizing the concept of 'congenital immunization' are justified by its benefits in terms of increasing disease resistance in a certain species. Following this route one has also to keep in mind that a given infectious pathogen will readily be able to escape the transgenic animal's immunity by changing its antigenic determinants. However, antibody-producing pigs could serve for the large-scale production of human monoclonal antibodies for therapeutic and diagnostic purposes (see below).

Production of recombinant proteins in transgenic swine

A variety of proteins required for diagnostic and therapeutic purposes are normally not available in sufficient quantity or quality when purified from natural sources or overproduced in bacterial expression systems. Transgene technologies have offered the possibility of producing recombinant proteins in the body fluids of farm animals (Logan, 1993; Jänne *et al.*, 1994; Wilmut and Whitelaw, 1994).

It is obviously more efficient to obtain milk from a ruminant than from swine; hence the transgenic conversion of the mammary gland is more frequently attempted in these species (reviewed by Bawden *et al.*, 1994; Brem and Müller, 1994; Maga and Murray, 1995). Nevertheless, due to higher efficiency in generating transgenic pigs, the production of proteins in the milk of swine has been tested. The regulatory elements required to direct gene expression into the mammary gland are reviewed in great detail elsewhere (Hennighausen, 1992; Bawden *et al.*, 1994). For the generation of transgenic swine, the regulatory sequences derived from the rodent whey acidic protein (WAP) were used. Wall *et al.* (1991) reported the expression of mouse WAP in the mammary gland up to a level of (1 g l^{-1}). However, the overproduction of this protein was found to impair the mammary development (Shamay *et al.*, 1992ab). In a second experiment a fusion gene consisting of the cDNA for human protein C (hPC) controlled by the mouse WAP promoter was used (Velander *et al.*, 1992). This gene construct was expressed at a high rate (up to 1 g l^{-1}) in the transgenic pigs. The recombinant hPC (zymogen of a serine protease) possessed an anticoagulant activity that was equivalent to that of plasma-derived hPC. The endoproteolytic processing of the hPC precursor to its mature form was found to be inefficient in the mammary gland (Lee *et al.*, 1995b). In transgenic mice, this problem could be solved by co-expressing of the processing enzyme (furin) in the milk (Drews *et al.*, 1995).

Expression of recombinant proteins in the blood is only feasible for proteins of which the biological activity or abnormally high levels are not detrimental to the health of the animal (Logan and Martin, 1994). This approach is applicable to the production of human antibodies (see above), human haemoglobin or inactive fusion proteins to be cleaved to their biologically active forms *ex vivo*. The production of human haemoglobin (Hb) in the erythrocytes of transgenic swine was achieved by combining the locus control region (LCR) of the human *globin beta* locus and the genes encoding the human α and β chains of haemoglobin (Swanson *et al.*, 1992). The human Hb purified from porcine proteins (including porcine haemoglobin) by ion-exchange chromatography exhibited similar oxygen equilibrium curves to haemoglobin derived from human serum. However, the level of expression was low with recoverable human Hb constituting only 10% of the total Hb in the porcine erythrocytes. The use of an isologous porcine promoter instead of the human regulatory sequences has resulted in swine with an expression level of 24% recoverable human Hb (Sharma *et al.*, 1994). Although more difficult than collecting milk, blood can be collected aseptically and at an earlier stage and is appropriate with a large number of animals. Thus transgenic swine might serve as an alternative to human erythrocytes as a source of human Hb.

Transgenic pigs as organ donors

Many groups are currently studying the transplantation of organs across species (xenografts) to eliminate the shortage of organs available for transplantation. For a variety of medical, practical, ethical and economic reasons the animal of choice in these studies is the pig (reviewed Dorling and Lechler, 1994; Lu *et al.*, 1994). According to the current understanding, the major barrier to successful xenotransplantation is hyperacute rejection, which is, in part, mediated by the deposition of high-titre preformed natural antibodies that activate the complement system. It has been established that these preformed naturally occurring anti-pig antibodies react with a single disaccharide, Galα-1,3-Gal (α-galactosyl epitope) (reviewed by Galili, 1993). This epitope is not expressed in Old World primates and humans because the α-1,3-galactosyltransferase is inactive in these species. The identification of this major xenogenic epitope led to the working hypothesis that genetic modification of the donor xenograft by stably downregulating the expression of the Galα-1,3-Gal epitope would significantly inhibit natural antibody binding and thus the activation of the complement cascade resulting in the hyperacute rejection. The a-1,3-galactosyltransferase would be an ideal target for producing 'knock-out pigs', which, however, depend on the availability of totipotential embryonic cells (see above). An alternative method involves the introduction of a gene for another enzyme (α-1,2-fucosyltransferase) competing with α-1,3-galactosyltransferase for adding a terminal sugar on to a carbohydrate chain. By overexpression of this enzyme, the synthesis of the xenogenic disaccharide is expected to be suppressed. Recently, tissue culture experiments and preliminary experiments in transgenic mice and pigs demonstrated convincingly the feasibility of this approach (Platt and Parker, 1995; Sandrin *et al.*, 1995; Sharma *et al.*, 1996). The complement cascade is downregulated at specific points by regulators such as decay-accelerating factor (DAF, CD55), membrane co-factor protein (MCP, CD46) and/or CD59. These proteins are collectively referred to as regulators of complement activation (RCAs) (reviewed by Cozzi and White, 1995). Transgenic swine expressing a particular RCA or combinations of RCAs have been generated. Transgenic pig organs perfused by human blood or transplanted into primates indicated the potential of this approach to trick the human immune response into seeing swine organs as human (Fodor *et al.*, 1994; Langford *et al.*, 1994; Carrington *et al.*, 1995; McCurry *et al.*, 1995; Rosengard *et al.*, 1995). The results reported up to date look promising, although prior to routine xenotransplantation lots of problems, not only in basic research but also in ethical aspects, remain to be solved (Bach *et al.*, 1995).

Conclusions

The transfer of genetic material by recombinant DNA technologies has a plethora of applications in the fields of medical basic research. The most prominent areas include laboratory animal disease models and studies of function and regulation of specific genes. Most gene transfer experiments in livestock have been performed with swine. This is due in part to the economic importance of pig production in many countries and in part to the reproductive nature of pigs. Current limitations of this technology for livestock are the unsatisfactory efficiency and the lack of knowledge in multigene interactions. The most promising research is focusing on the production of pharmaceutically useful proteins and the engineering of transgenic pig organs for cross-species transplantation. Although the transgene approaches show great potential to create beneficial changes in pigs, they are still at an early stage in development.

References

Anderson, G.B., Choi, S.J. and Bondurant, R.H. (1994) Survival of porcine inner cell masses in culture and after injection into blastocysts. *Theriogenology* 42, 204–212.

Archer, J.S., Kennan, W.S., Gould, M.N. and Bremel, R.D. (1994) Human growth hormone (hGH) secretion in milk of goats after direct transfer of the hGH gene into the mammary gland by using replication-defective retrovirus vectors. *Proceedings of the National Academy of Sciences, USA* 91, 6840–6844.

Bach, F.H., Robson, S.C., Winkler, H., Ferran, C., Stuhlmeier, K.M., Wrighton, C.J. and Hancock, W.W. (1995) Barriers to xenotransplantation. *Nature Medicine* 1, 869–873.

Baker, R.D. and Polge, C. (1976) Fertilization in swine and cattle. *Canadian Journal of Animal Science* 56, 105–119.

Bawden, W.S., Passey, R.J. and Mackinlay, A.G. (1994) The genes encoding the major milk-specific proteins and their use in transgenic studies and protein engineering. *Biotechnology and Genetic Engineering Reviews* 12, 89–137.

Besenfelder, U., Mödl, J., Müller, M. and Brem, G. (1997) Endoscopic embryo collection and transfer into oviduct and the uterus of pigs. *Theriogenology* 47, 1051–1060.

Blum-Reckow, B. and Holtz, W. (1991) Transfer of porcine embryos after 3 days of in vitro culture. *Journal of Animal Science* 69, 3335–3342.

Brem, G. (1993) Transgenic animals. In: Rehm, H.-J., Reed, G., Pühler, A. and Stadler, P. (eds) *Biotechnology.* VCH, Weinheim, Germany, pp. 745–832.

Brem, G. and Müller, M. (1994) Large transgenic animals. In: Maclean, N. (ed.) *Animals with Novel Genes.* Cambridge University Press, Cambridge, UK, pp. 179–244.

Brem, G., Brenig, B., Goodman, H.M., Selden, R.C., Graf, F., Kruff, B., Springman, K., Hondele, J., Meyer, J., Winnacker, E.-L. and Kräuszlich, H. (1985) Production of transgenic mice, rabbits and pigs by microinjection into pronuclei. *Zuchthygiene* 20, 251–252.

Brem, G., Besenfelder, U., Aigner, B., Müller, M., Liebl, I., Schütz, G. and Montoliu, L. (1996) YAC transgenesis in farm animals: rescue of albinism in rabbits. *Molecular Reproduction and Development* 44, 55–62.

Brinster, R.L. and Avarbook, M.R. (1994) Germline transmission of donor haplotype following spermatogonial transplantation. *Proceedings of the National Academy of Sciences, USA* 91, 11303–11307.

Brinster, R.L., Sandgren, E.P., Behringer, R.R. and Palmiter, R.D. (1989) No simple solution for making transgenic mice. *Cell* 59, 239–241.

Brüssow, K.-P. and Ratky, J. (1994) Repeated laparoscopical follicular puncture and oocyte aspiration in swine. *Reproduction in Domestic Animals* 29, 494–502.

Brüssow, K.-P., Ratky, J., Kanitz, W. and Becker, F. (1990) Determination of the duration of ovulation in gilts by means of laparoscopy. *Reproduction in Domestic Animals* 25, 184–190.

Bwanga, C.O., Ekwall, H. and Rodriguez Martinez, H. (1991a) Cryopreservation of boar sperm: III. Ultrastructure of boar spermatozoa frozen ultra-rapidly at various stages of conventional freezing and thawing. *Acta Veterinaria Scandinavica* 32, 463–471.

Bwanga, C.O., Hofmo, P.O., Grvle, I.S., Einarsson, S. and Rodriguez Martinez, H. (1991b) In vivo fertilizing capacity of deep frozen boar semen packaged in plastic bags and maxi-straws. *Zentralblatt der Veterinärmedizinischen Akademie* 38, 281–286.

Cameron, R.D.A., Durack, M., Fogarty, R., Putra, D.K.H. and McVeigh, J. (1989) Practical experience with commercial embryo transfer in pigs. *Australian Veterinary Journal* 66, 314–318.

Campbell, K.H.S., McWhir, J., Ritchie, W.A. and Wilmut, I. (1996) Sheep cloned by nuclear transfer from a cultured cell line. *Nature* 380, 64–66.

Carrington, C.A., Richards, A.C., Cozzi, E., Langford, G., Yannoutsos, N. and White, D.J. (1995) Expression of human DAF and MCP on pig endothelial cells protects from human complement. *Transplantation Proceedings* 27, 321–323.

Cozzi, E. and White, D.J.G. (1995) The generation of transgenic pigs as potential donors for humans. *Nature Medicine* 1, 964–966.

Donovan, P. (1994) Growth factor regulation of mouse primordial germ cell development. *Current Topics in Developmental Biology* 29, 189–225.

Dorling, A. and Lechler, R.I. (1994) Prospects for xenografting. *Current Opinion in Immunology* 6, 765–769.

Drews, R., Paleyanda, R.K., Lee, T.K., Chang, R.R., Rehemtulla, A., Kaufman, R.J., Drohan, W.N. and Lubon, H. (1995) Proteolytic maturation of protein C upon engineering in the mouse mammary gland to express furin. *Proceedings of the National Academy of Sciences, USA* 92, 10462–10466.

Ebert, K.M. and Schindler, J.E.S. (1993) Transgenic farm animals: progress report. *Theriogenology* 39, 121–135.

Ebert, K.M., Low, M.J., Overstrom, E.W., Buonomo, F.C., Baile, C.A., Roberts, T.M., Lee, A., Mandel, G. and Goodman, R.H. (1988) A moloney MLV-rat somatotropin fusion gene produces biologically active somatotropin in a transgenic pig. *Molecular Endocrinology* 2, 277–283.

Ebert, K.M., Smith, T.E., Buonomo, F.C., Overstrom, E.W. and Low, M.J. (1990) Porcine growth hormone gene expression from viral promoters in transgenic swine. *Animal Biotechnology* 1, 145–159.

Evans, M.J. and Kaufmann, M.H. (1981) Establishment in culture of pluripotent cells from mouse embryos. *Nature* 292, 154–156.

Feng, S., Zhang, Y., Li, S., Ma, Z., Wang, R. and Lu, D. (1990) Piglets from frozen (–20°C) embryos were born in China. *Theriogenology* 35, 199.

Fiser, P.S. and Fairfull, R.W. (1990) Combined effect of glycerol concentration and cooling velocity on motility and acrosomal integrity of boar spermatozoa frozen in 0.5 ml straws. *Molecular Reproduction and Development* 25, 123–129.

Fiser, P.S., Fairfull, R.W., Hansen, C., Panich, P.L., Shrestha, J.N.B. and Underhill, L. (1993) The effect of warming velocity on motility and acrosomal integrity of boar sperm as influenced by the rate of freezing and glycerol level. *Molecular Reproduction and Development* 34, 190–195.

Fodor, W.L., Williams, B.L., Matis, L.A., Madri, J.A., Rollins, S.A., Knight, J.W., Velander, W. and Squinto, S.P. (1994) Expression of a functional human complement inhibitor in a transgenic pig as a model for the prevention of xenogeneic hyperacute organ rejection. *Proceedings of National Academy of Sciences, USA* 91, 11153–11157.

Funahashi, H. and Day, B.N. (1993) Effects of different serum supplemental hormones on cytoplasmatic maturation of pig oocytes in vitro. *Journal of Reproduction and Fertility* 98, 177–185.

Funahashi, H., Cantley, T.C., Stumpf, T.T., Terlow, S.L. and Day, B.N. (1994a) In vitro development of in vitro-matured porcine oocytes following chemical activation or in vitro fertilization. *Biology of Reproduction* 50, 1072–1077.

Funahashi, H., Stumpf, T.T., Terlouw, S.L., Cantley, T.C., Rieke, A. and Day, B.N. (1994b) Development ability of porcine oocytes matured and fertilized in vitro. *Theriogenology* 41, 1425–1433.

Galili, U. (1993) Interaction of the natural anti-Gal antibody with α-galactosyl epitopes: a major obstacle for xenotransplantation in humans. *Immunology Today* 14, 480–482.

Glavin, J.M., Killian, D.B. and Stewart, A.N.V. (1994) A procedure for successful nonsurgical embryo transfer in swine. *Theriogenology* 41, 1279–1289.

Gordon, J.W. and Ruddle, F.H. (1981) Integration and stable germ line transmission of genes integrated into mouse pronuclei. *Science* 214, 1244–1246.

Hagen, D.R., Prather, R.S., Sims, M.M. and First, N.L. (1991) Development of one-cell porcine embryos to the blastocysts stage in simple media. *Journal of Animal Science* 69, 1147–1150.

Hammer, R.E., Pursel, V.G., Rexroad, C.E., Jr, Wall, R.J., Palmiter, R.D. and Brinster, R.L. (1985) Production of transgenic rabbits, sheep and pigs by microinjection. *Nature* 315, 680–683.

Harms, V.E. and Schmidt, D. (1970) Untersuchungen zur in-vitro-Befruchtung follikulärer und tubaler Eizellen vom Schwein. *Berliner und Münchner Tierärztliche Wochenschrift*, 269.

Hayashi, S., Kobayashi, K., Mizuno, J., Saitoh, K. and Hirano, S. (1989) Birth of piglets from frozen embryos. *Veterinary Record* 125, 43–44.

Haynes, J.R., McCabe, D.E., Swain, W.F., Widera, G. and Fuller, J.T. (1996) Particle-mediated nucleic acids immunization. *Journal of Biotechnology* 44, 37–42.

Hazeleger, W. and Kemp, B. (1994) Farrowing rate and litter size after transcervical embryo transfer in sows. *Reproduction in Domestic Animals* 29, 481–487.

Hennighausen, L. (1992) The prospects of domesticating milk protein genes. *Journal of Cellular Biochemistry* 49, 325–332.

Holtz, W., Schlieper, B., Stein-Stefani, J., Blum, B., Agrawala, P. and Rickert, J. (1987) Embryo transfer as a means to introduce new stock into spf pig herds. *Theriogenology* 27, 239.

Huff, B.G. and Esbenshade, K.L. (1992) Induction of follicular development and estrus in prepubertal gilts with gonadotropins. *Animal Reproduction Science* 27, 183–194.

Ishimori, H., Saeki, K., Inai, M., Itasaka, J., Miki, Y., Nozaki, N., Seike, N. and Kainuma, H. (1993) Direct transfer of vitrified bovine embryos. *Theriogenology* 39, 238.

Jaenisch, R. (1988) Transgenic animals. *Science* 240, 1468–1474.

Jänne, J., Hyttinen, J.M., Peura, T., Tolvanen, M., Alhonen, L., Sinervirta, R. and Halmekyto, M. (1994) Transgenic bioreactors. *International Journal of Biochemistry* 26, 859–870.

Kashiwazaki, N., Othani, S., Miyamoto, K. and Ogawa, S. (1991) Production of normal piglets from frozen hatched blastocysts –196°C. *Veterinary Record* 128, 256–257.

Kashiwazaki, N., Nakao, H., Othani, S. and Nakatsuji, N. (1992) Production of chimeric pigs by the blastocyst injection method. *Veterinary Record* 130, 186–187.

Kikuchi, K., Naito, K., Daen, F.P., Izaike, Y. and Toyoda, Y. (1995) Histone H1 kinase activity during in vitro fertilization of pig follicular oocytes matured in vitro. *Theriogenology* 43, 523–532.

Kramer, K.K. and Lamberson, W.R. (1991) Long-term effects of unilateral ovariectomy on ovarian function in gilts. *Animal Reproduction Science* 26, 137–150.

Langford, G.A., Yannoutsos, N., Cozzi, E., Lancaster, R., Elsome, K., Chen, P., Richards, A. and White, D.J. (1994) Production of pigs transgenic for human decay accelerating factor. *Transplantation Proceedings* 26, 1400–1401.

Lavitrano, M., Camaioni, A., Fazio, V.M., Dolci, S., Farace, M.G. and Sparafoda, C. (1989) Sperm cells as vectors for introducing foreign DNA into eggs: genetic transformation of mice. *Cell* 57, 717–723.

Lee, G.J., Ritchie, M., Thomson, M., MacDonald, A.A., Blasco, A., Santacreu, M.A., Argente, M.J. and Haley, C.S. (1995a) Uterine capacity and prenatal survival in Meishan and Large White pigs. *Animal Science* 60, 471–479.

Lee, T.K., Drohan, W.N. and Lubon, H. (1995b) Proteolytic processing of human protein C in swine mammary gland. *Journal of Biochemistry* 118, 81–87.

Lo, D., Pursel, V., Linto, P.J., Sandgren, E., Behringer, R., Rexroad, C., Palmiter, R.D. and Brinster, R.L. (1991) Expression of mouse IgA by transgenic mice, pigs and sheep. *European Journal of Immunology* 21, 25–30.

Logan, J.S. (1993) Transgenic animals: beyond 'funny milk'. *Current Opinion in Biotechnology* 4, 591–595.

Logan, J.S. and Martin, M.J. (1994) Transgenic swine as a recombinant production system for human hemoglobin. *Methods in Enzymology* 231, 435–445.

Lu, C.Y., Khair-El-Din, T.A., Dawidson, I.A., Butler, T.M., Brasky, K.M., Vazquez, M.A. and Sicher, S.C. (1994) Xenotransplantation. *FASEB Journal* 8, 1122–1130.

McCracken, K.J. (1993) Strategies for lean beef. *Food Sciences and Technologies Today* 7, 98–103.

McCurry, K.R., Kooyman, D.L., Diamond, L.E., Byrne, G.W., Martin, M.J., Logan, J.S. and Platt, J.L. (1995) Human complement regulatory proteins in transgenic animals regulate complement activation in xenoperfused organs. *Transplantation Proceedings* 27, 317–318.

Maga, E.A. and Murray, J.D. (1995) Mammary gland expression of transgenes and the potential for altering the properties of milk. *Bio/Technology* 13, 1452–1456.

Martin, M.J., Houtz, J., Adams, C., Thomas, D., Freeman, B., Keims, J. and Cottrill, F. (1996) Effect of pronuclear microinjection on the development of porcine ova in utero. *Theriogenology* 46, 695–701.

Mattioli, M., Bacci, M.L., Galeati, G. and Seren, E. (1988) Developmental competence of pig oocytes matured and fertilized in vitro. *Theriogenology* 31, 1201–1207.

Mödl, J., Reichenbach, H.-D., Wolf, E. and Brem, G. (1996) Development of frozen–thawed porcine blastocysts in vitro and in vivo. *Veterinary Record* 139, 208–210.

Morcom, C.B. and Dukelow, W.R. (1980) A research technique for the oviductal insemination of pigs using laparoscopy. *Laboratory Animal Science*, 1030–1031.

Motlik, J. and Fulka, J. (1974) Fertilization of pig follicular oocytes cultivated in vitro. *Journal of Reproduction and Fertility* 36, 235.

Müller, M. and Brem, G. (1991) Disease resistance in farm animals. *Experientia* 47, 923–934.

Müller, M. and Brem, G. (1994) Transgenic strategies to increase disease resistance in livestock. *Reproduction Fertility and Development* 6, 605–613.

Müller, M. and Brem, G. (1996) Intracellular, genetic or congenital immunisation – transgenic approaches to increase disease resistance of farm animals. *Journal of Biotechnology* 44, 233–242.

Müller, M., Brenig, B., Winnacker, E.L. and Brem, G. (1992) Transgenic pigs carrying cDNA copies encoding the murine Mx1 protein which confers resistance to influenza virus infection. *Gene* 121, 263–270.

Müller, M., Weidle, U.H. and Brem, G. (1996) Congenital immunisation of farm animals. In: Houdebine, L.M. (ed.) *Transgeneic animals – Generation and Use.* Harwood Academic Publishers GmbH, Chur, Switzerland.

Nagai, T. (1994) Current status and perspectives in IVM–IVF of porcine oocytes. *Theriogenology* 41, 73–78.

Nagai, T., Niwa, K. and Iritani, A. (1984) Effect of sperm concentration during preincubation in a defined medium on fertilization in vitro of pig follicular oocytes. *Journal of Reproduction and Fertility* 70, 271–275.

Nagai, T., Takahashi, T., Masuda, H., Shioya, Y., Kuwayama, M., Fukushima, M., Iwasaki, S. and Hanada, A. (1988) *In-vitro* fertilization of pig oocytes by frozen boar spermatozoa. *Journal of Reproduction and Fertility*, 585–591.

Nagashima, H., Kato, Y., Yamakawa, H. and Ogawa, S. (1988a) Survival of pig hatched blastocysts exposed below 15°C. *Theriogenology* 20, 280 (abstract).

Nagashima, H., Katoh, J., Shibata, K. and Ogawa, S. (1988b) Production of normal piglets from microsurgically split morulae and blastocysts. *Theriogeneology* 29, 485–495.

Nagashima, H., Kato, Y. and Ogawa, S. (1989) Microsurgical bisection of porcine morulae and blastocysts to produce monozygotic twin pregnancy. *Gamete Research* 23, 1–9.

Nagashima, H., Yamakawa, H. and Niemann, H. (1992) Freezability of porcine blastocysts at different peri-hatching stages. *Theriogenology* 37, 839–850.

Nagashima, H., Kashiwazaki, N., Ashman, R.J., Grupen, C.G. and Nottle, M.B. (1995) Cryopreservation of porcine embryos. *Nature* 374, 416.

Niemann, H. and Reichelt, B. (1993) Manipulating early pig embryos. *Journal of Reproduction and Fertility* (Suppl. 48), 75–94.

Onishi, A., Takeda, K., Komatsu, M., Akita, T. and Kojima, T. (1994) Production of chimeric pigs and the analysis of chimerism using mitochondrial desoxyribonucleic acid as a cell marker. *Biology of Reproduction* 51, 1069–1075.

Ortman, K. and Rodriguez Martinez, H. (1994) Membrane damage during dilution, cooling and freezing–thawing of boar spermatozoa packaged in plastic bags. *Zentralblatt der Veterinärmedizinischen Akademie* 41, 37–47.

Palmiter, R.D. and Brinster, R.L. (1985) Transgenic mice. *Cell* 41, 343–345.

Palmiter, R.D., Brinster, R.L., Hammer, R.E., Trumbauer, M.E., Rosenfeld, M.G., Birnberg, N.C. and Evans, R.M. (1982) Dramatic growth of mice that develop from eggs microinjected with metallotionein-growth hormone fusion genes. *Nature* 360, 611–615.

Pavlovic, J., Arzet, H.A., Hefti, H.P., Frese, M., Rost, D., Ernst, B., Kolb, E., Staeheli, P. and Haller, O. (1995) Enhanced virus resistance of transgenic mice expressing the human MxA protein. *Journal of Virology* 69, 4506–4510.

Pedersen, R.A. (1994) Studies of in vitro differentiation with embryonic stem cells. *Reproduction, Fertility and Development* 6, 543–552.

Piedrahita, J.A., Anderson, G.B., Martin, G.R., Bon Durant, R.H. and Pashen, R.L. (1988) Isolation of embryonic stem cell-like colonies from porcine embryos. *Theriogenology* 29, 286.

Pinkert, C.A., Galbreath, E.J., Yang, C.W. and Striker, L.J. (1994) Liver, renal and subcutaneous histopathology in PEPCK-bGH transgenic pigs. *Transgenic Research* 3, 401–405.

Platt, J.L. and Parker, W. (1995) Another step towards xenotransplantation. *Nature Medicine* 1, 1248–1250.

Polge, C. (1977) The freezing of mammalian embryos: perspectives and possibilities. In: Elliott, K. and Whelan, J. (eds), *The Freezing of Mammalian Embryos.* Elsevier Science Publishers, Amsterdam, pp. 3–13.

Polge, C. and Day, B.N. (1968) Pregnancy following non-surgical egg transfer in pigs. *Veterinary Record* 82, 712.

Polge, E.J.C., Barton, S.C., Surani, M.H.A., Miller, R., Wagner, T., Elsome, K., Davis, A.J., Goode, J.A., Foxcroft, G.R. and Heap, R.B. (1989) Induced expression of a bovine growth hormone construct in transgenic pigs. In: Heap, R.B., Prosser, C.G., and Lamming, G.E. (eds), *Biotechnology in Growth Regulation.* Butterworths, London, pp. 189–199.

Prather, R.S., Sims, M.M. and First, N.L. (1989) Nuclear transplantation in early pig embryos. *Biology of Reproduction* 41, 414–418.

Pursel, V.G. and Solomon, M.B. (1993) Alteration of carcass composition in transgenic swine. *Food Reviews International* 9, 423–439.

Pursel, V.G., Campbell, R.G., Miller, K.F., Behringer, R.B., Palmiter, R.D. and Brinster, R.L. (1988) Growth potential of transgenic pigs expressing a bovine growth hormone gene. *Journal of Animal Science* 66, 267.

Pursel, V.G., Pinkert, C.A., Miller, K.F., Bolt, D.J., Campbell, R.G., Palmiter, R.D., Brinster, R.L. and Hammer, R.E. (1989) Genetic engineering of livestock. *Science* 244, 1281–1288.

Pursel, V.G., Hammer, R.E., Bolt, D.J., Palmiter, R.D. and Brinster, R.L. (1990) Integration, expression and germ-line transmission of growth-related genes in pigs. *Journal of Reproduction and Fertility* (Suppl. 41), 77–87.

Pursel, V.G., Sutrave, P., Wall, R.J., Kelly, A.M. and Hughes, S.H. (1992) Transfer of c-ski gene into swine to enhance muscle development. *Theriogenology* 37, 278.

Pursel, V.G., Coleman, M.E., Wall, R.J., Elsasser, T.H., Haden, M., DeMayo, F. and Schwartz, R.J. (1996) Regulatory avian skeletal α-actin directs expression of insulin-like growth factor-I to skeletal muscle of transgenic pigs. *Theriogenology* 45, 348.

Rall, W.F. and Fahy, G.M. (1985) Ice-free cryopreservation of mouse embryos at −196°C by vitrification. *Nature* 313, 573–575.

Rath, D., Niemann, H. and Torres, C.R.L. (1995) In vitro development to blastocysts of early porcine embryos produced in vivo of in vitro. *Theriogenology* 43, 913–926.

Rath, D., Johnson, L.A., Dobrinsky, J.R., Welch, G.R. and Niemann, H. (1997) Production of piglets preselected for sex following *in vitro* fertilization with x and y chromosome-bearing spermatozoa sorted by flow cytometry. *Theriogenology* 47, 795–800.

Reichelt, B. and Niemann, H. (1994) Generation of identical twin piglets following bisection of embryos at the morula and blastocyst stage. *Journal of Reproduction and Fertility* 100, 163–172.

Reichenbach, H.-D., Mödl, J. and Brem, G. (1993) Piglets born after transcervical transfer of embryos into recipient gilts. *Veterinary Record* 133, 36–39.

Rosengard, A.M., Cary, N.R., Langford, G.A., Tucker, A.W., Wallwork, J. and White, D.J. (1995) Tissue expression of human complement inhibitor, decay-accelerating factor, in transgenic pigs. A potential approach for preventing xenograft rejection. *Transplantation* 59, 1325–1333.

Sandrin, M.S., Fodor, W.L., Mouhtouris, E., Osman, N., Chney, S., Rollins, S.C., Guilmette, E.R., Setter, E., Squinto, S.P. and McKenzie, I.F.C. (1995) Enzymatic remodelling of the carbohydrate surface of a xenogenic cell substantially reduces human antibody binding and complement-mediated cytolysis. *Nature Medicine* 1, 1261–1267.

Schellander, K. and Brem, G. (1997) The direct gene transfer through mammalian spermatozoa. In: Houdebine, L.M. (ed.) *Transgenic Animals – Generation and Use.* Harwood Academic Publishers, Amsterdam, The Netherlands, pp. 41–44.

Shamay, A., Pursel, V.G., Wall, R.J. and Hennighausen, L. (1992a) Induction of lactogenesis in transgenic virgin pigs: evidence for gene and integration site-specific hormonal regulation. *Molecular Endocrinology* 6, 191–197.

Shamay, A., Pursel, V.G., Wilkinson, E., Wall, R.J. and Hennighausen, L. (1992b) Expression of the whey acidic protein in transgenic pigs impairs mammary development. *Transgenic Research* 1, 124–132.

Sharma, A., Martin, M.J., Okabe, J.F., Truglio, R.A., Dhanjal, N.K., Logan, J.S. and Kumar, R. (1994) An isologous porcine promoter permits high level expression of human hemoglobin in transgenic swine. *Bio/Technology* 12, 55–59.

Sharma, A., Okabe, J., Birch, P., McClellan, S.B., Martin, M.J., Platt, J.L. and Logan, J.S. (1996) Reduction in the level of Gal(α1,3)Gal in transgenic mice and pigs by the expression of an α(1,2)fucosyltransferase. *Proceedings of the National Academy of Sciences, USA* 93, 7190–7195.

Sims, M.M. and First, N.L. (1987) Nonsurgical embryo transfer in swine. *Journal of Animal Sciences* 65, 386.

Smorag, Z., Wieczorek, B. and Jura, J. (1989) Stage-dependent viability of vitrified rabbit embryos. *Theriogenology* 31, 1227–1231.

Soede, N.M., Wetzels, C.C.H., Zondag, W., de Koning, M.A.I. and Kemp, B. (1995) Effects of time of insemination relative to ovulation, as determined by ultrasonography, on fertilization rate and accessory sperm count in sows. *Journal of Reproduction and Fertility* 104, 99–106.

Solomon, M.B., Pursel, V.G., Paroczay, E.W. and Bolt, D.J. (1994) Lipid composition of carcass tissue from transgenic pigs expressing a bovine growth hormone gene. *Journal of Animal Sciences* 72, 1242–1246.

Solter, D. (1996) Lambing by nuclear transfer. *Nature* 380, 24–25.

Springman, K. and Brem, G. (1989) Embryotransfer beim Schwein im Rahmen von gentransferprogrammen. *Tieraerztliche Praxis* Suppl. 4, 21–25.

Staeheli, P., Pitossi, F. and Pavlovic, J. (1993) Mx proteins: GTPases with antiviral activity. *Trends in Cell Biology* 3, 268–272.

Stein-Stefani, J. and Holtz, W. (1987) Surgical and endoscopic transfer of porcine embryos to different uterine sites. *Theriogenology* 27, 278.

Swanson, M.E., Martin, M.J., O'Donnell, J.K., Hoover, K., Lago, W., Huntress, V., Parsons, C.T., Pinkert, C.A., Pilder, S. and Logan, J.S. (1992) Production of functional human hemoglobin in transgenic swine. *Bio/Technology* 10, 557–559.

Velander, W.H., Johnson, J.L., Page, R.L., Russell, C.G., Subramanian, A., Wilkins, T.D., Gwazdauskas, F.C., Pittius, C. and Drohan, W.N. (1992) High-level expression of a heterologous protein in the milk of transgenic swine using the cDNA encoding human protein C. *Proceedings of the National Academy of Sciences, USA* 89, 12003–12007.

Vize, P.D., Michalska, A.E., Ashman, R., Lloyd, B., Stone, B.A., Quinn, P., Wells, J.R.E. and Seamark, R.F. (1988) Introduction of a porcine growth hormone fusion gene into transgenic pigs promotes growth. *Journal of Cell Science* 90, 295–300.

Waberski, D., Weitze, K.F., Gleumes, T., Schwarz, M., Willmen, T. and Petzoldt, R. (1994) Effect of time of insemination relative to ovulation on fertility with liquid and frozen boar semen. *Theriogenology* 42, 831–840.

Wall, R.J. (1996) Transgenic livestock: progress and prospects for the future. *Theriogenology* 45, 57–68.

Wall, R.J., Pursel, V.G., Shamay, A., McKnight, R.A., Pittius, C.W. and Hennighausen, L. (1991) High-level synthesis of a heterologous milk protein in the mammary glands of transgenic swine. *Proceedings of the National Academy of Sciences, USA* 88, 1696–1700.

Wall, R.J., Hawk, H.W. and Nel, N. (1992) Making transgenic livestock: genetic engineering on a large scale. *Journal of Cellular Biochemistry* 49, 113–120.

Wang, W.H., Niwa, K. and Okuda, K. (1991) In vitro penetration of pig oocytes matured in culture by frozen–thawed ejaculated spermatozoa. *Journal of Reproduction and Fertility* 93, 491–496.

Wanke, R., Wolf, E., Hermans, W., Folger, S., Buchmüller, T. and Brem, G. (1992) The GH-transgenic mouse as an experimental model for growth research: clinical and pathological studies. *Hormone Research* 37, 74–87.

Weidle, U.H., Lenz, H. and Brem, G. (1991) Genes encoding a mouse monoclonal antibody are expressed in transgenic mice, rabbits and pigs. *Gene* 98, 185–191.

Welch, G.R., Waldbieser, G.C., Wall, R.J. and Johnson, L.A. (1995) Flow cytometric sperm sorting and PCR to confirm separation of X- and Y-chromosome bearing bovine sperm. *Animal Biotechnology* 6, 131–139.

Wheeler, M.B. (1994) Development and validation of swine embryonic stem cells: a review. *Reproduction, Fertility and Development* 6, 563–568.

Wieghart, M., Hoover, J.L., McGrane, M.M., Hanson, R.W., Rottman, F.M., Holtzman, S.H., Wagner, T.E. and Pinkert, C.A. (1990) Production of transgenic pigs harbouring a rat phosphoenolpyruvate carboxykinase-bovine growth hormone fusion gene. *Journal of Reproduction and Fertility* 41, 89–96.

Wilmut, I. and Whitelaw, C.B.A. (1994) Strategies for production of pharmaceutical proteins in milk. *Reproduction, Fertility and Development* 6, 625–630.

Wu, G.M., Qin, P.C., Tan, J.H. and Wang, L.A. (1992) In vitro fertilization of in vitro matured pig oocytes. *Theriogenology* 37, 323.

Yoshino, J., Kojima, T., Shimizu, M. and Tomizuka, T. (1993) Cryopreservation of porcine blastocysts by vitrification. *Cryobiology* 30, 413–422.

Developmental Genetics 13

D. Pomp[1] and R. Geisert[2]
[1]Department of Animal Science, University of Nebraska-Lincoln, Lincoln, NE 68583, USA; [2]Department of Animal Science, Oklahoma State University, Stillwater, OK 74074, USA

Introduction

The genetics of development is by definition a very broad field, encompassing the control of genes regulating the life cycle from gamete formation, fertilization, embryonic development, postnatal growth, sexual maturity and ageing. Traditionally, however, developmental genetics has focused on regulation during the earliest stages of an animal's existence, emphasizing gene expression and its phenotypic consequences from fertilization through the early periods of embryogenesis. In the pig, major developmental events and their post-fertilization timing within this early period of ontogeny include: initial control by maternally derived messages and transcripts, followed by initiation of embryonic gene expression and transition from maternal to embryonic control during the four-cell stage; cell cleavage, blastocyst formation and the first differentiation event (inner cell mass versus trophectoderm) within the embryo by day 5; differentiation of the first cell lineages by day 10; a unique period of cellular remodelling leading to conceptus expansion and elongation between days 10 and 12; signalling for maternal recognition of pregnancy by conceptus oestrogen synthesis from days 12 to 16; and placental attachment by day 14 of embryonic development.

In addition to the basic desire to understand the genetic and physiological mechanisms underlying early embryonic development in the pig, such an understanding is of critical practical significance in pork production for several reasons. First, the ability to culture and nurture early porcine gametes and embryos *in vitro* without loss of viability or function is of fundamental importance in support of a growing array of reproductive biotechnologies such as *in vitro* fertilization, embryo storage and transfer, production of chimeras, cloning by nuclear transfer and genetic engineering by gene transfer or homologous recombination. While *in vitro* development from zygote to blastocyst has been successful in pigs (reviewed by Petters and Wells, 1993), difficulties resulting in embryo loss are often encountered during the period when genomic control of development is being transferred from maternal to embryonic genome, a phenomenon referred to as the 'four-cell block' to *in vitro* development (Davis, 1985). Perhaps of even greater significance is the need to decrease embryonic loss *in vivo*. One of the primary objectives for enhancing economic efficiency in pig production is to increase the number of pigs born per female per year. At present, reproduction (see Chapter 11) is an inefficient process, with substantial embryonic loss occurring prior to day 25 of pregnancy (Pope and First, 1985). Much of this inefficiency may be traced to the period between days 10 and 12 of gestation (see review by Pope, 1995) when the porcine conceptus is undergoing the unique period of cellular remodelling and trophoblastic elongation (see review by Geisert *et al.*, 1990).

Most knowledge on mammalian developmental genetics has originated from studies in the mouse, due to their ease of use and the availability of large quantities of embryonic material. It is not our intent in this chapter to fill in the many gaps that exist in our understanding of developmental genetic events in the pig by reviewing the current status in other species, but rather to focus on the current knowledge of porcine developmental genetics, targeting primarily the two critical periods encompassing transition of control from maternal to embryonic genome, and subsequent conceptus expansion and placental attachment. The latter period will be especially emphasized, given its direct relevance to embryo loss and the fact that it is a biological phenomenon that is unique to the pig and for which gene expression has been studied in some detail. Clearly, a greater understanding of the control of gene expression during these periods of early development may enable researchers to regulate better and eventually manipulate embryonic survival both *in vitro* and *in vivo*.

Early Gene Expression and Transition from Maternal to Embryonic Control

Early embryonic development in pigs (Figs. 13.1 and 13.2), as in other mammalian species, is characterized by a series of equal cell cleavages, followed by cell compaction, appearance of a blastocoele, formation of a blastocyst, and the initial differentiation steps leading to formation of the first cell lineages (see reviews by Cruz and Pederson, 1991, and Betteridge, 1995, for description of early developmental events, their timetable, and comparison within domesticated mammalian species). Upon fertilization, the porcine embryo possesses the entire complement of genetic material that will guide its development. However, developmental control during the first two cell cycles (through the one-cell and two-cell stages) of an embryo's life span is under the direct control of transcripts and peptides of maternal origin that had been synthesized and accumulated during oocyte development, with little or no contribution from the embryonic genome. It is not until the third cell cycle, or the four-cell stage, that the porcine embryonic genome is activated and begins to control its own developmental fate. Specific timing of this period of transition of control from maternal to embryonic genome is species dependent, and has clearly been best characterized in the mouse model (see Nothias *et al.*, 1995, and Schultz *et al.*, 1995, for recent reviews). However, the porcine story is becoming increasingly clear, largely due to the efforts of Randy Prather and his colleagues (see review by Prather, 1993).

Timing of the onset of RNA synthesis by the porcine embryo has been evaluated by many different experimental methods. Indirectly, the inability of embryos to survive *in vitro* culture conditions, referred to as the four-cell block (Davis, 1985), may be the result of a particular sensitivity at the time of initiation of transcriptional activity. Similar *in vitro* blocks have been observed in other species, and are also correlated temporally with the timing of initiation of transcription by the embryonic genome. Eyestone *et al.* (1986) have further defined the timing of the block in pigs to the gap phase 2 (G2) of the third cell cycle, when gene expression would be expected to occur. Further evidence for timing of the maternal to embryonic genome transition can be found in the appearance of definitive nucleoli, capable of transcribing ribosomal RNA of embryonic origin, at the four- and eight-cell stage (Tomanek *et al.*, 1989).

Perhaps one of the most powerful methods by which to evaluate embryonic transcription is the treatment of developing embryos with α-amanitin, which specifically inhibits RNA polymerase II. Protein profiles which are sensitive to this inhibitor must, therefore, be of embryonic as opposed to maternal origin. Jarrell *et al.* (1991) found that late four-cell embryos lacked specific proteins that were present at earlier stages and in unfertilized oocytes. Presence of these proteins caused resistance to treatment with α-amanitin, indicating messages of maternal origin that were subsequently degraded (Jarrell *et al.*, 1991). In contrast, the late four-cell embryo produced 26 kDa and 50 kDa proteins that were sensitive to α-amanitin. While others had shown that new RNA was being

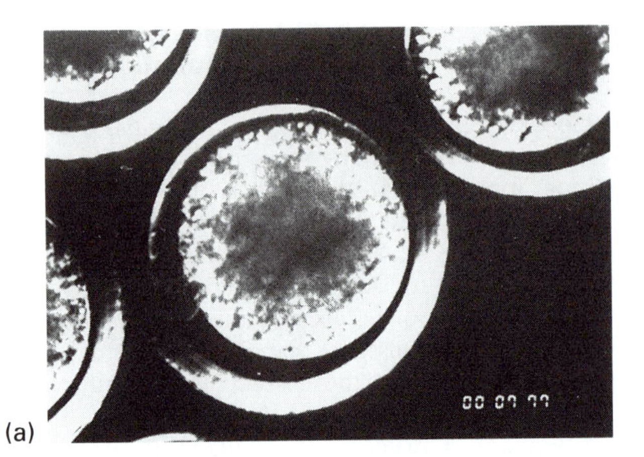
(a)

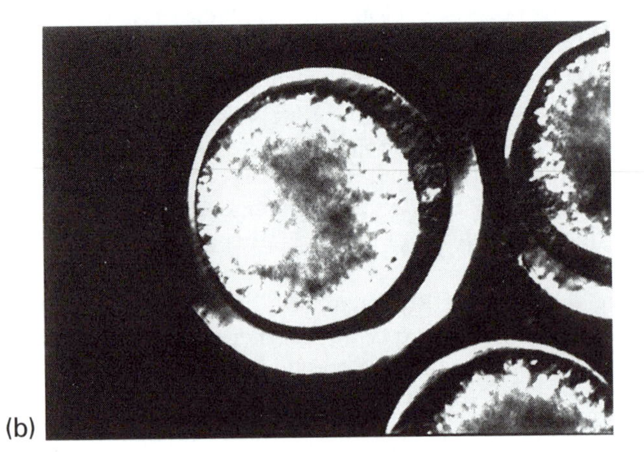
(b)

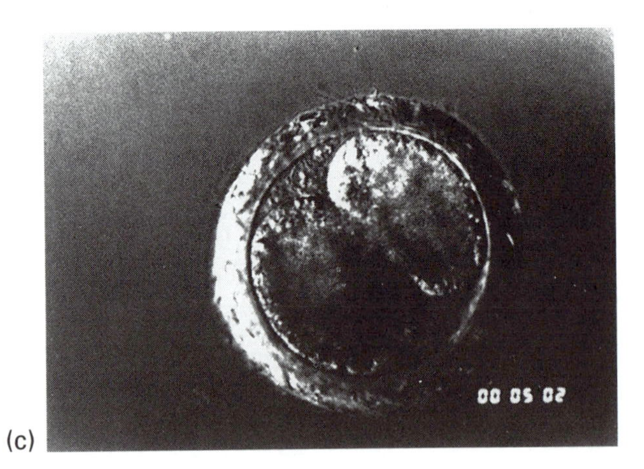
(c)

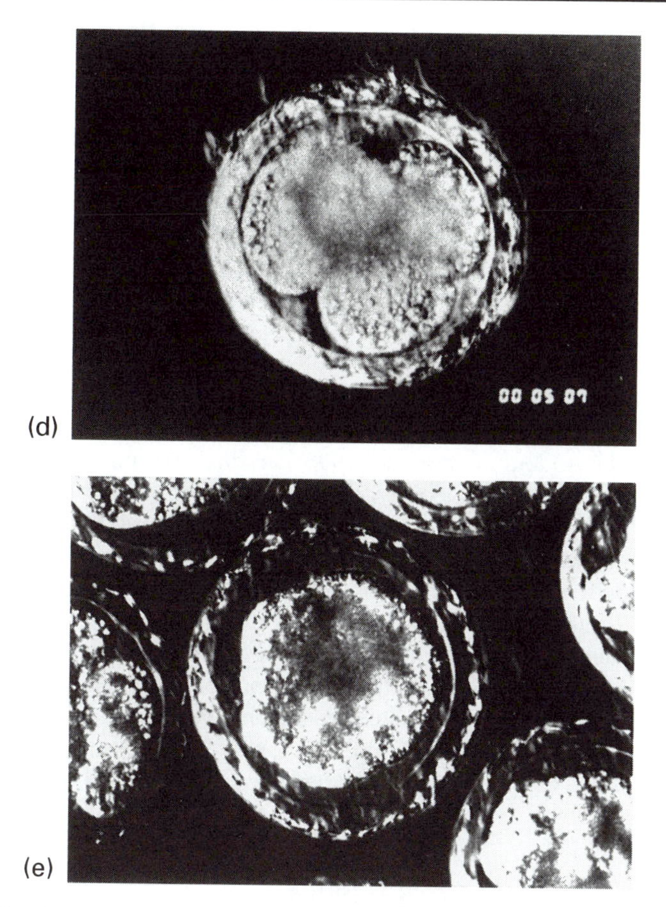

Fig. 13.1. Early pre-implantation development of pig embryos: (a) unfertilized oocyte at metaphase II of meiosis, with the first polar body at 8 o'clock; (b) zygote with two polar bodies. Note a clearing in the cytoplasm at 8 and 9 o'clock, where the pronuclei are located; (c) two-cell stage embryo. By this time, the zona pellucida has accumulated numerous additional spermatozoa; (d) four-cell stage embryo. One blastomere at 12 o'clock is out of focus; (e) compact morula. Distinct boundaries between individual cells have been lost.

synthesized in porcine four-cell embryos on the basis of incorporation of uridine (Freitag *et al.*, 1991), this was the first qualitative evidence demonstrating activation of the embryonic genome at the late four-cell stage in pigs. Jarell *et al.* (1991) also reported qualitative changes in protein profiles at later stages of pre-implantation development, indicating an expanding diversity of gene expression as ontogeny progresses. Prather and Rickords (1992) identified an additional α-amanitin sensitive event at the four-cell stage, namely the appearance of the Y12 antigen, which is correlated with mRNA production and processing. The Y12 antigen was further found to be present only at or beyond the four-cell stage (Prather and Rickords, 1992).

The specific timing when the embryonic genome is first activated was studied in detail by Schoenbeck *et al.* (1992), who cultured four-cell embryos *in vitro* and evaluated qualitative changes in protein profiles and their sensitivity to α-amanitin at various stages within the cell cycle. Significant qualitative alterations in protein synthesis were identified following 14 to 16 hours post-cleavage to the four-cell stage, corresponding with a G2 phase of the cycle. Furthermore, treatment with α-amanitin inhibited further cleavage *in vitro*, but only when exposure took place prior to 16 hours post-cleavage (Schoenbeck *et al.*, 1992). Embryos exposed after 24 hours post-cleavage to four-cells were able to continue development through the third cell cycle and cleave to an eight-cell embryo. Prather (1993) has suggested that this embryonic message necessary for cleavage to the eight-cell stage may be either the cyclin *B1* or *cdc*25 genes, and has recently found evidence for expression of these loci at the four-cell stage (R.S. Prather and J. Anderson, Missouri, 1996, personal communication).

Thus, maternal control of embryonic development in pigs predominates until the late four-cell stage, by which time maternal messages are mostly degraded and new RNA is synthesized by the embryonic genome. This period of transition of genetic control in the porcine embryo appears to be correlated with embryonic loss *in vitro*, and could potentially be a direct source of variation in embryo survival *in vivo* as well. Furthermore, it is possible that variation within a litter of embryos in timing of initiation of control of development by the embryonic genome may cascade to subsequent changes in protein profiles, and thus contribute to variation in embryonic development at later stages, which has been hypothesized as a potential cause for embryo loss in pigs (Pope, 1995).

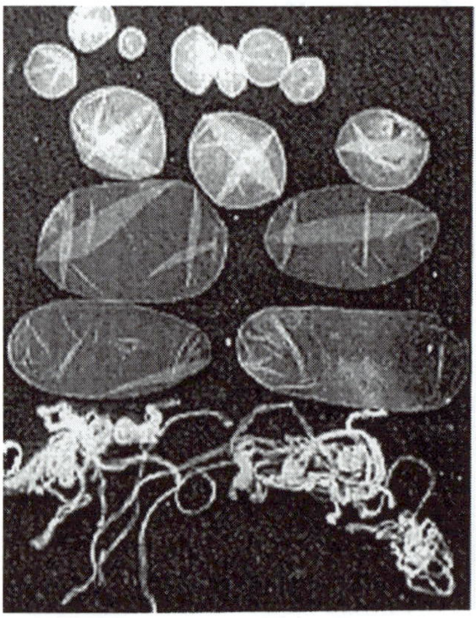

Fig. 13.2. Progression during rapid trophoblastic elongation in the porcine conceptus as morphology changes from spherical, ovoid, tubular and finally filamentous shape within 2 to 3 hours.

Unfortunately, little information exists regarding the identity of the gene products that result from early activation of the embryonic genome in porcine embryos, and which factors regulate initiation of transcription and early developmental events in pigs. Identification of these genes and regulatory mechanisms will contribute greatly to our understanding of the factors that control early embryonic development and survival in the pig. New methodologies such as reverse transcription-polymerase chain reaction (RT-PCR) and its utilization in differential display (Zimmerman and Schultz, 1994) should help to overcome the limitation of minute quantities of biological specimen in early embryos, and prove to be a powerful tool in attempts to characterize genes that regulate early development. Using this method, Schultz *et al.* (1995) identified an increase in expression of a gene corresponding to the protein synthesis translation initiation factor eIF-4C in mouse embryos, just subsequent to the transition from maternal to embryonic control of development. The presence of eIF-4c was found to be sensitive to α-amanitin, implicating the embryonic genome as the source of transcription. Abdelrahman *et al.* (1995) used RT-PCR methods to demonstrate the presence of transcripts for two transcription regulators, *OCT4* and *OCT6*, in early human embryos with more than ten cells. The findings that early embryonic gene expression events include specific regulators of transcription and translation may be indicative of the onset of autocrine initiation of gene regulation. In mice, regulation of early gene expression appears to be time-dependent and involves an initial chromatin-mediated repression of promoter activity, followed by acquisition of enhancer and TATA-box dependent transcriptional ability (Nothias *et al.*, 1995).

Genes Regulating Cleavage to Blastocyst Formation

There is a paucity of data regarding the specific expression of genes during the cleavage stages of porcine embryos that regulate rate of development and the ontogeny of the first differentiation events that occur. The availability of powerful methods utilizing PCR is likely to result in significant research advancement in this area in the near future. This has been the case in several other species, where relatively detailed glances into the profile of gene expression during this brief window of development have been documented in mice (Rappolee, 1992; Schultz *et al.*, 1992), cows (Schultz *et al.*, 1992; Watson *et al.*, 1992) and more recently in sheep (Watson *et al.*, 1994). These studies have naturally focused on the timing of appearance of a variety of growth factors and their receptors.

In the mouse, transcripts for *IGF-II*, *TGF-α*, *TGF-β1* and *-β2*, *kFGF/FGF-4*, *PDGF-α*, *LIF* and *IL-6* have been detected in the pre-implantation by RT-PCR embryo (see review by Rappolee, 1992), while no evidence for expression of the *IGF-I*, *Int-2/FGF-3*, *βFGF/FGF-2*, *EGF*, *NGF-β*, *IL-3*, *G-CSF* and *GM-CSF* loci was found (Rappolee, 1992). Furthermore, transcripts representing receptors for *IGF-I* and *-II*, insulin, *PDGF-α* and *-β*, *CSF-1*, *EGF* and mast cell growth factor were detected by the blastocyst stage of murine development (Rappolee, 1992). Relatively similar expression profiles were found in both bovine (Schultz *et al.*,

1992; Watson *et al.*, 1992) and ovine (Watson *et al.*, 1994) pre-implantation embryos, although some qualitative and temporal differences were suggested (see Watson *et al.*, 1994).

It is likely that the cleaving porcine embryo follows a relatively similar pattern of ontogeny in terms of growth factor ligand and receptor gene expression. Some supporting evidence was provided by Chastant *et al.* (1994), who used immunohistochemistry to detect IGF-II receptors but no IGF-I receptors on whole day 4 porcine embryos, while in a preliminary report, Li *et al.* (1996) used RT-PCR to identify low levels of *IGF-II* transcripts from four-cell to blastocyst stages and ubiquitous expression of the *IGF-I* and *-II* receptor loci throughout the pre-implantation period. The findings from Li *et al.* (1996) are of particular interest, because expression of the *IGF-I* receptor gene was detected in oocytes and electrically activated oocytes, indicating that this locus may not be maternally imprinted in pigs in contrast to reports in mice (Rappolee *et al.*, 1992). Transcripts for *IGF-II* were not found in porcine oocytes (Li *et al.*, 1996), potentially indicating conservation of maternal imprinting of this locus across mice (DeChiara *et al.*, 1990) and pigs.

It is clear that:

1. Various growth factors appear to represent a major element of transcriptional activity in the preimplantation embryo;
2. Various growth factors are likely to be essential for early development;
3. Embryos function to provide themselves with at least a subset of such factors in addition to those provided by the maternal oviductal and uterine milieux.

Growth factors are likely to be involved in regulation of gene expression at the levels of both transcription and translation (Schultz *et al.*, 1990), and through these influences help to determine the rate of development and differentiation of pre-implantation pig embryos. These effects do not appear to be dependent on sex of the embryo or on sex differences in gene expression, as several reports have found no evidence for effects of sex on size of early pig embryos (Pomp *et al.*, 1995; Kaminski *et al.*, 1996).

Breed differences have been established in cell number of pre-implantation (days 5.5 to 6.5) porcine embryos (Rivera *et al.*, 1996), primarily as a result of variation in the quantity of trophectoderm cells. While the *PED* (pre-implantation embryonic development) gene product, which has been shown to exert control over the rate of cleavage of pre-implantation mouse embryos, was recently characterized and identified as the Qa-2 antigen of the *Q9* gene within the *MHC* (Xu *et al.*, 1994), a porcine homologue with similar influences on development has not been identified. However, Ford *et al.* (1988) did demonstrate that genetic variation within the SLA can influence developmental rate of pre-implantation embryos in minature pigs. Rate of early embryonic development in pigs may be an important factor in determination of an embryo's fitness to survive to term (Pope, 1995). While several reports have linked *PED* genotype with embryonic survival in the mouse (Warner *et al.*, 1991, 1993), a recent study has shown that *PED* genotype of a female mouse has no influence on the survival rate of her embryos in a large, segregating population (D. Pomp and E.J. Eisen, 1996, unpublished data).

While it is important to understand developmental genetics and early embryonic differentiation events in order to regulate and limit embryonic loss both *in vitro* and *in vivo*, such an understanding should also prove beneficial towards the goal of producing multiple copies of genetically identical animals. Successful cloning by nuclear transfer (see Brem *et al.*, Chapter 12 this volume) requires that the donor nuclei be totipotent, or capable of directing the formation of fertile organisms that participate in the production of normal progeny (Prather, 1996). While these donor nuclei undergo processes of remodelling and reprogramming once fused to a recipient oocyte (see review by Prather, 1996), it is likely that totipotency may be regulated by the degree of differentiation of the cells that supply the nuclei for transfer. Understanding of early differentiation and the genes and gene products involved may allow for future extension of the developmental stages of embryos capable of providing totipotent cells for cloning, and/or enable regulation of the remodelling/reprogramming processes to include embryonic cells with a greater degree of differentiation. The recent birth of a lamb following cloning by nuclear transfer of an adult mammary cell (Wilmut *et al.*, 1997) indicates that at least some cell populations may not become irreversibly differentiated. It remains to be determined whether such results may be obtained in pigs and from adult cell populations other than mammary.

Genes Regulating Conceptus Expansion and Placental Attachment (Table 13.1)

Conceptus development

Following the stage of blastocyst formation and hatching, the unattached blastocysts have differentiated three developmental cell lineages. On days 10 to 11 of gestation, the 3 to 4 mm spherical porcine blastocyst consists of an outer covering of trophectoderm, inner lining of hypoblast (endoderm) and a cluster of loosely connected irregularly shaped cells forming the embryoblast or inner cell mass (Stroband *et al.*, 1984). Mesodermal outgrowth radiates from the embryoblast between the trophectoderm and hypoblast forming the third germ layer in embryonic development with continued growth and expansion of what is now called the conceptus (embryoblast and trophoblast) to an approximately 8 to 10 mm spherical morphology (Geisert *et al.*, 1982; Stroband and Van der Lende, 1990). Conceptus development between days 10 and 12 of gestation is a period of high early embryonic mortality in the pig (see review by Pope, 1995). This period of development corresponds to critical events involving conceptus intrauterine migration, trophoblastic elongation, steroidogenesis, establishment of pregnancy and placental attachment to the maternal uterine surface epithelial lining (see review by Geisert *et al.*, 1990).

Within the confines of the uterine lumen, porcine conceptuses undergo a phenomenal morphological change (Fig. 13.2) from 10 mm spherical to tubular (20 to 40 mm) and finally filamentous (100 mm in length) shape in less than 3 to 4 hours (Anderson, 1978; Geisert *et al.*, 1982; Stroband and Van der Lende, 1990). Rapid transition from spherical to filamentous thread-like morphology occurs through cellular remodelling and restructuring rather than an increase in mitotic activity (Geisert *et al.*, 1982). The drastic and rapid expansion of the porcine conceptuses allows them to establish independently a region of non-invasive, epitheliochorial placental contact for nutrient transfer throughout pregnancy. Conceptus elongation is regulated through developmental gene expression, as trophoblastic elongation is not necessarily synchronous as morphological diversity between conceptuses within a given litter often exists (Anderson, 1978; Geisert *et al.*, 1982; Stroband *et al.*, 1990, Pope, 1995). Geisert *et al.* (1982) suggested that a difference of as little as 1 mm in spherical diameter could result in 2 to 4 hours difference in the timing of trophoblast expansion between conceptuses. Elongation is therefore programmed through developmental cues by the conceptuses as they reach approximately 10 mm in diameter rather than through direct stimulation by uterine secretions (Geisert *et al.*, 1991), although uterine secretions play a significant role in conceptus growth and survival.

Rapid trophoblastic elongation, which sets the boundaries for placental attachment and allotment of uterine space available to each embryo, is essential for embryonic survival in the pig. It not only delineates the surface area for the placenta but also provides the transport vehicle to deliver oestrogen synthesized by the conceptuses throughout the uterus to maintain functional corpora lutea during pregnancy (see reviews by Bazer *et al.*, 1984; Geisert *et al.*, 1995). Conceptus oestrogen synthesis, which is the signal for maternal recognition of pregnancy on days 11 to 12 of gestation in the pig, is developmentally regulated and controls endometrial secretion and placental attachment. Pope (1995) proposed that a wider disparity in development among littermate blastocysts may contribute to subsequent embryonic losses. Competition between a group of developmentally diverse blastocysts could result in loss of less developed blastocysts through either limited space available for placental attachment following elongation, or alteration in uterine secretions following oestrogen release by more developed littermates and/or removal of synchrony for placental attachment to the uterine surface epithelium. Clearly, distinguishing the expression of genes involved with cellular growth, differentiation, steroidogenesis, immunoprotection and morphological change during the period of conceptus elongation, maternal recognition of pregnancy and placental attachment will provide an opportunity to regulate development and possibly improve embryonic survival. The following provides our current understanding of conceptus genes involved with early development and establishment of pregnancy in the pig.

Initiation of oestrogen synthesis is one of the first biosynthetic markers in conceptus development following blastocyst hatching in the pig. Conceptus oestrogen synthesis is detected in the uterine lumen (Ford *et al.*, 1982; Geisert *et al.*, 1982; Stone and Seamark, 1985) and *in vitro* culture (Fischer *et al.*, 1985; Pusateri *et al.*, 1990) when the conceptuses reach approximately a 5 mm spherical morphology. There is a large transient increase in oestrogen biosynthesis by the porcine conceptuses as they initiate elongation from 10 mm spherical to filamentous morphology (see review by Geisert *et al.*, 1990). Conceptus oestrogen secretion dramatically declines following completion of elongation until a second sustained increase occurs on day 16 of gestation (Bazer *et al.*, 1982; Geisert *et al.*, 1982; Ford *et al.*, 1982; Pusateri *et al.*, 1990). The two phases of conceptus oestrogen release are essential for establishment of pregnancy in the pig. Cytochromes P450 17α-hydroxylase ($P450_{17\alpha}$) and aromatase ($P450_{arom}$) are two key enzymes involved with the steroidogenic pathway for porcine conceptus oestrogen synthesis (Fig. 13.3). Increase and decline of conceptus oestrogen synthesis is directly correlated with the cellular content of $P450_{17\alpha}$ and $P450_{arom}$ (Conley *et al.*, 1992, 1994; Ko *et al.*, 1994). Expression of $P450_{17\alpha}$ is localized solely in the trophoblast whereas $P450_{arom}$ is present in the trophoblast and hypoblast, with greatest expression detected directly beneath the embryoblast (Conley *et al.*, 1994). Enzyme synthesis is cell specific as both embryoblast and mesoderm are negative for $P450_{17\alpha}$ and $P450_{arom}$. Developmental changes in steroidogenic enzyme gene expression, as analysed by RT-PCR, correlate with conceptus steroid synthesis. Gene expression of $P450_{17\alpha}$ is low in 10–12 day conceptuses, i.e. smaller than 6 mm spherical morphology, and greatly increases as conceptuses approach the initial stages (10 mm) of elongation (Green *et al.*, 1995; Yelich *et al.*, 1997b). Expression of the $P450_{arom}$ gene followed a similar pattern with a greatly enhanced expression detected at the time of conceptus elongation. A marked decline in both $P450_{17\alpha}$ and $P450_{arom}$ gene expression occurs immediately following elongation of the trophoblast (Conley *et al.*, 1992; Ko *et al.*, 1994; Green *et al.*, 1995).

Stromstedt *et al.* (1996) demonstrated that porcine conceptuses express the genes for NADPH-cytochrome P450 reductase, adrenodoxin, lanosterol 14-demethylase cytochrome P450 (*CYP51*), 17α-hydroxylase cytochrome P450 (*CYP17*), cholesterol side-chain cleavage cytochrome P450 (*CYP11A1*), sterol 27-hydroxylase cytochrome P450 (*CY27*) and aromatase cytochrome P450 (*CYP19*). Aromatase P450 is the product of the *CYP19* gene (Simpson *et al.*, 1994; Toda *et al.*, 1990). Tissue-specific regulation of *CYP19* is accomplished through alternative transcriptional start sites that are activated by tissue-specific promoters. Unique 5′-termini are present in mRNA from placental compared with ovarian tissues (Hinshelwood *et al.*, 1995) and placental $P450_{arom}$ cDNA is eightfold more active in transfected Cos1 cells compared with ovarian $P450_{arom}$ (Corbin *et al.*, 1995). Corbin *et al.* (1995) indicated that the distinct isoform of $P450_{arom}$ provides evidence that the pig conceptus not only contains an ovarian

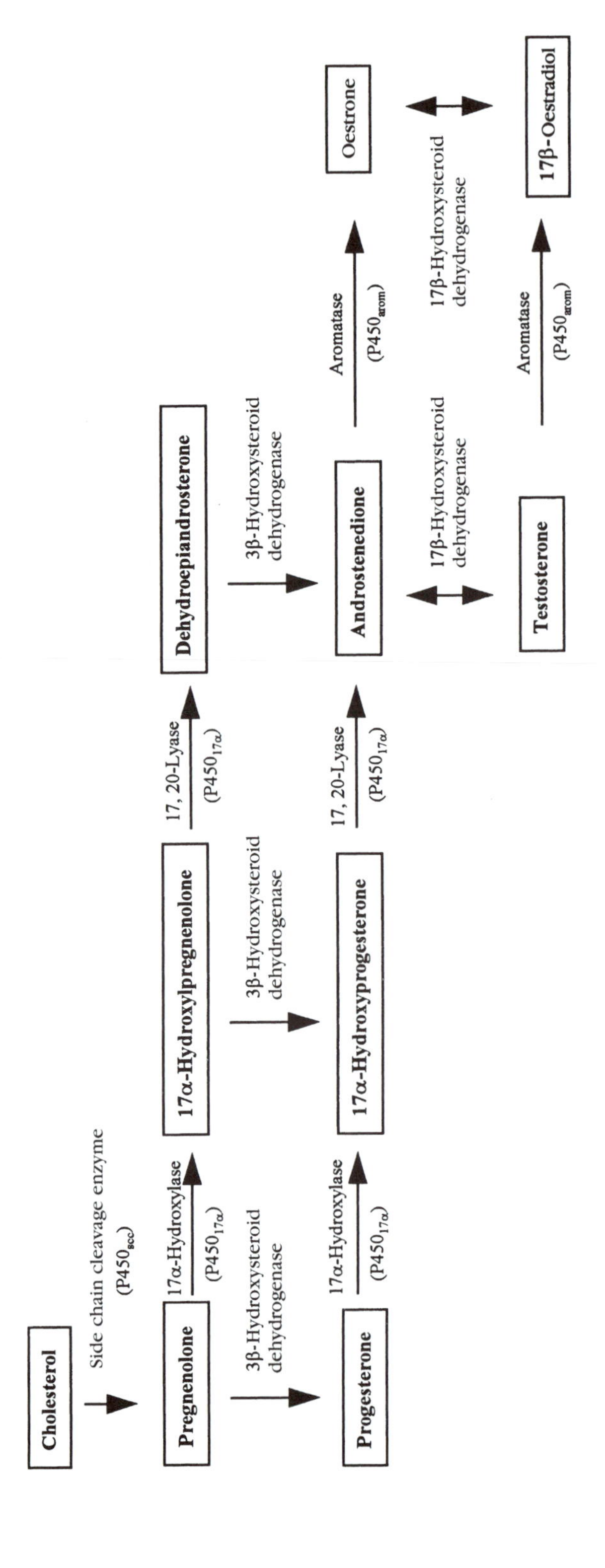

Fig. 13.3. Steroidogenic pathway of cholesterol metabolism to oestrogens in the porcine conceptus.

form of $P450_{arom}$ but also possesses a unique mechanism for regulation of androgen metabolism.

Since conceptus oestrogen synthesis plays a major role in subsequent embryo survival (see reviews by Geisert *et al.*, 1990; Pope 1995), regulation of gene expression for the key steroidogenic enzymes could provide a possible approach to improve the synchrony of the uterine environment when conceptuses are diverse in development. Conceptus oestrogen secretion has a direct effect on uterine function as oestrogen receptors are highly abundant in uterine epithelium at the time of rapid trophoblastic elongation (Geisert *et al.*, 1993). Oestrogen receptor (ESR) message has not been detected in the early conceptus by either immunocytochemistry or RT-PCR (Yelich *et al.*, 1997b). Conceptuses from Chinese Meishan pigs develop at a slower rate (Ford and Youngs, 1993) and have significantly lower gene expression for $P450_{17\alpha}$ (Kaminsky *et al.*, 1995) compared with domestic Yorkshire females. Breed differences in steroidogenic enzyme production would support the possible regulation of litter development through timing and quantity of conceptus oestrogen released into the uterine lumen. In addition, differences in steroidogenic enzyme receptor expression in the uterus may also contribute to variation in litter size both between and within breeds of pigs. Rothschild *et al.* (1996) found that a DNA polymorphism within the *ESR* locus was associated, either as a direct or as a linked effect, with litter size differences of 0.9 to 2.3 pigs/litter in Large White crosses or in Chinese breed crosses.

Conceptus oestrogen synthesis is regulated by the growth and development of the conceptuses within the uterine horns of the pig. The large increase in oestrogen synthesis from the conceptus is correlated with mesodermal outgrowth from the embryoblast of the 8 to 10 mm spherical conceptus (see reviews by Geisert *et al.*, 1990; Stroband and Van der Lende, 1990). *Brachyury* expression is a marker for mesoderm differentiation as it has been shown to have a direct role in early events in murine mesoderm formation (Wilkinson *et al.*, 1990). In the mouse, the *T* gene (*Brachyury*) is essential for mesodermal growth as homozygous mutations for the gene are lethal (Conlon *et al.*, 1995). *Brachyury* gene expression is detected in 5 mm spherical conceptuses when there is an initial increase in conceptus oestrogen production (Yelich *et al.*, 1997b). This first increase is prior to the appearance of mesoderm outgrowth from the embryoblast of 10 mm conceptuses. Thus, it appears that mesoderm formation in the embryoblast initiates oestrogen synthesis and the initial outgrowth is correlated with a rapid increase in aromatase production during elongation. The absence of $P450_{17\alpha}$ and $P450_{arom}$ within the mesoderm (Conley *et al.*, 1994) would appear to argue against mesoderm outgrowth as a mediator of oestrogen synthesis. However, the extracellular matrix (ECM) formed by this cell lineage may have an indirect role as ECM has been shown to direct cell shape, migration and differentiation (Luna and Hitt, 1992). Mesenchyme interactions with epithelial cells have also been demonstrated to serve an instructive role in epithelial differentiation and function (Pavlova *et al.*, 1994).

Insulin-like growth factors

Conceptus growth is affected by secretion of growth factors from both the uterine endometrium and conceptus (Pollard, 1990). A number of growth factors/cytokines, which may function through either autocrine or paracrine pathways to affect conceptus development, have been described for the porcine endometrium and conceptus (see reviews by Brigstock *et al.*, 1989; Simmen and Simmen, 1991). Insulin-like growth factors have been the most extensively studied system during early pregnancy in the pig. An increase of endometrial insulin-like growth factor-I (IGF-I) mRNA and IGF-I secretion into the uterine lumen (Tavakkol *et al.*, 1988; Simmen *et al.*, 1990) occurs at the time of rapid conceptus elongation and oestrogen release. Porcine conceptuses also express IGF-I mRNA but the level is significantly less than endometrial synthesis and is not altered by stage of development (Letcher *et al.*, 1989; Green *et al.*, 1995). IGF-I is a potent growth factor involved in regulation of cellular proliferation and differentiation (Sara and Hall, 1990) which, given the high content within the uterine lumen, could greatly influence conceptus development (Simmen *et al.*, 1993). Early conceptus IGF-I receptor mRNA expression appears to be constitutive as there is no effect of morphological development on gene expression (Green *et al.*, 1995). However, although the conceptus possesses mRNA for IGF-I receptor, receptor levels appear to be low as Chastant *et al.* (1994) failed to detect IGF-I receptors on early conceptuses through immunocytochemistry and autoradiography. It is possible that IGF-I plays a more important role in uterine growth as endometrial IGF-I receptors are abundant (Simmen *et al.*, 1992; Chastant *et al.*, 1994).

Development of murine stem cell technology has provided a powerful technique to selectively introduce mutations into genes of interest (Robertson, 1991). Targeted mutagenesis of the *IGF-I* and/or type 1 *IGF-I* receptor genes in mice causes prenatal growth deficiency but is not lethal during embryonic development (Liu *et al.*, 1993). However, null mutants for *IGF-IR* gene do not survive following birth. Thus, although IGF-I and IGF-I receptors can modulate embryonic growth, they are not essential for embryonic development. Trophectoderm concentration of IGF-II/mannose-6-phosphate receptors is abundant in early porcine blastocysts and therefore may serve to stimulate conceptus growth and differentiation (Chastant *et al.*, 1994). Murine embryonic growth has been demonstrated to involve IGF-II/mannose-6-phosphate receptors as gene mutation results in embryonic mortality (Barlow *et al.*, 1991). Disruption of the *IGF-II* gene impairs embryonic growth similar to disruption of *IGF-I* (Liu *et al.*, 1993). The *IGF-II* and *IGF-II* receptor genes are reciprocally imprinted genes. Only disruption of the paternal *IGF-II* gene will reduce murine embryonic growth (DeChiara *et al.*, 1990) while only the maternal *IGF2R* receptor gene is expressed during early development (Filson *et al.*, 1993). Baker *et al.* (1993) have indicated that an additional IGF-II receptor (XR) may serve to continue ligand–receptor activation to sustain development when IGF-I receptor is mutated. It is possible that IGFs modulate conceptus development in the pig. Increase of endometrial IGF-II mRNA occurs following elongation (day 15) and placental expression of *IGF-II* is increased in day 30 tissue (Simmen *et al.*, 1992). It is interesting that the mouse *Igf2r* gene is located in the proximal region of chromosome 17 which also includes the *T* gene that encodes for *Brachyury*. However in this region, only *Igf2r* is maternally imprinted. The IGF system is also regulated through binding proteins (see Simmen *et al.*, 1993). The porcine endometrium expresses IGFBP-2 compared with low or no expression detected in the placenta (Simmen *et al.*, 1992). Further studies will be necessary to elucidate the role of the IGFs and their receptors and binding proteins in porcine conceptus development and survival.

Other growth factors

The porcine uterus is a source of epidermal growth factor (EGF) as Kennedy *et al.* (1994) have detected EGF mRNA by RT-PCR and localized EGF protein in the glandular epithelium of the endometrium. Glandular secretion of EGF into the uterine lumen could stimulate uterine growth as well as conceptus growth and development. The gene for porcine EGF has been previously cloned and sequenced (Pascall *et al.*, 1991). Expression of mRNA for EGF, transforming growth factor-α (TGF-α) and EGF receptor have been demonstrated in the early porcine conceptus (Vaughan *et al.*, 1992). The TGF-α mRNA expression is maximal prior to and during conceptus elongation but declines post-elongation, whereas EGF mRNA expression is not detected until day 15 of gestation and was localized specifically in the embryo (Vaughan *et al.*, 1992). The EGF receptor is a membrane tyrosyl kinase receptor for which both EGF and TGF-α can serve as ligands (Burgess, 1989). Early spherical to elongated porcine conceptuses contain EGF receptors (Corps *et al.*, 1990; Zhang *et al.*, 1992) and gene expression was detected equally across all stages of development (Vaughan *et al.*, 1992). It is possible that increased gene expression of *TGF-α* could initiate events involved with rapid trophoblastic elongation while conceptus *EGF* gene activation regulates development of the embryo. Porcine conceptus mRNA expression of *TGF-*β_3 also increases during the period of conceptus elongation while gene expression of *TGF-*β_2 was not detected (Yelich *et al.*, 1997a). Expression of TGF-β_1, -β_2 and -β_3 is differentially regulated during the period of trophoblast elongation in the pig conceptus (Gupta *et al.*, 1996). The TGF-β family of growth factors can act as chemoattractants for fibroblasts, induce mesoderm formation with basic fibroblastic growth factor and regulate collagen and fibronectin gene expression (Roberts and Sporn, 1988). Changes in oncofetal fibronectin have been described in the early conceptus and chorioallantois of the pig throughout gestation (Tuo and Bazer, 1996). Conceptus activation of TGF-β could provide the stimulus for cellular remodelling of the trophoblast during elongation.

Retinoic acid receptors

TGF-β_2 synthesis and activation from its latent complex is stimulated by retinoic acid. Retinol is transported into the uterine lumen through retinol-binding protein (RBP) where the uterus and conceptus tissues can metabolize it to retinol and finally retinoic acid. Porcine endometrial mRNA expression of RBP is high and there is a dramatic increase in the uterine lumenal content of RBP during the period of rapid trophoblastic elongation and oestrogen release (Harney *et al.*, 1990, 1993; Trout *et al.*, 1991, 1992). The importance of retinoic acid for conceptus development is evident as RBP is the first major protein synthesized by the developing conceptus prior to trophoblastic elongation and is present throughout pregnancy (Godkin *et al.*, 1982; Harney *et al.*, 1990, 1994b). Gene expression of RBP is evident in early 3 mm spherical conceptuses and remains constant across all stages of development (Yelich *et al.*, 1997a). Delivery of retinol to the developing conceptus is critical as retinoic acid serves a major function in embryo morphogenesis (see review by De Luca, 1991). However, RBP also provides a protective role as retinoids must be tightly regulated since high cellular concentrations of retinol are teratogenic (Lammer *et al.*, 1985) and embryotoxic (Thompson *et al.*, 1993). Retinoic acid exerts its pleiotropic effects through binding directly to nuclear retinoic acid receptors (RAR) within the cell (Linney, 1992). Retinoic acid receptors belong to the family of steroid/thyroid hormone receptors of which there are three RAR subtypes (α,β,γ). Each subtype contains several isoforms with a distinct function in embryonic development (see De Luca, 1991; Glass *et al.*, 1994). Harney *et al.* (1994a) demonstrated the presence of RAR-α and RAR-γ mRNA and receptor protein in day 15 conceptus and late pregnant fetal tissues of the pig. Conceptus RAR-α and RAR-γ mRNA synthesis is evident during the early and post-elongation stages of conceptus development, although RAR-γ is more pronounced at the time of trophoblast elongation (Yelich *et al.*, 1997a). However, gene expression of *RAR*-β was distinctly activated immediately prior to and during elongation of the conceptus as was previously described for TGF-β_2. It is clearly evident that *RBP* and *RAR* play essential roles in conceptus development and survival (see Fig. 13.4).

Cytokines

Lymphohaematopoietic cytokines and their receptors have taken on a new perspective in reproductive biology beyond the strictly immunological aspects of maternal–fetal interactions necessary to prevent immune rejection of the fetus. Production of cytokines provides a rapid method for intercellular communication between the uterus and the developing conceptus (Robertson *et al.*, 1994). A broad spectrum of cytokines, once thought to be only produced in lymphocytes and macrophages, have been detected in conceptus and placental tissues throughout pregnancy in numerous species (Croy, 1994). Although the number of cytokines discovered to date is far from complete, gene expression of several important cytokines have been reported in the early conceptus and

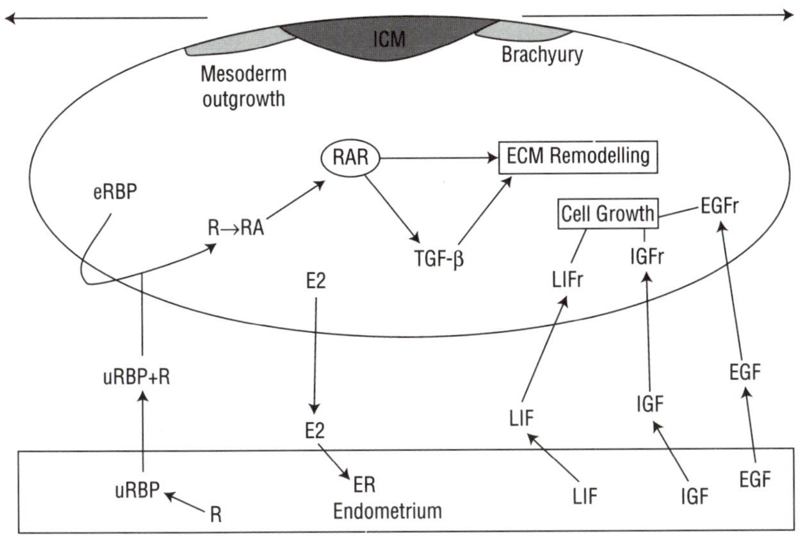

Fig. 13.4. Model depicting the possible involvement of retinoic acid (RA) stimulation of rapid cellular remodelling during trophoblastic elongation in the porcine conceptus. The model proposes that oestrogen (E2), synthesized by the conceptus, stimulates endometrial oestrogen receptors (ER) which initiate the release of uterine retinol-binding protein (uRBP) from the endometrium. Release of uRBP into the uterine lumen transfers retinol (R) either directly to the conceptus or through exchange with embryonic retinol-binding protein (eRBP). RBP serves to deliver R to the conceptus and protects the conceptus from elevated and toxic concentrations of free R. Release of R in the conceptus cytoplasm allows conversion to RA and activation of retinoic acid receptors (RAR) which serve as one of the initiators of extracellular matrix (ECM) remodelling for cell migration and trophoblast elongation through possible activation of the transforming growth factor-betas (TGF-β2,3) and other morphogens. Other cytokines like the insulin-like growth factors (IGF-I,II), epidermal growth factor (EGF) and leukaemia inhibitory factor (LIF) may function to regulate cell proliferation through receptors associated with the trophoblast prior to, during and following the period of elongation. (Model developed by Dr Joel Yelich, Oklahoma State University.)

placental tissue of the pig. Interleukin-6 (IL-6) and colony stimulating factor-1 (CSF-1) mRNA have been detected in porcine conceptuses and placental tissue (Mathialagan *et al.*, 1992; Anegon *et al.*, 1994; Tuo *et al.*, 1995). However, conceptus *IL-6* gene expression is much lower compared with levels in endometrial tissue (Anegon *et al.*, 1994). IL-6 stimulates acute-phase genes which may be involved with the local hyperaemic response of the endometrium to conceptus placental attachment (see Geisert *et al.*, 1990). Placental CSF-1 mRNA expression and protein detection are greatly increased after day 15 of gestation until shortly before term (Tuo *et al.*, 1995). Preliminary data have indicated that the trophoblast and endometrium express CSF-1 receptor (W. Tuo and F.W. Bazer, personal communication). The CSF-1 receptor, which is a product of the *c-fms* proto-oncogene, has been detected in the bovine placenta throughout pregnancy (Beauchamp and Croy, 1991). The close association of CSF-1 production with placental attachment, growth and water transport suggests that CSF-1 may contribute to placental development throughout pregnancy in the pig.

Leukaemia inhibitory factor (LIF) is another hematopoietic growth regulator whose gene expression has recently been demonstrated in porcine endometrium but not conceptus tissue (Anegon *et al.*, 1994). Production of LIF by the maternal endometrium is clearly involved with regulating embryonic differentiation and implantation in mice, as disruption of the *LIF* gene results in pregnancy failure (Stewart *et al.*, 1992; Stewart, 1994). However, early embryonic mortality is not related to direct effects on embryo development, but rather a lack of maternal endometrial LIF production prevents the invasive implantation in the mouse. It is possible that other cytokines can substitute for the absence of LIF as IL-6 does interact with the LIF receptor and mediates its cellular function through a common transmembrane protein gp130 (Gearing *et al.*, 1991). The LIF mRNA is first detected in the porcine endometrium on day 11 of gestation and a short surge of LIF occurs within the uterine lumen on day 12 of pregnancy (Anegon *et al.*, 1994). Endometrial secretion of LIF may affect uterine receptivity for placental attachment and/or conceptus differentiation, however gene expression and protein localization of LIF receptor await clarification.

Conceptus expression of IL-1β is also temporally associated with trophoblast elongation in the pig (Tuo *et al.*, 1996). High concentrations of IL-1β are detected in the porcine conceptus on days 11 and 12 of pregnancy, followed by a rapid decline in mRNA expression on day 15. Conceptus production of IL-1β may influence trophoblast elongation through its effects on fibroblast growth or regulate prostaglandin E production by the conceptus and/or endometrium to suppress maternal immunorejection until conceptus synthesis of interferons.

Trophoblastic interferons

The synthesis and secretion of trophoblastic interferons (IFN) by pre-implantation conceptuses of ungulate species has been extensively described (see Roberts *et al.*, 1992, 1996; La Bonnardiere, 1993). Interferons α, β and γ are well known for their involvement in initiating cellular mechanisms to block viral assault, but also serve as cytokines in regulation of cell mitosis, morphology, differentiation and expression of major histocompatibility antigens (Pestka *et al.*, 1987). Conceptus synthesis of trophoblastic IFN-τ is clearly involved with establishment and maintenance of pregnancy in ruminants (see Bazer *et al.*, 1994). The IFN-τ exerts its antiluteolytic effect through regulation of steroid receptors in endometrium of the ewe (Spencer *et al.*, 1995). Porcine conceptuses also secrete IFNs into the uterine lumen between days 12 and 20 of pregnancy (Cross and Roberts, 1989; Mirando *et al.*, 1990). However, although porcine conceptuses secrete Type I and II IFNs, maintenance of functional corpora lutea in the pig does not involve trophoblastic IFNs but rather conceptus oestrogen secretion (see Geisert *et al.*, 1994). The pig conceptus synthesizes two distinctly different IFNs compared with the IFN-τ of ruminant species (La Bonnardiere, 1993). The major IFN secreted by the conceptus has been identified as IFN-γ for which two mRNAs have been detected (Lefevre *et al.*, 1990). A second novel IFN designated as short porcine type I IFN (spI IFN), has been identified and is most likely the type I IFN secreted by the porcine conceptus (Lefevre and Boulay, 1993). Like other IFNs, this trophoblastic IFN has no introns but contains the greatest number of cysteine residues (seven) and is the smallest IFN presently known (La Bonnardiere, 1993). Conceptus gene expression of *IFN-γ* is unique in that this is the first demonstration of production of IFN-γ outside hematopoietic cells (La Bonnardiere, 1993). Gene expression of both porcine conceptus *IFN-γ* and *spI IFN* are developmentally regulated as maternal signals do not stimulate IFN secretion. Porcine trophoblastic IFNs appear to have no direct effects on the conceptus itself as no embryonic IFN receptors have been detected (D'Andrea *et al.*, 1994). The IFN receptors are present on the uterine epithelium (D'Andrea *et al.*, 1994) and thus suggest a direct effect of the cytokine on endometrial function.

The role of the porcine trophoblastic IFNs has not been completely defined. Conceptus IFNs would obviously act as antiviral agents to protect the developing embryo from viral attack. IFN-γ stimulates expression of MHC antigens in numerous cell types but, as stated earlier, porcine conceptuses lack IFN receptors (see La Bonnardiere, 1993). The IFN secretion by porcine conceptuses may recruit natural killer cells to the endometrium and thus influence cellular immunity at the placental–endometrial interface (Croy, 1994).

Pregnancy-associated glycoproteins (PAGs)

In order for eutherian mammalian species to successfully establish a pregnancy *in utero*, they must protect the developing allogenic conceptus from recognition and lysis by the maternal immune system (see Croy, 1994). Recently, the pregnancy-associated glycoproteins (PAG), which were previously identified in ruminant placenta (see Roberts *et al.*, 1996), have been shown to be expressed in the porcine trophoblast (Szafranska *et al.*, 1995). Placental PAG are members of the aspartyl proteinase gene family. Although their substrate-binding cleft is intact, PAG-1 has amino acid substitutions within the catalytic centre which interfere with its enzymatic activity (Roberts *et al.*, 1996). Roberts *et al.* (1996) have proposed that PAG-1 may interfere with the maternal immunological function of T cells by competing with MHC for peptides necessary for antigen presentation. The *PAG* gene expression is abundant in the invasive binucleate cells just prior to implantation and throughout pregnancy in sheep and cattle (Xie *et al.*, 1991). However, *PAG* is expressed in the non-invasive mononucleate cells of the porcine chorion as well (Szafranska *et al.*, 1995). Porcine *PAG* mRNA expression is not detected in the conceptus until the period of placental attachment on day 15 of pregnancy, however expression continues throughout pregnancy. Similar to the ruminant PAG, the porcine PAG-1 has amino acid substitutions within the catalytic centre that interfere with its enzymatic activity. Localization of PAG within the chorionic cells directly involved in adhesive contact to the uterine epithelium, and temporal relationship to the timing of placental attachment, suggest a role for PAG in immunoprotection as an attractive hypothesis.

Conceptus development, differentiation and survival obviously involve activation of a multitude of genes by the conceptus as well as the uterus. Although activation of key genes by the conceptus is essential for development and signalling to the maternal system, selective targeted mutagenesis studies in mice have clearly provided evidence that many genes previously thought essential for embryonic development can be deleted without lethal effects. Many cytokines and growth factors can interact with each other's receptors as previously described. Nature appears to have provided the conceptus with alternatives when gene defects do occur.

Table 13.1. Genes expressed by the porcine conceptus and placenta during early development and placentation.

Gene expressed	Developmental process affected	Reference
Brachyury (*T* gene)	Mesoderm formation Embryo and placental differentiation	Yelich et al., 1997b
Cytochromes P450 (17α, arom, scc)	Steroidogenesis Conceptus oestrogen synthesis	Green *et al.*, 1995 Yelich *et al.*, 1997b Corbin *et al.*, 1995
IGF-I	Regulation of cellular proliferation/differentiation	Letcher *et al.*, 1989 Green *et al.*, 1995
IGF-II	Regulation of cellular proliferation/differentiation	Simmen *et al.*, 1992
IGF-I receptor	Receptor activation for cellular proliferation and differentiation	Green *et al.*, 1995
IGFBP-2	Regulation of IGF-I and IGF-II activity	Simmen *et al.*, 1992
EGF	Regulation of cellular proliferation/ differentiation	Kennedy *et al.*, 1994
TGF-α	Regulation of cellular proliferation/ differentiation	Vaughan *et al.* 1992
TGF-β_1, β_2, β_3	Chemoattractant; regulation of cellular differentiation and morphogenesis; collagen and fibronectin gene expression	Yelich *et al.*, 1997a Gupta *et al.*, 1996
RBP	Retinol transport to conceptus Regulation of retinol concentration	Harney *et al.*, 1990
RAR-α,β,γ	Retinoic acid receptor for embryo morphogenesis	Harney *et al.*, 1994a,b Yelich *et al.*, 1997a
IL-6	Uterine immunological stimulation Uterine hyperaemic response	Mathialagan *et al.*, 1992 Anegon *et al.*, 1994
IL-1β	Uterine immunological stimulation	Tuo *et al.*, 1996
CSF-1	Hematopoietic growth factor for cellular proliferation and differentiation	Tuo *et al.*, 1995
IFN-γ, *spl IFN*	Immunological regulation of maternal system, antiproliferative	Lefevre *et al.*, 1990 Lefevre and Boulay, 1993
PAG	Possible role in placental adhesion to uterine surface, protease inhibitor	Szafranska *et al.*, 1995

Conclusions

While our knowledge base regarding developmental genetics in the pig is at an embryonic level, particularly at early stages of development up to formation of the blastocyst, there is a slowly growing database regarding which genes are expressed, how they are regulated and how their products contribute to early differentiation. In contrast, the groundwork for understanding regulation of gene expression in porcine conceptus growth, oestrogen synthesis, placental attachment and immunological interactions with the maternal system (see Table 13.1) is much further developed, albeit far from complete. Already, our current evidence has indicated that control of conceptus oestrogen synthesis and trophoblastic expansion may provide possible approaches to improve litter size in the pig. With continued research and discovery of genes involved in establishment and maintenance of pregnancy in pigs, we may be able to develop methods to regulate key developmental events during the periods of high conceptus mortality.

Acknowledgements

The authors gratefully acknowledge the work of Joel Yelich, who performed significant research on gene expression in early pig embryos, and contributed to our model of gene expression during embryo elongation, while working as a postdoctoral research associate in our laboratories at Oklahoma State University. We appreciate the critical reviews of this chapter by Randy Prather, William Trout and two anonymous referees. We thank Randy Prather for providing useful information and discussion, as well as the photographs reproduced in Fig. 13.1. Portions of this work were supported by Oklahoma State University Section 1433 Animal Health Funds awarded to R.G. and D.P.

References

Abdelrahman B., Fiddler, M., Rappolee, D. and Pergament, E. (1995) Expression of transcription regulating genes in human preimplantation embryos. *Human Reproduction* 10, 2787–2792.

Anderson, L.L. (1978) Growth, protein content and distribution of early pig embryos. *Anatomical Record* 190, 143–154.

Anegon, I., Cuturi, M.C. Godard, A., Moreau, M., Terqui, M., Martinat-Botte, F. and Soulillou, J.P. (1994) Presence of leukaemia inhibitory factor and interleukin 6 in porcine uterine secretions prior to conceptus attachment. *Cytokine* 6, 493–499.

Baker, J., Liu, J.-P., Robertson, E.J. and Efstratiadis, A. (1993) Role of insulin-like growth factors in embryonic and postnatal growth. *Cell* 75, 73–82.

Barlow, D.P., Stoger, R., Herrmann, B.G., Saito, K. and Schweifer, N. (1991) The mouse insulin-like growth factor type 2 receptor is imprinted and closely related to the Tme locus. *Nature* 349, 84–87.

Bazer, F.W., Geisert, R.D., Thatcher, W.W. and Roberts, R.M. (1982) The establishment and maintenance of pregnancy. In: Cole, D.J.A. and Foxcroft, G.R. (eds) *Control of Pig Reproduction.* Butterworth Scientific, London, pp. 227–252.

Bazer, F.W., Marengo, S.R., Geisert, R.D. and Thatcher, W.W. (1984) Exocrine versus endocrine secretion of prostaglandin $F_{2\alpha}$ in the control of pregnancy in swine. *Animal Reproductive Science* 7, 115–132.

Bazer, F.W., Spencer, T.E., Ott, T.L. and Johnson, H.M. (1994) Cytokines and pregnancy recognition. In: Hunt J.S. (ed.), *ImmunoBiology of Reproduction.* Springer-Verlag, New York, pp. 37–56.

Beauchamp, J.L. and Croy, B.A. (1991) Assessment of expression of the receptor for CSF-1 (fms) in bovine trophoblast. *Biology of Reproduction* 45, 811–817.

Betteridge, K. (1995) Phylogeny, ontogeny and embryo transfer. *Theriogenology* 44, 1061–1098.

Brigstock, D.R., Heap, R.B. and Brown, K.D. (1989) Polypeptide growth factors in uterine tissues and secretions. *Journal of Reproduction and Fertility* 85, 747–758.

Burgess, A.W. (1989) Epidermal growth factor and transforming growth factor α. *British Medical Bulletin* 45, 401–424.

Chastant, S., Monget, P. and Terqui, M. (1994) Localization and quantification of insulin-like growth factor-I (IGF-I) and IGF-II/mannose-6-phosphate (IGF-II/M6P) receptors in pig embryos during early pregnancy. *Biology of Reproduction* 51, 588–596.

Conley, A.J., Christenson, R.K., Ford, S.P., Geisert, R.D. and Mason, J.I. (1992) Steroidogenic enzyme expression in porcine conceptuses during and after elongation. *Endocrinology* 131, 896–902.

Conley, A.J., Christenson, R.K., Ford, S.P. and Christenson, R.K. (1994) Immunocytochemical localization of cytochrome P450 17α-hydroxylase and aromatase in embryonic cell layers of elongating porcine blastocysts. *Endocrinology* 135, 2248–2254.

Conlon, F.L., Wright, C.V. and Robertson, E.J. (1995) Effects of the Twis mutation on notochord formation and mesodermal patterning. *Mechanisms of Development* 49, 201–209.

Corbin, C.J., Khalil, M.W. and Conley, A.J. (1995) Functional ovarian and placental isoforms of porcine aromatase. *Molecular and Cellular Endocrinology* 113, 29–37.

Corps, A.N., Brigstock, D.R., Littlewood, C.J. and Brown, K.D. (1990) Receptors for epidermal growth factor and insulin-like growth factor-I on preimplantation trophectoderm of the pig. *Development* 110, 221–227.

Cross, J.C. and Roberts, R.M. (1989) Porcine conceptuses secrete an interferon during the preattachment period of early pregnancy. *Biology of Reproduction* 40, 1109–1118.

Croy, B.A. (1994) Embryo survival in domestic mammals: immunological aspects. In: Zavy, M.T. and Geisert, R.D. (eds) *Embryonic Mortality in Domestic Species.* CRC Press, Boca Raton, pp. 153–178.

Cruz, Y.P. and Pederson, R.A. (1991) Origin of embryonic and extraembryonic cell lineages in mammalian development. In: Pederson, R.A., McLaren, A. and First, N.L. (eds) *Animal Applications of Research in Mammalian Development.* Cold Spring Harbor Laboratory Press, Cold Spring Harbor, New York, pp. 147–204.

D'Andrea, S., Choustermanm S., Flechon, J.E. and La Bonnardiere, C. (1994) Paracrine activities of porcine trophoblastic interferons. *Journal of Reproduction and Fertility* 102, 185–194.

Davis, D.L. (1985) Culture and storage of pig embryos. *Journal of Reproduction and Fertility* (Suppl. 33), 115–124.

DeChiara, T.M., Efstratiadis, A. and Robertson, E.J. (1990) A growth-deficiency phenotype in heterozygous mice carrying an insulin-like growth factor II gene disrupted by targeting. *Nature* 345, 78–80.

De Luca, L.M. (1991) Retinoids and their receptors in differentiation, embryogenesis, and neoplasia. *FASEB Journal* 5, 2924–2933.

Eyestone, W.H., Lawyer, M.L., Critser, E.S. and Leibfried-Rutledge, M.L. (1986) Cell cycle stage of early porcine embryos during developmental arrest *in vitro*. *Biology of Reproduction* 34 (Suppl.), 98.

Filson, A.J., Louvi, A., Efstratiadis, A. and Robertson, E.J. (1993) Rescue of the T-associated maternal effect in mice carrying null mutations in Igf-2 and Igf2r, two reciprocally imprinted genes. *Development* 118, 731–736.

Fischer, H.E., Bazer, F.W. and Fields, M.J. (1985) Steroid metabolism by endometrial and conceptus tissues during early pregnancy and psuedopregnancy in gilts. *Journal of Reproduction and Fertility* 75, 69–78.

Ford, S.P. and Youngs, C.R. (1993) Early embryonic development in prolific Meishan pigs. *Journal of Reproduction and Fertility* (Suppl. 48), 271–278.

Ford, S.P., Christenson, R.K. and Ford, J.J. (1982) Uterine blood flow and uterine arterial, venous and luminal concentrations of oestrogens in Days 11, 13 and 15 after oestrus in pregnant and non-pregnant sows. *Journal of Reproduction and Fertility* 64, 185–190.

Ford, S.P., Schwartz, N.K., Rothschild, M.F., Conley, A.J. and Warner, C.M. (1988) Influence of SLA haplotype on preimplantation embryonic cell number in miniature pigs. *Journal of Reproduction and Fertility* 84, 99–104.

Freitag, M., Dopke, H.H., Niemann, H. and Elsaesser, F. (1991) Ontogeny of RNA synthesis in preimplantation pig embryos and the effect of antioestrogen on blastocyst formation in vitro. *Molecular Reproduction and Development* 29, 124–128.

Gearing, D.P., Thut, C.J., VandeBos, T., Gimpel, S.D., Delaney, P.B., King, J., Price, V., Cosman, D. and Beckmann, M.P. (1991) Leukemia inhibitory factor receptor is structurally related to the IL-6 signal transducer, pg130. *EMBO Journal* 10, 2839–2848.

Geisert, R.D., Renegar, R.H., Thatcher, W.W., Roberts, R.M. and Bazer, F.W. (1982) Establishment of pregnancy in the pig. I. Interrelationships between preimplantation development of the pig blastocyst and uterine endometrial secretions. *Biology of Reproduction* 27, 925–939.

Geisert, R.D., Zavy, M.T., Moffatt, R.J., Blair, R.M. and Yellin, T. (1990) Embryonic steroids and the establishment of pregnancy in pigs. *Journal of Reproduction and Fertility* (Suppl. 40), 293–305.

Geisert, R.D., Morgan, G.L., Zavy, M.T., Blair, R.M., Gries, L.K., Cox, A. and Yellin, T. (1991) Effect of asynchronous transfer and oestrogen administrations on survival and development of porcine embryos. *Journal of Reproduction and Fertility* 93, 475–481.

Geisert, R.D., Brenner, R.M., Moffatt, R.J., Harney, J.P., Yellin, T. and Bazer, F.W. (1993) Changes in estrogen receptor protein, mRNA expression and localization in the endometrium of cyclic and pregnant gilts. *Reproduction, Fertility and Development* 5, 247–260.

Geisert, R.D., Short, E.C. and Morgan, G.L. (1994) Establishment of pregnancy in domestic farm species. In: Zavy, M.T. and Geisert, R.D. (eds) *Embryonic Mortality in Domestic Species*. CRC Press, Boca Raton, pp. 23–52.

Glass, C.K., Devary, O.V. and Rosenfeld, M.G. (1994) Multiple cell type-specific proteins differentially regulate target sequence recognition by the α retinoic acid receptor. *Cell* 63, 729–738.

Godkin, J.D., Bazer, F.W., Lewis, G.S., Geisert, R.D. and Roberts, R.M. (1982) Synthesis and release of polypeptides by pig conceptuses during the period of blastocyst elongation and attachment. *Biology of Reproduction* 27, 977–987.

Green, M.L., Simmen, R.C.M. and Simmen, F.A. (1995) Developmental regulation of steroidogenic enzyme gene expression in the preimplantation porcine conceptus: a paracrine role for insulin-like growth factor-I. *Endocrinology* 136, 3961–3970.

Gupta, A., Bazer, F.W. and Jaeger, L.A. (1996) Differential expression of beta transforming growth factors (TGFβ1, TGFβ2, and TGFβ3) and their receptors (Type I and Type II) in peri-implantation porcine conceptuses. *Biology of Reproduction* 55, 796–802.

Harney, J.P., Mirando, M.A., Smith, L.C. and Bazer, F.W. (1990) Retinol-binding protein: a major secretory product of the pig conceptus. *Biology of Reproduction* 42, 523–532.

Harney, J.P., Ott, T.L. Geisert, R.D. and Bazer, F.W. (1993) Retinol-binding protein gene expression in cyclic and pregnant endometrium of pigs, sheep, and cattle. Biology of *Reproduction* 49, 1066–1073.

Harney, J.P., Ali, M., Vedeckis, W.V. and Bazer, F.W. (1994a) Porcine conceptus and endometrial retinoid-binding proteins. *Reproduction, Fertility and Development* 6, 211–219.

Harney, J.P., Smith, L.C., Simmen, L.C., Fliss, A.E. and Bazer, F.W. (1994b) Retinol-binding protein: Immunolocalization of protein and abundance of messenger ribonucleic acid in conceptus and maternal tissues during pregnancy in pigs. *Biology of Reproduction* 50, 1126–1135.

Hinshelwood, M.M., Liv, Z., Conley, A.J. and Simpson, E.R. (1995) Demonstration of tissue-specific promoters in non-primate species that express aromatase P450 in placentae. *Biology of Reproduction* 53, 1151–1159.

Jarrell, V.L., Day, B.N. and Prather, R.S. (1991) The transition from maternal to zygotic control of development occurs during the 4-cell stage in the domestic pig, *Sus scrofa*: quantitative and qualitative aspects of protein synthesis. *Biology of Reproduction* 44, 62–68.

Kaminsky, M.A., Conley, A.J. and Ford, S.P. (1995) Developmental expression of cytochrome P450 17α-hydroxylase (P450c17) in day 10.5–14 Meishan and Yorkshire conceptuses. *Biology Reproduction* 52 (Suppl. 1), 181.

Kennedy, T.G., Brown, K.D. and Vaughan, T.J. (1994) Expression of the genes for the epidermal growth factor receptor and its ligands in porcine oviduct and endometrium. *Biology of Reproduction* 50, 751–756.

Ko, Y., Choi, I., Green, M.L. Simmen, F.A. and Simmen R.C.M. (1994) Transient expression of the cytochrome P450 aromatase gene in elongating porcine blastocysts is correlated with uterine insulin-like growth factor levels during peri-implantation development. *Molecular Reproduction and Development* 37, 1–11.

La Bonnardiere, C. (1993) Nature and possible functions of interferons secreted by the preimplantation pig blastocyst. *Journal of Reproduction and Fertility* (Suppl. 48), 157–170.

Lammer, E.J., Chen, D.T., Hoar, R.M., Angish, N.D., Benke, P.J., Braun, J.T., Curry, C.J., Fernoff, P.M., Grix, A.W., Lott, I.T., Richard, J.M. and Sun, S.C. (1985) Retinoic acid and embryopathy. *New England Journal of Medicine* 313, 837–841.

Lefevre, F. and Boulay, V. (1993) A novel and atypical type one interferon gene expressed by trophoblast during early pregnancy. *Journal of Biological Chemistry* 268, 19760–19768.

Lefevre, F., Martinat-Botte, F., Guillomot, M., Zouari, K., Charley, B. and La Bonnardiere, C. (1990) Interferon-gamma gene and protein are spontaneously expressed by the porcine trophectoderm early in gestation. *European Journal of Immunology* 20, 2485–2490.

Letcher, R., Simmen, R.C.M., Bazer, F.W. and Simmen, F.A. (1989) Insulin-like growth factor-1 expression during early conceptus development in the pig. *Biology of Reproduction* 41, 1143–1151.

Li, J., Abeydeera, L.R., Matteri, R.L., Day, B.N. and Prather, R.S. (1996) Transcriptional activities of IGF-1R, IGF-2 and IGF-2R in porcine pre-implantation embryos. *Biology of Reproduction* 54 (Suppl. 1), 74.

Linney, E. (1992) Retinoic acid receptors: transcription factors modulating gene expression, development, and differentiation. *Current Topics in Developmental Biology* 27, 309–350.

Liu, J.-P., Baker, J., Perkins, A.S., Robertson, E.J. and Efstratiadis, A. (1993) Mice carrying null mutations of the genes encoding insulin-like growth factor I (Igf-1) and Type 1 IGF receptor (Igf1r). *Cell* 75, 59–72.

Luna, E.J. and Hitt, A.L. (1992) Cytoskeleton–plasma membrane interactions. *Science* 258, 955–964.

Mathialagan, N., Bixby, J.A. and Roberts, R.M. (1992) Expression of interleukin-6 in porcine, ovine, and bovine preimplantation conceptuses. *Molecular Reproduction and Development* 32, 324–330.

Mirando, M.A., Harney, J.P., Beers, S., Pontzer, C.H., Torres, B.A., Johnson, H.M. and Bazer, F.W. (1990) Onset of secretion of proteins with antiviral activity in pig conceptuses. *Journal of Reproduction and Fertility* 88, 197–203.

Nothias, J.-Y., Majumder, S., Kaneko, K.J. and DePamphilis, M.L. (1995) Regulation of gene expression at the beginning of mammalian development. *Journal of Biological Chemistry* 270, 22077–22080.

Pascall, J.C., Jones, D.S., Doel, S.M., Clements, J.M., Hunter, M., Fallon, T., Edwards, M. and Brown, K.D. (1991) Cloning and characterization of a gene encoding pig epidermal growth factor. *Journal of Molecular Endocrinology* 6, 63–70.

Pavlova, A., Boutin, E., Cunha, G. and Sassoon, D. (1994) Msx1 (Hox-7.1) in the adult mouse uterus: cellular interactions underlying regulation of expression. *Development* 120, 335–345.

Pestka, S., Langer, J.A., Zoon, K.C. and Samuel, C.E. (1987) Interferons and their actions. *Annual Reviews of Biochemistry* 56, 727–777.

Petters, R.M. and Wells, K.D. (1993) Culture of pig embryos. *Journal of Reproduction and Fertility* (Suppl. 48), 61–73.

Pollard, J.W. (1990) Regulation of polypeptide growth factor synthesis and growth factor-related gene expression in the rat and mouse uterus before and after implantation. *Journal of Reproduction and Fertility* 88, 721–731.

Pomp, D., Good, B.A., Geisert, R.D., Corbin, C.J. and Conley, A.J. (1995) Sex identification in mammals with polymerase chain reaction and its use to examine sex effects on diameter of day-10 or -11 pig embryos. *Journal of Animal Science* 73, 1408–1415.

Pope, W.F. (1995) Embryonic mortality in swine. In: Zavy M.T. and Geisert, R.D. (eds) *Embryonic Mortality in Domestic Species.* CRC Press, Boca Raton, pp. 53–78.

Pope, W.F. and First, N.L. (1985) Factors affecting the survival of pig embryos. *Theriogenology* 23, 91–105.

Prather, R.S. (1993) Nuclear control of early embryonic development. *Journal of Reproduction and Fertility* (Suppl. 48), 17–29.

Prather, R.S. (1996) Progress in cloning embryos from domesticated livestock. *Proceedings of the Society for Experimental Biology and Medicine* 121, 38–43.

Prather, E.S. and Rickords, L.F. (1992) Developmental regulation of an snRNP core protein epitope during pig embryogenesis and after nuclear transfer for cloning. *Molecular Reproduction and Development* 33, 119–123.

Pusateri, A.E., Rothschild, M.F., Warner, C.M. and Ford, S.P. (1990) Changes in morphology, cell number, cell size and cellular estrogen content of individual littermate pig conceptuses on days 9 to 13 of gestation. *Journal of Animal Science* 68, 3727–3735.

Rappolee, D.A. (1992) Endogenous insulin-like growth factor II mediates growth in preimplantation mouse embryos through the insulin-like growth factor I receptor: a case study on the role of growth factors in mammalian development. *Journal of Animal Science* 70 (Suppl. 2), 42–50.

Rivera, R.M., Youngs, C.R. and Ford, S.P. (1996) A comparison of the number of inner cell mass and trophectoderm cells of preimplantation Meishan and Yorkshire pig embryos at similar developmental stages. *Journal of Reproduction and Fertility* 106, 111–116.

Roberts, A.B. and Sporn, M.B. (1988) Transforming growth factor-betas: a large family of multifunctional regulatory proteins. *Journal of Animal Science* 66 (Suppl. 3), 67–75.

Roberts, R.M., Cross, J.C. and Leaman, D.W. (1992) Interferons as hormones of pregnancy. *Endocrine Reviews* 13, 432–452.

Roberts, R.M., Xie, S. and Mathialagan, N. (1996) Maternal recognition of pregnancy. *Biology of Reproduction* 54, 294–302.

Robertson, E.J. (1991) Using embryonic stem cells to introduce mutations into the mouse germ line. *Biology of Reproduction* 44, 238–245.

Robertson, S.A., Seamark, R.F., Guilbert, L.J. and Wegmann, T.G. (1994) The role of cytokines in gestation. *Critical Reviews in Immunology* 14, 239–292.

Rothschild, M., Jacobson, C., Vaske, D., Tuggle, C., Wang, L., Short, T., Eckardt, G., Sasaki, S., Vincent, A., McLaren, D., Southwood, O., van der Steen, H., Mileham, A. and Plastow, G. (1996) The estrogen receptor locus is associated with a major gene influencing litter size in pigs. *Proceedings of the National Academy of Sciences, USA* 93, 201–205.

Sara, V.R. and Hall K. (1990) Insulin-like growth factors and their binding proteins. *Physiological Reviews* 70, 591–614.

Schoenbeck, R.A., Peters, M.S., Rickords, L.F., Stumpf, T.T. and Prather, R.S. (1992) Characterization of DNA synthesis and the transition from maternal to embryonic control in the 4-cell porcine embryo. *Biology of Reproduction* 47, 1118–1125.

Schultz, G.A., Dean, W., Hahnel, A., Telford, N., Rappolle, D., Werb, Z. and Pederson, R. (1990) Changes in RNA and protein synthesis during the development of the preimplantation mouse embryo. In: Heyner, S. and Wiley, L. (eds.) *Early Embryo Development and Paracrine Relationships.* UCLA Symposium on Molecular and Cellular Biology, New Series 117. AR Liss, New York, pp. 27–46.

Schultz, G.A., Hogan A., Watson, A.J., Smith, R.M. and Heyner, S. (1992) Insulin-like growth factors and glucose transporters: temporal patterns of gene expression in early mouse and cow embryos. *Reproduction, Fertility and Development* 4, 361–371.

Schultz, R.M., Worrad, D.M., Davis, W. Jr, and De Sousa, P.A. (1995) Regulation of gene expression in the preimplantation mouse embryo. *Theriogenology* 44, 1115–1131.

Simmen, F.A. and Simmen R.C.M. (1991) Peptide growth factors and proto-oncogenes in mammalian conceptus development. *Biology of Reproduction* 44, 1–5.

Simmen, R.C.M., Simmen F.A., Hofig, A., Farmer, S.J. and Bazer, F.W. (1990) Hormonal regulation of insulin-like growth factor gene expression in pig uterus. *Endocrinology* 127, 2166–2174.

Simmen, R.C.M., Ko, Y. and Simmen, F.A. (1993) Insulin-like growth factors and blastocyst development. *Theriogenology* 39, 163–175.

Simmen, F.A., Simmen R.C.M., Geisert, R.D., Martinat-Botte, F., Bazer, F.W. and Terqui, M. (1992) Differential expression, during the estrous cycle and pre- and postimplantation conceptus development, of messenger ribonucleic acids encoding components of the pig uterine insulin-like growth factor system. *Endocrinology* 130, 1547–1556.

Simpson, E.R., Mahendroo, M.S., Means, G.D., Kilgore, M.W., Hinshelwood, M.M., Graham Lorence, S., Amarneh, B., Ito, Y., Fisher, C.R., Michael, M.D., Mendelson, C.R. and Bulun, S.E. (1994) Aromatase cytochrome P450, the enzyme responsible for estrogen biosynthesis. *Endocrine Reviews* 15, 342–355.

Spencer, T.E., Becker, W.C., George, P, Mirando, M.A., Ogle, T.F. and Bazer, F.W. (1995) Ovine interferon-τ regulates expression of endometrial receptors for estrogen and oxytocin but not progesterone. *Biology of Reproduction* 53, 732–745.

Stewart, C.L. (1994) Leukaemia inhibitory factor and the regulation of pre-implantation development of the mammalian embryo. *Molecular Reproduction and Development* 39, 233–238.

Stewart, C.L., Kaspar, P., Brunet, L.J., Bhatt, H., Gadi, I, Kontgen, F. and Abbondanzo, S. (1992) Blastocyst implantation depends on maternal expression of leukemia inhibitory factor. *Nature* 359, 76–79.

Stone, B.A. and Seamark, R.F. (1985) Steroid hormones in uterine washings and in plasma of gilts between days 9 and 15 after oestrus and between days 9 and 15 after coitus. *Journal of Reproduction and Fertility* 75, 209–221.

Stroband H.W.J. and Van der Lende, T. (1990) Embryonic and uterine development during early pregnancy in pigs. *Journal of Reproduction and Fertility* (Suppl. 40), 261–277.

Stroband, H.W.J., Taverne, N. and Van den Bogaard, M. (1984) The pig blastocyst: its ultrastructure and the absorption of protein macromolecules. *Cell and Tissue Research* 235, 347–356.

Stromstedt, M., Keeney, D.S., Waterman, M.R., Paria, B.C., Conley, A.J. and Dey, S.K. (1996) Preimplantation mouse blastocysts fail to express CYP genes required for estrogen biosynthesis. *Molecular Reproduction and Development* 43, 428–436.

Szafranska, B., Xie, S., Green, J. and Roberts, R.M. (1995) Porcine pregnancy-associated glycoproteins: new members of the aspartic proteinase gene family expressed in trophectoderm. *Biology of Reproduction* 53, 21–28.

Tavakkol, A., Simmen, F.A. and Simmen, R.C.M. (1988) Porcine insulin-like growth factor-I (pIGF-I): Complementary deoxyribonucleic acid cloning and uterine expression of messenger ribonucleic acid encoding evolutionarily conserved IGF-I peptides. *Molecular Endocrinology* 2, 674–681.

Thompson, J.N., Howell, J. and Pitt, G.A.J. (1993) Vitamin A and reproduction in rats. *Proceedings of the Royal Society London Series B.* 159, 510–535.

Toda, K., Terashima, M., Kawamoto, T., Sumimoto, H., Yokoyama, Y., Kuribayashi, Y., Mitsuuchi, Y., Maeda, T., Yamamoto, Y., Sagara, Y., Ikeda, H. and Shizuta, Y. (1990) Structural and functional characterizations of human aromatase P-450 gene. *European Journal of Biochemistry* 193, 559–565.

Tomanek, M., Kopency, V. and Kanka, J. (1989) Genome reactivation in developing early pig embryos: an ultrastructural and autoradiographic analysis. *Anatomy and Embryology* 180, 309–316.

Trout, W.E., McDonnell, J.J., Kramer, K.K., Baumbach, G.A. and Roberts, R.M. (1991) The retinol-binding protein of the expanding blastocyst: molecular cloning and expression in trophectoderm and embryonic disc. *Molecular Endocrinology* 5, 1533–1540.

Trout, W.E., Hall, J.A., Stallings-Mann, M.L., Galvin, J.M., Anthony, R.V. and Roberts, R.M. (1992) Steroid regulation of the synthesis and secretion of retinol-binding protein by the uterus of the pig. *Endocrinology* 130, 2557–2564.

Tuo, W. and Bazer, F.W. (1996) Expression of oncofetal fibronectin in porcine conceptuses and uterus throughout gestation. *Reproduction, Fertility, and Development* 8, 1207–1213.

Tuo, W., Harney, J.P. and Bazer, F.W. (1995) Colony-stimulating factor-1 in conceptus and uterine tissues in pigs. *Biology of Reproduction* 53, 133–142.

Tuo, W., Harney, J.P. and Bazer, F.W. (1996) Developmentally regulated expression of interleukin-1β by peri-implantation conceptuses in swine. *Journal of Reproductive Immunology* 31, 185–198.

Vaughan T.J., James, P.S., Pascall, J.C. and Brown, K.D. (1992) Expression of the genes for TGFα, EGF and the EGF receptor during early pig development. *Development* 116, 663–669.

Warner, C.M., Brownell, M.S. and Rothschild, M.F. (1991) Analysis of litter size and weight in mice differing in Ped gene phenotype and the Q region of the H-2 complex. *Journal of Reproductive Immunology* 19, 303–313.

Warner, C.M., Panda, P., Almquist, C.D. and Xu, Y. (1993) Preferential survival of mice expressing the Qa-2 antigen. *Journal of Reproduction and Fertility* 99, 145–147.

Watson, A.J., Hogan, A., Hahnel, A., Weimer, K.E. and Schultz, G.A. (1992) Expression of growth factor ligand and receptor genes in the preimplantation bovine embryo. *Molecular Reproduction and Development* 31, 87–95.

Watson, A.J., Watson, P.H., Arcellana-Panlilio, M., Warnes, D., Walker, S.K., Schultz, G.A., Armstrong, D.T. and Seamark, R.F. (1994) A growth factor phenotype map for ovine preimplantation development. *Biology of Reproduction* 50, 725–733.

Wilkinson, D.G., Bhatt, S. and Herrmann, B.G. (1990) Expression pattern of the mouse T gene and its role in mesoderm function. *Nature* 342, 657–659.

Wilmut, I., Schnieke, A.E., McWhir, J., Kind, A.J. and Campbell, K.H.S. (1997) Viable offspring derived from fetal and adult mammalian cells. *Nature* 385, 810–813.

Xie, S., Low, B.G., Kramer, K.K., Nagel, R.J., Anthony, R.V., Zoli, A.P., Beckers, J-F. and Roberts, R.M. (1991) Identification of the major pregnancy-specific antigens of cattle and sheep as inactive members of the aspartic proteinase family. *Proceedings of the National Academy of Sciences, USA* 88, 10247–10251.

Xu, Y., Jin, P., Mellor, A.L. and Warner, C.M. (1994) Identification of the Ped gene at the molecular level: The Q9 MHC class I transgene converts the Ped slow to the Ped fast phenotype. *Biology of Reproduction* 51, 695–699.

Yelich, J.V., Pomp, D. and Geisert, R.D. (1997a) Detection of transcripts for retinoic acid receptors, retinol binding protein, and transforming growth factors during rapid trophoblastic elongation in the porcine conceptus. *Biology of Reproduction* 57, 286–294.

Yelich, J.V., Pomp, D. and Geisert, R.D. (1997b) Ontogeny of elongation and gene expression in the early developing porcine conceptus. *Biology of Reproduction* 57, 1256–1265.

Zhang, Y., Paria, B.C., Dey, S.K. and Davis, D.L. (1992) Characterization of the epidermal growth factor receptor in preimplantation pig conceptuses. *Developmental Biology* 151, 617–621.

Zimmerman, J.W. and Schultz, R.M. (1994) Analysis of gene expression in the preimplantation mouse embryo: use of mRNA differential display. *Proceedings of the National Academy of Sciences, USA* 91, 5456–5460.

Genetic Resources and the Global Programme for their Management

14

K. Hammond[1] and H.W. Leitch[2]

[1]Animal Production and Health Division, Food and Agriculture Organization, Via delle terme di Caracalla, Rome 00100, Italy; [2]Centre for the Genetic Improvement of Livestock, Department of Animal and Poultry Science, University of Guelph, Guelph, Canada N1G 2W1

Introduction

At the current rate of population growth, during the second decade of the twenty-first century, consumption of food and agriculture products will be equivalent to that in the last 10,000 years. Domestic animals are a crucial element in meeting future global food requirements. The 40+ species which have been domesticated contribute directly and indirectly some 30–40% of the total value of food and agriculture production. Pigs are one of the most important species sustaining humankind. They are found in a broad spectrum of production environments, providing a range of important direct (meat and various other foods derived from the carcass and hides) and indirect benefits – income and foreign exchange generation, manure (fuel and organic fertilizer) as well as fulfilling important cultural needs. Pig production has been increasing in importance in both developing and developed countries. Pig numbers rank second only to poultry in terms of rate of increase: from 1960 to 1990 stocks increased by 45% and 201% in developed and developing countries, respectively (FAO, 1992). The increase is largely attributed to proliferation of a few breeds of pigs, and the erosion of important genetic diversity in several indigenous breeds of pig. A full 20% of all known breeds of pigs are at high risk of extinction.

Animal genetic diversity allows farmers to select stocks or develop new breeds in response to changes in the environment, threats of disease, new knowledge of human nutrition requirements, changing market conditions and societal needs, all of which are largely unpredictable. Animal genetic diversity is critical for achieving food security for the rapidly growing population, not only with respect to the local or national situation but also because countries are becoming increasingly interdependent for access to unique animal genetic resources. Furthermore, lifting trade barriers should create opportunities for developing countries to achieve more efficient and effective livestock sectors capable of increasing foreign exchange earnings and security for the world's majority. In addition, in developed countries consumer emphasis on product quality is increasing, markets are becoming more segmented, and efficiency of input resource use is being increasingly emphasized. These needs will intensify and should broaden efforts in the genetic development of animals for production, productivity, product quality and to help sustain primary national and regional agroecosystems.

The need for countries to develop and strengthen their capacity to benefit fully from their biological resources highlights the need for a global strategy for conservation of biological resources. The catalyst for such a global strategy was endorsement of the Convention on Biological Diversity (CBD) by representatives of 167 countries at the UN Conference on Environment and Development (UNCED) in Rio de Janeiro in June 1992. Following its ratification by 30 countries the CBD became law in December 1993. As of August 1996, 152 countries have ratified the CBD. The CBD specifically addresses agriculture and identifies the objectives: to conserve diversity, to use it in a sustainable manner and to share benefits arising from utilization of genetic resources.

In recognition of the importance of domestic animal genetic resources, and of the sizeable proportion which are currently at risk of loss, and in keeping with FAO's mandate under the CBD, a special action programme for the global management of domestic animal genetic resources was launched by FAO in 1992. This chapter focuses on the rationale and the operation of FAO's Global Programme for the Management of Farm Animal Genetic Resources being designed to assist countries characterize, develop, use and maintain some 5000 breeds comprising about 40+ livestock species. The state of pig genetic diversity is highlighted.

Achieving sustainability

The contribution of livestock production to global food security is generally undervalued, particularly by developed country communities. Sansoucy *et al.* (1995) provide a review of contributions of livestock in achieving sustainable production systems. They also suggest that as human needs for food and agriculture products double over the coming decades, global food production from animal products will increase more rapidly than from plants in most developing countries with increasing purchasing power. Over this same period and beyond, the large regional differences in human needs for food and agriculture and in production capacity will persist at the global level, with almost 75% of world agriculture remaining at the low- to medium-input levels. Indigenous livestock breeds often possess valuable traits such as disease resistance, high fertility, good maternal qualities, unique product qualities, longevity and adaptation to harsh conditions and poor quality feed. These are all desirable qualities for achieving sustainable agriculture under low-input conditions. In both developing and developed countries, food and agriculture production systems must increasingly utilize mixes of biological diversity, several plant and animal, and combinations of lines across species, with line development applying increasingly to particular primary production environments. With continual loss of agricultural land at alarming rates, the challenges of understanding, better measuring and introducing more sustainable systems must be accepted.

National strategies for resource use

The fitness traits associated with the adaptation complex in a particular environment are generally much more difficult to measure and change than are the 'production' traits. This seriously questions the common approach of introducing to low- to medium-input environments high-input/high-output exotic breeds developed primarily under comparatively benign environments and then relatively quickly expanding the exotic breeds expecting them to adapt rapidly in arriving at sustainable systems. Instead, a superior strategy for these low to medium environments will aim to develop the production traits in one or more indigenous breeds which are already adapted and which local farmers are still prepared to use. Remember, much of the very large amount of low- to medium input agriculture is likely to remain the same for much of the twenty-first century. Some medium- to even high-input pockets of agriculture are emerging in some developing countries particularly around large cities and to better utilize waste products. While these situations pose land use and rural development policy issues, they may enable use of some higher demanding and producing genetic resources particularly when disease, parasite and climatic stresses are substantially and permanently reduced by other changes to the particular production environment. However, even in many of these countries low-input rural production from a range of species will remain important and will require highly superior animal adaptation to achieve sustainable agriculture. It is also of interest that in these adapted types, under the low-input and otherwise high-stress environments, the relative variability between animals for the production traits is often very high, suggesting that rapid rates of improvement in these traits under these stresses will be feasible if an effective breeding strategy can be designed and maintained.

Genetic type by production environment differences
We have little understanding of the important ranking differences between most production environments of the genetic types developed under low-, medium- and high-input levels. Included among the low level category are the combination high-stress and scavenger production environments, relevant because of their continuing importance, particularly scavenger poultry, swine and goats in the rural communities in much of the developing world. The high selection intensities which have ruled for very long periods in these environments surely have resulted in types which are more likely to differ between different low environments, and also from lines developed under more benign high-input conditions. Most of the performance testing of these high input breeds takes place under conditions of *ad lib* feeding with genetic improvement directed to a breeding goal integrating soundness, growth, efficiency, carcass quality and reproductive performance. Studies to characterize this reranking of genetic types among environments are not easy to design and conduct well; however, they are very important and likely to be rewarding relatively quickly particularly for the short generation interval, prolific species.

Genetic erosion of pig genetic diversity

In 1990 FAO initiated, with the support of the United Nations Environmental Programme (UNEP) and the European Association of Animal Production, an inventory and basic description of the breeds of the domestic livestock species. As of December 1995, the Global Databank contained information on 3882 breeds of 28 domestic species in more than 180 countries. Represented in the Global Databank are over 350 breeds of swine. A comprehensive listing of these breeds together with country of ownership, most common breed name, population size and trends, morphology, performance and a range of other information can be accessed from the Domestic Animal Diversity Information System (DAD-IS) on the Internet at URL: http://www.fao.org/dad-is/. The Global Breeds Databank is constantly being updated by individual countries. Currently, there are population data on 265 breeds of pigs (Table 14.1). Based on the population data, 26% of pig breeds are classified as endangered and critical. FAO defines endangered as populations having less than 1000 breeding females and less than 20 breeding males; and critical populations having less than 100 breeding females and less than 5 breeding males. Of the breeds listed in these two categories, 25% are either managed through a conservation programme or maintained by an institute. This represents a far lower percentage conserved compared with the situation with all other breeds of livestock species identified at risk (Table 14.1), for which conservation management programmes are underway for 36% of the 873 breeds identified at risk.

Presumably, the risk of loss for these breeds actively managed or maintained is far lower than breed populations outside such management programmes. Endangered breeds which are not being managed under a conservation programme are at high risk of extinction. A full 20% of pig breeds fall in this category and their genetics may be irreversibly lost if action is not taken.

Swine genetic resources by global region are shown in Table 14.1. In Europe, where the situation is well documented, 40 of the 117 swine breeds with population data are at risk of loss. Although conservation management programmes are underway for 17 of these breeds, 23 swine breeds are in imminent danger of disappearance. Globally, more than 60% of genetic resources are in developing countries and due to lack of financial and human resources there is negligible information on many breeds and their status. Using this region as an indicator, the situation in the African and Latin America and Caribbean region, for example, is likely to be at least as desperate as in Europe.

The complete global analysis of breeds at risk of loss is published in the second edition of *World Watch List for Domestic Animal Diversity* (FAO/UNEP, 1995). The information contained in this second edition shows that globally 30% of the total number of breeds with population data comprising 285 species, are at risk of loss (Table 14.1).

Table 14.1. Regional distribution of swine breeds identified and at risk.[1,2]

Region	Total no. swine breeds on file	No. swine breeds with population data and per cent of total	No. of breeds categorized as critical or endangered and per cent of which are maintained	No. of breeds at high risk of loss[2] and per cent at risk
Africa	13	6	0 (0)	0
Asia and Pacific	157	113	15 (0)	15
Europe and former USSR	129	117	40 (43%)	23
Latin Am. and Caribbean	24	10	8 (0)	8
Near East	2	2	1 (0)	1
North America	28	17	5 (0)	5
Global total swine breeds	353	265 (75%)	69 (25%)	52 (20%)
Global total all breeds[3]	3882	2924 (75%)	873 (36%)	559 (19%)

[1]Source: Adapted from FAO/UNEP (1995).
[2]'At risk' determined based on breeds with population data having <1000 breeding females or <20 breeding males and for which there is no conservation programme in place.
[3]This includes 28 species of mammalian and avian species on Global Databank.

Rushing to one genetic resource per species

There are several factors which place breeds at risk of loss and threaten domestic animal diversity. These factors are listed in Table 14.2. By far the greatest cause of genetic erosion is the growing trend to global reliance on one or a few modern breeds suited for high input–output needs of industrial agriculture. This trend is of paramount concern as a cursory examination of the research literature suggests that about 50% of the total variation at the quantitative level is between-breed, the remainder being common to all breeds. Hence, a move to one breed would eliminate half this variation in the species, in addition to jeopardizing readily available gene combinations in other remaining unique genetic resources.

Conservation Imperatives for Animal Genetic Resources

With the projection that more than 1000 of the probably 5000 breeds of Earth's domestic animal species are currently at high risk of extinction, and with so little known about most, scarce international funds cannot at this point be concentrated on a small number of breed rescue projects. Emphasis must be on implementing a sound global management infrastructure and broad technical programme, which has the potential to help many countries design and implement national action strategies, as required under the CBD. With these considerations firmly in mind, the imperatives for conservation of domestic animal genetic diversity can be stated:

Table 14.2. List of causes for risk of loss or extinction of breeds.

Reason	Description
Aid	Lack of incentive to develop and use breeds, giving preference to those few developed for use in high-input, high-output relatively benign environments.
Product	Undue emphasis placed on a specific product or trait leading to the rapid dissemination of one breed of animal to exclusion and loss of others.
Crossbreeding	Indiscriminate crossbreeding which can quickly lead to the loss of original breeds.
Storage	Failure of the cryopreservation equipment and inadequate supply of liquid nitrogen to store samples of semen, ova or embryos; or inadequate maintenance of animal populations for breeds not currently in use.
Technology	Introduction of new machinery to replace animal draught and transport resulting in permanent change of farming system. Poorly interpreted international genetic evaluation.
Biotechnology	Artificial insemination and embryo transfer leading to rapid replacement of indigenous breeds.
Violence	Wars and other forms of socio-political instability.
Disaster	Natural disasters such as floods, drought or famine.

1. *Identify* and *understand* those unique genetic resources which collectively comprise the global gene pools for each of the important animal species domesticated and used to provide food and agriculture.
2. *Develop* and properly *utilize* the associated diversity, to increase production and productivity, achieve sustainable agricultural systems and meet demands for specific product types. Hence, the effective use of breeds is also an essential component of conservation, and perhaps the most cost effective, which is a further reason for enabling the development and use of more breeds. Development and use of genetic resources must be integral to an effective conservation effort for the domestic animal species.
3. *Monitor* particularly those resources which are currently represented by small populations of animals; or which are otherwise being displaced by one or other breed replacement strategies.
4. *Preserve* the unique resources for which sufficient current demand cannot be engendered.
5. *Train* and *involve* people in management of these resources, including their best use and development, and in the maintenance of diversity.
6. *Communicate* to the world community the importance of our domestic animal genetic resources and of the associated diversity, its current exposure to loss and its irreplaceability.

Working definitions

Clear terminology is necessary in the conservation effort, to advance understanding, facilitate education and training, communicate successfully with the wider public and realize a common purpose in application. Terms must accommodate all practical situations. In the case of domestic animals, these should include all genetic resources and diversity associated with each species, both resources currently in use and those not in use, the common and the rare, the long developed and the new, the commercial lines and the research stocks. The conservation literature includes a number of terms that are not well understood in the domestic animal context. Interpretation of these terms to form working definitions for domestic animal diversity (DAD) is required if meaningful involvement in conservation of DAD by the necessary range of people is to be realized. A minimum set of working definitions is presented in Table 14.3.

FAO's Global Management Programme: Structure and Work Elements

FAO is the sole intergovernmental organization with a broad international mandate for improving agriculture and food production for current and future populations – with particular emphasis on developing countries. Given the global nature of animal genetic resources (AnGR), few institutions have the capacity to coordinate the geographic, species, technical and intergovernmental issues which are necessarily involved in developing a successful global programme of management for the domestic animal sector. Based on the conservation imperatives, and in accordance with the CBD and Agenda 21 and the oversight provided by the International Commission on Sustainable Development, FAO's Global Programme for the Management of Farm Animal Genetic Resources was launched. The global strategy involves four components:

1. An intergovernmental support mechanism for enabling direct government involvement and ensuring continuity of policy advice.

Table 14.3. Working definitions proposed for conservation.[1]

Animal Genetic Resources (AnGR) At the breed level, the genetically unique breed populations formed throughout all domestication processes within each animal species used for the production of food and agriculture, together with their immediate wild relatives (here '**breed**' is accepted as a cultural rather than a technical term, i.e. to emphasize ownership, and also includes strains and research lines).

Domestic Animal Diversity (DAD) The genetic variation or genetic diversity existing among the species, breeds and individuals, for all animal species which have been domesticated and their immediate wild relatives.

Conservation (of domestic animal diversity) The sum total of all operations involved in the management of animal genetic resources, such that these resources are *best used and developed* to meet *immediate and short-term* requirements for food and agriculture, and to ensure the diversity they harbour remains available to meet possible *longer term* needs.

Conservation (in general) The management of human use of the biosphere so that it may yield the greatest sustainable benefit to *present* generations while maintaining its potential to meet the needs and aspirations of *future* generations. Thus conservation is positive, embracing preservation, maintenance, sustainable utilization, restoration and enhancement of the natural environment (IUCN-UNEP-WWF and FAO-UNESCO, 1980).

***In-situ* Conservation** Primarily the active breeding of animal populations for food production and agriculture, such that diversity is both best utilized in the short term and maintained for the longer term. Operations pertaining to *in-situ* conservation include performance recording schemes, and development (breeding) programmes. *In-situ* conservation also includes ecosystem management and use for the sustainable production of food and agriculture. For wild relatives *in-situ* conservation – generally called *in-situ* preservation – is the maintenance of live populations of animals in their adaptive environment or as close to it as practically possible.

***Ex-situ* Conservation** In the context of conservation of domestic animal diversity, *ex-situ* conservation means storage. It involves the preservation as animals of a sample of a breed in a situation removed from its normal production environment or habitat, and/or the collection and cryopreservation of resources in the form of living semen, ova, embryos or tissues, which can be used to regenerate animals.

Other methods of genetic manipulation, such as the use of various recombinant DNA techniques, may represent useful means of studying or improving breeds, but do not constitute *ex-situ* conservation, and may not serve conservation objectives. At present the technical capacity to regenerate whole organisms from isolated DNA does not exist.

[1]Source: FAO, 1995.

2. A geographically and country-based global structure, to assist and coordinate national actions.
3. A technical programme of activities grouped under seven elements.
4. A cadre of experts from throughout the world to guide the programme and maximize its cost effectiveness.

The global structure

Countries possess different subsets of the global total of breeds forming each domestic animal species. While the Convention on Biological Diversity clearly accepts each country's sovereignty over its genetic resources, countries are becoming increasingly interdependent in seeking to access unique animal genetic resources from elsewhere. Therefore, effective conservation programmes by nations provide the foundation for a successful global programme of management for each species. FAO has established the Global Programme for the Management of Farm Animal Genetic Resources, coordinated, led and facilitated by the Global Focal Point. The programme of management provides the structure for achieving country-based emphasis combined with the necessary regional and global coordination of policy and effort (Fig. 14.1). This essential primary global infrastructure will lead to the early implementation of the necessary range of activities to assist countries which have not done so, to design, implement and maintain comprehensive national strategies for the management of their animal genetic resources. Scarce financial resources are concentrated on initiating the key infrastructure required. Some aspects of the programme are already being implemented although complete implementation will take some years and will depend on strong collaborative support for the programme.

1. *National focus* for each country, comprising a coordinating institution strongly linked to the regional focus (Fig. 14.1) and a national technical coordinator nominated and supported by the member government. This coordinator will be the point of contact for the country's involvement in the FAO AnGR Programme and will assist in establishing and maintaining the essential in-country network. To date, 50 countries have established focal points: 38 in the European Community and 12 in Asia; 24 national focal points are now being established in Sub-Saharan Africa and the Americas regions.
2. *Regional focal point*, in each major genetic storehouse region of the world, provides support to the national coordinators. The regional focal point is responsible for operations within the region and provides critical support for country action planning, development of effective national coordinators and establishing strong networks with the region. The regional focal point will also trigger a range of most effective projects covering the conservation complex for domestic animals. Regional focal points are planned for Asia and the Pacific; Europe; the Americas and Caribbean; Africa; and the Near East and

Mediterranean. The regional focal point for Asia and the Pacific has now been initiated in Bangkok through funding from Japan.

3. *Global focus*, located at FAO headquarters in Rome, serves as the global management entity and is responsible for designing implementing and leading the programme, developing the necessary guidelines and procedures, and putting in place the global information system. The global focus has a demanding coordination role in order to involve the range of governmental, non-governmental and intergovernmental parties essential for the programme's success. The global focus also provides the secretariat for the Commission on Genetic Resources for Food and Agriculture and has been mandated to seek the essential extra-budgetary funding for the programme. The global focus is responsible for the global early warning system for AnGR, and maintains breed-related databases and the World Watch List for Domestic Animal Diversity.

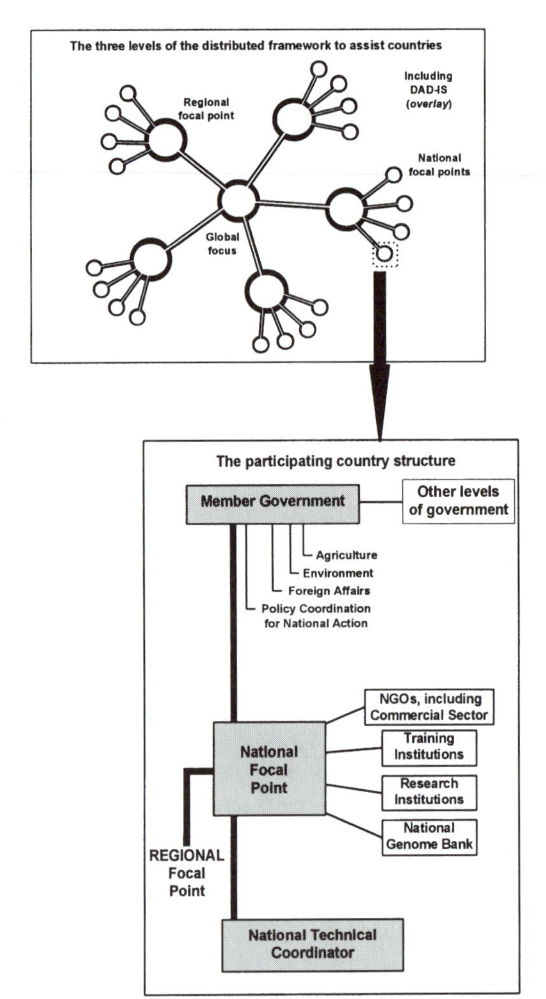

Fig. 14.1. Structure for FAO's Global Programme for Management of Farm Animal Genetic Resources.

The decentralized nature of the programme enables it to be responsive to the needs of countries and provide maximum assistance.

Work elements

The structure must be accompanied by a cost-effective programme of work if the global management is to be effective in achieving both intergovernmental and technical imperatives over time (Fig. 14.2).

Domestic Animal Diversity Information System (DAD-IS)
This system is designed to link all stockholders to the programme. These stockholders include a range of individuals from researchers, students, and farmers to conservationists, governmental and non-governmental organizations. DAD-IS is accessible via the Internet (http://www.fao.org/dad-is/) and in the future on CD-ROM. DAD-IS forms the information axis for all aspects of the programme and serves to:

1. Accommodate the range of essential and unique databases for the programme covering the global inventory and description and monitoring of resources for both species and genebanks, and for project MoDAD (see below) and to link other specific genetic databases.

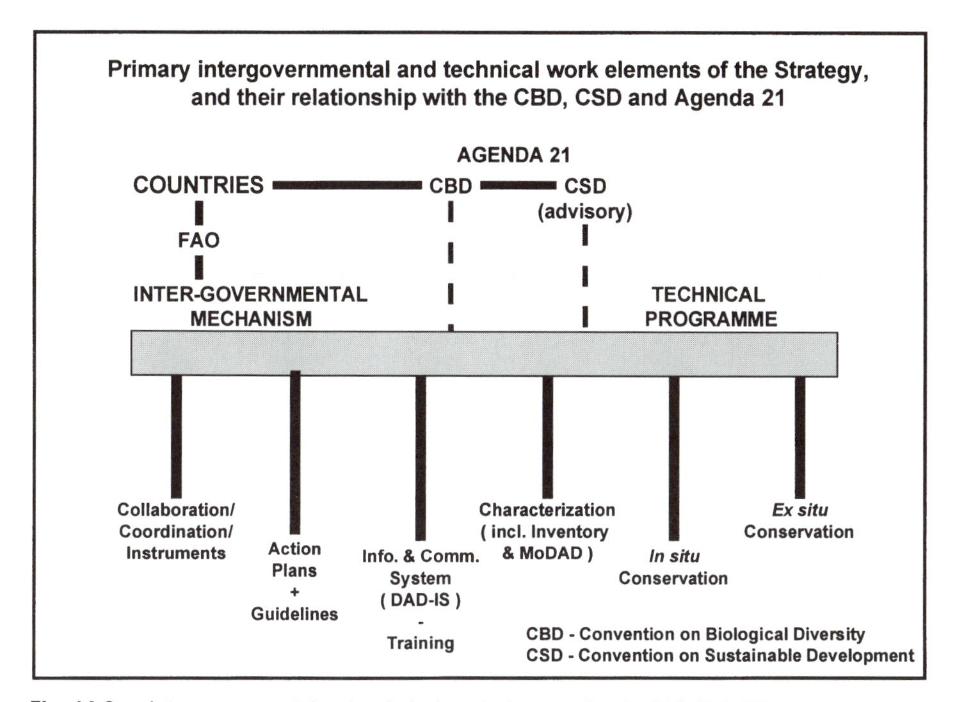

Fig. 14.2. Intergovernmental and technical work elements for the FAO Global Programme for Management of Farm Animals Resources.

2. Provide the global early warning system for AnGR and to facilitate continuous updating and ongoing access to this database.
3. Lower the cost and increase the amount and effectiveness of training and education in animal conservation genetics and procedures, through a system of shared expertise and information.
4. Provide, in the future, a central and reliable source of aids for experimental design and data analysis, in order to increase cost effectiveness of, and capacity for, research.
5. Provide the global bibliography for AnGR.
6. Assist in management of the programme, and execution of activities including the effective networking in project development and management.
7. Facilitate active involvement of the world community in the programme.

Characterization
Establishing the magnitude of existing animal genetic diversity and reliable rates of loss are corner stones for the programme. The enormity of this task is highlighted by the fact that globally there are some 5000 breeds, comprising the 40+ species with some in most countries and largely no systematic monitoring in place nor baseline information available. By the end of 1997 survey data contained in FAO's Global Databank for AnGR will include basic descriptive data on roughly 85% of these known breeds representing 28 species. The population data provided through surveys enable the monitoring of breeds at risk of extinction. This information was most recently summarized in the second edition of FAO's *World Watch List for DAD* (FAO, 1995). Information in the Databank will be regularly updated to incorporate missing data and as a mechanism to monitor breed genetic resources at risk and rates of loss. Each of these activities, together with information on the status of *ex situ* collections, forms the global early warning system for AnGR.

Characterizing animal genetic diversity

Comprehensive genetic evaluation at the breed level to cover all breeds for both current and future production potential for all primary production environments is neither feasible nor required. However, knowledge of the amount of the breed level variation in each species and of the size of each breed's contribution to this will assist priority setting for overall AnGR management. In order to better understand this relative uniqueness of animal genetic resources, a global research project is planned in genetic distancing (Barker *et al.*, 1993). This project referred to as the Global Project for the Measurement of Domestic Animal Genetic Diversity (Project MoDAD), will utilize microsatellite technology and initially focus on the analysis of genetic variation within some 14 species accounting for above 90% of food and agriculture production globally, pigs being among these most important species. The primary objective of Project MoDAD is substantially to increase the cost effectiveness of the total global programme by establishing the comparative uniqueness of genetic resources in each species to aid rationalization of the total management task for countries and globally. This will be achieved primarily by directly utilizing the results to objectively reduce the number of breeds that will need to be maintained. Secondary objectives of Project MoDAD are to establish global repositories for both AnGR microsatellite data and DNA, for enabling more effective research and use in capacity building; and to identify breed combinations having highest potential for heterosis for improvement in traits related to adaptation. Project MoDAD will benefit all nations. A detailed formulation document for the project, for use in assisting countries to be involved is available from FAO.

Mechanisms for conserving animal genetic resources

Conservation is not an end in itself but a means of ensuring that animal genetic resources are better understood and available and more effectively used by present and future generations. Once genetic resources have been identified and characterized, there are two basic conservation activities which may then be defined as either *in situ* or *ex situ.*

In situ conservation

Generation and loss of alleles is a dynamic process which should be maintained at about equilibrium through sound management. The strategy for the global *in situ* activity emphasizes 'wise use' of indigenous animal genetic resources by establishing and implementing breeding goals and strategies for sustainable production systems. Effective development of more of these adapted resources to meet the requirements is all important and will form the focus of FAO's *in situ* emphasis. Because *in situ* involves the maintenance of live populations of animals in their adaptive environment, animal populations continue to evolve and be developed for more effective use. Infrastructure for animal recording and breeding is well established for developed countries. But infrastructure which is appropriate to developing country systems remains scarce. Modalities for the simplified animal recording, genetic development and dissemination is needed for each species and a range of national livestock structures for developing countries.

Ex situ conservation

Ex situ conservation includes cryogenic preservation and the maintenance as animal populations of breeds of domesticated species in farm parks, zoos and locations away from the environment in which they are being developed; in effect this is storage of AnGR which farmers in a country are no longer interested in using. The biggest shortcoming of *ex situ* genebanks is that, once stored, animal genetic resources are removed from the evolutionary process they undergo in nature, and unless a concerted effort is made, the level of knowledge about them is also frozen. No selection pressure is imposed from changing natural conditions. Cryopreservation is technologically demanding and needs further development for handling pig semen, embryos and ova. A range of quarantine issues must be overcome before international storage and access can be effective.

The global programme's *ex situ* conservation strategy is still being developed but is based primarily on the use of live animal populations wherever possible, backed up by cryopreservation where technology exists or can be developed, combining within country genebanks with global gene repositories of last resort, in keeping with the CBD. Interested governments, NGOs (non-governmental organizations), research institutions, and private enterprises would also be encouraged to maintain *in vivo* samples of breeds at risk, with national inventories of these being established and maintained current, so that the genetic resources are directly available for use and study.

In situ and *ex situ* conservation are complementary, not mutually exclusive; their application for particular AnGR depending on farmers' current use of it and comparative uniqueness. Further, frozen germplasm can play an important role in the support of *in situ* breeding strategies. For example, the use of artificial insemination (AI) in *in situ* conservation of populations may enable much greater male selection differentials and dissemination than would be practical via natural mating using live adult males. The use of AI in a backcrossing systems enables efficient regeneration of a population and alternate use of male breeds in reciprocal crossbreeding systems may also be advantageous.

Global action plan and guidelines

The programme provides assistance to countries in the development and implementation of comprehensive and practical guidelines for designing national action strategies for the management of animal genetic resources to harmonize with the provisions of the CBD. The global action strategy will be further developed by integrating all national action plans, and will be continually updated as knowledge, technology, the negotiation of the CBD by member countries and implementation of policies progress.

A further early activity in the programme involves sub-regional project identification missions being carried out to establish a global portfolio of the most effective conservation activities, ready for formulation, funding and execution to add to the basic framework now being implemented. The specially designed missions are being trust-funded by various donors and focus on the following regions: (i) China; (ii) and (iii) Sub-Saharan Africa (Francophone and Anglophone); (iv) Central and Eastern Europe; (v) former USSR; (vi) and (vii) Latin America and the Caribbean (split into two sub-regions); (viii) Near East and the Mediterranean; (ix) the Indian sub-continent.

The programme for animal genetic resources will operate with technical input from FAO's intergovernmental Commission Genetic Resources on Food and Agriculture (CGRFA). The Commission has only recently expanded from being solely devoted to plant genetic resources, to include also animal genetic resources and eventually forestry and fish genetic resources. Many of the technical logistics and the policy and legal issues are fundamentally similar for both plants and animals. Further, there is a need to promote the development and maintenance of more sustainable agriculture systems. The first technical discussion related to animal genetic resources at the CGRFA will occur in 1997.

The global management entity

Successful management of AnGR involves a range of technical development, coordination, facilitation and operational activities for countries and for non-governmental and international parties involved. FAO established global focus as the management entity for this global programme with the mandate to:

1. Help countries requesting assistance and coordinate activities regionally and globally;
2. Develop uniform basic procedures, guidelines and protocols;
3. Maintain the global early warning system for AnGR;
4. Bolster capacity building and upgrade training;
5. Increase awareness and communicate issues globally;
6. Serve as the secretariat for the global intergovernmental mechanism for AnGR.

Balancing Biotechnology and Diversity

Molecular and reproductive biotechnologies afford great potential to both reduce the risks of loss and increase the wise use of animal genetic resources. An overview of positive and negative effects of biotechnology on development, use and maintenance of animal genetic resources, appears in Table 14.4.

Molecular

Rapid development of maps of the pig genome is enabling identification of some quantitative trait loci. Molecular biology is important in characterizing and conserving biodiversity. The global project proposing to measure the diversity within species (MoDAD) will enable more objective decision making in terms of where conservation efforts should be focused. The repositories of DNA will prove invaluable as additional molecular screening tools are refined and developed. Molecular markers can help to identify genes of interest to breeders. The ability to screen the DNA of live or cryopreserved germplasm for specific genes underscores the importance of ensuring that diversity is maintained. As such, DNA repositories will constitute an important component of genebanking; limited only by the current inability to regenerate populations of animals from DNA stores.

Although a powerful tool, molecular technology has been often wrongly heralded as a panacea for genetic erosion. Proponents of this argue that biodiversity can be genetically engineered in the laboratory. Such unrealistic expectations fail to consider the adaptive complex which determines animal productivity and sustainability in a given environment. Comprised of unique

Table 14.4. Some important positive and negative effects of biotechnology on development, use and maintenance of animal genetic resources.

Biotechnology	Positive	Negative
Molecular		
Marker assisted measurement of diversity	Quantify genetic variation to better target conservation efforts DNA stores as a 'backup' to other germplasm genebanks Repositories of DNA for research and training	Does not distinguish unique genetics for current or future needs. Requires good experimental design for integrity of results
Marker assisted selection	Identify marker genome sites to assist in selection of quantitative traits Potentially circumvents expensive performance recording; measurement prior to expression – age/sex	May require high level of pedigree and descriptive production information, i.e. linkage maps Markers required for substantial portion of variation in a trait
Transgenics	Transfer of useful genes improves production and productivity, i.e. major genes in sheep and pigs affecting litter size Animals producing unique gene products, e.g. human blood clotting factor in sheep milk Recombinant DNA involving bacteria for mass production of useful proteins, e.g. bovine somatotrophin	Technology needs further development Inefficient and costly for 'adaptive complex' Potential negative effect on displacing value of traditional stocks when useful qualities can be 'packaged' into few breeds or species. Potential consumer backlash
Reproductive		
Artificial insemination	Enables dissemination of superior genetics Important for *in-situ* conservation efforts *Ex-situ* conservation and regeneration of lost animal populations *In-situ* conservation when used for crossbreeding, combining local adapted with increased production characteristics of 'exotic' Affords more controlled breeding and mate selection and potentially decreases inbreeding Enables accurate comparative genetic evaluation for traits of low heritability or sex limited both within and across countries Reduces transmission of disease	Dilution/loss of indigenous adapted genetics with exotic semen Needs further development for many species Requires high level management for collection, processing of semen and use Requires access to technology When used for genebanking semen requires ongoing supply of liquid nitrogen

Embryo transfer	Effective for genebanking and regeneration of populations Passive immunity conferred on offspring by recipient female Evaluation of genetic potential in several different production environments Increased potential for dissemination of genetics from superior dams, and transboundary transport of genetic material	Technology developed for few species Costly and logistically complex Requires specialized training and expertise Requires ongoing supply of liquid nitrogen if used for genebanking Potential replacement of indigenous stock in one generation
Cloning	Mass production of crossbreed clones enables optimum combination of adapted indigenous with exotic genetics Accurate evaluation of genotype by environment interaction Enables evaluation for optimum genetic combinations (e.g. hybrid vigour)	Technology needs further development Potential cytoplasmic effect (may be viewed as a positive or negative) Expensive Requires specialized equipment and training Production of genetic 'copies' may erode diversity
Oocyte collection and *in vitro* embryo production	Ova can be collected from ovaries of slaughtered animals thus conserving breeds at risk, i.e. in emergency situation – drought, disease, war, etc. resulting in slaughter or death of populations Potential to reduce lengthy generation intervals by collection of ova from prepubertal females Efficient use of semen for *in-vitro* fertilization as few sperm required thus reducing the number of doses of semen to be gene banked Enables dissemination of genetics from superior females Reduced cost and increased flexibility of genebanking and embryo transfer	Technology needs further development for many species Requires specialized equipment and training Costly (but not as high as embryo transfer) Requires ongoing supply of liquid nitrogen if used for genebanking Potential replacement of indigenous stock in one generation

packages of genes formed over a millennia of evolution and selection, it is far more cost effective to rely on utilizing these already formed genetic types than attempting to create new ones. This is not to say that the ability to transfer genes within and between species does not hold promise for production, performance or morphological traits controlled by few or single genes. The ability to incorporate a transgene into the genome of a different breed, variety or species has been demonstrated in both avian and mammalian species.

Reproduction

Reproductive technology, which enables the widespread dissemination of semen from exotic boars using artificial insemination, has the potential to cause abrupt changes in the composition of indigenous breeds. Often the widespread use of one or a few breeds of exotics has been tied to international aid. Conversely, the use of reproduction technology when properly used plays a pivotal role in the management of animal genetic resources and conservation of domestic animal diversity. A most important current challenge for research in reproductive biology is the development of practical protocols for long-term semen and oocyte cryopreservation in order to markedly increase the cost effectiveness of genebanking.

Related to issues of 'genetic prospecting' and 'ownership', biotechnology is sometimes shrouded in controversy related to international property rights. Countries which have ratified the Convention on Biological Diversity recognize the sovereignty a country has over its genetic resources. But it is ironic that the developing countries which are rich in biodiversity, and stand to benefit the most through its development and use, often lack the necessary technology to enable exploitation of their genetic resources. Added to the issue of ownership are concerns for biosafety.

Strengthening Scientific and Technological Capacity

New paradigm for animal breeding

Traditional animal breeding has dwelt on quantitative genetic principles and theory. At the other extreme, conservation has been mainly focused on storage of genetic material. There can be no denying the enormous success of traditional animal breeding programmes in terms of improving production potential of modern livestock, particularly in developed countries (Smith, 1984a,b; Gibson and Smith, 1989). However, socio-political changes underway are emphasizing the need to account for the longer term costs and benefits; and for sustainability of production systems, including the breeding programmes involved. This broadens the domain which genetic development must address. The term conservation genetics perhaps better describes the new paradigm where the objective is now to manage wisely all of a country's genetic resources in terms of better understanding, developing, using and maintaining agrobiodiversity. Education and training programmes relating to management of animal genetic resources must address these needs. Malmfors *et al.* (1994) has identified areas where university teaching should focus.

Capacity building imperatives

The ability of countries to implement effective national action strategies for conservation of animal genetic resources will largely depend on human resources and institutional capacities. Countries will need to determine their training, technological and research needs. Countries may seek guidance and require support to complement local initiatives in order to strengthen their capacity and technical competence. FAO's programme will address these needs as part of, and in support of, the technical elements of the programme.

Implications

The outcomes being sought by FAO's Global Programme for Management of Farm Animal Genetic Resources are inventory control and early warning; establishing the comparative uniqueness of the breeds comprising each species; more effective development and use of a greater range of unique resources; a much higher level of public awareness; major reduction in the number of breeds at high risk of loss; maintenance of and ready access to the more unique AnGR; and helping countries to meet their obligations under the CBD. The DAD represents a resource that is critical for achieving food security for the rapidly growing human population, not only with respect to the local or national situation but also because countries are becoming increasingly interdependent for unique animal genetic resources. The country-based global structure of the programme is designed to recognize and emphasize responsibility and activities within countries, to involve countries and other essential parties, to maximize opportunities for conservation action and fully align with the CBD.

The programme introduces some changes of emphasis to and a broadening of our long-accepted approach to animal breeding principles and practice, to harmonize with the decisions of a large section of the world community. Much remains to be done if the vast majority of largely unique genetic resources in developing countries in particular are not to be lost under the increasingly intense pressures now being applied. Countries need assistance now.

This FAO programme has the potential to be effective for all species of domestic animals upon which humankind must rely for much of its existence. However, the programme's implementation is heavily dependent upon developed country funding. The primary reason being offered by a wide spectrum of developed countries for being unable currently to provide substantial support is budgetary constraints. The easiest short-term decision is to delay, but in so doing it must also be accepted that these domestic animal genetic resources are irreplaceable, and 'sharing of benefits' will inevitably also mean the granting of access to those providing assistance. There are several ways in which interested groups may become involved, including:

1. Learn more about your country's programme from your national focal point or international government representative.
2. Help advance the way genetics and animal breeding is taught, at all academic and applied levels.
3. Help address the new research and development needs.
4. Help develop DAD-IS and particularly its training and research modules.
5. Help communicate the issues to the public and commercial business to rethink the overemphasis on the development and sale of few AnGR.
6. Help your country to identify, better understand and monitor its AnGR.
7. Help establish national genebanks for each species.
8. Help develop more of the worlds AnGR in each of the species used for food and agriculture.
9. Help develop the range of detailed guidelines required to assist countries to upgrade their management of AnGR.
10. Contribute to the development and maintenance of your country's national action strategy for management of AnGR of each species.
11. Contribute to funding of your national programme and the global programme to conserve AnGR.

These recommendations form the basis for sound practices to maintain animal diversity.

References

Barker, J.S.F., Bradley, D.G., Fries, R., Hill, W.G., Nei, M. and Wayne, R.K. (1993) An integrated global programme to establish the genetic relationships among the breeds of each domestic animal species. FAO Animal Production and Health Paper. Rome, Italy.

FAO. (1992) The role of ruminant livestock in food security in developing countries. Committee on World Food Security, Seventeenth Session, 23–27 March 1992. Food and Agriculture Organization, Rome, Italy.

FAO/UNEP (1993) *World Watch List for Domestic Animal Diversity*, 1st edn. (Eds Loftus, R., and Scherf, B.) FAO, Rome, 369 pp.

FAO/UNEP (1995) *World Watch List for Domestic Animal Diversity*, 1st edn. (Ed. B. Scherf) FAO, Rome, 769 pp.

Gibson, J.P. and Smith, C. (1989) The incorporation of biotechnologies into animal breeding strategies. In: Moo-Young, M., Babiuk, L.A. and Phillips, J.P. (eds) *Animal Biotechnology: Comprehensive Biotechnology*, First Supplement. Pergamon, Elmsford, New York, pp. 203–231.

Hammond, K. and Leitch, H.W. (1996) The FAO Global Program for the Management of Farm Animal Genetic Resources. In: *Biotechnology's role in genetic improvement of farm animals.* American Society of Animal Science, pp. 24–42.

IUCN/UNEP/WWF/FAO/UNESCO (1980) *World Conservation Strategy. Living Resources Conservation for Sustainable Development.* IUCN, Switzerland.

Malmfors, B., Philipsson, J., and Haile-Mariam, M. (1994) Education and training in the conservation of domestic animal diversity – student needs and field experience. Symposium on Conservation of Domestic Animal Diversity. In: *Proceedings of the 5th World Congress on Genetics Applied to Livestock Production*, Vol 21, pp. 485–492.

Sansoucy, R., Jabbar, M.A., Ehui, S. and Fitzhugh, H. (1995) The Contribution of Livestock to Food Security and Sustainable Development. In: *Roundtable on Livestock Development Strategies for Low Income Countries* FAO, Rome and ILRI, Nairobi, 182 pp.

Smith, C. (1984a) Estimated costs of genetic conservation of farm animals. In: *Animal Genetic Resources Conservation and Management, Data Banks and Training.* FAO Animal Production and Health Paper no. 44/1. FAO, Rome, pp. 21–30.

Smith, C. (1984b) Genetic aspects of conservation in farm livestock. *Livestock Production Science* 11, 37–48.

UNEP (1992) Convention on Biological Diversity Final Text. UN Department of Public Information, New York.

Genetics of Performance Traits

A.C. Clutter[1] and E.W. Brascamp[2]

[1]*Department of Animal Science, Oklahoma State University, 206 Animal Science Building, Stillwater, OK 74078, USA;*
[2]*Wageningen Institute of Animal Science, Wageningen Agricultural University, PO Box 338, Wageningen 6700 AH, The Netherlands*

Introduction

Performance of an animal destined for market in a pork production system is defined by the efficiency with which it develops saleable product. That efficiency is largely determined by costs associated with feed and time, and by the amount of quality lean tissue produced. Genetic merit for market animal performance is improved in seedstock populations by performance testing and selection. For successful commercial pork production, a breeding system must be implemented which optimizes market pig performance and reproductive efficiency. Thus, the genetics of performance traits encompasses not only additive and non-additive genetic effects associated with feed intake and tissue growth, but also strategies for performance testing and selection, and genetic correlations between market pig performance and traits of animal soundness and reproduction.

This chapter begins with a discussion of traits of the postweaning period in the context of models of tissue growth and corresponding strategies for performance testing and selection. Summaries of reported estimates of relevant genetic parameters and of direct and correlated responses to selection for postweaning traits are presented, including reported effects on reproductive performance. Finally, knowledge regarding genetic aspects of structural soundness and stress susceptibility of the market pig is reviewed.

Performance Testing and Selection Strategies

Overview

In a typical selection programme an overall objective is defined and phenotypic measurements are chosen as criteria to estimate genetic merit for the selection objective. For objectives related to performance of the market pig, phenotypic measurements have traditionally included postweaning growth (days of age at an ideal market weight, typically 105 to 110 kg, or rate of gain from around 25 kg to market weight) and live-animal backfat thickness at market weight measured either with a ruler or more recently with an ultrasonic probe. Less frequently, phenotypic records may include individual feed intake and efficiency during the postweaning period and carcass information from relatives. A classical approach to this sort of multi-trait selection involves derivation of a linear index of the phenotypic measurements with weightings that maximize its correlation with the selection objective (Hazel, 1943).

The general structure of a livestock industry, with respect to the creation and dissemination of genetic improvement, can be described as a pyramid with nucleus, multiplier and commercial levels. Although selection may be practised at all levels, it is selection at the nucleus that determines the rate of permanent genetic improvement in the industry. Thus, the selection objectives addressed in nucleus herds must accurately reflect production goals at the commercial level.

Testing methods in nucleus populations are designed to provide unbiased estimates of genetic potential and generally result in relatively uniform performance and greater heritability. But the testing environment under which candidates for selection are evaluated in nucleus herds is often different from the commercial production environment. For example, boars tested in nucleus populations are typically penned individually or in small groups. They may receive feed *ad libitum* or some form of restricted or semi-restricted feeding such as scheduled feedings to appetite. Commercial market pigs, on the other hand, are usually penned in larger groups and while most commercial producers in the USA allow pigs free access to feed, restricted feeding in the latter part of the postweaning period is more common in European commercial units. Consequently, benefits of greater heritabilities from reduced phenotypic variance in nucleus testing may be offset by genotype × environment interactions, i.e. the extent to which the genetic correlation between performance in the nucleus testing environment and in the commercial production setting is less than unity (see Brascamp *et al.*, 1985). For growth and backfat, some estimates of the genetic correlation between environments have been less than 0.80 (e.g. Merks, 1988, 1989), a threshold value suggested by Robertson (1959) to indicate biological importance of a genotype × environment interaction. In contrast, there have been several other reports (e.g. Van Diepen and Kennedy, 1989; Crump *et al.*, 1990; Cameron, 1993) of correlations greater than 0.80 and near unity.

Variation in the testing environment can also change the effective selection objective for a given set of measurements. For example, single-trait selection for rate of gain among animals that have *ad libitum* access to feed puts emphasis on appetite and may result in increased fatness (Woltmann *et al.*, 1992, 1995), but selection solely for gain among animals limited to a standard amount of feed may emphasize lean gain by identifying those animals that partition the allotted feed to the relatively efficient process of lean deposition (Webster, 1977).

Several models of metabolizable energy utilization and resulting tissue growth in the pig have been described over the previous two decades (e.g. Whittemore and Fawcett, 1976; Whittemore, 1986; Moughan *et al.*, 1987). Performance testing environments and selection schemes based on these models have been proposed as alternatives with which to address selection objectives associated with postweaning performance. Ultimately the test environment must be designed to optimize heritability and the genetic correlation with a targeted commercial setting or range of commercial environments.

Classical index selection

For selection approaches within the framework of classical index theory, the selection objective is defined as a linear combination of breeding values for traits considered of economic importance. Phenotypic measurements are chosen as criteria with which to estimate genetic merit for the selection objective. The traits that define the selection objective may or may not be among the phenotypic measurements chosen, depending on the difficulty in measurement and the availability of correlated indicator traits. Estimates of genetic and phenotypic variation and covariation associated with the measured traits and those in the objective, and the relative economic values of traits in the objective, are used to derive the optimum combination or index of the phenotypic selection criteria. Thus, this classical selection index is sometimes also referred to as an economic index.

As discussed previously, the overriding objective of a pork enterprise is the efficient production of quality lean. Typical selection indices applied in the industry have reflected this purpose by relating to a breeding objective that includes genetic merit for leanness, growth rate and feed efficiency. The corresponding phenotypic index usually includes an ultrasonic measurement of backfat depth at market weight and average daily gain for the test period, but often does not include direct measurements of feed intake and efficiency. Hence, while one of the advantages of the classical index is the ability to select for an objective trait through correlated phenotypic measurements, the method demands estimates of genetic parameters and relative economic values. Also, the optimum index for one performance testing environment and production system may not apply if, for example, an alternative testing scheme or marketing option is employed.

Models of growth and alternative strategies

Whittemore (1986) described a relatively simple model based on a deductive approach in which the objective is to determine the causal forces that result in animal tissue growth. The model is best summarized in terms of daily rates of lean and fat tissue gain in response to daily feed intake (Fig. 15.1). The animal's genetic potential for maximum rate of lean tissue deposition is relatively constant beginning at an early age and over nearly the entire range of ages and weights associated with the postweaning production period. Therefore, as daily feed intake is increased, rate of lean gain increases linearly until it plateaus when the animal's maximum potential is reached. The area to the left of the point at which lean gain plateaus may be thought of as nutritionally limited and the area to the right, in which maximum lean gain is realized, as nutritionally unlimited.

During the nutritionally limited phase of growth, it is assumed that the animal will partition most of the available energy to lean growth while maintaining some physiologically normal, but minimal level of fat gain. In the nutritionally unlimited phase of growth, most of the feed consumed beyond what is needed for maximum lean gain is partitioned to fat deposition. Thus, when feed intake is allowed to increase into this range the animal begins to deposit fat rapidly, grow less efficiently and eventually more slowly due to the relatively greater energy requirements of fat deposition.

The principles described in this model are the basis for postweaning production environments in which *ad libitum* feed intake is restricted, a practice common in European systems. The objective of restricting access to feed by pigs

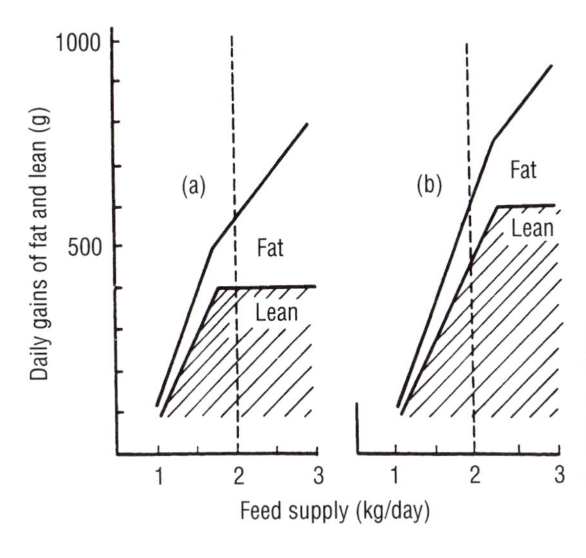

Fig. 15.1. A hypothesis for the relationship between daily gains of lean and fat and the daily feed supply. (From Whittemore (1986, p. 619) with permission from *Journal of Animal Science*.)

in finishing facilities is to limit daily feed intake to the minimum level at which maximum lean gain is realized, thereby avoiding unnecessary and costly fat gains. In some populations, feed intake in the latter stages of the postweaning period can be restricted by as much as 75% of *ad libitum* without inhibiting lean growth potential (Fowler *et al*., 1976). Restricting postweaning feed intake is sometimes referred to as scale feeding because systems are typically scaled so that each pig's daily allotment of feed is increased as the postweaning period progresses according to either time or body weight.

As depicted in graphs (a) and (b) of Fig. 15.1, a given amount of daily feed intake can result in the nutritionally unlimited growth of one animal, but in nutritionally limited growth of another, depending on their relative genetic merit for maximum lean gain. Differences in lean gain potential between sexes of pigs and between some genetic strains of pigs have been well quantified (Campbell and Taverner, 1988). Maximum lean gain potential is also assumed to vary among individuals of the same sex within a breed or genetic strain and is an important component of a breeding objective. Consequently, the principles outlined here also become useful in the development of hypotheses regarding expected response to selection, and the design of testing and selection schemes to improve genetic potential for lean gain and the efficiency of lean gain.

Fowler *et al.* (1976) attempted to model interactions between selection objectives, nutritional environments during performance testing and nutrition provided during commercial production. The model of growth that they considered (Fig. 15.2) was similar to that of Whittemore (1986). Metabolizable energy intake not lost as heat, i.e. energy retained as product (PE), is partitioned into either skeletal muscle and essential accessory tissues, categorized together as lean tissue (LT), or into fat tissue. In an environment in which the pig has *ad libitum* access to feed, energy will be partitioned to deposition of lean tissue until the maximum potential for lean growth rate is met, after which energy will

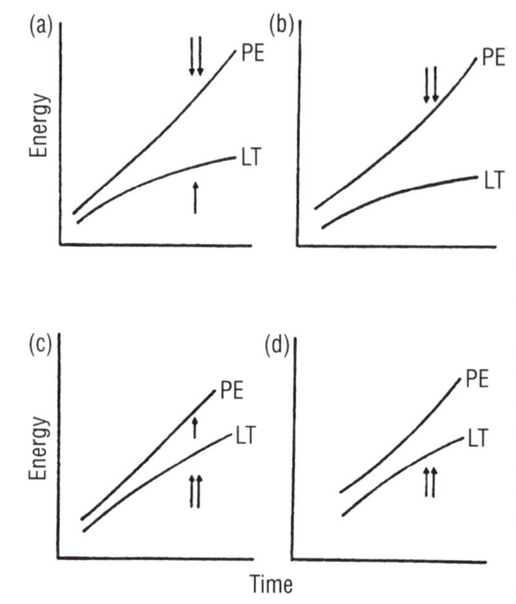

Fig. 15.2. Schematic models showing different ways in which improvement of lean tissue feed conversion (LTFC) may occur. PE = product energy and LT = lean tissue: (a) simultaneous increase in lean tissue growth rate (LTGR) and decrease in mean rate of feed intake; (b) decrease in mean rate of feed intake; (c) increased LTGR with rate of feed intake unconstrained; (d) increased LTGR with rate of intake kept constant. (From Fowler *et al.* (1976, p. 379) with permission from the *British Society of Animal Science.*)

be deposited as fat. The nutritional environments considered, both for performance testing and for the commercial production system, were scale feeding and *ad libitum* access to feed. In the context of performance testing, the time- or weight-based feeding scale would be designed to limit intake to a proportion of *ad libitum* such that variation in appetite is not expressed, but restraint of lean growth is minimized. Discussion of *ad libitum* access also applies, to some extent, to systems in which animals are allowed to eat freely for a set period of time twice each day, i.e. semi-*ad libitum.*

As an alternative to the classical selection index objective, expressed as a linear function of breeding values for traits deemed to have economic importance, Fowler *et al.* (1976) suggested a more biological definition of the objective based on the physiological factors related to the market value of the pig. Because the primary product of the industry is lean pork and the largest costs to the system are those associated with feed and time, the two 'biological' objectives suggested were lean tissue growth rate (LTGR) and the feed required per unit of lean produced or lean tissue feed conversion (LTFC).

The simplest of these biological indices would use either LTGR or LTFC as the selection objective and as the selection criterion. In practice, selection for the objective LTGR would be based on the difference between estimates of lean content at the onset of the test and at the end of test, in each case based on a function of live weight and ultrasonic measurements of subcutaneous fat depth and loin muscle area or depth. In a testing environment in which candidates are given free access to feed, selection for the objective LTFC would require direct measurement or estimation of individual feed intake. When performance testing is conducted with a time-scale feeding system in which pigs start the test at a standard weight, are tested for a standard length of time, and are given an amount of feed based on time on test, variation in feed intake and in days on test is zero by design. Consequently, the biological indices LTGR and LTFC are perfectly correlated under this system.

Three primary testing/selection scenarios were considered by Fowler *et al.* (1976): Case 1, testing environment – *ad libitum*, objective – LTFC; Case 2, testing environment – scale feeding, objective – LTFC (LTGR); Case 3, testing environment – *ad libitum*, objective – LTGR.

Case 1. Response to selection for LTFC is expected to be through a reduction in genetic potential for feed intake (appetite) and an increase in lean tissue gain (Fig. 15.2a). The relative emphasis on decreased feed intake will be greater when full expression of appetite is allowed and emphasis may shift almost entirely towards reduced feed intake if the estimate of lean gain is poor, or if a great amount of consideration is given to leanness rather than lean gain (Fig. 15.2b). Decreased appetite may presently be desirable in many industry populations in which lean gain potential is not limited by intake. However, as lean gain potential nears the limit presented by *ad libitum* energy intake, selection emphasis on feed intake would need to be reversed to improve LTFC (Fig. 15.2c).

Case 2. In this case the testing environment of scale feeding is aimed at removing variation in feed intake without limiting expression of lean gain potential, improvement in LTFC is entirely through selection emphasis on LTGR with genetic potential for feed intake unconstrained (Fig. 15.2c). As lean gain potential is increased, additional improvements in LTGR and LTFC under this scenario will eventually require increased emphasis on appetite. However, unlike Case 1, a reversal in the selection pressure on feed intake would not be necessary.

Case 3. This is typical of many on-farm testing situations in which individual feed intake is not measured. Feed intake is expected to remain unchanged under this scenario (Fig. 15.2d) until lean gain potential becomes limited, after which improvements in LTGR must be accompanied by increased feed intake (Fig. 15.2c). This method avoids the cost of measuring individual feed intake, but only addresses one component of LTFC.

Based on the principles of the model, the authors also determined expectations for performance of animals resulting from the testing/selection scenarios when used in commercial systems that provided either *ad libitum* or restricted access to feed. They concluded that pigs from Case 1 would be most likely to produce acceptable carcasses in systems allowing *ad libitum* feed intake. Pigs from Cases 2 and 3 would be expected to perform better in systems with restricted intake, than with *ad libitum* intake and a lean-based marketing system.

Summary

Understanding the genetics of postweaning performance in pork production must include knowledge of the important biological components and associated strategies for genetic improvement through performance testing and selection. Expectations for response to selection depend not only on the phenotypic criteria applied but also on the conditions of the performance test. The literature provides reports of experiments to determine direct and correlated responses to selection for individual component traits of postweaning performance, classical or economic indices of postweaning traits, and more recently for the biological objectives LTGR or LTFC under different performance testing environments. The following section will include, in addition to summaries of estimates of heritabilities and genetic correlations associated with performance traits, discussion of results from these selection experiments.

Additive (Co)variation and Selection Response

Parameter estimates

There have been many reported estimates of heritabilities and genetic correlations associated with postweaning performance traits based on covariation among relatives. A summary of estimates from several of these studies is presented in Tables 15.1 and 15.2 for environments of *ad libitum* or semi-*ad libitum* feed intake and restricted feeding, respectively. The estimates are listed in the order of the citations in each case, along with the range and a simple average. It is important to mention that in addition to differences in the way feed was provided to the animals, there were also differences in the breeds studied and some variation in the methods implemented, such as the testing interval used and location of backfat measurements. For some traits (h^2) or trait combinations (r_g), most of the estimates were fairly similar across experiments. But for those cases in which there were conflicting results, possible interactions with experimental differences or fixed effects within experiments will be discussed.

Table 15.1 is a summary of estimates from animals either allowed *ad libitum* access to feed or fed by hand to appetite (semi-*ad libitum*). Heritability estimates for the basic postweaning traits of average daily gain (ADG) and backfat thickness under those conditions were mostly moderate and clearly indicate that these traits will respond to selection. Estimates for backfat thickness tended to be higher than those for ADG. Heritability estimates for daily feed intake on test (DFI) and feed conversion (feed/gain) were also generally moderate. There were fewer estimates for LTGR and LTFC reported, each based on an ultrasonic prediction of lean content of the animal at the completion of test, but the reports were similar and indicate that these traits are also moderately heritable.

Estimates of the genetic correlations of ADG and backfat with DFI were all positive and most were moderate to high. Estimates tended to be higher with *ad libitum* than with semi-*ad libitum* consumption (for ADG/DFI, the average estimate was 0.75 and 0.40 for *ad libitum* and semi-*ad libitum*, respectively), probably reflecting less than full expression of appetite in the latter setting.

Although, based on these estimates, DFI would be expected to increase with greater genetic merit for ADG, the genetic correlation between ADG and

Table 15.1. Estimates of heritabilities and genetic correlations for pigs with *ad libitum* or semi-*ad libitum* access to feed.

References[a]	Parameter/trait(s)	Estimates[b]	Range	Average
	Heritabilities			
1, 2, 3, 4, 5, 6, 7, 8, 9, 10, 11, 12, 13, 14	ADG	0.41*; 0.41*; 0.20*; 0.40; 0.07; 0.36; 0.41; 0.41, 0.03; 0.28; 0.49; 0.43; 0.37, 0.14; 0.17, 0.33; 0.39	0.03–0.49	0.31
1, 2, 3, 6, 8, 9, 10, 11, 12, 13, 14, 15, 16	Backfat (BF)	0.66*; 0.74*; 0.56*; 0.47; 0.60, 0.12; 0.34; 0.57; 0.59; 0.46, 0.25; 0.56, 0.50; 0.43; 0.56; 0.38	0.12–0.74	0.49
1, 2, 3, 6, 7, 8, 10, 11, 12, 17, 18	Feed intake (DFI)	0.34*; 0.41*; 0.13*; 0.62; 0.23; 0.17, 0.26; 0.16; 0.45; 0.29, 0.19; 0.29; 0.18	0.13–0.62	0.29
1, 2, 3, 6, 7, 8, 10, 11, 12, 14	Feed/gain (FCR)	0.58*; 0.48*; 0.23*; 0.27; 0.18; 0.41, 0.12; 0.51; 0.28; 0.19, 0.15; 0.20	0.12–0.58	0.30
11, 17, 18	LTGR	0.39; 0.38; 0.25	0.25–0.39	0.34
11, 17, 18	LTFC	0.34; 0.35; 0.25	0.25–0.35	0.31
	Genetic correlations[c]			
1, 2, 3, 6, 7, 8, 11, 12, 19	ADG/DFI	0.32*; 0.50*; 0.37*; 0.83; 0.89; 0.69;0.73; 0.80; 0.76, 0.41; 0.89	0.32–0.89	0.65
1, 2, 3, 6, 11, 12	BF/DFI	0.38*; 0.08*; 0.35*; 0.59; 0.42; 0.51;0.24	0.08–0.59	0.37
1, 2, 3, 6, 7, 8, 11, 12, 19	ADG/FCR	–0.57*; –0.71*; –0.86*; 0.34; –0.05; –0.69; –1.24; –0.28; –0.52; –0.52; –0.78	–1.24–0.34	–0.53
1, 2, 3, 6, 11, 12, 19	BF/FCR	0.10*; 0.44*; 0.24*; 0.41; 0.24; 0.28, 0.36; 0.33	0.10–0.44	0.30
1, 2, 3, 6, 9, 11, 12, 13, 19	ADG/BF	0.02*; –0.26*; –0.06*; 0.55; 0.35; 0.32; 0.26, –0.25; 0.09, –0.12; 0.37	–0.26–0.55	0.12
11	ADG/LTGR	0.96		
11	ADG/LTFC	–0.09		
11	BF/LTGR	0.02		
11	BF/LTFC	0.52		
17, 18	DFI/LTGR	0.23; 0.31		0.27
17, 18	DFI/LTFC+	–0.45; –0.36		–0.41
17, 18	LTGR/LTFC+	0.76; 0.87		0.82

[a]References: 1, Smith *et al.* (1962); 2, Smith and Ross (1965); 3, Standal and Vangen (1985); 4, Nordskog *et al.* (1944); 5, Fahmy and Bernard (1970); 6, McPhee *et al.* (1979); 7, Wyllie *et al.* (1979); 8, Cameron *et al.* (1988); estimates presented for Large White and Landrace populations; estimates for backfat are an average of those for shoulder, mid-back and loin measurements; 9, McPhee *et al.* (1988); 10, DeHaer and DeVries (1993); 11, Mrode and Kennedy (1993); 12, Cameron and Curran (1994a); estimates presented for Large White and Landrace populations; 13, Hetzer and Miller (1972a); estimates presented for Duroc and Yorkshire populations; 14, Cameron and Curran (1995a), restriction: 75% of *ad libitum*, all estimates presented are averaged across selection lines in Large White population, estimates presented for backfat are an average of those for shoulder, mid-back and loin measurement sites; 15, Gray *et al.* (1968); 16, Berruecos *et al.* (1970); 17, Cameron (1994); 18, Cameron and Curran (1994b); 19, Robison and Berruecos (1973).

[b] * = Semi-*ad libitum* (e.g. hand-fed to appetite).

[c] + = LTFC based on an index of backfat and feed/gain; increasing values of LTFC correspond to greater lean efficiency (less feed/lean).

feed conversion was moderate to high and negative in most of the reports cited. The exceptions (McPhee *et al.*, 1979 [0.34]; Wyllie *et al.*, 1979 [–0.05]) were from studies in which pigs had continuous access to feed. Thus, the majority of reports suggest that the cost of greater DFI accompanying increased ADG would be more than offset by fewer days on feed. Estimates of the genetic correlation between backfat thickness and feed conversion revealed that selection for improved (less) backfat should also improve efficiency.

Reports of the correlation between ADG and backfat thickness were quite variable, ranging from moderate and favourable (–0.26) to moderate and unfavourable (0.55). The previously mentioned differences among experiments including differences in the methods and technicians used for backfat measurement, and the typically large sampling errors associated with estimates of genetic correlations in experimental populations (Koots and Gibson, 1994), make interpretation of the variation among estimates difficult. However, it appears that some of the variation may be due to breed differences. For example, Cameron and Curran (1994a) reported that the correlation was 0.26 in Large White selection lines, but –0.25 in Landrace selection lines. Perhaps the genetic correlation between ADG and backfat in a population depends on how tightly coupled the traits are with DFI versus the ability to partition energy intake to lean tissue growth. In the report by Cameron and Curran (1994a), the estimated genetic correlations of DFI with ADG and BF were 0.76 and 0.51, respectively in the Large White lines, and 0.41 and 0.24, respectively in the Landrace lines. The extent to which free access to feed is actually available may determine this link between growth and backfat with DFI. The average estimate of the correlation between ADG and backfat was 0.20 and –0.10 under *ad libitum* and semi-*ad libitum* intake, respectively.

Mrode and Kennedy (1993) included LTGR and LTFC (feed/lean), based on ultrasonic predictions of lean content of Yorkshire, Landrace and Duroc boars at the completion of test, in their study of DFI, ADG and backfat. They reported that genetic merit for ADG and LTGR was closely correlated (0.96), but that the correlation between ADG and LTFC was small (–0.09). Conversely, the genetic correlation between backfat and LTGR was near zero (0.02), but the relationship between backfat and LTFC was moderate and positive (0.52). Cameron and Curran (1994a) and Cameron (1994) reported genetic correlations of DFI with LTGR and LTFC. Notice that LTGR and LTFC were based on ultrasonic backfat measurements and ADG and feed/gain, respectively, and that decreasing values for LTFC corresponded with more feed per unit of lean. These estimates indicate that under an environment of *ad libitum* feed intake, LTGR and LTFC are positively but not perfectly correlated, and selection for improved LTFC would be expected to result in decreased genetic potential for DFI – both results are consistent with the suggestions of Fowler *et al.* (1976) as discussed above.

Estimates of genetic parameters for pigs restricted in their access to feed are available from evaluations of data from Danish test stations and from relatively new investigations of selection based on performance testing with scale feeding (Table 15.2). For studies in which parameters from environments of *ad libitum*

Table 15.2. Estimates of heritabilities and genetic correlations for pigs with restricted feed intake.[a]

References[b]	Parameters/trait(s)	Estimates	Range	Average
	Heritabilities			
1, 2, 3, 4, 5, 6, 7, 8	ADG	0.24; 0.39(0.20); 0.22, 0.14; 0.76(0.41), 0.14(0.03); 0.41(0.28); 0.30, 0.35; 0.17, 0.16; 0.26 (0.39)	0.14–0.76	0.30
1, 2, 3, 4, 5, 6, 7, 8	Backfat (BF)	0.47; 0.41(0.56); 0.26, 0.29; 0.06(0.60), 0(0.12); 0.60(0.34); 0.32, 0.36; 0.29, 0.28; 0.35 (0.43)	0–0.60	0.31
2	Feed intake(DFI)	0.20 (0.13)		
2, 3, 4, 8	Feed/gain (FCR)	0.35(0.23); 0.23, 0.19; 0.56(0.41), 0.16(0.12); 0.24(0.20)	0.16–0.56	0.29
7	LTGR	0.34; 0.28		0.31
	Genetic correlations			
2	ADG/DFI	0.28 (0.37)		
2	BF/DFI	0.29 (0.35)		
2, 3, 4	ADG/FCR	–0.93 (–0.86); –1.07, –1.02; –1.03 (–0.69); –0.99 (–1.24)	–1.07–-0.93	–1.0
2, 3	BF/FCR	0.30 (0.24); 0.16; 0.23	0.16–0.30	0.23
2, 3, 5, 6, 7	ADG/BF	–0.21(–0.06); –0.07, –0.31; –0.22 (0.35); 0.08, –0.39; –0.10; –0.02	–0.39–0.08	–0.16

[a]Estimates in parentheses are for (semi-) *ad libitum* intake; see Table 15.1.

[b]References: 1, Lush (1936); 2, Standal and Vangen (1985); 3, Merks (1987); estimates presented are for Landrace and Yorkshire boars; 4, Cameron *et al.* (1988); estimates presented are for Large White and Landrace populations; estimates presented for backfat are an average of those for shoulder, mid-back and loin measurements; 5, McPhee *et al.* (1988); 6, Gu *et al.* (1989), fed to a time-based scale; male and female averages presented; 7, Cameron *et al.* (1994), restriction: 75% of *ad libitum*, estimates presented are for Large White and Landrace populations; 8, Cameron and Curran (1995a), restriction: 75% of *ad libitum*, all estimates presented are averaged across selection lines in Large White population, estimates presented for backfat are an average of those for shoulder, mid-back and loin measurement sites.

and restricted intake were compared directly, the comparable estimate from Table 15.1 is presented in parentheses in Table 15.2.

As might be expected, phenotypic variation in ADG tended to be less with restricted than with *ad libitum* intake (Cameron *et al.*, 1988; McPhee *et al.*, 1988; Cameron and Curran, 1995a). Restricted feeding tended to reduce phenotypic variation in feed conversion ratio (Cameron *et al.*, 1988) and backfat thickness (McPhee *et al.*, 1988) in some studies, but there have also been reports that observable variation in feed conversion (Cameron and Curran, 1995a) and backfat (Cameron *et al.*, 1988) was unaffected by feed restriction. Restriction of feed would be expected to have the greatest effects on phenotypic variances in backfat and feed conversion in populations in which *ad libitum* energy intake greatly exceeds lean growth potential. The associated variation in fat deposition and partitioning of energy to lean versus fat would be reduced as feed restriction moves energy intake closer to lean growth potential.

Effects of feeding environment on heritabilities (Table 15.2) are difficult to summarize from literature reports. In none of the studies in which (semi-) *ad libitum* and restricted intake were directly compared (Standal and Vangen, 1985; Cameron *et al.*, 1988; McPhee *et al.*, 1988; Cameron and Curran, 1995a) was precision adequate to conclude that the heritabilities were truly different.

Most estimates of the genetic correlation between ADG and backfat in pigs with restricted access to feed were negative (Table 15.2), consistent with the trend noted when comparing estimates of the same correlation in pigs with semi-*ad libitum* versus *ad libitum* access (Table 15.1). McPhee *et al.* (1988) reported that the correlation changed from 0.35 with *ad libitum* to −0.22 with restricted intake, and Standal and Vangen (1985) estimated that the correlation was −0.06 and −0.21 for pigs with semi-*ad libitum* and restricted intake, respectively. Although Gu *et al.* (1989) reported values of 0.08 and −0.39 in boars and gilts, respectively, under an environment of scale feeding, they noted that boars were fed more liberally than gilts and attributed the difference in the estimates to the relatively greater restriction of intake in the gilts. As pointed out by Gu *et al.* (1989), the change in sign of this correlation is also consistent with the suggestions of Fowler *et al.* (1976) and Webster (1977). Greater merit for ADG with *ad libitum* intake is associated with greater daily consumption and fat deposition, but superior merit for ADG when intake is restricted is due to the ability to partition available energy to the relatively efficient process of lean growth.

The genetic correlation between ADG and feed conversion is affected by the feeding regime. Reported correlations under restricted feeding are close to −1.0, but with freer access to feed generally differ from −1.0. This result is consistent with the mathematical characteristics of a correlation between a ratio and its denominator (Sutherland, 1965). If heritabilities of gain and feed intake are similar, and the genetic coefficient of variation is much smaller for feed intake than for gain, the genetic correlation between gain and feed conversion will always be highly negative. When the genetic coefficients of variation for feed intake and gain are more similar, the genetic correlation between gain and feed conversion moves towards zero.

The potential to exploit additive genetic variation and covariation associated with postweaning performance traits has been studied in numerous selection experiments. Discussion of correlated responses in this section will be limited to postweaning traits; correlated changes in reproduction from selection for postweaning performance are discussed on page 445–447. Unless otherwise indicated, *ad libitum* access to feed was allowed during performance testing and for all measurements of responses in postweaning performance.

Single-trait selection

Consistent with moderate heritability estimates based on covariation among relatives (Table 15.1), the fundamental postweaning traits of body growth and fatness exhibit significant response to selection. In an early study (Krider *et al.*, 1946), divergent selection on weight at three postweaning ages was applied in a Hampshire population. After nine generations of selection (Baird *et al.*, 1952; Craig *et al.*, 1956), average weight at 180 days of age was 27.7 kg greater in the high line than in the low line. Although weight at a given age includes variation in birth weight and preweaning growth, most of the divergence between the lines occurred after weaning. Subsequently, there have been several additional reports of significant direct responses to single-trait selection for weight at a given age (Kuhlers and Jungst, 1990, 1991a,b) or postweaning ADG (Rahnefeld, 1971a,b; Rahnefeld and Garnett, 1976; Fredeen and Mikami, 1986a,b; Woltmann *et al.*, 1992, 1995; Clutter *et al.*, 1995).

Selection for total body growth with free access to feed has been accompanied by changes in DFI. High-line Hampshires consumed 0.64 kg per day more feed than those from the low line during a 72-day postweaning test that followed the nine generations of selection (Baird *et al.*, 1952), and Clutter and Buchanan (1997) reported a 0.71 kg per day difference in DFI between lines after eight generations of divergent selection for postweaning ADG. When the divergent lines in the latter study were restricted to standard amount of feed intake at generation four, differences in ADG between the lines were not significant (Woltmann *et al.*, 1992). Thus, most of the response in ADG was attributed to changes in feed intake. Direct responses to selection for growth rate have been great enough to offset the correlated changes in feed intake, resulting in improved feed conversion from upward selection for growth (Rahnefeld, 1973) and less feed/gain in upward than in downward divergent lines (Baird *et al.*, 1952; Clutter and Buchanan, 1997).

Changes in body composition in response to single-trait selection for growth have varied somewhat among experiments. Divergent selection for ADG resulted in greater backfat thickness at 105 kg body weight, but also greater LTGR, in the fast line than in the slow line (Woltmann *et al.*, 1992, 1995). Selection for greater weight at 200 days of age in a Duroc population was accompanied by an increase in backfat thickness, a decrease in percentage of muscle and no change in LTGR (Kuhlers and Jungst, 1992b), but selection on the same criterion in a Landrace herd resulted in increased LTGR without significant changes in backfat and percentage of muscle (Kuhlers and Jungst, 1993). Differences in correlated responses may be due to the specific criterion applied, the selection differential achieved, random drift and the effects of sampling (Hill, 1971). However, the change in body composition from selection for growth may also depend on the genetic potential for DFI relative to lean growth in the base populations in which selection is applied or, as depicted by the estimates of the genetic correlation between ADG and backfat in Tables 15.1 and 15.2, the real degree of access to feed. In general, it appears that direct response to selection for rate of total body growth with complete access to feed results largely through increased DFI, and may be associated with an increased rate of fat as well as lean tissue growth.

Due to its high correlation with leanness (Robison *et al.*, 1960; multiple correlation of backfat at shoulder and loin with per cent lean cuts = 0.73) and relative ease of measurement (Hazel and Kline, 1952), average backfat thickness is a valuable indicator of body composition. Responses to selection for backfat thickness have been reported in three primary studies. Divergent, controlled selection for backfat in Duroc and Yorkshire populations resulted in greater than four phenotypic standard deviations of total direct response (Hetzer and Harvey, 1967; Hetzer and Miller, 1972a,b). Corresponding realized heritabilities were similar for upward and downward selection in both breeds (0.47 and 0.48, respectively in Durocs; 0.38 and 0.43, respectively in Yorkshires). Similar direct responses were also reported from two independent studies in which five generations of downward selection for backfat were practised (Gray *et al.*, 1968; Berruecos *et al.*, 1970).

Berruecos *et al.* (1970) reported that there was no correlated change in pig growth associated with selection for decreased fat, but Hetzer and Miller (1972a) concluded that the correlated response in growth rate may vary by breed. In Durocs, ADG increased significantly in both divergent lines; in Yorkshires, ADG did not change significantly in the high fat line, but decreased in the low fat line. The variable correlated responses in ADG or backfat to direct selection on the other trait is consistent with the variability in estimates of the genetic correlation between the two traits (Table 15.1).

Because of costs associated with feed intake, the potential for genetic improvement through selection of the efficiency with which feed is converted to body weight gain has been of interest for some time. Although estimates of heritability (Table 15.1) suggest that feed/gain will respond to direct selection, results from selection experiments have been mostly discouraging. Dickerson and Grimes (1947) reported that divergent selection for feed/gain resulted in significant direct response and a corresponding realized heritability of approximately 24%. But in two other studies in which downward selection for feed/gain was practised (Jungst *et al.*, 1981; Webb and King, 1983), realized heritabilities were not significantly different from zero. Dickerson and Grimes (1947) determined, based on estimates of the relevant parameters in their population, that selection for ADG would be as effective at improving genetic merit for feed/gain as would direct selection based on individual feed records. Their expectation is supported by the correlated responses in feed conversion to selection for ADG discussed previously.

The experiments described thus far have provided information regarding the effectiveness of single-trait selection for components of postweaning performance and some measurements of correlated responses. However, modern demands for the efficient production of lean pork and unfavourable relationships between some of the defining components have led to the evaluation of alternative selection criteria, including classical and biological indices, as means for improving LTGR and LTFC.

Classical index selection

Several experiments have been conducted to test responses to selection for a classical index of ADG and backfat thickness (e.g. Vangen, 1979, 1980a,b; Cleveland *et al.*, 1982, 1983a,b, 1988; Fredeen and Mikami, 1986a,b,c; McKay, 1990, 1992). Although the objective of each of the studies was to evaluate the same two-trait index, there was variation in the breeds used, the specific design employed, and the amounts of intended and effective selection emphasis given to ADG and backfat.

In general, index selection resulted in significant response in the component traits of ADG and backfat thickness. In a rather comprehensively designed experiment (Fredeen and Mikami, 1986a), index selection improved ADG and backfat as much as each component was improved by single-trait selection in two contemporary lines. However, consistent with the generally unfavourable relationship between these components observed under *ad libitum* intake (Table 15.1), neither of the single-trait lines was as effective as the index at improving overall merit for the two traits combined. Only in the experiment reported by McKay (1990) was there little response in ADG, but that result could be explained by the relatively greater selection emphasis applied to backfat.

The most detailed evaluation of correlated changes in LTGR and LTFC in response to the index selection was reported by Cleveland *et al.* (1983a,b) with an index = 100 + 286.6 (ADG, kg/day) −39.4 (backfat, cm). In a study of barrows from the select and control lines at three levels of feed intake, the select line was superior in rate of protein growth and feed required per unit of edible lean at each of the intake levels. Although specific responses in LTGR and LTFC in the other studies were not reported, the simultaneous improvements in ADG, backfat and total body feed conversion reported by Vangen (1980a) and Sather and Fredeen (1978) indicate that correlated improvements were made. Even with selection against backfat thickness, improvements in efficiency were achieved without significant reductions in genetic potential for DFI (Sather and Fredeen, 1978; Vangen, 1980a; Cleveland *et al.*, 1983a).

Selection indices that included measurements of feed conversion, in addition to ADG and backfat, have also been evaluated as a means of improving LTFC (McPhee, 1981; Ellis *et al.*, 1988). Both McPhee (1981) and Ellis *et al.*

(1988) reported improvements in backfat and feed conversion, but not significant improvements in ADG, resulting from index selection. Both groups also reported that reduced *ad libitum* DFI accompanied selection. The lack of improvement in ADG was explained by the relatively greater negative emphasis on feed conversion (feed/gain) and backfat, and the tendency for a positive genetic correlation between backfat and ADG, and possibly between feed conversion and ADG, as feeding becomes more liberal (Table 15.1). Even though LTGR was increased by selection on the indices, the improvements in LTFC were primarily through decreased appetite.

Alternative methods of selection

Leymaster *et al.* (1979a,b) reported significant direct responses to selection based on ultrasonic estimates of either LTGR (weight of lean cuts at 160 days of age) or percentage of carcass lean (percentage of lean cuts at 81.6 kg live weight), but only selection on the estimate of LTGR improved both traits simultaneously. While selection for LTGR resulted in less carcass fat and greater ADG, response to selection for a greater percentage of carcass lean was accompanied by decreased ADG. These results are consistent with the positive estimates of the genetic correlation between ADG and backfat presented in Table 15.1, and with the correlated decrease in ADG reported by Hetzer and Miller (1972a) when selection was for lower backfat thickness.

There have been a limited number of reports of experiments designed to evaluate restricted feeding as a performance testing environment and to test the theories of Fowler *et al.* (1976) regarding interactions of LTGR and LTFC with testing regime. McPhee *et al.* (1988) reported that selection on an index that estimated LTGR in a testing environment of scale feeding resulted in significant response in ADG, backfat thickness, feed conversion and ham lean, and that responses were greater when offspring from the line were allowed *ad libitum* intake than when they were restricted. Appetite (*ad libitum* DFI) was also increased by selection on the index under scale feeding. Although the study did not include a line in which selection on the index was practiced with *ad libitum* feed intake, the authors speculated that response would be greater when testing is with scale feeding since it was under that environment that estimates of heritabilities were greater and the genetic correlation between ADG and backfat was negative (Table 15.2).

A comprehensive study was conducted in Landrace and Large White populations to test divergent, controlled selection for either DFI, LTGR or LTFC in pigs allowed *ad libitum* feed intake, and LTGR in pigs with restricted feed intake (Cameron, 1994; Cameron and Curran, 1994b; Cameron *et al.*, 1994). When the testing environment allowed *ad libitum* feed intake and performance was measured in pigs given free access to feed, the greatest amount of improvement in LTGR was from direct selection. In addition, correlated response in LTFC from selection on LTGR was similar to (in Landrace) or greater than (in Large White) direct response from selection on LTFC. Correlated response in DFI was zero or slightly positive from selection on LTGR, but selection on LTFC caused a reduction in DFI.

Progeny from the line in which selection was for LTGR in a testing environment of scale feeding, and from the lines in which selection was with *ad libitum* consumption, were compared in both feeding environments to determine the best overall selection strategy (Cameron and Curran, 1995a). When pigs from the Large White lines were allowed *ad libitum* intake, ADG was greatest in the line selected for high LTGR under scale feeding followed, respectively, by those selected under *ad libitum* intake for high LTGR and improved LTFC. The lines ranked the same for DFI as for ADG, but were similar to one another in backfat and feed conversion. Performance in the lines was more similar when progeny were scale-fed, but pigs from the line selected for LTGR with scale feeding still tended to have faster gains and less backfat than those from the lines in which selection was for LTGR or LTFC with *ad libitum* access. In Landrace pigs, response in the same traits was similar in the lines selected for either high LTGR or improved LTFC with *ad libitum* intake, or for high LTGR with scale feeding, regardless of whether progeny were allowed *ad libitum* or restricted feed intake.

Cameron and Curran (1995b) reported results from an evaluation of carcass traits in which boar and gilt progeny from the three lines (LTGR-*ad libitum*, LTFC-*ad libitum* and LTGR-scale) were tested under the feeding conditions of selection in their respective lines. The most extensive results were presented for the Large White population. Selection for LTGR-*ad libitum* increased LTGR, but did not change rate of fat growth. Rate of fat growth was reduced in the line selected for LTFC-*ad libitum*, but the change in LTGR was not significant. Selection for LTGR-scale significantly increased LTGR and decreased rate of fat growth.

Summary

Great opportunities exist for genetic improvement of postweaning performance through within-line selection. Various selection experiments have shown that predicted responses in the components of LTGR and LTFC based on models of growth (Fowler *et al.*, 1976) are remarkably accurate. In a testing environment of *ad libitum* intake, selection for LTFC was successful primarily through a reduction in appetite accompanied by a lesser rate of fat growth, rather than through increased LTGR. Selection for LTGR by testing with *ad libitum* intake improved LTGR and LTFC without reducing appetite, but did not reduce rate of fat growth. As pointed out by Cameron and Curran (1995b), implementation of a scale-fed testing environment to select for LTGR seems to combine the best features of selection for LTGR and LTFC under *ad libitum* intake: LTFC is improved through increased LTGR and a decreased rate of fat growth without a decrease in *ad libitum* DFI. When Cameron and Curran (1995a) compared progeny from these three selection strategies on *ad libitum* and restricted intake, a genotype × feeding regime interaction was largely due to an increased advantage of the line selected for LTGR on scale feeding when progeny were allowed *ad libitum* versus restricted intake. Thus, as suggested by Fowler *et al.* (1976) and supported by Cameron and Curran (1995b), this testing and selection approach may be an effective strategy for either commercial feeding environment.

Selection for Postweaning Performance: Effects on Reproduction

To determine the optimum emphasis on lean gain and efficiency in selection objectives for sire and dam lines contributing to a commercial breeding system, the genetic relationships of the components of LTGR and LTFC (i.e. ADG, body composition and feed conversion) with traits of reproductive performance must be known. Expectations for changes in reproductive performance from selection for market animal performance can be derived from estimates of genetic correlations between the relevant traits and their heritabilities. As mentioned previously, however, acceptable levels of precision are rarely achieved when genetic correlations are estimated from covariances among relatives in experimental populations or relatively small samples from industry populations. Estimates of correlated responses in reproductive traits in experiments of selection for postweaning performance may provide more useful information, but often have limitations as well. The following is a summary of information available in the literature, including a discussion of shortcomings in the present base of knowledge and needs for future research.

Estimates of genetic correlations

Two publications in 1981 included summaries of estimated genetic correlations between postweaning performance and traits of puberty or reproduction based on covariances among relatives (Hutchens and Hintz, 1981; Johansson, 1981). A review by Brien (1986) also included a discussion of literature estimates of genetic relationships between growth and reproduction in the pig and other mammalian species. Vogt *et al.* (1963) first reported that the genetic correlation between postweaning growth rate and litter size was near zero, but in subsequent studies in which first and second parities were considered separately (Morris, 1975; Johannson, 1981), the estimated correlation was positive for second litters. Young *et al.* (1977) were unable to estimate correlations between ADG and litter size because of a negative estimate of sire variance, but they reported a positive genetic correlation between ADG and ovulation rate, and a tendency for a negative relationship between ADG and embryo survival (number of ovulations/number of embryos). These results for ADG and ovulation rate seem to be consistent with the positive genetic correlation between body size and ovulation rate observed in other species (Brien, 1986).

Reported estimates of genetic correlations between traits of reproduction and feed conversion or backfat depth have been inconsistent and generally imprecise. For example, Morris (1975) and Bereskin (1984) each reported genetic correlations between backfat and litter size that were negative, but not significantly different from zero. Positive estimates reported by Johannson and Kennedy (1983; $r_g = 0.13$ to 0.22 for backfat and number born alive) were not accompanied by standard errors, but were probably also not different from zero.

In attempts to draw overall conclusions regarding the genetic relationship between postweaning performance and litter characteristics, genetic correlations between indices of the relevant traits have been estimated. Bereskin (1984) reported that genetic correlations of a performance index of ADG and backfat thickness with two indices of sow productivity (litter sizes and weights) were not different from zero (average estimate = 0.07) and Morris (1975) reported that litter sizes and litter weight were not genetically correlated with a total point score that included ADG, feed conversion and carcass traits (estimates ranged from −0.04 to 0.02). In general, these studies have not detected significant genetic relationships between performance and reproduction traits, and imply that selection for lean gain and efficiency would not harm reproduction.

However, significant estimates of genetic correlations in two studies may have implications for selection LTGR and LTFC in some situations. Rydhmer *et al.* (1994) reported a genetic correlation of 0.40 ± 0.10 between lean percentage and age of puberty. In addition, Kerr and Cameron (1996) reported positive genetic correlations between litter weights at birth and weaning with DFI (0.48 ± 0.15 and 0.42 ± 0.12, respectively). These results suggest that selection for a greater percentage of lean will delay sexual maturity and that selection strategies that decrease genetic potential for DFI will have detrimental effects on litter characteristics.

Correlated responses to selection

Many studies of selection for postweaning performance have included measurements of those traits of reproduction which are routinely available in the production system, primarily litter sizes and litter weights from birth to weaning. In a few studies, traits such as age of puberty, breeding performance and sow weight changes have also been recorded. Unfortunately, data on reproductive performance in experimental populations selected for traits of market pig performance typically have several limitations:

1. The need/desire to turn generations rapidly results in data that are exclusively from gilts and first litters. As noted earlier, correlations between performance and reproduction may differ by parity. If selection for lean gain and efficiency affects reproductive ability, it is possible that the effects are manifested after the gilt weans her first litter. Reports of correlated responses in rebreeding performance and later parity litter traits, for example, are rare.

2. Because data on reproduction are usually collected only for females kept for breeding in the selection study, measurements of correlated responses are often based on the reproduction of selected females and consequently subject to bias.
3 Many important traits of reproduction are quite variable and inherently difficult to measure with precision. Thus, data collected from the relatively small number of females maintaining a given selection line each generation are usually not adequate to allow meaningful conclusions.
4. Most experimental selection lines of pigs are unreplicated and of relatively small effective size. Therefore, measurements of correlated (and direct) responses to selection are likely confounded with random drift in gene frequencies.

Nonetheless, some researchers have given considerable emphasis to the study of correlated responses in traits of reproduction, and the literature does contain useful information and food for thought. Unless otherwise noted, results discussed are for gilts and first litters only.

Among experiments in which selection was for some measurement of growth, significant changes in litter weights or pig weaning weights reflecting changes in the pig's direct genetic merit for preweaning growth and/or the maternal ability of the sow were only observed by Craig *et al.* (1956) with selection for weight at a postweaning age and by Garnett and Rahnefeld (1976) with selection for ADG. In both cases a positive impact of upward selection for growth was indicated. Although trends in litter sizes were not significant in most of the studies in which selection was for growth, a positive effect of selection for weight at 70 days of age on 21-day litter size was reported by Kuhlers and Jungst (1992b). Conversely, selection for weight at 200 days of age in Durocs decreased total litter size at birth (Kuhlers and Jungst, 1992a), and preweaning mortality increased due to selection for postweaning ADG (Garnett and Rahnefeld, 1976). Number born alive did not change significantly in the former study (Kuhlers and Jungst, 1992a).

Litter traits did not change significantly from eight generations of divergent selection for ADG (Clutter and Buchanan, 1997), but correlated response in appetite seems to be reflected in weight changes of gilts during gestation and first lactation. Gilts from the line selected for slow growth and expressing relatively lesser appetite weighed less at farrowing, but lost a significantly greater amount of weight during lactation than gilts from the line selected for fast growth. Although size and weight of the first litter was not affected, weight loss in the female was great enough to suggest that future reproduction might be inhibited. Unfortunately, females were marketed after the first litter, hence rebreeding performance and subsequent litter characteristics were not measured.

For studies in which selection was for divergent or decreased backfat thickness, only Berruecos *et al.* (1970) reported significant changes in reproduction. Selection for increased leanness resulted in a significant decline in litter size at birth and weaning, but surprisingly by 130 days of age the difference in litter size was nearly gone. Although, based on the difference between sire and dam components of variance for feed conversion, Dickerson and Grimes (1947) concluded that females with greater efficiency for total body weight gain expressed poorer nursing ability, correlated responses in reproduction were not reported from studies in which single-trait selection for feed conversion was practised.

As with most of the experiments implementing single-trait selection, many of the studies of direct or alternative criteria to improve LTGR or LTFC revealed no significant trends in traits of reproduction. However, some exceptions suggest the potential for detrimental relationships. Although the regressions of first parity litter sizes on generation number in a line selected for weight of lean cuts at 160 days (LTGR) were negative but not different from zero (DeNise *et al.*, 1983), second parity litter sizes and corresponding litter weights were significantly decreased by selection. Second parity litter characteristics were also significantly decreased by selection for percentage of lean cuts at approximately 80 kg body weight. Kerr and Cameron (1995) did not detect correlated responses in litter traits to selection for either LTGR with *ad libitum* intake or LTGR with scale feeding, but litter sizes and weights at birth and weaning were significantly decreased by selection for LTFC with *ad libitum* intake. This result is consistent with their estimate of positive genetic correlations between litter traits and DFI (Kerr and Cameron, 1996), given the decrease in DFI when selection was for LTFC with *ad libitum* intake.

Summary

Estimates of genetic correlations between postweaning performance and traits of puberty and reproduction based on covariances among relatives have not revealed consistent relationships, suggesting that selection to improve genetic merit for both within a single line could be effective. However, some estimates of unfavourable relationships have been reported (Rydhmer *et al.*, 1994; Kerr and Cameron, 1996) and in general estimates of genetic correlations have been plagued by insufficient precision.

The selection experiments discussed here encompass a variety of criteria, objectives and methods. Even among those studies of a classical index including ADG and backfat, selection varied due to differences in specific index weightings, selection differentials achieved and the population in which selection was applied. Based on these factors alone, it is not surprising that results for correlated changes in reproductive traits are mixed. In addition, it is difficult to generate the number of observations necessary for powerful tests of correlated responses in these inherently variable traits of reproduction. This lack of precision adds to the inconsistency in results and leads to precarious interpretation of non-significant trends.

Several studies reported significant responses in LTGR or LTFC with small or non-significant changes observed in traits of reproduction, but some unfavourable correlated responses were reported. The potential for adverse effects seems most likely when selection is for decreased fat (Berruecos *et al.*, 1970; DeNise *et al.*, 1983) or improved LTFC (Kerr and Cameron, 1995) in a testing environment of *ad libitum* intake, but was also evident in second parity litters when selection was for LTGR with *ad libitum* intake (DeNise *et al.*, 1983).

The literature does not provide conclusive evidence that selection for leanness, LTGR or LTFC will have detrimental effects on reproduction, but the indication of antagonistic relationships or negative effects in some studies and the general lack of precision in estimation of covariances among relatives and line means for many of the reproductive traits measured are causes for concern. In addition, many of the studies cited here were in populations with performance and carcass characteristics no longer acceptable in today's industry. It is possible that genetic correlations between components of LTFC and reproduction are not linear and that antagonistic relationships may develop as performance reaches new thresholds. This may be particularly true for selection approaches that take body fat to extremely low levels and/or depress *ad libitum* energy intake. Thus, while experiments such as those reported by Kerr and Cameron (1995, 1996) have begun to evaluate relevant selection objectives in present industry populations, much remains to be learned regarding future targets for pork product and interactions with lifetime reproductive efficiency.

Non-additive Genetic Effects and Breeding Systems

The most efficient commercial production of pork occurs in a breeding system that optimizes genetic merit for market pig performance and for reproduction in the contributing lines, and maximizes exploitation of non-additive genetic effects through heterosis. To determine the efficiency of a breeding system from an overall industry perspective, however, the cost of maintaining the pureline and multiplier herds that supply the commercial system must also be considered. In this section, estimates of maximum heterosis available for traits of postweaning performance will be briefly reviewed, and effective breeding systems for the pork industry will be discussed.

Heterosis estimates for postweaning performance traits are included in several reviews (e.g. Sellier, 1976; Johnson, 1980, 1981; Buchanan, 1987). Individual, maternal and paternal heterosis, the difference in performance due to a crossbred versus purebred individual, dam or sire, respectively, are considered. Consistent with theoretical expectations, heterotic effects are greatest for those traits that have relatively low heritability, such as litter sizes and preweaning litter weights. Accordingly, for components of postweaning performance which are in general moderately heritable (see Tables 15.1. and 15.2.), relatively moderate to small proportions of heterosis have been reported.

Average estimates of individual heterosis, expressed as a percentage of the purebred mean, ranged from 6.0% (Sellier, 1976) to 8.8% (Johnson, 1981) for ADG. Favourable individual heterosis has also been documented for feed conversion ratio. Average estimates of 5.9% for gain/feed and −3.0% for feed/gain were reported by Johnson (1981) and Sellier (1976), respectively. A slight unfavourable individual heterosis may exist for backfat, i.e. crossbred pigs may have a tendency to be fatter than the average of their purebred contemporaries, but estimates of individual heterosis for carcass traits have generally not been significantly different from zero (Johnson, 1981). As Sellier (1976) points out, the amount of heterosis for performance traits, especially feed conversion, may depend on the feeding system (restricted versus *ad libitum*).

Effects of maternal heterosis, while important for litter size and litter weights until weaning, were small and unimportant for postweaning performance (Johnson, 1981). Paternal heterosis, i.e. the advantage of a crossbred boar, may be important for breeding efficiency of young boars, but was also negligible for postweaning performance (Buchanan, 1987).

All three types of heterosis (individual, maternal and paternal) can be maximized in a static (terminal) cross of F_1 sires and dams of unrelated breed background (e.g. AB sires × CD dams). Static crosses also offer the best opportunity to match the complementary strengths of lines as sires and dams to fill the paternal and maternal roles in the breeding system. As described by Sellier (1976), differences in maternal effects are the basis for most of the benefit from complementarity in a breeding system, and the appropriate choice of a maternal line is most obvious for traits of reproduction. However, reported differences in feed efficiency (Kuhlers *et al.*, 1972; Johnson *et al.*, 1973) and carcass composition (Bereskin *et al.*, 1971; Johnson *et al.*, 1973) between reciprocal crosses indicate that complementarity may also affect postweaning performance traits in the system.

Fortunately, the static cross is also the highest ranking breeding system for pork production in terms of overall industry efficiency (Dickerson, 1973). The reproductive rate of the pig results in a relatively small proportion of herds needed for seedstock and multiplier production to maintain the static commercial cross and, thus, a large proportion of animals that realize the maximum benefits of heterosis and complementarity.

In some situations, rate of genetic improvement in the industry can be maximized by implementing specialized selection objectives for each line based on its intended role in the breeding system. The objective in paternal lines might focus exclusively on LTFC. Since maternal lines not only produce the commercial sow herd, but also contribute half the genes of market pigs, maternal line selection objectives must give relatively equal emphasis to LTFC and traits of reproduction. Smith (1964) concluded that pursuit of specialized objectives in sire and dam lines would result in at least as much overall genetic improvement as selection on a general objective, and significantly greater improvement if genetic antagonism exists between market pig performance and traits of reproduction.

Bereskin (1984) estimated that the genetic correlation between an index of postweaning performance traits and a sow productivity index was near zero and suggested that benefits from selection on specialized sire and dam objectives may be marginal. However, he also pointed out that the crossing of lines resulting from specialized selection may enhance heterosis, and Smith (1964) notes that the diversity in lines resulting from specialized selection will guarantee flexibility of the industry as management practices and market demands evolve. In addition, as discussed on page 448, situations may presently exist or may arise in which there is significant genetic antagonism between LTFC and reproduction, thus demanding that different relative emphasis be applied to each in sire and dam lines.

Structural Soundness and Stress Susceptibility

In order to realize their genetic potential for feed intake, growth and reproduction, animals must possess adequate structural soundness and the ability to cope with stresses of the production system and marketing process. In this section, genetic variation and covariation associated with structural soundness and performance in the pig will be discussed, and rapid advancements regarding genetic stress susceptibility will be briefly described.

Structural soundness and leg weakness

The pig's ability to move and function is the culmination of a complex of mechanisms involving skeletal and muscular anatomy. Problems with this ability are most evident in the feet and legs of the animal. Consequently, structural integrity and ability for movement and function is usually termed 'structural soundness' in the United States, while the term 'leg weakness' is more often used in Europe when addressing concerns in this area.

Several non-genetic factors including type of housing can affect structural soundness and associated leg weakness (Grøndalen, 1974a) and the industry's evolution towards more intensive housing facilities over the past 25 years has been accompanied by an increased incidence of leg weakness. Individual breeders or producers are likely to consider varying degrees of leg weakness to be justification for culling; culling rates as high as 35% have been reported (Sabec *et al.*, 1980). As pointed out by Webb *et al.* (1983), rejection of potential breeding animals because of leg weakness can cause an unnecessary reduction in selection differential and add to the cost of genetic improvement. Knowledge of genetic variation associated with leg weakness and its relationships with traits of production and reproduction is needed to determine the optimum approach to leg weakness in genetic programmes.

The genetics of leg weakness in pigs has been studied using various scoring systems in an attempt to quantify variation in anatomical structure and the animal's ability to move. For example, Smith (1966) and Webb *et al.* (1983) used scores of 0 (satisfactory), 1 (mild fault) or 2 (serious fault) to evaluate 19 specific traits of the hind- and front-ends of pigs (e.g. sickle-hocked, bow-legged, over at the knee). Wider scoring scales, for example ranging from 1 to 9 where 9 is associated with the greatest soundness and least leg weakness (Rothschild and Christian, 1988), have been used to quantify overall hind- and front-end structure and movement.

Some estimates of phenotypic correlations have revealed a weak to moderate, favourable relationship between leg weakness and ADG, but a weak, unfavourable relationship between leg weakness and backfat (Bereskin, 1979; Woltmann *et al.*, 1995). Pigs with below average leg weakness tended to gain faster, but had more backfat at market weight. A small, unfavourable phenotypic relationship between leg weakness and backfat was also reported by Webb *et al.* (1983), but they estimated that the phenotypic correlation between leg weakness and ADG was near zero.

Estimates of the heritability of structural soundness/leg weakness and genetic correlations between leg weakness and postweaning performance traits have been derived based on covariances among relatives. Smith (1966) reported the estimated heritability of an overall leg and action score was near zero, and that estimates for more specific characteristics (e.g. sickle hocks, knock-knees) were also generally low. However, others have reported heritabilities for leg weakness scores ranging from 0.16 to 0.30 (Bereskin, 1979; Wilson *et al.*, 1980; Webb *et al.*, 1983).

Estimates of the genetic correlation of leg weakness score with ADG have been variable. Smith (1966), Drewry (1979) and Webb *et al.* (1983) reported that genetic correlation between leg weakness and ADG was zero, but there have been other reports that the correlation is either favourable (Bereskin, 1979; Sather and Fredeen, 1982) or unfavourable (Grondalen, 1974b; Lundeheim, 1987). Although Drewry (1979) and Sather and Fredeen (1982) did not detect a genetic relationship between leg weakness and backfat, some estimates of the genetic correlation suggest an unfavourable relationship (Bereskin, 1979; Webb *et al.*, 1983; Lundeheim, 1987): pigs with genetic merit for below average backfat have greater genetic potential for leg weakness. Webb *et al.* (1983) reported no genetic correlation between leg weakness and feed conversion ratio. Most recently, Van Steenbergen *et al.* (1990) estimated genetic correlations between postweaning performance traits and linear scores of several exterior traits of the legs and movement, but not overall leg weakness. They considered the score for gait pattern to be most closely indicative of leg weakness and reported undesirable genetic correlations of gait with ADG, backfat and feed conversion.

Rothschild and Christian (1988) and Rothschild *et al.* (1988) reported direct and correlated response, respectively, to divergent, single-trait selection for front-end leg score in Duroc pigs. Possible scores were from 1 (unable to stand) to 9 (best structure, no observed weakness). Realized heritabilities for leg weakness were moderate in both directions, but tended to be slightly greater in the downward (greater weakness; $h^2 = 0.42$) than in the upward (less weakness; $h^2 = 0.29$) direction. The five generations of divergent selection completed did not affect postweaning ADG, but there was an undesirable trend for increased backfat in the line selected for less leg weakness. The absence of a correlated response in ADG and the unfavourable trend in backfat are consistent with reports of zero genetic correlations between ADG and leg weakness (Smith, 1966; Drewry, 1979; Webb *et al.*, 1983) and of an unfavourable genetic correlation between backfat and leg weakness (Bereskin, 1979; Webb *et al.*, 1983; Lundeheim, 1987).

The absence of a change in ADG from selection on leg score was also supported by a report of divergent selection for postweaning ADG (Woltmann *et al.*, 1995). After approximately 5 standard deviations of cumulative, divergent selection ADG was 27% greater in the high line than in the low (0.84 vs. 0.66 kg-day averaged across sex and farrowing group), but front-end leg weakness score did not differ between the lines. In contrast to the upward trend in backfat associated with less leg weakness reported by Rothschild *et al.* (1988), Woltmann *et al.* (1995) reported that high line pigs had greater backfat thickness than low line pigs without a difference in leg weakness.

Response to selection for leg weakness (Rothschild and Christian, 1988) confirms that structural soundness can be improved and problems with leg weakness reduced by direct selection. But in practice, breeders have a limited amount of selection effort to spend and increasing selection pressure against leg weakness or for greater structural soundness decreases the emphasis given to other traits. The optimum approach to leg weakness in a breeding programme is determined not only by its heritability, but also by its relative economic value and genetic correlations with other traits. The results of Woltmann *et al.* (1995), supported by those of Rothschild *et al.* (1988), indicate that aggressive selection for ADG will not have an adverse effect on leg weakness. Although in the population studied by Woltmann *et al.* (1995), high line pigs had greater LTGR than low line pigs, they were also fatter at market weight. Thus, the effects of decreased fatness on leg weakness in scenarios of selection for LTGR and LTFC remain unclear. Until results become available from studies of leg weakness in lines selected for LTGR/LTFC, practical emphasis on leg weakness in breeding programmes should probably be limited to culling based on observed limitations in function.

Stress susceptibility

Although pigs naturally respond to stimuli within the production setting with varying degrees of stress, Topel *et al.* (1968) were the first to report a syndrome in pigs that exhibit abnormal response to some stimuli. Porcine stress syndrome (PSS) usually occurs in susceptible pigs when subjected to extreme excitement during the marketing process or other management procedures that require restraint or movement of groups of animals. Pigs afflicted with PSS first exhibit muscle and tail tremors followed by laboured breathing, reddened and blotched skin, increased body temperature and extreme acidosis if stress continues. The final stages of the syndrome are marked by hyperthermia, total collapse and extreme muscle rigidity, usually followed by death.

Pigs that are susceptible to PSS exhibit pale, soft and exudative (PSE) musculature postmortem, even in the absence of extreme stress prior to slaughter, resulting in a negative economic impact of PSS that goes far beyond death loss in the production unit. The incidence of PSS and PSE tends to be greatest in breeds that express extreme muscling, such as the Piétrain, and has increased in association with relatively intense selection for leanness. Several comprehensive reviews concerning PSS and related PSE are available in the literature (e.g. Webb *et al.*, 1982; Christian and Lundstrom, 1992).

Administration of the anaesthetic gas halothane will elicit symptoms of PSS in susceptible pigs (Allen *et al.*, 1970) and detection tests using halothane have been a vital tool in efforts to determine mode of inheritance. Christian (1972) first hypothesized that PSS is inherited by a simple recessive allele with incomplete penetrance. Reports of several subsequent breeding studies have supported that hypothesis (see Christian and Lundstrom, 1992). Identification of normal carrier (heterozygous) pigs was previously only possible through breeding tests, but the discovery of the specific mutation in the ryanodine receptor gene (*RYR*) responsible for PSS (Fujii *et al.*, 1991; Otsu *et al.*, 1991) has allowed accurate identification of all three genotypes with DNA from small samples of tissue and use of polymerase chain reaction (Rempel *et al.*, 1993).

Because accurate determination of homozygous normal (*NN*) and heterozygous normal (*Nn*) genotypes has only recently become possible, much of the information regarding the effects associated with this locus on postweaning performance only distinguishes between halothane positive (*nn*) and negative (*N–*) pigs. In a summary of the literature, Webb *et al.* (1982) reported that rate of growth was not significantly different between positive and negative groups, but that halothane positive pigs had superior feed conversion ratio and LTFC, leaner carcasses, and a greater incidence of PSE than halothane negative pigs.

Even though results have not been completely consistent, reports from some studies in which all three genotypes were compared, or *NN* and *Nn* pigs were compared, have suggested an additive effect on some traits of postweaning performance and body composition. Differences between *NN* and *Nn* for ADG have generally been small and non-significant (Simpson and Webb, 1989; Sather *et al.*, 1991a; Pommier *et al.*, 1992; NPPC, 1995; Leach *et al.*, 1996), but advantages of *Nn* over *NN* pigs have been reported for feed conversion ratio (McPhee *et al.*, 1994; Leach *et al.*, 1996), dressing percentage and lean content of the carcass (Sather *et al.*, 1991b, Pommier *et al.*, 1992; NPPC, 1995; Leach *et al.*, 1996). The NPPC (1995) study also revealed sire line by genotype interactions for several traits in that heterozygotes differed from homozygous negative pigs in some sire lines but not in others.

As a result of the observed advantage in efficiency and leanness for carrier pigs, at least in some genetic backgrounds, systems using stress susceptible (*nn*) terminal sires to produce carrier (*Nn*) market hogs have been implemented by some breeders. However, reports of a higher incidence of PSE and generally poorer meat quality in carrier (*Nn*) versus homozygous negative (*NN*) pigs (NPPC, 1995; Leach *et al.*, 1996) have led to recommendations to the industry for total eradication of the recessive allele through DNA testing and selection (e.g. NPPC, 1995). (See Chapter 16 for a complete discussion of the effects associated with this locus on body composition and tissue quality.)

The protein encoded by *RYR* regulates calcium permeability of the cell and is likely involved in many physiological processes. Even so, it is not completely clear if differences between genotype groups for performance, carcass and muscle quality traits are due to pleiotropic effects of the *RYR* mutation or to linkage disequilibrium with alleles at a nearby locus. If more than one gene is involved, selection for recombinants could remove the halothane mutation from a population while exploiting the positive effects of a linked gene(s). However, evaluation of this chromosomal region in a resource family produced from a cross of wild and domestic pigs failed to detect a similar effect on muscle quality independent of the *RYR* mutation (Lundstrom *et al.*, 1995).

Conclusions

The future competitiveness of pork in the food market depends on continued genetic improvement in the efficiency of quality lean production. From the perspective of postweaning performance of the market pig, great opportunities exist for improvement through selection within seedstock lines. Market pig performance can be characterized by several component traits that respond to selection, but the biological index LTFC has been proposed as the most appropriate expression of the industry's objective for this phase of production.

Investigations of models describing tissue growth have yielded strategies for performance testing and selection to improve LTFC. It has been hypothesized that LTGR and LTFC are perfectly correlated in a time-based scale feeding system and that direct response to selection for LTGR (LTFC) among scale-fed candidates will be through increased LTGR and decreased fat growth without depressed *ad libitum* feed intake. Recently reported experimental evidence supports the hypothesis and suggests this testing and selection scheme as an effective approach in seedstock lines supplying commercial systems that allow either *ad libitum* or restricted feed intake.

Knowledge of the genetic relationships of LTFC with the animal's ability to function and reproduce is required to determine the optimum selection emphasis for LTFC in the various lines of a breeding system. Although significant correlated responses in reproductive traits have not been detected in most studies of selection for postweaning performance, detrimental effects on reproduction have been reported in some cases. The greatest risk for negative effects on reproduction from selection for LTFC probably exists for approaches that result in decreased genetic potential for *ad libitum* feed intake. Research to determine genetic correlations between components of LTFC and traits of reproduction, especially at levels of performance desired for future competitiveness, should remain a priority. Similarly, antagonistic genetic relationships between performance and structural soundness have not been established and intense selection for ADG has been shown experimentally not to have an effect on leg weakness score. However, much remains to be learned regarding the effects on structural soundness of long-term selection for LTFC.

This chapter has focused on the genetics of efficiency in lean tissue growth. The economic importance of LTFC is largely driven by the modern consumer's demand for lean cuts of meat, but the quality of tissue in the final product is also of increasing importance. As a result, the genetics of tissue quality is an area of great interest (see Chapter 16) and genetic correlations of components of LTFC with quality characteristics must be determined. In addition, the genetics of performance has been presented here in terms of quantitative variation and covariation, and the approaches to genetic improvement reviewed are based on traditional animal breeding principles. With the development and application of genetic linkage maps (Chapter 9), much will likely be learned about the contribution of specific loci to quantitative variation in performance traits. Consequently, methods of selection for postweaning performance will evolve to incorporate supplemental marker information and in response to changing global market demands.

Acknowledgements

The authors gratefully acknowledge Dr Lauren Christian for his helpful suggestions. A.C.C. acknowledges research support from the Oklahoma Agricultural Experiment Station. The authors are also grateful for permission to reproduce the figures from Fowler *et al.* (1976) and Whittemore (1986).

References

Allen, W.M., Berrett, S. and Harding, J.D.J. (1970) Experimentally induced acute stress syndrome in Piétrain pigs. *Veterinary Record* 87, 64–69.

Baird, D.M., Nalbandov, A.V. and Norton, H.W. (1952) Some physiological causes of genetically different rates of growth in swine. *Journal of Animal Science* 11, 292–300.

Bereskin, B. (1979) Genetic aspects of feet and leg soundness in swine. *Journal of Animal Science* 48, 1322–1328.

Bereskin, B. (1984) Genetic correlations of pig performance and sow productivity traits. *Journal of Animal Science* 59, 1477–1487.

Bereskin, B., Shelby, C.E. and Hazel, L.N. (1971) Carcass traits of purebred Durocs and Yorkshires and their crosses. *Journal of Animal Science* 32, 413–419.

Berruecos, J.M., Dillard, E.U. and Robison O.W. (1970) Selection for low backfat thickness in swine. *Journal of Animal Science* 30, 844–848.

Brascamp, E.W., Merks, J.W.M. and Wilmink, J.B.M. (1985) Genotype environment interaction in pig breeding programmes: methods of estimation and relevance of the estimates. *Livestock Production Science* 13, 135–146.

Brien, F.D. (1986) A review of the genetic and physiological relationships between growth and reproduction in mammals. *Animal Breeding Abstracts* 12, 975–997.

Buchanan, D.S. (1987) The crossbred sire: experimental results for swine. *Journal of Animal Science* 65, 117–127.

Campbell, R.G. and Taverner, M.R. (1988) Genotype and sex effects on the relationship between energy intake and protein deposition in growing pigs. *Journal of Animal Science* 66, 676–686.

Cameron, N.D. (1993) Methodologies for estimation of genotype with environment interaction. *Livestock Production Science* 35, 237–249.

Cameron, N.D. (1994) Selection for components of efficient lean growth rate in pigs 1. Selection pressure applied and direct responses in a Large White herd. *Animal Production* 59, 251–262.

Cameron, N.D. and Curran, M.K. (1994a) Selection for components of efficient lean growth rate in pigs 4. Genetic and phenotypic parameter estimates and correlated responses in performance test traits with ad-libitum feeding. *Animal Production* 59, 281–291.

Cameron, N.D. and Curran, M.K. (1994b) Selection for components of efficient lean growth rate in pigs 2. Selection pressure applied and direct responses in a Landrace herd. *Animal Production* 59, 263–269.

Cameron, N.D. and Curran, M.K. (1995a) Genotype with feeding regime interaction in pigs divergently selected for components of efficient lean growth rate. *Animal Science* 61, 123–132.

Cameron, N.D. and Curran, M.K. (1995b) Responses in carcass composition to divergent selection for components of efficient lean growth rate in pigs. *Animal Science* 61, 347–359.

Cameron, N.D., Curran, M.K. and Thompson, R. (1988) Estimation of sire with feeding regime interaction in pigs. *Animal Production* 46, 87–95

Cameron, N.D., Curran, M.K. and Kerr, J.C. (1994) Selection for components of efficient lean growth rate in pigs 3. Responses to selection with a restricted feeding regime. *Animal Production* 59, 271–279.

Campbell, R.G. and Taverner, M.R. (1988) Genotype and sex effects on the relationship between energy intake and protein deposition in growing pigs. *Journal of Animal Science* 66, 676–686.

Christian, L.L. (1972) A review of the role of genetics in animal stress susceptibility and meat quality. In: *Proceedings of the Pork Quality Symposium.* University of Wisconsin, Madison, pp. 91–115.

Christian, L.L. and Lundstrom, K. (1992) Porcine stress syndrome. In: Leman, A.D., Straw, B.E., Mengeling, W.L., D'Allaire, S. and Taylor, D.J. (eds) *Diseases of Swine.* Iowa State University Press, Ames, pp. 763–771.

Cleveland, E.R., Cunningham, P.J. and Peo, E.R., Jr (1982) Selection for lean growth in swine. *Journal of Animal Science* 54, 719–727.

Cleveland, E.R., Johnson, R.K. and Mandigo, R.W. (1983a) Index selection and feed intake restriction in swine. I. Effect on rate and composition of growth. *Journal of Animal Science* 56, 560–569.

Cleveland, E.R., Johnson, R.K., Mandigo, R.W. and Peo, E.R., Jr (1983b) Index selection and feed intake restriction in swine. II. Effect on energy utilization. *Journal of Animal Science* 56, 570–578.

Cleveland, E.R., Johnson, R.K. and Cunningham, P.J. (1988) Correlated responses of carcass and reproductive traits to selection for rate of lean growth in swine. *Journal of Animal Science* 66, 1371–1377.

Clutter, A.C. and Buchanan, D.S. (1998) Effects of divergent selection for postweaning average daily gain on body weight changes of gilts. *Oklahoma Agricultural Experiment Station Research Report* P-958, 1–5.

Clutter, A.C., Spicer, L.J., Woltmann, M.D., Grimes, R.W., Hammond, J.M. and Buchanan, D.S. (1995) Plasma growth hormone, insulin-like growth factor I, and insulin-like growth factor binding proteins in pigs with divergent genetic merit for postweaning average daily gain. *Journal of Animal Science* 73, 1776–1783.

Craig, J.V., Norton, H.W. and Terrill, S.W. (1956) A genetic study of weight at five ages in Hampshire swine. *Journal of Animal Science* 14, 242–256.

Crump, R.E., Thompson, R. and Mercer, J.T. (1990) The genetic relationship between sexes for performance traits as recorded in a commercial nucleus herd. *Proceedings of the 4th World Congress on Genetics Applied to Livestock Production* 15, 454–457.

De Haer, L.C.M. and De Vries, A.G. (1993) Effects of genotype and sex on the feed intake pattern of group housed growing pigs. *Livestock Production Science* 36, 223–232.

DeNise, R.S., Irvin, K.M., Swiger, L.A. and Plimpton, R.F. (1983) Selection for increased leanness of Yorkshire swine. IV. Indirect responses of the carcass, breeding efficiency and preweaning litter traits. *Journal of Animal Science* 56, 551–559.

Dickerson, G.E. (1973) Inbreeding and heterosis in animals. In: *Proceedings of the Animal Breeding and Genetics Symposium in Honor of Dr Jay L. Lush.* Virginia Polytechnic Institute and State University, Blacksburg, pp. 54–77.

Dickerson, G.E. and Grimes, J.C. (1947) Effectiveness of selection for efficiency of gain in Duroc swine. *Journal of Animal Science* 6, 265–287.

Drewry, K.J. (1979) Production traits and visual scores of tested boars. *Journal of Animal Science* 48, 723–734.

Ellis, M., Chadwick, J.P., Smith, W.C. and Laird R. (1988) Index selection for improved growth and carcass characteristics in a population of Large White pigs. *Animal Production* 46, 265–275.

Fahmy, M.H. and Bernard, C. (1970) Genetic and phenotypic study of pre- and post-weaning weights and gains in swine. *Canadian Journal of Animal Science* 50, 593–599.

Fowler, V.R., Bichard, M. and Pease, A. (1976) Objectives in pig breeding. *Animal Production* 23, 365–387.

Fredeen, H.T. and Mikami, H. (1986a) Mass selection in a pig population: experimental design and responses to direct selection for rapid growth and minimum fat. *Journal of Animal Science* 62, 1492–1508.

Fredeen, H.T. and Mikami, H. (1986b) Mass selection in a pig population: realized heritabilities. *Journal of Animal Science* 62, 1509–1522.

Fredeen, H.T. and Mikami, H. (1986c) Mass selection in a pig population: correlated changes in carcass merit. *Journal of Animal Science* 62, 1546–1554.

Fujii, J., Otsu, K., Zorzato, F., DeLeon, S., Khanna, V.K., Weiler, J.E., O'Brien, P. and MacLennan, D.H. (1991) Identification of a mutation in porcine ryanodine receptor associated with malignant hyperthermia. *Science* 253, 448–451.

Garnett, I. and Rahnefeld, G.W. (1976) Mass selection for post-weaning growth in swine V. Correlated response of reproductive traits and pre-weaning growth. *Canadian Journal of Animal Science* 56, 791–801.

Gray, R.C., Tribble, L.F., Day, B.N. and Lasley, J.F. (1968) Results of five generations of selection for low backfat thickness in swine. *Journal of Animal Science* 27, 331–335.

Grøndalen, T. (1974a) Leg weakness in pigs. I. Incidence and relationship to skeletal lesions, feed level, protein and mineral supply, exercise and exterior conformation. *Acta Veterinario Scandinavica* 15, 555–573.

Grøndalen, T. (1974b) Leg weakness in pigs. II. Litter differences in leg weakness skeletal lesions, joint shape and exterior conformation. *Acta Veterinario Scandinavica* 15, 574–586.

Gu, Y., Haley, C.S. and Thompson, R. (1989) Estimates of genetic and phenotypic parameters of growth and carcass traits from closed lines of pigs on restricted feeding. *Animal Production* 49, 467–475.

Hazel, L.N. (1943) The genetic basis for constructing selection indexes. *Genetics* 28, 476–490.

Hazel, L.N. and Kline, E.A. (1952) Mechanical measurement of fatness and carcass value on live hogs. *Journal of Animal Science* 11, 313–318.

Hetzer, H.O. and Harvey, W.R. (1967) Selection for high and low fatness in swine. *Journal of Animal Science* 26, 1244–1251.

Hetzer, H.O. and Miller, R.H. (1972a) Rate of growth as influenced by selection for high and low fatness in swine. *Journal of Animal Science* 35, 730–742.

Hetzer, H.O. and Miller, R.H. (1972b) Correlated responses of various body measurements in swine selected for high and low fatness. *Journal of Animal Science* 35, 743–751.

Hill, W.G. (1971) Design and efficiency of selection experiments for estimating genetic parameters. *Biometrics* 27, 293–311.

Hutchens, L.K. and Hintz, R.L. (1981) A summary of genetic and phenotypic statistics for pubertal and growth characteristics in swine. *Technical Bulletin* T-155, Oklahoma State University, Stillwater, 37 pp.

Johansson, K. (1981) Some notes concerning the genetic possibilities of improving sow fertility. *Livestock Production Science* 8, 431–447.

Johansson, K. and Kennedy, B.W. (1983) Genetic and phenotypic relationships of performance test measurements with fertility in Swedish Landrace and Yorkshire sows. *Acta Agriculturæ Scandinavica* 33, 195–199.

Johnson, R.K. (1980) Heterosis and breed effects in swine. *North Central Regional Publication No. 262*, University of Nebraska, Lincoln, 51 pp.

Johnson, R.K. (1981) Crossbreeding in swine: Experimental results. *Journal of Animal Science* 52, 906–923.

Johnson, R.K., Omtvedt, I.T. and Walters, L.E. (1973) Evaluation of purebreds and two-breed crosses in swine: Feedlot performance and carcass merit. *Journal of Animal Science* 37, 18–26.

Jungst, S.B., Christian, L.L. and Kuhlers, D.L. (1981) Response to selection for feed efficiency in individually fed Yorkshire boars. *Journal of Animal Science* 53, 323–331.

Kerr, J.C. and Cameron, N.D. (1995) Reproductive performance of pigs selected for components of efficient lean growth. *Animal Science* 60, 281–290.

Kerr, J.C. and Cameron, N.D. (1996) Genetic and phenotypic relationships between performance test and reproduction traits in Large White pigs. *Animal Science* 62, 531–540.

Koots, K.R. and Gibson, J.P. (1994) How precise are genetic correlation estimates? *Proceedings of the 5th World Congress on Genetics Applied to Livestock Production* 18, 353–360.

Krider, J.L., Carroll, B.W. and Roberts, E. (1946) Effectiveness of selecting for rapid and slow growth in Hampshire swine. *Journal of Animal Science* 5, 3–15.

Kuhlers, D.L. and Jungst, S.B. (1990) Mass selection for increased 70-day weight in a closed line of Landrace pigs. *Journal of Animal Science* 68, 2271–2278.

Kuhlers, D.L. and Jungst, S.B. (1991a) Mass selection for increases 200-day weight in a closed line of Landrace pigs. *Journal of Animal Science* 69, 977–984.

Kuhlers, D.L. and Jungst, S.B. (1991b) Mass selection for increases 200-day weight in a closed line of Duroc pigs. *Journal of Animal Science* 69, 507–516.

Kuhlers, D.L. and Jungst, S.B. (1992a) Correlated responses in reproductive and carcass traits to selection for 200-day weight in Duroc swine. *Journal of Animal Science* 70, 2707–2713.

Kuhlers, D.L. and Jungst, S.B. (1992b) Correlated responses in reproductive and carcass traits to selection for 70-day weight in Landrace swine. *Journal of Animal Science* 70, 372–378.

Kuhlers, D.L. and Jungst, S.B. (1993) Correlated responses in reproductive and carcass traits to selection for 200-day weight in Landrace pigs. *Journal of Animal Science* 71, 595–601.

Kuhlers, D.L., Chapman, A.B. and First, N.L. (1972) Estimates of genotype-environment interactions in production and carcass traits in swine. *Journal of Animal Science* 35, 1–7.

Leach, L.M., Ellis, M., Sutton, D.S., McKeith, F.K. and Wilson, E.R. (1996) The growth performance, carcass characteristics, and meat quality of halothane carrier and negative pigs. *Journal of Animal Science* 74, 934–943.

Leymaster, K.A., Swiger, L.A. and Harvey, W.R. (1979a) Selection for increased leanness of Yorkshire swine. I. Experimental procedures and selection applied. *Journal of Animal Science* 48, 789–799.

Leymaster, K.A., Swiger, L.A. and Harvey, W.R. (1979b) Selection for increased leanness of Yorkshire swine. II. Population parameters, inbreeding effects and response to selection. *Journal of Animal Science* 48, 800–809.

Lundeheim, N. (1987) Genetic analysis of osteochondrosis and leg weakness in the Swedish Pig Progeny Testing Scheme. *Acta Agriculturæ Scandinavica* 37, 159–173.

Lundstrom, K., Karlsson, A., Hakansson, J., Hansson, I., Johansson, M., Andersson, L. and Andersson, K. (1995) Production, carcass and meat quality traits of F_2-crosses between European Wild pigs and domestic pigs including halothane gene carriers. *Animal Science* 61, 325–332.

Lush, J.L. (1936) Genetic aspects of the Danish system of progeny-testing swine. *Iowa Agricultural Experiment Station Research Bulletin* (204) 105–196.

McKay, R.M. (1990) Responses to index selection for reduced backfat thickness and increased growth rate in swine. *Canadian Journal of Animal Science* 70, 973–977.

McKay, R.M. (1992) Effect of index selection for reduced backfat thickness and increased growth rate on sow weight changes through two parities in swine. *Canadian Journal of Animal Science* 72, 403–408.

McPhee, C.P. (1981) Selection for efficient lean growth in a pig herd. *Australian Journal of Agricultural Research* 32, 681–690.

McPhee, C.P., Brennan, P.J. and Duncalfe, F. (1979) Genetic and phenotypic parameters of Australian Large White and Landrace boars performance-tested when offered food ad libitum. *Animal Production* 28, 79–85.

McPhee, C.P., Rathmell, G.A., Daniels, L.J. and Cameron, N.D. (1988) Selection in pigs for increased lean growth rate on a time-based feeding scale. *Animal Production* 47, 149–156.

McPhee, C.P., Daniels, L.J., Kramer, H.L., Macbeth, G.M., and Noble, J.W. (1994) The effects of selection for lean growth and the halothane allele on growth performance and mortality of pigs in a tropical environment. *Livestock Production Science* 38, 117.

Merks, J.W.M. (1987) Genotype × environment interactions in pig breeding programmes. II. Environmental effects and genetic parameters in central test. *Livestock Production Science* 16, 215–227.

Merks, J.W.M. (1988) Genotype × environment interactions in pig breeding programmes. IV. Sire × herd interaction in on-farm test results. *Livestock Production Science* 30, 325–336.

Merks, J.W.M. (1989) Genotype × environment interactions in pig breeding programmes. VI. Genetic relations between performances in central test, on-farm test and commercial fattening. *Livestock Production Science* 22, 325–339.

Morris, C.A. (1975) Genetic relationships of reproduction with growth and with carcass traits in British pigs. *Animal Production* 20, 31–44.

Moughan, P.J., Smith, W.C. and Pearson, G. (1987) Description and validation of a model simulating growth in the pig (20–90 kg liveweight). *New Zealand Journal of Agricultural Research* 30, 481–489.

Mrode, R.A. and Kennedy, B.W. (1993) Genetic variation in measures of food efficiency in pigs and their genetic relationships with growth rate and backfat. *Animal Production* 56, 225–232.

National Pork Producers Council, USA (1995) *Genetic Evaluation: Terminal Line Program Results.* Des Moines, Iowa, 312 pp.

Nordskog, A.W., Comstock, R.E. and Winters, L.M. (1944) Hereditary and environmental factors affecting growth rate in swine. *Journal of Animal Science* 3, 257–272.

Otsu, K., Khanna, V.K., Archiband, A. and MacLennan, D.H. (1991) Cosegregation of porcine malignant hyperthermia and a probable causal mutation in the skeletal muscle ryanodine receptor gene in backcross families. *Genomics* 11, 744–750.

Pommier, S.A., Houde, A., Rousseau, F. and Savoie, Y. (1992) The effect of the malignant hyperthermia genotype as determined by a restriction endonuclease assay on carcass characteristics of commercial crossbred pigs. *Canadian Journal of Animal Science* 72, 973–976.

Rahnefeld, G.W. (1971a) Mass selection for post-weaning growth in swine I. The value of a pedigreed control population. *Canadian Journal of Animal Science* 51, 481–496.

Rahnefeld, G.W. (1971b) Mass selection for post-weaning growth in swine II. Response to selection. *Canadian Journal of Animal Science* 51, 497–502.

Rahnefeld, G.W. (1973) Mass selection for post-weaning growth in swine III. Correlated response in weaning weight and feed efficiency to recurrent selection for postweaning average daily gain in swine. *Canadian Journal of Animal Science* 53, 173–178.

Rahnefeld, G.W. and Garnett, I. (1976) Mass selection for post-weaning growth in swine IV. Selection response and control population stability. *Canadian Journal of Animal Science* 56, 783–790.

Rempel, W.E., Lu, M.Y., El Kandelgy, S., Kennedy, C.F.H., Irvin, L.R., Mickelson, J.R. and Louis, C.F. (1993) Relative accuracy of the halothane challenge test and a molecular genetic test in detecting the gene for porcine stress syndrome. *Journal of Animal Science* 71, 1395–1399.

Robertson, A. (1959) The sampling variance of the genetic correlation coefficient. *Biometrics* 15, 469–485.

Robison, O.W. and Berruecos, J.M. (1973) Feed efficiency in swine. II. Prediction of efficiency and genetic correlations with carcass traits. *Journal of Animal Science* 37, 650–657.

Robison, O.W., Cooksey, J.H., Chapman, A.B. and Self, H.L. (1960) Estimation of carcass merit of swine from live animal measurements. *Journal of Animal Science* 19, 1013–1023.

Rothschild, M.F. and Christian, L.L. (1988) Genetic control of front-leg weakness in Duroc swine. I. Direct response to five generations of divergent selection. *Livestock Production Science* 19, 459–471.

Rothschild, M.F., Christian, L.L. and Jung, Y.C. (1988) Genetic control of front-leg weakness in Duroc swine. II. Correlated responses in growth rate, backfat and reproduction from five generations of divergent selection. *Livestock Production Science* 19, 473–485.

Rydhmer, L., Eliasson-Selling, L., Johansson, K., Stern, S. and Andersson, K. (1994) A genetic study of estrus symptoms at puberty and their relationship to growth and leanness in gilts. *Journal of Animal Science* 72, 1964–1970.

Sabec, D., Zagozen, F., Urbas, J. and Subelj, J. (1980) Locomotory disturbances in testing boars. *Proceedings of the 6th International Pig Veterinary Society Congress*, Copenhagen, Denmark, p.331.
Sather, A.P. and Fredeen, H.T. (1978) Effect of selection for lean growth rate upon feed utilization of market hogs. *Canadian Journal of Animal Science* 58, 285–289.
Sather, A.P. and Fredeen, H.T. (1982) The effect of confinement upon the incidence of leg weakness in swine. *Canadian Journal of Animal Science* 62, 1119–1128.
Sather, A.P., Jones, S.D.M., Tong, A.K.W. and Murray, A.C. (1991a) Halothane genotype by weight interactions on lean yield from pork carcasses. *Canadian Journal of Animal Science* 71, 633–643.
Sather, A.P., Jones, S.D.M., Tong, A.K.W. and Murray, A.C. (1991b) Halothane genotype by weight interactions on pig meat quality. *Canadian Journal of Animal Science* 71, 645–658.
Sellier, P. (1976) The basis of crossbreeding in pigs; a review. *Livestock Production Science* 3, 203–226.
Simpson, S.P. and Webb, A.J. (1989) Growth and carcass performance of British Landrace pigs heterozygous at the halothane locus. *Animal Production* 49, 503–509.
Smith, C. (1964) The use of specialized sire and dam lines in selection for meat production. *Animal Production* 6, 337–344.
Smith, C. (1966) A note on the heritability of leg weakness scores in pigs. *Animal Production* 8, 345–348.
Smith, C. and Ross, G.J.S. (1965) Genetic parameters of British Landrace bacon pigs. *Animal Production* 7, 291–301.
Smith, C., King, J.W.B. and Gilbert, N. (1962) Genetic parameters of British Large White bacon pigs. *Animal Production* 4, 128–143.
Standal, N. and Vangen, O. (1985) Genetic variation and covariation in voluntary feed intake in pig selection programmes. *Livestock Production Science* 12, 367–377.
Sutherland, T.M. (1965) The correlation between feed efficiency and rate of gain, a ratio and its denominator. *Biometrics* 21, 739–749.
Topel, D.G., Bicknell, E.J., Preston, K.S., Christian, L.L. and Matsushima, C.Y. (1968) Porcine stress syndrome. *Modern Veterinary Practice* 49, 40.
Van Diepen, T.A. and Kennedy, B.W. (1989) Genetic correlations between test station and on-farm performance for growth rate and backfat in pigs. *Journal of Animal Science* 67, 1425–1431.
Vangen, O. (1979) Studies on a two trait selection experiment in pigs. II. Genetic changes and realized genetic parameters in the traits under selection. *Acta Agriculturæ Scandinavica* 29, 305–319.
Vangen, O. (1980a) Studies on a two trait selection experiment in pigs. III. Correlated responses in daily feed intake, feed conversion and carcass traits. *Acta Agriculturæ Scandinavica* 30, 125–141.
Vangen, O. (1980b) Studies on a two trait selection experiment in pigs. V. Correlated responses in reproductive performance. *Acta Agriculturæ Scandinavica* 30, 309–319.
Van Steenbergen, E.J., Kanis, E. and Van Der Steen, H.A.M. (1990) Genetic parameters of fattening performance and exterior traits of boars tested in central stations. *Livestock Production Science* 24, 65–82.
Vogt, D.W., Comstock, R.E. and Rempel, W.E. (1963) Genetic correlations between some economically important traits in swine. *Journal of Animal Science* 22, 214–217.
Webb, A.J. and King, J.W.B. (1983) Selection for improved food conversion ratio on ad libitum group feeding in pigs. *Animal Production* 37, 375–385.

Webb, A.J., Carden, A.E., Smith, C. and Imlah, P. (1982) Porcine stress syndrome in pig breeding. In: *Proceedings of the 2nd International Congress on Genetics Applied to Livestock Production* vol. 5, pp. 588–608.

Webb, A.J., Russell, W.S. and Sales, D.I. (1983) Genetics of leg weakness in performance-tested boars. *Animal Production* 36, 117–130.

Webster, A.J.F. (1977) Selection for leanness and the energetic efficiency of growth in meat animals. In: *Proceedings of Nutrition Society*, vol. 36, pp. 53–59.

Whittemore, C.T. (1986) An approach to pig growth modeling. *Journal of Animal Science* 63, 615–621.

Whittemore, C.T. and Fawcett, R.H. (1976) Theoretical aspects of a flexible model to simulate protein and lipid growth in pigs. *Animal Production* 22, 87–96.

Wilson, R.D., Christian, L.L. and Schneider, J.F. (1980) The effects of sire soundness classification and feed restriction on performance and leg scores of pigs. *Journal of Animal Science* 51 (Suppl. 1), 132 (abstract).

Woltmann, M.D., Clutter, A.C., Buchanan, D.S. and Dolezal, H.G. (1992) Growth and carcass characteristics of pigs selected for fast or slow gain in relation to feed intake and efficiency. *Journal of Animal Science* 70, 1049–1059.

Woltmann, M.D., Clutter, A.C. and Buchanan, D.S. (1995) Effect of divergent selection for postweaning average daily gain on front-end soundness of market-weight pigs. *Journal of Animal Science* 73, 1940–1947.

Wyllie, D., Morton, J.R. and Owen, J.B. (1979) Genetic aspects of voluntary food intake in the pig and their association with gain and food conversion efficiency. *Animal Production* 28, 381–390.

Young, L.D., Johnson, R.K. and Omtvedt, I.T. (1977) An analysis of the dependency structure between a gilt's prebreeding and reproductive traits. I. Phenotypic and genetic correlations. *Journal of Animal Science* 44, 557–564.

Genetics of Meat and Carcass Traits

16

P. Sellier
INRA, Station de Génétique quantitative et appliquée, 78352 Jouy-en-Josas Cedex, France

Introduction

Domestic pigs are raised for the human consumption of fresh meat and processed meat products, and traits referring to meat are of particular importance for the pig industry as a whole. Increasing the lean to fat ratio of pig carcasses has been, for several decades, a major objective of pig breeding programmes in many countries, and a very large number of genetic studies have been devoted to body composition traits. More recently, emphasis began to be put also on qualitative properties of muscle and fat tissues, and research has increasingly been conducted on this topic. This chapter aims at presenting the current state of knowledge about genetics of carcass traits and meat and fat quality attributes both from technological and sensory points of view.

Traits of Interest

A list of carcass and meat quality traits of current or potential interest for pig breeding purposes is given in Table 16.1. Regarding body composition, the most frequently measured traits comprise killing out percentage, carcass length, carcass lean percentage, loin muscle area and backfat thickness. The latter can be accurately and easily recorded on the live pig using ultrasound scanning equipment (e.g. Wilson, 1992), and has a good predictive value for assessing the total amount of carcass fat separable by dissection because the dorsal subcutaneous region is the preferential site of fat deposition in the pig. Ultrasonic backfat depth has been a widely used selection criterion for improving carcass lean to fat ratio in the last 30 years (see Chapter 15).

As pointed out by Cameron (1993) and Moller and Iversen (1993), meat quality unlike meat quantity is a composite concept and as such is hard to define and to measure in a simple and unique manner. This complexity is magnified in the pig because pork is used in many different ways which do not necessarily require the same qualitative characteristics from the raw matter. In most European countries, roughly half the pigmeat is consumed as fresh meat whereas the other half is processed into meat products such as cured–cooked or dry-cured hams, bacon and sausages (Moller *et al.*, 1993). So, a single term 'meat quality' covers a mixture of instrumentally measured compositional and physiochemical characteristics of fresh meat, various processing yields, traits relating to human health, safety or nutritional needs and sensory properties contributing to the consumer's acceptability of fresh pork or processed pork products.

Technological quality of pork classically refers to a set of properties comprising water-holding capacity (e.g. drip loss during storage), intensity and homogeneity of colour, firmness, shelf-life, cooking loss and various processing yields. Sensory quality of meat primarily involves appearance (essentially colour and marbling) and eating quality as assessed by trained taste panels or consumer surveys. Usual components of eating quality consist of texture (including tenderness and juiciness) and flavour-liking or pork flavour intensity (including taste

Table 16.1. A list of meat and carcass traits of interest in the pig.

Attributes of quality	Main component or predictive traits
Body composition	Killing out percentage
	Carcass length
	Carcass lean percentage or lean to fat ratio
	Backfat thickness (live and carcass measurements)
	Loin muscle area (or depth)
Technological quality of pork (as fresh meat or as raw matter for processing)	Water-holding capacity (drip loss, 'filter paper' methods, free water content)
	Firmness
	Cooking loss
	Technological yield (e.g. 'Napole' yield for cured–cooked products)
	Muscle glycogen content (glycolytic potential)
	Muscle protein content and solubility
	Electrical conductivity of meat
	Seasoning loss (dry-cured hams)
Technological quality of pork; Sensory quality of fresh pork and processed pork products	pH_1 (at 45–60 min postmortem)
	pH_u (ultimate pH at 24 h post mortem)
	Colour parameters (e.g. *CIE* L*-value for light reflectance, two-toning aspect, myoglobin level)
	Muscle fibre characteristics (contractile/metabolic type, cross-sectional areas)
Sensory quality of fresh pork and processed pork products	Tenderness (instrumental determination or sensory panel score)
	Juiciness (sensory panel score)
	Pork flavour (sensory panel score or volatile compound profile)
	Overall acceptability (sensory panel score)
	Intramuscular fat (content and fatty acid composition, visible marbling)
	Postmortem proteolysis
Quality of fatty depots	Firmness (subjective score or penetrometer value)
	Keepability (rancidness)
	Moisture, lipid and connective tissue contents
	Fatty acid composition (e.g. ratio of polyunsaturated over saturated fatty acids)
	Boar taint (androstenone, skatole)

and smell). Some attention is also increasingly paid in pig breeding to quality of fatty depots (e.g. firmness and keepability) owing to the greater proneness to the 'soft fat' defect in leaner pigs (Metz, 1985; Wood and Enser, 1989).

Two widely used meat quality indicators, i.e. pH at 45–60 min postmortem (pH_1) and pH on the day after slaughter (ultimate pH or pH_u), are of value for predicting technological as well as eating quality of pork (e.g. Jacquet *et al.*, 1984; Warriss and Brown, 1987; Bendall and Swatland, 1988; Berger *et al.*, 1994; Buscailhon *et al.*, 1994; Fernandez *et al.*, 1994; Van Laack *et al.*, 1994; Van der Wal *et al.*, 1995). A similar 'dual-purpose' interest could also prevail regarding muscle fibre characteristics (Lefaucheur, 1989; Essén-Gustavsson, 1993).

Although there exists a continuum in the phenotypic variability of meat quality and kinetics of postmortem pH fall (both rate and extent), reference is commonly made to three distinct pH-related abnormalities. These are namely the PSE (pale, soft, exudative) meat condition (associated with pH_1 values lower than, say, 5.9–6.1 depending on the muscle), the DFD (dark, firm, dry) meat condition (associated with pH_u values higher than 6.0–6.2) and the 'acid meat' condition (associated with pH_u values lower than 5.4–5.5). The DFD defect is essentially due to adverse environmental circumstances (e.g. long transportation and/or fasting times causing muscle glycogen depletion), and genetic factors are of minor importance in the occurrence of DFD meat (e.g. Nielsen, 1981). In contrast, the PSE and 'acid meat' defects are strongly related to genetic factors as two single major genes (*HAL* and *RN*, respectively) are primarily involved (see below).

Intramuscular fat content is generally considered as having a favourable influence on eating quality of pork, particularly tenderness and juiciness (e.g. Bejerholm and Barton-Gade, 1986; DeVol *et al.*, 1988; Gandemer *et al.*, 1990; Essén-Gustavsson *et al.*, 1994). However, this association is somewhat controversial (Tornberg *et al.*, 1993; Casteels *et al.*, 1995). Several authors have failed to find a significantly positive phenotypic correlation between intramuscular fat content and tenderness (e.g. Cameron, 1990b; Purchas *et al.*, 1990; Lan *et al.*, 1993). In addition, an excess of visible marbling could be detrimental to the acceptability of fresh meat by consumers (Touraille *et al.*, 1989; Fernandez *et al.*, 1996).

Heritabilities

Body composition traits

Since the pioneer work of Lush (1936) on the Danish Landrace, additive genetic variation of body composition traits has been investigated by many authors. Published estimates of heritability (h^2) of the most thoroughly studied carcass traits were reviewed by Stewart and Schinckel (1989) and Ducos (1994). These traits are moderately to highly heritable, as average h^2 values range from 0.30 to 0.35 for killing out percentage to 0.55 to 0.60 for carcass length (Table 16.2). Available h^2 estimates were averaged by Ducos (1994) according to breed, testing environment (central stations vs. breeding herds) and feeding system (*ad libitum* vs. scale feeding). Heritabilities of backfat thickness and lean percentage are higher in the heavily-muscled Piétrain and Belgian Landrace breeds than in the Duroc and Hampshire breeds, with the Large White and Landrace breeds being intermediate. Heritability of ultrasonic backfat thickness is increased by some 30% in central station testing ($h^2 = 0.49$) as compared with on-farm testing ($h^2 = 0.36$). The influence of feeding system is smaller, but body composition traits are slightly more heritable on restricted feeding than on *ad libitum* feeding (0.52 vs. 0.46 in backfat thickness and 0.60 vs. 0.56 in lean percentage).

Table 16.2. Average values of heritability (h^2) for body composition traits.

Trait	Reference	
	Stewart and Schinckel (1989)	Ducos (1994)[1]
Ultrasonic backfat thickness	0.41	0.45 (143)
Fat depth over the 10th rib	0.52	—
Loin muscle area	0.47	0.48 (35)
Lean percentage	0.48	0.54 (77)
Killing out percentage	0.30	0.36 (16)
Carcass length	0.56	0.57 (43)

[1]In brackets, number of estimates taken into account.

Meat and fat quality traits

Heritability estimates for meat quality traits were first published in the early 1960s (Duniec *et al.*, 1961; Jonsson, 1963; Ollivier and Meslé, 1963). Since then, meat colour, pH, water-holding capacity and intramuscular fat content were among the most often studied traits. Information is also available on additive genetic variation of eating quality traits and compositional characteristics of fat depots. A survey of average h^2 values is given in Table 16.3 for a number of meat and fat quality criteria.

Meat (muscle) quality as a whole is slightly to moderately heritable as heritability values fall in the range 0.10–0.30 for most meat quality traits. However, intramuscular lipid content is well outside of this usual range of h^2 values and shows a markedly higher heritability (around 0.50). Among eating quality traits, tenderness (h^2 = 0.25–0.30) appears to be more heritable than juiciness or flavour (h^2 around 0.10). Meat colour (h^2 close to 0.30) is also more heritable than pH or water-holding capacity measurements (h^2 = 0.15–0.20). As for the high average value of heritability reported for Napole yield (see Table 16.10 for definition) and muscle glycogen content, it should be pointed out that all the involved studies but one have dealt with lines which were in segregation at the *RN* locus, known for exerting a very large effect on these traits (see below). For example, the heritability of muscle glycolytic potential is strongly increased in such lines (h^2 larger than 0.80; Le Roy *et al.*, 1994b) as compared with a Large White selection line presumably devoid of the RN^- allele (h^2 = 0.29; Larzul, 1997).

Heritabilities of compositional traits (water, stearic and linoleic acid content) and firmness of subcutaneous fat depots range from 0.35 to 0.65, indicating that additive genetic variance of fat quality is similar to that of carcass fatness. The fatty acid composition of intramuscular lipids shows the same heritability as that of adipose tissue: h^2 = 0.40–0.70 for percentages of the main fatty acids in loin muscle (Berger *et al.*, 1994). Regarding eating quality of meat from intact male pigs, androstenone (5α-androst-16-en-3-one) is a testicular steroid implied in the occurrence of the so-called boar taint, and fat androstenone level is highly heritable (e.g. Jonsson and Wismer-Pedersen, 1974; Sellier and Bonneau, 1988;

Table 16.3. Average values of heritability (h^2) for meat and fat quality traits.[1]

Trait	Average h^2	No. and range of estimates	
pH_1	0.16	14	(0.04–0.41)
pH_u	0.21	33	(0.07–0.39)
Colour (light reflectance, *CIE* L* value)	0.28	29	(0.15–0.57)
Water-holding capacity[2]	0.15	15	(0.01–0.43)
Drip loss	0.16	10	(0.01–0.31)
Cooking loss	0.16	9	(0.00–0.51)
Technological yield (cooked ham processing)	0.27	3	(0.09–0.40)
Napole yield[3]	0.45	5	(0.26–0.78)
Meat quality index[4]	0.20	13	(0.11–0.33)
Visual score of meat quality	0.20	8	(0.10–0.37)
Tenderness (instrumental determination)	0.26	10	(0.17–0.46)
Tenderness (sensory panel score)	0.29	9	(0.18–0.70)
Flavour (sensory panel score)	0.09	6	(0.01–0.16)
Juiciness (sensory panel score)	0.08	8	(0.00–0.28)
Overall acceptability (sensory panel score)	0.25	2	(0.16–0.34)
Muscle composition traits			
% water	0.25	7	(0.14–0.52)
% lipid	0.50	19	(0.26–0.86)
% protein	0.22	1	(—)
% glycogen (glycolytic potential)[3]	0.69	3	(0.29–0.90)
Fat composition traits (backfat)			
% water	0.44	2	(0.27–0.59)
% lipid	0.26	1	(—)
% stearic acid (C18:0)[5]	0.51	3	(0.42–0.57)
% linoleic acid (C18:2)[5]	0.58	3	(0.47–0.67)
androstenone level (entire males)	0.56	5	(0.25–0.88)
Intensity of boar taint (score)	0.54	1	(—)
Firmness of backfat (instrumental determination)	0.43	1	(—)

[1]Updated and completed from Sellier (1988a, 1995) and Sellier and Monin (1994).
[2]As assessed by various 'filter paper' methods.
[3]See text and Table 16.10.
[4]Meat quality index IQV ('indice de qualité de la viande') used in French central test stations and constructed for predicting the technological yield of cured–cooked ham processing from pH_u, reflectance and water-holding capacity measurements.
[5]Expressed in % of total fatty acids.

Willeke, 1993). Evidence for a major gene influencing fat androstenone content in boars was recently provided in two distinct populations (Bidanel *et al.*, 1996; Fouilloux *et al.*, 1997). Deposition of skatole (3-methylindole) in adipose tissue, another factor giving rise to boar taint (Claus *et al.*, 1994), would also be under some genetic control, and a recessive gene having a pronounced effect, especially under triggering environmental conditions, has been hypothesized by Lundström and Malmfors (1993).

Genetic Correlations

Genetic correlations among body composition traits

Genetic correlations (r_A) among major body composition traits have been reviewed by Stewart and Schinckel (1989) and Ducos (1994) (Table 16.4). The strong genetic relationship between live measurement of backfat thickness and carcass lean percentage is of special interest for breeding purposes and has largely been exploited by pig breeders since the advent of performance testing of candidates to selection in the 1960s. The magnitude of this genetic relationship appears to remain unchanged in very lean populations, e.g. the Piétrain breed (r_A = –0.79; Bidanel and Ducos, 1995), as compared with the Large White or Landrace breeds (r_A = –0.70 to –0.90; Ducos *et al.*, 1993; Bidanel and Ducos, 1996; Labroue *et al.*, 1997). A higher lean to fat ratio is genetically associated with an increased carcass length (r_A near 0.20). The sign of the genetic correlation of carcass lean content with killing out percentage might depend on the population involved. This genetic correlation is zero or slightly negative in certain breeds (e.g. Ducos *et al.*, 1993; Labroue *et al.*, 1997) whereas it is

Table 16.4. Average values of genetic correlations among body composition traits.

	Reference	
Traits	Stewart and Schinckel (1989)	Ducos (1994)
Lean percentage		
× ultrasonic backfat thickness	—	–0.65
× 10th rib fat depth	–0.87	—
× loin muscle area	0.65	—
× killing out percentage	–0.10	0.20
× carcass length	0.18	—
Tenth rib fat depth		
× loin muscle area	–0.38	
× killing out percentage	0.19	
× carcass length	–0.21	
Ultrasonic backfat thickness		
× killing out percentage	—	0.18
Loin muscle area		
× killing out percentage	0.50	—
× carcass length	–0.18	—
Killing out percentage		
× carcass length	–0.32	—

Table 16.5. Average values of genetic correlations among meat quality traits.

Traits[1]	Genetic correlation		References[2]
	Mean	Range of estimates	
Drip loss			
× pH_1	–0.27	–0.55–0.01	9,11
× pH_u	–0.71	–0.99––0.50	5,9,11,12
× reflectance	0.49	0.49–0.49	11,12
× WHC	–0.94	–0.99––0.90	11,12
× IMF	–0.08	–0.23–0.05	5,11,12
× myofibre diameter	0.73	—	7
WHC			
× pH_1	–0.65	—	11
× pH_u	0.45	0.26–0.92	11–13
× reflectance	–0.39	–0.66––0.18	11–13
× IMF	0.12	0.02–0.22	11,12
Cooking loss			
× pH_1	–0.14	–0.23–0.04	9,11
× pH_u	–0.68	–0.82––0.45	8,9,11
× reflectance	0.26	0–0.47	2,8,11,12
× WHC	–0.25	–0.30––0.21	11,12
× IMF	0.07	–0.03–0.23	8,11,12
× drip loss	0.66	0.45–0.80	2,9,11,12
Napole yield or technological yield			
× pH_u	0.70	0.26–0.99	1,4,14
× GPIV	–0.83	–1––0.5	10,14
Reflectance			
× pH_1	–0.38	—	11
× pH_u	–0.53	–0.66––0.38	3,8,11–13
× IMF	0.01	–0.12–0.15	3,8,11,12
Tenderness			
× pH_1	0.27	—	11
× pH_u	0.49	0.40–0.68	3,8,11,12
× reflectance	–0.16	–0.22––0.08	3,8,11,12
× WHC	0.23	0.08–0.41	6,11,12
× IMF	0.15	–0.08–0.53	3,6,8,11,12
× drip loss	–0.16	–0.19––0.14	11,12
× cooking loss	–0.46	–0.57––0.40	8,11,12
Overall acceptability			
× pH_u	0.59	—	3
× reflectance	–0.02	—	3
× WHC	0.46	—	6
× IMF	0.61	0.54–0.68	3,6

significantly positive in other breeds (e.g. Johansson *et al.*, 1987; Bidanel and Ducos, 1995).

Genetic correlations among meat and fat quality traits

Table 16.5 reports average values of genetic correlations of the main components of technological and sensory quality of pig meat (drip loss, cooking loss, processing yield, colour, tenderness, overall acceptability) with predictive criteria such as pH_1, pH_u, muscle glycolytic potential and intramuscular fat content. Ultimate pH shows a marked genetic association with all components of meat quality, especially cooking yield and cured-cooked ham processing yield (r_A close to 0.70). Muscle glycolytic potential as measured on the live animal is genetically correlated with Napole yield or technological yield of cured-cooked ham processing in composite lines segregating for the major RN^- allele (r_A very close to –1) as well as in Large White lines presumably free of this allele (r_A = –0.50). In halothane-negative Yorkshire and Duroc populations studied by Hovenier *et al.* (1992) and De Vries *et al.* (1994), drip loss during storage of fresh meat was found to be genetically correlated with both pH_1 and pH_u (r_A close to –0.60). A positive genetic correlation between drip loss and average myofibre diameter was reported by Dietl *et al.* (1993). Score for overall acceptability of pork would be equally correlated at the genetic level (r_A = 0.60) with pH_u and intramuscular fat content (Cameron, 1990b; Lo *et al.*, 1992b). However, cooked meat tenderness assessed by a mechanical measurement or a sensory panel score appears to be genetically correlated with ultimate pH and cooking yield (r_A close to 0.50) rather than with intramuscular fat content (r_A less than 0.20).

Regarding fat quality, subcutaneous fat firmness showed strong genetic relationships with compositional traits of adipose tissue in the study of Cameron (1990b). These genetic correlations were positive (around 0.60) for saturated fatty acid content (C16:0 and C18:0) and negative for per cent moisture (r_A = –0.60) and per cent C18:2 (r_A = –0.85).

---

[1]WHC = water-holding capacity, as assessed by percentage of free water content (reference 6) or by 'filter paper' tests (references 11–13); IMF = intramuscular fat content; GPIV = muscle glycolytic potential as measured *in vivo*; tenderness = instrumental determination or sensory panel score; overall acceptability = sensory panel score.

[2]1 = Ollivier and Meslé (1963), Large White; 2 = Malmfors and Nilsson (1979), Yorkshire/Landrace; 3 = Cameron (1990b), Duroc/Landrace; 4 = Le Roy *et al.* (1990b), composite lines; 5 = Hovenier *et al.* (1992), Duroc/Yorkshire; 6 = Lo *et al.* (1992b), Duroc/Landrace; 7 = Dietl *et al.* (1993), several German breeds; 8 = Berger *et al.* (1994), several US breeds; 9 = Bidanel *et al.* (1994b), Piétrain; 10 = Le Roy *et al.* (1994b), composite lines; 11 = de Vries *et al.* (1994), Yorkshire; 12 = NPPC (1995), several US genetic types; 13 = Tribout *et al.* (1996), Landrace/Large White; 14 = Larzul (1997). Large White.

Genetic correlations of carcass composition traits with meat and fat quality traits

Average values of genetic correlations with body composition traits are reported in Table 16.6 for meat and eating quality traits and in Table 16.7 for fat quality traits. Regarding the usual criteria of meat quality (pH, reflectance, water-holding capacity, drip loss), a moderate genetic antagonism with carcass lean to fat ratio (up to ±0.25) is generally found. The genetic correlation of meat quality index (IQV) is, on average, −0.23 with carcass lean percentage and 0.18 with ultrasonic backfat thickness. Available estimates for cooking yield surprisingly indicate a slight tendency to a favourable genetic relationship between this trait and carcass lean to fat ratio. However, muscle glycolytic potential shows a positive genetic correlation (about 0.30) with lean to fat ratio. As could

Table 16.6. Average values of genetic correlations of meat and eating quality traits with body composition traits.[1]

	Genetic correlation with:				
	carcass leanness[2]		carcass fatness[3]		
Trait	Mean	Range	Mean	Range	References[4]
pH_1	0.10	—	0.26	—	9,11
pH_u	−0.13	−0.50–0.08	0.15	−0.05–0.45	3–5,8,9,11–13,15, 17,20
Reflectance	0.16	−0.16–0.42	−0.21	−0.48–0.07	3,8,11–13,15,17
Water-holding capacity	−0.19	−0.57–0.24	0.02	−0.25–0.24	6,11–13,17
Drip loss	0.05	−0.10–0.13	−0.10	−0.20–−0.01	5,9,11,12
Cooking loss	−0.07	−0.16–−0.06	0.12	−0.04–0.39	8,9,11,12
Napole yield	—	—	0.15	−0.12–0.41	4
Muscle glycolytic potential	0.40	—	−0.21	−0.34–−0.10	10,14
Meat quality index (IQV)	−0.23	−0.44–0.06	0.18	0.01–0.39	13,17,19, 21–23
Intramuscular fat content	−0.34	−0.55–−0.07	0.30	0.04–0.60	3,5,6,8,11,12,16,18
Tenderness	−0.20	−0.48–0.12	0.24	0.11–0.48	3,6,8,11,12
Juiciness	−0.18	−0.47–0.08	0.29	−0.19–0.85	3,6,8,12
Pork flavour	−0.27	−0.60–0.02	0.35	−0.03–0.72	3,6,8,12
Overall acceptability	−0.48	−0.71–−0.32	0.34	−0.04–0.70	3,6

[1]This table refers to estimates published in the years 1987–1996. For earlier studies, see, e.g. Sellier (1988a).
[2]Lean percentage, lean weight, loin muscle area, % premium cuts.
[3]Live or carcass backfat thickness, backfat weight.
[4]See Table 16.5 for references 3 to 14. Other references: 15 = Johansson (1987), Landrace/Yorkshire; 16 = Schwörer *et al.* (1987), Landrace/Large White; 17 = Cole *et al.* (1988), Belgian Landrace/Landrace/Large White; 18 = Bout *et al.* (1989b), Landrace/Large White; 19 = Ducos *et al.* (1993), Landrace/Large White; 20 = Bidanel *et al.* (1994a), Landrace/Large White; 21 = Bidanel and Ducos (1995), Piétrain; 22 = Bidanel and Ducos (1996), Landrace/ Large White; 23 = Labroue *et al.* (1997), Landrace/Large White.

Table 16.7. Average values of genetic correlations of qualitative characteristics of subcutaneous backfat with body composition traits.[1]

	Genetic correlation with:	
Backfat characteristic	carcass leanness[2]	carcass fatness[3]
% Water	0.35	–0.65
% Lipid	–0.70	0.70
% C18:0	–0.40	0.40
% C18:2	0.60	–0.70
Firmness (instrumental)	–0.40	0.70

[1]Data from Schwörer *et al.* (1988), Bout *et al.* (1989a) and Cameron (1990b).
[2]Lean weight or percentage, or % premium cuts.
[3]Backfat thickness or weight.

be expected, intramuscular fat content is genetically associated with carcass fatness, but the average genetic correlation between the two traits does not exceed 0.30 showing that part of the genetic variation in lipid content of the muscle is independent of the genetic variation in overall lipid content of the carcass. The same pattern is observed when considering the between-breed covariation of the two traits (see below). Eating quality traits consistently show unfavourable genetic correlations with carcass lean to fat ratio (around –0.25 for tenderness and juiciness of meat and around –0.35 for pork flavour intensity and overall acceptability of meat). So, the 'quantity–quality' antagonism would be more pronounced, at the genetic level, for eating than for technological quality of pig meat. This genetic antagonism is even stronger regarding qualitative characteristics of subcutaneous backfat. Average genetic correlations shown in Table 16.7 range from ± 0.35–0.40 up to ± 0.70–0.75. Higher carcass lean to fat ratio is genetically associated with lower lipid or stearic acid content and softer fat as well as with higher moisture or linoleic acid content of fat depots. The close positive genetic correlation between carcass leanness and ratio of polyunsaturated over saturated fatty acids of adipose tissue mainly originates from the relatively smaller contribution of the *de novo* fatty acid synthesis and the relatively larger contribution of dietary fatty acids to total fat deposition in genetically leaner pigs (Scott *et al.*, 1981; Metz, 1985).

Inbreeding

As reviewed by Johnson (1989), few estimates of the effects of inbreeding on carcass traits have been reported, but available information indicates that body composition traits are essentially unaffected by inbreeding (King and Roberts, 1959; Mikami *et al.*, 1977; Leymaster and Swiger, 1981).

Major Genes and QTL

A number of single major genes affecting meat and/or carcass traits are known (*HAL*, *RN*) or postulated (*MU*, *MI*) in the pig, whereas the recent development of marker-QTL studies has enabled researchers to detect significant associations between these traits and anonymous microsatellite polymorphisms. Within Table 16.8 is a summarized view of the current information on major genes and quantitative trait loci (QTL).

Halothane sensitivity gene

The occurrence of 'soft' and 'colourless' pig meat was mentioned in the literature more than a century ago, as noted by Goutefongea (1963). Earlier studies showed that this serious meat defect is related to an abnormal rate of post-mortem change in muscle pH (Wismer-Pedersen, 1959). It became clear in the early 1960s that the liability to pale, soft, exudative (PSE) meat condition had a

Table 16.8. A list of single genes affecting meat and/or carcass traits in the pig.

Gene	Chromosome mapped	Main effect(s) and reference(s)
Halothane sensitivity gene (*HAL*)/ryanodine receptor gene (*RYR*)/calcium release channel gene (*CRC*)	6	Malignant hyperthermia syndrome (MHS), pale, soft, exudative (PSE) meat condition, muscular hypertrophy phenomenon (Eikelenboom and Minkema, 1974; Ollivier *et al.*, 1975; Smith and Bampton, 1977; Ollivier, 1980; Webb, 1981; Christian and Mabry, 1989; Fujii *et al.*, 1991)
Acid meat gene (*RN*)	15	'Rendement Napole' (*RN*), muscle glycogen and protein content (Naveau, 1986; Le Roy *et al.*, 1990b, 1996b; Lundström *et al.*, 1996)
Muscle content gene (*MU*)	?	Backfat thickness (Le Roy *et al.*, 1990a)
Meishan intramuscular fat gene (*MI*)	?	Muscle lipid content (Janss *et al.*, 1997)
Swine leucocyte antigen (*SLA*) genes and microsatellite markers (*S0064*, *S0066*, *S0102*) or genes (*TNFα*) linked to *SLA*	7	Backfat thickness, loin muscle area, meat quality traits, boar taint (Capy *et al.*, 1981; Jung *et al.*, 1989; Renard *et al.*, 1992; Rothschild *et al.*, 1995; Bidanel *et al.*, 1996)
Adenosine deaminase (*ADA*)	?	Meat quality (Hyldgaard-Jensen, 1986)
Microsatellite markers (*S0001* and *S0175*)	4	Backfat thickness, abdominal fat percentage (Andersson *et al.*, 1994)
Pituitary-specific transcription factor (*PIT1*)	13	Backfat thickness (Yu *et al.*, 1995)
Meishan cooking loss gene (*MC*)	?	Cooking loss, ultimate pH (Janss *et al.*, 1997)
Meishan backfat gene	?	Backfat thickness (Janss *et al.*, 1997)

marked genetic component (Briskey, 1964). It was reported at that time that certain breeds (Piétrain, Poland China), or certain strains within breeds (Landrace), contained a large proportion of PSE-prone animals, whereas other breeds or strains were practically free of this defect. Moreover, PSE-proneness was shown to be closely associated with porcine stress syndrome (Judge, 1972), and evidence was presented by Ollivier (1968) for a monogenic inheritance of the muscular hypertrophy phenomenon in the Piétrain breed. Christian (1972) was the first to suggest that the inheritance of stress susceptibility is monogenic autosomal recessive with some variation in penetrance. A major step in this field of research was the finding that porcine stress syndrome (PSS), or malignant hyperthermia syndrome (MHS), can be triggered by a short exposure to the halothane gas, an anaesthetic agent, and halothane reactors recover if the halothane administration is stopped as soon as the first PSS symptoms are observed (Christian, 1974; Eikelenboom and Minkema, 1974). Studies based on the all-or-none reaction to the halothane challenge test confirmed the recessive inheritance of PSS/MHS (Ollivier *et al.*, 1975; Smith and Bampton, 1977). The locus responsible for halothane sensitivity was called *HAL* (Andresen and Jensen, 1977), with two alleles *N* (normal, dominant) and *n* (halothane sensitivity, recessive)(Minkema *et al.*, 1977).

Associations between certain alleles at several blood type loci, *S* (*A-O*) and *H* blood group systems, glucose phosphate isomerase (*GPI*, formerly *PHI*), 6-phosphogluconate dehydrogenase (*PGD*) and alpha-1-B-glycoprotein (*A1BG*, formerly *PO2*) and lower meat quality were reported in the Danish Landrace breed in the late 1970s (Jorgensen, 1979) and then confirmed in various pig populations. As discussed by Andresen (1987), these alleles are very likely to exert no direct effect on meat quality. Their apparent effect originates from the location of the corresponding loci in the same chromosomal region as the *HAL* locus, the most likely order of the loci being *S* (*A-O*)–*HAL*–*GPI*–*H*–*A1BG*–*PGD*. In several populations, the HAL^n allele is preferentially associated with the H^a, GPI^B, $A1BG^S$ and PGD^B (or PGD^A) alleles due to linkage disequilibria. This property has been successfully used in selection against HAL^n by means of blood typing (e.g. Vögeli *et al.*, 1988; Saugère *et al.*, 1989). The 'halothane' linkage group is mapped to the proximal part (centromere to q2.5) of the long arm of pig chromosome 6 (Davies *et al.*, 1988; Harbitz *et al.*, 1990; Yerle *et al.*, 1990; Chowdhary *et al.*, 1991).

Studies taking benefit from the homologies between segments of human chromosome 19 (q13.1–q13.2) and porcine chromosome 6 (p1.1–q2.1) have dealt with the ryanodine receptor, the calcium release channel of the skeletal muscle sarcoplasmic reticulum (Mickelson and Louis, 1993). The linkage study carried out by MacLennan *et al.* (1990) indicated that the basic defect in human MHS results from mutations in the ryanodine receptor gene (*RYR*). Similarly, a single point mutation (C to T) at nucleotide 1843 of that gene was identified by Fujii *et al.* (1991) and Otsu *et al.* (1991) as being correlated with, and probably being causative of, halothane-induced MHS in six different pig breeds. This finding has rapidly resulted in the development of DNA-based methods to detect non-carriers, single and double carriers of the halothane sensitivity allele by means of an accurate and non-invasive test.

Numerous comparisons between the two halothane phenotypes (halothane-positive = HP or halothane-negative = HN) concerning meat and carcass traits have been published since the pioneer work of Eikelenboom and Minkema (1974)(for reviews, see Webb, 1981; Christian and Mabry, 1989). It is well documented that HP pigs give heavier, shorter and leaner carcasses than HN pigs. Halothane sensitivity primarily induces an acceleration of the post-mortem fall in muscle pH (lower pH_1 at 45–60 min after death). This feature results in a much increased incidence of PSE meat among HP pigs, although the figures concerning the latter vary from one study to another. The difference in PSE frequency between HP and HN pigs from the same population can be strongly influenced by handling, fasting and slaughtering conditions as shown by Monin *et al.* (1981). Preslaughter stress, such as that caused by long trucking distances from the farm to the abattoir, tends to reduce the PSE incidence among HP pigs (McPhee and Trout, 1995). Increasing the length of time off feed prior to slaughter could also decrease the PSE incidence among HP pigs (Murray *et al.*, 1989).

Earlier comparisons between the three halothane genotypes (*NN*, *Nn* and *nn*) were most often based on the above-mentioned blood markers linked to the *HAL* locus, and were reviewed by Sellier (1987, 1988a) and Sellier and Monin (1994). More recently, studies involving the use of DNA tests for halothane genotyping have been carried out, and results from most of them are summarized in Table 16.9. It is established that the HAL^n allele acts in an approximately additive manner regarding lean content, killing out percentage and carcass length. For lean to bone ratio of primal cuts, *nn*, *Nn* and *NN* pigs average 5.7, 5.1 and 4.6, respectively, according to Aalhus *et al.* (1991). The advantage of *nn* over *NN* pigs in carcass lean content generally falls in the range 2–5% and is, on average, of the order of one phenotypic standard deviation (SD) of the trait. Regarding meat quality, the difference between *NN* and *nn* pigs amounts to 3 SD for pH_1, whereas it is close to one SD for meat colour ($L*$ value), drip loss and tenderness of fresh meat, and is around 0.5 SD for technological yield of cured–cooked ham processing and seasoning yield of dry-cured ham processing. For most meat quality traits, the heterozygotes lie in an intermediate position between the two homozygotes, though being closer to *NN* than to *nn* in pH_1, colour, cooking loss and tenderness. The halothane gene does not affect muscle glycolytic potential (Monin and Sellier, 1985; Sellier *et al.*, 1988; Kocwin-Podsiadla *et al.*, 1995) and ultimate pH of meat.

According to Sather *et al.* (1991b,c), the effect of the HAL^n allele in the heterozygous state may depend on live weight. The effect of this allele appears to be recessive in light weight pigs (80 kg) and becoming dominant in heavy weight pigs (130 kg). However, this halothane genotype by slaughter weight interaction was not found by Garcia-Macias *et al.* (1996), Leach *et al.* (1996) and Larzul *et al.* (1997).

As discussed by Sellier and Monin (1994), halothane sensitivity probably does not affect eating quality of meat primarily through muscle composition.

Higher toughness of pork from *nn* animals is more likely to be due to an impaired proteolysis during the ageing process, as the degradation of the myofibrillar structure is much less marked after several days of storage in HP than in HN pigs (Estrade *et al.*, 1991; Boles *et al.*, 1992). This can be related to the findings of Fernandez *et al.* (1994) and Minelli *et al.* (1995) showing from rheological data that PSE meat from Piétrain pigs tenderizes at a lower rate and to a lesser extent during ageing, in accordance with previous observations of Buchter and Zeuthen (1971).

As to the mechanism involved in the effect of the halothane sensitivity gene on carcass leanness and muscular hypertrophy phenomenon (Ollivier, 1980), it remains to be elucidated as pointed out by Reiner (1993). According to

Table 16.9. Estimates of differences between halothane genotypes (as determined by a DNA-based assay) for meat and carcass traits.

Trait	References[1]	Differences between genotypes[2]			
		nn – *NN*		*Nn* – *NN*	
Killing out percentage	1–3,6–8,10,12,13	1.2	*(0.8)*	0.5	*(0.3)*
Carcass length (cm)	1–3,6–8,10,12,13	–2.7	*(–1.1)*	–0.5	*(–0.2)*
Backfat thickness (mm)	1–8,10,12,13	–3.0	*(–1.0)*	–0.7	*(–0.3)*
Carcass lean percentage	2,4,6,7,10,12,13	3.5	*(1.1)*	1.2	*(0.4)*
Loin muscle area (cm^2)	2,3,6–8,10,13	5.9	*(1.2)*	2.0	*(0.4)*
pH_1 (LD)	2,6,10,12,13	–0.73	*(–3.0)*	–0.24	*(–1.0)*
pH_u (LD)	1–3, 6–8, 10,12,13	0.01	*(0.1)*	–0.03	*(–0.2)*
L* (LD)[3]	1,3,6,7,10,12,13	5.0	*(1.2)*	1.9	*(0.4)*
Drip loss (LD) (%)	6,7,13	1.2	*(0.8)*	1.0	*(0.6)*
Technological yield (%)[4]	6	–1.3	*(–0.4)*	–1.1	*(–0.3)*
Napole yield (%)[5]	6	–3.0	*(–0.8)*	–0.7	*(–0.2)*
Seasoning loss (%)[6]	9,11	1.7	*(0.7)*	0.5	*(0.2)*
Juiciness score[7]	3,7,11–13	—	*(–0.3)*	—	*(–0.2)*
Flavour score[7]	3,7,11	—	*(–0.1)*	—	*(0.0)*
Tenderness score[7]	3,7,11–13	—	*(–1.0)*	—	*(–0.3)*
Marbling score	3,4,7,8,11,13	—	*(–0.8)*	—	*(–0.4)*
Intramuscular fat content (%)	3,7,10,13	—	*(—)*	–0.2	*(–0.3)*
% C18:2 in backfat lipids	10	—	*(—)*	0.5	*(0.4)*
% C18:2 in LD muscle lipids	3	—	*(—)*	0.9	*(0.3)*

[1]1 = Pommier *et al.* (1992); 2 = Wittmann *et al.* (1993); 3 = Berger *et al.* (1994); 4 = O'Brien *et al.* (1994); 5 = Webb *et al.* (1994); 6 = Guéblez *et al.* (1995); 7 = NPPC (1995); 8 = Rempel *et al.* (1995); 9 = Russo and Nanni Costa (1995); 10 = Garcia-Macias *et al.* (1996); 11 = Guéblez *et al.* (1996); 12 = Larzul *et al.* (1996); 13 = Leach *et al.* (1996).
[2]In brackets, differences expressed in standard deviation units of the trait.
[3]L*: *CIE* L*value for light reflectance; LD: Longissimus dorsi muscle.
[4]Curing–cooking yield of ham processing. [5]See Table 16.10.
[6]Weight loss during 7-month ageing period in dry-cured ham processing.
[7]As assessed by a sensory panel on cooked roasts or loin chops.

Lundström *et al.* (1995), the effect of the halothane sensitivity gene on meat quality and lean meat content is primarily due to pleiotropy and not to closely linked genes.

Acid meat gene

The hypothesis was put forward by Naveau (1986) that a single gene affecting meat quality, and different from the halothane sensitivity gene, was segregating in two French composite lines (Penshire and Laconie). These two lines shared the characteristic of having been built from Hampshire blood (at a rate of one-half and one-third, respectively). In that study, meat quality was assessed using the 'Napole yield', which is an indicator of the technological yield of cured–cooked ham processing (weight of cooked ham/weight of raw deboned–defatted ham). The dominant allele assumed to decrease the Napole yield (Rendement Napole in French) was called RN^-, the favourable recessive allele being rn^+, and the corresponding defect was designated 'acid meat' by Naveau (1986). The dominant gene hypothesis was substantiated by Le Roy *et al.* (1990b, 1994a) by use of segregation analysis methods on several independent data sets, and then definitively confirmed by Le Roy *et al.* (1996b) from direct comparisons between animals of proven *RN* genotype. These studies agree in showing that the disadvantage in Napole yield of the RN^- carriers as compared with the rn^+rn^+ homozygotes amounts approximately to 8% (i.e. 3 phenotypic standard deviations of the trait). At the time of the investigation of Le Roy *et al.* (1990b), the frequency of the RN^- allele was around 0.60 in both lines studied, and the *RN* locus accounted for 75–80% of the total additive genetic variance of Napole yield. It should also be mentioned that Wassmuth *et al.* (1991) have established the presence in the Hampshire breed of a dominant allele, called HF^- ('Hampshirefaktor') and affecting meat quality in the same way as RN^-: *HF* and *RN* are very likely to be the same locus.

Besides the Napole yield, i.e. the trait involved in the original discovery of the *RN* gene, this gene also exerts a major effect on other meat quality traits (Table 16.10). The carriers of the RN^- allele, as compared with non-carriers, show a larger extent of the postmortem fall in muscle pH, resulting in lower ultimate pH, as well as paler colour and decreased water-holding capacity of fresh meat (Le Roy *et al.*, 1996b; Lundström *et al.*, 1996). For eating quality traits, Le Roy *et al.* (1996a) found that RN^- carriers give meat of inferior texture (lower sensory panel scores for tenderness, juiciness and mellowness) but of increased pork flavour intensity. Higher taste and smell intensity scores were similarly reported by Lundström *et al.* (1996) in RN^-rn^+ pigs, compared with rn^+rn^+ pigs, but there was no difference between the two genotypes in sensory panel score for tenderness, in spite of a lower shear-force value in the heterozygous pigs.

There seems to exist a favourable influence of the RN^- allele on carcass composition as RN^- carriers, when compared with non-carriers, exhibit decreased backfat thickness and increased loin muscle area (Le Roy *et al.*,

1996b). The advantage of RN^-RN^- over rn^+rn^+ pigs would be of the order of 1 percentage point in carcass lean content.

With regard to how RN^- affects meat quality, there is very strong evidence that it primarily acts by increasing the *intra vitam* glycogen content of muscle. Glycogen content, assessed on the live animal or after slaughter through glycolytic potential, is higher by some 70% in the white-type, longissimus dorsi (LD), muscle of RN^- carrier pigs (Table 16.10). In the experiment carried out by Le Roy *et al.* (1996b), the heterozygous RN^- carriers were slightly below the

Table 16.10. Estimates of differences between RN genotypes for meat and carcass traits.

		Differences between genotypes			
Trait	References[1]	$RN^-RN^- - rn^+rn^+$		$RN^-rn^+ - rn^+rm^+$	
Napole yield (%)[3]	1,3,5,6	−8.1	*(−3.0)*	−7.5	*(−2.8)*
Muscle GP (μmol/g)[4]					
LD muscle	5,6	125	*(5.1)*	92	*(3.8)*
SC muscle	5	24	*(1.1)*	19	*(0.9)*
Killing out percentage	4,5	−0.1	*(−0.1)*	0.2	*(0.1)*
Carcass length (cm)	5	2.0	*(0.6)*	1.6	*(0.5)*
Backfat thickness (mm)	5	−2.1	*(−0.9)*	−1.3	*(−0.5)*
Carcass lean percentage	4,5	1.1	*(0.4)*	0.6	*(0.2)*
Loin muscle area (cm^2)	5	2.9	*(0.5)*	3.3	*(0.6)*
pH_1 (LD)[5]	5	0.01	*(0.1)*	0.02	*(0.1)*
pH_u (LD)	5,6	−0.20	*(−1.7)*	−0.17	*(−1.4)*
pH_u (SM)	2,5	−0.24	*(−1.8)*	−0.21	*(−1.6)*
pH_u (AF)	5	−0.37	*(−2.5)*	−0.35	*(−2.3)*
L* (LD)	5	2.5	*(0.7)*	3.4	*(0.9)*
WHC (LD)	5,6	—	*(−0.7)*	—	*(−0.8)*
Muscle composition traits					
% Water	5	1.0	*(2.0)*	0.9	*(1.8)*
% Protein	2,5,6	−1.4	*(−1.5)*	−1.1	*(−1.2)*
% Lipid	5	−0.1	*(−0.3)*	−0.1	*(−0.2)*

[1]1 = Le Roy *et al.* (1990b); 2 = Wassmuth *et al.* (1991); 3 = Le Roy *et al.* (1994a); 4 = Lundström *et al.* (1994); 5 = Le Roy *et al.* (1996b); 6 = Lundström *et al.* (1996).
[2]In brackets, differences expressed in standard deviation units of the trait.
[3]The 'Napole' technique consists of curing and cooking a 100 g meat sample taken from the Semimembranosus muscle, and Napole yield is the ratio of cooked to raw weight (Naveau *et al.*, 1985).
[4]Muscle glycolytic potential (GP) is defined as the potential of lactate formation during the postmortem glycogenolysis and is expressed in mol lactate equivalents per g fresh muscle weight (Monin and Sellier, 1985); LD: Longissimus dorsi muscle; SC: Semispinalis capitis muscle.
[5]L*: *CIE* L*value for light reflectance; WHC: water-holding capacity assessed by drip loss measurement (reference 6) or by a 'filter paper' test (reference 5); SM: Semimembranosus muscle; AF: Adductor femoris muscle.

homozygous ones (287 vs. 314 μmol/g) for *in vivo* glycolytic potential of LD muscle. It should also be pointed out that the glycolytic potential of a typical 'red-type' muscle (semispinalis capitis) is much less affected by the *RN* genotype. The biochemical mechanism underlying the glycogen hyperaccumulation in 'white-type' muscles from RN^- carriers is not fully understood so far. The causal metabolic defect is very likely to be located in muscle cells (Monin *et al.*, 1992), and the glycogen granules accumulate in the sarcoplasmic compartment of the white myofibres (Estrade *et al.*, 1993b). The activity of the glycogen branching enzyme is higher in the LD muscle from RN^- carriers (Estrade *et al.*, 1994).

Another feature of interest associated with the *RN* gene is that muscle protein content is lower (by 1–1.5 percentage points) and, conversely, muscle water content is higher (by around 1 percentage point) in RN^- carriers than in non-carriers (Wassmuth *et al.*, 1991; Monin *et al.*, 1992; Estrade *et al.*, 1993a; Le Roy *et al.*, 1996b; Lundström *et al.*, 1996). This increase in water to protein ratio (3.7 vs. 3.4) could also contribute to the extra release of water during cooking of meat from RN^- carriers, as pointed out by Sellier and Monin (1994).

The *RN* locus has been mapped to pig chromosome 15 (Milan *et al.*, 1995, 1996; Looft *et al.*, 1996; Mariani *et al.*, 1996). (Additional details may be found in Chapter 9.)

Muscle content gene

It was postulated by Le Roy *et al.* (1990a) that a two-allele single locus (*MU*) exerting a major effect on backfat thickness was segregating in a French composite line (Laconie). Maximum likelihood estimates of parameters under the hypothesis of mixed (monogenic + polygenic) inheritance indicate that the difference between the means of MU^+MU^+ and mu^-mu^- homozygotes is about 3 mm (i.e. 2 phenotypic standard deviations of the trait), whereas the mean value of heterozygotes is very close to that of MU^+MU^+ homozygotes. It can be noted that a recessive allele, causing an increased backfat thickness and likely present in the Chinese Meishan breed, was similarly found by Janss *et al.* (1997).

Intramuscular fat gene

By modelling a mixed mode of inheritance, i.e. one single gene and polygenic background genes, for intramuscular fat content (IMF) of pigs from F_2 crosses between the Chinese Meishan breed and Dutch dam lines, Janss *et al.* (1997) detected a major IMF locus called *MI*. Double carriers of the MI^+ allele would exhibit approximately 3.9% IMF vs. 1.8% IMF in non-carriers or single carriers. This difference corresponds to more than 3 phenotypic standard deviations of the trait. This MI^+ allele is likely to originate from the Meishan breed in which it would be at a frequency of about 0.5.

Other genes

In the earlier studies reviewed by Ollivier and Sellier (1982, 1985), associations of carcass traits with polymorphisms of various erythrocyte antigen or blood protein genes have occasionally been reported, but scarcely in a repeated manner. The most consistent data in this respect dealt with effects of the loci belonging to the halothane linkage group (chromosome 6), e.g. the effect of *GPI* on visual score for muscular hypertrophy (Guérin *et al.*, 1979) and the effect of H blood group system (*EAH* locus) on carcass length (Andresen and Jensen, 1979) and meat colour (Imlah and Thomson, 1979). Our increasing knowledge of the proper influence of the *HAL* gene (Table 16.9) provides new evidence that these apparent effects result from linkage disequilibria in the corresponding chromosomal region. However, Clamp *et al.* (1992) have suggested that their finding of a significant effect of *PGD* variants on firmness score of meat could imply a chromosome 6 gene other than *HAL*.

Since the preliminary investigation of Capy *et al.* (1981), a number of studies have presented some evidence for associations of the swine leucocyte antigen (*SLA*) complex with carcass traits. The *SLA* class I polymorphisms involved in these studies relied either on serological typing (Renard *et al.*, 1988; Vaiman *et al.*, 1988) or on restriction fragment length polymorphism (RFLP) analyses (Jung *et al.*, 1989; Rothschild *et al.*, 1990). Significant associations between certain *SLA* haplotypes and enzyme activities (malic enzyme in muscle and acetyl coenzyme A carboxylase in fat) were found by Renard *et al.* (1992, 1996). The *SLA* complex is located on chromosome 7 (Geffrotin *et al.*, 1984), and results reported by Rothschild *et al.* (1995) show that *TNFα* (a gene located within the *SLA* region) and several chromosome 7 microsatellite markers are associated with carcass and meat quality traits. Some QTLs for these traits would therefore exist in or near the pig major histocompatibility complex. The *SLA* region has also been found to have a major influence on fat androstenone levels in entire male pigs (Bidanel *et al.*, 1996).

In pigs from the F_2 generation of a cross between Large White and European Wild boar, Andersson *et al.* (1994) and Marklund *et al.* (1996) have identified chromosome 4 microsatellite markers of a major locus affecting carcass fatness and length. According to Yu *et al.* (1995), fat deposition would also be associated with RFLP of PIT1, a pituitary-specific transcription factor which is a regulatory factor of several hormones involved in the developmental processes and which maps to chromosome 13. Two RFLPs of the growth hormone gene (*GH* locus, chromosome 12) have been found to affect carcass composition traits in both Meishan × Piétrain and Wild boar × Piétrain F_2 families (Geldermann *et al.*, 1996). A significant effect of a chromosome 14 microsatellite marker (*S0058*) has been reported by Bidanel *et al.* (1996) for backfat thickness of pigs from a Meishan × Large White F_2 population. In a similar F_2 population, Janss *et al.* (1997) have found evidence for a major Meishan cooking loss gene (*MC*), with the recessive allele decreasing cooking loss and increasing ultimate pH. The locus responsible for red cell adenosine deaminase (*ADA*) might offer another genetic polymorphism of interest with respect to meat quality, as discussed by

Sellier and Monin (1994). In particular, the recessive silent allele ADA^0, producing no or barely detectable enzyme activity, would be related to lower meat quality in Danish Landrace pigs (Hyldgaard-Jensen, 1986).

Among the rare familial abnormalities for which a genetic aetiology is suspected, some of them refer to carcass tissues (e.g. asymmetric hindquarter syndrome, acute back muscle necrosis, myositis ossificans, obesity). For references on these defects, see Ollivier and Sellier (1982, 1985) and Robinson (1991).

Breed Variation

In the last four decades, innumerable comparisons between pure breeds and/or breed crosses have been achieved for meat and carcass traits. A worldwide survey of earlier breed comparison trials dealing with these traits was performed by Sutherland *et al.* (1985). Reviews specifically devoted to meat quality traits were published by Sellier (1988a,b, 1995), Schwörer *et al.* (1989) and Sellier and Monin (1994) whereas breed differences in fat quality were reviewed by Sellier (1983) and Wood and Enser (1989).

Contribution of major genes to breed differences

Part of the breed variation in meat quality and, to a lesser extent, in body composition traits can be ascribed to large breed differences in allelic frequencies for some of the major genes known in the pig.

The most illustrative example is offered by the halothane sensitivity allele for which recent estimates of frequency (q_n) in major pure breeds ranged from zero to near unity (Table 16.11). 'Stress-resistant' breeds (with q_n lower than, say, 0.10) include Chester White, Duroc, Hampshire and most of Landrace and Large White (or Yorkshire) national varieties. Poland China and certain Landrace populations are among the breeds showing intermediate values of q_n (around 0.40). 'Stress-sensitive' breeds are essentially the Belgian Landrace breed (q_n = 0.60–0.90) and the Piétrain breed (q_n greater than 0.90). As reported by Guéblez *et al.* (1995) and Hanset *et al.* (1995), the difference between the Piétrain and the Large White for q_n would contribute to the difference between these breeds for a proportion of 30–70% in carcass traits (killing out percentage, carcass length, lean percentage) and for a still higher proportion in meat quality traits strongly influenced by halothane genotype (e.g. pH_1).

Likewise, more and more convincing evidence supports the earlier hypothesis put forward by Sellier (1987) on the central role played by the 'acid meat' RN^- allele as responsible for the 'Hampshire effect' (Monin and Sellier, 1985) on several components of meat quality. A high proportion of RN^- carriers, 90 % or so, following Enfält *et al.* (1994), are found in Hampshire purebreds whereas the proportion of RN^- carriers is presumably zero or close to zero in other major pure breeds.

Finally, the postulated MI^+ recessive allele would be at a medium frequency in the Chinese Meishan breed (Janss *et al.*, 1997), and could also be present in other breeds (e.g. Duroc) exhibiting high levels of muscle lipids.

Body composition traits

The general pattern of differences between major pig breeds is summarized in Table 16.12 for body composition. As a general rule, Large White (Yorkshire) and 'conventional' Landrace populations (i.e. with a low frequency of the

Table 16.11. Estimates of halothane sensitivity gene frequency in pig breeds.

	Reference (country)[1,2]					
Breed	*a* (Belgium)	*b,c* (Canada)	*d* (France)	*e,f* (Germany)	*g* (UK)	*h* (USA)
Belgian Landrace	0.56			0.88		
Berkshire						0.13
Chester White						0
Duroc		0.03				0.03
Hampshire						0.04
Landrace		0.13	0	0.44	0.33	0.06
Large White or Yorkshire		0.10	0.03		0.11	0.02
Piétrain			0.97	0.91		
Poland China						0.42

[1]*a* = de Smet *et al.* (1995); *b* = Hubbard *et al.* (1990); *c* = Houde *et al.* (1993); *d* = Amigues *et al.* (1994); *e* = Reinecke and Kalm (1988); *f* = Wittmann *et al.* (1993); *g* = Southwood *et al.* (1988); *h* = Berger *et al.* (1994).
[2]Estimates are based on results of the DNA-based test for the ryanodine receptor gene in references *c, d, f* and *h*, and are derived from results of halothane testing in references *a, b, e* and *g*.

Table 16.12. A summary of differences between major pig breeds in body composition traits.

	Trait				
Breed	Lean percentage	Backfat thickness	Loin muscle area	Killing out percentage	Carcass length
Large White[1]	=	=	=	=	=
Landrace	=	=	=	–/=	+
Duroc	–/+	–/+	–/+	–/=	–
Hampshire	+	–	+	=/+	–/=
Belgian Landrace	++	–	++	+	=
Piétrain	+++	– –	+++	++	––
Meishan	–––––	+++++	–––––	––	––

[1]Large White (or Yorkshire): taken as a 'control' breed.

halothane sensitivity allele) are close to each other in carcass traits except for carcass length which is greater by around 1.5 cm in the Landrace (e.g. Johansson *et al.*, 1987; Sather *et al.*, 1991a; Hammell *et al.*, 1993; Berger *et al.*, 1994). There is also a general trend towards a higher killing out percentage in Large Whites or Yorkshires than in Landraces (e.g. Johansson *et al.*, 1987; Smith *et al.*, 1990; Sather *et al.*, 1991a; Hammell *et al.*, 1993).

Duroc purebreds or crossbreds give rather variable results when compared with Large White or Landrace purebreds or crossbreds for carcass lean percentage and backfat thickness. A higher lean to fat ratio of Duroc is found in certain studies (e.g. Barton-Gade, 1987; Bout *et al.*, 1990; Smith *et al.*, 1990; Hammell *et al.*, 1993) whereas other studies show opposite results (e.g. Langlois and Minvielle, 1989; Cameron, 1990a; Hovenier *et al.*, 1992; Berger *et al.*, 1994). Rather consistently, Durocs give slightly shorter and lighter carcasses than Large Whites.

Results reported by Barton-Gade (1987), Johansson *et al.* (1987), Gunsett and Robison (1989), Smith *et al.* (1990), Hammell *et al.* (1993), Berger *et al.* (1994) and NPPC (1995) indicate that carcasses from Hampshire purebreds or crossbreds when compared with those from Large White (Yorkshire) purebreds or crossbreds are leaner while being slightly shorter and heavier in most cases. This favourable ranking of the Hampshire breed in carcass lean to fat ratio is perhaps partly due to the high frequency of the RN^- allele in that breed. The RN^- allele has indeed been shown to be associated with slightly better carcass composition (Le Roy *et al.*, 1996b).

The Piétrain breed and to a lesser extent the Belgian Landrace breed exhibit the muscular hypertrophy (or double-muscling) phenomenon (Ollivier, 1980) which results in markedly leaner, shorter and heavier carcasses. According to Guéblez *et al.* (1993a), the advantage of Piétrain over Large White or Landrace purebreds amounts to 9% in carcass lean content and 2.5% in killing out percentage. As stated above, this advantage mainly, but not completely, originates from the high frequency of the halothane sensitivity allele in the Piétrain breed.

Several comparisons of highly prolific Chinese breeds (Meishan, Jiaxing, Fengjing) with European or North-American breeds have been carried out for carcass composition traits in the last 15 years (Legault *et al.*, 1985; Poilvet *et al.*, 1990; Haley *et al.*, 1992; Serra *et al.*, 1992; Young, 1992; Bidanel *et al.*, 1993; Lan *et al.*, 1993). These Chinese breeds showed very poor carcass performance. The considerable range of between-breed variation in carcass lean to fat ratio is illustrated by Table 16.13 reporting results of a comparison between the 'extremely fat' Meishan and the 'extremely lean' Piétrain (Bidanel *et al.*, 1991).

Meat quality traits

The average differences between five major porcine breeds are reported in Table 16.14 for pH_1, pH_u and chemical composition of muscle. Most studies agree that Large White (or Yorkshire) and 'conventional' Landrace populations are very close to each other for these traits and, more generally, for technological and eating qualities of pork (see Malmfors and Nilsson, 1979; Touraille and Monin, 1984; Sellier *et al.*, 1985; Sather *et al.*, 1991a; Guéblez *et al.*, 1993a; Berger *et al.*, 1994; Russo and Nanni Costa, 1995). These two breeds are therefore put together as a 'control' in Table 16.14.

As seen above, pH is among the prominent sources of variation in pork quality, postmortem changes in muscle pH being greatly influenced by two major genes (*HAL* and *RN* for the rate and the extent of pH fall, respectively); the most important breed differences in meat quality are 'governed' by these major genes.

Table 16.13. Carcass performance of purebred Meishan and Piétrain female pigs slaughtered at 97 kg liveweight.

Trait	Meishan	Piétrain
Killing out percentage	72	78
Carcass length (cm)	94	93
Backfat thickness at rump (mm)	33	8
Proportion of ham + loin (%)[1]	42	63
Proportion of backfat + leaf fat (%)[1]	21	7
Carcass lean content (%)[2]	34	61
Carcass fat content (%)[2]	43	15

Source: From Bidanel *et al.*, 1991.
[1]Expressed as percentage of the half-carcass.
[2]Estimated from the proportions of loin, ham, belly, backfat and leaf fat.

Table 16.14. Average differences between major pig breeds in meat pH and muscle compositional traits (water, protein, lipid, glycogen).

Trait	Mean value of Large White–Landrace 'control' breeds	Deviation from the 'control' (range of estimates in brackets)		
		Duroc	Hampshire	Piétrain
pH_1 (LD)	6.3 (6.1–6.4)	0.05 (–0.06–0.16)	0.01 (–0.03–0.05)	–0.63 (–0.83––0.44)
pH_u (LD)	5.5 (5.4–5.7)	0.05 (0.01–0.09)	–0.08 (–0.16–0.02)	–0.02 (–0.10–0.04)
pH_u (ham)	5.7 (5.5–5.9)	–0.01 (–0.09–0.04)	–0.16 (–0.28––0.07)	–0.03 (–0.07––0.01)
% water	74.5 (73.1–75.6)	–0.9 (–1.2––0.7)	0.0 (–0.1–0.1)	–0.3 (–0.8–0.5)
% protein	23.0 (22.2–23.9)	–0.5 (—)	–1.5 (–2.1––0.9)	–0.1 (–0.5–0.6)
% lipid	1.5 (1.1–2.8)	1.4 (0.2–2.2)	0.2 (0–0.4)	0.0 (–0.3–0.4)
% glycogen[1]	1.2 (—)	0.1 (—)	0.8 (—)	0.0 (–0.2–0.1)
References[2]	1–21	4–6,9–11,13,15, 16,18–20	1–6,8,9,12,18,20,21	2,3,7,12,14,17–19

[1]Estimated from muscle glycolytic potential values (see Sellier and Monin, 1994).
[2]1 = Monin and Sellier (1985); 2 = Fjelkner-Modig and Persson (1986); 3 = Monin *et al.* (1986); 4 = Schwörer *et al.* (1987); 5 = Smith and Pearson (1987); 6 = Barton-Gade (1988); 7 = Bout *et al.* (1989b); 8 = Lundström *et al.* (1989); 9 = Barton-Gade (1990); 10 = Bout *et al.* (1990); 11 = Cameron *et al.* (1990); 12 = Krieter *et al.* (1990); 13 = Maassen-Francke *et al.* (1991); 14 = Hartmann *et al.* (1992); 15 = Hovenier *et al.* (1992); 16 = Lo *et al.* (1992a); 17 = Guéblez *et al.* (1993a); 18 = Kallweit *et al.* (1993); 19 = Oliver *et al.* (1993); 20 = Berger *et al.* (1994); 21 = Schwörer *et al.* (1994).

The very high frequency of the halothane sensitivity allele (HAL^n) in the Piétrain is the primary cause of the frequently occurring PSE meat condition in heavily-muscled Piétrain purebreds. This results in well-established deficiencies of the Piétrain for many components of meat quality such as drip loss, two-toning aspect and technological yield of cured–cooked ham processing (Sellier *et al.*, 1988; Guéblez *et al.*, 1993a), seasoning loss of dry-cured hams (Russo and Nanni Costa, 1995) and eating quality of fresh meat, especially tenderness (Dumont, 1974; Touraille and Monin, 1982). Although the Belgian Landrace breed shows an incidence of halothane sensitivity similar to that observed in the Piétrain breed, the adverse effects of halothane sensitivity on technological yields of cured–cooked or dry-cured ham processing are partly counterbalanced by the relatively limited extent of pH fall in Belgian Landraces which are intermediate between Large Whites and Piétrains in those traits (Sellier *et al.*, 1988; Russo and Nanni Costa, 1995).

The 'acid meat' RN^- allele is very likely to be at the origin of most features of the particular pattern usually shown by Hampshire purebred pigs, i.e. normal pH_1 but higher glycogen content, lower ultimate pH and protein content and higher cooking loss (Sayre *et al.*, 1963; Monin and Sellier, 1985; Monin *et al.*, 1986; Fjelkner-Modig and Persson, 1986). However, this 'Hampshire effect' on meat quality is of variable magnitude in different studies and/or countries, which could be ascribed to varying frequencies of RN^- carriers in the involved Hampshire samples and/or populations. Whatever the magnitude of the effects, results found on Hampshire pigs closely resemble the above-mentioned results found when comparing RN^- carriers with non-carriers (Table 16.10). The respective roles played by the increased glycogen content and the decreased protein to water ratio in muscle from Hampshires are discussed by Fernandez *et al.* (1991) and Sellier and Monin (1994) in order to explain the extra release of water during cooking of meat from these pigs. Regarding eating quality, meat from Hampshires was judged to be more juicy and more tender than that from Swedish Yorkshires in the studies of Fjelkner-Modig and Persson (1986) and Fjelkner-Modig and Tornberg (1986). Similarly, Berger *et al.* (1994) reported that Hampshires have a substantial advantage over Yorkshires and Landraces for tenderness score while showing lower ultimate pH of meat. That meat from Hampshire pigs is favourably ranked in eating quality may be partly related to its slightly higher lipid content (Table 16.14). However, other mechanisms are likely to be implied, e.g. water distribution in raw and fried meat (Fjelkner-Modig and Tornberg, 1986) and higher percentage of 'red-type' fibres and oxidative capacity of muscle (Essén-Gustavsson and Fjelkner-Modig, 1985; Monin *et al.*, 1986; Ruusunen and Puolanne 1997; Feddern *et al.*, 1995).

The Duroc breed is, on the whole, close to the Large White and Landrace breeds in pH_1, pH_u and technological quality of meat. The outstanding specificity of Duroc pigs lies in the approximately twofold increase in intramuscular fat content. This increase in lipid content is accompanied by higher concentrations of saturated and monounsaturated fatty acids and lower concentrations of polyunsaturated fatty acids in muscle lipids (Honkavaara, 1989; Bout *et al.*, 1990;

Cameron and Enser, 1991). Because of its higher level of marbling, the Duroc breed has often been claimed to improve eating quality of fresh pork and processed products such as dry-cured hams (e.g. Bejerholm and Barton-Gade, 1986; Oliver *et al.*, 1994). However, this recommendation is not substantiated by the results from certain breed comparisons dealing with the Duroc. In the study of Cameron *et al.* (1990), meat from Duroc pigs was scored by taste and consumer panels as being more juicy but less tasty, less tender and less acceptable than that from halothane-negative British Landrace pigs. In the large-scale comparison reported by Berger *et al.* (1994), sensory scores for juiciness, tenderness, chewiness and flavour of cooked meat did not differ between Duroc, Landrace and Yorkshire pigs. Excess of marbling of meat from Duroc purebred or Duroc-sired pigs could also be frequently unacceptable regarding the consumer's willingness to buy fresh meat (Fernandez *et al.*, 1996) and for processing of cooked hams (Barton-Gade, 1988) or dry-cured hams (Gou *et al.*, 1995; Russo and Nanni Costa, 1995).

The results obtained so far on the Chinese breeds newly introduced into Europe and North America indicate that the Meishan is similar or slightly superior to the Large White in technological quality of meat (Legault *et al.*, 1985; Serra *et al.*, 1992; Bidanel *et al.*, 1993; Lan *et al.*, 1993). The Meishan shows high levels of intramuscular fat while being probably slightly below the Duroc in that respect (Poilvet *et al.*, 1990; Bidanel *et al.*, 1991; Lan *et al.*, 1993). Regarding eating quality, conclusions are rather unclear. The fresh meat from Chinese purebred or crossbred pigs was judged by sensory panels as more tender, more juicy and more tasty than that from European controls in certain studies (e.g. Touraille *et al.*, 1989) but no benefit from using Chinese breeds was found by Lan *et al.* (1993) and Ellis *et al.* (1995). In a consumer survey, Touraille *et al.* (1989) found no difference in overall acceptability of fresh pork from Large White and Large White × Chinese pigs. Meat of the latter obtained significantly better scores for 'in the mouth' quality, but this advantage was counterbalanced by an unfavourable score for meat appearance, the amount of visible fat being judged as excessive.

Fat quality traits

The between-breed variation in backfat quality closely matches that in backfat thickness, in the same way as the within-breed genetic correlation between the two traits is strongly positive (Table 16.7). That breeds with thicker backfat generally exhibit better fat quality is illustrated by the comparison of 11 breeds performed by Warriss *et al.* (1990) and dealing with scores for subcutaneous fat firmness and the degree of separation of the fat layer from the underlying lean tissue. Regarding chemical characteristics of fatty depots (water and lipid contents, fatty acid profile, iodine value), breeds of similar carcass fatness (e.g. Large White vs. Landrace, or Piétrain vs. Belgian Landrace) are generally close to each other (Pascal *et al.*, 1975; Bonneau *et al.*, 1979; Bout *et al.*, 1989a; Guéblez *et al.*, 1993b). The lean Piétrain breed exhibits subcutaneous dorsal adipose tissue containing more water and fewer lipids and presenting a higher ratio of polyunsaturated over saturated fatty acids (Wood, 1973; Bout *et al.*, 1989a; Warriss *et al.*, 1990; Hartmann *et al.*, 1992). The magnitude of these differences between Piétrains and Large Whites or Landraces depends on the energy concentration of the diet (Pascal *et al.*, 1975) and is larger in males than in castrated males (Guéblez *et al.*, 1993b). Higher water to lipid ratio and degree of unsaturation of fat have been found in Hampshire purebreds or crossbreds by Lea *et al.* (1970), Villegas *et al.* (1973) and Barton-Gade (1987), which is probably related to the relatively low fatness of the Hampshire. Results concerning the fat quality of Duroc pigs are fairly variable, but it can be noted that even in the studies where it showed similar or higher backfat thickness than the Large White, the Duroc had softer and more unsaturated fat (Barton-Gade, 1987; Bout *et al.*, 1990; Cameron *et al.*, 1990; Edwards *et al.*, 1992). There could be, in the case of Duroc, a proper breed effect on fat quality, independently of overall carcass fatness. The Chinese Meishan tends to have greater fat firmness, less fat separation and lower iodine value than the Large White (Gandemer *et al.*, 1992; Serra *et al.*, 1992), but the breed difference in fat quality remains smaller than could be expected from the considerable breed difference in carcass fat deposition.

Regarding breed differences in fatty tissue levels of the 'boar taint' compounds, available results are fairly variable. Fat skatole content is little affected by breed effects (Patterson *et al.*, 1990; Bonneau *et al.*, 1992; Squires *et al.*, 1992; Xue *et al.*, 1996). The influence of breed on fat androstenone content of intact males appears to be more important with higher levels in Meishans, Durocs and possibly Hampshires and Piétrains than in Landraces and Yorkshires (Prunier *et al.*, 1987; Willeke, 1993; Xue *et al.*, 1996). The 'Meishan effect' is likely to be related to the much earlier sexual maturity of this breed, but this explanation does not hold for the Duroc according to Xue *et al.* (1996).

Heterosis and Maternal Effects in Breed Crosses

Crossbreeding is today extensively used in the pig industry, and knowing the 'behaviour' of the above-mentioned breed effects in crossing is required for implementing optimal crossbreeding strategies. It is therefore needed for all breeds and traits of interest to know whether: (i) breed effects are additively inherited (i.e. no heterosis) or not when crossing breeds; and (ii) the difference between estimates of breed of sire and breed of dam effects for any breed is zero (i.e. no maternal effect, assuming that sex-linked inheritance or genomic imprinting are of minor importance) or not.

Body composition traits

Regarding direct (or individual) heterosis effect (defined as the deviation from the mean of the reciprocal F_1 crosses from the mean of the two parent breeds), reported experimental estimates have in general been small and not significantly different from zero for body composition traits, as reviewed by Sellier (1976), Johnson (1981) and Gunsett and Robison (1989). The general lack of direct heterosis in carcass traits is exemplified by the results referring to F_1 crosses between two breeds largely divergent in that respect, the Large White (or Yorkshire) and the Chinese Meishan (Poilvet *et al.*, 1990; Serra *et al.*, 1992; Lan *et al.*, 1993). No maternal heterosis effect on carcass traits was found by Bidanel *et al.* (1993) for the same breed combination and by Schneider *et al.* (1982) for other breed combinations. However, it may be noted that there exists a trend towards a slightly positive (i.e. unfavourable) direct heterosis effect on backfat thickness. This trend has reached statistical significance in certain crossbreeding experiments conducted with *ad libitum* fed pigs (McLaren *et al.*, 1987; Lo *et al.*, 1992a). This is probably associated with the positive direct heterosis in average daily gain (more than 10% of the parental mean in the most 'heterotic' crosses) and daily feed intake.

In a complete diallel cross involving four breeds (Duroc, Landrace, Spotted and Yorkshire), significant differences between reciprocal F_1 crosses were found in the crosses involving the Duroc for backfat thickness, loin muscle area and weight of lean cuts (McLaren *et al.*, 1987). These results showed a substantial advantage of using the Duroc as the sire breed in such crosses, particularly in the Duroc by Yorkshire or Landrace crosses in accordance with findings of other authors (Table 16.15). The mechanism by which such maternal breed effects operate remains unclear. Prenatal influences on embryos or fetuses might result in developmental differences which carried over up to the market weight (Wilken *et al.*, 1992). This negative 'Duroc maternal effect' on carcass lean to fat ratio could contribute to the fact that Duroc-sired pigs when compared with Large White or Landrace-sired pigs generally show better carcass merit than could be expected from the differences between pure breeds (McGloughlin *et al.*, 1988; Smith *et al.*, 1988, 1990; Hammell *et al.*, 1993; Blasco *et al.*, 1994; NPPC, 1995).

Table 16.15. Estimates of reciprocal cross differences in Duroc by Yorkshire or Landrace crosses for carcass composition traits.[1]

	Average difference[3] between F_1 from Duroc dams and:	
Trait[2]	F_1 from Yorkshire dams	F_1 from Landrace dams
Backfat thickness (mm)	4.0 (2.2–7.4)	3.0 (2.1–3.9)
Loin muscle area (cm^2)	–3.5 (–7.0––1.8)	–4.6 (–5.9––3.4)

[1]Data from Bereskin *et al.* (1971), Young *et al.* (1976), Bereskin and Steele (1986), McLaren *et al.* (1987) and Lo *et al.* (1992a).
[2]Residual standard deviations were about 3.0 mm for carcass backfat thickness and 2.5 cm^2 for loin muscle area.
[3]Range of estimates in brackets.

Meat quality traits

The need for reconsidering the classical view, i.e. additive inheritance of breed effects on meat quality in crossing, was put forward by Sellier (1987). That direct and maternal heterosis effects on these traits are negligible is supported by experimental results for most breed combinations and/or most meat and eating quality traits (e.g. Young *et al.*, 1976; Schneider *et al.*, 1982; McLaren *et al.*, 1987; Lo *et al.*, 1992a; Serra *et al.*, 1992; Bidanel *et al.*, 1993; Lan *et al.*, 1993; Ellis *et al.*, 1995). However, this assumption does not hold for certain traits (pH_1, pH_u and related traits) and certain crosses. Such 'specific' direct heterosis effects can be essentially ascribed to the influence of the above-mentioned *HAL* and *RN* major genes, as jointly shown by results reported by Krieter *et al.* (1990) for the Piétrain by Hampshire cross (Table 16.16).

Regarding the rate of pH fall, the crosses involving the 'stress-sensitive' Piétrain generally show favourable direct heterosis effects on pH_1, and most of the Piétrain F_1 crosses reviewed by Sellier (1987) are rather close to the 'stress-resistant' partner breed in that respect. This can be interpreted in terms of halothane genotypes and related to the partial recessiveness of the HAL^n gene for pH_1 (see Table 16.9). Favourable heterosis was also found in Piétrain crosses for other PSE indicators such as muscle protein solubility and transmission value, but results are less consistent for meat colour. Piétrain F_1 crosses have exhibited the same meat colour as the 'stress-resistant' parent breed in some studies but have been intermediate between the parent breeds in other studies. There is some evidence that the paler meat colour of the Piétrain results not only from the detrimental HAL^n allele effect but also from a presumably additive 'Piétrain breed effect' which is independent of the *HAL* locus. Meat from halothane-positive (HP) Piétrains has indeed higher reflectance values than that from HP Belgian or French Landraces (Sellier *et al.*, 1984, 1988). The effect of varying proportions of Piétrain blood (0, 0.25, 0.5 and 1) on eating quality scores given to cooked roasts by a sensory panel was studied by Pellois and Runavot (1991).

Table 16.16. Heterosis effects on meat quality traits in the Hampshire (Ha) by Pietrain (Pi) cross.[1]

	Breed group means			Parameters[2]	
Trait	Pi	Ha	Pi × Ha	a	H
pH_1 LD	5.70	6.19	6.03	1.2	0.4
LF_1 LD [3]	6.45	3.73	3.79	−0.6	−0.5
pH_1 LD	5.53	5.45	5.44	−0.4	−0.4
Glycolytic potential LD (μmol g^{-1} fresh tissue)	163	209	221	0.9	1.4

[1]Data from Krieter *et al.* (1990).
[2]Expressed in standard deviation units of the trait: 2a = Hampshire–Piétrain; H = direct heterosis effect.
[3]LF_1 = electrical conductivity at 45 min postmortem (the threshold value above which meat is usually considered as being PSE is 4.5–5.0).

Results supported the expectation of a favourable heterosis effect on meat tenderness in Piétrain crosses, in accordance with the partial recessiveness of the main genetic factor (*HAL*^{n}) responsible for the increased toughness of meat from Piétrain purebreds (see Table 16.9).

Regarding the extent of pH fall, it has been found that ultimate pH is not affected by heterosis except in Hampshire crosses. The 'acid meat' condition usually exhibited by the Hampshire appears to be inherited as a more or less completely dominant trait in the crosses involving that breed. Combining the results of a series of French experiments dealing with Hampshire purebreds or crossbreds, Sellier (1987) concluded that the difference in ultimate pH between Hampshire-sired and Piétrain-sired pigs (–0.17 to –0.24 depending on the ham muscle and the experiment) is of the same magnitude as that found between Hampshire and Piétrain purebreds (–0.16 to –0.25). Paler colour of meat is associated with lower ultimate pH in the Hampshire, and Hampshire crosses follow a particular pattern for colour score, as pointed out by Johnson (1981). In the studies of Young *et al.* (1976) and Schneider *et al.* (1982), four out of five estimates of direct heterosis effect on colour score were significantly negative (i.e. unfavourable) for Hampshire crosses whereas none of the heterosis estimates concerning various crosses between Duroc, Yorkshire, Chester White, Landrace and Spotted breeds differed from zero for the same trait, though being also negative in most cases (Young *et al.*, 1976; Schneider *et al.*, 1982; McLaren *et al.*, 1987). On average, Hampshire-sired pigs are therefore fairly close to Hampshire purebreds in ultimate pH and meat colour, suggesting that the 'Hampshire effect' on meat quality is partially to completely dominant. This statement is also supported by reports evaluating the Hampshire as a sire breed (Sellier and Jacquet, 1973; Barton-Gade, 1987; Purchas *et al.*, 1990; Hammell *et al.*, 1993; NPPC, 1995), and it can be interpreted as primarily resulting from the high frequency of the dominant *RN*$^{-}$ allele in the Hampshire.

Estimates of direct heterosis effect on marbling score are close to zero according to Young *et al.* (1976), Schneider *et al.* (1982) and McLaren *et al.* (1987). However, significant negative heterosis in this trait was reported in females, but not in males, by Serra *et al.* (1992) in the Large White by Meishan cross, which is to be related with the recessiveness of the high intramuscular fat allele (*MI*$^{+}$) present at a medium frequency in the Meishan (Janss, *et al.*, 1997). Similar intramuscular fat levels were found by Poilvet *et al.* (1990) for Large White and Meishan × Large White pigs, which were markedly below Meishan purebreds in that respect. In addition, Wood *et al.* (1987) and Lo *et al.* (1992a) found significant direct heterosis effects (–20% and –13% of the parental mean, respectively) on muscle lipid content in the Duroc by Landrace cross.

Fat quality traits

As far as fat quality is closely related to the 'non-heterotic' carcass fatness, it may be anticipated that fat quality traits are essentially unaffected by heterosis. However, experimental information on this topic is still scarce, and available data come from studies of rather small scale. There is no evidence for noticeable heterosis in fatty acid composition, iodine value or firmness of subcutaneous adipose tissue in Yorkshire by Duroc or Hampshire crosses (Kellogg *et al.*, 1977) and in the Meishan by Large White cross (Gandemer *et al.*, 1992; Serra *et al.*, 1992). Reported fat quality differences between crossbred pigs from various sire breeds (Duroc, Hampshire, Large White, Piétrain) are in fairly good agreement with the expectations based on halving the above-mentioned differences between pure breeds (Barton-Gade, 1987; Edwards *et al.*, 1992; Affentranger *et al.*, 1996; Ellis *et al.*, 1996).

Prospects for Breeding on Carcass and Meat Traits

Increasing carcass lean to fat ratio has been for 30–40 years a major goal in pig breeding. Because of the high heritability of this trait, substantial genetic gains have been achieved in numerous pig populations. Reported annual rates of genetic change in 'national' populations range from −0.1 to −0.5 mm in backfat thickness, from 0.2 to 0.5 cm^2 in loin muscle area and from 0.3 to 0.6 percentage points in carcass lean content, as reviewed by Sellier and Rothschild (1991).

Most pig breeders today concur in the opinion that the objective of removing backfat has been largely reached and they should pause in their efforts to reduce fat levels even further. The weighting of carcass leanness in the overall breeding goal is already diminishing in several European countries. Moreover, selection on the sole subcutaneous backfat depth could induce some correlated genetic trends in the cranio-caudal and/or dorso-ventral distribution of carcass fat (Godfrey *et al.*, 1991; Cameron and Curran, 1995). New techniques (e.g. Akridge *et al.*, 1992) and more sophisticated ultrasonic scanning devices using computerized image analysis (e.g. Liu and Stouffer, 1995) are now appearing and offer means for assessing the whole carcass value in a 'deeper' way. It remains to evaluate the economic balance between extra accuracy of carcass assessment and greater cost of measurement, as pointed out by Webb (1995).

The inclusion of meat quality traits among breeding goals dates back to the years 1970–1980. Several different approaches have been followed in various countries for genetic improvement of meat quality, as surveyed by Kallweit (1985), Ianssen and Sehested (1989), Lundström *et al.* (1989), Monin (1991), Cameron (1993), Hovenier *et al.* (1993b), Le Roy and Sellier (1994) and Schwörer *et al.* (1994). These breeding strategies have recourse to both classical (polygenic) selection methods and utilization of major genes. Polygenic selection, based on different forms of the selection index theory (including restricted or desired-gains selection indices) and nowadays on the best linear unbiased predictions (BLUP) methodology, usually dealt with a limited number of meat quality traits. Traits of interest (colour, pH, intramuscular fat) were given varying emphasis depending on the country, but meat colour (assessed either subjectively or instrumentally) has been the trait mainly involved in past breeding for meat quality. Regarding major genes, much breeding work has been done since the late 1970s to eradicate the halothane sensitivity allele in numerous pig populations (especially specialized dam lines) by successively using halothane challenge test, blood typing and, today, DNA-based test for the ryanodine receptor gene. Efforts have been made since the early 1990s to eliminate the *RN*⁻ allele in certain pig lines by use of *in vivo* assessment of muscle glycolytic potential. The frequency distributions of this trait for *RN*⁻ carriers and non-carriers are essentially non-overlapping.

For the foreseeable future, advances are expected from techniques which are being developed for assessing meat quality on the live animal using measurements performed on small biopsy samples of muscle or fat (e.g. Lengerken *et al.*, 1993; Le Roy *et al.*, 1994b; Cheah *et al.*, 1995; Larzul, 1997). Sensory quality of meat will be paid greater attention, and research is needed for appraising the predictive value of traits such as muscle fibre types and/or areas and for better understanding of the variable influence of intramuscular fat content (e.g. 'within the fibres' vs. 'between the fibres' deposition and fatty acid composition of muscle lipids). Evaluation of available genetic resources (including exotic or local unimproved breeds) could also participate in the search for 'high-quality' pork products, particularly in extensive management situations. More generally, as most meat quality traits can be defined by optimal ranges, the most appropriate approach could consist of selecting for the intermediate optimum of each involved trait (e.g. Hovenier *et al.*, 1993a). In addition to changing the mean levels, one could also try to reduce the phenotypic variability of meat quality traits even though the expected impact of reducing genetic variances is fairly limited in this area (de Vries and van der Wal, 1993).

Apart from the well-known *HAL* and *RN* genes, there might be other single genes with significant effects on pork quality. Large-scale research programmes are currently in progress for detecting molecular markers of quantitative trait loci (QTL) and should permit identification of new genes of interest through closely linked DNA-based markers. The use of blood typing for predicting an animal's halothane status was the first example of marker-assisted selection in pigs, and other applications resulting from our rapidly increasing knowledge of the porcine genome are to be expected in the near future (e.g. Sellier, 1994; Visscher and Haley, 1995).

Acknowledgements

The author would like to thank G. Monin, L. Ollivier and the reviewers for their comments as well as Marie-Laure Le Paih for her assistance in preparing the manuscript.

References

Aalhus, J.L., Jones, S.D.M., Robertson, W.M., Tong, A.K.W. and Sather, A.P. (1991) Growth characteristics and carcass composition of pigs with known genotypes for

stress susceptibility over a weight range of 70 to 120 kg. *Animal Production* 52, 347–353.
Affentranger, P., Gerwig, C., Seewer, G.J.F., Schwörer, D. and Künzi, N. (1996) Growth and carcass characteristics as well as meat and fat quality of three types of pigs under different feeding regimens. *Livestock Production Science* 45, 187–196.
Akridge, J.T., Brorsen, B.W., Whipker, L.D., Forrest, J.C., Kuei, C.H. and Schinckel, A.P. (1992) Evaluation of alternative techniques to determine pork carcass value. *Journal of Animal Science* 70, 18–28.
Amigues, Y., Runavot, J.P. and Sellier, P. (1994) Evaluation de la fréquence du gène de la sensibilité à l'halothane dans les principales races porcines françaises en 1993. *Techni-porc* 17(3), 23–28.
Andersson, L., Haley, C.S., Ellegren, H., Knott, S.A., Johansson, M., Andersson, K., Andersson-Eklund, L., Edfors-Lijla, I., Fredholm, M., Hansson, I., Hakansson, J. and Lundström, K. (1994) Genetic mapping of quantitative trait loci for growth and fatness in pigs. *Science* 263, 1771–1774.
Andresen, E. (1987) Selection against PSS by means of blood typing. In: Tarrant, P.V., Eikelenboom, G. and Monin, G. (eds.) *Evaluation and Control of Meat Quality in Pigs.* Martinus Nijhoff Publishers, Dordrecht, The Netherlands, pp. 317–328.
Andresen, E. and Jensen, P. (1977) Close linkage established between the HAL locus for halothane sensitivity and the PHI (phosphohexose isomerase) locus in pigs of the Danish Landrace breed. *Nordisk Veterinaer Medicine* 29, 502–504.
Andresen, E. and Jensen, P. (1979) Functional additive effect of H blood group alleles on carcass length in Danish Landrace pigs. *Nordisk Veterinaer Medicine* 31, 229–232.
Barton-Gade, P.A. (1987) Meat and fat quality in boars, castrates and gilts. *Livestock Production Science* 16, 187–196.
Barton-Gade, P.A. (1988) The effect of breed on meat quality characteristics in pigs. In: *Proceedings of the 34th International Congress of Meat Science and Technology*, 29 August–2 September 1988, Brisbane, Australia, part B, pp. 568–570.
Barton-Gade, P.A. (1990) Danish experience in meat quality measurements. In: Hill, W.G., Thompson, R. and Woolliams, J.A. (eds.) *Proceedings of the 4th World Congress on Genetics Applied to Livestock Production*, 23–27 July 1990, Edinburgh, UK, vol. 15, pp. 511–520.
Bejerholm, C. and Barton-Gade, P. (1986) Effect of intramuscular fat level on eating quality of pig meat. In: *Proceedings of the 32nd European Meeting of Meat Research Workers*, 24–29 August 1986, Ghent, Belgium, part II, pp. 389–391.
Bendall, J.R. and Swatland, H.J. (1988) A review of the relationships of pH with physical aspects of pork quality. *Meat Science* 24, 85–126.
Bereskin, B. and Steele, N.C. (1986) Performance of Duroc and Yorkshire boars and gilts and reciprocal breed crosses. *Journal of Animal Science* 62, 918–926.
Bereskin, B., Shelby, C.E. and Hazel, L.N. (1971) Carcass traits of purebred Durocs and Yorkshires and their crosses. *Journal of Animal Science* 32, 413–419.
Berger P.J., Christian, L.L., Louis, C.F. and Mickelson, J.R. (1994) Estimation of genetic parameters for growth, muscle quality, and nutritional content of meat products for centrally tested purebred market pigs. In: *Research Investment Report 1994.* National Pork Producers Council, Des Moines, Iowa, USA, pp. 51–63.
Bidanel, J.P. and Ducos, A. (1995) Variabilité et évolution génétique des caractères mesurés dans les stations publiques de contrôle de performances chez les porcs de race Piétrain. *Journées de la Recherche Porcine en France* 27, 149–154.

Bidanel, J.P. and Ducos, A. (1996) Genetic correlations between test station and on-farm performance traits in Large White and French Landrace pig breeds. *Livestock Production Science* 45, 55–62.

Bidanel, J.P., Bonneau, M., Pointillart, A., Gruand J., Mourot, J. and Demade, I. (1991) Effects of exogenous porcine somatotropin (pST) administration on growth performance, carcass traits, and pork meat quality of Meishan, Piétrain, and crossbred gilts. *Journal of Animal Science* 69, 3511–3522.

Bidanel, J.P., Caritez, J.C., Gruand, J. and Legault, C. (1993) Growth, carcass and meat quality performance of crossbred pigs with graded proportions of Meishan genes. *Genetics, Selection, Evolution* 25, 83–99.

Bidanel, J.P., Ducos, A., Guéblez, R. and Labroue, F. (1994a) Genetic parameters of backfat thickness, age at 100 kg and ultimate pH in on-farm tested French Landrace and Large White pigs. *Livestock Production Science* 40, 291–301.

Bidanel, J.P., Ducos, A., Labroue, F., Guéblez, R. and Gasnier, C. (1994b) Genetic parameters of backfat thickness, age at 100 kg and meat quality traits in Piétrain pigs. *Annales de Zootechnie* 43, 141–149.

Bidanel, J.P., Milan, D., Chevalet, C., Woloszyn, N., Caritez, J.C., Gruand, J., Le Roy, P., Bonneau, M., Renard, C., Vaiman, M., Gellin, J. and Ollivier, L. (1996) Mapping quantitative trait loci in a Meishan × Large White F_2 population. *47th Annual Meeting of the European Association for Animal Production*, 26–29 August 1996, Lillehammer, Norway, paper G3.6 (5 pp.).

Blasco, A., Gou, P., Gispert, M., Estany, J., Soler, Q., Diestre, A. and Tibau, J. (1994) Comparison of five types of pig crosses. I. Growth and carcass traits. *Livestock Production Science* 40, 171–178.

Boles, J.A., Parrish, F.C., Huiatt, T.W. and Robson, R.M. (1992) Effect of porcine stress syndrome on the solubility and degradation of myofibrillar/cytoskeletal proteins. *Journal of Animal Science* 70, 454–464.

Bonneau, M., Desmoulin, B. and Dumont, B.L. (1979) Qualités organoleptiques des viandes de porcs mâles entiers ou castrés: composition des graisses et odeurs sexuelles chez les races hypermusclées. *Annales de Zootechnie* 28, 53–72.

Bonneau, M., Le Denmat, M., Vaudelet, J.C., Veloso Nunes, J.R., Mortensen, A.B. and Mortensen, H.P. (1992) Contributions of fat androstenone and skatole to boar taint: I. Sensory attributes of fat and pork meat. *Livestock Production Science* 32, 63–80.

Bout, J., Girard, J.P., Runavot, J.P. and Sellier, P. (1989a) Genetic variation in chemical composition of fat depots in pigs. *40th Annual Meeting of the European Association for Animal Production*, 27–31 August 1989, Dublin, Ireland, paper GP3.12 (7 pp.).

Bout, J., Girard, J.P., Runavot, J.P. and Sellier, P. (1989b) Genetic variation of intramuscular lipid content and composition in pigs. *40th Annual Meeting of the European Association for Animal Production*, 27–31 August 1989, Dublin, Ireland, paper GP3.15 (6 pp.).

Bout, J., Girard, J.P., Sellier, P. and Runavot, J.P. (1990) Comparaison de porcs Duroc et Large White pour la composition chimique du gras de bardière et du muscle Long dorsal. *Journées de la Recherche Porcine en France* 22, 29–34.

Briskey, E.J. (1964) Etiological status and associated studies of pale, soft, exudative porcine musculature. *Advances in Food Research* 13, 89–178.

Buchter, L. and Zeuthen, P. (1971) The effect of ageing on the organoleptic qualities of PSE and normal pork loins. In: *Proceedings of the Second International Symposium on Condition and Meat Quality in Pigs.* Pudoc, Wageningen, pp. 247–254.

Buscailhon, S., Berdagué, J.L., Gandemer, G., Touraille, C. and Monin, G. (1994) Effects of initial pH on compositional changes and sensory traits of French dry-cured hams. *Journal of Muscle Foods* 5, 257–270.

Cameron, N.D. (1990a) Comparison of Duroc and British Landrace pigs and the estimation of genetic and phenotypic parameters for growth and carcass traits. *Animal Production* 50, 141–153.

Cameron, N.D. (1990b) Genetic and phenotypic parameters for carcass traits, meat and eating quality traits in pigs. *Livestock Production Science* 26, 119–135.

Cameron, N.D. (1993) Selection for meat quality: objectives and criteria. *Pig News and Information* 14, 161N–168N.

Cameron, N.D. and Curran, M.K. (1995) Responses in carcass composition to divergent selection for components of efficient lean growth rate in pigs. *Animal Science* 61, 347–359.

Cameron, N.D. and Enser, M.B. (1991) Fatty acid composition of lipid in *Longissimus dorsi* muscle of Duroc and British Landrace pigs and its relationship with eating quality. *Meat Science* 29, 295–307.

Cameron, N.D., Warriss, P.D., Porter, S.J. and Enser, M.B. (1990) Comparison of Duroc and British Landrace pigs for meat and eating quality. *Meat Science* 27, 227–247.

Capy, P., Renard, C., Sellier, P. and Vaiman, M. (1981) Etude préliminaire des relations entre le complexe majeur d'histocompatibilité (SLA) et des caractères de production chez le porc. *Annales de Génétique et de Sélection Animale* 13, 441–446.

Casteels, M., van Oeckel, M.J., Boschaerts, L., Spincemaille, G. and Boucqué, Ch.V. (1995) The relationship between carcass, meat and eating quality of three pig genotypes. *Meat Science* 40, 253–269.

Cheah, K.S., Cheah, A.M. and Krausgrill, D.I. (1995) Variations in meat quality in live halothane heterozygotes identified by biopsy samples of *M. longissimus dorsi*. *Meat Science* 39, 293–300.

Chowdhary, B.P., Harbitz, I., Davies, W. and Gustavsson, I. (1991) Mapping of genes belonging to the halothane linkage group using *in situ* hybridization. *Genetics, Selection, Evolution* 23 (Suppl. 1), 96S–99S.

Christian, L.L. (1972) A review of the role of genetics in animal stress susceptibility and meat quality. In: Cassens, R.G., Giesler, F. and Kolb, Q. (eds.) *Proceedings of the Pork Quality Symposium*. University of Wisconsin, Madison, pp. 91–115.

Christian, L.L. (1974) Halothane test for PSS-field application. In: *Proceedings of American Association of Swine Practitioners Conference*, Des Moines, Iowa, pp. 6–13.

Christian, L.L. and Mabry, J.W. (1989) Stress susceptibility of swine. In: Young, L.D. (ed.) *Genetics of Swine*. USDA-ARS, Clay Center, Nebraska, pp. 49–57.

Clamp, P.A., Beever, J.E., Fernando, R.L., McLaren, D.G. and Schook, L.B. (1992) Detection of linkage between genetic markers and genes that affect growth and carcass traits in pigs. *Journal of Animal Science* 70, 2695–2706.

Claus, R., Weiler, U. and Herzog, A. (1994) Physiological aspects of androstenone and skatole formation in the boar. A review with experimental data. *Meat Science* 38, 289–305.

Cole, G., Le Hénaff, G. and Sellier, P. (1988) Paramètres génétiques de quelques caractères de qualité de la viande dans les races porcines Large White, Landrace Français et Landrace Belge. *Journées de la Recherche Porcine en France* 20, 249–254.

Davies, W., Harbitz, I., Fries, R., Stranzinger, G. and Hauge, J.G. (1988) Porcine malignant hyperthermia carrier detection and chromosomal assignment using a linked probe. *Animal Genetics* 19, 203–212.

De Smet, S., Pauwels, H., Vervaeke, I., Demeyer, D., de Bie, S., Eeckhout, W. and Casteels, M. (1995) Meat and carcass quality of heavy muscled Belgian slaughter pigs as influenced by halothane sensitivity and breed. *Animal Science* 61, 109–114.

DeVol, D.L., McKeith, F.K., Bechtel, P.S., Novakofski, J., Shanks, R.D. and Carr, T.R. (1988) Variation in composition and palatability traits and relationships between muscle characteristics and palatability in a random sample of pork carcasses. *Journal of Animal Science* 66, 385–395.

De Vries, A.G. and van der Wal, P.G. (1993) Breeding for pork quality. In: Puolanne, E. and Demeyer, D.I. (eds.) *Pork Quality: Genetic and Metabolic Factors.* CAB International, Wallingford, Oxon, UK, pp. 58–75.

De Vries, A.G., van der Wal, P.G., Long, T., Eikelenboom, G. and Merks, J.W.M. (1994) Genetic parameters of pork quality and production traits in Yorkshire populations. *Livestock Production Science* 40, 277–289.

Dietl, G., Groeneveld, E. and Fiedler, I. (1993) Genetic parameters of muscle structure traits in pig. *44th Annual Meeting of the European Association for Animal Production,* 16–19 August 1993, Aarhus, Denmark, paper P1.8 (5 pp.).

Ducos, A. (1994) Paramètres génétiques des caractères de production chez le porc. Mise au point bibliographique. *Techni-porc* 17(3), 35–67.

Ducos, A., Bidanel, J.P., Ducrocq, V., Boichard, D. and Groeneveld, E. (1993) Multivariate restricted maximum likelihood estimation of genetic parameters for growth, carcass and meat quality traits in French Large White and French Landrace pigs. *Genetics, Selection, Evolution* 25, 475–493.

Dumont, B.L. (1974) Propriétés sensorielles et qualités technologiques de la viande de trois races (Landrace Belge, Landrace Français et Piétrain). *Journées de la Recherche Porcine en France* 6, 233–239.

Duniec, H., Kielanowski, J. and Osinska, Z. (1961) Heritability of chemical fat content in the loin muscle of baconers. *Animal Production* 3, 195–198.

Edwards, S.A., Wood, J.D., Moncrieff, C.B. and Porter, S.J. (1992) Comparison of the Duroc and Large White as terminal sire breeds and their effect on pigmeat quality. *Animal Production* 54, 289–297.

Eikelenboom, G. and Minkema, D. (1974) Prediction of pale, soft, exudative muscle with a non-lethal test for the halothane-induced porcine malignant hyperthermia syndrome. *Tijdschrift voor Diergeneeskunde* 99, 421–426.

Ellis, M., Lympany, C., Haley, C.S., Brown, I. and Warkup, C.C. (1995) The eating quality of pork from Meishan and Large White pigs and their reciprocal crosses. *Animal Science* 60, 125–131.

Ellis, M., Webb, A.J., Avery, P.J. and Brown, I. (1996) The influence of terminal sire genotype, sex, slaughter weight, feeding regime and slaughter-house on growth performance and carcass and meat quality in pigs and on the organoleptic properties of fresh pork. *Animal Science* 62, 521–530.

Enfält, A.C., Lundström, K., Lundkvist, L., Karlsson, A. and Hansson, I. (1994) Technological meat quality and the frequency of the RN-gene in purebred Swedish Hampshire and Yorkshire pigs. In: *Proceedings of the 40th International Congress of Meat Science and Technology,* The Hague, The Netherlands, paper S.IVA.08.

Essén-Gustavsson, B. (1993) Muscle-fiber characteristics in pigs and relationships to meat-quality parameters – Review. In: Puolanne, E. and Demeyer, D.I. (eds) *Pork Quality: Genetic and Metabolic Factors.* CAB International, Wallingford, Oxon, UK, pp. 140–155.

Essén-Gustavsson, B. and Fjelkner-Modig, S. (1985) Skeletal muscle characteristics in different breeds of pigs in relation to sensory properties of meat. *Meat Science* 13, 33–47.

Essén-Gustavsson, B., Karlsson, A., Lundström, K. and Enfält A.C. (1994) Intramuscular fat and muscle fiber lipid contents in halothane-gene-free pigs fed high or low protein diets and its relation to meat quality. *Meat Science* 38, 269–277.

Estrade, M., Rock, E. and Vignon, X. (1991) Evolution *post mortem* de la structure myofibrillaire chez des porcs normaux et des porcs sensibles à l'halothane. *Journées de la Recherche Porcine en France* 23, 365–368.

Estrade, M., Vignon, X. and Monin, G. (1993a) Effect of the RN^- gene on ultrastructure and protein fractions in pig muscle. *Meat Science* 35, 313–319.

Estrade, M., Vignon, X., Rock, E. and Monin, G. (1993b) Glycogen hyperaccumulation in white muscle fibres of RN^- carrier pigs. A biochemical and ultrastructural study. *Comparative Biochemistry and Physiology* 104B, 321–326.

Estrade, M., Ayoub, S., Talmant, A. and Monin G. (1994) Enzyme activities of glycogen metabolism and mitochondrial characteristics in muscles of RN^- carrier. *Comparative Biochemistry and Physiology* 108B, 295–301.

Feddern, E., Krieter, J. and Kalm, E. (1994) Verlauf der postmortalen Glykogenolyse und Merkmale der Fleischbeschaffenheit bei Hampshire-Reinzuchttieren und verschiedenen Kreuzungskombinationen. *Archiv für Tierzucht, Dummerstorf* 37, 229–243.

Feddern, E., Wegner, J., Ender, K. and Kalm, E. (1995) Untersuchung von Muskelstrukturmerkmalen bei Hampshire-Reinzuchttieren und verschiedenen Kreuzungskombinationen. *Archiv für Tierzucht, Dummerstorf* 38, 43–56.

Fernandez, X., Lefaucheur, L., Guéblez, R. and Monin, G. (1991) Paris ham processing: technological yield as affected by residual glycogen content of muscle. *Meat Science* 29, 121–128.

Fernandez, X., Culioli, J. and Guéblez, R. (1994) Relationship between rate of *post mortem* pH fall and ageing of *Longissimus* muscle in Piétrain pigs. *Journal of Science of Food and Agriculture* 65, 215–222.

Fernandez, X., Monin, G., Talmant, A., Mourot, J., Lebret, B., Bernard, P., Gilbert, S., Sirami, J. and Malter, D. (1996) Influence de la teneur en lipides intramusculaires sur l'acceptabilité, par les consommateurs, de la viande de porc et du jambon cuit. *Journées de la Recherche Porcine en France* 28, 163–170.

Fjelkner-Modig, S. and Persson, J. (1986) Carcass properties as related to sensory properties of pork. *Journal of Animal Science* 63, 102–113.

Fjelkner-Modig, S. and Tornberg, E. (1986) Water distribution in porcine *M. Longissimus dorsi*, as related to sensory properties. *Meat Science* 17, 213–231.

Fouilloux, M.N., Le Roy, P., Gruand, J., Renard, C., Sellier, P. and Bonneau, M. (1997) Support for single major genes influencing fat androsterone level and development of bulbo-urethral glands in young boars. *Genetics, Selection, Evolution* 29, 357–366.

Fujii, J., Otsu, K., Zorzato, F., de Leon, S., Khanna, V.K., Weiler, J.E., O'Brien, P.J. and MacLennan, D.H. (1991) Identification of a mutation in porcine ryanodine receptor associated with malignant hyperthermia. *Science* 253, 448–451.

Gandemer, G., Pichou, D., Bouguennec, B., Caritez, J.C., Berge, P., Briand, E. and Legault, C. (1990) Influence du système d'élevage et du génotype sur la composition chimique et les qualités organoleptiques du muscle Long dorsal chez le porc. *Journées de la Recherche Porcine en France* 22, 101–110.

Gandemer, G., Viau, M., Caritez, J.C. and Legault, C. (1992) Lipid composition of adipose tissue and muscle in pigs with an increasing proportion of Meishan genes. *Meat Science* 32, 105–121.

Garcia-Macias, J.A., Gispert, M., Oliver, M.A., Diestre, A., Alonso, P., Munoz-Luna, A., Siggens, K. and Cuthbert-Heavens, D. (1996) The effects of cross, slaughter weight and halothane genotype on leanness and meat and fat quality in pig carcasses. *Animal Science* 63, 487–496.

Geffrotin, C., Popescu, C.P., Cribiu, E.P., Boscher, J., Renard, C., Chardon, P. and Vaiman, M. (1984) Assignment of MHC in swine to chromosome 7 by *in situ* hybridization and serological typing. *Annales de Génétique* 27, 213–219.

Geldermann, H., Müller, E., Beeckmann, P., Knorr, C., Yue, G. and Moser, G. (1996) Mapping of quantitative-trait loci by means of marker genes in F_2 generations of Wild boar, Piétrain and Meishan pigs. *Journal of Animal Breeding and Genetics* 113, 381–387.

Godfrey, N.W., Frapple, P.G., Paterson, A.M. and Payne, H.G. (1991) Differences in the composition and tissue distribution of pig carcasses due to selection and feeding level. *Animal Production* 53, 97–103.

Gou, P., Guerrero, L. and Arnau, J. (1995) Sex and crossbreed effects on the characteristics of dry-cured ham. *Meat Science* 40, 21–31.

Goutefongea, R. (1963) Les viandes exsudatives. *Annales de Zootechnie* 12, 297–357.

Guéblez, R., Sellier, P., Fernandez, X. and Runavot, J.P. (1993a) Comparaison des caractéristiques physico-chimiques et technologiques des tissus maigre et gras de trois races porcines françaises (Large White, Landrace Français et Piétrain). 1 – Caractéristiques du tissu maigre. *Journées de la Recherche Porcine en France* 25, 5–12.

Guéblez, R., Sellier, P. and Runavot, J.P. (1993b) Comparaison des caractéristiques physico-chimiques et technologiques des tissus maigre et gras de trois races porcines françaises (Large White, Landrace Français et Piétrain). 2 – Caractéristiques de la bardière. *Journées de la Recherche Porcine en France* 25, 23–28.

Guéblez, R., Paboeuf, F., Sellier, P., Bouffaud, M., Boulard, J., Brault, D., Le Tiran M.H. and Petit, G. (1995) Effet du génotype halothane sur les performances d'engraissement, de carcasse et de qualité de la viande du porc charcutier. *Journées de la Recherche Porcine en France* 27, 155–164.

Guéblez, R., Bouyssière, M. and Sellier, P. (1996) Evaluation sensorielle de différents produits issus de porcs de génotype halothane connu. *Journées de la Recherche Porcine en France* 28, 45–52.

Guérin, G., Ollivier L. and Sellier, P. (1979) Effet d'entraînement d'un gène sélectionné et association gamétique (déséquilibre de 'linkage') : l'exemple de deux locus étroitement liés chez le porc, Hal (sensibilité à l'halothane) et PHI (phosphohexose isomérase). *Compte Rendus Hebdomadaires des Séances de l'Académie des Sciences (Paris)* 289, 153–156.

Gunsett, F.C. and Robison, O.W. (1989) Crossbreeding effects on reproduction, growth and carcass traits. In: Young, L.D. (ed.) *Genetics of Swine.* USDA-ARS, Clay Center, Nebraska, pp. 110–118.

Haley, C.S., D'Agaro, E. and Ellis, M. (1992) Genetic components of growth and ultrasonic fat depth traits in Meishan and Large White pigs and their reciprocal crosses. *Animal Production* 54, 105–115.

Hammell, K.L., Laforest, J.P. and Dufour, J.J. (1993) Evaluation of the growth performance and carcass characteristics of commercial pigs produced in Quebec. *Canadian Journal of Animal Science* 73, 495–508.

Hammell, K.L., Laforest, J.P. and Dufour, J.J. (1994) Evaluation of the lean meat colour of commercial pigs produced in Quebec. *Canadian Journal of Animal Science* 74, 443–449.

Hanset, R., Dasnois, C., Scalais, S., Michaux, C. and Grobet, L. (1995) Génotypes au locus de la sensibilité à l'halothane et caractères de croissance et de carcasse dans une F_2 Piétrain × Large White. *Genetics, Selection, Evolution* 27, 63–76.

Harbitz, I., Chowdhary, B., Thomsen, P.D., Davies, W., Kaufmann, U., Kran, S., Gustavsson, I., Christensen, K. and Hauge, J.G. (1990) Assignment of the porcine calcium release channel gene, a candidate for the malignant hyperthermia locus, to the 6p11→q21 segment of chromosome 6. *Genomics* 8, 243–248.

Hartmann, S., Otten, W., Kratzmair, M., Berrer, A. and Eichinger, H.M. (1992) Effects of breed, halothane genotype and sex on the lipid composition of two skeletal muscles and adipose tissue in swine. In: *Proceedings of the 38th International Congress of Meat Science and Technology*, 23–28 August 1992, Clermont-Ferrand, France, vol. 2, pp. 77–80.

Honkavaara, M. (1989) Influence of porcine stress and breed on the fatty acid profiles of subcutaneous and intramuscular total lipids. *Fleischwirtschaft* 69, 1429–1432.

Houde, A., Pommier, S.A. and Roy, R. (1993) Detection of the ryanodine receptor mutation associated with malignant hyperthermia in purebred swine populations. *Journal of Animal Science* 71, 1414–1418.

Hovenier, R., Kanis, E., van Asseldonk, Th. and Westerink, N.G. (1992) Genetic parameters of pig meat quality traits in a halothane negative population. *Livestock Production Science* 32, 309–321.

Hovenier, R., Brascamp, E.W., Kanis, E., van der Werf, J.H.J. and Wassenberg A.P.A.M. (1993a) Economic values of optimum traits: the example of meat quality in pigs. *Journal of Animal Science* 71, 1429–1433.

Hovenier, R., Kanis, E., van Asseldonk, Th. and Westerink, N.G. (1993b) Breeding for pig meat quality in halothane negative populations – a review. *Pig News and Information* 14, 17N–25N.

Hubbard, D.J., Southwood, O.I. and Kennedy, B.W. (1990) Estimation of the frequency of the halothane gene and its effects in Landrace and Yorkshire pigs in Ontario. *Canadian Journal of Animal Science* 70, 73–79.

Hyldgaard-Jensen J. (1986) Adenosine and meat quality in pigs. Role of an antilipolytic action of adenosine. *Israel Journal of Veterinary Medicine* 42, 278–285.

Ianssen, K. and Sehested, E. (1989) Intramuscular fat in the Norwegian pig breeding program. *40th Annual Meeting of the European Association for Animal Production*, 27–31 August 1989, Dublin, Ireland, paper GP3.20 (6 pp.).

Imlah, P. and Thomson, S.R.M. (1979) The H blood group locus and meat colour, and using blood groups to predict halothane reactors. *Acta Agriculturae Scandinavica* Suppl. 21, 403–410.

Jacquet, B., Sellier, P., Runavot, J.P., Brault, D., Houix, Y., Perrocheau, C., Gogué, J. and Boulard, J. (1984) Prediction of the technological yield of 'Paris ham' processing by using measurements at the abattoir. In: *Proceedings of the Scientific Meeting 'Biophysical PSE-muscle Analysis'*, 26–27 April 1984, Technical University, Vienna, Austria, pp. 143–153.

Janss, L.L.G., Van Arendonk, J.A.M. and Brascamp, E.W. (1997) Bayesian statistical analyses for presence of single genes affecting meat quality traits in a crossed pig population. *Genetics* 145, 395–408.

Johansson, K. (1987) Evaluation of station testing of pigs. II. Multiple trait versus single trait estimation of genetic parameters for meat quality measurements. *Acta Agriculturae Scandinavica* 37, 108–119.

Johansson, K., Andersson, K. and Sigvardsson, J. (1987) Evaluation of station testing of pigs. III. Genetic parameters for carcass measurements of partially dissected pigs. *Acta Agriculturae Scandinavica* 37, 120–129.

Johnson, R.K. (1981) Crossbreeding in swine: experimental results. *Journal of Animal Science* 52, 906–923.

Johnson, R.K. (1989) Inbreeding effects on reproduction, growth and carcass traits. In: Young, L.D. (ed.) *Genetics of Swine.* USDA-ARS, Clay Center, Nebraska, pp. 107–109.

Jonsson, P. (1963) Danish pig progeny testing results. *Zeitschrift für Tierzüchtung und Zuchtüngsbiologie* 78, 205–252.

Jonsson, P. and Wismer-Pedersen, J. (1974) Genetics of sex odour in boars. *Livestock Production Science* 1, 53–66.

Jorgensen, P.F. (1979) Polymorphic systems in blood. Associations with porcine halothane sensitivity and meat quality. *Acta Agriculturae Scandinavica* Suppl. 21, 386–395.

Judge, M.D. (1972) A review of possible methods to detect stress susceptibility and potential low quality pork. In: Cassens, R.G., Giesler, F. and Kolb, Q. (eds) *Proceedings of the Pork Quality Symposium.* University of Wisconsin, Madison, pp. 91–115.

Jung, Y.C., Rothschild, M.F., Flanagan, M.P., Christian, L.L. and Warner, C.M. (1989) Associations of restriction fragment length polymorphisms of swine leukocyte antigen class I genes and production traits of Duroc and Hampshire boars. *Animal Genetics* 20, 79–91.

Kallweit, E. (1985) Selection for stress resistance in pigs in various West-European countries. In: Ludvigsen, J.B. (ed.), *Stress Susceptibility and Meat Quality in Pigs,* EAAP Publication no. 33, pp. 60–67.

Kallweit, E., Baulain, U. and Hoppenbrock, K.H. (1993) Intramuscular fat in some German pig breeds and crosses. *44th Annual Meeting of the European Association for Animal Production,* 16–19 August 1993, Aarhus, Denmark, paper P1.5 (5 pp.).

Kellogg, T.F., Rogers, R.W. and Miller, H.W. (1977) Differences in tissue fatty acids and cholesterol of swine from different genetic backgrounds. *Journal of Animal Science* 44, 47–52.

King, J.W.B. and Roberts, R.C. (1959) The effects of inbreeding on carcass traits in the bacon pig. *Animal Production* 1, 123–127.

Kocwin-Podsiadla, M., Przybylski, W., Kuryl, J., Talmant, A. and Monin, G. (1995) Muscle glycogen level and meat quality in pigs of different halothane genotypes. *Meat Science* 40, 121–125.

Krieter, J., Lang, J.J., Looft, C. and Kalm, E. (1990) Verlauf der postmortalen Glykogenolyse beim Schwein. *Fleischwirtschaft* 70, 1097–1098.

Labroue, F., Guéblez, R. and Sellier, P. (1997) Genetic parameters of feeding behaviour and performance traits in group-housed Large White and French Landrace growing pigs. *Genetics, Selection, Evolution* 29 (in press).

Lan, Y.H., McKeith, F.K., Novakofski, J. and Carr, T.R. (1993) Carcass and muscle characteristics of Yorkshire, Meishan, Yorkshire × Meishan, Meishan × Yorkshire, Fengjing × Yorkshire, and Minzhu × Yorkshire pigs. *Journal of Animal Science* 71, 3344–3349.

Langlois, A. and Minvielle, F. (1989) Comparisons of three-way and backcross swine: II. Wholesale cuts and meat quality. *Journal of Animal Science* 67, 2025–2032.

Larzul, C. (1997). Variabilité génétique d'une mesure *in vivo* du potentiel glycolytique musculaire chez le porc. Thèse de doctorat INA-PG, Paris, 138 pp.

Larzul, C., Rousset-Akrim, S., Le Roy, P., Gogué, J., Talmant, A., Vernin, P., Touraille, C., Monin, G. and Sellier, P. (1996) Effet du génotype halothane sur la texture de la viande de porc. *Journées de la Recherche Porcine en France* 28, 39–44.

Larzul, C., Le Roy, P., Guéblez, R., Talmant, A., Gogué, J., Sellier, P. and Monin, G. (1997) Effect of halothane genotype (*NN,Nn,nn*) on growth, carcass and meat quality traits of pigs slaughtered at 95 kg or 125 kg live weight. *Journal of Animal Breeding and Genetics* 114, 309–320.

Lea, C.H., Swoboda, P.A.T. and Gatherum, D.P. (1970) A chemical study of soft fat in cross-bred pigs. *Journal of Agricultural Science, Cambridge* 74, 279–289.

Leach, L.M., Ellis, M., Sutton, D.S., McKeith, F.K. and Wilson, E.R. (1996) The growth performance, carcass characteristics, and meat quality of halothane carrier and negative pigs. *Journal of Animal Science* 74, 934–943.

Lefaucheur, L. (1989) Les différents types de fibres musculaires chez le porc. Conséquences sur la production de viande. *INRA Productions Animales* 2, 205–213.

Legault, C., Sellier, P., Caritez, J.C., Dando, P. and Gruand, J. (1985) L'expérimentation sur le porc chinois en France. 2. Performances de production en croisement avec les races européennes. *Génétique Sélection Evolution* 17, 133–152.

Lengerken, G., Wicke, M. and Maak, S. (1993) Suitability of structural and functional traits of skeletal muscle for the genetic improvement of meat quality in pigs. *44th Annual Meeting of the European Association for Animal Production*, 16–19 August 1993, Aarhus, Denmark, paper P1.3 (9 pp.).

Le Roy, P. and Sellier, P. (1994) The current French breeding programme for improvement of meat quality in pigs. In: *Proceedings of the 2nd International Conference 'The influence of genetic and non genetic traits on carcass and meat quality'*, 7–8 November 1994, Siedlce, Poland, pp. 1–13.

Le Roy, P., Elsen, J.M. and Naveau, J. (1990a) Etude de la variabilité génétique de l'adiposité dans la lignée Laconie. *Journées de la Recherche Porcine en France* 22, 11–16.

Le Roy, P., Naveau, J., Elsen, J.M. and Sellier, P. (1990b) Evidence for a new major gene influencing meat quality in pigs. *Genetical Research* 55, 33–40.

Le Roy, P., Caritez, J.C., Elsen, J.M. and Sellier, P. (1994a) Pigmeat quality: experimental study on the RN major locus. In: *Proceedings of the 5th World Congress on Genetics Applied to Livestock Production*, 7–12 August 1994, University of Guelph, Canada, vol. 19, pp. 473–476.

Le Roy, P., Przybylski, W., Burlot, T., Bazin, C., Lagant H. and Monin, G. (1994b) Etude des relations entre le potentiel glycolytique du muscle et les caractères de production dans les lignées Laconie et Penshire. *Journées de la Recherche Porcine en France* 26, 311–314.

Le Roy, P., Juin, H., Caritez, J.C., Billon, Y., Lagant, H., Elsen, J.M. and Sellier, P. (1996a) Effet du génotype RN sur les qualités sensorielles de la viande de porc. *Journées de la Recherche Porcine en France* 28, 53–56.

Le Roy, P., Monin, G., Elsen, J.M., Caritez, J.C., Talmant, A., Lebret, B., Lefaucheur, L., Mourot J., Juin, H. and Sellier, P. (1996b) Effect of the RN genotype on growth and carcass traits in pigs. *47th Annual Meeting of the European Association for Animal Production*, 26–29 August 1996, Lillehammer, Norway, paper AG7.9 (8 pp.).

Leymaster, K.A. and Swiger, L.A. (1981) Selection for increased leanness of Yorkshire swine. III. Inbreeding effects on secondary traits. *Journal of Animal Science* 53, 620–628.

Liu, Y. and Stouffer, J.R. (1995) Pork carcass evaluation with an automated and computerized ultrasonic system. *Journal of Animal Science* 73, 29–39.

Lo, L.L., McLaren, D.G., McKeith, F.K., Fernando, R.L. and Novakofski, J. (1992a) Genetic analysis of growth, real-time ultrasound, carcass, and pork quality traits in Duroc and Landrace pigs: I. Breed effects. *Journal of Animal Science* 70, 2373–2386.

Lo, L.L., McLaren, D.G., McKeith, F.K., Fernando, R.L. and Novakofski, J. (1992b) Genetic analyses of growth, real-time ultrasound, carcass, and pork quality traits in Duroc and Landrace pigs: II. Heritabilities and correlations. *Journal of Animal Science* 70, 2387–2396.

Looft, C., Reinsch, N., Rudat, I. and Kalm, E. (1996) Mapping the porcine RN gene to chromosome 15. *Genetics, Selection, Evolution* 28, 437–442.

Lundström, K. and Malmfors, B. (1993) Genetic influence on skatole deposition in entire male pigs. In: Bonneau, M. (ed.) *Measurement and Prevention of Boar Taint in Entire Male Pigs.* INRA, Paris (Les Colloques, no. 60), pp. 159–165.

Lundström, K., Karlsson, A., Lundeheim, N., Rydhmer, L. and Vestergaard, T. (1989) Current methods and perspectives in breeding for meat quality in pigs. *40th Annual Meeting of the European Association for Animal Production*, 27–31 August 1989, Dublin, Ireland, paper GP3.4 (16 pp.).

Lundström, K., Andersson, A., Maerz, S. and Hansson, I. (1994) Effect of the RN-gene on meat quality and lean meat content in crossbred pigs with Hampshire as terminal sire. In: *Proceedings of the 40th International Congress of Meat Science and Technology*, The Hague, The Netherlands, paper S.IVA.07 (5 pp.).

Lundström, K., Karlsson, A., Hakansson, J., Hansson, I., Johansson, M., Andersson, L. and Andersson, K. (1995) Production, carcass and meat quality traits of F_2-crosses between European Wild Pigs and domestic pigs including halothane gene carriers. *Animal Science* 61, 325–331.

Lundström, K., Andersson, A. and Hansson, I. (1996) Effect of the RN gene on technological and sensory meat quality in crossbred pigs with Hampshire as terminal sire. *Meat Science* 42, 145–153.

Lush, J.L. (1936) Genetic aspects of the Danish system of progeny-testing swine. *Iowa Agricultural Experiment Station Research Bulletin* no. 204.

Maassen-Francke, B., Krieter, J. and Kalm, E. (1991) Vergleichende Untersuchungen über Wachstum, Fleischbeschaffenheit und postmortale Glykogenolyse bei verschiedenen Schweinerassen. *Züchtungskunde* 63, 366–374.

McGloughlin, P., Allen, P., Tarrant, P.V., Joseph, R.L., Lynch, P.B. and Hanrahan, T.J. (1988) Growth and carcass quality of crossbred pigs sired by Duroc, Landrace and Large White boars. *Livestock Production Science* 18, 275–288.

McLaren, D.G., Buchanan, D.S. and Johnson, R.K. (1987) Individual heterosis and breed effects for postweaning performance and carcass traits in four breeds of swine. *Journal of Animal Science* 64, 83–98.

MacLennan, D.H., Duff, C., Zorzato, F., Fujii, J., Phillips, M., Korneluk, R.G., Frodis, W., Britt, B.A. and Worton, R.G. (1990) Ryanodine receptor gene is a candidate for predisposition to malignant hyperthermia. *Nature* 343, 559–561.

McPhee, C.P. and Trout, G.R. (1995) The effects of selection for lean growth and the halothane allele on carcass and meat quality of pigs transported long and short distances to slaughter. *Livestock Production Science* 42, 55–62.

Malmfors, B. and Nilsson, R. (1979) Meat quality traits in Swedish Landrace and Yorkshire pigs with special emphasis on genetics. *Acta Agriculturae Scandinavica* Suppl. 21, 81–90.

Mariani, P., Lundström, K., Gustafsson, U., Enfält, A.C., Juneja, R.K. and Andersson, L. (1996) A major locus (RN) affecting muscle glycogen content is located on pig chromosome 15. *Mammalian Genome* 7, 52–54.

Marklund, L., Nyström, P.E., Stern, S. and Andersson, L. (1996) Further characterization of a major QTL for fatness on pig chromosome 4. *Animal Genetics* 27 (Suppl. 2), 114 (abstract).

Metz, S.H.M. (1985) Genetic effects on fat deposition and fat quality in the growing pig. *Pig News and Information* 6, 291–294.

Mickelson, J.R. and Louis, C.F. (1993) Calcium (Ca^{2+}) regulation in porcine skeletal muscle – Review. In: Puolanne, E. and Demeyer, D.I. (eds) *Pork Quality: Genetic and Metabolic Factors.* CAB International, Wallingford, Oxon, UK, pp. 160–182.

Mikami, H., Fredeen, H.T. and Sather, A.P. (1977) Mass selection in a pig population. 2. The effects of inbreeding within the selected populations. *Canadian Journal of Animal Science* 57, 627–634.

Milan, D., Le Roy, P., Woloszyn, N., Caritez, J.C., Elsen, J.M., Sellier P. and Gellin, J. (1995) The RN locus for meat quality maps to pig chromosome 15. *Genetics, Selection, Evolution* 27, 195–199.

Milan, D., Woloszyn, N., Yerle, M., Le Roy, P., Bonnet, M., Riquet, J., Lahbib-Mansais, Y., Caritez, J.C., Robic, A., Sellier, P., Elsen, J.M. and Gellin, J. (1996) Accurate mapping of the 'acid meat' RN gene on genetic and physical maps of pig chromosome 15. *Mammalian Genome* 7, 47–51.

Minelli, G., Culioli, J., Vignon, X. and Monin, G. (1995) *Postmortem* changes in the mechanical properties and ultrastructure of the *Longissimus* in two porcine breeds. *Journal of Muscle Foods* 6, 313–326.

Minkema, D., Eikelenboom, G. and Van Eldik, P. (1977) Inheritance of MHS – susceptibility in pigs. In: *Proceedings of the 3rd International Conference on Production Disease in Farm Animals.* Pudoc, Wageningen, The Netherlands, pp. 203–207.

Moller, A.J. and Iversen, P. (1993) Elements in the concept of pig meat quality. *44th Annual Meeting of the European Association for Animal Production,* 16–19 August 1993, Aarhus, Denmark, paper P1.1 (10 pp.).

Moller, A.J., Bertelsen, G. and Olsen, A. (1993) Processed pork: technological parameters related to type of raw material – Review. In: Puolanne, E. and Demeyer, D.I. (eds) *Pork Quality: Genetic and Metabolic Factors.* CAB International, Wallingford, Oxon, UK, pp. 217–237.

Monin, G. (1991) Techniques used in Europe for evaluation of pig meat quality: current situation and research needs. *Conference 'Qualita delle carni: problematiche relative alla valutazione strumentale'*, 30 April 1991, Reggio Emilia, Italy (11 pp.).

Monin, G. and Sellier, P. (1985) Pork of low technological quality with a normal rate of muscle pH fall in the immediate *post-mortem* period: the case of the Hampshire breed. *Meat Science* 13, 49–63.

Monin, G., Sellier, P., Ollivier, L., Goutefongea, R. and Girard, J.P. (1981) Carcass characteristics and meat quality of halothane negative and halothane positive Piétrain pigs. *Meat Science* 5, 413–423.

Monin, G., Talmant, A., Laborde, D., Zabari, M. and Sellier, P. (1986) Compositional and enzymatic characteristics of the *Longissimus dorsi* muscle from Large White, halothane-positive and halothane-negative Piétrain, and Hampshire pigs. *Meat Science* 16, 307–316.

Monin, G., Mejenes-Quijano, A., Talmant, A. and Sellier, P. (1987) Influence of breed and muscle metabolic type on muscle glycolytic potential and meat pH in pigs. *Meat Science* 20, 149–158.

Monin, G., Brard, C., Vernin, P. and Naveau, J. (1992) Effects of the RN⁻ gene on some traits of muscle and liver in pigs. In: *Proceedings of the 38th International Congress of Meat Science and Technology*, 23–28 August 1992, Clermont-Ferrand, France, vol. 3, pp. 391–394.

Murray, A.C., Jones, S.D.M. and Sather, A.P. (1989) The effects of preslaughter feed restriction and genotype for stress susceptibility on pork lean quality and composition. *Canadian Journal of Animal Science* 69, 83–91.

National Pork Producers Council (1995) *Genetic Evaluation/Terminal Line Program Results* (Goodwin, R. and Burroughs, S. (ed.)). NPPC, Des Moines, Iowa, USA, 312 pp.

Naveau, J. (1986) Contribution à l'étude du déterminisme génétique de la qualité de la viande porcine. Héritabilité du rendement technologique Napole. *Journées de la Recherche Porcine en France* 18, 265–276.

Naveau, J., Pommeret, P. and Lechaux, P. (1985) Proposition d'une méthode de mesure du rendement technologique: la 'méthode Napole'. *Techni-porc* 8(6), 7–13.

Nielsen, N.J. (1981) The effects of environmental factors on meat quality and on deaths during transportation and lairage before slaughter. In: Froystein, T., Slinde, E. and Standal, N. (eds), *Porcine Stress and Meat Quality: Causes and Possible Solutions to the Problems.* Agricultural Food Research Society, As, Norway, pp. 287–297.

O'Brien, P.J., Ball, R.O. and MacLennan, D.H. (1994) Effects of heterozygosity for the mutation causing porcine stress syndrome on carcass quality and live performance characteristics. In: *Proceedings of the 13th International Pig Veterinary Science Congress,* 26–30 June 1994, Bangkok, Thailand, p. 481.

Oliver, M.A., Gispert, M. and Diestre, A. (1993) The effects of breed and halothane sensitivity on pig meat quality. *Meat Science* 35, 105–118.

Oliver, M.A., Gou, P., Gispert, M., Diestre, A., Arnau, J., Noguera, J.L. and A. Blasco (1994) Comparison of five types of pig crosses. II. Fresh meat quality and sensory characteristics of dry cured ham. *Livestock Production Science* 40, 179–185.

Ollivier, L. (1968) Etude du déterminisme génétique de l'hypertrophie musculaire du porc de Piétrain. *Annales de Zootechnie* 17, 393–407.

Ollivier, L. (1980) Le déterminisme génétique de l'hypertrophie musculaire chez le porc. *Annales de Génétique et de Sélection Animale* 12, 383–394.

Ollivier, L. and Meslé, L. (1963) Résultats d'un contrôle de descendance portant sur la qualité de la viande chez le porc. *Annales de Zootechnie* 12, 173–179.

Ollivier, L. and Sellier, P. (1982) Pig genetics: a review. *Annales de Génétique et de Sélection animale* 14, 481–544.

Ollivier, L. and Sellier, P. (1985) *La Génétique du Porc: Mise au Point.* Institut Technique du Porc, Paris, 70 pp.

Ollivier, L., Sellier, P. and Monin, G. (1975) Déterminisme génétique du syndrome d'hyperthermie maligne chez le porc de Piétrain. *Annales de Génétique et de Sélection Animale* 7, 159–166.

Otsu, K., Khanna, V.K., Archibald, A.L. and MacLennan, D.H. (1991) Cosegregation of porcine malignant hyperthermia and a probable causal mutation in the skeletal muscle ryanodine receptor gene in backcross families. *Genomics* 11, 744–750.

Pascal, G., Macaire, J.P., Desmoulin, B. and Bonneau, M. (1975) Composition des graisses de porcs femelles: influence du type génétique (LF, LB ou PP) et évolution au cours de la croissance (entre 40 et 100 kg). *Journées de la Recherche Porcine en France* 7, 203–214.

Patterson, R.L.S., Elks, P.K., Lowe, D.B. and Kempster, A.J. (1990) The effects of different factors on the levels of androstenone and skatole in pig fat. *Animal Production* 50, 551 (abstract).

Pellois, H. and Runavot, J.P. (1991) Comparaison des performances d'engraissement, de carcasse et de qualité de la viande de 4 types de porcs ayant une proportion variable de sang Piétrain. *Journées de la Recherche Porcine en France* 23, 369–376.

Poilvet, D., Bonneau, M., Caritez, J.C. and Legault, C. (1990) Carcass tissue composition in Meishan (MS), Large White (LW) and F_1 (MS × LW) pigs. In: Molénat, M. and Legault, C. (eds) *Symposium sur le Porc Chinois,* 5–6 July 1990, Toulouse, France. INRA, Jouy en Josas, pp. 237–238.

Pommier, S.A., Houde, A., Rousseau, F. and Savoie, Y. (1992) The effect of the malignant hyperthermia genotype as determined by a restriction endonuclease assay on carcass characteristics of commercial crossbred pigs. *Canadian Journal of Animal Science* 72, 973–976.

Prunier, A., Caritez, J.C. and Bonneau, M. (1987) Développement de l'appareil génital des porcs mâles et femelles et évolution de la teneur en androsténone du tissu adipeux des verrats de race européenne ou chinoise. *Annales de Zootechnie* 36, 49–58.

Purchas, R.W., Smith, W.C. and Pearson, G. (1990) A comparison of the Duroc, Hampshire, Landrace, and Large White as terminal sire breeds of crossbred pigs slaughtered at 85 kg liveweight. 2. Meat quality. *New Zealand Journal of Agricultural Research* 33, 97–104.

Reinecke, S. and Kalm, E. (1988) Zusammenhänge zwischen Halothanreaktion, Markergenen, CK-Aktivität und Leisungseigenschaften bei Schweinen der Rassen Piétrain und Landrasse B. *Züchtungskunde* 60, 330–344.

Reiner, G. (1993) A new physiological pathway controlling muscle growth and its potential relevance for pig production. *Pig News and Information* 14, 123N–125N.

Rempel, W.E., Ming Yu, L., Mickelson, J.R. and Louis, C.F. (1995) The effect of skeletal muscle ryanodine receptor genotype on pig performance and carcass quality traits. *Animal Science* 60, 249–257.

Renard, C., Bidanel, J.P., Palovics, A., Vaiman, M., Guérin, G. and Runavot, J.P. (1988) Relations entre des marqueurs génétiques et les caractères de production chez le porc. *Journées de la Recherche Porcine en France* 20, 315–320.

Renard, C., Mourot, J., Götz, K.U., Caritez, J.C., Bidanel, J.P. and Vaiman, M. (1992) Analyse des liaisons génétiques entre les marqueurs SLA et les caractères de croissance et d'adiposité chez le porc. *Journées de la Recherche Porcine en France* 24, 9–16.

Renard, C., Mourot, J., Nunes, M., Lahbib-Mansais, Y., Geffrotin, C., Bourgeaux, N., Caritez, J.C., Götz, K.U., Bidanel, J.P. and Vaiman, M. (1996) Association between the swine MHC region and malic enzyme activity in muscle. *Animal Genetics* 27 (Suppl. 2), 115–116 (abstract).

Robinson, R. (1991) Genetic defects in the pig. *Journal of Animal Breeding and Genetics* 108, 61–65.

Rothschild, M.F., Hoganson, D.L., Warner, C.M. and Schwartz, N.K. (1990) The use of major histocompatiblity complex class I restriction fragment length polymorphism analysis to predict performance in the pig. In: Hill, W.G., Thompson, R. and Woolliams, J.A. (eds) *Proceedings of the 4th World Congress on Genetics Applied to Livestock Production,* 23–27 July 1990, Edinburgh, vol. 13, pp. 125–128.

Rothschild, M.F., Liu, H.C., Tuggle, C.K., Yu, T.P. and Wang, L. (1995) Analysis of pig chromosome 7 genetic markers for growth and carcass performance traits. *Journal of Animal Breeding and Genetics* 112, 341–348.

Russo, V. and Nanni Costa, L. (1995) Suitability of pig meat for salting and the production of quality processed products. *Pig News and Information* 16, 17N–26N.

Ruusunen, M. and Puolanne, E. (1997) Comparison of biochemical properties of different pig breeds. *Meat Science* 45, 119–125.

Sather, A.P., Jones, S.D.M. and Joyal, S. (1991a) Feedlot performance, carcass composition and pork quality from entire male and female Landrace and Large White market-weight pigs. *Canadian Journal of Animal Science* 71, 29–42.

Sather, A.P., Jones, S.D.M. and Tong, A.K.W. (1991b) Halothane genotype by weight interactions on lean yield from pork carcasses. *Canadian Journal of Animal Science* 71, 633–643.

Sather, A.P., Jones, S.D.M., Tonk, A.K.W. and Murray, A.C. (1991c) Halothane genotype by weight interactions on pig meat quality. *Canadian Journal of Animal Science* 71, 645–658.

Saugère, D., Runavot, J.P. and Sellier, P. (1989) Un premier bilan du programme de sélection contre le gène de la sensibilité à l'halothane chez le porc Landrace Français. *Journées de la Recherche Porcine en France* 21, 335–344.

Sayre, R.N., Briskey, E.J. and Hoekstra, W.G. (1963) Comparison of muscle characteristics and *post mortem* glycolysis in three breeds of swine. *Journal of Animal Science* 23, 1012–1020.

Schneider, J.F., Christian, L.L. and Kuhlers, D.L. (1982) Crossbreeding in swine: genetic effects on pig growth and carcass merit. *Journal of Animal Science* 54, 747–756.

Schwörer, D., Morel, P. and Rebsamen, A. (1987) Selektion auf intramuskuläres Fett beim Schwein. *Der Tierzüchter* 39, 392–394.

Schwörer, D., Morel, P., Prabucki, A. and Rebsamen, A. (1988) Genetic parameters of fatty acids of pork fat. In: *Proceedings of the 24th International Congress of Meat Science and Technology*, 29 August–2 September 1988, Brisbane, Australia, part B, pp. 598–600.

Schwörer, D., Morel, P. and Rebsamen, A. (1989) Genetic variation in intramuscular fat content and sensory properties of pork. *40th Annual Meeting of the European Association for Animal Production*, 27–31 August 1989, Dublin, Ireland, paper GP3.2 (12 pp.).

Schwörer, D., Rebsamen, A. and Lorenz, D. (1994) Twenty years of selection for meat quality in Swiss pig breeding. *Pig News and Information* 15, 63N–66N.

Scott, R.A., Cornelius, S.G. and Mersmann, H.J. (1981) Fatty acid composition of adipose tissue from lean and obese swine. *Journal of Animal Science* 53, 977–981.

Sellier, P. (1976) The basis of crossbreeding in pigs: a review. *Livestock Production Science* 3, 203–226.

Sellier, P. (1983) Effets de la sélection sur l'adiposité chez le porc. *Revue Française des Corps Gras* 30, 103–111.

Sellier, P. (1987) Crossbreeding and meat quality in pigs. In: Tarrant, P.V., Eikelenboom, G. and Monin, G. (eds) *Evaluation and Control of Meat Quality in Pigs.* Martinus Nijhoff Publishers, Dordrecht, The Netherlands, pp. 329–342.

Sellier, P. (1988a) Aspects génétiques des qualités technologiques et organoleptiques de la viande chez le porc. *Journées de la Recherche Porcine en France* 20, 227–242.

Sellier, P. (1988b) Meat quality in pig breeds and in crossbreeding. In: *Proceedings of the Meeting 'Pig Carcass and Meat Quality'*, 2–3 June 1988, Reggio Emilia, Italy, pp. 145–164.

Sellier, P. (1994) The future role of molecular genetics in the control of meat production and meat quality. *Meat Science* 36, 29–44.

Sellier, P. (1995) Genetics of pork quality. In: *Proceedings of the Conference on Pork Science and Technology*, 24–26 April 1995, Campinas SP, Brazil, pp. 1–35.

Sellier, P. and Bonneau, M. (1988) Genetic relationships between fat androstenone level in males and development of male and female genital tract in pigs. *Journal of Animal Breeding and Genetics* 105, 11–20.

Sellier, P. and Jacquet, B. (1973) Comparaison de porcs Hampshire × Large White et Piétrain × Large White. *Journées de la Recherche Porcine en France* 5, 173–180.

Sellier, P. and Monin, G. (1994) Genetics of pig meat quality: a review. *Journal of Muscle Foods* 5, 187–219.

Sellier, P. and Rothschild, M.F. (1991) Breed identification and development in pigs. In: Maijala, K. (ed.) *Genetic Resources of Pig, Sheep and Goat, World Animal Science, B8.* Elsevier Science Publishers, Amsterdam, The Netherlands, pp. 125–143.

Sellier, P., Monin, G., Houix, Y. and Dando, P. (1984) Qualité de la viande de quatre races porcines: relations avec la sensibilité à l'halothane et l'activité créatine phosphokinase plasmatique. *Journées de la Recherche Porcine en France* 16, 65–74.

Sellier, P., Monin, G. and Talmant, A. (1985) Aptitude de divers types génétiques de porcs à la fabrication du jambon sec. In: *Proceedings of the 31st European Meeting of Meat Research Workers*, Albena, Bulgaria, vol. 1, pp. 7–10.

Sellier, P., Mejenes-Quijano, A., Marinova, P., Talmant, A., Jacquet, B. and Monin, G. (1988) Meat quality as influenced by halothane sensitivity and ultimate pH in three porcine breeds. *Livestock Production Science* 18, 171–186.

Serra, J.J., Ellis, M. and Haley, C.S. (1992) Genetic components of carcass and meat quality traits in Meishan and Large White pigs and their reciprocal crosses. *Animal Production* 54, 117–127.

Smith, C. and Bampton, P.R. (1977) Inheritance of reaction to halothane anaesthesia in pigs. *Genetical Research* 29, 287–292.

Smith, W.C. and Pearson, G. (1986) Comparative voluntary feed intakes, growth performance, carcass composition, and meat quality of Large White, Landrace, and Duroc pigs. *New Zealand Journal of Experimental Agriculture* 14, 43–50.

Smith, W.C. and Pearson, G. (1987) Comparative voluntary feed intakes, growth performance, carcass measurements, and some meat quality characteristics of Duroc, Hampshire and Large White pigs. *New Zealand Journal of Experimental Agriculture* 15, 39–43.

Smith, W.C., Pearson, G. and Garrick, D.J. (1988) Evaluation of the Duroc in comparison with the Landrace and Large White as a terminal sire of crossbred pigs slaughtered at 85 kg liveweight. *New Zealand Journal of Agricultural Research* 31, 421–430.

Smith, W.C., Pearson, G. and Purchas, R.W. (1990) A comparison of the Duroc, Hampshire, Landrace, and Large White as terminal sire breeds of crossbred pigs slaughtered at 85 kg liveweight. 1. Performance and carcass characteristics. *New Zealand Journal of Agricultural Research* 33, 89–96.

Southwood, O.I., Simpson, S.P., Curran, M.K. and Webb, A.J. (1988) Frequency of the halothane gene in British Landrace and Large White pigs. *Animal Production* 46, 97–102.

Squires, E.J., Deng, H. and Wu, L. (1992) Taint testing of intact male pigs from swine breeding herds in Ontario. *Journal of Animal Science* 70 (Suppl. 1), 223 (abstract).

Stewart, T.S. and Schinckel, A.P. (1989) Genetic parameters for swine growth and carcass traits. In: Young, L.D. (ed.) *Genetics of Swine.* USDA-ARS, Clay Center, Nebraska, pp. 77–79.

Sutherland, R.A., Webb, A.J. and King, J.W.B. (1985) A survey of world pig breeds and comparisons. *Animal Breeding Abstracts* 53, 1–22.

Tornberg, E., Andersson, A., Göransson, A. and Von Seth, G. (1993) Water and fat distribution in pork in relation to sensory properties. In: Puolanne, E. and Demeyer, D.I. (eds) *Pork Quality: Genetic and Metabolic Factors.* CAB International, Wallingford, Oxon, pp. 239–258.

Touraille, C. and Monin, G. (1982) Qualités organoleptiques de la viande de porc en relation avec la sensibilité à l'halothane. *Journées de la Recherche Porcine en France* 14, 33–36.

Touraille, C. and Monin, G. (1984) Comparaison des qualités organoleptiques de la viande de porcs de trois races: Large White, Landrace français, Landrace belge. *Journées de la Recherche Porcine en France* 16, 75–80.

Touraille, C., Monin, G. and Legault, C. (1989) Eating quality of meat from European × Chinese crossbred pigs. *Meat Science* 25, 177–186.

Tribout, T., Garreau, H. and Bidanel, J.P. (1996) Paramètres génétiques de quelques caractères de qualité de la viande dans les races porcines Large White et Landrace Français. *Journées de la Recherche Porcine en France* 28, 31–38.

Vaiman, M., Renard, C. and Bourgeaux, N. (1988) SLA, the major histocompatibility complex in swine: its influence on physiological and pathological traits. In: Warner, C.M., Rothschild, M.F. and Lamont, S. (eds) *The Molecular Biology of the Major Histocompatibility Complex of Domestic Animal Species.* Iowa State Press, Ames, Iowa, pp. 23–38.

Van der Wal, P.G., de Vries, A.G. and Eikelenboom, G. (1995) Predictive value of slaughterhouse measurements of ultimate pork quality in seven halothane negative Yorkshire populations. *Meat Science* 40, 183–191.

Van Laack, R.L.J.M., Kauffman, R.G., Sybesma, W., Smulders, F.J.M., Eikelenboom, G. and Pinheiro, J.C. (1994) Is color brightness (L-value) a reliable indicator of water-holding capacity in porcine muscle ? *Meat Science* 38, 193–201.

Villegas, F.J., Hedrick, H.B., Veum, T.L., McFate, K.L. and Bailey, M.E. (1973) Effect of diet and breed on fatty acid composition of porcine adipose tissue. *Journal of Animal Science* 36, 663–668.

Visscher, P.M. and Haley, C.S. (1995) Utilizing genetic markers in pig breeding programmes. *Animal Breeding Abstracts* 63, 1–8.

Vögeli, P., Kühne, R., Gerwig, C., Kaufmann, A., Wysshaar, M. and Stranzinger, G. (1988) Bestimmung des Halothangenotyps (HAL) mit Hilfe der S, PHI, HAL, H, PO2, PGD Haplotypen von Eltern und Nachkommen beim Schweizerischen Veredelten Landschwein. *Züchtungskunde* 60, 24–37.

Warriss, P.D. and Brown, S.N. (1987) The relationships between initial pH, reflectance and exudation in pig muscle. *Meat Science* 20, 65–74.

Warriss, P.D., Brown, S.N., Franklin, J.G. and Kestin, S.C. (1990) The thickness and quality of backfat in various pig breeds and their relationship to intramuscular fat and the setting of joints from the carcasses. *Meat Science* 28, 21–29.

Wassmuth, R., Surmann, H. and Glodek, P. (1991) Untersuchungen zum 'Hampshire-faktor' in der Fleischbeschaffenheit von Schweinen. 1. Mitteilung: Fleischbeschaffenheit bei Schweinen mit unterschiedlichen Hampshire-Anteilen und Segregationsanalyse ausgesuchter Merkmale. *Züchtungskunde* 63, 445–455.

Webb, A.J. (1981) The halothane sensitivity test. In: Froystein, T., Slinde, E. and Standal, N. (ed.) *Porcine Stress and Meat Quality: Causes and Possible Solutions to the Problems.* Agricultural Food Research Society, As, Norway, pp. 105–124.

Webb, A.J. (1995) Future challenges in pig genetics. *Animal Breeding Abstracts* 63, 731–736.

Webb, A.J., Grundy, B. and Kitchin, P. (1994) Within-litter effect of the Hal-1843 heterozygote on lean growth in pigs. In: *Proceedings of the 5th World Congress on Genetics Applied to Livestock Production,* 7–12 August 1994, University of Guelph, Canada, vol. 17, pp. 421–424.

Wilken, T.M., Lo, L.L., McLaren, D.G., Fernando, R.L. and Dziuk, P.J. (1992) An embryo transfer study of reciprocal cross differences in growth and carcass traits of Duroc and Landrace pigs. *Journal of Animal Science* 70, 2349–2358.

Willeke, H. (1993) Possibilities of breeding for low 5α-androstenone content in pigs. *Pig News and Information* 14, 31N–33N.

Wilson, D.E. (1992) Application of ultrasound for genetic improvement. *Journal of Animal Science* 70, 973–983.

Wismer-Pedersen, J. (1959) Quality of pork in relation to rate of pH change *post mortem. Food Research* 24, 711–727.

Wittmann, W., Peschke, W., Littmann, E., Behringer, J., Birkenmaier, St., Dovc, P. and Förster, M. (1993) Mast- und Schlachtleistungen von DL-Kastraten in Abhängigkeit vom MHS-Genotyp. *Züchtungskunde* 65, 197–205.

Wood, J.D. (1973) The fatty acid composition of backfat from Piétrain and Large White pigs. *Animal Production* 17, 281–285.

Wood, J.D. and Enser, M. (1989) Fat quality in pigs with special emphasis on genetics. *40th Annual Meeting of the European Association for Animal Production,* 27–31 August 1989, Dublin, Ireland, paper GP3.3 (8 pp.).

Wood, J.D., Kempster, A.J., David, P.J. and Bovey, M. (1987) Observations on carcass and meat quality in Duroc, Landrace and Duroc × Landrace pigs. *Animal Production* 44, 488 (abstract).

Xue, J.L., Dial, G.D., Holton, E.E., Vickers, Z., Squires, E.J., Lou, Y., Godbout, D. and Morel, N. (1996) Breed differences in boar taint: relationship between tissue levels of boar taint compounds and sensory analysis of taint. *Journal of Animal Science* 74, 2170–2177.

Yerle, M., Gellin, J., Dalens, M. and Galman, O. (1990) Localization on pig chromosome 6 of markers GPI, APOE and ENO1, carried by human chromosomes 1 and 19, using *in situ* hybridization. *Cytogenetics and Cell Genetics* 54, 86–91.

Young, L.D. (1992) Effects of Duroc, Meishan, Fengjing and Minzhu boars on carcass traits of first-cross barrows. *Journal of Animal Science* 70, 2030–2037.

Young, L.D., Johnson, R.K., Omtvedt, I.T. and Walters, L.E. (1976) Postweaning performance and carcass merit of purebred and two-breed cross pigs. *Journal of Animal Science* 42, 1124–1132.

Yu, T.P., Tuggle C.K., Schmitz, C.B. and Rothschild, M.F. (1995) Association of PIT1 polymorphisms with growth and carcass traits in pigs. *Journal of Animal Science* 73, 1282–1288.

Genetic Improvement of the Pig

17

L. Ollivier
INRA, Station de Génétique Quantitative et Appliquée, 78352 Jouy-en-Josas cedex, France

Introduction

The first attempts at domesticating the pig date back to the Neolithic age, i.e. circa 5000 BC, and remains of domestic pig dating as early as 7000 BC have been found also in Asia (see Chapter 2). The process of domestication, as described for instance by Jonsson (1991), includes selective breeding for specific characteristics, and may be considered as a first step in genetic improvement. An idea of the distance covered from the beginning may be gained by comparing the modern domestic pig to its wild counterpart. The improvement of the pig, as with other farm animals, has for a long time been the result of empirical methods employed by individual farmers. A long period of 'unconscious selection', as termed by Darwin (1859), elapsed before selection became a technique scientifically based, progressing with the knowledge of the biological processes involved, and particularly with the development of genetics. A history of pig improvement methods has been attempted by Ollivier (1976).

The development of breeding plans for pig improvement has been an important step. An early example of this is the programme presented by P.A. Moerkeberg to the Royal Danish Agricultural Society in Copenhagen, on 5 February 1896. The programme, as described by Jonsson (1965), aimed at developing a new breed, the Danish Landrace, to be crossed with imported Yorkshire boars for the production of bacon. Recognized 'breeding centres', or farms, were then established and official performance recording for carcass traits started in 1907 when the first progeny-testing station was opened. In such a scheme, one can already find the essential ingredients of modern pig improvement, i.e. definition of a breeding objective, selection of the best breed combination and within-breed improvement by selection.

In this chapter on modern genetic improvement, the *organization and genetic structure* of 'artificial pig populations' will first be reviewed. The definition of breeding goals, their weighting in an overall *breeding objective*, and the choice of optimal *selection criteria* will then be considered. The principles applied to the design of efficient *breeding programmes* will be outlined in the following section. In a final section, the assistance that *genetic markers* are expected to provide for enhancing genetic progress in the future will be discussed.

The Organization of Genetic Improvement

Breed societies

The creation of distinct breeds has been an important step in the history of genetic improvement of farm animals and, in this respect, the pig has followed the example of the other domestic species. Purebreeding, as we define it now, started in Britain in the second half of the 18th century. However, breed societies as official organizations only appeared about one century later in pig breeding.

Details on the time of formation of breed societies relevant to our present-day breeds are given by Sellier and Rothschild (1991). They also explain the changes which have progressively occurred in the role and activities of those societies. From an initial concern essentially limited to breed purity and commercial promotion, breed societies have moved towards greater attention to economic merit of their breeding stock, performance recording and also education of their members. Breed societies thus became more and more constructive forces in pig breeding, and increasingly contributed to overall genetic improvement.

Breeding structures: breeding companies and national improvement plans

In order to be fully efficient, any performance recording needs an adequate population structure. This was pointed out as early as 1965 in a historical survey of pig breeding in Canada (see the analysis of Fredeen, 1980). Pig farms have indeed specific functions in genetic improvement, according to whether they actually generate genetic changes (nucleus), or disseminate these (multiplier), or just take advantage of them in a production system (producer). This classical three-tier breeding pyramid has been the basic structure for pig genetic improvement in most countries for the last three decades, together with the extension of crossbreeding. The general context has thus become dominated by crossbreeding systems calling for specialized sire and dam lines and the development of a stratified supply system, in which commercial farms get their female stock from multiplier herds, usually as F_1 gilts, and their terminal boars, either pure or crossbred, from different nucleus or multiplier herds (see Fig. 17.1a). A typical breeding pyramid, in the case of a four-way crossing system, without artificial insemination, would show sow numbers in the approximate ratio 1:7:70 for the respective nucleus, multiplier and commercial herds (e.g. Sellier, 1986, p.217).

Such an organization usually requires a more or less complete integration of the two upper tiers of the breeding pyramid. This has been attempted by independent breeding organizations, either private or cooperative, which appeared in a number of countries during the 1960s, a notable early exception being Farmers' Hybrid created in 1945 in the USA. Several of those breeding companies now operate internationally. In many countries, national schemes have also been implemented, following the early Danish example mentioned in the introduction to this chapter. In such structures, usually more dispersed than a single company nucleus, animals from several breeding farms are evaluated at central testing facilities, and the selected animals may then be used on the national herd by artificial insemination (see Fig. 17.1b). The role played by breeding companies varies greatly between countries, as they can provide from over 50% of the replacements in some countries of Western Europe to hardly any in Scandinavian countries. An overview of the situation in the late 1980s has been presented by Sellier and Rothschild (1991).

(a)

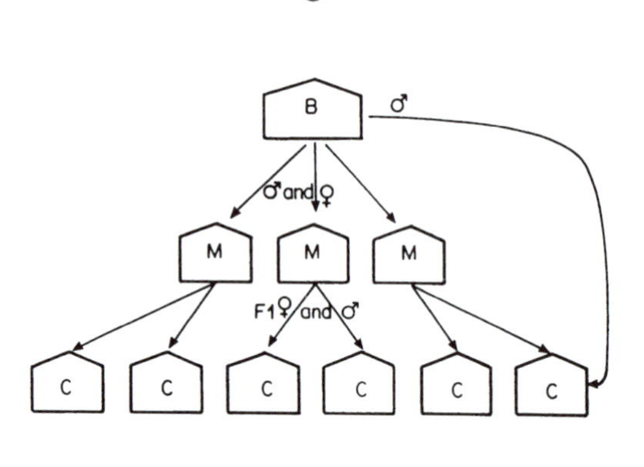

(b)

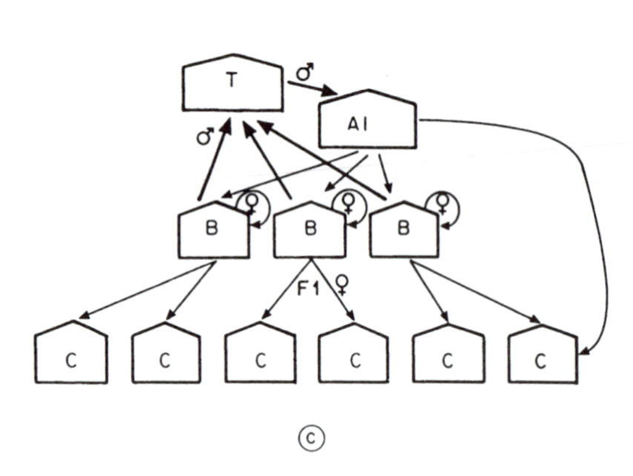

(c)

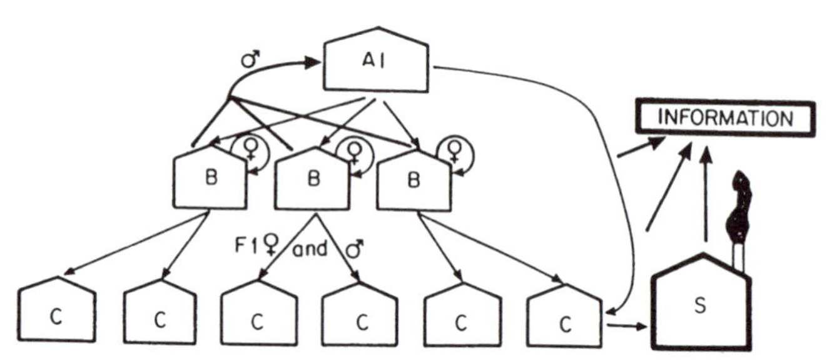

Fig. 17.1. Structures of pig breeding programmes: (a) standard breeding company pyramid, made up of a breeding nucleus (B), multiplier (M) and commercial (C) farms; (b) a national breeding structure, with breeding (B) and commercial (C) farms, central testing (T) and artificial insemination centres (AI); (c) a potential future structure encompassing breeding companies and a national programme, with breeding farms (B) genetically linked through AI, and information collected from B, commercial farms (C) and slaughterhouses (S) for genetic evaluations. (Graph adapted from Figure 4 in Haley (1991, p. 316) with the kind permission of the author. M. Weber is gratefully acknowledged for drawing the figure.)

The situation has more recently been reviewed by Brascamp (1994), who stressed the role that national breeding programmes have had in the past and that they still have in many countries, in contrast with the poultry situation for instance, where hardly any national involvement exists. Various routes have been taken in the countries which have tried to maintain national programmes. The extremes are on one side a *closed selection* system, where a breeding company system is, so to speak, extended to the whole country, as in Denmark for instance, and on the other end an *open selection* system, allowing for instance genetic evaluation to be carried across breeding organizations of different status. In that respect, France may be an extreme case, where most breeding companies are incorporated in the national programme. An overview of the relative market shares of breeding companies and national programmes has been given by Brascamp (1994), covering several European countries, Canada and Australia.

Artificial insemination (AI)

The advantages of AI regarding health control, herd management and genetic improvement have been recognized for a long time, and AI has played an important role in several breeding programmes. In their review of the place of pig AI in various countries, Sellier and Rothschild (1991) concluded that, prior to 1991, it had remained on average at a low level. Recently, because of the development of semen delivery services, which enable the individual farmer to carry out his own inseminations, an increase in artificial breeding has been observed in many countries. In France for instance, the percentage of AI, which had remained at a low level of about 3% of the matings until 1987, has been increasing very rapidly since and AI now contributes to more than 50% of the litters produced.

As pointed out by Haley (1991), the structure of national and company breeding programmes could converge in the future, with genetic links between herds being increasingly provided by AI. Evaluation would then take place centrally, with appropriate methodologies (see below) and data coming from a number of sources, including breeding farms, commercial farms and slaughterhouses (see Fig. 17.1c).

The Definition of an Overall Breeding Objective

Choice of breeding goals

The choice of breeding goals has always been a major preoccupation in pig breeding. Since the early goal of preserving breed type in traditional breed societies, the evolution has been towards increased attention being paid to the economic considerations relevant to the pig producers. This has developed along with the evolution and diversification of the breeding structures, as outlined above. International exchanges of breeding stock have now become common practice and worldwide perspectives tend to be increasingly considered by breeding organizations.

There will usually be a number of breeding goals in any pig production system. They classically fall into two categories, i.e. *reproduction* and *production* traits, which are of interest to weaner (feeder pigs) producing farms or to slaughter pig producers respectively. The first group of traits includes the main components of sow productivity, such as age at puberty, conception rate, number born alive per litter, number weaned and weaning to oestrus interval. Production traits include production costs, such as traits associated with growth rate and food conversion, as well as components of product value, such as lean content and quality of the lean and fat tissues. Other traits such as leg structure may also play a role.

Derivation of economic weights

Once the breeding goals have been chosen, the question arises of giving to each of them its proper weight in an overall objective, also termed aggregate genotype (H), because the targets are the breeding values (A) of each element. A common practice is to use profit equations of the form:

$$P(\text{profit}) = R(\text{returns}) - C(\text{costs}) \quad (17.1)$$

By taking the partial derivatives of P with respect to the n traits included as goals in R and C, economic weights are obtained and an aggregate genotype, linear in the objectives, can be established, i.e. $H = a_1A_1 + a_2A_2 + \ldots + a_nA_n$. This has been the approach in the classic paper of Hazel (1943) on selection index theory. However, as pointed out by Moav (1973), the weights so derived depend on the perspective taken in defining P. This dilemma of different economic weights for different perspectives in production, under unchanged economic conditions, has been resolved by Smith *et al.* (1986), who suggested imposing two conditions. The first is that fixed costs, incurred in running the production enterprise, should be included with other costs and expressed per unit of output. The second condition is that any extra profit from genetic change that could also be obtained by altering the size of the enterprise should not be counted. The application of those two conditions also shows that the economic weights are those obtained in considering cost per unit return, i.e. C/R instead of $R - C$. More complex economic models have recently been advocated as possible alternatives (see Amer *et al.*, 1994).

Economic values may also be derived from models describing life-cycle efficiency as a function of production costs and outputs, which arise from both the sow herd and the pigs marketed. This is the approach of the bioeconomic model of Tess *et al.* (1983), as well as of the model of De Vries (1989a), to estimate efficiency of production. Change in efficiency as a response to a given change of any trait may be computed directly in such models, yielding the economic weights needed to establish an aggregate genotype. A comparison among various approaches used to define aggregate genotypes for reproduction and production traits is given in Table 17.1.

The use of a profit equation such as (17.1) assumes a well-defined relationship between P (profit) and each of the traits included in the right hand side of the equation. If this is usually the case for C (costs), the same does not always hold for R (returns), as shown by the difficulty of properly integrating meat quality traits in the overall objective. The 'real' economic value of such traits at the producer's level is indeed zero in most countries, because of the lack of proper online measurement systems on which a payment system could be based. Arbitrary choices have then to be made, such as, for instance, applying a constraint of zero genetic change for such traits, which assumes a present satisfactory level of the trait, a system applied in several European countries for meat quality (see Sellier, 1988, and also Chapter 16). An alternative is to rely on surveys, whereby customers are asked to indicate their willingness to pay different prices for different quality classes. Such a method has been suggested for determining economic values of various meat and fat quality traits in Switzerland (von Rohr *et al.*, 1996).

Adaptation of breeding objectives to the breeding–production system

The need to develop a variety of breeding stocks in order to satisfy the widest possible range of potential customers has been stressed by Smith (1986). The first level at which diversity of breeding objectives has to be considered is between lines in the system of specialized sire and dam lines, nowadays extensively adopted. Smith (1964) laid out the principles on which to base the choice of a breeding objective in each line, by considering the profit made in the slaughter generation obtained in the crossing system, which involves both sow productivity and market pig performances. As shown in Table 17.2, when three breeds are available and allow six crossbreeding schemes combining them in various manners, four different breeding objectives should be defined. In such a situation, a breeding company wishing to provide optimal breeding stock for the six systems considered would have to select seven different lines.

When different production–marketing systems are considered, additional variation has to be taken into account. The relative emphasis on production and reproduction traits, i.e. the value of a in Table 17.2, may vary considerably. Relative emphasis may also vary within a group of traits. Tess *et al.* (1983) give an extreme example of such variation, showing that increased lean growth may have adverse effect on economic efficiency when increased feed costs are not compensated by an adequate payment for lean content. Given that the number of lines which can be selected by a breeding organization meets financial limitations, some strategy is required in order to maximize genetic gains over a range of production conditions at a reasonable cost. Two possibilities for defining the relative weights of reproduction and production traits, i.e. an average versus an extreme strategy, have been simulated by Ollivier *et al.* (1990), who show that an extreme strategy may in some situations yield substantial benefits.

An example of extreme strategy is the 'hyperprolific' scheme proposed by Legault and Gruand (1976), which has attracted the attention of several breeding organizations. Basically, this scheme implies a two-stage selection of boars, i.e.

Table 17.1. Relative importance of reproduction and production traits.

Trait	Relative importance[1]		
	Tess[2]	de Vries[3]	Bidanel[4]
Reproduction			
Age at puberty	6	3	6
Conception rate	28	36	14
Number born alive/litter	35	34	48
Piglet viability	31	27	32
Total	100	100	100
Production			
Growth rate	28	38	15
Food conversion	—	20	23
Carcass percent of live weight	—	12	16
Lean per cent of carcass	72	30	46
Total	100	100	100

[1]Relative increase (%) in profitability expected from an increase of one phenotypic standard deviation of each trait.
[2]Tess *et al.*, 1983 (USA: cost per unit of lean meat).
[3]De Vries, 1989a (The Netherlands: return per slaughter pig).
[4]Bidanel, 1988, for reproduction; Ducos, 1994, for production (France: profit function).

Table 17.2. Breeding objectives for reproduction (H_1) and production (H_2) traits combined according to Smith (1964) in crossbreeding systems using three different breeds.

Crossbreeding system (dam × sire)		Breed		
		A[1]	B	C
Single cross	A × C	$aH_1 + 0.5H_2$	—	H_2
Two-breed rotation	[(A × B) × A] × B . . .	$aH_1 + H_2$	$aH_1 + H_2$	—
Three-breed rotation	{[(A × B) × C] × A} × B . .	$aH_1 + H_2$	$aH_1 + H_2$	$aH_1 + H_2$
Back-cross	(A × B) × B	$aH_1 + 0.5H_2$	$aH_1 + 1.5H_2$	—
Three-way cross	(A × B) × C	$aH_1 + 0.5H_2$	$aH_1 + 0.5H_2$	H_2
Four-way cross	(A × B) × (B × C)	$aH_1 + 0.5H_2$	$aH_1 + H_2$	H_2
Number of lines (breeding objectives) per breed		2	3	2

[1]a = the value of one standard deviation change in H_1 relative to the value of n standard deviations change in H_2, n being the number of offspring marketed per litter (see Smith, 1964, p. 337).

a very intense first-stage selection on dam litter size averaged over four successive parities, followed by a milder second-stage selection on individual lean growth. Given the respective selection intensities of about 3 and 2 on the two traits, assuming phenotypic independence and the usual genetic parameters of litter size and lean growth, it can be shown that such a selection is equivalent to using the optimal selection index corresponding to a value of 9 for the *a* parameter defined in Table 17.2, i.e the ratio of economic weights of litter size relative to lean growth.

Profit of the breeding organizations should not be disregarded, because in a competitive environment saleability of the breeding stock may actually become their primary objective. When several breeding companies compete in the same market, differences in performances play a major role in determining the market share of each of them. They tend to put more emphasis on traits for which they are behind their competitors, thus implicitly assuming that a higher performance in one trait does not compensate for a lower performance in the other. This is often done rather empirically, but it can be done in a more elaborate way by manipulating the selection index system (Brascamp, 1984). An objective way of incorporating competitive position in the breeding goal has been proposed by De Vries (1989b), using a marketing approach. Given an acceptance level normally distributed over the customers, the performance level of a given line determines a percentage p of this distribution which is below the acceptance level, p being also the percentage of customers satisfied. In order to optimize saleability, the weight to be given to the trait in the breeding objective should be the economic value of the trait for the producer, as for instance derived from a profit function, multiplied by 1.25 z/p (where z is the ordinate of the standardized normal curve at the cut-off point corresponding to p). When there is compensation between traits, the mean acceptance level has to be corrected, knowing the average performance of the competitors and a compensation factor going from 0 (no compensation) to 1 (full compensation). In addition to the difficulties of application of the method, as discussed by De Vries (1989b), it should be realized that the economic weights then deviate from their economic optimum, and consequently the overall response is reduced. The method can only be considered as a short-term strategy to correct for weaknesses of a particular breed or strain, and it would need periodical revision.

Adaptation of breeding objectives to changing economic conditions

Past experience shows that re-evaluation of the economic weights applied to each trait is needed continuously. In spite of the efforts made to foresee the changes to be expected in the economic conditions, this exercise can never be entirely satisfactory, because of inevitable discrepancies between present objectives and future production conditions. However, this should be of little concern as long as the evolution of these conditions is slow and gradual, because using slightly 'false' economic weights has a limited impact on overall selection efficiency in most cases (Vandepitte and Hazel, 1977).

Present and future trends in breeding objectives have been discussed in detail by Ollivier *et al.* (1990) and Haley (1991), among others. Traits such as feed intake capacity, quality of fat and lean tissues, and various components of the sow reproductive ability are likely to receive more and more attention. An overall view on past and future evolution of breeding objectives is given in Table 17.3. The prospects on carcass and meat traits are also discussed in Chapter 16.

The Choice of Selection Criteria and Their Use in Breeding Value Estimation

A distinction is classically made between traits considered as *objectives* for improvement, and traits actually used in ranking the male and female candidates, which are termed *criteria* of selection. In pig breeding, the two sets only partly coincide, and each set includes a fairly large number of traits. It is the purpose of performance recording programmes to define the measurements to be used as selection criteria and to organize the collection and processing of the corresponding data. Such programmes will now be reviewed briefly, as more details can be found in several textbooks on pig breeding, e.g. in Sellier (1986) or Glodek (1992) among others. A historical presentation of performance recording and genetic evaluation since the beginning of the century has also been given in Sellier and Rothschild (1991).

Reproduction performance has for a long time been assessed through on-farm litter-recording systems, including litter size at birth and at weaning, sometimes completed by litter weights, these traits being viewed as the most important reproduction traits (see Chapter 11 for more details). *Production*

Table 17.3. Breeding objectives in pigs: past, present and near future (from Ollivier *et al.*, 1990).

	Breeding objective by year[1]		
	1980	1990	2000
Reproduction traits			
Litter size	+	++	+++
Sow productivity	0	+	++
Other (longevity, uterine capacity . . .)	0	0	+
Production traits			
Growth rate	++	++	++
Food conversion	+++	++	++
Lean content	+++	++	+
Meat quality	+	++	+++
Fat quality	0	0	+

[1]0, +, ++, +++ : increasing emphasis on the trait in the aggregate genotype.

traits, namely growth rate, feed efficiency and carcass measurements (see Chapter 15), were initially recorded almost exclusively in central testing stations, built on the Danish model, as progeny testing stations became progressively available to pig breeders in most countries from the early 1920s to the late 1950s. An important step has been the advent of techniques allowing a fairly accurate evaluation of body composition on the live pig, starting with the metal ruler of Hazel and Kline (1952), followed by the development of ultrasonic machines and more sophisticated technologies such as computer tomography. This has opened the way to central performance testing stations of young boars, replacing progeny-testing stations, and also to on-farm testing programmes. Boar testing stations usually record average daily gain over a given live weight interval, food conversion ratio over the same interval and backfat thickness at the end of test, whereas on-farm tests generally include only daily gain (expressed as days to a given weight) and backfat thickness at a given weight. More recently, meat and fat quality traits have been introduced into testing programmes, based on measurements made in slaughterhouses, such as ultimate pH, colour and water-holding capacity. More details on the meat and carcass traits of interest can be found in Chapter 16, and especially on the advances to be expected regarding live animal measurements based on tissue biopsies.

As shown by Hazel in his classical paper of 1943, pig improvement is essentially a multiple-trait selection problem, which can be solved by combining performance records into a linear index (I). This index is a *predictor of breeding value*, defined so as to maximize the correlation between I and the aggregate genotype (H), described in the previous section. This principle has eventually become almost universally applied in pig breeding, use being made of the necessary genetic parameters (heritabilities and genetic correlations), as presented in other chapters of this book. The performance records entering the selection indices, collected either on farm or in central test stations, are generally expressed as deviations from contemporary group means. In the meantime, selection index theory was being refined through the development of mixed model methodologies taking account of unequal information among candidates and unknown means, and providing best linear unbiased predictions (BLUP) of breeding values. Application of the theory could then be made in a two-step procedure, first the estimation of individual trait breeding values and, in a second step, the application of relative weights to those breeding values for deriving the estimation of H. As noted by Hazel *et al.* (1994), such a procedure brings more flexibility in the adaptation of the economic weights to any breeding system, without the need to recalculate individual breeding values.

However, BLUP methodology has shortcomings due to the heavy computations involved, and particularly so in pigs, where selection decisions have to be taken rapidly and repeatedly, and where records are available for a large number of traits as well as large numbers of relatives. Such situations make it difficult to provide breeding farms with their proper BLUP of breeding values in a timely manner. This has been done for the first time in Canada in 1985 for growth and backfat on-farm genetic evaluations (Hudson and Kennedy, 1985), later extended to maternal traits. In the USA, a cooperative project between

Purdue University and purebred associations, termed Swine Testing and Genetic Evaluation System (STAGES), was also initiated in 1985 and reported its first evaluations in 1986 (Stewart *et al.*, 1991). STAGES and the Canadian programme are designed as on-farm evaluation systems with the added capability of across-herd analysis performed at longer time intervals. Progress in speed of data transmission and computing has now made it possible to shorten these intervals and to make centrally-produced BLUP available to the breeding farms within a few days. Such systems are currently applied by several European countries (Ducos, 1994), in most cases with implementation of the multivariate prediction–estimation software (PEST) of Groeneveld *et al.* (1990). Most breeding companies also apply similar procedures.

The Design of Breeding Programmes

General principles

The first objective of a breeding programme is to produce the 'most improvement per unit of time', as stated by Dickerson and Hazel (1944). They showed that selection response depends on three parameters, which may differ between the sexes, i.e. selection accuracy (ρ, defined as the correlation between the aggregate genotype and the selection criterion), selection intensity (i) or standardized selection differential, and generation interval (t). The expected *annual response* (R_a), expressed in genetic standard deviation units, is:

$$R_a = (i_1 \rho_1 + i_2 \rho_2)/(t_1 + t_2) \quad (17.2)$$

where the indices refer to dams and sires respectively. R_a is seen to be a function of both genetics (ρ) and demography (i,t), and an efficient breeding programme thus depends on a proper choice of evaluation methods and replacement policies. As demographic parameters are largely determined by the reproductive rate of the species, an upper limit of R_a exists for any given selection method. This limit corresponds to an optimal replacement policy for breeding males and females (Ollivier, 1974).

A second aspect to consider is the *dissemination* of the genetic changes from the nucleus down to the multiplier and commercial farms. This process creates genetic lags (Bichard, 1971), which can be minimized by acting on the structure of the breeding pyramid (using AI for instance), as well as on the genetic level of the individuals migrating from one tier to the next, using the gene flow techniques proposed by Elsen and Mocquot (1974) and Hill (1974). By introducing cost–benefit considerations at all levels of the breeding pyramid, as Elsen and Sellier (1978) did for determining an optimal selection policy in dam lines, an economic optimization of the whole system may also be attempted.

Chance is another important factor to consider in most cases, as intense selection in small closed populations is known to generate genetic drift and to increase the inbreeding coefficient. Stochastic models can take into account the variability expected in genetic responses, as well as the changes in genetic variances and covariances under selection in populations of limited size. Models such as those simulated by Belonsky and Kennedy (1988) and by De Roo (1988) are now increasingly relied upon in the study of pig breeding schemes.

Breeding for production traits

Production traits in pigs are measurable on both sexes before breeding (growth rate, food conversion and backfat thickness), or after slaughter (lean content, lean and fat characteristics). Individual and sib information therefore provide the essential criteria of selection in most breeding programmes, as additional culling on progeny performance has been known for a long time to be ineffective (Dickerson and Hazel, 1944; King, 1955). The expected annual response in equation (17.2) is then maximized when $i/t = (i_1 + i_2)/(t_1 + t_2)$, i.e. the annual selection intensity, is at its maximum. With the common demographic parameters of the pig, such as one year of age at first offspring, six candidates per litter at breeding age and a mating ratio (sow : boar) of 15 under conditions of natural mating, it can be shown that the optimal t_1 and t_2 in individual selection are close to one year, which yields a maximum i/t of 1.75 (Ollivier, 1974). For a trait of medium heritability, $h^2 = 0.30$, a maximum annual response of nearly 1 genetic standard deviation can be obtained with such a method of selection.

When information on slaughtered sibs is used in selection, the value of i/t has to be adapted to the ensuing reduction in the number of candidates remaining and change of sex ratio among them (Ollivier, 1988a). The loss in i/t incurred in any sib, or combined sib–individual, testing scheme can then be set against the increased accuracy of evaluation.

It should be borne in mind that the responses predicted in such a theory refer to an idealized situation of a large population of sows, farrowing simultaneously at fixed intervals of six months, and that selection is assumed to be carried out among a large number of independent observations. In practice, however, those assumptions are not fulfilled, as farrowings do generally occur quasi-continuously in breeding herds, and comparisons between candidates are made within 'batches' containing a limited number of full-sib groups. The resulting reduction in selection intensity for the nested family structure typical of the pig has been worked out by Meuwissen (1991). It can be shown that in a herd producing 100 gilt litters per year in 17 batches, the expected maximum of i/t in individual selection is reduced by approximately 15%, to a value of 1.5, with the same demographic parameters as above (Ollivier, 1988b). The latter objective will even rarely be achieved in practice, because of incomplete testing or of culling for reasons other than performances. Actual values of i/t achieved in individual on-farm tests rarely exceed 1. Considerably lower values have been reported in national programmes emphasizing either on-farm tests or central station family selection (Table 17.4). It can be noted that the individual selection schemes simulated by De Roo (1988) and Belonsky and Kennedy (1988) do not

Table 17.4. Retrospective evaluation of selection intensities and generation intervals in breeding schemes for production traits.

Breeding structure	Selection criterion	Selection intensity (i)	Generation interval (t)	i/t[1]
Breeding company	Individual[2]	1.28	1.25	1.02
National programme	Individual[3]	0.48	1.92	0.25
	Sib and progeny[4]	1.46	2.00	0.73
	Progeny[5]	0.91	2.00	0.46

[1]For i/t = 0.5, using an individual selection index, expected annual responses of 5 g in average daily gain and –0.4 mm in backfat thickness have been given by Sellier (1986, p.201). The ranges of annual genetic trends reported by Sellier and Rothschild (1991), 3–6 g and –0.1–0.4 mm based on central testing records, indicate that the corresponding i/t would generally be below 0.5.
[2]Bichard *et al.* (1986) : 1966–1985; growth and backfat index.
[3] Kennedy *et al.* (1986) : 1977–1983; growth and backfat index in retrospect (Canadian Yorkshire and Landrace).
[4]Christensen *et al.* (1986) : 1980–1985; index on growth, lean and meat quality (Danish Landrace and Yorkshire).
[5]Lundeheim *et al.* (1994) : 1982–1986; growth and lean index in retrospect (Swedish Landrace and Yorkshire).

exploit fully the potential of such schemes, as the values of i/t realized are generally well below 1.

With the advent of BLUP (see above), records from all relatives, such as sibs, cousins and ancestors, can be used to predict breeding values. Genetic response is then expected to increase, following the increases in average accuracies ρ_1 and ρ_2 of equation (17.2), predicted from standard index selection theory, and knowing that those increases are partially offset by the decreases expected in selection intensities, because of the high correlation between BLUP of relatives. A further increase of response, more difficult to predict because it depends on the breeding structure, results from a better estimation of the fixed effects which allows across-farm (or station) evaluation and also from taking into account genetic trends (Sorensen, 1988). Another advantage of BLUP is to allow sequential culling, whereby sows and boars are culled on the basis of their estimated breeding values, the best remaining longest in the herd. The simulation of Belonsky and Kennedy (1988) shows that this culling scheme reduces generation intervals from 16 to 22% relative to strictly individual selection. This advantage actually depends on the herd replacement policy, which in this study implied culling sows after a maximum of five litters. Hagenbuch and Hill (1978) have shown that with a more stringent policy of keeping sows only for a maximum of two litters, the gain from sequential culling is much reduced, to about 2–3%. Overall, the advantage of BLUP evaluation over individual selection is in a range of 10–30% for traits of high to moderate heritability, i.e. for most production traits, as shown by various simulations (Belonsky and Kennedy, 1988; Sorensen, 1988; Röhe *et al.*, 1990).

The gains from BLUP selection are to some extent counterbalanced in small populations by a higher increase in inbreeding than under individual selection, a tendency which is also enhanced by sequential culling, as shown by Belonsky and Kennedy (1988). Various methods to restrict inbreeding without significant loss of response have been proposed by Toro and Perez-Enciso (1990). Another strategy, recommended by Brisbane and Gibson (1995), is to include genetic relationships in selection decisions, assuming a given value of a unit of inbreeding relative to a unit of genetic gain.

Breeding for production and reproduction traits

As shown in Table 17.2, reproduction traits have to be included in the breeding objective in crossbreeding systems with specialized lines, apart from the breeds or lines which only serve to produce the terminal boars. However, until recently, little attention was paid to such traits in most breeding programmes. The situation is now progressively changing, as a combination of theoretical (see Table 17.5) and experimental results (see Chapter 11) have shown that litter size can be successfully improved by selection, and given also that economic conditions will tend to make such selection increasingly worth-while (Haley *et al.*, 1988; Ollivier, 1988b).

The efficiency of simultaneous selection for reproduction and production traits has been extensively investigated, either by using the index selection approach of Smith (1964), or by more elaborate methods based on the economic returns of the entire crossbreeding system (Elsen and Sellier, 1978), life-cycle economic efficiency (Smith *et al.*, 1983), or stochastic models in closed dam lines (De Vries *et al.*, 1989). The general conclusion of those studies is that some benefit is expected from the inclusion of the reproduction objective (H_1) in addition to the production objective (H_2) in specialized dam lines. The benefit will increase with the relative importance (a) of H_1 in the overall breeding objective, as defined in Table 17.2, with the accuracy (r_1) of evaluating H_1

Table 17.5. Predicted annual selection response in litter size (number born per litter).

	Reference[1]			
	a	*b*	*c*	*d*
Annual response	0.25	0.47	0.34	0.17
Selection criterion	Dam	←	Dam and family	→
Population size	Large	Large	100 Sows	400 Sows

[1] References:
a Ollivier (1973)
b Avalos and Smith (1987)
c Toro *et al.* (1988)
d De Vries *et al.* (1989)

{ Theoretical predictions based on dam and family selection indices of prolificacy, with two litters recorded. (*a*–*c*)

Simulated selection on an index combining reproduction and production traits. (*d*) relative to H_2 (r_2), and with an increasingly unfavourable genetic correlation (r) between H_1 and H_2. A useful approximation of the relative benefit (*RB*) in terms of those parameters has been proposed by W.G. Hill (in Webb and Bampton, 1987):

$$RB = [(x^2 + xr + 0.5)^{0.5} + 0.5]/(x^2 + 2rx + 1) \quad (17.3)$$

in which $x = ar_1/r_2$.

By comparing the simulations of De Roo (1988) and De Vries *et al.* (1989), who use the same stochastic models and the same economic parameters, the benefit from including reproduction traits among the goals of the dam line appears to be about 25%, in a situation of genetic independence between H_1 and H_2. This is markedly above the predictions of Webb and Bampton (1987) in their projection to 1992. Gains of 5 to 15% with $r = 0$ and of 6 to 22% with $r = -0.2$ were anticipated. Though most studies have so far concluded that litter size is genetically uncorrelated with growth and carcass traits (see the review of Haley *et al.*, 1988), there are indications that this might not be a general rule, as shown by the indirect responses observed in the selection experiment analysed by Kerr and Cameron (1994), or by recent estimations of genetic correlations between reproduction and production traits (see Chapter 11).

Breeding based on crossbred performance

The breeding programmes discussed so far assume that the selection criteria exclusively rely on purebred information, which is the common practice in pigs. Therefore predicted responses in a crossbreeding scheme implicitely assume a genetic correlation of 1 between purebred and crossbred performance. The use of crossbred information, as in reciprocal recurrent selection schemes, offers some advantages compared to pure line selection especially for traits showing large non-additive genetic variation, as reviewed by Sellier (1982) and more recently by Wei and van der Steen (1991). Two specific parameters are needed to evaluate the efficiency of such a type of selection, namely the genetic correlation between purebred and crossbred performance (r_{pc}) and the crossbred heritability (h_c^2). Purebred and crossbred information should in fact be combined in an optimal way in order to maximize genetic response in crossbreds, which can be done within the general framework of index selection theory (Wei and van der Werf, 1994). For instance, the crossbred paternal half-sib family mean may be combined in such a way to the purebred information on paternal half-sibs, full-sibs and own performance. As emphasized by Wei and van der Werf (1994), such a combined selection is always superior to pure-line selection when testing of crossbreds is not at the expense of testing purebreds, as for instance when a crossbreeding structure exists, through which crossbred information is collected for management purposes. In such situations, which apply to pig breeding (see for instance Fig. 17.1c), crossbred information may indeed come as an addition to purebred data and the increase that can be expected in genetic response will then depend on a proper evaluation of the two genetic parameters, r_{pc} and h_c^2. Precise information on these parameters, however, is so far lacking in pigs.

Breeding for particular objectives

The breeding programmes which have been described above are in principle able to address most objectives of present-day pig production, the relevant traits being amenable to a treatment in the framework of classical quantitative genetics theory (Falconer, 1989). Exceptions to this general rule are however worth mentioning, such as the elimination of genetic abnormalities or breeding for disease resistance. The existence of chromosomal aberrations responsible for drastic reductions in litter size, as in the case of reciprocal translocations, makes it worth-while to screen the boars of paternal and maternal lines for karyotype abnormalities. The screening may be systematic, although it will usually be more profitable to restrict cytogenetical investigations to boars known for having sired litters well below average size in the population, so-called hypoprolific boars (Popescu *et al.*, 1984; Popescu and Legault, 1988). Another field of some importance is resistance to diseases, especially when genes of resistance are known, as for various forms of diarrhoea (Gibbons *et al.*, 1977; Bertschinger *et al.*, 1993). Breeding strategies for resistance to neonatal diarrhoea due to various strains of K88 *Escherichia coli* have, in particular, been discussed by Ollivier and Renjifo (1991), who showed the advantage of selecting sire lines for the genes of resistance, while disregarding their frequency in dam lines. As will be seen in the next section, genetic markers are expected to enhance the efficiency of breeding for such specific traits.

Marker-assisted Breeding

In the late 1980s, it became apparent that progress in molecular genetics, and particularly the discovery of new classes of DNA polymorphisms, could contribute to more efficient breeding programmes. This prospect has now materialized for most farm animals including the pig, with the linkage maps developed in recent years providing abundant and highly polymorphic markers evenly spaced over the 19 chromosomes of the species (see Chapters 8 and 9). This situation has given a new impetus to the theoretical investigations initiated by Neimann-Sorensen and Robertson (1961) on the role that individual gene-marker identification could play in breeding schemes, compared to classical quantitative genetics methods relying only on measurements of the traits of interest. The use of genetic markers to enhance genetic progress in the pig has been recently discussed by Visscher and Haley (1995).

The assistance provided by genetic markers had already been illustrated in the 1980s by the use of biochemical markers closely linked to the halothane gene in order to eliminate the susceptibility allele from maternal lines. Such a scheme was applied successfully in several Landrace breeds in Europe (see

Amigues *et al.*, 1994, for a report on the French case). The objective was to guarantee obtaining of a halothane-negative slaughter generation, protected from the deleterious effects of the recessive susceptibility gene on liveability and PSS-related meat characteristics, even though terminal boars carrying the susceptibility gene might happen to be employed. This example enters a first category of marker-assisted breeding, whereby gene frequency at a locus of interest can be acted upon by using markers, sometimes more efficiently than through a direct approach based on the phenotypic effects of the gene itself. This type of use of markers applies to *monogenic traits*, usually with an all-or-none pattern of effect. A second category of traits consists of *polygenic traits* resulting from the combined effects of several quantitative trait loci (QTL) and the environment. Major genes, with large phenotypic effects against a polygenic background, are intermediate, but they can in most cases be assimilated to single-locus situations. A third category of application is the use of markers to characterize the *whole genome*. In the following presentation, it will be assumed that the genes of interest are precisely located on a genetic map and that markers closely linked to them are available, the relevant mapping methods being presented elsewhere (see Chapter 9).

Markers for single-locus traits, or major genes

A first objective is to change gene frequency at a given locus by selecting this gene (when possible) as well as marker genes at one or several other linked loci. The efficiency of such a *marker-assisted selection* (MAS) programme depends heavily on the linkage disequilibria existing between the loci implied. Linkage disequilibrium (D) is a population genetics parameter measuring associations between genes at two loci in the process of their transmission to the progeny within a given population. The fate of the loci of interest under MAS depends on the marker genes 'hitchhiking' those of interest, in a manner that can be quantified by standard two-locus selection theory (see, for example, Karlin, 1975). The basic parameters in a biallelic situation are the initial gametic frequencies x_1, x_2, x_3, x_4, which determine the initial linkage disequilibrium $D = x_1x_4 - x_2x_3$, the recombination fraction between the two loci and the selective values of the nine genotypes. Neglecting unlikely epistatic effects, selective values can be derived from the selection procedures considered at both loci. At the marker locus the selection procedure is usually fixed by the breeder, whereas at the locus of interest selection may be natural, e.g. against a lethal recessive gene, or artificial, e.g. if the candidates expressing the trait are eliminated, or even neutral if selection is assumed to be limited to the marker locus. This extreme case is envisaged in Table 17.6, with two different recombination fractions and two different values of the initial linkage disequilibrium between the marker locus and the locus of interest. It can be seen that a strong initial linkage disequilibrium is more important than a tight linkage in such a situation.

When the initial linkage disequilibrium is large, selection on the marker locus may be more advantageous than direct selection, which is obviously the

Table 17.6. Hitchiking effect of a marker locus in the elimination of a neutral recessive gene linked to the marker, with complete or incomplete initial linkage disequilibrium (*D*).

		Frequency of the hitchhiked gene[1]	
Generation	(*)[2]	*D* = 1	*D* = 0.05
0	(0.20)	0.20	0.20
		0.20	0.20
1	(0.17)	0.10	0.15
		0.10	0.15
2	(0.14)	0.05	0.13
		0.05	0.13
5	(0.10)	0.01	0.10
		0.01	0.11
10	(0.07)	0.00	0.10
		0.01	0.11
100	(0.01)	0.00	0.10
		0.01	0.11

[1]First line, 1 cM between loci; second line, 10 cM between loci; D = standardized linkage disequilibrium; marker selection: elimination of the homozygotes and half of the heterozygotes for the marker allele associated to the undesirable gene.
[2](*) Frequency of the recessive gene when directly selected against.

case in the elimination of a rare deleterious recessive gene, as in Table 17.6 and in the halothane example mentioned previously. It should be noted that this advantage is relative to the knowledge at hand about the locus of interest, as a marker locus obviously cannot be of any help when the alleles at the neighbouring locus are easy to identify. In such a case the locus of interest may be considered as being itself marked.

In practical breeding situations, other advantages of MAS come from the possibilities it offers of selecting both sexes (for sex-limited traits), of selecting earlier (for reproduction traits or traits measured on carcass) and often at lower costs. The present situation of the pig linkage map offers several possibilities which could be exploited in a not too distant future. Meat quality trait major genes, such as for yield of cured–cooked meat (Milan *et al.*, 1996), or for intramuscular fat content (Janss *et al.*, 1994) and genetic resistances to various forms of diarrhoea for which single genes are precisely mapped on chromosome 18 (Guérin *et al.*, 1993; Edfors-Lilja *et al.*, 1995) and chromosome 6 (Meijerink *et al.*, 1996), are also areas of interest for future MAS in pigs.

Whatever the situation, the efficiency of such a type of selection is expected to decrease progressively in time because of the recombinations occurring between linked loci at each generation (see Table 17.6). The decrease may be minimized if the locus of interest can be bracketed by two marker loci. The relevant linkage disequilibrium is then between the segment bracketed (or marker haplotype) and the trait locus included in the segment, with a rate of decay corresponding approximately to the probability of double recombination, e.g. 0.25 r^2 for a recombination proportion r between the two markers, i.e. about 1% for a 20 centimorgan (cM) long segment.

Another use of markers is for introgressing individual genes from one breed into another. This can be done more efficiently when markers are available around the gene to be introgressed. The starting point of an introgression scheme is usually a cross between two breeds, a donor and a recipient, a situation likely to yield the strong linkage disequilibria needed for successful *marker-assisted introgression* (MAI). This has been exploited in introgressing the halothane-resistance allele into the Piétrain breed in Belgium. By typing the closely linked *GPI* (glucose phosphate isomerase) locus, Hanset *et al.* (1995) were able to fix the Large White normal allele and to obtain a halothane-negative Piétrain strain after three backcrosses. The possibility of relying on marker brackets for increasing the efficiency of introgressing unidentified genes had also been previously demonstrated by Soller and Plotkin-Hazan (1977). They showed, for instance, that the frequency of the introgressed gene could be maintained at 0.94 among homozygotes for two markers 20 cM apart, as against only 0.53 among homozygotes for a single marker 10 cM apart from the gene introgressed, after five backcross generations and an intercross generation necessary for recovering the homozygotes. During the many generations needed for successful introgression, the rest of the genome introduced from the donor breed cannot be ignored, and a quick recovery of the full genome of the recipient breed is usually looked for, leading to another way in which markers can be useful, as we shall see in the next section.

Markers for polygenic traits

Referring to the general principles of efficient selection discussed above, the advantage of marker-assisted selection (MAS) for polygenic traits will depend on the way the three essential parameters, i, t, and ρ of equation (17.2), may be affected, as stressed by Smith (1967). First, selection on markers may allow a *higher selection intensity*, because of the increase in the number of candidates, when typing is made at an early age, and more so for traits expressed only in one sex or traits that can only be assessed on carcass. Secondly, earlier selection and consequently *shorter generation intervals* are made possible, as markers can be typed early in life and long before the trait can be measured. However, in the pig, where short generation intervals are usually achieved, this early selection may not be of great advantage. Thirdly, a *more accurate selection* can be performed due to the additional information brought in by the markers. Lande and Thompson (1990) have investigated the relative efficiency of various MAS schemes, combining classical selection indices (I_Q) for quantitative traits, based on individual and family information, and marker information summarized by a 'molecular score' or index (I_M). The MAS criterion is then a linear combination $I = b_1 I_Q + b_2 I_M$ which maximizes the expected response. The accuracy R of I is a function of the accuracy ρ of I_Q and of the proportion m of additive genetic variance associated with the marker loci contributing to I_M:

$$R = \rho[1 + m(1-\rho^2)^2/\rho^2(1-m\rho^2)]^{0.5} \quad (17.4)$$

This expression generalizes to index selection the formula derived by Neimann-Sorensen and Robertson (1961, Appendix III) for individual selection. However, it assumes that marker information included in I_M does only refer to the individual candidates to selection. When the index I_Q includes family information and when marker information on relatives can also be included in I_M, an additional gain in accuracy is obtained, as shown by Lande and Thompson (1990) for the case of individual and family (i.e. full or half-sib) index. Equation (17.4) is applied in Table 17.7 to three typical situations in pig breeding. The case of meat production and quality traits in pigs has also been reviewed by Sellier (1994). A general feature is that expected gains from MAS are higher for traits of low heritability. However, as shown by Lande and Thompson (1990), QTL for low-heritability traits will be more difficult to detect because of their expected smaller effects, and so lower values of m are likely to be attained for reproduction than for production traits.

The theoretical predictions of Table 17.7 rest on the basic assumption that the associations detected between markers and quantitative traits are the result of linkage disequilibria. The need for close linkages if MAS is to be successful at all should therefore be emphasized (Smith and Smith, 1993). For that reason, theoretical predictions, as noted by Lande and Thompson (1990), can only be strictly valid for one generation of selection. A gradual decrease in efficiency is to be expected, as shown in the simulation study of Zhang and Smith (1993), although it is difficult to quantify the decrease in the absence of knowledge on the disequilibria involved at each marker locus. As suggested by Gimelfarb and Lande (1994), periodical re-examination of marker–QTL associations would be required in order to maintain the advantage of MAS over phenotypic selection.

When linkage disequilibria are rare, which would be the case in populations kept closed for several generations, and with the present relatively low density of markers over the pig genome, marker–QTL associations can only be detected within families. Knowledge of marker transmission within families can be used to determine more exact co-ancestry coefficients than calculations solely based on pedigree information. In the absence of any knowledge about the transmission of actually identified genes, the covariances between relatives needed to establish selection indices are based on co-ancestries estimated over the whole genome. If markers close to QTL can be followed in their transmission, more accurate co-ancestries for segments of the genome including QTL can be estimated. Chevalet *et al.* (1984) introduced the concept of *conditional co-ancestry* for a marked region of the genome, from which conditional covariances between relatives can be obtained. This additional information allows a better prediction of breeding values, as shown by Fernando and Grossman (1989) with BLUP methodology using marker genotypes and trait phenotypes. Wang *et al.* (1995) presented a theory and algorithms to construct the matrix of conditional covariances between relatives, given marker genotypes, and to obtain its inverse

Table 17.7. Relative efficiency of one generation of marker-assisted selection (MAS) compared with standard phenotypic selection for various traits in pigs, according to the proportion *m* of additive genetic variance associated with the marker loci.

Trait	Litter size	Meat quality	Carcass lean
Heritability[1] (h^2)	0.10	0.30	0.50
Accuracy of standard selection[2] (ρ)	0.31	0.36	0.78
Annual selection intensity[3] (i/t)			
Standard selection	1.43	1.23	1.75
MAS	1.66	1.56	1.75
Relative efficiency of MAS[4]			
$m = 0.10$	1.40	1.30	1.03
	1.62	1.65	1.03
$m = 0.30$	1.98	1.76	1.11
	2.29	2.23	1.11
$m = 0.50$	2.44	2.14	1.21
	2.83	2.71	1.21

[1]Fraction of variance due to common litter environment and dominance assumed to be zero.
[2]Litter size: 2 litters on dam + dam and sire family, with records only on females selected (case 2 of Avalos and Smith, 1987).
Meat quality: 2 sibs of different sexes slaughtered per litter.
Carcass lean: BLUP of individual + sibs (simulation of Ruane and Colleau, 1995).
[3]Litter size standard: one male candidate per litter (i.e. 1/10 selected).
MAS: males selected across litters (i.e. 1/30 selected).
Meat quality standard: one male and one female candidate per litter (value of i/t assuming 1 candidate of each sex per litter and a mating ratio of 15, as given in Table 1 of Ollivier, 1988b).
MAS: males and females selected across litters (value of i/t assuming 2 candidates of each sex per litter and a mating ratio of 15, as given in Table 1 of Ollivier, 1988b).
Carcass lean: selection across litters for both sexes in both types of selection (value of i/t assuming 3 candidates of each sex per litter and a mating ratio of 15, as given in Table 1 of Ollivier, 1988b).
[4]First line: relative accuracy. Second line: relative annual response, i.e. relative accuracy × relative i/t.

efficiently, and also to accommodate situations where marker information is incomplete.

However, information on relatives has so far played a relatively minor role in breeding evaluation of pigs, and fewer gains are to be expected than in species such as dairy cattle which rely heavily on progeny testing for sire evaluation. Selection for a trait measured in both sexes with a population structure of 8 sires and 64 dams, similar to a closed pig line situation, has been simulated by Ruane and Colleau (1995), when one-eighth of the additive genetic variance is associated with one QTL located in the middle of a 20 cM marker bracket and with linkage equilibrium. They show that the gain in accuracy due to MAS over conventional BLUP is indeed small. The gain actually increases from a maximum value, over six generations of selection, of 1.3% to 6.0% when heritability changes from 0.5 to 0.1.

In situations of linkage equilibrium, another route could be to extend the granddaughter design proposed in dairy cattle breeding to a trait such as litter size in pigs, as suggested by Visscher and Haley (1995). In such a scheme, linkages between markers and litter size loci could be detected and then exploited by preselecting grand-offspring of sons of widely used boars on the basis of the appropriate marker genotypes. However, the present structure of pig breeding programmes, with a relatively large number of boars used in any breed, makes it unlikely that extra responses of the same order as those obtained in dairy cattle can be reached.

Markers for the whole genome

The assistance that markers can provide in introgressing a gene from one breed into another has been previously discussed. When the donor breed is inferior for all but the gene introgressed, there is a need to recover as quickly as possible the genomic background of the recipient breed during the repeated backcrossing scheme. Hillel *et al.* (1990) proposed that DNA fingerprints could be used to accelerate such a recovery, by selecting individuals for maximum genomic similarity to the recipient line, a process they called *genomic selection* (GS). They found that moderate selection applied in two backcross generations would be sufficient to recover a high percentage of the recipient genome. Less optimistic predictions were made by Hospital *et al.* (1992), who showed that GS should be carried out at least during three backcross generations. They also showed a marked advantage when GS was performed on evenly spaced markers across the genome, such as microsatellites, compared to markers of unknown chromosomal location, such as DNA fingerprints. They recommended an optimal density of two or three markers per morgan, which would mean between 50 and 80 markers for the pig genome. Under such a scheme, in order to recover 98% of the recipient genome, a gain of about two generations is achieved by using GS, when a proportion of 10% is selected on the markers at each generation. It should be noted that a similar breeding programme could be implemented for re-establishing a breed from a panel of frozen semen, if no female of the breed were available, as for instance if the breed had become extinct. GS would thus have some potential also in genetic conservation (see Chapter 14).

Heterozygosity can be measured directly for a set of marker loci and used as an indicator of the heterozygosity of the whole genome. As heterosis in a cross is proportional to the degree of heterozygosity, it should in theory be possible to predict heterosis on such a basis. One would expect the various crosses among a set of parental lines or breeds to rank in accordance with the corresponding genetic distances, for traits showing heterosis. This was confirmed in an early German study showing significant correlations between crossbred performances for heterotic traits such as litter size and survival, among a set of pig lines, and genetic distances based on 12 polymorphic biochemical loci (Glodek, 1974). Similar investigations would certainly deserve being pursued using the molecular markers recently developed.

In a similar way, *homozygosity* can be assessed directly from markers and used to estimate realized inbreeding at the individual level, as discussed by Visscher and Haley (1995). Genetic drift in a population of limited size might thus be counteracted by selecting the least inbred animals at each generation. In a theoretical simulation, Chevalet (1992) showed that three markers per morgan would allow significant gains in maintaining heterozygosity over the whole genome in populations of small size (effective size $N < 100$), provided an optimal selection pressure of about 85% selected would be applied at each generation. For instance, with $N = 50$ the homozygosity normally reached after 25 generations could be delayed until 40 generations under such a selection scheme. However, it remains to be seen how selection against inbreeding would interfere with response to selection for a quantitative trait.

Conclusions

The efficiency of breeding programmes can be assessed by measuring genetic changes occurring over time. The situation in pigs is well documented for the most important traits, based on the central station testing and on-farm testing records available. From their review on genetic trends estimated in large national purebred populations, Sellier and Rothschild (1991) conclude that appreciable genetic gains have been obtained for growth and body composition traits, of the order of 0.5–1.5% of the mean annually over periods of time ranging from 5 to 10 years in most studies. Less evidence exists for genetic trends in meat quality traits, but most available results point towards mildly unfavourable trends, as could be expected from the generally negative genetic correlations between lean content and meat quality (see Chapter 16). In contrast to production traits, no change of appreciable magnitude has occurred for litter size at birth up to now in most countries. Annual genetic gains of about 0.01–0.02 piglet born have for instance been reported by Bidanel and Ducos (1994), over the period 1975–1991.

It can be concluded that, even though numerous examples of very successful improvement plans exist, there is still scope for designing more efficient breeding schemes than those generally implemented so far. Opportunities will undoubtedly derive from combining the knowledge which has accumulated on classical, quantitative and molecular genetics, as covered in the various chapters of this book. In this chapter, attention has been drawn, in various places, to the role that the structure of the breeding system can also play and its adaptation to the ever evolving conditions of production and performance recording, or to the gains that may be realized by a proper management of the selection herds.

Specific challenges have to be faced for some traits, such as reproductive ability and quality of lean and fat tissue, and here much hope is founded on a better knowledge of the genome and in particular the identification of useful genetic markers. However, many uncertainties still remain as to the real value of marker-assisted breeding in future pig improvement programmes, as emphasized by Visscher and Haley (1995) in their conclusions. They point out difficulties arising from a lack of precision in the essential parameters, such as marker effects and their distribution in various breeds. They also draw attention to the fact that the prospects depend on the state of marker technology. An evolution towards high density genetic maps makes it possible that markers sufficiently close to any gene of interest will eventually be found and then used successfully for genetic improvement. In any case, the building up of linkage maps in the pig is a well-engaged enterprise, from which various kinds of benefits, some unknown at present, may be expected. The applications can be foreseen as complementary to present conventional breeding plans, which will essentially continue to be based on the optimal exploitation of the pig *reproductive* capacity (see also Chapter 12), the efficient use of *breeding value evaluation* tools and the adequate management of *genetic variability*. Among these levers of genetic improvement, the latter two are expected to benefit most from a better knowledge of the genome. A well-balanced approach, taking into account all opportunities, will remain essential in any genetic improvement scheme.

Acknowledgements

Comments and suggestions by P. Sellier and three anonymous referees are gratefully acknowledged.

References

Amer, P.R., Fox, G.C. and Smith, C. (1994) Economic weights from profit equations: appraising their accuracy in the long run. *Animal Production* 58, 11–18.

Amigues, Y., Runavot, J.P. and Sellier, P. (1994) Evaluation de la fréquence du gène de sensibilité à l'halothane dans les principales races porcines françaises en 1993. *Techni-Porc* 17, 23–28.

Avalos, E. and Smith, C. (1987) Genetic improvement of litter size in pigs. *Animal Production* 44, 153–164.

Belonsky, G.M. and Kennedy B.W. (1988) Selection on individual phenotype and best linear unbiased predictor of breeding value in a closed swine herd. *Journal of Animal Science* 66, 1124–1131.

Bertschinger, H.U., Stamn, M. and Vögeli, P. (1993) Inheritance of resistance to oedema disease in the pig: Experiments with an *Escherichia coli* strain expressing fimbriae 107. *Veterinary Microbiology* 35, 79–89.

Bichard, M. (1971) Dissemination of genetic improvement through a livestock industry. *Animal Production* 13, 401–411.

Bichard, M., David, P.J. and Bovey, M. (1986) Selection between and within lines and crossbreeding strategies for worldwide production of hybrids. *Proceedings of the 3rd World Congress on Genetics Applied to Livestock Production*, vol. 10, pp. 130–142.

Bidanel, J.P. (1988) Bases zootechniques et génétiques de l'utilisation en élevage intensif des races prolifiques chinoises. Cas du porc Meishan. Thèse de Docteur-Ingénieur. Institut National Agronomique Paris-Grignon, 194 pp.

Bidanel, J.P. and Ducos, A. (1994) Utilisation du BLUP modèle animal pour l'évaluation génétique des porcs de race Large White et Landrace français sur la prolificité. *Journée de la Recherche Porcine en France* 26, 321–326.

Brascamp, E.W. (1984) Selection indices with constraints. *Animal Breeding Abstracts* 52, 645–654.

Brascamp, E.W. (1994) Current status and future of national breeding programmes. *Proceedings of the 5th World Congress on Genetics Applied to Livestock Production*, vol. 17, pp. 371–372.

Brisbane, J.R. and Gibson, J.P. (1995) Balancing selection response and rate of inbreeding by including genetic relationships in selection decisions. *Theoretical and Applied Genetics* 91, 421–431.

Christensen, A., Sorensen, D.A., Vestergaard, T. and van Kemenade, P. (1986) The Danish pig breeding program: current system and future developments. *Proceedings of the 3rd World Congress on Genetics Applied to Livestock Production*, vol. 10, pp. 143–148.

Chevalet, C. (1992) Apports actuels et futurs des marqueurs génétiques dans l'amélioration des populations animales. Utilisation de marqueurs pour la sauvegarde de la variabilité génétique des populations. *INRA Productions Animales*, Suppl. *Génétique Quantitative*, 295–298.

Chevalet, C., Gillois, M. and Vu Tien Khang, J. (1984) Conditional probabilities of identity of genes at a locus linked to a marker. *Génétique Sélection Evolution* 16, 431–444.

Darwin, C. (1859) *On the Origin of Species by Means of Natural Selection, or the Preservation of Favoured Races in the Struggle for Life.* John Murray, London, 605 pp.

De Roo, G. (1988) Studies on breeding schemes in a closed pig population. I. Population size and selection intensities. *Livestock Production Science* 19, 417–441.

De Vries, A. (1989a) A model to estimate economic value of traits in pig breeding. *Livestock Production Science* 21, 49–66.

De Vries, A. (1989b) A method to incorporate competitive position in the breeding goal. *Animal Production* 48, 221–227.

De Vries, A., van der Steen, H.A.M. and De Roo, G. (1989) Optimal population size and sow/boar ratio in a closed dam line of pigs. *Livestock Production Science* 22, 305–325.

Dickerson, G.E. and Hazel, L.N. (1944) Effectiveness of selection on progeny performance as a supplement to earlier culling in livestock. *Journal of Agricultural Research* 69, 459–476.

Ducos, A. (1994) Evaluation génétique des porcs contrôlés dans les stations publiques à l'aide d'un modèle animal multicaractère. Thèse de Docteur de l'Institut National Agronomique Paris-Grignon, 230 pp.

Edfors-Lilja, I., Gustafsson, U., Duval-Iflah, Y., Ellegren, H., Johansson, M., Juneja, R.K., Marklund, L. and Andersson, L. (1995) The porcine intestinal receptor for *Escherichia coli* K88*ab*, K88*ac*: regional localization on chromosome 13 and influence on IgG response to the K88 antigen. *Animal Genetics* 26, 237–242.

Elsen, J.M. and Mocquot, J.C. (1974) Méthode de prévision de l'évolution du niveau génétique d'une population soumise à une opération de sélection et dont les générations se chevauchent. *Bulletin Technique du Département de Génétique Animale* 17, 30–54.

Elsen, J.M. and Sellier, P. (1978) Etude conjointe de l'intérêt de la sélection sur la prolificité et de l'utilisation d'une lignée mâle spécialisée chez le Porc. *Annales de Génétique et Sélection Animale* 10, 403–411.

Falconer, D.S. (1989) *Introduction to Quantitative Genetics*, 3rd edn. Longman, Harlow, 438 pp.
Fernando, R.L. and Grossman, M. (1989) Marker assisted selection using best linear unbiased prediction. *Genetics, Selection, Evolution* 21, 467–477.
Fredeen, H.T. (1980) Pig breeding: current programs vs. future production requirements. *Canadian Journal of Animal Science* 60, 241–251.
Gibbons, R.A., Sellwood, R., Burrows, M. and Hunter, P.A. (1977) Inheritance of resistance to neonatal *Escherichia coli* diarrhoea in the pig: examination of the genetic system. *Theoretical and Applied Genetics* 51, 65–70.
Gimelfarb, A. and Lande, R. (1994) Simulation of marker assisted selection in hybrid populations. *Genetical Research* 63, 39–47.
Glodek P. (1974) Specific problems of breed evaluations and crossing in pigs. In: *Proceedings of the Working Symposium on Breed Evaluation and Crossing Experiments with Farm Animals*, Zeist, The Netherlands, pp. 267–281.
Glodek, P. (1992) Züchtung und Genetik. In: P. Glodek (ed.) *Schweinezucht.* Ulmer, Stuttgart, pp. 55–107.
Groeneveld, E., Kovac, M. and Wang, T. (1990) PEST, a general purpose BLUP package for multivariate prediction and estimation. *Proceedings of the 4th World Congress on Genetics Applied to Livestock Production* vol. 13, pp. 488–491.
Guérin, G., Duval-Iflah, Y., Bonneau, M., Bertrand, M., Guillaume, P. and Ollivier, L. (1993) Evidence for linkage between K88ab, K88ac intestinal receptors to *Escherichia coli* and transferrin loci in pigs. *Animal Genetics* 24, 393–396.
Hagenbuch, P. and Hill, W.G. (1978) Effectiveness of sequential selection in pig improvement. *Animal Production* 27, 21–27.
Haley, C.S. (1991) Considerations in the development of future pig breeding programmes. *Asian–Australasian Journal of Animal Science* 4, 305–328.
Haley, C.S., Avalos, E. and Smith, C. (1988) Selection for litter size in the pig. *Animal Breeding Abstracts* 56, 317–332.
Hanset, R., Dasnois, C., Scalais, S., Michaux, C. and Grobet, L. (1995) Effets de l'introgression dans le génome Piétrain de l'allèle normal au locus de sensibilité à l'halothane. *Genetics, Selection, Evolution* 27, 77–88.
Hazel, L.N. (1943) The genetic basis for constructing selection indexes. *Genetics* 28, 476–490.
Hazel, L.N. and Kline, E.A. (1952) Mechanical measurement of fatness and carcass value on live hogs. *Journal of Animal Science* 11, 318.
Hazel, L.N., Dickerson, G.E. and Freeman, A.E. (1994) The selection index – then, now, and for the future. *Journal of Dairy Science* 77, 3236–3251.
Hill, W.G. (1974) Prediction and evaluation of response to selection with overlapping generations. *Animal Production* 18, 117–139.
Hillel, J., Schaap, T., Haberfeld, A., Jeffreys, A.J., Plotzky, Y., Cahaner, A. and Lavi, U. (1990) DNA fingerprints applied to gene introgression in breeding programs. *Genetics* 124, 783–789.
Hospital, F., Chevalet, C. and Mulsant, P. (1992) Using markers in gene introgression breeding programs. *Genetics* 132, 1199–1210.
Hudson, G.F.S. and Kennedy, B.W. (1985) Genetic evaluation of swine for growth rate and backfat thickness. *Journal of Animal Science* 61, 83–91.
Janss, L.L.G., van Arendonk, J.A.M. and Brascamp, E.W. (1994) Identification of a single gene affecting intramuscular fat in Meishan crossbreds using Gibbs sampling. *Proceedings of the 5th World Congress on Genetics Applied to Livestock Production*, vol. 18, pp. 361–364.

Jonsson, P. (1965) Analyse af egenskaber hos svin af Dansk Landrace med en historik indledning. *350 Beretning fra Forsoegslaboratoriet.* Copenhagen.

Jonsson, P. (1991) Evolution and domestication, an introduction. In: K. Maijala (ed.) *Genetic Resources of Pig, Sheep and Goat. World Animal Science,* vol. 12, Elsevier, Amsterdam, pp. 1-10.

Karlin, J. (1975) General two-locus selection models: some objectives, results and interpretations. *Theoretical Population Biology* 7, 364–398.

Kennedy, B.W., Hudson, G.F.S. and Schaeffer, L. (1986) Evaluation of genetic change in performance tested pigs in Canada. *Proceedings of the 3rd World Congress on Genetics Applied to Livestock Production,* vol. 10, pp. 149–154.

Kerr, J.C. and Cameron, N.D. (1994) Reproductive performance of pigs selected for components of efficient lean growth. *Proceedings of the 5th World Congress on Genetics Applied to Livestock Production,* vol. 17, pp. 339–342.

King, J.W.B. (1955) The use of testing stations for pig improvement. *Animal Breeding Abstracts* 23, 347–356.

Lande, R. and Thompson, R. (1990) Efficiency of marker-assisted selection in the improvement of quantitative traits. *Genetics* 124, 743–756.

Legault, C. and Gruand, J. (1976) Amélioration de la prolificité du porc par la création d'une lignée 'hyperprolifique' et l'usage de l'insémination artificielle: principe et résultats expérimentaux préliminaires. *Journées de la Recherche Porcine en France* 8, 201–206.

Lundeheim, N., Johansson, K., Rydhmer, L. and Andersson, K, (1994) Realized generation intervals, selection differentials and predicted genetic progress. *Proceedings of the 5th World Congress on Genetics Applied to Livestock Production* 17, 378–381.

Meijerink, E., Voegeli, P., Fries, R., Bertschinger, H.U. and Stranzinger, G. (1996) Genetic mapping in Swiss Landrace pigs of the gene specifying receptors for F18 fimbriated *Escherichia coli* strains causing oedema disease and post-weaning diarrhoea. *25th International Conference on Animal Genetics,* 21–25 July 1996, Tours (France) 128 (abstract).

Meuwissen, T.H.E. (1991) Reduction of selection differentials in finite populations with a nested full-half sib family structure. *Biometrics* 47, 195–203.

Milan, D., Woloszyn, N., Yerle, M., Le Roy, P., Bonnet, M., Riquet, J., Lahbib-Mansais, Y., Caritez, J.C., Robic, A., Sellier, P., Elsen, J.M. and Gellin, J. (1996) Accurate mapping of the 'acid meat' RN gene on genetic and physical maps of pig chromosome 15. *Mammalian Genome* 7, 47–51.

Moav, R. (1973) Economic evaluation of genetic differences. In: R. Moav (ed.) *Agricultural Genetics.* Wiley, New York, pp. 319–352.

Neimann-Sorensen, A. and Robertson, A. (1961) The association between blood groups and several production characteristics in three Danish cattle breeds. *Acta Agriculturae Scandinavica* 11, 163–196.

Ollivier, L. (1973) Five generations of selection for increasing litter size in swine. *Genetics* 74, (Suppl. 2), 202–203.

Ollivier, L. (1974) Optimum replacement rates in animal breeding. *Animal Production* 19, 257–271.

Ollivier, L. (1976) Evolution des méthodes de sélection du porc. *Bulletin de l'Institut Technique du Porc* 6, 11–27

Ollivier, L. (1988a) Current principles and future prospects in selection of farm animals. In: Weir, B.S., Eisen, E.J., Goodman, M.M. and Namkoong, G. (eds): *Proceedings of the Second International Conference on Quantitative Genetics.* Sinauer Associates, Sunderland, Massachusets, pp. 438–450.

Ollivier, L. (1988b) Future breeding programmes in pig. In: Korver, S., van der Steen, H.A.M., van Arendonk, J.A.M., Bakker, H., Brascamp, E.W. and Dommerholt, J. (eds) *Advances in Animal Breeding*. Pudoc, Wageningen, pp. 90–106.

Ollivier, L. and Renjifo, X. (1991) Utilisation de la résistance génétique à la colibacillose K88 dans les schémas d'amélioration génétique du porc. *Genetics, Selection, Evolution* 23, 235–248.

Ollivier, L., Guéblez, R., Webb, A.J. and van der Steen, H.A.M. (1990) Breeding goals for nationally and internationally operating pig breeding organizations. *Proceedings of the 4th World Congress on Genetics Applied to Livestock Production*, vol. 15, pp. 383–394.

Popescu, C.P. and Legault, C. (1988) Anomalies chromosomiques et hypoprolificité chez le porc. *Journées de la Recherche Porcine en France* 20, 297–303.

Popescu, C.P., Bonneau, M., Tixier, M., Bahri, I. and Boscher, J. (1984) Reciprocal translocations in pigs. Their detection and consequences on animal performance and economic loss. *Journal of Heredity* 75, 448–452.

Röhe, R., Krieter, J. and Kalm, E. (1990) Efficiency of selection in closed nucleus herds of pigs using an animal model. A simulation study. *Proceedings of the 4th World Congress of Genetics applied to Livestock Production*, vol. 15, pp. 469–472.

von Rohr, P., Hofer, A. and Künzi, N. (1996) Economic values for meat quality traits in pigs. *47th Annual Meeting of EAAP*, 25–29 August 1996, Lillehammer, Norway, 4 pp.

Ruane, J. and Colleau, J.J. (1995) Marker assisted selection for genetic improvement of animal populations when a single QTL is marked. *Genetical Research* 66, 71–83.

Sellier, P. (1982) Selecting populations for use in crossbreeding. *Proceedings of the 2nd World Congress on Genetics Applied to Livestock Production*, vol. 6, pp. 15–49.

Sellier, P. (1986) Amélioration génétique. In: Perez, J.M., Mornet, P. and Rérat, A. (eds) *Le Porc et Son Élevag.*, Maloine, Paris, pp. 159–230.

Sellier, P. (1988) Aspects génétiques des qualités technologiques et organoleptiques de la viande chez le porc. *Journées de la Recherche Porcine en France* 20, 227–242.

Sellier, P. (1994) The future role of molecular genetics in the control of meat production and meat quality. *Meat Science* 36, 29–44.

Sellier, P. and Rothschild, M.F. (1991) Breed identification and development. In: Maijala, K. (ed.) *Genetic Resources of Pig, Sheep and Goat. World Animal Science*, vol. 12. Elsevier, Amsterdam, pp. 125–143.

Smith, C. (1964) The use of specialised sire and dam lines in selection for meat production. *Animal Production* 6, 337–344.

Smith, C. (1967) Improvement of metric traits through specific genetic loci. *Animal Production* 9, 349–358.

Smith, C. (1986) Variety of breeding stocks for the production-marketing range, and for flexibility and uncertainty. *Proceedings of the 3rd World Congress on Genetics Applied to Livestock Production*, vol. 10, pp. 14-18.

Smith, C. and Smith, D.B. (1993) The need for close linkages in marker-assisted selection for economic merit in livestock. *Animal Breeding Abstracts* 61, 197–204.

Smith, C., Dickerson, G.E., Tess, M.W. and Bennett, G.L. (1983) Expected relative responses to selection for alternative measures of life cycle economic efficiency of pork production. *Journal of Animal Science* 56, 1306–1314.

Smith, C., James, J.W. and Brascamp, E.W. (1986) On the derivation of economic weights in livestock improvement. *Animal Production* 43, 545–551.

Soller, M. and Plotkin-Hazan, J. (1977) The use of marker alleles for the introgression of linked quantitative alleles. *Theoretical and Applied Genetics* 51, 133–137.

Sorensen, D.A. (1988) Effect of selection index versus mixed model methods of prediction of breeding value on response to selection in a simulated pig population. *Livestock Production Science* 20, 135–148.

Stewart, T.S., Lofgren, D., Harris, D.L., Einstein, M.E. and Schinckel, A.P. (1991) Genetic improvement programs in livestock: swine testing and genetic evaluation system (STAGES). *Journal of Animal Science* 69, 3882–3890.

Tess, M.W., Bennett, G.L. and Dickerson, G.E. (1983) Simulation of genetic changes in life cycle efficiency of pork production. I. A bioeconomic model. II. Effects of components on efficiency III. Effect of management systems and feed prices on importance of genetic components. *Journal of Animal Science* 56, 336–379.

Toro, M.A. and Perez-Enciso, M. (1990) Optimization of selection response under restricted inbreeding. *Genetics, Selection, Evolution* 22, 93–107.

Toro, M.A., Silio, L., Rodriganez, J. and Dobao, M.T. (1988) Inbreeding and family index selection for prolificacy in pigs. *Animal Production* 46, 79–85.

Vandepitte, W.M. and Hazel, L.N. (1977) The effect of errors in economic weights on accuracy of selection indexes. *Annales de Génétique et de Sélection Animale* 9, 87–103.

Visscher, P.M. and Haley, C.S. (1995) Utilizing genetic markers in pig breeding programmes. *Animal Breeding Abstracts* 63, 1–8.

Wang, T., Fernando, R.L., van der Beck, S., Grossman, M. and van Arendonk, J.A.M. (1995) Covariances between relatives for a marked quantitative trait locus. *Genetics, Selection, Evolution* 27, 251–274.

Webb, J. and Bampton, P.R. (1987) Choice of selection objectives in specialised sire and dam lines for commercial crossbreeding. *38th Annual Meeting of EAAP*, 28 September–1st October, Lisbon, 14 pp.

Wei, M. and van der Steen, H.A.M. (1991) Comparison of pure-line selection with reciprocal recurrent selection systems in animal breeding (a review). *Animal Breeding Abstract* 59, 281–298.

Wei, M. and van der Werf, J.H.J. (1994). Maximizing genetic response in crossbreds using both purebred and crossbred information. *Animal Production* 59, 401–413.

Zhang, W. and Smith, C. (1993) The use of marker assisted selection with linkage disequilibrium: the effects of several additional factors. *Theoretical and Applied Genetics* 86, 492–496.

Standard Nomenclature and Pig Genetic Glossary

L.D. Young[1]
USA-ARS, U.S. Meat Animal Research Center, PO Box 166, Clay Center, NE 68933-0166, USA

Introduction

This chapter addresses nomenclature issues for individual genes and phenotypes. The entire area of gene nomenclature is undergoing revision and the guidelines presented here are those in use at the time of the publication. In addition a limited glossary is included for some of the genetic terms used in this book.

Guidelines for Gene Nomenclature

These guidelines for nomenclature are adapted and abbreviated from the COGNOSAG *ad hoc* committee 1995 Revised guidelines for gene nomenclature in ruminants 1993, *Genetique, Selection, Evolution* 27, 89–93 and those from the HUGO Nomenclature meeting March 1997 (White *et al.*, 1997).

[1]Now deceased.

Locus name and symbol

The name of a locus in English should be as brief as possible but should not be a single letter. The name should convey as accurately as possible the character affected or the function by which the locus is recognized. The name may indicate a morphological character, disease or metabolic condition, a body system or body function, a biochemical property, or a nucleotide segment. All Greek symbols should be replaced by Latin type and placed after the name, e.g. β *Haemoglobin* becomes *Haemoglobin beta.*

The locus name should be in Latin letters or a combination of Latin letters and Arabic numerals. The initial letter of the locus name should be a capital Latin character. If the locus name is two or more words, each word after the first word may begin with either a capital or a lower case Latin character.

The locus symbol should consist of as few Latin letters as possible or a combination of Latin letters and Arabic numerals and should not exceed six characters in length. The characters of the symbol should always be capital Latin characters which, if possible, should include the initial letter of the name of the locus. If the locus name is two or more words, and the initial letters are used in the locus symbol, then these letters should be in Latin capitals. All characters in a locus symbol should be written on the same line with no superscripts or subscripts and no Roman numbers used. The locus name and symbol should be printed in italics wherever possible or otherwise be underlined.

New symbols must not duplicate existing gene symbols and every effort to check genome databases should be made to ensure no duplicate symbols. Rules for naming genes predicted solely from sequence data are regarded as putative and are designated by the chromosome of origin and arbitrary number (for example *C2ORF1* where ORF stands for ortholog). For homologous genes first discovered in other species (orthologs), the same gene nomenclature should be used whenever possible. Some letters have been reserved for special usage as the last letter in a symbol to represent a specific meaning and include P for pseudogene (but also note BP for binding protein), L for like, R for receptor or regulator and N for inhibitor.

Allele name and symbol

The name of the allele should be as brief as possible, but should convey the variation associated with the allele. Alleles do not have to be named, but should be given symbols as described later. If a new allele is similar to one that is already named, it should be named according to the breed, geographic location, or population of origin. The names of new alleles at a recognized locus should conform to nomenclature established for that locus. A lower first letter of the allele name is preferred. This does not apply when the allele has only a symbol and no name.

The allele symbol should be as brief as possible and consist of a combination of Latin letters and/or Arabic numerals. As far as possible, the allele symbol should be an abbreviation of the allele name, and should start with the same letter. The allele name and symbol may be identical for a locus detected by biochemical, serological, or nucleotide methods. The symbol + can be used alone for identification of the standard or 'wild type' allele which has no visible effects. Neither + nor – symbols should be used in alleles detected by biochemical, serological or nucleotide methods. Null alleles should be designated by the number zero. The initial letter of the symbol of the top dominant allele should be a capital. All alleles that are codominant should have an initial capital letter. The initial letter of all other alleles should be lower case. Numbers may also be used for alleles.

The allele symbol should always be written with the locus symbol. It was previously written as a superscript (PGM^1) following the locus symbol but now the preferred manner is following an asterisk on the same line as the locus symbol (e.g. *PGM*1*). The allele symbol should be printed immediately adjacent to the locus symbol, i.e. with no gaps.

The allele name and symbol should be printed in italics whenever possible or otherwise be underlined.

Genotype terminology

The genotype of an individual should be shown by printing the relevant locus and allele symbols for the two homologous chromosomes concerned, separated by a slash, e.g. *PGM*1/PGM*2*. Unlinked loci should be separated by semicolons. Linked or syntenic loci should be separated by a space and listed in linkage order or in alphabetical order if the linkage order is not known. For X-linked loci, the hemizygous case should have a /Y following the locus and allele symbol. Y-linked loci should be designated by /X following the locus and allele symbol.

Phenotype terminology

The phenotype symbol should be in the same characters as genotype and allele symbols. The difference is that the characters should not be underlined or in italics and should be written with a space between locus characters and allele characters instead of an asterisk. Square brackets [ ] may also be used.

Genetic Glossary

This glossary is limited in scope but contains many useful terms. A dictionary of animal production terms in English, French, German, Spanish and Latin has been published (Straszewska and Ollivier, 1993) and also may be of use.

Additive gene effects. The average effect on phenotype when one allele is replaced by another.
Adenine. A purine base found in RNA and DNA.
Allele. One of a pair, or series of alternative forms of a gene that can occur at a given locus on homologous chromosomes.
Amino acids. Any one of a class of organic compounds containing the amino (NH_2) group and the carboxyl (COOH) group. Amino acids are combined to form proteins.
Ancestor. Any individual from which an animal is descended.
Assortive mating. Assigning animals as mates based on phenotypic or genetic likeness. Positive assortive mating is mating animals that are more similar than average. Negative assortive mating is mating animals that are less similar than average.
Atresia ani. The condition when an animal is born without an external rectal opening.
Autosome. Any chromosome that is not a sex chromosome.
Backcross. The cross produced by mating a first-cross animal back to one of its parent lines or breeds.
Breeding value. The mean genetic value of an individual as a parent. It can be estimated as the average superiority of an individual's progeny relative to all other progeny under conditions of random mating.
Categorical trait. A trait whose measurements fall into discrete classes or values, e.g. number of pigs born.
Centromere. Spindle-fiber attachment region of a chromosome.
Chromosome. Microscopically observable linear arrangement of DNA in the nucleus of a cell. Chromosomes carry the genes responsible for the determination and transmission of hereditary characteristics.
Codominant alleles. Alleles, each of which produces an independent effect in heterozygotes.
Combining ability. The mean performance of a line when involved in a crossbreeding system. General combining ability is the average performance when a breed or line is crossed with two or more other breeds or lines. Specific combining ability is the degree to which the performance of a specific cross deviates from the average general combining ability of two lines.
Composite. A line developed from a crossbred foundation.
Congenital. A condition present at birth but not hereditary.
Control line. A line that is randomly selected and randomly mated. Usually used in selection experiments to monitor environment effects in order to estimate genetic change in a selected line.
Covariance. The degree to which two measurements vary together. A positive covariance is when two measurements tend to increase together. A negative covariance is when one measurement increases and the other measurement tends to decrease.
Crossbreeding. Matings between animals of different breeds or lines.
Crossover. The process during meiosis when chromosomal segments from different members of a homologous pair of chromosomes break and part of one

will join a part of the other so that two gametes form possessing new combinations of genes. The frequency of crossover between two loci is proportional to the physical distance between the loci.

Crossover unit. Each unit is equal to a 1% frequency of crossover gametes.

Cytoplasm. The protoplasm outside a cell nucleus.

Cytosine. A pyrimidine base found in RNA and DNA.

Descendant. An individual descended from other individuals.

Diallele cross. When both males and females from each breed (or line) in a set of breeds (or lines) are mated to males and females of each breed (or line) in the set including their own breed (or line).

DNA. Deoxyribonucleic acid, the chemical material which forms a gene.

Dominant. Applied to one member of an allelic pair of genes which has the ability to express itself wholly or largely at the exclusion of the expression of the other member.

Economic trait loci. Loci that have effects on traits of economic importance.

Economic value. A measure of the contribution an individual trait makes to the overall economic value of an animal.

Environment. The aggregate of all the external conditions and influences affecting the life and development of the organism.

Environmental correlation. When two traits tend to change in the same or different direction as a result of environmental effects.

Environmental variance. Variation in phenotype which results from variation in environmental effects.

Epistasis. When genes at one locus affect the expression of genes at another locus.

F_1. Animals resulting from crossing parents from different lines or breeds.

F_2. Animals resulting from matings among F_1 parents.

F_3. Animals resulting from matings among F_2 parents.

Family size. The mean number of offspring per parent that successfully reproduce.

Fullsibs. Individuals having the same male and female parents.

Gamete. A sperm or egg cell containing the haploid ($1n$) number of chromosomes.

Gene. A functional hereditary unit that occupies a fixed location on a chromosome, has a specific influence on phenotype, and is capable of mutation to various allelic forms.

Generation interval. The average age of the parents when the progeny that will replace them are born.

Genetic correlation. When two traits tend to change in the same or different direction as a result of genetic effects.

Genetic drift. Changes in gene frequency in small breeding populations due to chance fluctuations.

Genetic variance. Variation in phenotype which results from variation in genetic composition among individuals.

Genome. A complete set of chromosomes (hence genes).

Genotype. The genetic constitution of an organism.

Genotype–environment interaction. When the difference in performance between two genotypes differs depending upon the environment in which performance is measured. This may be a change in the magnitude of the difference or a change in rank of the genotypes.

Germplasm. The germinal material or physical basis of heredity. The sum total of the genes.

Grade-up. The process of repeated backcrossing to one parental line to produce a population that is nearly purebred.

Guanine. A purine base found in DNA and RNA.

Halfsibs. Individuals that share one common parent.

Hardy–Weinberg law. A population is in genotypic equilibrium if p and q are the frequencies of alleles A and a, respectively, and p^2, $2pq$ and q^2 are the genotypic frequencies of AA, Aa, and aa under the condition of random mating.

Heritability. Degree to which a given trait is controlled by inheritance. Proportion of total phenotypic variation that is attributable to genetic variation.

Heterosis. The degree to which the performance of a crossbred animal is better or worse than the average performance of the parents.

Heterozygote (adj. **heterozygous).** An organism with unlike members of any given pair or series of alleles, which consequently produces unlike gametes.

Homologous chromosomes. Chromosomes which occur in pairs and are similar in size and shape, one having come from the male and one from the female parent.

Homozygote (adj. **homozygous).** An organism whose chromosomes carry identical members of a given pair of genes. The gametes are therefore all alike with respect to this locus.

Inbreeding. Matings among related individuals which results in progeny that have more homozygous gene pairs than the average of the population.

Inbreeding coefficient. A measurement of the increase in homozygosity and each unit is equal to a 1% increase in homozygosity relative to the average homozygosity in the base population.

Inbreeding depression. The decreased performance normally associated with accumulation of inbreeding. Many recessive genes result in undesired traits or decreased performance when they are expressed. Inbred animals have more recessive genes in the homozygous condition that are expressed and result in reduced performance or undesired traits.

Independent culling. When animals are culled if they do not meet all of the minimum levels of performance for a set of traits.

Karyotype. The appearance of the metaphase chromosomes of an individual or species which shows the comparative size, shape and morphology of the different chromosomes.

Lethal gene. A gene that results in the death of the animal.

Liability. Both internal (e.g. genetic merit) and external (e.g. nutrition, disease, exposure) forces that influence the expression of a threshold character (e.g. disease, conception, abnormalities, etc.).

Linebreeding. Mating of selected individuals from successive generations to produce animals with a high relationship to one or more selected ancestors. It is a form of inbreeding.

Linkage. Association of genes that are physically located on the same chromosome. A group of linked genes is called a linkage group.

Locus (pl. **Loci).** A fixed position on a chromosome occupied by a given gene or one of its alleles.

Major gene. A gene that has an easily recognizable and measurable effect on a characteristic.

Marker. Specific and identifiable sequences of the DNA molecule. These markers may or may not be functional genes.

Mating systems. The rules which describe how selected breeds and/or individuals will be paired at mating.

Meiosis. The process by which the chromosome number of a reproductive cell becomes reduced to half the diploid ($2n$) or somatic number and results in the formation of eggs or sperm.

Migration. Movement of animals, and consequently genes, from one population to another.

Mitochondria. Small bodies in the cytoplasm of most plant and animal cells.

Mitosis. Cell division in which there is first a duplication of chromosomes followed by migration of chromosomes to the ends of the spindle and a dividing of the cytoplasm resulting in the formation of two cells with diploid ($2n$) number of chromosomes.

Multiple alleles. Three or more alternative forms of a gene representing the same locus in a given pair of chromosomes.

Mutation. A sudden change in the genotype of an organism. The term is most often used in reference to point mutations (changes in base sequence within a gene), but can refer to chromosomal changes.

Natural selection. Natural processes favouring reproduction by individuals that are better adapted and tending to eliminate those less adapted to their environment.

Nucleus. Part of a cell containing chromosomes and surrounded by cytoplasm.

Outcrossing. Mating of individuals that are less closely related than the average of the population.

Overdominant. The property shown by two alleles when the heterozygote lies outside the range of the homozygotes in genotypic value with respect to a specific character.

Pedigree. The ancestors of an individual.

Penetrance. The proportion of individuals with a particular gene combination that express the corresponding trait.

Permanent environmental effects. Environmental effects that result in permanent effects on the phenotypic expression of a trait. For example, severe mastitis during a lactation may have a permanent effect on milk production and litter weaning weight in subsequent litters.

Phenotype. Characteristics of an individual that are observable such as size, shape, colour, or performance.

Phenotypic correlation. When two traits tend to change in the same or different direction as a net result of genetic and environmental effects.

Phenotypic variation. Variation in phenotype which results from variation in genetic and environment effects on the individuals.

Pleiotropy. The property of a gene whereby it affects two or more characters, so that if the gene is segregating it causes simultaneous variation in the characters it affects.

Population. Entire group of organisms of a kind that interbreed.

Population genetics. The branch of genetics which deals with frequencies of alleles in groups of individuals.

Porcine stress syndrome (PSS). A syndrome, commonly initiated by extreme physical stress such as fighting, marked by difficult breathing, increased rate of respiration, blanching and reddening of the skin, followed by cyanosis and acidosis. Total collapse, muscle rigidity and extreme hyperthermia accompany these changes which are generally followed by death. Caused by a single autosomal recessive gene, called ryanodine receptor gene.

Progeny. Offspring or individuals resulting from specific matings.

Protein. Any of a group of complex nitrogenous organic compounds that contain amino acids as their basic structural units and that occur in all living matter and are essential for the growth and repair of animal tissue.

PSE. Refers to pork that is pale, soft and exudative (watery surface).

Qualitative trait. A trait that can generally be classified into a limited number of categories and the animal can be said to 'possess' the quality or not. Examples include hair colour, skin colour and ear stature.

Quantitative trait. A trait that is represented by an almost continuous distribution of measurements. Examples include average daily gain, backfat thickness and litter size.

Random mating. A mating system in which animals are assigned as breeding pairs at random without regard to genetic relationship or performance.

Recessive. Applied to one member of an allelic pair which lacks the ability to manifest itself when the other or dominant member is present.

Reciprocal cross. When males of breed A are mated with females of breed B and males of breed B are mated with females of breed A.

Reciprocal recurrent selection. A method of selection for combining ability or heterosis. Selection within two lines is based on the performance of crossbred progeny produced by crossing the two lines.

Recombination. The observed new combinations of traits different from those combinations exhibited by the parents.

Recurrent selection. A method of selection for combining ability or heterosis. Selection within one line is based on performance of crossbred progeny from matings with a 'tester' line.

Repeatability. The proportion of total phenotypic variation that is attributable to variation caused by genetic and permanent environmental effects. It is a

measure of the degree to which early measures of a trait can predict later records of the same trait.
RNA. Ribonucleic acid which is involved in the transcription of genetic information from DNA.
Scrotal hernia. A condition caused by protruding of the intestines through the inguinal canal and into the scrotum of male pigs.
Selection. Any natural or artificial process favouring the survival and propagation of certain individuals in population.
Selection criteria. The character(s) upon which selection decisions are based with the intent of changing the character(s) in the selection objective.
Selection differential. The difference in mean performance of the selected group of animals relative to the mean performance of all animals available for selection.
Selection index. The combining of measurements from several sources into an estimate of genetic value. When more than one measurement on trait, and/or measurements of the trait on relatives, and/or the measurements on more than one trait are combined into a single estimate of overall genetic value.
Selection intensity. The proportion of animals selected to be parents relative to the total number available for selection. The smaller the proportion selected the higher the selection intensity.
Selection objective. The character(s) which are intended to be modified by selection.
Sex chromosomes. The X or Y chromosomes.
Sex-influenced. Traits in which the expression depends on the sex of the individual.
Sex-limited. A trait that can be expressed only in one sex, such as milk production.
Sex linked. Genes that are located on the sex (X or Y) chromosomes.
Splay legged. A condition occurring at birth in pigs that generally affects the rear legs. The rear legs are splayed to each side and the pig cannot use them to walk.
Synthetic. See Composite.
Thymine. A pyrimidine base found in DNA but replaced by uracil in RNA.
Umbilical hernia. A condition of males and females in which the umbilical area is weak and the intestines protrude through the weak area and are carried between the skin and abdominal muscles.
Uracil. A pyrimidine base found in RNA but replaced by thymine in DNA.
Zygote. The cell produced by the union of mature gametes (egg and sperm) in reproduction.

References

COGNOSAG *ad hoc* Committee. (1995) Revised guidelines for gene nomenclature in ruminants 1993. *Genetique, Selection,Evolution* 27, 89–93.

Straszewska, S. and Ollivier, L. (1993) *Dictionary of Animal Production Terminology.* Elsevier, Amsterdam, 660 pp.

White, J.A., McAlpine, P.J., Antonarakis, S., Cann, H., Frazer, K., Frezal, J., Lancet, D., Nahmias, J., Pearson, P., Peters, J., Scott, A., Scott, H., Spurr, N., Talbot C. Jr and Povey, S. 1997. *Guidelines for Human Gene Nomenclature.* Proceedings of HUGO Nomenclature meeting, Toronto, Canada.

Appendix 1: [1]Genetic Linkage Maps of Pig

The following linkage maps independently constructed by three groups supported by EU, Scandinavian and US governments, reflect current progress in gene mapping in the pig. It is very likely that in the near future these maps will be substituted by the unified map of the pig genome. These maps were downloaded from PiGBASE in June 1997. For each chromosome, maps from the PiGMaP collaboration, Nordic collaboration and the USDA-MARC group are shown in that order. Loci found on all three maps are linked by solid lines, those on two maps by dotted lines. Several databases around the world provide information about studied loci and gene maps in the pig:

http://www.ri.bbsrc.ac.uk/pigmap/pigbase/pigbase.html

http://www.public.iastate.edu/~pigmap/

http://sol.marc.usda.gov/genome/swine/swine.html

http://probe.nalusda.gov:8300/animal/index.html

[1]Compiled by A. Ruvinsky from http://www.ri.bbsrc.ac.uk/pigmap/pigbase/pigbase.html

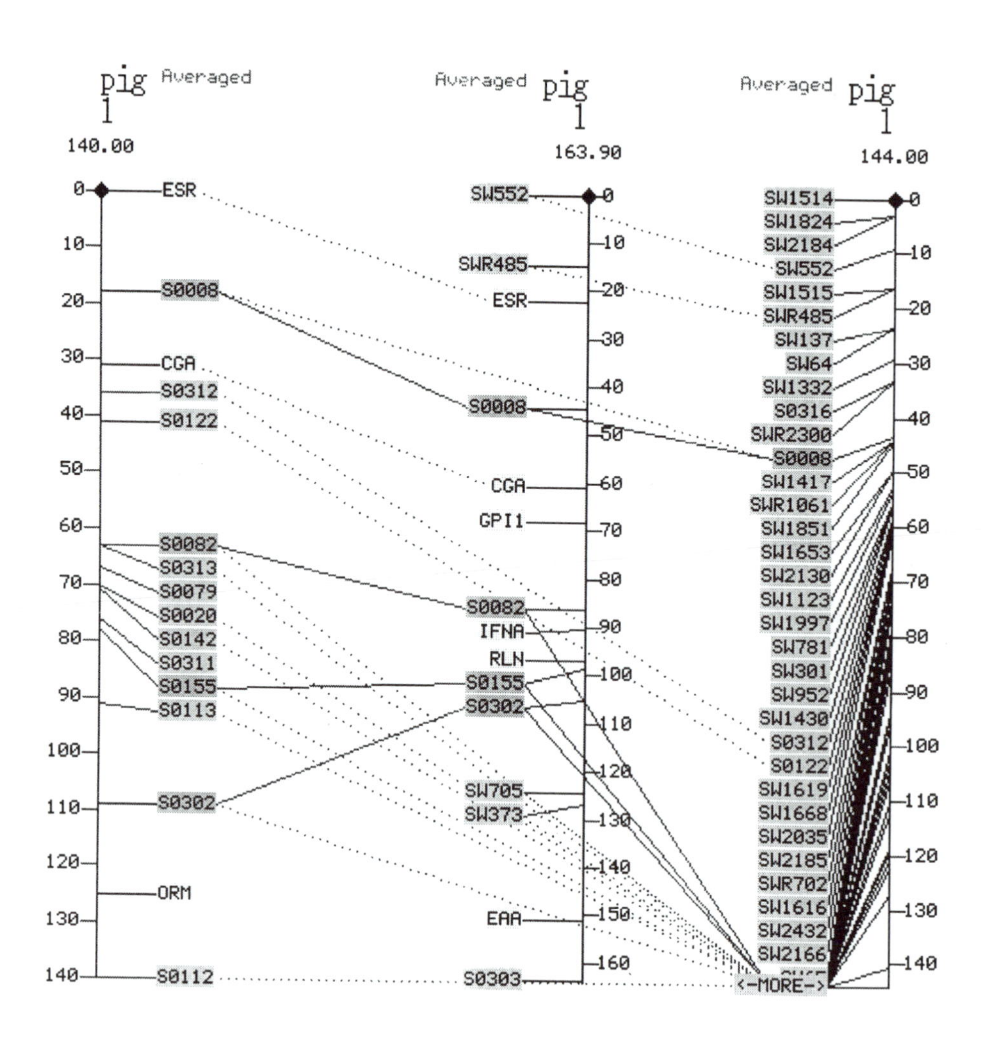

The pig linkage maps of chromosome 1 from three independent sources (June 1997). Obtained from Anubis Map Screen, http://www.ri.bbsrc.ac.uk/cgi-bin/anubis

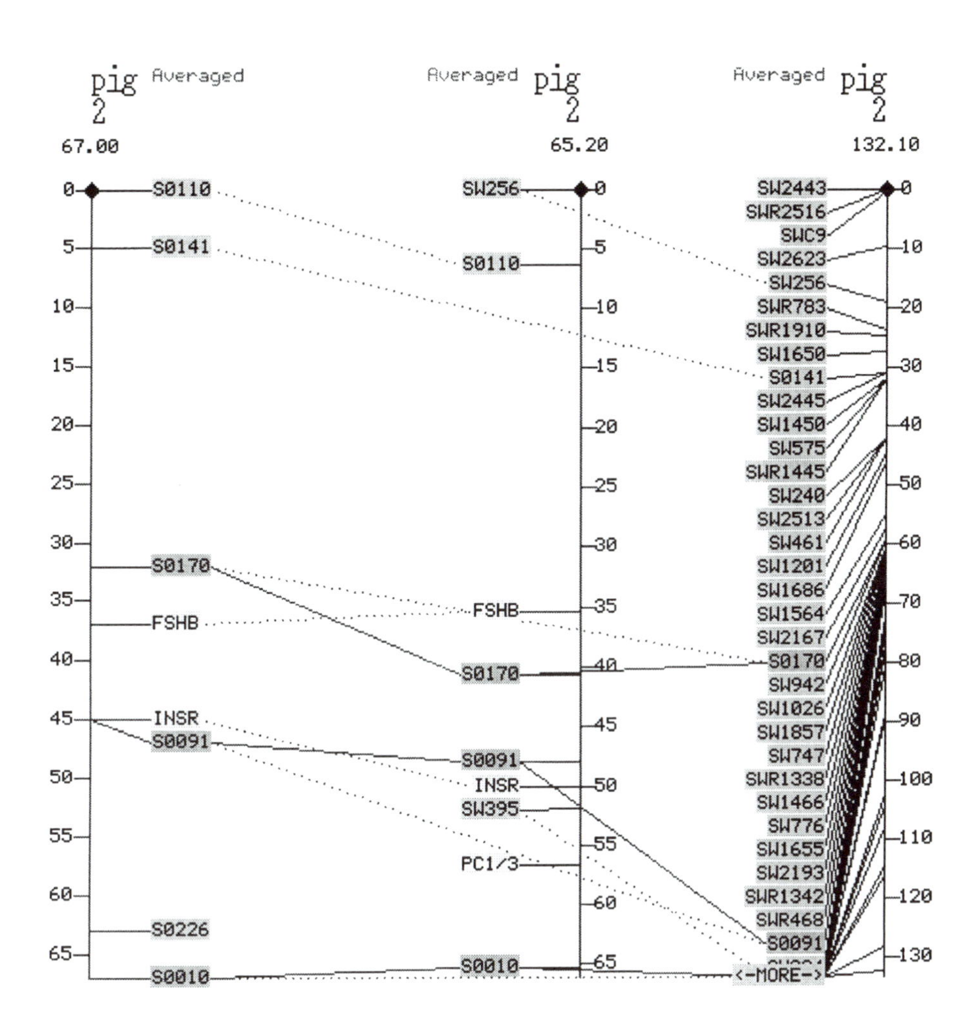

The pig linkage maps of chromosome 2 from three independent sources (June 1997). Obtained from Anubis Map Screen, http://www.ri.bbsrc.ac.uk/cgi-bin/anubis

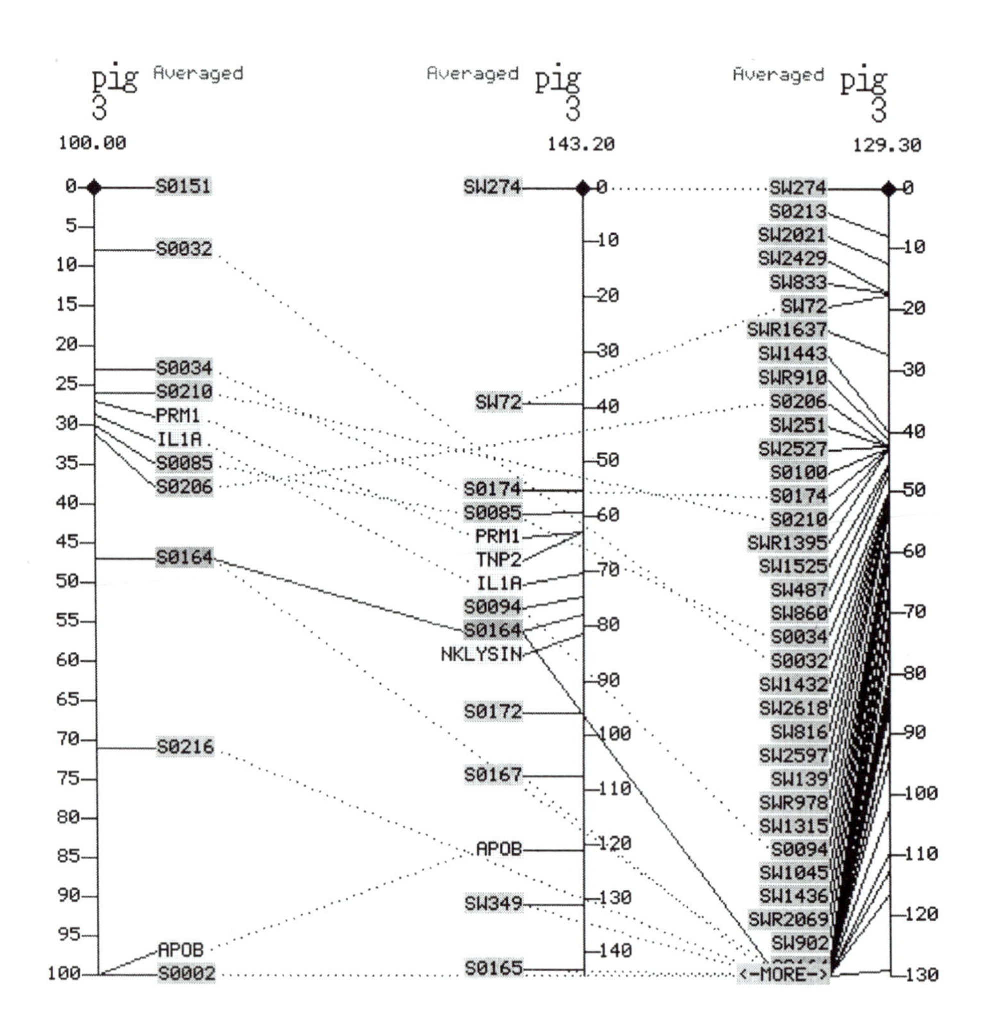

The pig linkage maps of chromosome 3 from three independent sources (June 1997). Obtained from Anubis Map Screen, http://www.ri.bbsrc.ac.uk/cgi-bin/anubis

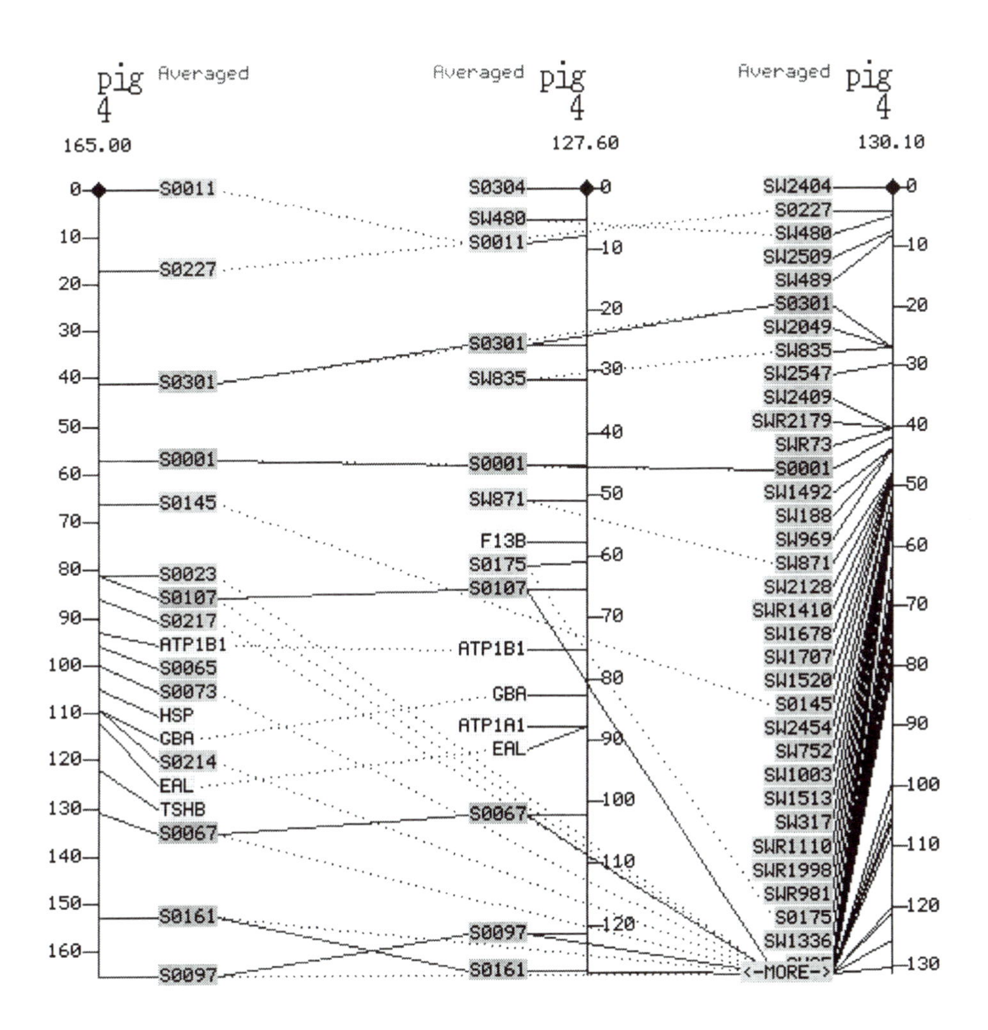

PiGMaP.1 Nordic.2 USDA-MARC.2

The pig linkage maps of chromosome 4 from three independent sources (June 1997). Obtained from Anubis Map Screen, http://www.ri.bbsrc.ac.uk/cgi-bin/anubis

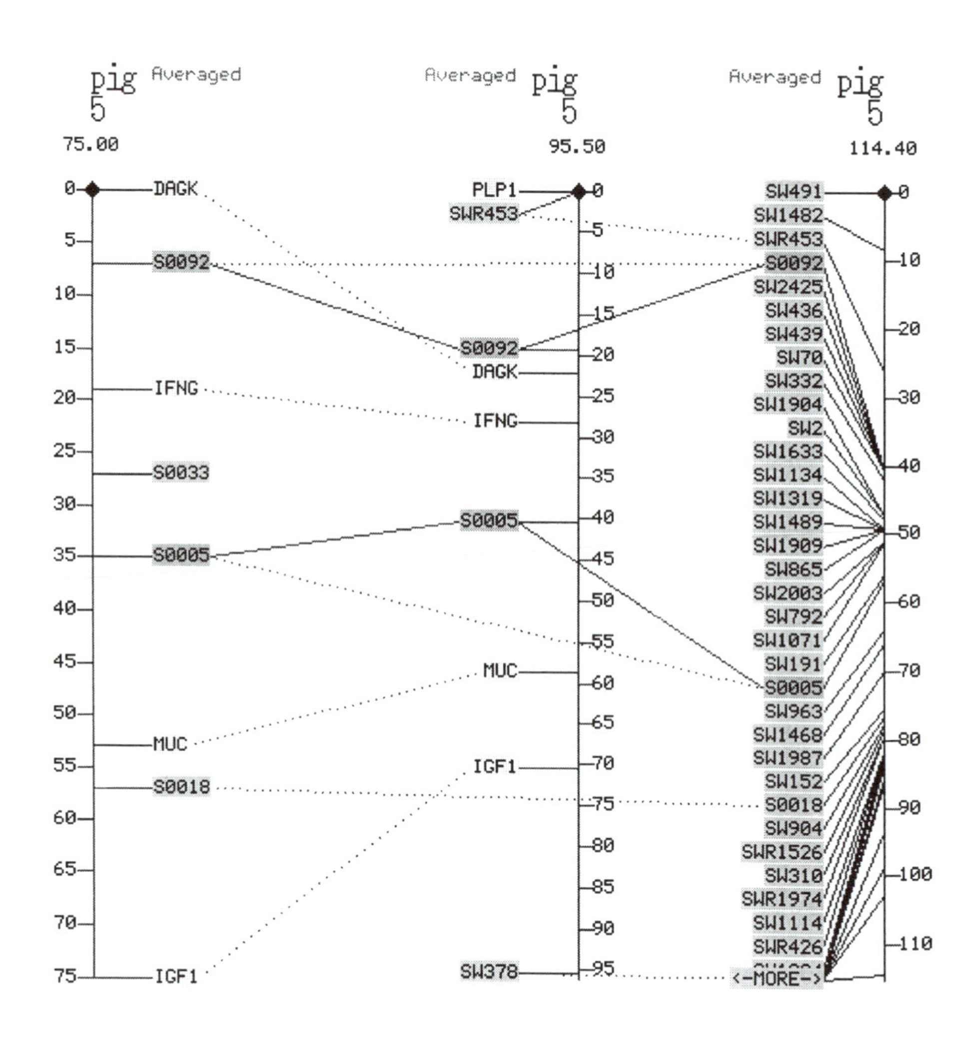

The pig linkage maps of chromosome 5 from three independent sources (June 1997). Obtained from Anubis Map Screen, http://www.ri.bbsrc.ac.uk/cgi-bin/anubis

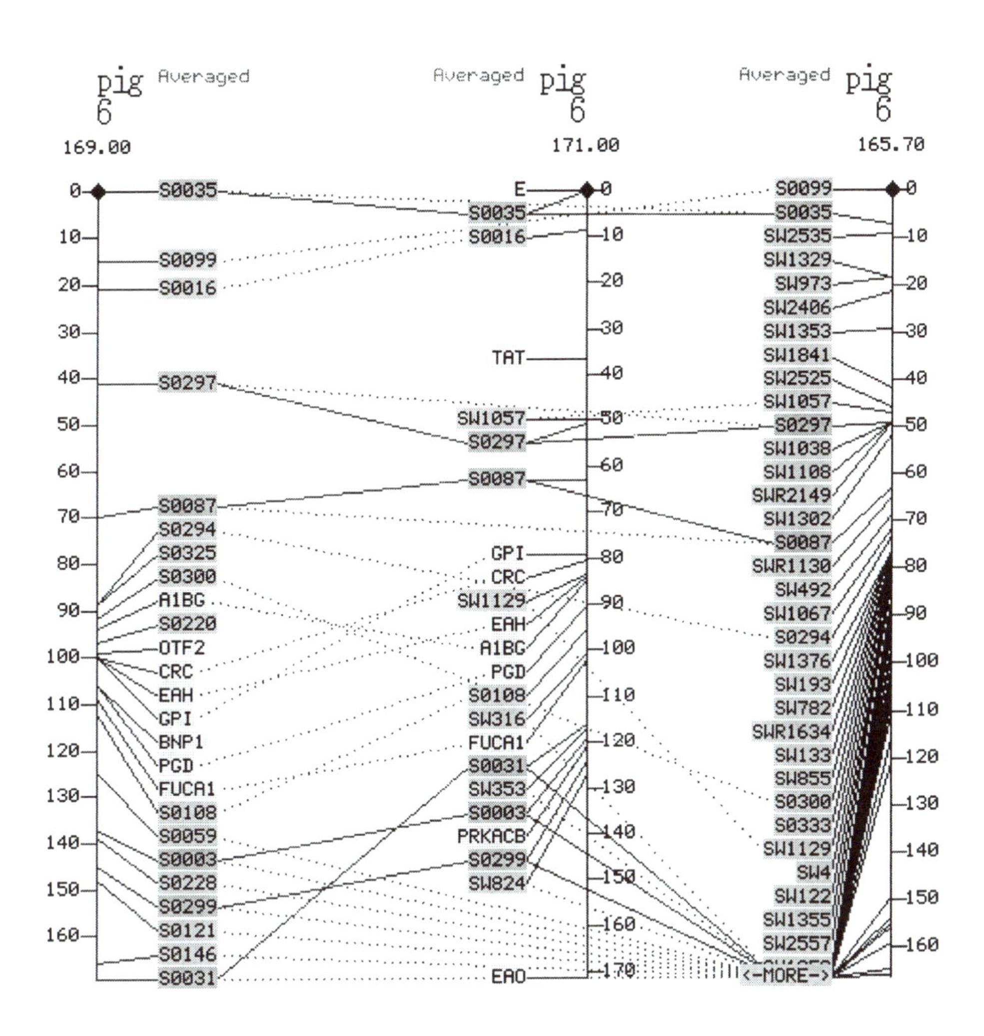

The pig linkage maps of chromosome 6 from three independent sources (June 1997). Obtained from Anubis Map Screen, http://www.ri.bbsrc.ac.uk/cgi-bin/anubis

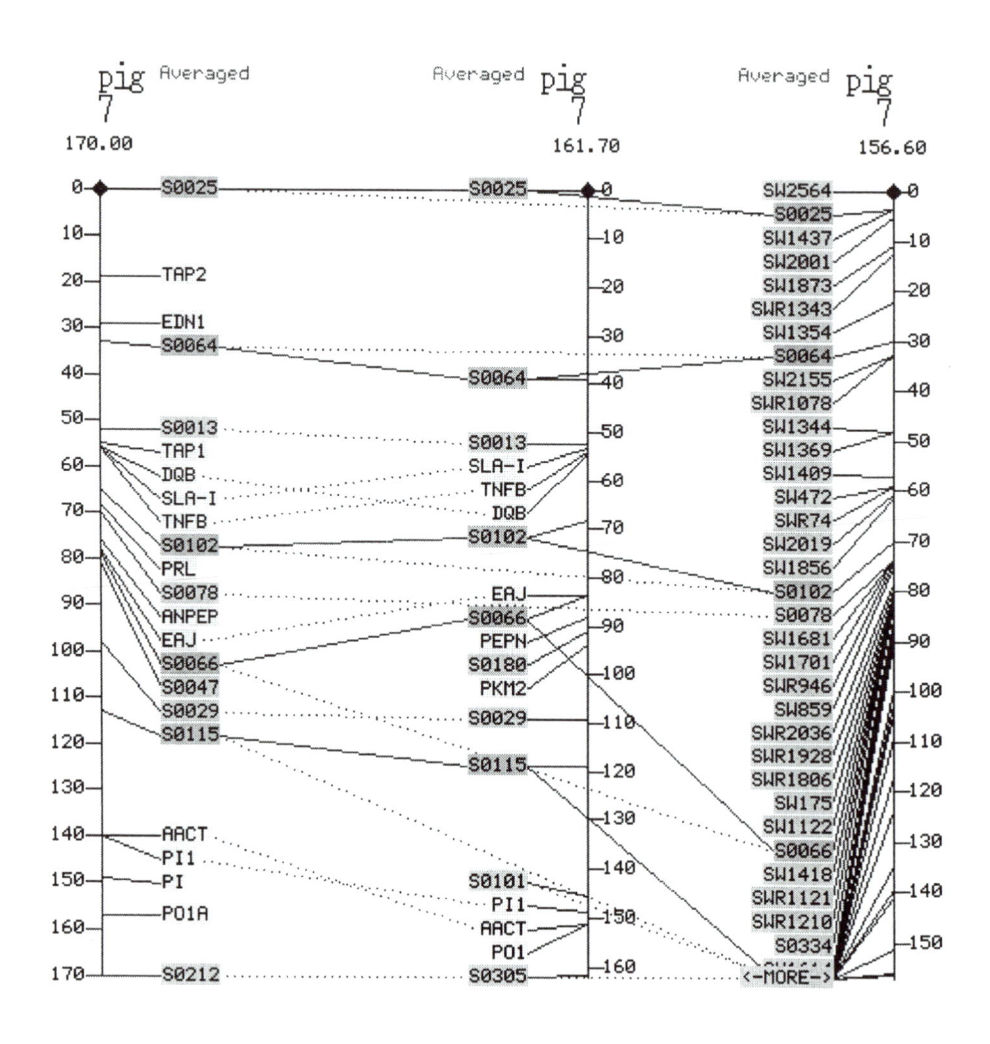

PiGMaP.1 Nordic.2 USDA-MARC.2

The pig linkage maps of chromosome 7 from three independent sources (June 1997). Obtained from Anubis Map Screen, http://www.ri.bbsrc.ac.uk/cgi-bin/anubis

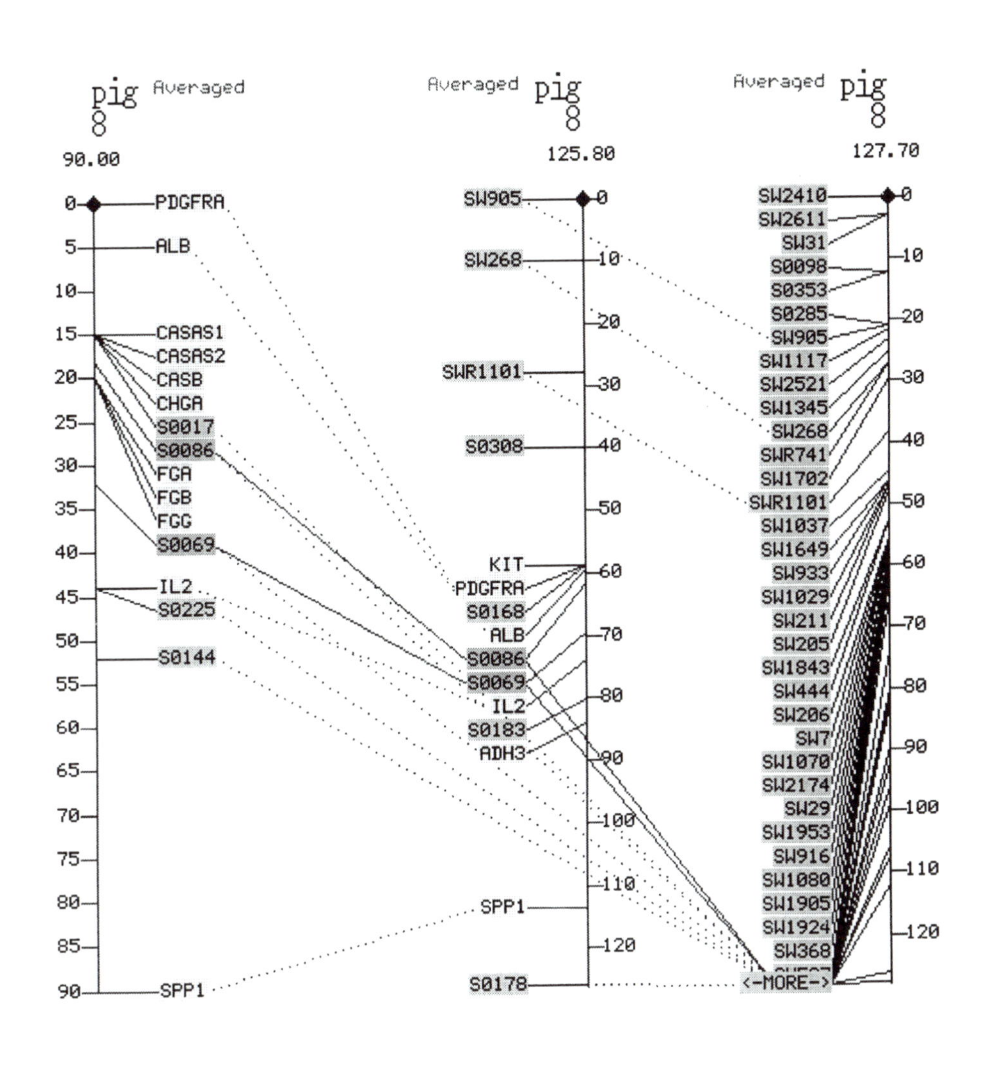

The pig linkage maps of chromosome 8 from three independent sources (June 1997). Obtained from Anubis Map Screen, http://www.ri.bbsrc.ac.uk/cgi-bin/anubis

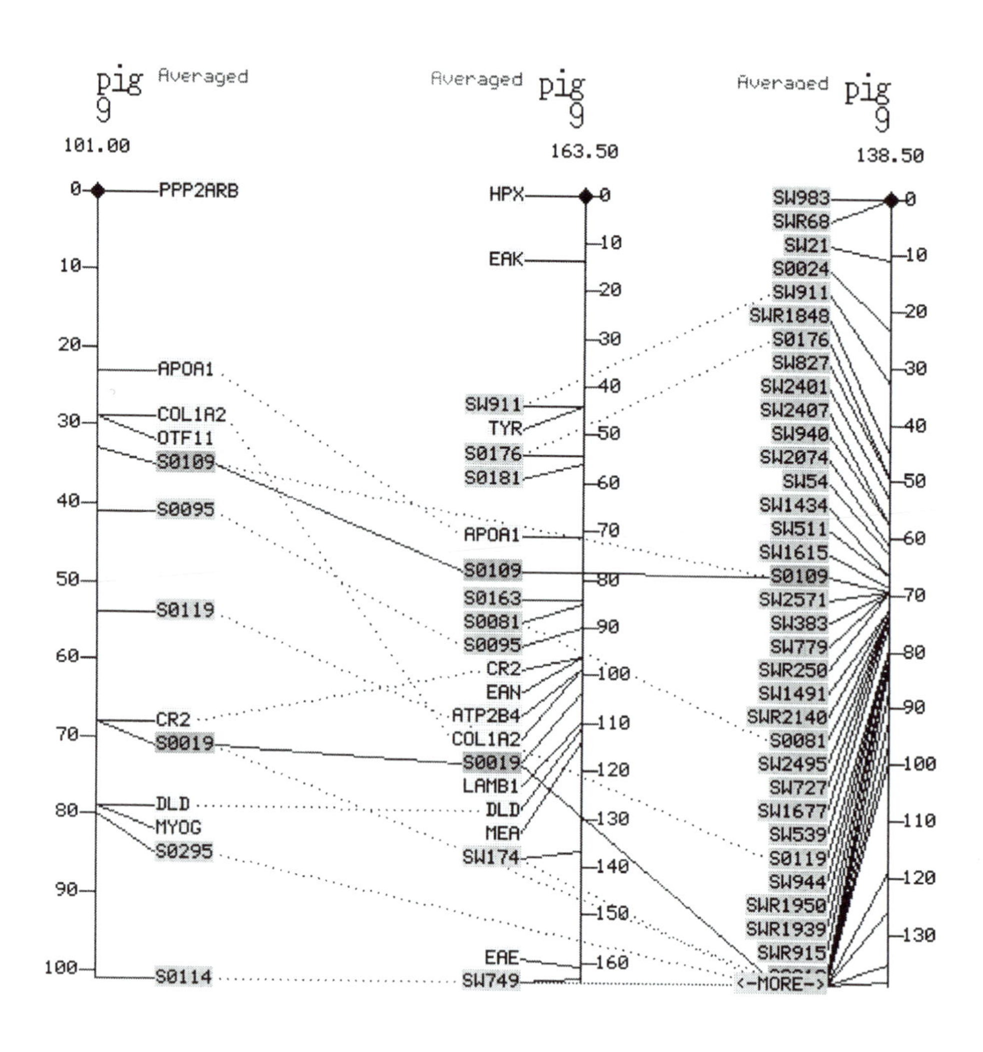

The pig linkage maps of chromosome 9 from three independent sources (June 1997). Obtained from Anubis Map Screen, http://www.ri.bbsrc.ac.uk/cgi-bin/anubis

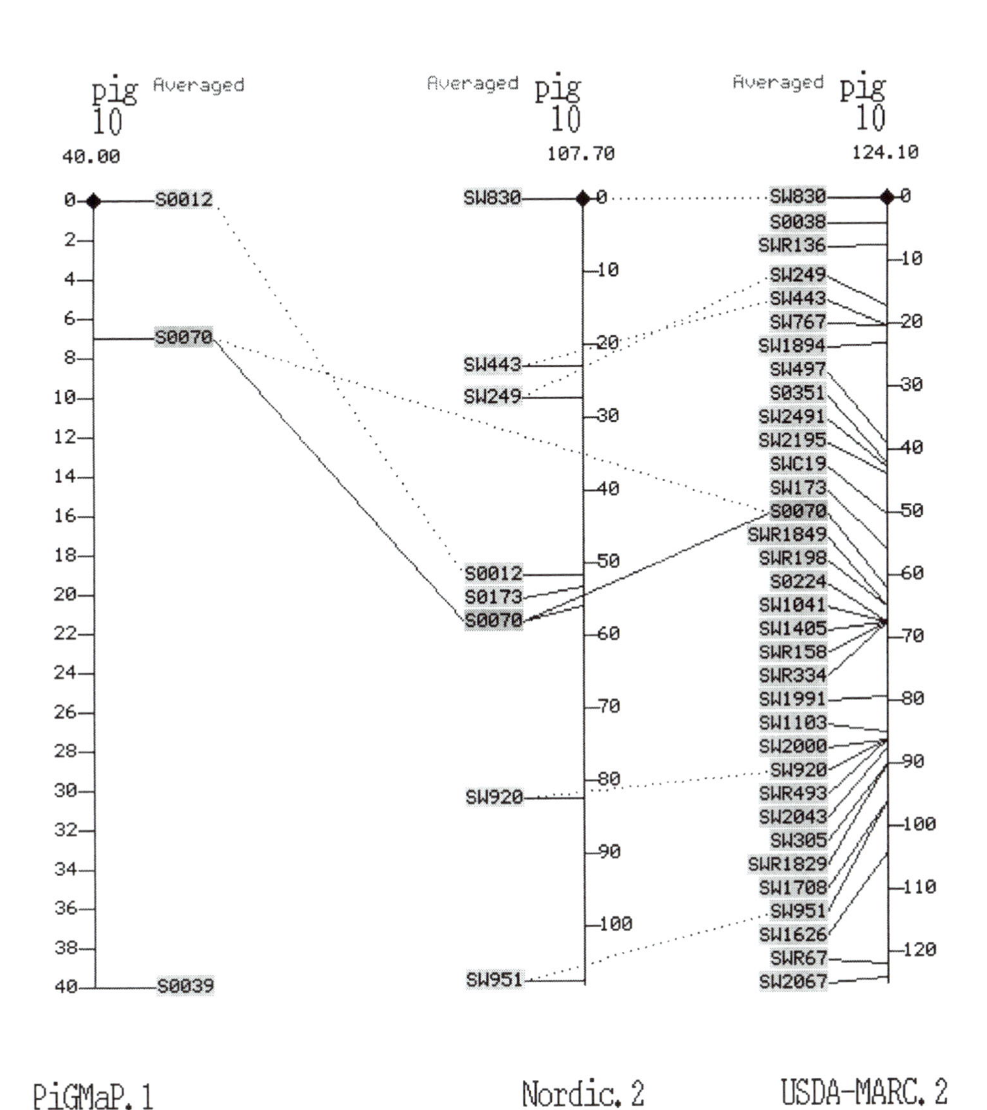

The pig linkage maps of chromosome 10 from three independent sources (June 1997). Obtained from Anubis Map Screen, http://www.ri.bbsrc.ac.uk/cgi-bin/anubis

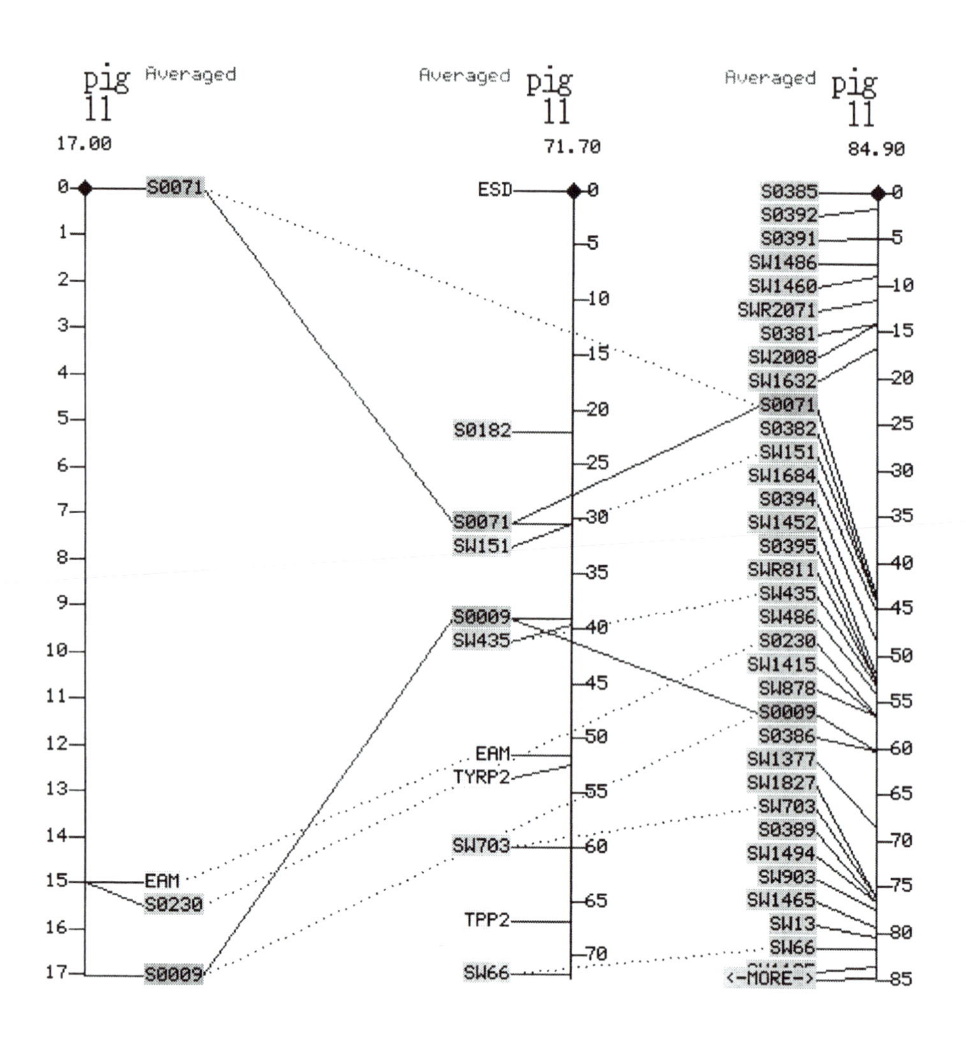

The pig linkage maps of chromosome 11 from three independent sources (June 1997). Obtained from Anubis Map Screen, http://www.ri.bbsrc.ac.uk/cgi-bin/anubis

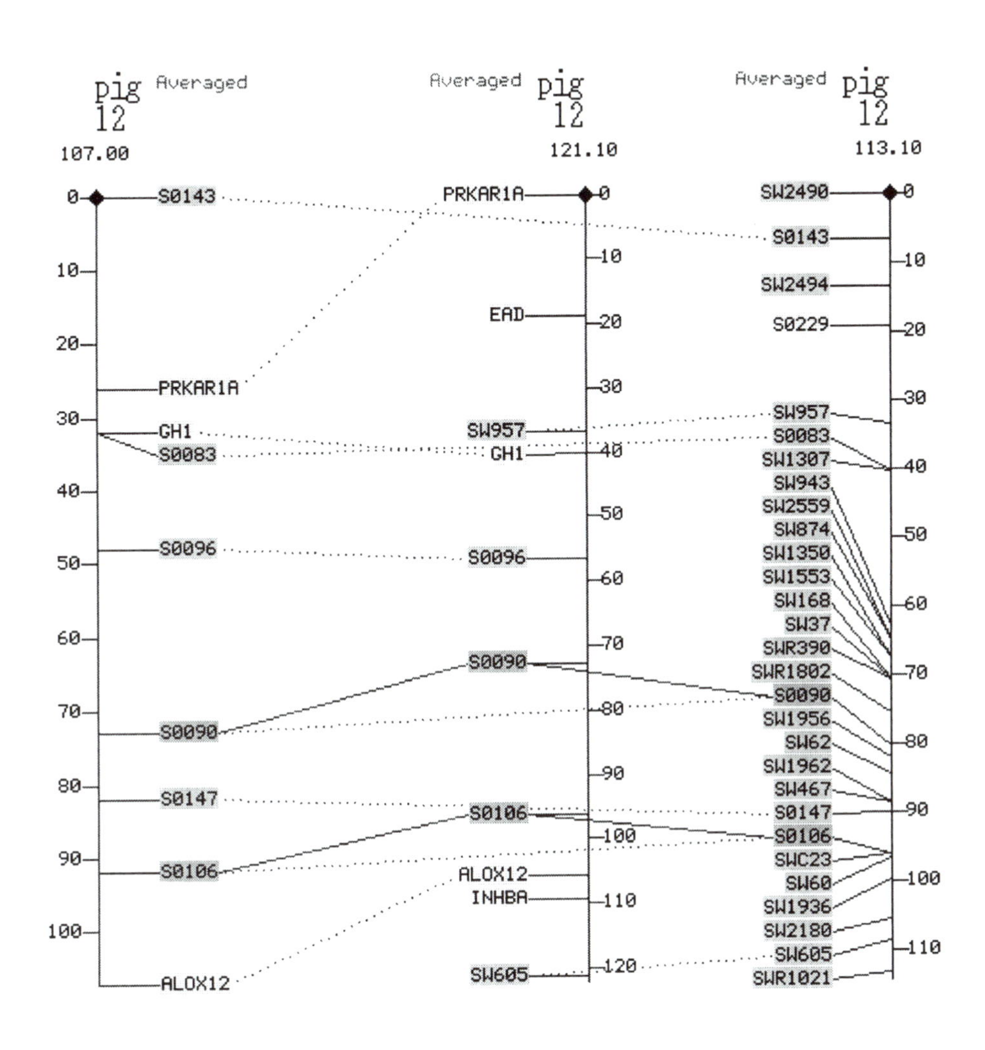

The pig linkage maps of chromosome 12 from three independent sources (June 1997). Obtained from Anubis Map Screen, http://www.ri.bbsrc.ac.uk/cgi-bin/anubis

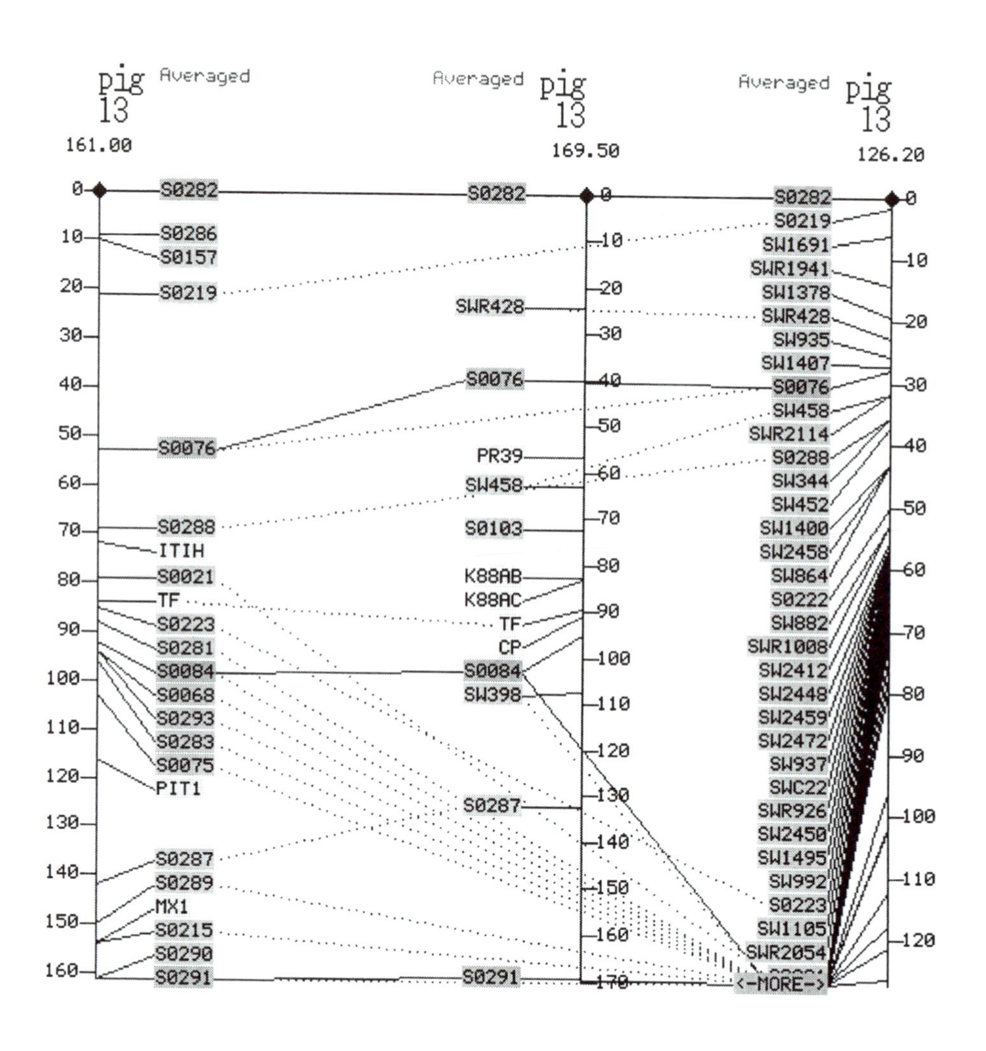

The pig linkage maps of chromosome 13 from three independent sources (June 1997). Obtained from Anubis Map Screen, http://www.ri.bbsrc.ac.uk/cgi-bin/anubis

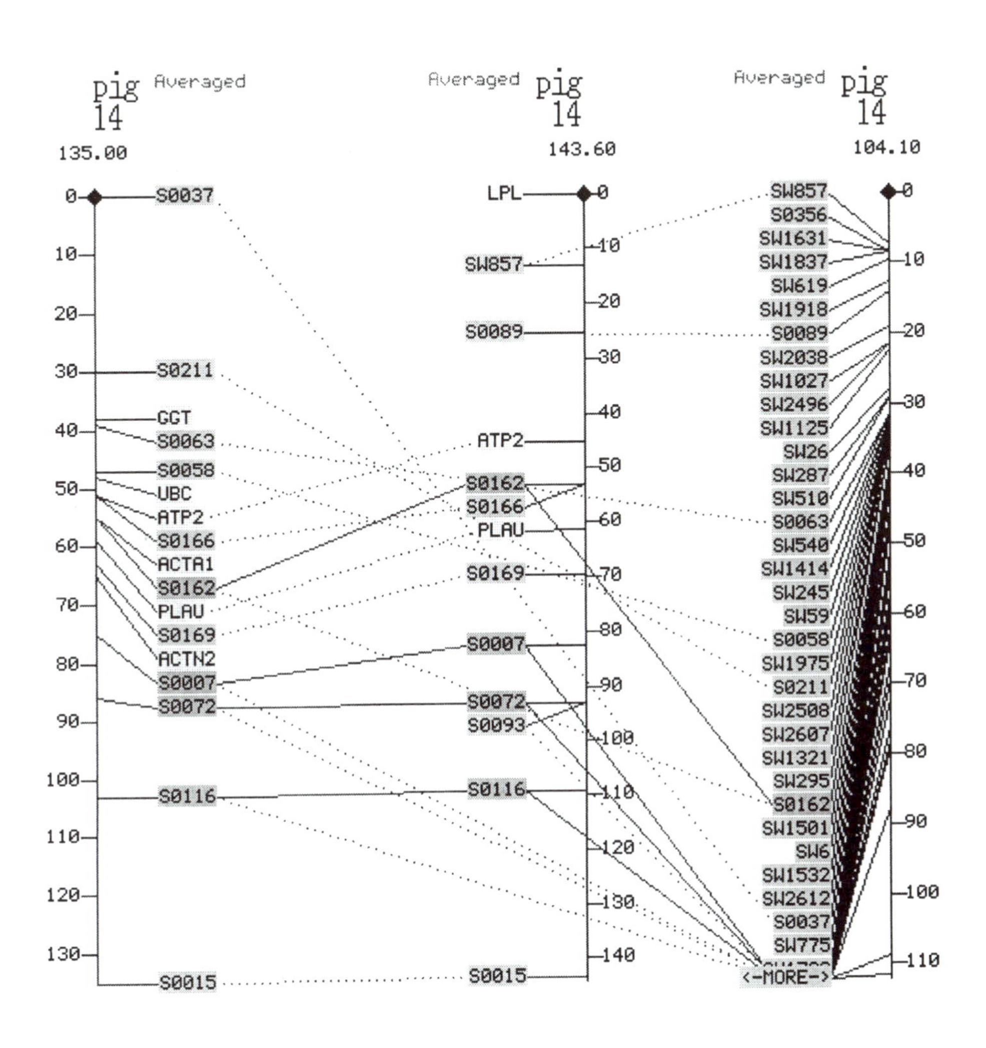

PiGMaP.1 Nordic.2 USDA-MARC.2

The pig linkage maps of chromosome 14 from three independent sources (June 1997). Obtained from Anubis Map Screen, http://www.ri.bbsrc.ac.uk/cgi-bin/anubis

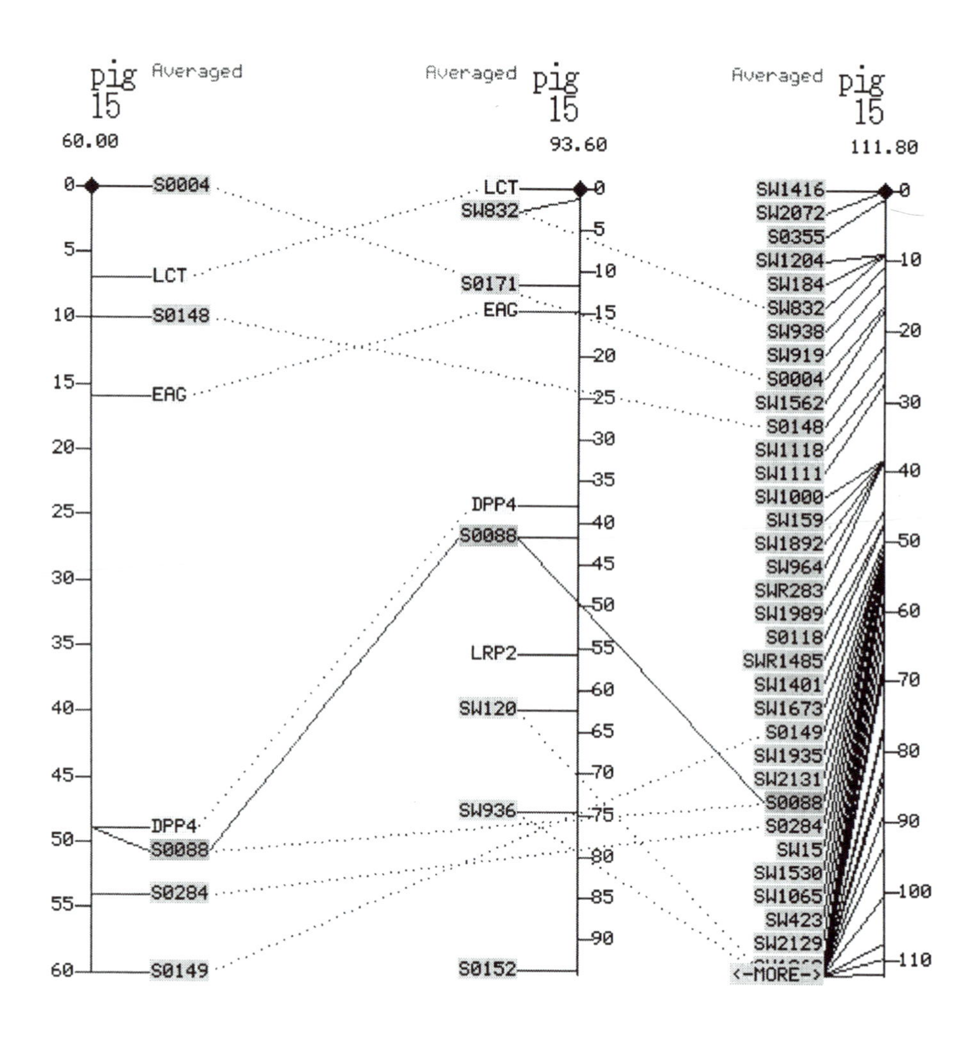

PiGMaP.1 Nordic.2 USDA-MARC.2

The pig linkage maps of chromosome 15 from three independent sources (June 1997). Obtained from Anubis Map Screen, http://www.ri.bbsrc.ac.uk/cgi-bin/anubis

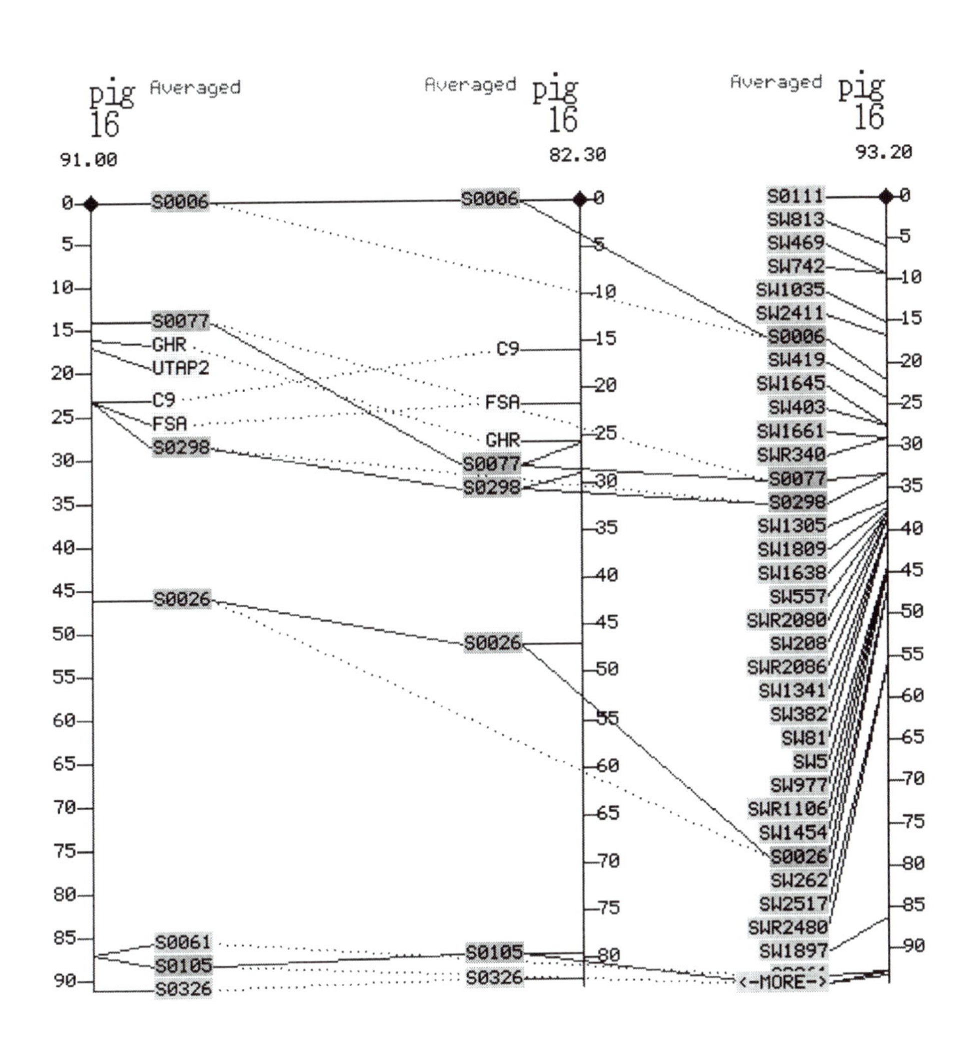

The pig linkage maps of chromosome 16 from three independent sources (June 1997). Obtained from Anubis Map Screen, http://www.ri.bbsrc.ac.uk/cgi-bin/anubis

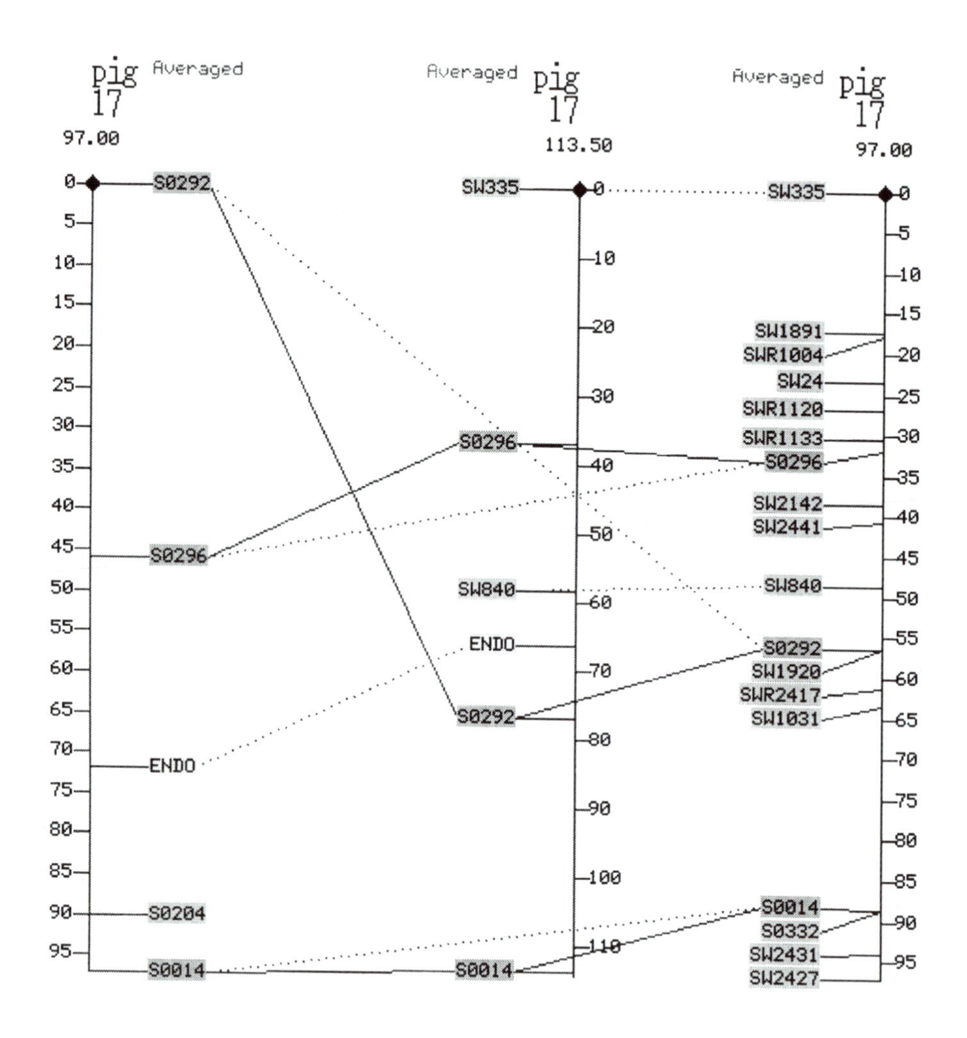

The pig linkage maps of chromosome 17 from three independent sources (June 1997). Obtained from Anubis Map Screen, http://www.ri.bbsrc.ac.uk/cgi-bin/anubis

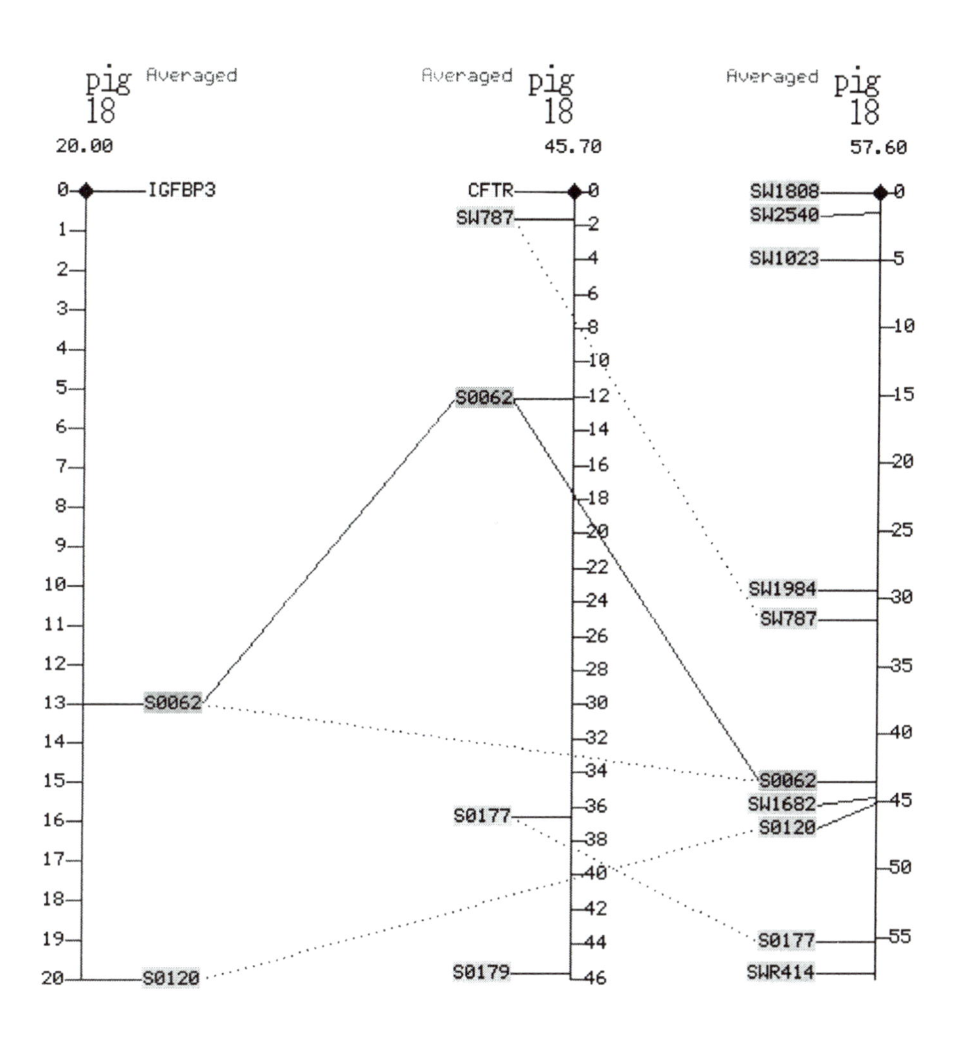

The pig linkage maps of chromosome 18 from three independent sources (June 1997). Obtained from Anubis Map Screen, http://www.ri.bbsrc.ac.uk/cgi-bin/anubis

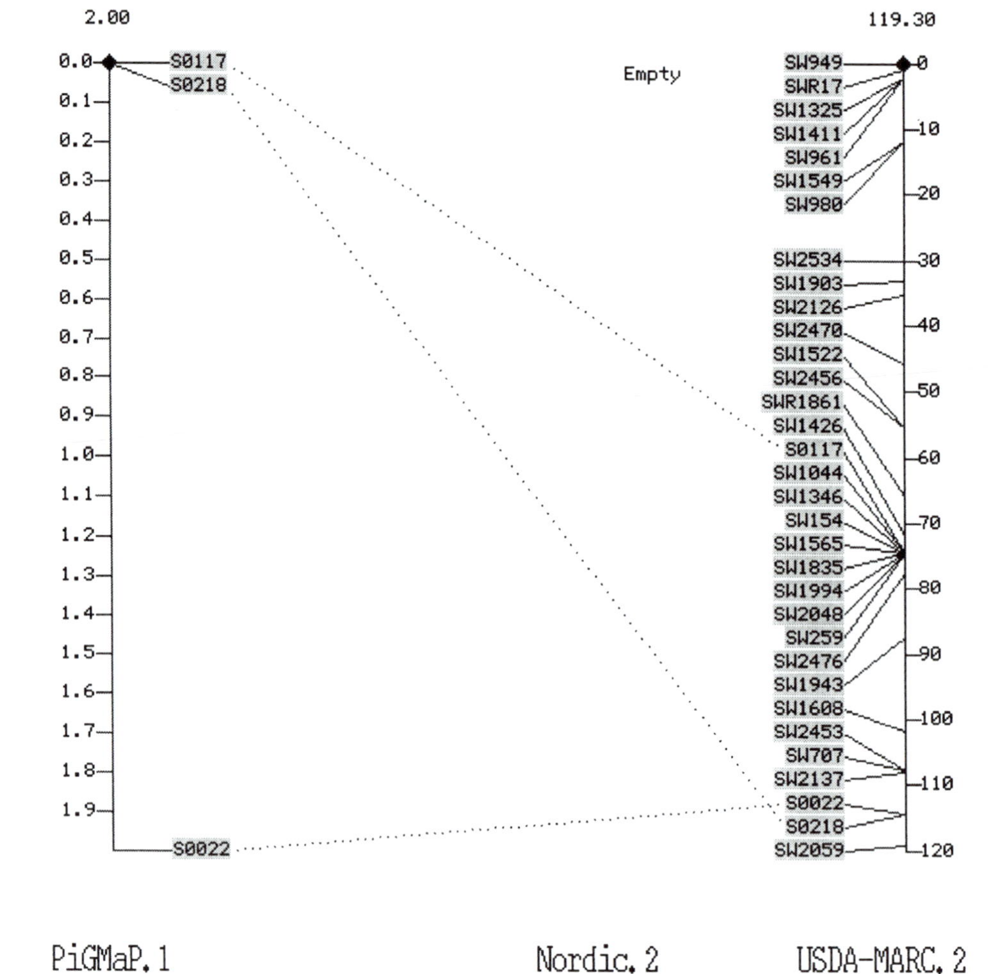

The pig linkage maps of chromosome X from three independent sources (June 1997). Obtained from Anubis Map Screen, http://www.ri.bbsrc.ac.uk/cgi-bin/anubis

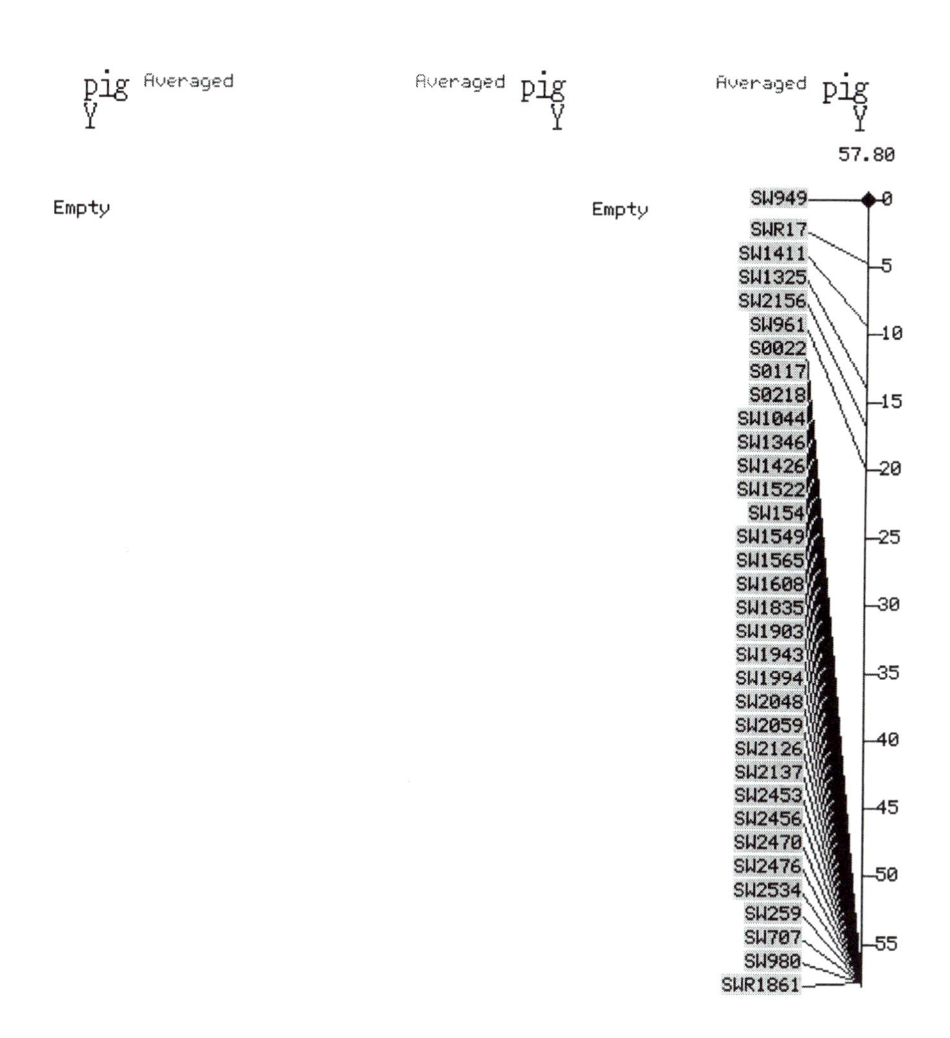

The pig linkage maps of chromosome Y from three independent sources (June 1997). Obtained from Anubis Map Screen, http://www.ri.bbsrc.ac.uk/cgi-bin/anubis

Appendix 2: List of Identified Loci in the Pig

This working list was downloaded from http://www.ri.bbsrc.ac.uk/pigmap/pig-base/ loclist.html in June 1997. Given the nature of this list, some errors may exist. Using tools in the following databases you may obtain more information about the locus you are interested in:

http://www.ri.bbsrc.ac.uk/pigmap/pigbase/pigbase.html

http://www.public.iastate.edu/~pigmap/

http://sol.marc.usda.gov/genome/swine/swine.html

http://probe.nalusda.gov:8300/animal/index.html

List of identified loci in the pig (June 1997).

Sorted by: uppersymbol

Chromosome	Symbol	Name
17	*13R43U*	anonymous DNA sequence, microsatellite
6	*6-PGD*	6-phosphogluconate dehydrogenase
6	*A1BG*	alpha1B-glycoprotein
14	*A47*	anonymous DNA sequence, contains microsatellite
7	*AACT*	alpha-1-antichymotrypsin
16	*ABP1*	angiotensin binding protein 1, microsatellite
5	*ACO2*	aconitase 2, mitochondrial
2	*ACP5*	acid phosphatase 5, tartrate resistant
5	*ACR*	acrosin

14	*ACTA1*	actin, alpha 1
14	*ACTN2*	actinin, alpha 2
8	*ADH3*	alcohol dehydrogenase 3
8	*ALB*	albumin
12	*ALOX12*	arachidonate 12-lipoxygenase
6	*ALPL*	bone, liver, kidney alkaline phosphatase
7	*ANPEP*	alanyl (membrane) aminopeptidase N, CD13
9	*APOA1*	apolipoprotein AI
3	*APOB*	apolipoprotein B
UN	*APOC3*	apolipoprotein CIII
6	*APOE*	apolipoprotein E
4	*ATP1A1*	ATPase, Na$^+$/K$^+$ transporting, alpha 1 polypeptide
4	*ATP1B1*	ATPase, Na$^+$/K$^+$ transporting, beta 1 polypeptide
14	*ATP2*	Ca^{2+}-transporting ATPase
9	*ATP2B4*	ATPase, Ca^{2+} transporting, plasma membrane 4
7	*BAT1*	*BAT1*
7	*BAT2*	*BAT2*
7	*BAT3*	*BAT3*
7	*BAT6*	*BAT6*
7	*BF*	factor B
1	*BHT49*	DNA segment, cosmid (contains microsatellite)
UN	*BMP5*	bone morphogenetic protein 5
UN	*BMP6*	bone morphogenetic protein 6
6	*BNP*	brain natriuretic peptide
6	*BNP1*	brain natriuretic peptide[1]
7	*C2*	complement component 2
7	*C4*	complement component 4
UN	*C6*	complement component 6
6	*C6014*	anonymous DNA sequence, C6014
6	*C6045*	anonymous DNA sequence, C6045
16	*C9*	complement component 9
9	*CALCR*	calcitonin receptor
1	*CAPN*	calpain, skeletal muscle
UN	*CASA*	alpha casein
8	*CASAS1*	casein, alpha s1
8	*CASAS2*	casein, alpha s2
8	*CASB*	casein, beta
UN	*CASB1*	beta-1 casein
8	*CASK*	casein, kappa
UN	*CAST*	calpastatin
13	*CCK*	cholecystokinin
15	*CEN(15)*	centromere 15 size polymorphism
18	*CFTR*	cystic fibrosis transmembrane conductance regulator
1	*CGA*	glycoprotein hormones, alpha polypeptide

17	*CH13*	anonymous DNA sequence, microsatellite
14	*CHAT*	choline acetyltransferase
7	*CHGA*	chromogranin A
8	*CHGA*	chromogranin A, (*Eco*RV RFLP)
9	*COL1A2*	collagen, type 1, alpha 2
13	*CP*	ceruloplasmin
6	*CR101*	anonymous DNA sequence, CR101
6	*CR103*	anonymous DNA sequence, CR103
9	*CR2*	complement component (3d/Epstein–Barr virus) receptor[2]
6	*CRC*	calcium release channel (skeletal muscle)
5	*CS*	citrate synthase
7	*CYP21*	cytochrome P450, steroid 21-hydroxylase
UN	*CYP7*	pig cholesterol 7 alpha-hydroxylase locus
5	*DAGK*	diacylglycerol kinase (includes microsatellite)
14	*DAO*	D-amino acid oxidase
9	*DLD*	lipoamide dehydrogenase
7	*DOB*	MHC class II antigen, DO beta chain
7	*DPA*	MHC class II antigen, DP alpha chain
7	*DPB*	MHC class II antigen, DP beta chain
15	*DPP4*	dipeptidylpeptidase IV
7	*DQA*	MHC class II antigen, DQ alpha chain
7	*DQB*	MHC class II DQ beta locus
7	*DRA*	MHC class II antigen, DR alpha chain
7	*DRB*	MHC class II antigen, DR beta chain
7	*DRB1*	MHC class II DRB1
7	*DRB2*	MHC class II pseudogene DRB2
Y	*DYZ1*	male-specific repetitive DNA sequence
Y	*DYZ2*	male-specific repetitive DNA sequence
7	*DZA*	MHC class II antigen, DZ alpha chain
6	*E*	extension, coat colour locus
1	*EAA*	erythrocyte antigen A
UN	*EAB*	erythrocyte antigen B
7	*EAC*	erythrocyte antigen C
12	*EAD*	erythrocyte antigen D
9	*EAE*	erythrocyte antigen E
15	*EAG*	erythrocyte antigen G
6	*EAH*	erythrocyte antigen H
UN	*EAI*	erythrocyte antigen I
7	*EAJ*	erythrocyte antigen J
9	*EAK*	erythrocyte antigen K
4	*EAL*	erythrocyte antigen L
11	*EAM*	erythrocyte antigen M
9	*EAN*	erythrocyte antigen N
6	*EAO*	erythrocyte antigen O
6	*EAS*	erythrocyte antigen S – inhibition of A–O antigens

6	*ECF107*	receptor for *E. coli* expressing fimbriae 107
6	*ECF18R*	receptor for F18 fimbriated *E. coli*
7	*EDN1*	endothelin-1
10	*EGGT*	porcine repeat family (GGT)10
UN	*ELF*	early lethal factor
17	*ENDO*	dynorphin
6	*ENO1*	enolase 1
11	*ESD*	esterase D
1	*ESR*	oestrogen receptor
6	*ETH5001*	microsatellite at *RYR1* locus
4	*F13B*	coagulation factor XIII, B polypeptide
UN	*F8VWF*	von Willebrand factor
X	*F9*	coagulation factor IX
8	*FGA*	fibrinogen, alpha
8	*FGB*	fibrinogen, beta
8	*FGG*	fibrinogen, gamma
16	*FSA*	follistatin
2	*FSHB*	follicle stimulating hormone, beta polypeptide
3	*FSHR*	follicle stimulating hormone receptor
6	*FUCA1*	fucosidase, alpha
UN	*G2*	anonymous DNA sequence, contains microsatellite
X	*G6PD*	glucose-6-phosphate dehydrogenase
UN	*G9,G16*	serum protein antigens (perhaps alpha-globulins)
15	*GAD1*	glutamate decarboxylase 1
5	*GAPD*	glyceraldehyde-3-phosphate dehydrogenase
4	*GBA*	glucosidase, beta; acid
14	*GGT*	gamma-glutamyl transpeptidase
1	*GGTA1*	alpha 1,3 galactosyltransferase
12	*GH1*	growth hormone
16	*GHR*	growth hormone receptor
X	*GLA*	galactosidase, alpha
7	*GLO1*	glyoxylase 1
UN	*GNRHR*	gonadotrophin-releasing hormone receptor
6	*GPI*	glucose phosphate isomerase
1	*GPI1*	glucose phosphate isomerase, psuedogene
1	*GRP78*	glucose-regulated protein 78-kDa
UN	*GSN*	gelsolin
6	*HAL*	halothane sensitivity
3	*Hbx24*	porcine homeobox gene 24
UN	*HMGCR*	3-hydroxy-3-methylglutaryl coenzyme A reductase
18	*HOXA11*	homeobox gene A11
12	*HOXB6*	homeobox gene B6
5	*HOXC8*	homeobox gene C8

15	*HOXD4*	homeobox gene D4
X	*HPRT*	hypoxanthine-guanine phosphoribosyltransferase
9	*HPX*	haemopexin
4	*HSP*	heat shock protein
7	*HSPA1*	heat-shock 70 protein 1
7	*HSPA2*	heat shock 70-kDa protein 2
7	*HSPA3*	heat shock 70-kDa protein 3
14	*HSPA4*	heat shock 70-kDa protein 4
14	*HSPA6*	heat-shock 70-kDa protein HC4.2
1	*HXB*	tenascin (also called hexabrachion or cytotactin)
8	*I*	dominant white mutation
1	*IFNA*	interferon, alpha
1	*IFNB1*	interferon, beta 1
5	*IFNG*	interferon, gamma (includes microsatellite)
5	*IGF1*	insulin-like growth factor I (includes microsatellite)
1	*IGF1R*	insulin-like growth factor 1 receptor
18	*IGFBP3*	insulin-like growth factor binding protein 3
7	*IGH*	immunoglobulin heavy chain
UN	*IGH1*	Immunoglobulin heavy chain 1
UN	*IGH2*	immunoglobulin heavy chain 2
UN	*IGH3*	immunoglobulin heavy chain 3
UN	*IGH4*	immunoglobulin heavy chain 4
7	*IGHG*	immunoglobulin gamma heavy chain
3	*IGKC*	immunoglobulin kappa constant region
14	*IGLV*	immunoglobulin lambda, variable region
7	*IKBA*	nuclear factor of kappa light polypeptide gene *enh*
UN	*IL-6*	interleukin 6
3	*IL1A*	interleukin 1, alpha
3	*IL1B*	interleukin 1, beta
8	*IL2*	interleukin 2
9	*IL6*	interleukin 6
UN	*INHA*	inhibin, alpha
12	*INHBA*	inhibin, beta A
12	*INHBB*	inhibin, beta subunit
2	*INSL3*	Leydig insulin-like hormone
2	*INSR*	insulin receptor
UN	*ITGA7*	alpha7-integrin
13	*ITIH*	inter-alpha-trypsin inhibitor
4	*IVL*	involucrin
13	*K88AB*	colibacillose resistance b
13	*K88ABR*	K88ab small intestinal receptors to *E. coli* adhesin
13	*K88AC*	colibacillose resistance c
13	*K88ACR*	K88ac small intestinal receptors to *E. coli* adhesin

UN	*K88E.coli*	receptors mediating adhesion of K88 *E. coli* pilus a
8	*KIT*	mast/stem cell growth factor receptor
6	*KRT10L*	keratin-like locus, homologue to rat K51
UN	*LALBA*	lactalbumin, alpha
9	*LAMB1*	laminin, B1 polypeptide
15	*LCT*	lactase phlorizin hydrolase
2	*LDHA*	lactate dehydrogenase A
5	*LDHB*	lactate dehydrogenase B
UN	*LDLR*	low density lipoprotein receptor
18	*LEP*	porcine obese (leptin)
6	*LHB*	luteinizing hormone, beta polypeptide
3	*LHCGR*	luteinizing hormone/choriogonadotrophin receptor
6	*LIPE*	hormone-sensitive lipase
UN	*LPB*	lipoprotein B
14	*LPL*	lipoprotein lipase
9	*LPR*	lipoprotein R
UN	*LPS*	lipoprotein S
UN	*LPT*	lipoprotein T
UN	*LPU*	lipoprotein U
15	*LRP2*	calcium sensing protein
7	*LTBETA*	lymphotoxin beta
3	*MDH1*	malate dehydrogenase, NAD (soluble)
1	*ME1*	cytosolic malic enzyme
9	*MEA*	male-enhanced antigen
UN	*MIC1*	anonymous DNA sequence, microsatellite
UN	*MIC15*	anonymous DNA sequence, microsatellite
UN	*MIC27*	anonymous DNA sequence, microsatellite
UN	*MIC3*	anonymous DNA sequence, microsatellite
UN	*MP35*	anonymous DNA sequence, microsatellite, MP35
UN	*MP47*	anonymous DNA sequence, microsatellite, MP47
UN	*MP66*	anonymous DNA sequence, microsatellite, MP66
UN	*MP7*	anonymous DNA sequence, microsatellite, MP7
UN	*MP75*	anonymous DNA sequence, microsatellite, MP75
UN	*MP77*	anonymous DNA sequence, microsatellite, MP77
7	*MPI*	mannose phosphate isomerase
5	*MUC*	apomucin
13	*MX1*	myxovirus (influenza) resistance 1
UN	*MX2A*	Mx2 protein, SINE poly (A) sequence
UN	*MYBPH*	myosin binding protein H
UN	*MYF6*	myogenic factor 6
UN	*MYHCB*	myosin heavy chain, cardiac muscle, beta
9	*MYOG*	myogenin
UN	*NEB*	nebulin
7	*NG*	DNA-binding factors

4	*NGFB*	nerve growth factor, beta polypeptide
3	*NKLYSIN*	natural killer cells, lysin
7	*NP*	nucleoside phosphorylase
UN	*OCT11*	octamer-binding transcription factor 11
Y	*OPF.07*	Y-specific RAPD using operon primer OPF.07
Y	*OPF.20*	Y-specific RAPD using operon primer OPF.20
Y	*OPH.08*	Y-specific RAPD using operon primer OPH.08
Y	*OPH.18*	Y-specific RAPD using operon primer OPH.18
1	*ORM*	orosomucoid (alpha-1-acid glycoprotein)
7	*OSG*	opposite strand gene, tenascin-like
UN	*OSG/X*	opposite strand gene
4	*OTF1*	octamer transcription factor 1
9	*OTF11*	octamer binding transcription factor 11
6	*OTF2*	octamer binding transcription factor 2
UN	*P10*	PRE-1 sequence from cosmid P10
UN	*P14*	PRE-1 sequence from cosmid P14
UN	*P18*	PRE-1 sequence from cosmid P18
1	*P1_44M18*	anonymous DNA sequence, P1 clone
UN	*P20*	PRE-1 sequence from cosmid P20
UN	*P25*	PRE-1 sequence from cosmid P25
UN	*P27*	PRE-1 sequence from cosmid P27
6	*P3*	uncharacterized allotype P3
UN	*P38*	PRE-1 sequence from cosmid P38
2	*PC1/3*	prohormome convertase 1/3
X	*pCMS14x*	anonymous DNA, VNTR
Y	*pCMS14y*	anonymous DNA, VNTR
7	*PD1A*	porcine SLA class I gene
7	*PD6*	microsatellite
8	*PDGFRA*	platelet-derived growth factor receptor, alpha
2	*PDGFRB*	platelet-derived growth factor receptor, beta poly
5	*PEPB*	peptidase B
7	*PEPN*	withdrawn, = ANPEP
2	*PGA*	pepsinogen A
6	*PGD*	6-phosphogluconate dehydrogenase
X	*PGK*	phosphoglycerate kinase
X	*PGK1*	phosphoglycerate kinase 1
6	*PGM1*	phosphoglucomutase 1
7	*PI*	protease inhibitors (includes PI1, PI2, PO1A and P)
7	*PI1*	protease inhibitor 1
UN	*PI3*	protease inhibitor 3
7	*PI4*	alpha-protease inhibitor-4
13	*PIT1*	PIT-1 (a member of the POU-domain gene familiy)
7	*PKM2*	pyruvate kinase
14	*PLAU*	plasminogen activator urokinase type

5	*PLP1*	unidentified plasma protein
7	*PO1*	post albumin 1 (i.e 1A + 1B)
7	*PO1A*	postalbumin 1A
UN	*PO1B*	postalbumin 1B
12	*POLR2*	large polypeptide RNA polymerase II
9	*PPP2ARB*	65-kDa regulatory subunit of protein phosphatase 2
13	*PR39*	peptide antibiotic PR-39
UN	*PRKACA*	protein kinase catalytic alpha subunit
6	*PRKACB*	protein kinase, cAMP-dependent catalytic, beta
12	*PRKAR1A*	protein kinase, cAMP-dependent, regulatory, type I
UN	*PRKCE*	protein kinase C, epsilon polypeptide
7	*PRL*	prolactin
3	*PRM1*	protamine 1
UN	*PRM2*	protamine 2
5	*RARG*	retinoic acid receptor, gamma
14	*RBP4*	retinol-binding protein 4
UN	*RELXA*	relaxin
1	*RLN*	relaxin
2	*RLNCE*	pro-relaxin converting enzyme
15	*RN*	Napole, meat quality
14	*RN5S@*	RNA, 5S cluster
UN	*RNR*	ribosomal RNA
8	*RNR1*	ribosomal RNA (NOR)
10	*RNR2*	ribosomal RNA (NOR)
16	*RNR3*	ribosomal RNA (NOR)
14	*RYR2*	Ryanodine receptor 2 (cardiac)
4	*S0001*	anonymous DNA sequence, microsatellite (ctg1)
3	*S0002*	microsatellite locus (CGT2) (S0002)
6	*S0003*	microsatellite (ctg3)
15	*S0004*	microsatellite (CGT4) (EMBL: M97231)
5	*S0005*	microsatellite (ctg5)
16	*S0006*	microsatellite (CGT6) (EMBL: M97233)
14	*S0007*	microsatellite (ctg7)
1	*S0008*	microsatellite (CGT8) (EMBL: M97235)
11	*S0009*	anonymous DNA sequence, microsatellite (ctg9)
2	*S0010*	microsatellite (ctg10)
4	*S0011*	anonymous DNA sequence
10	*S0012*	VNTR, minisatellite (pS3)
7	*S0013*	anonymous DNA sequence, microsatellite
17	*S0014*	anonymous DNA sequence, microsatellite
14	*S0015*	anonymous DNA sequence, VNTR
6	*S0016*	anonymous DNA sequence, microsatellite
8	*S0017*	anonymous DNA sequence, microsatellite S0017
5	*S0018*	anonymous DNA sequence, microsatellite S0018

9	*S0019*	anonymous DNA sequence
1	*S0020*	anonymous DNA sequence
13	*S0021*	anonymous DNA sequence, microsatellite
Y	*S0022*	anonymous DNA sequence, microsatellite
X	*S0022*	anonymous DNA sequence, microsatellite
4	*S0023*	anonymous DNA sequence, microsatellite (DA1E6)
9	*S0024*	anonymous DNA sequence, microsatellite
7	*S0025*	anonymous DNA sequence, microsatellite
16	*S0026*	anonymous DNA sequence, microsatellite
UN	*S0027*	anonymous DNA sequence, microsatellite
7	*S0029*	anonymous DNA sequence, microsatellite
6	*S0031*	anonymous DNA sequence, microsatellite
3	*S0032*	anonymous DNA sequence, microsatellite
5	*S0033*	anonymous DNA sequence, microsatellite
3	*S0034*	anonymous DNA sequence, microsatellite
6	*S0035*	anonymous DNA sequence, microsatellite
2	*S0036*	anonymous DNA sequence, microsatellite
14	*S0037*	anonymous DNA sequence, microsatellite
10	*S0038*	microsatellite from P1 clone P1_34P14
10	*S0039*	microsatellite from P1 clone P1_12A1
UN	*S0040*	anonymous DNA sequence, microsatellite
1	*S0041*	anonymous DNA sequence, cosmid K12
8	*S0042*	anonymous DNA sequence, cosmid R2
6	*S0044*	anonymous DNA sequence, cosmid KG13
10	*S0045*	anonymous DNA sequence, cosmid PB6
14	*S0046*	anonymous DNA sequence, cosmid KC21
7	*S0047*	anonymous DNA sequence, cosmid PRC3
11	*S0048*	anonymous DNA sequence, cosmid KC19
2	*S0049*	anonymous DNA sequence, cosmid P3, contains microsatellite
2	*S0050*	anonymous DNA sequence, cosmid M9, contains microsatellite
6	*S0051*	anonymous DNA sequence, cosmid M3, contains microsatellite
6	*S0052*	anonymous DNA sequence, cosmid P4, contains microsatellite
7	*S0053*	anonymous DNA sequence, cosmid H3, contains microsatellite
13	*S0054*	anonymous DNA sequence, cosmid E13, contains microsatellite
14	*S0055*	anonymous DNA sequence, cosmid E3, contains microsatellite
1	*S0056*	anonymous DNA sequence (E2), contains microsatellite

14	*S0058*	anonymous DNA sequence (E18), contains microsatellite
6	*S0059*	anonymous DNA sequence (E22), contains microsatellite
16	*S0061*	anonymous DNA sequence, microsatellite (ctg11)
18	*S0062*	anonymous DNA sequence, microsatellite (ctg12)
14	*S0063*	anonymous DNA sequence, microsatellite (ctg13)
7	*S0064*	anonymous DNA sequence, microsatellite (ctg14)
4	*S0065*	anonymous DNA sequence, microsatellite (ctg15)
7	*S0066*	anonymous DNA sequence, microsatellite (ctg16)
4	*S0067*	anonymous DNA sequence, microsatellite (ctg17)
13	*S0068*	anonymous DNA sequence, microsatellite (ctg18)
8	*S0069*	anonymous DNA sequence, microsatellite (ctg19)
10	*S0070*	anonymous DNA sequence, microsatellite (ctg20)
11	*S0071*	anonymous DNA sequence, microsatellite (ctg21)
14	*S0072*	anonymous DNA sequence, microsatellite (ctg22)
4	*S0073*	anonymous DNA sequence, microsatellite (ctg23)
13	*S0075*	anonymous DNA sequence, microsatellite
13	*S0076*	anonymous DNA sequence, microsatellite
16	*S0077*	anonymous DNA sequence, microsatellite (cgt27)-
7	*S0078*	anonymous DNA sequence, microsatellite
1	*S0079*	anonymous DNA sequence, microsatellite
9	*S0081*	anonymous DNA sequence, microsatellite
1	*S0082*	anonymous DNA sequence, microsatellite
12	*S0083*	anonymous DNA sequence, microsatellite
13	*S0084*	anonymous DNA sequence, microsatellite
3	*S0085*	anonymous DNA sequence, microsatellite
8	*S0086*	anonymous DNA sequence, microsatellite
6	*S0087*	anonymous DNA sequence, microsatellite
15	*S0088*	anonymous DNA sequence, microsatellite
14	*S0089*	anonymous DNA sequence, microsatellite
12	*S0090*	anonymous DNA sequence, microsatellite
2	*S0091*	anonymous DNA sequence, microsatellite locus
5	*S0092*	anonymous DNA sequence, microsatellite
14	*S0093*	anonymous DNA sequence, microsatellite
3	*S0094*	microsatellite
9	*S0095*	anonymous DNA sequence, contains microsatellite
12	*S0096*	anonymous DNA sequence, contains microsatellite
4	*S0097*	anonymous DNA sequence, microsatellite
8	*S0098*	anonymous DNA sequence, microsatellite
UN	*S0098*	anonymous DNA sequence, microsatellite
6	*S0099*	anonymous DNA sequence, microsatellite
3	*S0100*	anonymous DNA sequence, microsatellite
7	*S0101*	anonymous DNA sequence, microsatellite

7	*S0102*	anonymous DNA sequence, microsatellite
13	*S0103*	anonymous DNA marker – SINE poly (A) sequence
UN	*S0104*	anonymous DNA marker – SINE poly (A) sequence
16	*S0105*	anonymous DNA, microsatellite [CA]n*[GT]n
12	*S0106*	anonymous DNA segment, contains compound microsatellite
4	*S0107*	anonymous DNA segment, contains (CA)n microsatellite
6	*S0108*	anonymous DNA segment, contains (AC)n microsatellite
9	*S0109*	anonymous DNA segment contains compound microsatellite
2	*S0110*	anonymous DNA segment, containing a (CA)n microsatellite
16	*S0111*	anonymous DNA, microsatellite [TG]n
1	*S0112*	anonymous DNA, microsatellite [TG]n[AG]nGG[AG]n
1	*S0113*	anonymous DNA, microsatellite [TG]n
9	*S0114*	anonymous DNA, microsatellite [TG]n[TA]nN10[TG]nTA
7	*S0115*	anonymous DNA, microsatellite [GT]n[GA]n
14	*S0116*	anonymous DNA sequence, microsatellite
X	*S0117*	anonymous DNA sequence, microsatellite
Y	*S0117*	anonymous DNA sequence, microsatellite
15	*S0118*	anonymous DNA sequence, microsatellite
9	*S0119*	anonymous DNA sequence, microsatellite
18	*S0120*	anonymous DNA sequence, microsatellite
6	*S0121*	anonymous DNA sequence (A30), microsatellite
1	*S0122*	anonymous DNA sequence (A16), microsatellite
2	*S0141*	anonymous DNA sequence, microsatellite
1	*S0142*	anonymous DNA sequence, microsatellite
12	*S0143*	anonymous DNA sequence, microsatellite
8	*S0144*	anonymous DNA sequence, microsatellite
4	*S0145*	anonymous DNA sequence, microsatellite
6	*S0146*	anonymous DNA sequence, microsatellite
12	*S0147*	anonymous DNA sequence, microsatellite
15	*S0148*	anonymous DNA sequence, microsatellite
15	*S0149*	anonymous DNA sequence, microsatellite
3	*S0151*	anonymous DNA sequence, microsatellite
15	*S0152*	anonymous DNA sequence, microsatellite
1	*S0155*	anonymous DNA sequence, microsatellite
13	*S0157*	anonymous DNA sequence, microsatellite
UN	*S0158*	anonymous DNA sequence, microsatellite
4	*S0161*	anonymous DNA sequence, microsatellite

14	*S0162*	anonymous DNA segment, contains (CA)*n* microsatellite
9	*S0163*	anonymous DNA segment, contains complex microsatellite
3	*S0164*	anonymous DNA sequence, microsatellite
3	*S0165*	anonymous DNA segment, contains (CA)*n* microsatellite
14	*S0166*	anonymous DNA sequence, cosmid Q14
3	*S0167*	anonymous DNA segment, contains compound microsatellite
8	*S0168*	anonymous DNA segment, contains complex microsatellite
8	*S0168M*	microsatellite S0168m
14	*S0169*	anonymous DNA segment
2	*S0170*	anonymous DNA segment, contains SINE 3′ poly(A) anonymous microsatellite marker
15	*S0171*	anonymous DNA segment
3	*S0172*	anonymous DNA segment
10	*S0173*	anonymous DNA segment (S0168 derived from this clone)
3	*S0174*	anonymous DNA sequence, microsatellite
4	*S0175*	anonymous DNA sequence, microsatellite
9	*S0176*	anonymous DNA sequence, microsatellite
18	*S0177*	anonymous DNA sequence, microsatellite
8	*S0178*	anonymous DNA sequence, microsatellite
18	*S0179*	anonymous DNA sequence, microsatellite
7	*S0180*	anonymous DNA sequence, microsatellite
9	*S0181*	anonymous DNA sequence, microsatellite
11	*S0182*	anonymous DNA sequence, microsatellite
8	*S0183*	anonymous DNA sequence, microsatellite
10	*S0201*	anonymous cosmid (cos PC37)
6	*S0202*	anonymous cosmid (cos KG3)
2	*S0203*	anonymous cosmid (cos KL8)
17	*S0204*	anonymous DNA sequence, microsatellite
3	*S0206*	anonymous DNA sequence (E6), contains microsatellite
7	*S0207*	anonymous DNA sequence (E7), contains microsatellite
13	*S0208*	anonymous cosmid (cos PF26)
3	*S0210*	anonymous DNA sequence (C69), microsatellite
14	*S0211*	anonymous DNA sequence (A34), microsatellite
7	*S0212*	anonymous DNA sequence (H37), microsatellite
3	*S0213*	anonymous DNA sequence (A13), microsatellite
4	*S0214*	anonymous DNA sequence (E14), contains microsatellite
13	*S0215*	anonymous DNA sequence (G41), microsatellite

3	*S0216*	anonymous DNA sequence (E16), contains microsatellite
4	*S0217*	anonymous DNA sequence (C46), microsatellite
X	*S0218*	anonymous DNA sequence (A18), microsatellite
Y	*S0218*	anonymous DNA sequence, microsatellite
13	*S0219*	anonymous DNA sequence (C49), microsatellite
6	*S0220*	anonymous DNA sequence (C37), microsatellite
UN	*S0221*	anonymous DNA sequence (C62), microsatellite
13	*S0222*	anonymous DNA sequence (C64), microsatellite
13	*S0223*	anonymous DNA sequence (C67), microsatellite
10	*S0224*	anonymous DNA sequence (G40), microsatellite
8	*S0225*	anonymous DNA sequence (G25), microsatellite
2	*S0226*	anonymous DNA sequence (H26), microsatellite
4	*S0227*	anonymous DNA sequence (B49), microsatellite
6	*S0228*	anonymous DNA sequence (C66), microsatellite
12	*S0229*	anonymous DNA sequence, microsatellite
11	*S0230*	anonymous DNA sequence (D47), microsatellite
13	*S0281*	anonymous DNA sequence, microsatellite (13N03R)
13	*S0282*	anonymous DNA sequence, microsatellite (13N04R)
13	*S0283*	anonymous DNA sequence, microsatellite (13N05R)
15	*S0284*	anonymous DNA sequence, microsatellite (13N06R)
8	*S0285*	anonymous DNA sequence, microsatellite
UN	*S0285*	anonymous DNA sequence, microsatellite (13N12U)
13	*S0286*	anonymous DNA sequence, microsatellite (13N14F)
13	*S0287*	anonymous DNA sequence, microsatellite (13N17R)
13	*S0288*	anonymous DNA sequence, microsatellite (13N18R)
13	*S0289*	anonymous DNA sequence, microsatellite (13N32R)
13	*S0290*	anonymous DNA sequence, microsatellite (13N33R)
13	*S0291*	anonymous DNA sequence, microsatellite (13R33U)
17	*S0292*	anonymous DNA sequence, microsatellite (13R43U)
13	*S0293*	anonymous DNA sequence, microsatellite (13R44R)
6	*S0294*	anonymous DNA sequence, contains microsatellite

9	*S0295*	anonymous DNA sequence, contains microsatellite
17	*S0296*	anonymous DNA sequence, contains microsatellite
6	*S0297*	anonymous DNA sequence, contains microsatellite
16	*S0298*	anonymous DNA sequence, contains microsatellite
6	*S0299*	anonymous DNA sequence (BHA248), contains microsatellite
6	*S0300*	anonymous DNA sequence (BHA254), contains microsatellite
4	*S0301*	anonymous DNA sequence, contains microsatellite
1	*S0302*	anonymous DNA sequence, microsatellite
1	*S0303*	anonymous DNA sequence, VNTR
4	*S0304*	minisatellite, VNTR
7	*S0305*	minisatellite, VNTR
18	*S0306*	anonymous DNA sequence
8	*S0308*	anonymous DNA, RFLP
1	*S0311*	anonymous DNA sequence, microsatellite
1	*S0312*	anonymous DNA, microsatellite
1	*S0313*	anonymous DNA sequence, microsatellite
1	*S0316*	anonymous DNA sequence, microsatellite S0316
UN	*S0317*	anonymous DNA sequence, microsatellite S0317
UN	*S0319*	anonymous DNA sequence, microsatellite S0319
1	*S0320*	anonymous DNA sequence, microsatellite S0320
5	*S0321*	anonymous DNA sequence, VNTR (pCMS7)
UN	*S0322*	anonymous DNA sequence, VNTR (pCMS12)
14	*S0323*	anonymous DNA sequence, VNTR (pCMS)23
Y	*S0324*	anonymous DNA sequence, VNTR, (pCMS14)
6	*S0325*	anonymous DNA sequence, VNTR (pCMS5)
16	*S0326*	anonymous DNA sequence, VNTR (pCMS16)
1	*S0331*	anonymous DNA sequence, microsatellite
17	*S0332*	anonymous DNA sequence, microsatellite
6	*S0333*	anonymous DNA sequence, microsatellite
7	*S0334*	anonymous DNA sequence, microsatellite
UN	*S0335*	anonymous DNA sequence, microsatellite
UN	*S0341*	anonymous DNA sequence, microsatellite S0341
10	*S0351*	anonymous DNA sequence (A12), microsatellite
3	*S0352*	anonymous DNA sequence (C52), microsatellite
8	*S0353*	anonymous DNA sequence (B53), microsatellite
1	*S0354*	anonymous DNA sequence (D49), microsatellite
15	*S0355*	anonymous DNA sequence (G22), microsatellite
14	*S0356*	anonymous DNA sequence (G31), microsatellite
1	*S0357*	anonymous DNA sequence (G35), microsatellite

3	*S0358*	anonymous DNA sequence (G48), microsatellite
17	*S0359*	anonymous DNA sequence, contains microsatellite
14	*S0360*	anonymous DNA sequence, contains microsatellite
9	*S0361*	anonymous DNA sequence, contains microsatellite
10	*S0362*	anonymous DNA sequence, contains microsatellite
15	*S0364*	anonymous DNA sequence, contains microsatellite
14	*S0365*	anonymous DNA sequence, contains microsatellite
10	*S0366*	anonymous DNA sequence, contains microsatellite
1	*S0367*	anonymous DNA sequence, contains microsatellite
11	*S0381*	anonymous DNA sequence, microsatellite
11	*S0382*	anonymous DNA sequence, microsatellite
11	*S0385*	anonymous DNA sequence, microsatellite
11	*S0386*	anonymous DNA sequence, microsatellite
11	*S0388*	anonymous DNA sequence, microsatellite
11	*S0389*	anonymous DNA sequence, microsatellite
11	*S0391*	anonymous DNA sequence, microsatellite
11	*S0392*	anonymous DNA sequence, microsatellite
11	*S0394*	anonymous DNA sequence, microsatellite
11	*S0395*	anonymous DNA sequence, microsatellite
14	*S0400*	anonymous DNA sequence, containing microsatellite
UN	*S0411*	anonymous DNA – 3′ SINE-PCR (SINE58 × oligo35)
UN	*S0412*	anonymous DNA – 5′ SINE-PCR (SINE74 × oligo39)
X	*S0511*	anonymous DNA sequence, microsatellite
1	*S0531*	anonymous DNA, microsatellite
13	*S0532*	anonymous DNA, microsatellite
7	*SLA-DQA*	DQ alpha class II genes
7	*SLA-DRA*	DR alpha class II genes
7	*SLA-I*	major histocompatibility complex class I genes
7	*SLA-II*	major histocompatibility complex class II genes
UN	*SLB*	pig alloantigen
6	*SLC2A1*	solute carrier family 2 (facilitated glucose trans)
9	*SOD1*	superoxide dismutase 1
UN	*SOD2*	superoxide dismutase 2, mitochondrial
X	*SPL*	splay leg

8	*SPP1*	secreted phosphoprotein 1 (includes microsatellite)
Y	*SRY*	sex-determining region
7	*SUDOI*	RFLPs sudo class I
9	*SW1*	anonymous DNA sequence, microsatellite
15	*SW1000*	anonymous DNA sequence, microsatellite
4	*SW1003*	anonymous DNA sequence, microsatellite
9	*SW1006*	anonymous DNA sequence, microsatellite
3	*SW102*	microsatellite
1	*SW1020*	anonymous DNA sequence, microsatellite
18	*SW1023*	anonymous DNA sequence, microsatellite
2	*SW1026*	anonymous DNA sequence, microsatellite
14	*SW1027*	anonymous DNA sequence, microsatellite
8	*SW1029*	anonymous DNA sequence, microsatellite
13	*SW1030*	anonymous DNA sequence, microsatellite
17	*SW1031*	anonymous DNA sequence, microsatellite
14	*SW1032*	microsatellite
16	*SW1035*	anonymous DNA sequence, microsatellite
8	*SW1037*	anonymous DNA sequence, microsatellite
6	*SW1038*	microsatellite
14	*SW104*	microsatellite
4	*SW1040*	anonymous DNA sequence, microsatellite
10	*SW1041*	anonymous DNA sequence, microsatellite
X	*SW1044*	anonymous DNA sequence, microsatellite
Y	*SW1044*	anonymous DNA sequence, microsatellite
3	*SW1045*	anonymous DNA sequence, microsatellite
6	*SW1053*	anonymous DNA sequence, microsatellite
6	*SW1055*	anonymous DNA sequence, microsatellite
13	*SW1056*	anonymous DNA sequence, microsatellite
6	*SW1057*	anonymous DNA sequence, microsatellite
6	*SW1059*	anonymous DNA sequence, microsatellite
15	*SW1065*	anonymous DNA sequence, microsatellite
3	*SW1066*	anonymous DNA sequence, microsatellite
6	*SW1067*	anonymous DNA sequence, microsatellite
6	*SW1069*	anonymous DNA sequence, microsatellite
8	*SW1070*	anonymous DNA sequence, microsatellite
5	*SW1071*	anonymous DNA sequence, microsatellite
4	*SW1073*	anonymous DNA sequence, microsatellite
8	*SW1080*	anonymous DNA sequence, microsatellite
14	*SW1081*	anonymous DNA sequence, microsatellite
14	*SW1082*	anonymous DNA sequence, microsatellite
7	*SW1083*	anonymous DNA sequence, microsatellite
8	*SW1085*	anonymous DNA sequence, microsatellite
4	*SW1089*	anonymous DNA sequence, microsatellite
1	*SW1092*	anonymous DNA sequence, microsatellite
5	*SW1094*	anonymous DNA sequence, microsatellite

8	*SW1101*	anonymous DNA sequence, microsatellite
10	*SW1103*	anonymous DNA sequence, microsatellite
13	*SW1105*	anonymous DNA sequence, microsatellite
6	*SW1108*	anonymous DNA sequence, microsatellite
14	*SW1109*	anonymous DNA sequence, microsatellite
15	*SW1111*	anonymous DNA sequence, microsatellite
5	*SW1114*	anonymous DNA sequence, microsatellite
8	*SW1117*	anonymous DNA sequence, microsatellite
15	*SW1118*	anonymous DNA sequence, microsatellite
15	*SW1119*	anonymous DNA sequence, microsatellite
7	*SW1122*	anonymous DNA sequence, microsatellite
1	*SW1123*	anonymous DNA sequence, microsatellite
14	*SW1125*	anonymous DNA sequence, microsatellite
6	*SW1129*	anonymous DNA sequence, microsatellite
5	*SW1134*	anonymous DNA sequence, microsatellite
11	*SW1135*	anonymous DNA sequence, microsatellite
15	*SW120*	anonymous DNA sequence, microsatellite
5	*SW1200*	anonymous DNA sequence, microsatellite
2	*SW1201*	anonymous DNA sequence, microsatellite
6	*SW1202*	anonymous DNA sequence, microsatellite
15	*SW1204*	anonymous DNA sequence, microsatellite
15	*SW1217*	anonymous DNA sequence, microsatellite
6	*SW122*	anonymous DNA sequence, microsatellite
15	*SW1262*	anonymous DNA sequence, microsatellite
15	*SW1263*	anonymous DNA sequence, microsatellite
13	*SW129*	anonymous DNA sequence, microsatellite
11	*SW13*	anonymous DNA sequence, microsatellite
1	*SW1301*	anonymous DNA sequence, microsatellite
6	*SW1302*	anonymous DNA sequence, microsatellite
7	*SW1303*	anonymous DNA sequence, microsatellite
16	*SW1305*	anonymous DNA sequence, microsatellite
12	*SW1307*	anonymous DNA sequence, microsatellite
15	*SW1309*	anonymous DNA sequence, microsatellite
1	*SW1311*	anonymous DNA sequence, microsatellite
8	*SW1312*	anonymous DNA sequence, microsatellite
3	*SW1315*	anonymous DNA sequence, microsatellite
15	*SW1316*	anonymous DNA sequence, microsatellite
5	*SW1319*	anonymous DNA sequence, microsatellite
2	*SW1320*	anonymous DNA sequence, microsatellite
14	*SW1321*	anonymous DNA sequence, microsatellite
X	*SW1325*	anonymous DNA sequence, microsatellite
Y	*SW1325*	anonymous DNA sequence, microsatellite
3	*SW1327*	anonymous DNA sequence, microsatellite
6	*SW1328*	anonymous DNA sequence, microsatellite
6	*SW1329*	anonymous DNA sequence, microsatellite
6	*SW133*	anonymous DNA sequence, microsatellite

1	*SW1332*	anonymous DNA sequence, microsatellite
14	*SW1333*	anonymous DNA sequence, microsatellite
4	*SW1336*	anonymous DNA sequence, microsatellite
8	*SW1337*	anonymous DNA sequence, microsatellite
15	*SW1339*	anonymous DNA sequence, microsatellite
16	*SW1341*	anonymous DNA sequence, microsatellite
7	*SW1344*	anonymous DNA sequence, microsatellite
8	*SW1345*	anonymous DNA sequence, microsatellite
X	*SW1346*	anonymous DNA sequence, microsatellite
Y	*SW1346*	anonymous DNA sequence, microsatellite
9	*SW1349*	anonymous DNA sequence, microsatellite
12	*SW1350*	anonymous DNA sequence, microsatellite
6	*SW1353*	anonymous DNA sequence, microsatellite
7	*SW1354*	anonymous DNA sequence, microsatellite
6	*SW1355*	anonymous DNA sequence, microsatellite
3	*SW1359*	anonymous DNA sequence, microsatellite
4	*SW1364*	anonymous DNA sequence, microsatellite
7	*SW1369*	anonymous DNA sequence, microsatellite
1	*SW137*	anonymous DNA sequence, microsatellite
2	*SW1370*	anonymous DNA sequence, microsatellite
6	*SW1376*	anonymous DNA sequence, microsatellite
11	*SW1377*	anonymous DNA sequence, microsatellite
13	*SW1378*	anonymous DNA sequence, microsatellite
7	*SW1380*	anonymous DNA sequence, microsatellite
5	*SW1383*	anonymous DNA sequence, microsatellite
13	*SW1386*	anonymous DNA sequence, microsatellite
3	*SW139*	anonymous DNA sequence, microsatellite
2	*SW14*	anonymous DNA sequence, microsatellite
13	*SW1400*	anonymous DNA sequence, cosmid containing microsatellite
15	*SW1401*	anonymous DNA sequence, microsatellite
10	*SW1405*	anonymous DNA sequence, microsatellite
13	*SW1407*	anonymous DNA sequence, microsatellite
2	*SW1408*	anonymous DNA sequence, microsatellite
7	*SW1409*	anonymous DNA sequence, microsatellite
X	*SW1411*	anonymous DNA sequence, microsatellite
Y	*SW1411*	anonymous DNA sequence, microsatellite
14	*SW1414*	anonymous DNA sequence, microsatellite
11	*SW1415*	anonymous DNA sequence, microsatellite
15	*SW1416*	anonymous DNA sequence, microsatellite
1	*SW1417*	anonymous DNA sequence, microsatellite
7	*SW1418*	anonymous DNA sequence, microsatellite
3	*SW142*	anonymous DNA sequence, microsatellite
14	*SW1425*	anonymous DNA sequence, microsatellite
X	*SW1426*	anonymous DNA sequence, microsatellite
Y	*SW1426*	anonymous DNA sequence, microsatellite

1	*SW1430*	anonymous DNA sequence, microsatellite
1	*SW1431*	anonymous DNA sequence, microsatellite
3	*SW1432*	anonymous DNA sequence, microsatellite
9	*SW1434*	anonymous DNA sequence, microsatellite
9	*SW1435*	anonymous DNA sequence, microsatellite
3	*SW1436*	anonymous DNA sequence, microsatellite
7	*SW1437*	anonymous DNA sequence, microsatellite
3	*SW1443*	anonymous DNA sequence, microsatellite
2	*SW1450*	anonymous DNA sequence, microsatellite
11	*SW1452*	anonymous DNA sequence, microsatellite
16	*SW1454*	anonymous DNA sequence, microsatellite
11	*SW1460*	anonymous DNA sequence, microsatellite
4	*SW1461*	anonymous DNA sequence, microsatellite
1	*SW1462*	anonymous DNA sequence, microsatellite
11	*S1465*	anonymous DNA sequence, microsatellite
2	*SW1466*	anonymous DNA sequence, microsatellite
5	*SW1468*	anonymous DNA sequence, microsatellite
7	*SW147*	anonymous DNA sequence, microsatellite
6	*SW1473*	anonymous DNA sequence, microsatellite
4	*SW1475*	anonymous DNA sequence, microsatellite
5	*SW1482*	anonymous DNA sequence, microsatellite
11	*SW1486*	anonymous DNA sequence, microsatellite
5	*SW1489*	anonymous DNA sequence, microsatellite
8	*SW149*	anonymous DNA sequence, microsatellite
9	*SW1491*	anonymous DNA sequence, microsatellite
4	*SW1492*	anonymous DNA sequence, microsatellite
14	*SW1493*	anonymous DNA sequence, microsatellite
11	*SW1494*	anonymous DNA sequence, microsatellite
13	*SW1495*	anonymous DNA sequence, microsatellite
15	*SW15*	anonymous DNA sequence, microsatellite
14	*SW1501*	anonymous DNA sequence, microsatellite
11	*SW151*	anonymous DNA sequence, microsatellite
15	*SW1510*	anonymous DNA sequence, microsatellite
4	*SW1513*	anonymous DNA sequence, microsatellite
1	*SW1514*	anonymous DNA sequence, microsatellite
1	*SW1515*	anonymous DNA sequence, microsatellite
2	*SW1517*	anonymous DNA sequence, microsatellite
5	*SW152*	anonymous DNA sequence, microsatellite
4	*SW1520*	anonymous DNA sequence, microsatellite
X	*SW1522*	anonymous DNA sequence, microsatellite
Y	*SW1522*	anonymous DNA sequence, microsatellite
3	*SW1525*	anonymous DNA sequence, microsatellite
14	*SW1527*	anonymous DNA sequence, microsatellite
15	*SW1530*	anonymous DNA sequence, microsatellite
14	*SW1532*	anonymous DNA sequence, microsatellite
14	*SW1536*	anonymous DNA sequence, microsatellite

Y	*SW154*	anonymous DNA sequence, microsatellite
X	*SW154*	anonymous DNA sequence, microsatellite
X	*SW1549*	anonymous DNA sequence, microsatellite
Y	*SW1549*	anonymous DNA sequence, microsatellite
13	*SW1550*	anonymous DNA sequence, microsatellite
8	*SW1551*	anonymous DNA sequence, microsatellite
14	*SW1552*	anonymous DNA sequence, microsatellite
12	*SW1553*	anonymous DNA sequence, microsatellite
14	*SW1556*	anonymous DNA sequence, microsatellite
14	*SW1557*	anonymous DNA sequence, microsatellite
15	*SW1562*	anonymous DNA sequence, microsatellite
2	*SW1564*	anonymous DNA sequence, microsatellite
X	*SW1565*	anonymous DNA sequence, microsatellite
Y	*SW1565*	anonymous DNA sequence, microsatellite
1	*SW157*	anonymous DNA sequence, microsatellite
15	*SW159*	anonymous DNA sequence, microsatellite
8	*SW16*	anonymous DNA sequence, microsatellite
3	*SW160*	anonymous DNA sequence, microsatellite
2	*SW1602*	anonymous DNA sequence, microsatellite
X	*SW1608*	anonymous DNA sequence, microsatellite
Y	*SW1608*	anonymous DNA sequence, microsatellite
7	*SW1614*	anonymous DNA sequence, microsatellite
9	*SW1615*	anonymous DNA sequence, microsatellite
1	*SW1616*	anonymous DNA sequence, microsatellite
1	*SW1619*	anonymous DNA sequence, microsatellite
1	*SW1621*	anonymous DNA sequence, microsatellite
10	*SW1626*	anonymous DNA sequence, microsatellite
2	*SW1628*	anonymous DNA sequence, microsatellite
13	*SW163*	anonymous DNA sequence, microsatellite
14	*SW1631*	anonymous DNA sequence, microsatellite
11	*SW1632*	anonymous DNA sequence, microsatellite
5	*SW1633*	anonymous DNA sequence, microsatellite
16	*SW1638*	anonymous DNA sequence, microsatellite
16	*SW1645*	anonymous DNA sequence, microsatellite
6	*SW1647*	anonymous DNA sequence, cosmid containing microsatellite
8	*SW1649*	anonymous DNA sequence, microsatellite
2	*SW1650*	anonymous DNA sequence, microsatellite
9	*SW1651*	anonymous DNA sequence, microsatellite
1	*SW1653*	anonymous DNA sequence, microsatellite
2	*SW1655*	anonymous DNA sequence, microsatellite
2	*SW1658*	anonymous DNA sequence, microsatellite
16	*SW1661*	anonymous DNA sequence, microsatellite
7	*SW1667*	anonymous DNA sequence, microsatellite
1	*SW1668*	anonymous DNA sequence, microsatellite
8	*SW1671*	anonymous DNA sequence, microsatellite

15	*SW1673*	anonymous DNA sequence, microsatellite
9	*SW1677*	anonymous DNA sequence, microsatellite
4	*SW1678*	anonymous DNA sequence, microsatellite
8	*SW1679*	anonymous DNA sequence, microsatellite
12	*SW168*	anonymous DNA sequence, microsatellite
6	*SW1680*	anonymous DNA sequence, microsatellite
7	*SW1681*	anonymous DNA sequence, microsatellite
18	*SW1682*	anonymous DNA sequence, microsatellite
15	*SW1683*	anonymous DNA sequence, microsatellite
11	*SW1684*	anonymous DNA sequence, microsatellite
2	*SW1686*	anonymous DNA sequence, microsatellite
13	*SW1691*	anonymous DNA sequence, microsatellite
3	*SW1693*	anonymous DNA sequence, microsatellite
2	*SW1695*	anonymous DNA sequence, microsatellite
7	*SW1701*	microsatellite SW1701
8	*SW1702*	anonymous DNA sequence, microsatellite
4	*SW1707*	anonymous DNA sequence, microsatellite
10	*SW1708*	anonymous DNA sequence, microsatellite
14	*SW1709*	anonymous DNA sequence, microsatellite
8	*SW171*	anonymous DNA sequence, microsatellite
4	*SW1710*	anonymous DNA sequence, microsatellite
10	*SW173*	anonymous DNA sequence, microsatellite
9	*SW174*	anonymous DNA sequence, microsatellite
7	*SW175*	anonymous DNA sequence, microsatellite
16	*SW18*	anonymous DNA sequence, microsatellite
14	*SW1804*	anonymous DNA sequence, microsatellite
18	*SW1808*	anonymous DNA sequence, microsatellite
16	*SW1809*	anonymous DNA sequence, microsatellite
7	*SW1816*	anonymous DNA sequence, microsatellite
6	*SW1818*	anonymous DNA sequence, microsatellite
6	*SW1823*	anonymous DNA sequence, microsatellite
1	*SW1824*	anonymous DNA sequence, microsatellite
11	*SW1827*	anonymous DNA sequence, microsatellite
1	*SW1828*	anonymous DNA sequence, microsatellite
13	*SW1833*	anonymous DNA sequence, microsatellite
X	*SW1835*	anonymous DNA sequence, microsatellite
Y	*SW1835*	anonymous DNA sequence, microsatellite
14	*SW1837*	anonymous DNA sequence, microsatellite
15	*SW184*	anonymous DNA sequence, microsatellite
6	*SW1841*	anonymous DNA sequence, microsatellite
8	*SW1843*	anonymous DNA sequence, microsatellite
2	*SW1844*	anonymous DNA sequence, microsatellite
1	*SW1846*	anonymous DNA sequence, microsatellite
1	*SW1851*	anonymous DNA sequence, microsatellite
7	*SW1856*	anonymous DNA sequence, microsatellite
2	*SW1857*	anonymous DNA sequence, microsatellite

2	*SW1860*	anonymous DNA sequence, microsatellite
13	*SW1864*	anonymous DNA sequence, microsatellite
15	*SW1865*	anonymous DNA sequence, microsatellite
7	*SW1873*	anonymous DNA sequence, microsatellite
13	*SW1876*	anonymous DNA sequence, microsatellite
2	*SW1879*	anonymous DNA sequence, microsatellite
4	*SW188*	anonymous DNA sequence, microsatellite
6	*SW1881*	anonymous DNA sequence, microsatellite
2	*SW1883*	anonymous DNA sequence, microsatellite
17	*SW1891*	anonymous DNA sequence, microsatellite
15	*SW1892*	anonymous DNA sequence, microsatellite
10	*SW1894*	anonymous DNA sequence, microsatellite
16	*SW1897*	anonymous DNA sequence, microsatellite
13	*SW1898*	anonymous DNA sequence, microsatellite
10	*SW19*	anonymous DNA sequence, microsatellite
13	*SW1901*	anonymous DNA sequence, microsatellite
1	*SW1902*	anonymous DNA sequence, microsatellite
X	*SW1903*	anonymous DNA sequence, microsatellite
Y	*SW1903*	anonymous DNA sequence, microsatellite
5	*SW1904*	anonymous DNA sequence, microsatellite
8	*SW1905*	anonymous DNA sequence, microsatellite
5	*SW1909*	anonymous DNA sequence, microsatellite
5	*SW191*	anonymous DNA sequence, microsatellite
14	*SW1918*	anonymous DNA sequence, microsatellite
17	*SW1920*	anonymous DNA sequence, microsatellite
8	*SW1924*	anonymous DNA sequence, microsatellite
14	*SW1925*	anonymous DNA sequence, microsatellite
6	*SW193*	anonymous DNA sequence, microsatellite
13	*SW1930*	anonymous DNA sequence, microsatellite
15	*SW1935*	anonymous DNA sequence, microsatellite
12	*SW1936*	anonymous DNA sequence, microsatellite
8	*SW194*	anonymous DNA sequence, microsatellite
X	*SW1943*	anonymous DNA sequence, microsatellite
Y	*SW1943*	anonymous DNA sequence, microsatellite
15	*SW1945*	anonymous DNA sequence, microsatellite
8	*SW1953*	anonymous DNA sequence, microsatellite
5	*SW1954*	anonymous DNA sequence, microsatellite
12	*SW1956*	anonymous DNA sequence, microsatellite
1	*SW1957*	anonymous DNA sequence, microsatellite
7	*SW1959*	anonymous DNA sequence, microsatellite
12	*SW1962*	anonymous DNA sequence, microsatellite
4	*SW1967*	anonymous DNA sequence, microsatellite
1	*SW1970*	anonymous DNA sequence, microsatellite
14	*SW1975*	anonymous DNA sequence, microsatellite
3	*SW1978*	anonymous DNA sequence, microsatellite
13	*SW1979*	anonymous DNA sequence, microsatellite

8	*SW1980*	anonymous DNA sequence, microsatellite
5	*SW1982*	anonymous DNA sequence, microsatellite
15	*SW1983*	anonymous DNA sequence, microsatellite
18	*SW1984*	anonymous DNA sequence, microsatellite
5	*SW1987*	anonymous DNA sequence, microsatellite
15	*SW1989*	anonymous DNA sequence, microsatellite
10	*SW1991*	anonymous DNA sequence, microsatellite
X	*SW1994*	anonymous DNA sequence, microsatellite
Y	*SW1994*	anonymous DNA sequence, microsatellite
4	*SW1996*	anonymous DNA sequence, microsatellite
1	*SW1997*	anonymous DNA sequence, microsatellite
5	*SW2*	anonymous DNA sequence, microsatellite
10	*SW2000*	anonymous DNA sequence, microsatellite
7	*SW2001*	anonymous DNA sequence, microsatellite
7	*SW2002*	anonymous DNA sequence, microsatellite
5	*SW2003*	anonymous DNA sequence, microsatellite
13	*SW2007*	anonymous DNA sequence, microsatellite
11	*SW2008*	anonymous DNA sequence, microsatellite
7	*SW2019*	anonymous DNA sequence, microsatellite
3	*SW2021*	anonymous DNA sequence, microsatellite
5	*SW2034*	anonymous DNA sequence, microsatellite
1	*SW2035*	anonymous DNA sequence, microsatellite
14	*SW2038*	anonymous DNA sequence, microsatellite
7	*SW2040*	anonymous DNA sequence, microsatellite
10	*SW2043*	anonymous DNA sequence, microsatellite
3	*SW2047*	anonymous DNA sequence, microsatellite
X	*SW2048*	anonymous DNA sequence, microsatellite
Y	*SW2048*	anonymous DNA sequence, microsatellite
4	*SW2049*	anonymous DNA sequence, microsatellite
8	*SW205*	anonymous DNA sequence, microsatellite
6	*SW2052*	anonymous DNA sequence, microsatellite
15	*SW2053*	anonymous DNA sequence, microsatellite
14	*SW2057*	anonymous DNA sequence, microsatellite
X	*SW2059*	anonymous DNA sequence, microsatellite
Y	*SW2059*	anonymous DNA sequence, microsatellite
8	*SW206*	anonymous DNA sequence, microsatellite
4	*SW2066*	anonymous DNA sequence, microsatellite
10	*SW2067*	anonymous DNA sequence, microsatellite
13	*SW207*	anonymous DNA sequence, microsatellite
15	*SW2072*	anonymous DNA sequence, microsatellite
1	*SW2073*	anonymous DNA sequence, microsatellite
9	*SW2074*	anonymous DNA sequence, microsatellite
16	*SW208*	anonymous DNA sequence, microsatellite
15	*SW2083*	anonymous DNA sequence, microsatellite
9	*SW2093*	anonymous DNA sequence, microsatellite
13	*SW2097*	anonymous DNA sequence, microsatellite

6	*SW2098*	anonymous DNA sequence, microsatellite
9	*SW21*	anonymous DNA sequence, microsatellite
14	*SW210*	anonymous DNA sequence, microsatellite
7	*SW2108*	anonymous DNA sequence, microsatellite
8	*SW211*	anonymous DNA sequence, microsatellite
9	*SW2116*	anonymous DNA sequence, microsatellite
13	*SW2118*	anonymous DNA sequence, microsatellite
14	*SW2122*	anonymous DNA sequence, microsatellite
X	*SW2126*	anonymous DNA sequence, microsatellite
Y	*SW2126*	anonymous DNA sequence, microsatellite
4	*SW2128*	anonymous DNA sequence, microsatellite
15	*SW2129*	anonymous DNA sequence, microsatellite
1	*SW2130*	anonymous DNA sequence, microsatellite
15	*SW2131*	anonymous DNA sequence, microsatellite
2	*SW2134*	anonymous DNA sequence, microsatellite
X	*SW2137*	anonymous DNA sequence, microsatellite
Y	*SW2137*	anonymous DNA sequence, microsatellite
3	*SW2141*	anonymous DNA sequence, microsatellite
17	*SW2142*	anonymous DNA sequence, microsatellite
7	*SW2155*	anonymous DNA sequence, microsatellite
Y	*SW2156*	anonymous DNA sequence, microsatellite
1	*SW216*	anonymous DNA sequence, microsatellite
8	*SW2160*	anonymous DNA sequence, microsatellite
1	*SW2166*	anonymous DNA sequence, microsatellite
2	*SW2167*	anonymous DNA sequence, microsatellite
6	*SW2173*	anonymous DNA sequence, microsatellite
8	*SW2174*	anonymous DNA sequence, microsatellite
12	*SW2180*	anonymous DNA sequence, microsatellite
1	*SW2184*	anonymous DNA sequence, microsatellite
1	*SW2185*	anonymous DNA sequence, microsatellite
2	*SW2192*	anonymous DNA sequence, microsatellite
2	*SW2193*	anonymous DNA sequence, microsatellite
10	*SW2195*	anonymous DNA sequence, microsatellite
13	*SW2196*	anonymous DNA sequence, microsatellite
13	*SW225*	anonymous DNA sequence, microsatellite
3	*SW236*	anonymous DNA sequence, microsatellite
17	*SW24*	anonymous DNA sequence, microsatellite
2	*SW240*	anonymous DNA sequence, microsatellite
9	*SW2401*	anonymous DNA sequence, cosmid containing microsatellite
4	*SW2404*	anonymous DNA sequence, cosmid containing microsatellite
6	*SW2406*	anonymous DNA sequence, cosmid containing microsatellite
9	*SW2407*	anonymous DNA sequence, cosmid containing microsatellite

3	*SW2408*	anonymous DNA sequence, cosmid containing microsatellite
4	*SW2409*	anonymous DNA sequence, cosmid containing microsatellite
8	*SW2410*	anonymous DNA sequence, cosmid containing microsatellite
16	*SW2411*	anonymous DNA sequence, microsatellite
13	*SW2412*	anonymous DNA sequence, cosmid containing microsatellite
11	*SW2413*	anonymous DNA sequence, cosmid containing microsatellite
6	*SW2415*	microsatellite SW2415
1	*SW2416*	anonymous DNA sequence, cosmid containing microsatellite
6	*SW2419*	microsatellite SW2419
5	*SW2425*	anonymous DNA sequence, cosmid containing microsatellite
17	*SW2427*	anonymous DNA sequence, cosmid containing microsatellite l
7	*SW2428*	anonymous DNA sequence, microsatellite
3	*SW2429*	anonymous DNA sequence, microsatellite
13	*SW2430*	anonymous DNA sequence, microsatellite
17	*SW2431*	anonymous DNA sequence, cosmid containing microsatellite
1	*SW2432*	anonymous DNA sequence, microsatellite
4	*SW2435*	anonymous DNA sequence, cosmid containing microsatellite
14	*SW2439*	anonymous DNA sequence, cosmid containing microsatellite
13	*SW2440*	anonymous DNA sequence, cosmid containing microsatellite
17	*SW2441*	anonymous DNA sequence, cosmid containing microsatellite
2	*SW2442*	anonymous DNa sequence, cosmid containing microsatellite
2	*SW2443*	anonymous DNA sequence, cosmid containing microsatellite
2	*SW2445*	anonymous DNA sequence, cosmid containing microsatellite
13	*SW2448*	anonymous DNA sequence, cosmid containing microsatellite
14	*SW245*	microsatellite
13	*SW2450*	anonymous DNA sequence, microsatellite
X	*SW2453*	anonymous DNA sequence, cosmid containing microsatellite
Y	*SW2453*	anonymous DNA sequence, microsatellite

4	*SW2454*	anonymous DNA sequence, microsatellite
X	*SW2456*	anonymous DNA sequence, cosmid containing microsatellite
Y	*SW2456*	anonymous DNA sequence, microsatellite
13	*SW2458*	anonymous DNA sequence, cosmid containing microsatellite
13	*SW2459*	anonymous DNA sequence, cosmid containing microsatellite
6	*SW2466*	anonymous DNA sequence, microsatellite
X	*SW2470*	anonymous DNA sequence, cosmid containing microsatellite
Y	*SW2470*	anonymous DNA sequence, microsatellite
13	*SW2472*	anonymous DNA sequence, cosmid containing microsatellite
14	*SW2474*	anonymous DNA sequence, microsatellite
X	*SW2476*	anonymous DNA sequence, microsatellite
Y	*SW2476*	anonymous DNA sequence, microsatellite
14	*SW2478*	anonymous DNA sequence, microsatellite
3	*SW248*	microsatellite
14	*SW2488*	anonymous DNA sequence, microsatellite
10	*SW249*	microsatellite
12	*SW2490*	anonymous DNA sequence, cosmid containing microsatellite
10	*SW2491*	anonymous DNA sequence, cosmid containing microsatellite
12	*SW2494*	anonymous DNA sequence, cosmid containing microsatellite
9	*SW2495*	anonymous DNA sequence, containing microsatellite
14	*SW2496*	anonymous DNA sequence, microsatellite
14	*SW2504*	anonymous DNA sequence, microsatellite
6	*SW2505*	anonymous DNA sequence, cosmid containing microsatellite
14	*SW2507*	anonymous DNA sequence, cosmid containing microsatellite
14	*SW2508*	anonymous DNA sequence, cosmid containing microsatellite
4	*SW2509*	anonymous DNA sequence, cosmid containing microsatellite
3	*SW251*	microsatellite
1	*SW2512*	anonymous DNA sequence, microsatellite
2	*SW2513*	anonymous DNA sequence, cosmid containing microsatellite
2	*SW2514*	anonymous DNA sequence, cosmid containing microsatellite

14	*SW2515*	anonymous DNA sequence, cosmid containing microsatellite
16	*SW2517*	anonymous DNA sequence, cosmid containing microsatellite
14	*SW2519*	anonymous DNA sequence, microsatellite
7	*SW252*	anonymous DNA sequence, microsatellite
14	*SW2520*	anonymous DNA sequence, cosmid containing microsatellite
8	*SW2521*	anonymous DNA sequence, microsatellite
6	*SW2525*	anonymous DNA sequence, cosmid containing microsatellite
3	*SW2527*	anonymous DNA sequence, cosmid containing microsatellite
3	*SW2532*	anonymous DNA sequence, microsatellite
X	*SW2534*	anonymous DNA sequence, microsatellite
Y	*SW2534*	anonymous DNA sequence, microsatellite
6	*SW2535*	anonymous DNA sequence, cosmid containing microsatellite
7	*SW2537*	anonymous DNA sequence, cosmid containing microsatellite
18	*SW2540*	anonymous DNA sequence, cosmid containing microsatellite
4	*SW2547*	anonymous DNA sequence, microsatellite
7	*SW255*	anonymous DNA sequence, microsatellite
1	*SW2551*	anonymous DNA sequence, cosmid containing microsatellite
6	*SW2557*	anonymous DNA sequence, cosmid containing microsatellite
12	*SW2559*	anonymous DNA sequence, microsatellite
2	*SW256*	microsatellite
7	*SW2564*	anonymous DNA sequence, microsatellite
3	*SW2570*	anonymous DNA sequence, microsatellite
9	*SW2571*	anonymous DNA sequence, cosmid containing microsatellite
2	*SW2586*	anonymous DNA sequence, cosmid containing microsatellite
Y	*SW259*	anonymous DNA sequence, microsatellite
X	*SW259*	anonymous DNA sequence, microsatellite
14	*SW2591*	anonymous DNA sequence, microsatellite
14	*SW2593*	anonymous DNA sequence, cosmid containing microsatellite
3	*SW2597*	anonymous DNA sequence, cosmid containing microsatellite
14	*SW26*	anonymous DNA sequence, microsatellite
14	*SW2604*	anonymous DNA sequence, cosmid containing microsatellite

14	*SW2607*	anonymous DNA sequence, microsatellite
15	*SW2608*	anonymous DNA sequence, microsatellite
8	*SW2611*	anonymous DNA sequence, cosmid containing microsatellite
14	*SW2612*	anonymous DNA sequence, cosmid containing microsatellite
3	*SW2618*	anonymous DNA sequence, cosmid containing microsatellite
16	*SW262*	anonymous DNA sequence, microsatellite
2	*SW2623*	anonymous DNA sequence, cosmid containing microsatellite
7	*SW263*	anonymous DNA sequence, microsatellite
8	*SW268*	anonymous DNA sequence, microsatellite
4	*SW270*	anonymous DNA sequence, microsatellite
3	*SW271*	anonymous DNA sequence, microsatellite
3	*SW274*	anonymous DNA sequence, microsatellite
6	*SW280*	anonymous DNA sequence, microsatellite
4	*SW286*	anonymous DNA sequence, microsatellite
14	*SW287*	anonymous DNA sequence, microsatellite
14	*SW288*	anonymous DNA sequence, microsatellite
8	*SW29*	anonymous DNA sequence, microsatellite
14	*SW295*	anonymous DNA sequence, microsatellite
1	*SW301*	anonymous DNA sequence, microsatellite
7	*SW304*	anonymous DNA sequence, microsatellite
10	*SW305*	anonymous DNA sequence, microsatellite
1	*SW307*	anonymous DNA sequence, microsatellite
8	*SW31*	anonymous DNA sequence, microsatellite
5	*SW310*	anonymous DNA sequence, microsatellite
3	*SW314*	anonymous DNA sequence, microsatellite
6	*SW316*	anonymous DNA sequence, microsatellite
4	*SW317*	anonymous DNA sequence, microsatellite
6	*SW322*	anonymous DNA sequence, microsatellite
14	*SW328*	anonymous DNA sequence, microsatellite
14	*SW330*	anonymous DNA sequence, microsatellite
5	*SW332*	anonymous DNA sequence, microsatellite
17	*SW335*	anonymous DNA sequence, microsatellite
14	*SW342*	anonymous DNA sequence, microsatellite
13	*SW344*	anonymous DNA sequence, microsatellite
3	*SW349*	anonymous DNA sequence, microsatellite
4	*SW35*	anonymous DNA sequence, microsatellite
7	*SW352*	anonymous DNA sequence, microsatellite
6	*SW353*	anonymous DNA sequence, microsatellite
2	*SW354*	anonymous DNA sequence, microsatellite
14	*SW361*	anonymous DNA sequence, microsatellite
8	*SW368*	anonymous DNA sequence, microsatellite
12	*SW37*	anonymous DNA sequence, microsatellite

1	*SW373*	anonymous DNA sequence, microsatellite
8	*SW374*	anonymous DNA sequence, microsatellite
5	*SW378*	anonymous DNA sequence, microsatellite
13	*SW38*	anonymous DNA sequence, microsatellite
16	*SW382*	anonymous DNA sequence, microsatellite
9	*SW383*	anonymous DNA sequence, microsatellite
2	*SW395*	anonymous DNA sequence, microsatellite
13	*SW398*	anonymous DNA sequence, microsatellite
6	*SW4*	anonymous DNA sequence, microsatellite
16	*SW403*	anonymous DNA sequence, microsatellite
16	*SW419*	anonymous DNA sequence, microsatellite
15	*SW423*	anonymous DNA sequence, microsatellite
11	*SW435*	anonymous DNA sequence, microsatellite
5	*SW436*	anonymous DNA sequence, microsatellite
5	*SW439*	anonymous DNA sequence, microsatellite
10	*SW443*	anonymous DNA sequence, microsatellite
8	*SW444*	anonymous DNA sequence, microsatellite
4	*SW445*	anonymous DNA sequence, microsatellite
6	*SW446*	anonymous DNA sequence, microsatellite
4	*SW45*	anonymous DNA sequence, microsatellite
13	*SW452*	anonymous DNA sequence, microsatellite
5	*SW453*	anonymous DNA sequence, microsatellite
13	*SW458*	anonymous DNA sequence, microsatellite
3	*SW460*	anonymous DNA sequence, microsatellite
2	*SW461*	anonymous DNA sequence, microsatellite
12	*SW467*	anonymous DNA sequence, microsatellite
16	*SW469*	anonymous DNA sequence, microsatellite
7	*SW472*	anonymous DNA sequence, microsatellite
4	*SW480*	anonymous DNA sequence, microsatellite
13	*SW482*	anonymous DNA sequence, microsatellite
1	*SW485*	anonymous DNA sequence, microsatellite
11	*SW486*	anonymous DNA sequence, microsatellite
3	*SW487*	anonymous DNA sequence, microsatellite
4	*SW489*	anonymous DNA sequence, microsatellite
5	*SW491*	anonymous DNA sequence, microsatellite
6	*SW492*	anonymous DNA sequence, microsatellite
10	*SW497*	anonymous DNA sequence, microsatellite
16	*SW5*	anonymous DNA sequence, microsatellite
1	*SW501*	anonymous DNA sequence, microsatellite
14	*SW510*	anonymous DNA sequence, microsatellite
9	*SW511*	anonymous DNA sequence, microsatellite
4	*SW512*	anonymous DNA sequence, microsatellite
13	*SW520*	anonymous DNA sequence, microsatellite
4	*SW524*	anonymous DNA sequence, microsatellite
8	*SW527*	anonymous DNA sequence, microsatellite
9	*SW539*	anonymous DNA sequence, microsatellite

9	*SW54*	anonymous DNA sequence, microsatellite
14	*SW540*	anonymous DNA sequence, microsatellite
14	*SW55*	anonymous DNA sequence, microsatellite
1	*SW552*	anonymous DNA sequence, microsatellite
16	*SW557*	anonymous DNA sequence, microsatellite
3	*SW57*	anonymous DNA sequence, microsatellite
2	*SW575*	anonymous DNA sequence, microsatellite
4	*SW58*	anonymous DNA sequence, microsatellite
7	*SW581*	anonymous DNA sequence, microsatellite
4	*SW589*	anonymous DNA sequence, microsatellite
14	*SW59*	anonymous DNA sequence, microsatellite
3	*SW590*	anonymous DNA sequence, microsatellite
14	*SW6*	anonymous DNA sequence, microsatellite
12	*SW60*	anonymous DNA sequence, microsatellite
15	*SW604*	anonymous DNA sequence, microsatellite
12	*SW605*	anonymous DNA sequence, microsatellite
6	*SW607*	anonymous DNA sequence, microsatellite
8	*SW61*	anonymous DNA sequence, microsatellite
6	*SW617*	anonymous DNA sequence, microsatellite
14	*SW619*	anonymous DNA sequence, microsatellite
12	*SW62*	anonymous DNA sequence, microsatellite
14	*SW63*	anonymous DNA sequence, microsatellite
7	*SW632*	anonymous DNA sequence, microsatellite
1	*SW64*	anonymous DNA sequence, microsatellite
1	*SW65*	anonymous DNA sequence, microsatellite
11	*SW66*	anonymous DNA sequence, microsatellite
14	*SW69*	anonymous DNA sequence, microsatellite
13	*SW698*	anonymous DNA sequence, microsatellite
8	*SW7*	anonymous DNA sequence, microsatellite
5	*SW70*	anonymous DNA sequence, microsatellite
11	*SW703*	anonymous DNA sequence, microsatellite
1	*SW705*	anonymous DNA sequence, microsatellite
Y	*SW707*	anonymous DNA sequence, microsatellite
X	*SW707*	anonymous DNA sequence, microsatellite
6	*SW709*	anonymous DNA sequence, microsatellite
6	*SW71*	anonymous DNA sequence, microsatellite
4	*SW714*	anonymous DNA sequence, microsatellite
3	*SW717*	anonymous DNA sequence, microsatellite
3	*SW72*	anonymous DNA sequence, microsatellite
4	*SW724*	anonymous DNA sequence, microsatellite
9	*SW727*	anonymous DNA sequence, microsatellite
3	*SW730*	anonymous DNA sequence, microsatellite
7	*SW732*	anonymous DNA sequence, microsatellite
16	*SW742*	anonymous DNA sequence, microsatellite
1	*SW745*	anonymous DNA sequence, microsatellite
2	*SW747*	anonymous DNA sequence, microsatellite

8	*SW748*	anonymous DNA sequence, microsatellite
9	*SW749*	anonymous DNA sequence, microsatellite
4	*SW752*	anonymous DNA sequence, microsatellite
14	*SW761*	anonymous DNA sequence, microsatellite
8	*SW763*	anonymous DNA sequence, microsatellite
7	*SW764*	anonymous DNA sequence, microsatellite
2	*SW766*	anonymous DNA sequence, microsatellite
10	*SW767*	anonymous DNA sequence, microsatellite
13	*SW769*	anonymous DNA sequence, microsatellite
14	*SW77*	anonymous DNA sequence, microsatellite
14	*SW775*	anonymous DNA sequence, microsatellite
2	*SW776*	anonymous DNA sequence, microsatellite
9	*SW779*	anonymous DNA sequence, microsatellite
1	*SW780*	anonymous DNA sequence, microsatellite
1	*SW781*	anonymous DNA sequence, microsatellite
6	*SW782*	anonymous DNA sequence, microsatellite
18	*SW787*	anonymous DNA sequence, microsatellite
8	*SW790*	anonymous DNA sequence, microsatellite
5	*SW792*	anonymous DNA sequence, microsatellite
1	*SW80*	anonymous DNA sequence, microsatellite
1	*SW803*	anonymous DNA sequence, microsatellite
16	*SW81*	anonymous DNA sequence, microsatellite
16	*SW813*	anonymous DNA sequence, microsatellite
3	*SW816*	anonymous DNA sequence, microsatellite
4	*SW818*	anonymous DNA sequence, microsatellite
6	*SW824*	anonymous DNA sequence, microsatellite
9	*SW827*	anonymous DNA sequence, microsatellite
3	*SW828*	anonymous DNA sequence, microsatellite
10	*SW830*	anonymous DNA sequence, microsatellite
15	*SW832*	anonymous DNA sequence, microsatellite
3	*SW833*	anonymous DNA sequence, microsatellite
2	*SW834*	anonymous DNA sequence, microsatellite
4	*SW835*	anonymous DNA sequence, microsatellite
3	*SW836*	anonymous DNA sequence, microsatellite
4	*SW839*	anonymous DNA sequence, microsatellite
17	*SW840*	anonymous DNA sequence, microsatellite
4	*SW841*	anonymous DNA sequence, microsatellite
8	*SW853*	anonymous DNA sequence, microsatellite
6	*SW855*	anonymous DNA sequence, microsatellite
4	*SW856*	anonymous DNA sequence, microsatellite
14	*SW857*	anonymous DNA sequence, microsatellite
7	*SW859*	anonymous DNA sequence, microsatellite
3	*SW860*	anonymous DNA sequence, microsatellite
13	*SW864*	anonymous DNA sequence, microsatellite
5	*SW865*	anonymous DNA sequence, microsatellite
9	*SW866*	anonymous DNA sequence, microsatellite

4	*SW871*	anonymous DNA sequence, microsatellite
13	*SW873*	anonymous DNA sequence, microsatellite
12	*SW874*	anonymous DNA sequence, microsatellite
11	*SW878*	anonymous DNA sequence, microsatellite
13	*SW882*	anonymous DNA sequence, microsatellite
14	*SW886*	anonymous DNA sequence, microsatellite
3	*SW902*	anonymous DNA sequence, microsatellite
11	*SW903*	anonymous DNA sequence, microsatellite
5	*SW904*	anonymous DNA sequence, microsatellite
8	*SW905*	anonymous DNA sequence, microsatellite
15	*SW906*	anonymous DNA sequence, microsatellite
9	*SW911*	anonymous DNA sequence, microsatellite
8	*SW916*	anonymous DNA sequence, microsatellite
6	*SW917*	anonymous DNA sequence, microsatellite
15	*SW919*	anonymous DNA sequence, microsatellite
10	*SW920*	anonymous DNA sequence, microsatellite
8	*SW933*	anonymous DNA sequence, microsatellite
13	*SW935*	anonymous DNA sequence, microsatellite
15	*SW936*	anonymous DNA sequence, microsatellite
13	*SW937*	anonymous DNA sequence, microsatellite
15	*SW938*	anonymous DNA sequence, microsatellite
9	*SW940*	anonymous DNA sequence, microsatellite
2	*SW942*	anonymous DNA sequence, microsatellite
12	*SW943*	anonymous DNA sequence, microsatellite
9	*SW944*	anonymous DNA sequence, microsatellite
1	*SW947*	anonymous DNA sequence, microsatellite
Y	*SW949*	anonymous DNA sequence, microsatellite
X	*SW949*	anonymous DNA sequence, microsatellite
10	*SW951*	anonymous DNA sequence, microsatellite
1	*SW952*	anonymous DNA sequence, microsatellite
13	*SW955*	anonymous DNA sequence, microsatellite
12	*SW957*	anonymous DNA sequence, microsatellite
13	*SW960*	anonymous DNA sequence, microsatellite
X	*SW961*	anonymous DNA sequence, microsatellite
Y	*SW961*	anonymous DNA sequence, microsatellite
1	*SW962*	anonymous DNA sequence, microsatellite
5	*SW963*	anonymous DNA sequence, microsatellite
15	*SW964*	anonymous DNA sequence, microsatellite
5	*SW967*	anonymous DNA sequence, microsatellite
4	*SW969*	anonymous DNA sequence, microsatellite
1	*SW970*	anonymous DNA sequence, microsatellite
6	*SW973*	anonymous DNA sequence, microsatellite
1	*SW974*	anonymous DNA sequence, microsatellite
16	*SW977*	anonymous DNA sequence, microsatellite
Y	*SW980*	anonymous DNA sequence, microsatellite
X	*SW980*	anonymous DNA sequence, microsatellite

9	*SW983*	anonymous DNA sequence, microsatellite
5	*SW986*	anonymous DNA sequence, microsatellite
9	*SW989*	anonymous DNA sequence, microsatellite
13	*SW992*	anonymous DNA sequence, microsatellite
5	*SW995*	anonymous DNA sequence, microsatellite
10	*SWC19*	anonymous DNA sequence, cosmid containing microsatellite
13	*SWC22*	anonymous DNA sequence, cosmid containing microsatellite
12	*SWC23*	anonymous DNA sequence, cosmid containing microsatellite
14	*SWC27*	anonymous DNA sequence, cosmid containing microsatellite
2	*SWC9*	anonymous DNA sequence, cosmid containing microsatellite
15	*SWR1002*	anonymous DNA sequence, microsatellite
17	*SWR1004*	anonymous DNA sequence, microsatellite
13	*SWR1008*	anonymous DNA sequence, microsatellite
12	*SWR1021*	anonymous DNA sequence, microsatellite
14	*SWR1042*	anonymous DNA sequence, microsatellite
1	*SWR1061*	anonymous DNA sequence, microsatellite
1	*SWR1063*	anonymous DNA sequence, microsatellite
7	*SWR1078*	anonymous DNA sequence, microsatellite
8	*SWR1101*	anonymous DNA sequence, microsatellite
16	*SWR1106*	anonymous DNA sequence, microsatellite
4	*SWR1110*	anonymous DNA sequence, microsatellite
5	*SWR1112*	anonymous DNA sequence, microsatellite
14	*SWR1113*	anonymous DNA sequence, microsatellite
17	*SWR1120*	anonymous DNA sequence, microsatellite
7	*SWR1121*	anonymous DNA sequence, microsatellite
6	*SWR1130*	anonymous DNA sequence, microsatellite
17	*SWR1133*	anonymous DNA sequence, microsatellite
7	*SWR1210*	anonymous DNA sequence, microsatellite
6	*SWR1211*	anonymous DNA sequence, microsatellite
13	*SWR1218*	anonymous DNA sequence, microsatellite
9	*SWR123*	anonymous DNA sequence, microsatellite
13	*SWR1306*	anonymous DNA sequence, microsatellite
2	*SWR1338*	anonymous DNA sequence, microsatellite
2	*SWR1342*	anonymous DNA sequence, microsatellite
7	*SWR1343*	anonymous DNA sequence, microsatellite
9	*SWR1357*	anonymous DNA sequence, microsatellite
10	*SWR136*	anonymous DNA sequence, microsatellite
4	*SWR1367*	anonymous DNA sequence, microsatellite
6	*SWR1384*	microsatellite SWR1384
3	*SWR1395*	anonymous DNA sequence, microsatellite
4	*SWR1410*	anonymous DNA sequence, microsatellite

1	*SWR1427*	anonymous DNA sequence, microsatellite
2	*SWR1445*	anonymous DNA sequence, microsatellite
15	*SWR1485*	anonymous DNA sequence, microsatellite
2	*SWR1512*	anonymous DNA sequence, microsatellite
3	*SWR1521*	anonymous DNA sequence, microsatellite
5	*SWR1526*	anonymous DNA sequence, microsatellite
4	*SWR153*	anonymous DNA sequence, microsatellite
15	*SWR1533*	anonymous DNA sequence, microsatellite
10	*SWR158*	anonymous DNA sequence, microsatellite
13	*SWR1627*	anonymous DNA sequence, microsatellite
6	*SWR1634*	anonymous DNA sequence, microsatellite
3	*SWR1637*	anonymous DNA sequence, microsatellite
Y	*SWR17*	anonymous DNA sequence, microsatellite
X	*SWR17*	anonymous DNA sequence, microsatellite
8	*SWR1801*	anonymous DNA sequence, microsatellite
12	*SWR1802*	anonymous DNA sequence, microsatellite
7	*SWR1806*	anonymous DNA sequence, microsatellite
10	*SWR1829*	anonymous DNA sequence, microsatellite
9	*SWR1848*	anonymous DNA sequence, microsatellite
10	*SWR1849*	anonymous DNA sequence, microsatellite
X	*SWR1861*	anonymous DNA sequence, microsatellite
Y	*SWR1861*	anonymous DNA sequence, microsatellite
2	*SWR1910*	anonymous DNA sequence, microsatellite
8	*SWR1921*	anonymous DNA sequence, microsatellite
6	*SWR1923*	anonymous DNA sequence, microsatellite
7	*SWR1928*	anonymous DNA sequence, microsatellite
9	*SWR1939*	anonymous DNA sequence, microsatellite
13	*SWR1941*	anonymous DNA sequence, microsatellite
9	*SWR1950*	anonymous DNA sequence, microsatellite
5	*SWR1974*	anonymous DNA sequence, microsatellite
10	*SWR198*	anonymous DNA sequence, microsatellite
4	*SWR1998*	anonymous DNA sequence, microsatellite
3	*SWR201*	anonymous DNA sequence, microsatellite
7	*SWR2036*	anonymous DNA sequence, microsatellite
13	*SWR2054*	anonymous DNA sequence, microsatellite
14	*SWR2063*	anonymous DNA sequence, microsatellite
3	*SWR2069*	anonymous DNA sequence, microsatellite
11	*SWR2071*	anonymous DNA sequence, microsatellite
16	*SWR2080*	anonymous DNA sequence, microsatellite
16	*SWR2086*	anonymous DNA sequence, microsatellite
3	*SWR2096*	anonymous DNA sequence, microsatellite
13	*SWR2114*	anonymous DNA sequence, microsatellite
15	*SWR2121*	anonymous DNA sequence, microsatellite
9	*SWR2140*	anonymous DNA sequence, microsatellite
6	*SWR2149*	anonymous DNA sequence, microsatellite
7	*SWR2152*	anonymous DNA sequence, microsatellite

2	*SWR2157*	anonymous DNA sequence, microsatellite
4	*SWR2179*	anonymous DNA sequence, microsatellite
1	*SWR2182*	anonymous DNA sequence, microsatellite
3	*SWR2188*	anonymous DNA sequence, microsatellite
13	*SWR2189*	anonymous DNA sequence, microsatellite
3	*SWR2204*	anonymous DNA sequence, microsatellite
1	*SWR2300*	anonymous DNA sequence, microsatellite
17	*SWR2417*	anonymous DNA sequence, cosmid containing microsatellite
16	*SWR2480*	anonymous DNA sequence, microsatellite
9	*SWR250*	anonymous DNA sequence, microsatellite
2	*SWR2516*	anonymous DNA sequence, cosmid containing microsatellite
15	*SWR283*	anonymous DNA sequence, microsatellite
2	*SWR308*	anonymous DNA sequence, microsatellite
15	*SWR312*	anonymous DNA sequence, microsatellite
10	*SWR334*	anonymous DNA sequence, microsatellite
1	*SWR337*	anonymous DNA sequence, microsatellite
16	*SWR340*	anonymous DNA sequence, microsatellite
2	*SWR345*	anonymous DNA sequence, microsatellite
14	*SWR346*	anonymous DNA sequence, microsatellite
4	*SWR362*	anonymous DNA sequence, microsatellite
2	*SWR389*	anonymous DNA sequence, microsatellite
12	*SWR390*	anonymous DNA sequence, microsatellite
18	*SWR414*	anonymous DNA sequence, microsatellite
5	*SWR426*	anonymous DNA sequence, microsatellite
13	*SWR428*	anonymous DNA sequence, microsatellite
5	*SWR453*	anonymous DNA sequence, microsatellite
2	*SWR468*	anonymous DNA sequence, microsatellite
1	*SWR485*	anonymous DNA sequence, microsatellite
10	*SWR493*	anonymous DNA sequence, microsatellite
10	*SWR67*	anonymous DNA sequence, microsatellite
9	*SWR68*	anonymous DNA sequence, microsatellite
1	*SWR702*	anonymous DNA sequence, microsatellite
6	*SWR726*	anonymous DNA sequence, microsatellite
4	*SWR73*	anonymous DNA sequence, microsatellite
7	*SWR74*	anonymous DNA sequence, microsatellite
8	*SWR741*	anonymous DNA sequence, microsatellite
8	*SWR750*	anonymous DNA sequence, microsatellite
3	*SWR756*	anonymous DNA sequence, microsatellite
7	*SWR773*	anonymous DNA sequence, microsatellite
2	*SWR783*	anonymous DNA sequence, microsatellite
11	*SWR811*	anonymous DNA sequence, microsatellite
1	*SWR817*	anonymous DNA sequence, microsatellite
6	*SWR823*	anonymous DNA sequence, microsatellite
14	*SWR84*	anonymous DNA sequence, microsatellite

3	*SWR85*	anonymous DNA sequence, microsatellite
3	*SWR910*	anonymous DNA sequence, microsatellite
9	*SWR915*	anonymous DNA sequence, microsatellite
14	*SWR925*	anonymous DNA sequence, microsatellite
13	*SWR926*	anonymous DNA sequence, microsatellite
7	*SWR946*	anonymous DNA sequence, microsatellite
13	*SWR950*	anonymous DNA sequence, microsatellite
3	*SWR978*	anonymous DNA sequence, microsatellite
4	*SWR981*	anonymous DNA sequence, microsatellite
1	*SWR982*	anonymous DNA sequence, microsatellite
6	*SWR987*	anonymous DNA sequence, microsatellite
14	*SYK*	spleen tyrosine kinase
UN	*T11*	anonymous DNA sequence, contains microsatellite
14	*T20*	anonymous DNA sequence, cosmid T20
7	*TAP1*	transporter associated with antigen processing 1
7	*TAP2*	transporter associated with antigen processing 2
6	*TAT*	tyrosine aminotransferase
UN	*TCP1*	*t* complex polypeptide 1
13	*TF*	transferrin
6	*TGFB1*	transforming growth factor beta-1
10	*TGFB2*	transforming growth factor beta-2
12	*TK1*	thymidine kinase 1, soluble
7	*TNFA*	tumour necrosis factor, alpha
7	*TNFB*	tumour necrosis factor, beta
15	*TNP1*	transition protein 1
3	*TNP2*	transition protein 2
12	*TP53*	tumour protein p53
11	*TPP2*	tripeptidyl peptidase II
X	*TRAIII*	congenital tremor type A III
4	*TSHB*	thyroid stimulating hormone, beta subunit
6	*TTR*	transthyretin, prealbumin
9	*TYR*	tyrosinase
11	*TYRP2*	tyrosinase-related protein 2
14	*UBB*	ubiquitin
14	*UBC*	polyubiquitin C
12	*UMPH2*	uridine 5′-monophosphate phosphohydrolase 2
6	*UOX*	urate oxidase
16	*UTAP2*	TAP2-like sequences
UN	*VCAM1*	vascular cellular adhesion molecule 1
UN	*VWD*	von Willebrand disease
7	*WARS*	tryptophanyl-tRNA synthetase
11	*X11001*	anonymous DNA sequence, microsatellite
11	*X11002*	anonymous DNA sequence, microsatellite
8	*X11003*	anonymous DNA sequence, microsatellite
12	*X11004*	anonymous DNA sequence, microsatellite

11	*X11005*	anonymous DNA sequence, microsatellite
11	*X11006*	anonymous DNA sequence, microsatellite
11	*X11008*	anonymous DNA sequence, microsatellite
11	*X11011*	anonymous DNA sequence, microsatellite
11	*X11012*	anonymous DNA sequence, microsatellite
11	*X11014*	anonymous DNA sequence, microsatellite
11	*X11015*	anonymous DNA sequence, microsatellite
UN	*ZFX*	zinc finger protein (X-specific)
UN	*ZFY*	zinc finger protein (Y specific)
3	*ZP3*	zona pellucida glycoprotein 3

This list obtained from Database in Roslin Institute (Edinburgh),
http://www.ri.bbsrc.ac.uk/pigmap/pigbase/loclist.html

Index